The Physics and Technology of Amorphous SiO_2

The Physics and Technology of Amorphous SiO_2

Edited by

Roderick A. B. Devine

National Center for Telecommunications Studies
Meyland, France

Plenum Press • New York and London

Library of Congress Cataloging in Publication Data

Symposium on the Physics and Technology of Amorphous SiO_2 (1987: Les Arcs, France)

The physics and technology of amorphous SiO_2 / edited by Roderick A. B. Devine.

p. cm.

"Proceedings of a Symposium on the Physics and Technology of Amorphous SiO_2, held June 29–July 3, 1987, in Les Arcs, France"—T.p. verso.

Includes bibliographical references and indexes.

ISBN-13: 978-1-4612-8301-0 e-ISBN-13: 978-1-4613-1031-0

DOI: 10.1007/ 978-1-4613-1031-0

1. Silica—Congresses. 2. Surfaces (Physics)—Congresses. 3. Surface chemistry—Congresses. I. Devine, Roderick A. B. II. Title.

QC585.75.S55S95 1987

620.1′93—dc19 88-17464

CIP

Proceedings of a symposium on the Physics and Technology of Amorphous SiO_2, held June 29–July 3, 1987, in Les Arcs, France

Softcover reprint of the hardcover 1st edition 1988

A Division of Plenum Publishing Corporation
233 Spring Street, New York, N.Y. 10013

FOREWORD

The contents of this volume represent most of the papers presented either orally or as posters at the international conference held in Les Arcs, Savoie, from June 29th to July 3rd 1987. The declared objective of the conference was to bring together specialists working in various fields, both academic and applied, to examine the state of our understanding of the physics of amorphous SiO_2 from the point of view of its structure, defects (both intrinsic and extrinsic), its ability to transport current and to trap charges, its sensitivity to irradiation, etc. For this reason, the proceedings is divided, as was the conference schedule, into a number of sections starting from a rather academic viewpoint of the internal structure of idealized SiO_2 and progressing towards subjects of increasing technological importance such as charge transport and trapping and breakdown in thin films. The proceedings terminates with a section on novel applications of amorphous SiO_2 and in particular, buried oxide layers formed by ion implantation. Although every effort was made at the conference to ensure that each presentation occured in its most obvious session, in editing the proceedings we have taken the liberty of changing the order where it seems that a paper was in fact more appropriate to an alternative section. In any event, because of the natural overlap of subjects, many papers could have been suitably placed in several different sections.

The successful operation of any conference is always the result of a group effort and Les Arcs was no exception. The groundwork for the conference resulted from discussions with Jörg Arndt, Akos Revesz, Guy Ghibaudo and Jean-Jacques Niez whilst the on-site smooth running was assured by Julia, Bob and Neil Devine and Bernadette Bonnefond - sincere thanks go to all of these people for their unflagging help. A special word of thanks goes also to the session chairmen who, without recourse to buzzers, bells or flashing lights, steered the sessions to a timely end so that coffee breaks were not delayed and nobody suffered from malnutrition.

Finally, the conference could not have been held were it not for the kind support, both financial and moral, obtained from the following organisations to whom we are sincerely grateful:

Bull S. A.
Centre National d'Etudes des Télécommunications
Department of the Army (European Research Office)
Department of the Navy (European Research Office)
Direction des Recherches, Etudes et Techniques
Heraeus (France)
IBM (France)

Laboratoire d'Electronique et de Physique Appliquée
Laboratoire d'Electronique et de Technologie de l'Informatique
Thomson CSF
Universität Tübingen (Federal Republic of Germany)

R. A. B. Devine
Grenoble, February 1988

CONTENTS

STRUCTURE: THEORY AND EXPERIMENT

INTRINSIC AND EXTRINSIC DEFECTS: THEORY

INTRINSIC AND EXTRINSIC DEFECTS: EXPERIMENT

INTERFACES

OXIDATION, OXYNITRIDES AND DEPOSITED FILMS

TRANSPORT, TRAPPING AND BREAKDOWN

RADIATION EFFECTS

BURIED DIELECTRIC LAYERS AND NOVEL APPLICATIONS

CURRENT MODELS FOR AMORPHOUS SiO_2

Frank L. Galeener

Department of Physics
Colorado State University
Fort Collins, CO 80523, USA

ABSTRACT

I discuss central questions of glass structure and define four ranges of order in an amorphous solid : short, intermediate, long and global. These concepts are used to catalog all of the elements of order in the Zachariasen-Warren model for a-SiO_2, in preparation for a later deduction of the logically possible improvements in that model. Six existing approximate forms of the ZW model are discussed using the language developed in the paper. This provides a fundamental background regarding structure for the rest of the papers in this volume.

INTRODUCTION

The description and determination of the structure of amorphous materials is a ubiquitous and fundamental problem for those who wish to understand and predict their properties. In this paper I will discuss conceptual aspects of the problem, using a-SiO_2 as a vehicle for the presentation. The content will build on ideas recently reported elsewhere[1].

SOME CENTRAL QUESTIONS OF GLASS STRUCTURE

It is clear that glasses contain sufficient disorder that their structure must ultimately be defined statistically. There are too many slightly different structural units for a complete enumeration, so an approximate and statistical description of structure is needed, just as one is needed for treating the properties of gases. How completely can one specify the structure of a real glass ? What is the best "language" for doing this ? How precise must be the description - for adequate understanding of each of various physical properties ? What are useful simplifications of these structural models, for purposes of facilitating calculations or conceptual reasoning ? How does structure vary with the conditions and mode of preparation[2] of the amorphous solid ? Previous reports[3] and several papers at this conference show that numerous properties of a-SiO_2 vary importantly with preparation conditions ; the statistical description of structure must be sensitive enough to account for these changes. Most of the questions just raised are not yet well-answered.

THREE KINDS OF STRUCTURAL MODELS

Occasionally an author will present evidence against a particular model and then infer erroneously that the parent class of models has been disproven. Thus, a specific inadequacy of the Bell and Dean model[4] for $a\text{-}SiO_2$ does not[5] necessarily invalidate[6] the parent continuous random network hypothesis. To help avoid this logical error, I will classify models as conceptual, specific or approximate.

A conceptual model consists of conceptual statements only, is very general in nature, typically stating the topology, connectivity or order with no quantitative detail. It is insufficiently specified for quantitative comparison with experiment. Examples are models that simply assert that an amorphous solid is "not crystalline", is a "continuous random network" or is an "array of microcrystals".

A specific model is characterized by conceptual statements plus sufficient quantitative specifications to be compared quantitatively with one or more experiments. To the extent that there is disagreement, the model may be wrong only in the quantitative specifications, or it may also be wrong conceptually if the conceptual aspects require quantitative specifications in the range tested[7] : this is, of course, simple logic, but it is all too frequently violated.

An approximate model is a specific model which is simplified from its pristine form in order to enable physical or mathematical construction of the model or to enable or facilitate mathematical calculations of properties that will be compared with experiment. A Bethe lattice[8] may be viewed as an approximate form of some specific continuous random network, and is used to calculate properties like the vibrational spectra of $a\text{-}SiO_2$ [9]; if those predictions cannot be made to agree with experiment the fault may lie in the Bethe lattice approximation, not in the specific structure it is used to model, or in the conceptual model that lies underneath. In more subtle situations this distinction is also too often overlooked. In fact, most structural models used to compare with experiment are approximate in the present sense : the Bell and Dean models are highly approximate because of the small size and large surface area of the clusters actually built and used for calculations.

SOME IMPORTANT FACTS ABOUT STRUCTURAL STUDIES

It is important to note some general facts about experimental studies of the structure of amorphous solids.

While the diffraction of x-rays, neutrons or electrons gives definite information about the structure of crystalline solids, such measurements give much less information about the structure of amorphous solids[10]. The diffraction pattern of an a-solid consists of diffuse rings whose radial variation of intensity provides a merely 1-dimensional representation of the 3-dimensional glass structure. Clearly some information about the structure is "averaged out" in the experiment, and the structure is surely not 1-1 with the diffraction data. In practice, then, one imagines a likely a-structure, predicts experiment and compares with observation. If comparison is poor, the specific model is rejected ; however, good comparison merely qualifies the model as an acceptable candidate and leaves open the possibility that some other model would do as well, or better.

This illustrates the fact that statistical structural models are almost impossible to prove uniquely. All probes will involve averages over statistical structural models. A poor fit to data with one probe implies rejection of the specific model used for calculation, but the conceptual form of the specific model may still be correct. A good fit is encouraging, but may in principle be had with other models not yet considered.

This discussion illustrates the need for the use of many different experimental probes of structure. Diffraction, NMR, vibrational spectroscopy and other techniques emphasize different aspects of structure, such as interatomic distances, or angles, or local symmetry, or range of order, and the like. Each technique provides an alternate view of the structure and generally averages the structure differently, thus revealing aspects that may be de-emphasized or obscured by another technique. Observation of ordered rings of bonds in a-SiO_2 by Raman spectroscopy is a vivid example[11]. Although each probe gives limited information, the results of several different probes can dramatically reduce the number of possible models, leading to a conviction of truth (if not its proof).

RANGES OF ORDER IN AMORPHOUS SOLIDS

This logical process of elimination of possible models would also be greatly helped by the development of clearer concepts and more certain theoretical principles concerning the structure of a-solids. To this end I have tried to develop a more precise set of terms for categorizing the scales (or ranges) of order in amorphous solids[1]. The resultant four ranges of order are as follows.

Short Range Order

Short range order (SRO) describes the nearest neighbor (nn) bonding environment of each atomic species. It involves bond types, distances, and angles, extends over a volume whose dimension is about twice the typical nn

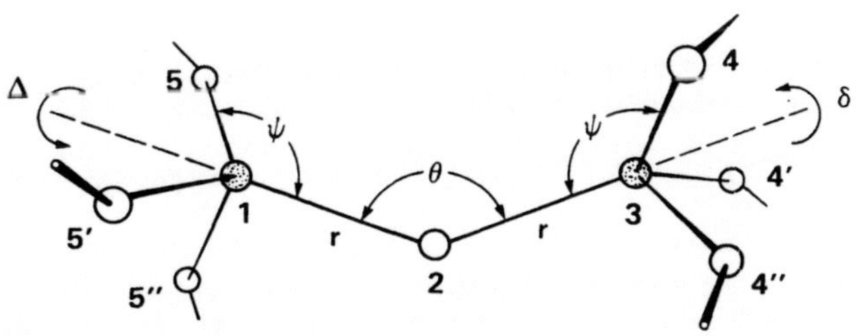

Fig. 1. The relative orientation of two corner-sharing tetrahedra in a-SiO_2, showing the short range order and one aspect of the intermediate range order. Each Si atom (dark) is surrounded tetrahedrally at distance r by four O atoms and each O atom bridges between two Si atoms with an angle θ that varies from site to site ; these two molecule-like arrangements specify the SRO in its usual "simplified" form. The dihedral angles δ and Δ, giving the angular orientation of the tetrahedra about their bridging O-Si bonds (relative to the Si-O-Si plane) are defined in this paper as elements of IRO.

structure of B_2O_3 glass[17]. The model is based on regular planar "boroxol" rings within which the dihedral angles of the BO_3 units on either side of an O atom are equal and thus completely correlated. On the other hand the dihedral angles of BO_3 units on either side of an O atom bridging between two rings are not equal, are much less correlated and are generally assumed to be random and independent in recent studies[18] (which are quantitatively consistent with that assumption).

Similarly, the assumption of random dihedral angles in fig. 1, used in the Zachariasen-Warren model[19-26] for a-SiO_2 is also a specification of an element of IRO.

SRO and IRO will specify the relative atomic positions over a volume whose dimension is several nearest neighbor distances, and is therefore in the range 5-10 Å, or larger.

Long Range Order

At larger distances we must consider the possibility of crystal-like arrangements of atoms which we will call long range order (LRO). We will define two mutually exclusive kinds of order on this next larger scale of distances, namely crystalline LRO and morphological LRO.

Crystalline Long Range Order

Crystalline LRO recognizes the possibility of periodic repetition of an IRO over several repeat lengths, and thus accounts for the possible existence of microcrystals in the structure of amorphous solids. Its existence shows clearly as sharp features in a diffraction pattern at positions related closely to those for more macroscopic crystals.

Roughly speaking, microcrystallinity over dimensions of 10 Å or more will be revealed as diffraction features that are sharper than those seen in glasses which are not microcrystalline. In a-SiO_2 it has been argued that microcrystals larger than about 8 Å would produce diffraction lines narrower than those seen[5]. Clearly, microcrystalline LRO specifies, SRO, IRO and other elements of structure - over the range of the particle size. (Note, as I have shown elsewhere[1], that the surface or interface layers between such particles comprise a very large fraction of the entire sample).

The important idea here is that we operationally define a distance d_{LRO}, which is the dimension of a microcrystal that would give sharp enough diffraction lines to be revealed by the narrowness of those lines. Clearly, d_{LRO} varies from material to material, and for different crystal structures of the same material.

Order on this scale or larger will be called LRO, thus defining a lower limit for the scale of LRO and an upper limit for the scale of IRO. We will use the same d_{LRO} to define a scale for LRO in amorphous materials that are not microcrystalline.

Morphological Long Range Order

This category accounts for order in non-microcrystalline structures on a longer range scale than that of IRO, ie on a scale larger than the minimum dimension d_{LRO} just defined. This is useful in order to catalog structures that are clearly not covered by the previously defined notions of IRO. Morphological LRO includes extended voids, channels, spherulites, distance, and is therefore in the range 3-5 Å for most materials, 3 Å for $a\text{-}SiO_2$. It is usually well determined from the radial distribution function (rdf) obtained from x-ray or neutron diffraction.

In $a\text{-}SiO_2$, the SRO is commonly specified by stating that each Si atom is surrounded almost tetrahedrally by four oxygen atoms at bonding distance r of 1.61 Å, while each O atom bridges between two Si atoms at the same distance. There is a small spread in the bond distance r, a small spread in the O-Si-O angles Φ, and a much larger spread in the Si-O-Si angles θ. These elements of SRO in $a\text{-}SiO_2$ can be visualized with the help of fig. 1, where one can see schematically the SRO around the Si and that around the O atomic species.

For the present definition of SRO, all neighbors of a species are treated as if they are spherically symmetric. This allows logical separation of SRO from intermediate range order (which recognizes that the neighbors are not spherically symmetric, and have orientations associated with them).

Intermediate Range Order

Intermediate range order (IRO) involves specification of relative atomic positions over several nn distances, given the SRO. Most fundamentally, it recognizes that nearest neighbors are not spherically symmetric. IRO may take the form of specification of the dihedral angles[12] δ, Δ shown in fig. 1 for the two corner sharing tetrahedra, distributions of the sizes of rings of completed bonds, properties of network connectivity[13] or some presently unformulated measure. There can be distributions of dihedral angles[12,14] and correlations [15] among dihedral angles if there are regular structures such as rings [16] embedded in the a-solid.

An example of IRO involving both correlated and uncorrelated dihedral angles is illustrated in fig. 2, which shows elements of the likely

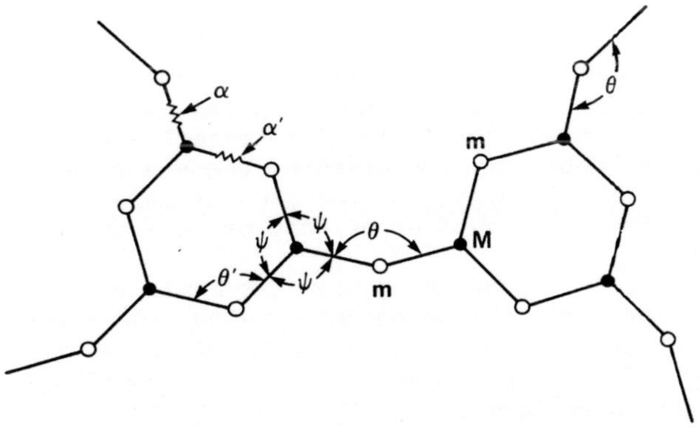

Fig. 2. Schematic (planarized) drawing of some elements of IRO in B_2O_3 glass, where • = boron and ○ = oxygen. The hexagonal B_3O_3 "boroxol" θ varies from site to site. The dihedral angles of the BO_3 units at each end of this bridging O atom are assumed random (the rings are generally not in the plane of the figure), while those within the rings all have the same value ($\delta = 0$), and are thus highly correlated.

amorphous microphases, island structures in thin films, hillocks, and other microscopic growth induced structures. These are identifiable non-crystalline elements of structure which are often called "extended" and are certainly not "point" in their nature. Nevertheless, they are usually seen only with the microscope and do not extend over the entire sample - and this raises the need for a fourth range of order.

Global Range Order

For a microscopically amorphous sample, neither crystalline nor morphological LRO extends over the entire sample or major portions of it. Crystalline LRO extending over an entire sample would produce a crystal, not a glass ! To cover this largest scale in amorphous solids we define a longest range of order, as follows.

Global range order (GRO) accounts for structural order which exists or is defined over macroscopic sample distances, often the entire sample. Specific examples are perfect crystallinity, macroscopic isotropy (often attributed to glasses), network connectivity, chemical order, chemical or structural homogeneity or heterogeneity. Idealized models are often assigned aspects of order on the global range for conceptual or mathematical simplicity. These global aspects of order may or may not be realized.

ELEMENTS OF ORDER IN THE ZACHARIASEN-WARREN MODEL FOR a-SiO_2

Elsewhere [1], I have briefly described five famous models for a-SiO_2 in terms of their SRO, IRO, LRO and GRO. These are : (1), the 1921 microcrystallite model of Lebedev[27] ; (2), the 1936 extension of this by Valenkov and Porai-Koshits[28] ; (3), the 1932 continuous random network (crn) model of Zachariasen[19] ; (4), the 1936 extension of this by Warren, et al.[20] ; and (5), the 1982 "paracrystallite" model of J.C. Phillips[6]. Strong arguments against all but the Zachariasen-Warren model were presented in ref. 1, many of which have previously been given by others. In this paper I shall therefore proceed by giving a more detailed characterization of the Zachariasen-Warren (ZW) model.

The ZW model is properly defined by the papers of Zachariasen[19], Warren, Krutter and Mornigstar[20], Warren[21] and Mozzi and Warren[22]. Although its overall features have been described in numerous textbooks[23-26] and articles, I shall here attempt to list exactly and comprehensively all elements of order that were specified by the above authors. The resultant list follows, in the language of the present paper and using the logical framework of the four ranges of order that we have defined.

The asterisks (*) mark the only statements which distinguish the ZW model from at least one of the perfect crystalline forms of SiO_2. In this list, distributions are continuous, not discrete (as they are in a crystal). Many of the statements were not mentioned by the original authors, but seem to be implicit in the present context.

The SRO of the Zachariasen-Warren Model

•Only Si and O atoms (no impurities).
•Only Si-O bonds (chemically ordered).
*A narrow unimodal distribution of bond lengths r, peaked at 1.61 Å.
•Each Si bonded almost tetrahedrally to four O atoms (4-coordinated Si).
*A narrow unimodal distribution of O-Si-O angles Φ, peaked at 109.5°.

•Each O bridges between two Si atoms (2-coordinated O).
*A broad unimodal distribution of Si-O-Si angles θ, peaked at θ = 144°.
*The r, θ, Φ are uncorrelated, amongst themselves and with each other.

The IRO of the Zachariasen-Warren Model

•The tetrahedra share corners only (no 2-rings).
*A random (flat) distribution of dihedral angles δ (or Δ).
*The δ (or Δ) are uncorrelated, with each other and with all r, θ, Φ.

The LRO of the Zachariasen-Warren Model

•There is no morphological LRO.
*There is no crystalline LRO (not microcrystalline).

The GRO of the Zachariasen-Warren Model

•The material is chemically ordered everywhere.
•The network is "continuous", in that there are no broken bonds.
•The network is topologically 4-connected everywhere.
•The structural parameters are homogeneous statistically (all ranges).
*The network is microscopically isotropic.
*The macroscopic density of the model is that of a real glass sample.

SOME ADVANTAGES AND SHORTCOMINGS OF THE ZW MODEL

A glance at the preceding list indicates that the ZW model : (1), makes many specifications of SRO ; (2), assumes several elements of GRO ; (3), specifies very little IRO, and (4) effectively states that LRO is merely an extension of the IRO into longer ranges, since there is no microcrystallinity and no morphological LRO. This observation suggests that efforts at improvement of the model might well concentrate on better determination and specification of IRO (and LRO) in a-SiO_2.

Some advantages of the ZW model are as follows. The several assumptions of GRO (such as statistical homogeneity), lead to mathematical advantages and simplifications (like use of the Bethe lattice). This is in contrast to the inhomogeneity of the microcrystalline models. The ZW model is a specific model, as we have defined the term, and is therefore subject to quantitative tests and subsequent improvement ; failures do not necessarily reject the conceptual crn model of Zachariasen that lies underneath. The ZW model fits diffraction data well[5] (although not perfectly) so that it is a promising starting point for improvements. The list of structural elements just presented reveals the model is well-defined for cataloging possible improvements.

Some obvious shortcomings of the ZW model are as follows. In real life, chemical order may occasionally be broken[29]. Point defects are known to exist in a-SiO_2 [30]. Structural parameters are not uncorrelated : r and θ are known from chemical theory to be correlated[31-32] ; furthermore, δ must be correlated with r, θ, Φ and other δ's in order for the rings of bonds to close on themselves[12,33].

Significant numbers of regular rings (planar 3-fold and puckered 4-fold) are now believed to exist in a-SiO_2 [34], and this implies special non-random values of δ. Furthermore, there are density fluctuations in real glass, hence some aspects of statistical inhomogeneity.

None of these shortcomings appears to be fatal for the ZW model in the case of a-SiO_2. It should be possible to account for them in perturbed forms of the model. A logically complete list of possible perturbations will be published elsewhere.

For some other glasses, elements of IRO and LRO not included in the ZW model may be too strongly involved to be viewed merely as a "perturbation". Examples may include the GeS_2 and $GeSe_2$ tetrahedral glasses studied by Boolchand and coworkers[35] ; these nominally SiO_2 like glasses appear to have large amounts of chemical disorder (wrong bonds) and special IRO. Other glasses are known to be microcrystalline or submicroscopically phase separated, and these heterogeneous structures are more than a perturbation away from the homogeneous ZW model.

SOME APPROXIMATE FORMS OF THE ZW MODEL

In the beginning of this paper, I defined a logical heirarchy of three kinds of models : conceptual, specific, and approximate. The ZW model is a specific model based upon the crn as its conceptual parent. Numerous specific models have been developed that can be considered as approximate forms of the ZW model. They involve simplifications designed to facilitate construction or calculations of properties of the ZW model of a glass. I shall briefly discuss six examples.

The Isolated Molecule Model

The isolated molecule model of Lucovsky and Martin[36] was introduced in an effort to provide for easy interpretation of the vibrational spectra of simple glasses like a-SiO_2. It is an extreme simplification, taking only the SRO depicted in fig. 1, and therefore treating the glass as if it were a collection of uncoupled SiO_4 and OSi_2 molecules. It has utility in those cases where some modes of vibration in the real glass are effectively confined to the structural units depicted and are therefore decoupled from vibrations of the neighboring units. When this model is unsucessful, the fault must first be sought in its excessive approximations, not with the ZW or crn models which it attempts to approximate.

The Bell and Dean Large Cluster Models

The "Bell and Dean" models for a-SiO_2 consist of ball and stick structures constructed by hand and containing ~1000 atoms[4]. They are intended as approximations to the ZW model, but contain some specifications not included in the ZW model. Because there are numerous completed rings of bonds in the structures, there are correlations among r, θ, ϕ, δ not specified in the ZW mode, etc. The model is "anthropomorphic" in that it takes the form unconsciously given to it by its builder : thus, no 3-rings were built-in, presumably because they put too much stress on the sticks representing bonds. Also, a large fraction of the atoms are on the surface and improperly bonded relative to a ZW model. In spite of these and other shortcomings, the model has several important successes in interpreting diffraction and vibrational data ; its failures are not necessarily to be attributed to the ZW or crn models, which it is intended to approximate.

The Large Amorphous Unit Cell Models

The problem of finite size and large numbers of surface atoms has been circumvented by many workers by building large amorphous structures (~3000 atoms) which are then extended periodically in 3-dimensions. Various techniques for doing this have been used[37,38]. The resultant isotropic crystals with large amorphous unit cells are then used to calculate physical properties. The models contain elements of IRO (rings, dihedral angles, correlations) not specified in the ZW model, but also of dubious physical origin, since the methods used to develop the unit cell structures are highly approximate or ad-hoc. For example, molecular dynamic generation of structures for a-B_2O_3 produces none of the 3-rings of which the real material is composed almost entirely[39]. This is most likely because molecular dynamic methods do not yet include the intrinsically 3-body forces that would model the chemical forces (which force special bond angles and cause 3-rings in real life). The models also contain unrealistic LRO, because of their long range periodicity.

The Sen and Thorpe Model

Another kind of approximation to the ZW model is that of Sen and Thorpe[40], which was developed to treat vibrational excitations of glasses and which has been interpreted and elaborated by others[41,42]. Their model is infinite, so has no trouble with surface atoms. It assumes the usual SRO but with simple specific values of r, θ, Φ, the same everywhere. The model makes no specifications regarding dihedral angles, or rings ; this is acceptable because the vibrations are treated in central forces only and in that case the results are completely independent of any possible values of δ. Again, shortcomings of this model do not necessarily undermine the ZW or crn models, but must first raise questions about the use of central forces only, and the non-specification of values of δ.

The Bethe Lattice Model

Still another approximation to the ZW model has been developed that is infinite, but has no rings of bonds and has an infinite surface ! This is the Bethe lattice[8], which has shown much promise as a mathematical simplification (no return paths, ie rings) enabling quantitative calculation of the electronic[43] and vibrational[9,44] excitations of a-SiO_2. The success of this apparently bizare approximation to the ZW network is illustrated by the results of vibrational calculations shown in figs. 3 and 4.

Figure 3 compares Bethe lattice calculations for a-SiO_2 with exact calculations[45] of the vibrational density of states of a Bell and Dean cluster, in (c). Panel (b) uses the most reasonable set of parameters for the Bethe lattice calculation and compares excellently with the Bell and Dean result in panel (c). This means that the Bethe lattice is capable of modeling the vibrational response of a large cluster model even though there are no completed rings or correlated angles in the Bethe lattice. It does not mean[46] that the authors[5] believe that SiO_2 has the structure of a Bethe lattice, nor would its failure necessarily invalidate the ZW or crn models.

Fig. 4 compares an experimental Raman spectrum of a-SiO_2 in (a) with a calculation in (b) of the Raman response of a Bethe lattice, chosen to have realistic structural parameters and force constants (adjusted to give an optimum fit)[9]. The agreement is remarkable, when it is realiszed that the major discrepancies are the sharp lines in experiment at $495 cm^{-1}$ and $606 cm^{-1}$, and that these are due[11,16] to regular rings in the real glass, rings that are non-existent in the Bethe lattice. Again, the authors[9] do not conclude that the structure of a-SiO_2 is a Bethe lattice, but rather that the Bethe lattice is an approximate form of the ZW model which enables

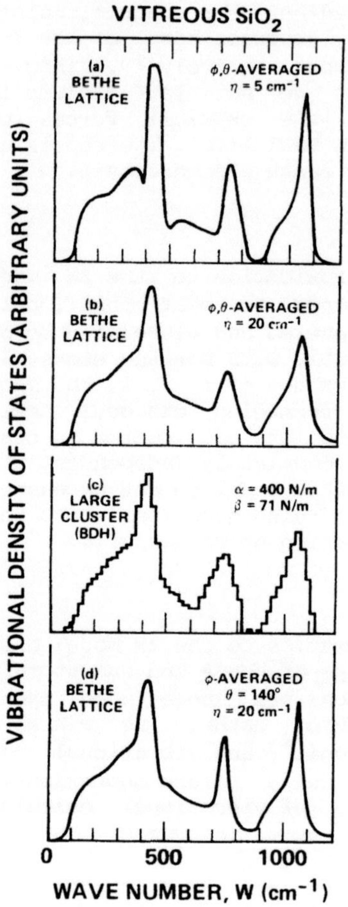

Fig. 3. Comparison of the density of states for a-SiO_2 obtained from a Bethe lattice structure with results obtained from numerical calculations on a hand-built large cluster model. (a) Averaged over a flat distribution of ϕ and a triangular distribution of θ. (b) Same as (a) but with a large imaginary part (η) added to the frequency. (c) Histogram obtained by Bell, Bird and Dean in ref. 45. (d) Theoretical result with θ fixed.

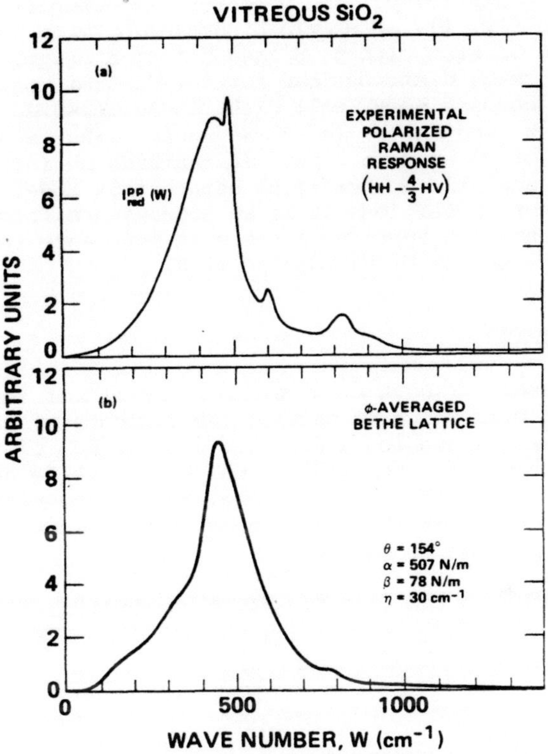

Fig. 4. Comparison between the experimental (a) and theoretical (b) polarized Raman spectra of a-SiO_2. The experimental curve is from ref. 47. The sharp peaks in (a) are not reproduced in (b) because they arise from defects in the real glass (ref. 11) which are not included in the Bethe lattice used to calculate (b).

reasonably accurate and economical calculations of the vibrational properties of a-SiO_2. This is supported by the agreement with the large cluster results.

The Oxygen Steric Hindrance Model

The final approximate form of the ZW model which I shall mention is the oxygen steric hindrance model, treated statistically by Thathachari and Tiller[33] and analytically by Galeener[12]. This structure is also infinite, has the usual "simplified" local order and considers neighboring dihedral angles δ and Δ explicity. The geometry is exemplified in fig. 1. On the assumption (in fig. 1) of steric hindrance between the oxygen atoms 4, 4' and 4'' on one Si atom with 5, 5' and 5'' on a neighboring Si atom, one is able to deduce exact distributions for the Si-O-Si angle θ and the dihedral angle δ that compare very well with those measured on the Bell and Dean ball and stick models[14]. In this sense, the model produces good distributions of θ and δ, yet it contains no information about rings. Because of its analytic mathematical nature this model may prove useful for calculating other properties. It is an approximate form of the ZW model but also goes beyond, in specifying a steric hindrance and thereby deducing a non-random distribution of dihedral angles.

CONCLUDING REMARKS

In this paper I have developed a framework of terms and defined four ranges of order with which to catalog the elements of order in an amorphous solid. This lexicon enabled a detailed description of the structure of the Zachariasen-Warren model for a-SiO_2, which will serve as the starting point for a future analysis of all possible improvements of the model. Six important approximations to the ZW model were discussed briefly, using the language developed in the paper.

I am grateful to the Colorado State University, for support of this work.

REFERENCES

1. F.L. Galeener, in "Proceedings of the Second International Conference on the Effects of Modes of Formation on the Structure of Glasses", edited by R.A. Weeks and D.L. Kinser Trans. Techn. Publications, Aedermansdorf, (1988).
2. F.L. Galeener in "Effects of Modes of Formation on the Structure of Glass", Eds. R.A. Weeks, D.L. Kinser and G. Kordas North Holland, Amsterdam, (1985) ; or, J. Non-Cryst. Solids, 71:373 (1985).
3. See numerous papers in proceedings (1) and (2).
4. R.J. Bell and P. Dean, Philos. Mag. 25:1381 (1972).
5. F.L. Galeener and A.C. Wright, Solid. State Commun. 57:677 (1986).
6. J.C. Phillips, Solid. State Phys. 37:93 (1982).
7. For example, the fact that the specific Bell and Dean model for a-SiO_2 predicts certain diffraction lines with incorrect widths may only point to incorrect details of the model built, not to incorrectness of the continuous random network conceptual model on which it is based.
8. M.F. Thorpe, in "Excitations in Disordered Systems", ed. M.F. Thorpe Plenum, New York, (1981), p. 85.
9. R.A. Barrio, F.L. Galeener and E. Martinez, Phys. Rev. B 31:7779 (1985).
10. B.E. Warren, "X-ray Diffraction" Addison-Wesley, Reading, (1969), Chapter 10.

11. F.L. Galeener, Solid State Commun. 44:1037 (1982). Also, F.L. Galeener, in "The Physics of Disordered Materials", eds. D. Adler, H. Fritzsche and S.R. Ovshinsky Plenum, New York, (1985), p. 159.
12. F.L. Galeener, Phil. Mag. B 51:L1 (1985).
13. M.F. Thorpe and F.L. Galeener, Phys. Rev. B 22:3078 (1980).
14. P.H. Gaskell and I.D. Tarrant, Phil. Mag. B 42:265 (1980).
15. F.L. Galeener, J. Non-Cryst. Solids 75:399 (1985).
16. F.L. Galeener, in "The Physics of Disordered Materials", eds. D. Adler, H. Fritzsche and S.R. Ovshinsky Plenum, New York, (1985).
17. D.L. Griscom in "Borate Glasses", edited by L.D. Pye, V.D. Frechette and N.J. Kreidl Plenum, New York, (1978), p. 11.
18. F.L. Galeener and M.F. Thorpe, Phys. Rev. B 28:5802 (1983).
19. W.H. Zachariasen, J. Am. Chem. Soc. 54:3841 (1932).
20. B.E. Warren, H. Krutter and O. Morningstar, J. Am. Ceram. Soc. 19:202 (1936).
21. B.E. Warren, J. Appl. Phys. 8:645 (1937).
22. R.L. Mozzi and B.E. Warren, J. Appl. Crystallogr. 2:164 (1969).
23. R.H. Doremus, Glass Science Wiley, New York (1973).
24. W.D. Kingery, H.K. Bowen and D.R. Uhlman, "Introduction to Ceramics, 2nd ed". Wiley, New York (1976).
25. R. Zallen, "The Physics of Amorphous Solids", Wiley, New York (1983).
26. S.R. Elliott, "Physics of Amorphous Materials" Longman, London (1984).
27. N. Lebedev, Proc. State Opt. Inst. Leningr. 2:10 (1921).
28. N. Valenkov and E.A. Porai-Koshits, Z. Kristallogr. A 95:195 (1936) ; also E.A. Porai-Koshits, J. Non-Cryst. Solids 73:79 (1985).
29. P. Boolchand, in "Defects in Glasses", Ref. 30, p. 57.
30. F.L. Galeener, D.L. Griscom and M.J. Weber, eds. "Defects in Glasses" Mater. Res. Soc. Proc. 61, PA, 1986).
31. M.D. Newton and G.V. Gibbs, Phys. Chem. Minerals 6:221 (1980). See also ref. 11.
32. A.K. Revesz and G.V. Gibbs, in "The Physics of MOS Insulators" edited by G. Lucovsky, S.T. Pantelides and F.L. Galeener Pergamon, New York, (1980), p. 92.
33. Y.T. Thathachari and W.A. Tiller, J. Appl. Phys. 53:8615 (1982).
34. F.L. Galeener, R.A. Barrio, E. Martinez and R.J. Elliot, Phys. Rev. Lett. 53:2429 (1984).
35. P. Boolchand, J. Grothaus, M. Tenhover, M.A. Hazle and R.K. Grasselli, Phys. Rev. B 33:5421 (1986).
36. G. Lucovsky and R.M. Martin, J. Non-Cryst. Solids 8-10:185 (1972).
37. F. Wooten, K. Winer and D. Weaire, Phys. Rev. Lett. 54:1392 (1985).
38. M.F. Thorpe and S.W. de Leeuw, Phys. Rev. B 33:8490 (1986).
39. T.F. Soules, J. Non-Cryst. Solids 49:29 (1982) ; see esp. p. 40, and earlier work on B_2O_3 by Soules.
40. P.N. Sen and M.F. Thorpe, Phys. Rev. B 15:4030 (1977).
41. F.L. Galeener, Phys. Rev. B 19:4292 (1979).
42. M.F. Thorpe and F.L. Galeener, Phys. Rev. B 22:3078 (1980).
43. R. Barrio, J. Taguena-Martinez, E. Martinez and F. Yndurain, J. Non-Cryst. Solids 72:181 (1985).
44. R.B. Laughlin and J.D. Joannopoulos, Phys. Rev. B 16:2942 (1977).
45. R.J. Bell, N.F. Bird and P. Dean, J. Phys. C 1:299 (1968).
46. J.C. Phillips, Phys. Rev. B 33:4443 (1986) ; also, Phys. Rev. B 32:5350 (1985).
47. F.L. Galeener, A.J. Leadbetter and M.W. Stringfellow, Phys. Rev. B 27:1052 (1983).

STRUCTURAL SIMILARITIES AND DISSIMILARITIES BETWEEN SiO_2 AND H_2O

Friedrich Liebau

Mineralogisches Institut
der Universität Kiel
Kiel, West Germany

INTRODUCTION

To date the atomistic structure of a crystalline substance can be determined with high accuracy. Due to lack of long-range order (distances r > 10Å) and the rather low degree of intermediate-range order (5Å ≲ r ≲10Å), the structure of amorphous substances can only be determined with considerably lower precision. For example, in crystalline SiO_2 the bond lengths d(Si-O) and the bond angles ∢ O-Si-O and ∢ Si-O-Si can be determined with standard deviations of approximately 0.003Å, 0.2° and 0.2° respectively, and the bond topology can be precisely given. In contrast, for amorphous silica corresponding values for the standard deviations of bond lengths and angles are considerably higher and only speculations can be made about the ratio between the numbers of rings containing 3, 4, 5, 6, etc.. Si atoms. Therefore, in order to derive reliable information on the intermediate range order of an amorphous phase this phase should be studied not only by diffraction (X-ray, neutron) and spectroscopic (IR, Raman, NMR etc.) methods but these results should also be compared in light of the structural knowledge of the crystalline phases of the same composition and, if possible, of crystal chemically related compounds. In the case of amorphous SiO_2, the structures of the crystalline silica polymorphs and of the various crystalline phases having H_2O frameworks should be considered.

CRYSTALLINE SILICA POLYMORPHS

To date (August 1987) more than 35 crystalline phases of silica are known. Of these only high-temperature (hT) cristobalite, hT-tridymite*, low-temperature (lT) quartz, hT-quartz, coesite, and stishovite have a temperature-pressure field of thermodynamic stability.

---

* Tridymite always contains small amounts of Al^{3+} plus corresponding amounts of alkali and/or alkaline earth cations which replace Si^{4+}. They seem to be essential to stabilize tridymite which, therefore, is to be considered as silica only if the presence of these cations is ignored.

Within the more than 35 crystalline silica polymorphs known, there are several groups with topologically identical structures ; for example, 1T-, hT- and incommensurate [1] quartz form one such group, 1T- and hT-cristobalite another, and the various tridymites a third. Not distinguishing between the members of these groups, there are to date 25 topologically non-identical crystalline SiO_2 polymorphs known. Sixteen of them have been discovered within the last ten years. In Table 1, the 25 topologically non-identical crystalline polymorphs are listed together with their framework densities d_f (number of SiO_2 per 1000Å^3). They cover the range between 43.0 SiO_2/1000Å^3 (stishovite) and 17.4 SiO_2/1000Å^3 (silica-sodalite).

Table 1. Survey of crystalline phases which have topologically distinct SiO_2 frameworks. D_{pore} : pore dimensionality, Abbr. : abbreviated name, d_f : framework density in n(SiO_2)/1000Å^3, M^f : guest molecule located in a cage that has $f = \Sigma m_i$ faces. Reference is made to the most recent publication only.

		Name	Abbr. (D_{pore})	d_f	Formula, unit cell content	Ref.
Pyknosils		Stishovite	Sti	43.0	SiO_2	2
		SiO_2(Fe_2N-type)	---	42.8	SiO_2	3
		Coesite	Coe	29.3	SiO_2	4
		Quartz	Qu	26.6	SiO_2	5
		Moganite	Mog	26.3	SiO_2	6
		Keatite	Kea	25.1	SiO_2	7
		Cristobalite	Cr	23.2	SiO_2	8
		Tridymite	Tr	22.9	SiO_2	9
Porosils	Clathrasils	Nonasils	Non	19.2	$88SiO_2 \cdot 8M^8 \cdot 8M^9 \cdot 4M^{20}$	10
		Melanophlogites	Mel	19.0	$46SiO_2 \cdot 2M^{12} \cdot 6M^{14}$	11
		Dodecasils 3C (Silica-ZSM-39)	D3C	18.6	$136SiO_2 \cdot 16M^{15} \cdot 8M^{16}$	12
		Dodecasil 1H	D1H	18.5	$34SiO_2 \cdot 3M^{12} \cdot 2M'^{12} \cdot 1M^{20}$	13
		SIGMA-2	---	17.8	$64SiO_2 \cdot 8M^9 \cdot 4M^{20}$	14
		Silica-sodalites	Sod	17.4	$12SiO_2 \cdot 2M^{14}$	15
	Zeosils	Deca-dodecasils 3R	DD3R (3)	17.6	$120SiO_2 \cdot 6M^{10} \cdot 9M^{12} \cdot 6M^{19}$	16
		Deca-dodecasils 3H	DD3H (3)	17.6	$120SiO_2 \cdot 6M^{10} \cdot 9M^{12} \cdot 1M^{15} \cdot 4M^{19} \cdot 1M^{23}$	16
		Silica-ZSM-23	(1)	20.0	$24SiO_2 \cdot (CH_3)_2N(CH_2)_7N(CH_3)_2$	17
		Silica-ZSM-48	(1)	19.9	$48SiO_2 \cdot H_2N(CH_2)_8NH_2$	18
		Silica-ZSM-22	(1)	19.7	$24SiO_2 \cdot HN(C_2H_5)_2$	19
		Silica-ferrierite	(2)	19.3	$36SiO_2 \cdot 2H_2N(CH_2)_2NH_2$ (en)	20
		Silica-ZSM-12	(1)	18.5	$28SiO_2 \cdot N(C_2H_5)_3$	21
		Silica-ZSM-50	(1)	18.2	$112SiO_2 \cdot n[(CH_3)_3N(CH_2)_6N(CH_3)_3](OH)_2$	22
		Silicalite II (Silica-ZSM-11)	(3)	17.9	$96SiO_2 \cdot n[N(C_4H_9)_4]OH$	23
		Silicalite I (Silica-ZSM-5)	(3)	17.8	$96SiO_2 \cdot 4[N(C_3H_7)_4]F$	24
Other		Fibrous silica		19.6	SiO_2	25

Syntheses of crystalline SiO_2 polymorphs

All known crystalline silica polymorphs can be crystallized from aqueous solutions of silicic acids with the exception of fibrous silica which is extremely hygroscopic and moganite which has not yet been synthesized. If the solution does not contain suitable "guest molecules" in sufficiently high concentrations, monomeric silicic acid, $Si(OH)_4$, condenses to form dense SiO_2 frameworks. However, in the presence of suitable guest molecules the silicic acid monomers have to condense around these molecules enclosing them in pores of the host framework. It is obvious that the size and shape of these pores is controlled by the size and shape of the guest molecules[26,27].

The guest molecules can be expelled from the host framework by thermal treatment without changing the framework topology. The degassed products are crystalline silica polymorphs of remarkable stability. Although they are not thermo-dynamically stable, most of them can be heated to more than 1250 K for weeks without transformation into a topologically different phase.

Silica polymorphs with frameworks too dense to enclose guest molecules that are larger than He and Ne are called pyknosils. Phases with SiO_2 frameworks which have pores wide enough to accommodate larger guest molecules are called porosils independent of whether their pores are filled or empty[28].

Structures of dense SiO_2 polymorphs (pyknosils)

Two pyknosils in which the coordination number of silicon towards oxygen CN(Si/O), is 6 are known. For these the term pyknosil-[6] is proposed, in contrast to the pyknosil-[4] which then contain four-coordinated silicon.

Pyknosils with six-coordinated silicon (pyknosils-[6])

Stishovite[2], the densest silica polymorph known, is isotypic with rutile, TiO_2. Its structure can be described as a distorted hexagonally close-packed arrangement of oxygen atoms in which 50 percent of the octahedral interstices are occupied by silicon atoms in an ordered way. It can be deduced from fig. 1, that the (SiO_6) octahedra share edges and corners.

SiO_2 (Fe_2N-type)[3] : Another very high pressure silica polymorph has recently been synthesized. Like stishovite, it also contains six-coordinated silicon. Its structure may be described as having a nearly undistorted hexagonally close-packed arrangement of oxygen atoms ($c/a = 1.606$ compared with the ideal value 1.633). Silicon atoms are statistically distributed over 50 percent of the octahedral interstices. This structure can either be derived from that of niccolite, NiAs, in which all the octahedral interstices of the hexagonally close-packed anion arrangement are occupied by cations, or from Fe_2N by interchanging cations and anions.

In both, stishovite and Fe_2N-type SiO_2 polymorph, each silicon atom has CN(Si/O) = 6, however, there is a difference in their coordination numbers of oxygen towards silicon. In stishovite each oxygen atom has (CN(O/Si) = 3 ; in the Fe_2N-type polymorph of SiO_2, in contrast, CN(O/Si) = 3 on average only due to the statistical silicon atom distribution.

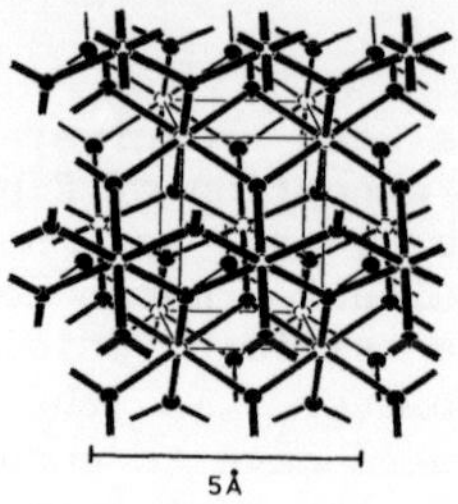

Fig. 1. The crystal structure of stishovite.
o : Silicon atoms,
o : oxygen atoms.

The very high framework density of the pyknosils-[6], when compared to that of the other silica phases, is also due to factors other than the higher values of the coordination numbers CN(Si/O) and CN(O/Si). More important is the coordination of oxygen by other oxygen atoms : The two above examples have CN(O/O) = 12 due to the (more or less distorted) close-packed oxygen arrangement whereas CN(O/O) is lower than 12 in the pyknosils-[4].

Pyknosils with four-coordinated silicon (pyknosils-[4]).

With the exception of the pyknosils-[6] in the remaining crystalline SiO_2 structures, pyknosils-[4] and porosils, silicon is tetrahedrally coordinated by oxygen and each oxygen atom is bonded to two silicon atoms : CN(Si/O) = 4, CN(O/Si) = 2. With the exception of fibrous silica, these structures contain only (SiO_4) tetrahedron thus forming 3-dimensional 4-connected tetrahedral frameworks[29].

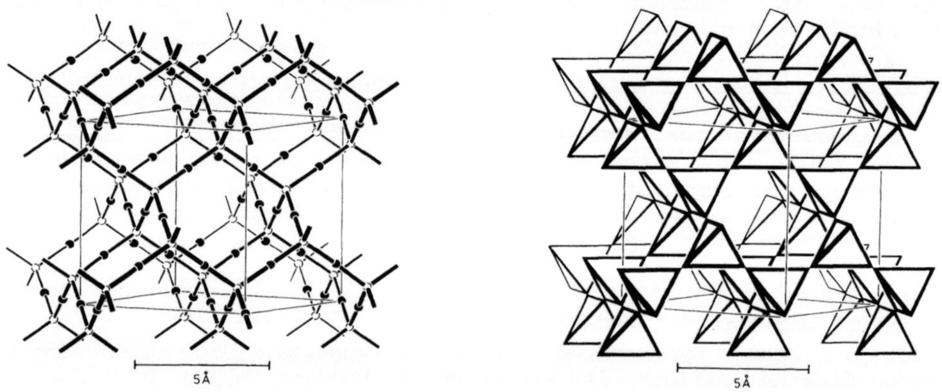

Fig.2. Schematic drawings of the crystal structures of cristobalite and ice I_c (a) showing the atoms and bonds (o : Si in cristobalite, O in ice I_c, o : O in cristobalite, H in ice I_c) ; (b) showing the linking between (SiO_4) and (OH_4) tetrahedra respectively.

In Fig. 2 the framework of cristobalite, a typical example of a pyknosil-[4], is illustrated in two ways, once by linking Si and O atoms and once by representing each (SiO_4) group by a solid tetrahedron. In the first presentation the oxygen atoms are shown as black dots at the midpoints between the silicon atoms although, in the actual structure, they deviate from the lines linking two Si atoms. The latter method (fig.2b) is more suitable to depict portions of a structure such as rings or chains from a thicker slice of the structure. This method is used to show the structure of the high-pressure phase coesite, another dense silica polymorph (fig.3).

In the pyknosil-[4] structures the voids between the oxygen atoms which are not occupied by Si are either too small to enclose other atoms or ions or they are just wide enough to accommodate small species such as He or Li^+, Mg^{2+} (in the quartz and keatite structure) or Na^+, K^+, Ca^{2+} (in the tridymite structure) to compensate for the charge deficit produced by some Al^{3+}/Si^{4+} replacement. The pyknosils-[4] have framework densities $d_f \gtrsim 21$ SiO_2/1000Å^3.

Structures of porous SiO_2 polymorphs (porosils)

As the framework density of the silica polymorphs decreases beyond the value of 21 SiO_2/1000Å^3, the voids between the (SiO_4) tetrahedra become large enough to accommodate large atoms (Kr, Xe) or polyatomic molecules.

In some porous SiO_2 polymorphs the voids are closed pores, that is, they have only finite extension. In others the voids are channel-like open pores of infinite extension. Porosils with closed pores are called clathrasils, those with open pores zeosils[28]. Molecules located in closed pores are trapped as in cages. Molecules located in open pores have a higher degree of freedom to migrate within the channels. The degree of mobility depends on size and shape of both the molecules and the channels.

Table 2 gives a selection of molecules successfully incorporated in the various cages of clathrasils ; some typical guest molecules of zeosils are given in Table 1. The largest of these are 1-aminomethyladamantane, $C_{11}NH_{19}$, and 1, 7-bis (dimethylamino)-heptane, $C_{11}N_2H_{28}$, respectively.

Clathrasils. The cage-like pores found in seven of the eight known clathrasil families (Table 1) are schematically shown in fig. 4. Here silicon atoms are located at the corners of the polyhedra ; oxygen atoms bridging two Si atoms are near the midpoints of the lines linking them. The resulting polyhedra are circumscribed by polygons which represent rings

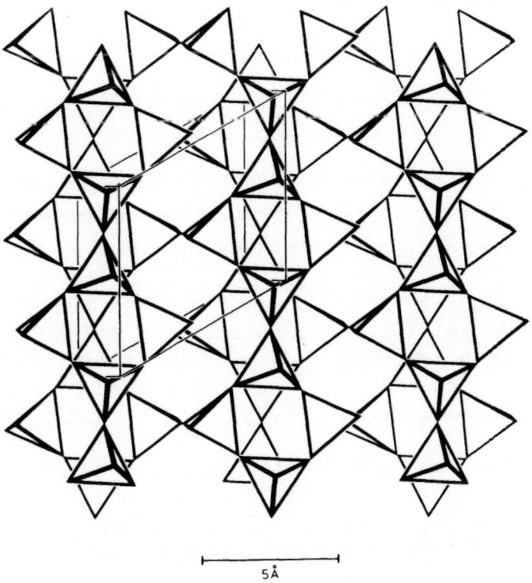

Fig.3. The crystal structure of coesite.

Table 2. A selection of guest molecules M^f which have successfully been enclathrated into the various clathrasil cages with $f = \Sigma m_i$.

Clathrasil	Cage $[n_i^{m_i}]$	Guest molecules M^f
Nonasils	$[5^4 6^4]$, $[4^1 5^8]$	—
	$[5^8 6^{12}]$	$CH_3CH(NH_2)C_2H_5$, 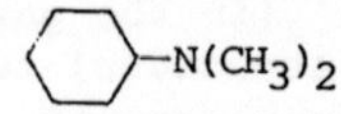
Melanophlogites	$[5^{12}]$	CH_4, N_2, Kr
	$[5^{12} 6^2]$	N_2, Kr, Xe, CO_2, CH_3NH_2
Dodecasils 3C	$[5^{12}]$	N_2
	$[5^{12} 6^4]$	SF_6, $N(CH_3)_3$, 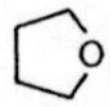, , 
Dodecasils 1H	$[5^{12}]$, $[4^3 5^6 6^3]$	N_2
	$[5^{12} 6^8]$	$(C_2H_5)_2NH$, 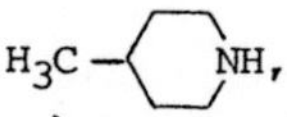, 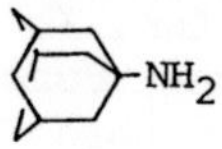, 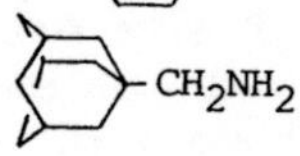
SIGMA-2	$[4^3 5^6]$	—
	$[5^{12} 6^8]$	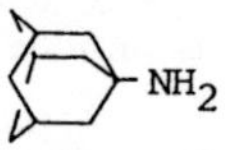
Silica-sodalite	$[4^6 6^8]$	$HOCH_2CH_2OH$
Deca-dodecasil 3R	$[4^3 5^6 6^1]$, $[5^{12}]$	?
	$[4^3 5^{12} 6^1 8^3]$	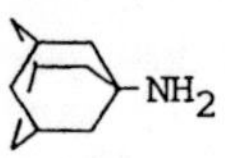
Deca-dodecasil 3H	$[4^3 5^6 6^1]$, $[5^{12}]$	?
	$[4^6 5^6 8^3]$	?
	$[4^3 5^{12} 6^1 8^3]$, $[5^{18} 6^2 8^3]$	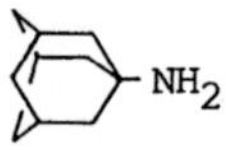

containing n silicon atoms. The polyhedra are described as $(n_i^{m_i})$ where m_i is the number of faces (rings) which have n_i corners (Si atoms).

The polyhedral pores share corners, edges and/or faces thus forming 3-dimensional frameworks. fig. 5 represents the melanophlogite framework as it is obtained from a structure analysis[11]. It contains $[5^{12}]$ and $[5^{12}6^2]$ polyhedra. For comparison, in fig. 6 the framework structure of the nonasil family is shown to be built from $[5^4T^4]$, $[4^1 5^8]$, and $[5^8 6^{12}]$ cages[10].

The frameworks of the typical clathrasils known (Table 1) contain only cages which are circumscribed by small rings (n = 4, 5, 6). The frameworks of the deca-dodecasils[16], however, contain some polyhedra with a small number of octagons (n = 8). The free aperture of the octagons is too small to let larger enclosed guest molecules pass, it is, however, large enough to let smaller guests pass. Therefore, with regard to larger guest molecules, the deca-dodecasils are considered as clathrasils, however, with regard to smaller guests that are located in their large polyhedra they are considered as zeosils.

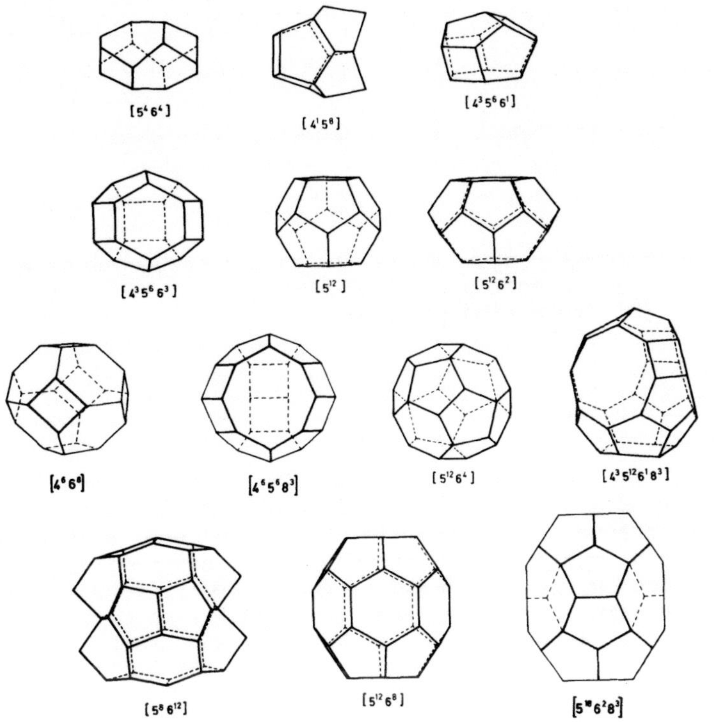

Fig. 4. Schematic drawings of the polyhedral pores (cages) found in clathrasils. (For designation of the cages compare with Table 2.)

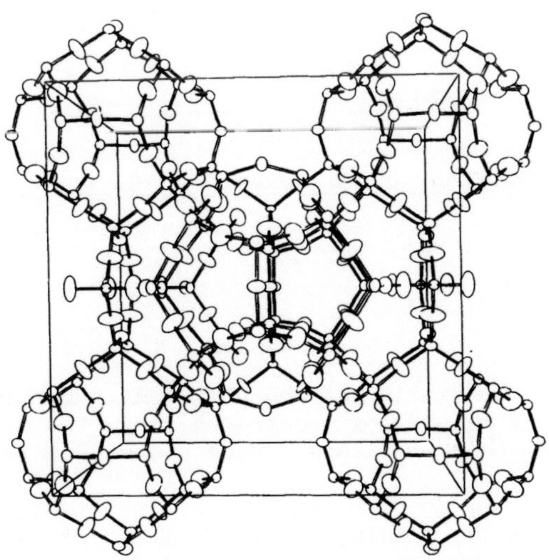

Fig. 5. The structure of the silica framework of the clathrasil melanophlogite. The atoms are represented by their thermal ellipsoids derived from X-ray diffraction data[11].

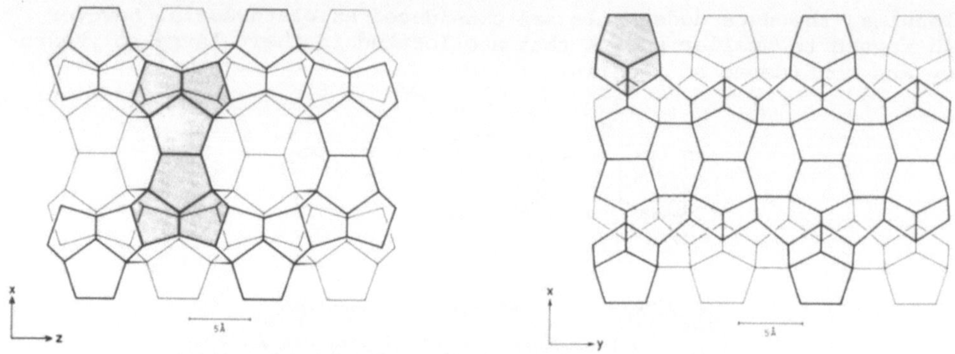

Fig. 6. Schematic drawings of the silica framework of nonasils.
(a) Projection parallel [010],
(b) Projection parallel [001].

Zeosils. If polyhedra share at least two rings with n ⩾ 8 with adjacent polyhedra, channel-like pores can be formed like the one shown in fig.7. Such channels of infinite extension are either separated or intersect each other. The corresponding channel systems have pore dimensionalities D_{pore} = 1, 2 and 3, respectively, compared with D_{pore} = 0 for the cage-like pores of the clathrasils. In fig. 8 the tetrahedral framework of silica-ferrierite is shown as an example. It contains two kinds of channels, one runs perpendicular to the plane of projection, the

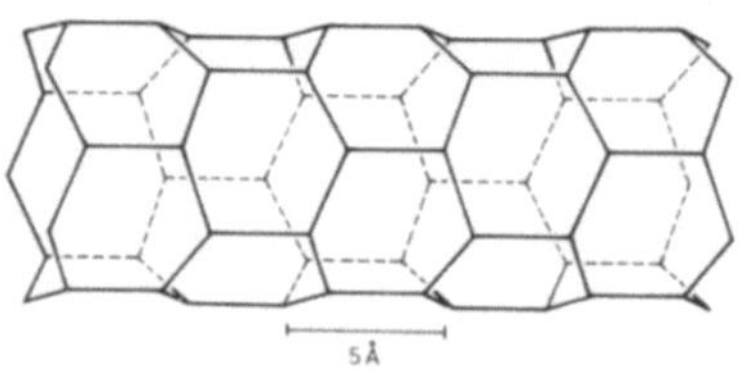

Fig. 7. Schematic diagram of a channel-like pore found in the zeosil silica-ZSM-48.

other parallel to the a-axis of the unit cell. Channels of the two kinds intersect each other by sharing 8-membered rings two of which are enhanced in the figure by thicker lines, thus forming a 2-dimensional channel system.

The zeosils can be considered to be aluminum-free endmembers of the zeolites. A survey of the presently known zeosils is given in Table 1.

Fibrous silica. Another way to achieve a low-density silica polymorph is observed in fibrous silica[25]. It contains chains of edge-sharing [SiO_4] tetrahedra which are held together by weak Van der Waals forces (fig. 9).

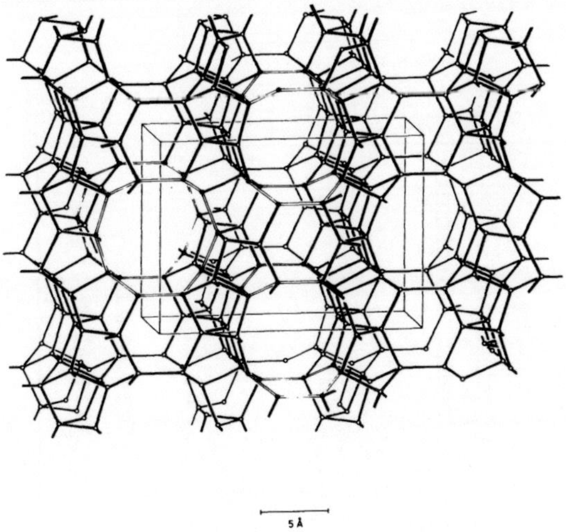

Fig. 8. Schematic representation of the tetrahedral framework of the zeosil silica-ferrierite.

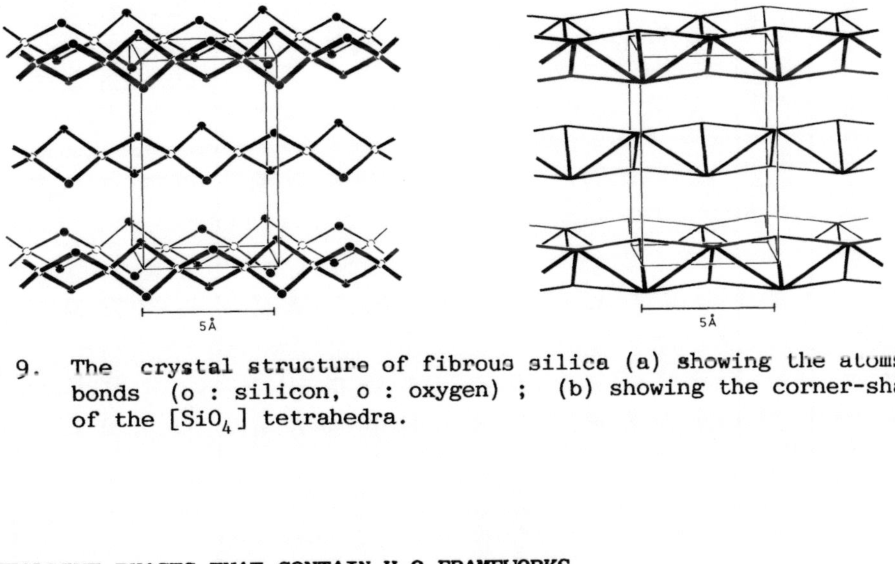

Fig. 9. The crystal structure of fibrous silica (a) showing the atoms and bonds (o : silicon, o : oxygen) ; (b) showing the corner-sharing of the $[SiO_4]$ tetrahedra.

CRYSTALLINE PHASES THAT CONTAIN H_2O FRAMEWORKS

Structural relationships between SiO_2 and H_2O

To date 17 structure types are known which contain 3-dimensional 4-connected frameworks of water molecules (Tab.3). Eight of them are isotypic with silica polymorphs, several others have structural relationships with silica polymorphs. These structural relationships are based on the fact that each water oxygen atom is linked to four others by hydrogen bonds O-H-O in much the same way as each silicon atom is linked to four others by Si-O-Si bonds. The $[SiO_4]$ tetrahedra in the silica frameworks correspond to $[OH_4]$ tetrahedra in water frameworks.

Table 3 Survey of crystalline phases which contain 3-dimensional 4-connected H_2O frameworks. d_f : framework density in $n(H_2O)/1000Å^3$, O : ordered distribution, D : disordered distribution, M^f : guest molecule located in a cage that has $f = \Sigma m_i$ faces.

Class	Order	Phase	d_f	Proton order	Formula, unit cell content	Ref.	Silica analogue
Pyknohydrates	Isoclathrates	Ice X	80	O	H_2O	30	Interpenetrating cristobalite
		Ice VIII	53.6	O	H_2O	31	
		Ice VII	53.0	D	H_2O	32, 33	
		Ice VI	45.9	D	H_2O	31	interp. edingtonite
	autoclathrates	Ice IV	42.6	D	H_2O	34	—
		Ice V	41.2	D	H_2O	35	—
	non-clathrates	Ice II	39.5	O	H_2O	36, 37	—
		Ice IX	39.1	O	H_2O	37	Keatite
		Ice III	38.8	D	H_2O	38	
		Ice I_c	31.2	D	H_2O	39	Cristobalite
		Ice I_h	31.2	D	H_2O	40	Tridymite
Porohydrates		Type I	27.0	D	$46H_2O \cdot 2M^{12} \cdot 6M^{14}$	41	Melanophlogites
		Type II	26.8	D	$136H_2O \cdot 16M^{12} \cdot 8M^{16}$	41	Dodecasils 3C
		Type VII	26.5	D	$12H_2O \cdot 2M^{14}$	41	Silica-sodalites
		Type H	25.9	D	$34H_2O \cdot 3M^{12} \cdot 2M'^{12} \cdot 1M^{20}$	42	Dodecasils 1H
		Type III	24.9	D	$172H_2O \cdot 10M^{12} \cdot 16M^{14} \cdot 4M^{15}$	41	—
		Type VI	23.4	D	$156H_2O \cdot 12M^{8} \cdot 16M^{17}$	41	—

With the exception of the recently discovered high-pressure phase ice X[30] all other of these water frameworks contain asymmetric hydrogen bonds in which a hydrogen atom occupies one out of two energetically equivalent positions between two oxygen atoms. The distribution of the hydrogen atoms over the energetically equivalent positions can either be ordered or disordered. In ice X, in contrast, the hydrogen bond is symmetrical with the hydrogen atom midway between the two oxygen atoms.

In Table 4 the phases which contain H_2O frameworks are classified on the basis of the degree of porosity (framework density) and the pore dimensionality, which is analogous to the classification of the corresponding silica phases. It can be seen that no pyknohydrates-[6] and zeohydrates are known which would be the H_2 analogues to the pyknosils-[6] and the zeosils respectively.

Phases which differ only in the symmetry of the hydrogen bond and/or in the distribution of hydrogen atoms, such as ices III and IX or ices VII, VIII and X (Table 3) have topologically identical frameworks. Without distinguishing between such phases, there are to date (August 1987) phases with 14 topologically non-identical 4-connected H_2O frameworks known. They are equivalent to the 25 topologically non-identical SiO_2 frameworks listed in Table 1.

By analogy to the silica polymorphs, one can distinguish between ices (which may systematically be called pyknohydrates) and porohydrates (Table 4). Whereas clathrate-hydrates, the water analogues of the clathrasils, have been known for a long time no zeohydrate has been reported which would be the analogue to zeosils and zeolites.

In contrast to the porosils which can be degassed without breaking the Si-O-Si bonds, the water frameworks of the clathrate hydrates are disrupted when the guest molecules are expelled. This is due to the weakness of the hydrogen bond in comparison with the stronger Si-O bond.

In the following section the structures of crystalline H_2O frameworks are treated in the order of increasing framework density.

Structures of clathrate hydrates

The known clathrate hydrates[41] that have 3-dimensional 4-connected water frameworks are listed in Table 3, which also gives their silica analogues. According to this table the type I clathrate hydrates, for instance, are isotypic with the melanophlogites (fig. 5). No silica analogues to the type III and VI clathrate hydrates are known. In addition, no water analogues to the nonasils, SIGMA-2 and the deca-dodecasils have been observed.

Structures of pyknohydrates (ice polymorphs)

Non-clathrates. From the survey of phases given in Table 3 it follows that the diagrams shown in fig. 2 equally represent both the structure of cristobalite and that of cubic ice I_c. In the same way, hexagonal ice I_h is isotypic with tridymite, and the ices III and IX with keatite. No silica analogue to ice II has been reported.

The structures of the ice phases I_h, I_c, II, III, and IX, each consist of one "normal" 3-dimensional 4-connected water framework. To distinguish them from the denser ice polymorphs described in the next two paragraphs the term non-clathrates has been introduced[43].

TABLE 4 Classification, based on porosity and pore dimensionality, of crystalline phases which have SiO_2 and H_2O frameworks. Within each group the known phases are arranged in the order of decreasing framework density.

a) Ice VI may alternatively be considered as a zeohydrate with two identical interpenetrating zeohydrate frameworks.

SiO_2				H_2O			
P y k n o s i l s				P y k n o h y d r a t e s (ices)			
Pyknosils-[6]	Pyknosils-[4]			Pyknohydrates-[6]	Pyknohydrates-[4]		
	Non-clathr.	Auto-clathr.	Iso-clathr.		Non-clathr.	Auto-clathr.	Iso-clathr.
stishovite,	Coe, Qu, Mog, Kea, Cr, TR	unknown	unknown	unknown	II, IX, III, I_c, I_h	IV, V	X, VIII, VII, VI[a]
P o r o s i l s				P o r o h y d r a t e s			
Clathrasils (Heteroclathr.)	Zeosils			Clathrate hydrates (Heteroclathrates)	Zeohydrates		
	Zeosils-1D	Zeosils-2D	Zeosils-3D				
Non, Mel, D3C, D1H, SIGMA-2, Sod	silica-ZSM-23, -48, -22, -12, -50	silica-ferrierite	silica-ZSM-11, -5	CI, CVII, CII, H, CIII, CVI	unknown[a]		
D D 3 R D D 3 H							

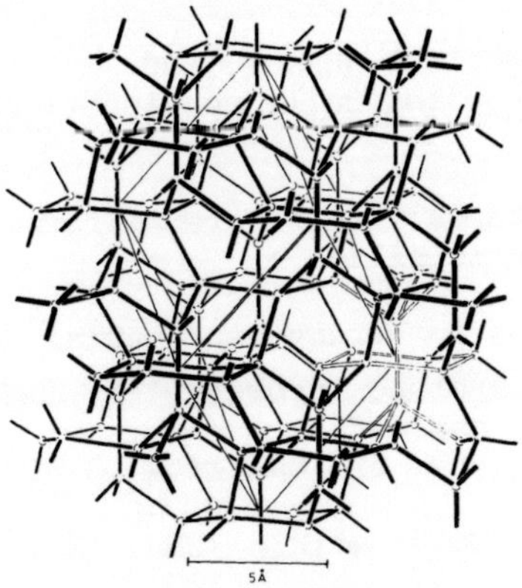

Fig. 10. Schematic representation of the tetrahedral framework of autoclathrate ice IV.

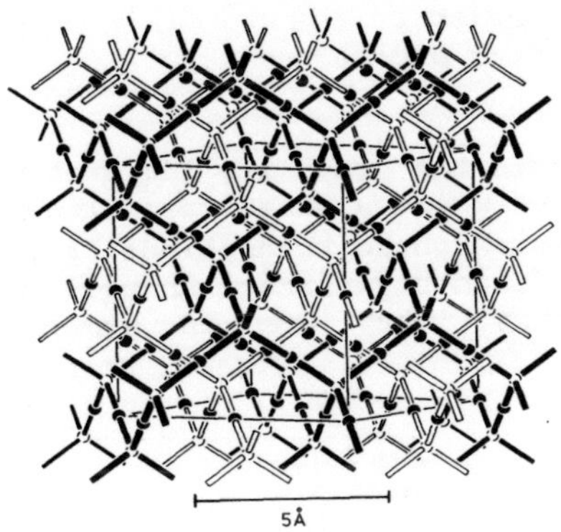

Fig. 11. Schematic representation of the tetrahedral frameworks of the isoclathrates ice VII, VIII, and probably X.

Autoclathrates. The structures of the somewhat denser ice polymorphs IV and V each consist also of one 3-dimensional 4-connected H_2 framework. However, in contrast to the "normal" frameworks of the non-clathrates, autoclathrates have frameworks which contain O-H-O bonds which interpenetrate rings of the same framework. In ice IV the interpenetrated rings are formed by six water molecules (fig. 10), in ice V by eight H_2O.

Isoclathrates. The densest ice polymorphs have structures in which two identical water frameworks interpenetrate each other. There are no hydrogen bonds between the two interpenetrating frameworks. Fig. 11 demonstrates that in the topologically identical polymorphs ice VII, VIII, and probably X, two ice I_h frameworks are interpenetrating each other. The two identical interpenetrating frameworks of ice VI have the same topology as the zeolite edingtonite although no zeohydrate (Table 4) with non-interpenetrating framework has been discovered. The term isoclathrates has been introduced[43] for phases with interpenetrating identical frameworks.

DENSIFICATION OF SiO_2

Structure changes that are accompanied by volume changes

Increase of pressure and decrease of temperature cause changes of structure that aim to reduce the mole volume of a substance. Such structure changes take place

(a) within the stability field of a phase,
(b) by phase transition between topologically identical phases, and
(c) by reconstructive phase transition between topologically non-identical phases.

As in other polyhedral framework structures[44] there are different mechanisms to bring about these structure changes in silica :

(i) rotation of (SiO_4) tetrahedra away from an arrangement having or approaching maximum topological symmetry (aristotype symmetry[45]). Such rotation is accompanied by a reduction of Si-O-Si angles* ;
(ii) distortion of (SiO_4) tetrahedra by way of changing O-Si-O angles* ; and
(iii) reduction of the mean Si-O distances* within the (SiO_4) tetrahedra.

The energy change required to achieve a volume change of, for example, 3 % by one of these mechanisms increases in the order :
change of Si-O-Si angles —— change of O-Si-O angles ——
—— change of Si-O distances.

As a consequence, structure changes within the stability field of a phase can be brought about by mechanism (i) alone whereas phase transitions between topologically identical phases (displacive phase transitions[44]) in general involve both mechanisms (i) and (ii). Since significant changes of Si-O distances require much higher energies, they are in general

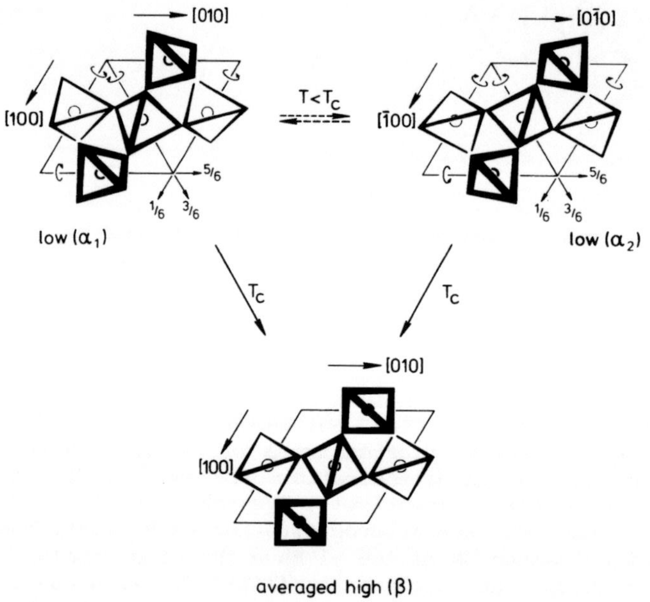

Fig. 12. Structure changes during the phase transition from low-temperature to high-temperature quartz[47].

*Si-O distances and Si-O-Si and O-Si-O angles discussed in this review are distances and angles between the mean positions of the atoms as determined by diffraction methods, averages over space and time. These distances and angles have to be corrected for thermal and, if necessary, static disorder effects in order to obtain true bond angles and bond lengths[46].

accompanied by breaking of Si-O bonds, for example, by change of the bond topology (reconstructive phase transition[44]).

Structure changes without bond breaking

Quartz is a good example to demonstrate the structural changes which do not involve breaking of Si-O bonds. At room temperature and 0.1 MPa hydrostatic pressure, the (SiO_4) tetrahedra have within experimental errors the symmetry 222. They are rotated away from the aristotype orientation of ideal hT-quartz by an angle $|\varphi| = 16°$ (fig. 12). As temperature increases the rotation angle $|\varphi|$ approaches 0°. In contrast, increasing the pressure gives rise to further increase of $|\varphi|$ which reaches 26° at 10.2 GPa and room temperature. This change of the angle of tetrahedron rotation from about 0° at 875 K and 0.1 MPa to 26° at room temperature and 10.2 GPa is accompanied by a decrease of the Si-O-Si angle from 153° to 130° and by a volume change of 19 %[48].

In quartz the transition from the high-temperature β phase to the low-temperature α phase on cooling does not take place in one step but an incommensurate phase exists in a small temperature range of ca 1.4 K around 846 K.[1]

An even larger change of the Si-O-Si angle than in quartz is observed in tridymite. In hexagonal hT-tridymite, which is stable above 653 K at normal pressure (0.1 MPa), the Si-O-Si angles are 180° (averaged over space and time)[49]. As temperature is lowered, pressure is increased and/or other experimental parameters are changed, there is a monotonic decrease of the the mean Si-O-Si angle. This structure change is not continuous but interrupted by successive displacive phase transitions from the phase with the maximum topological symmetry $P6_3/mmc$ to phases with successively lower subgroup symmetries[50]. For a natural 1T-tridymite, for example, symmetry P1 and a mean value 148° of the Si-O-Si angle has been found[9].

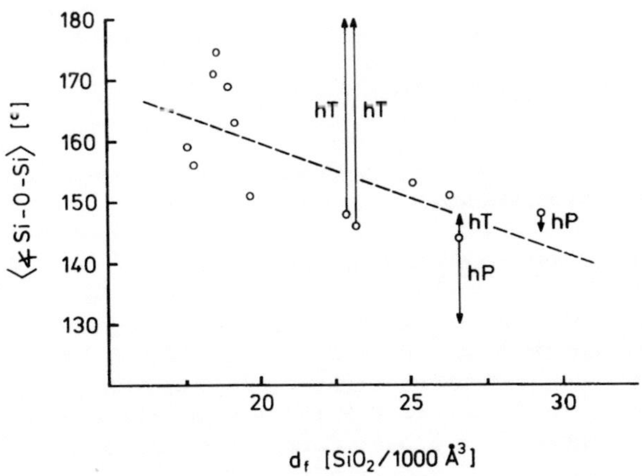

Fig. 13. Correlation between framework density, d_f, and mean Si-O-Si angles of crystalline silica polymorphs. o: mean angles at room temperatures. The arrows indicate the ranges of mean Si-O-Si angles at elevated temperatures (hT) and elevated pressures (hT) respectively.

A similar decrease of the mean Si-O-Si angle from 180° to 146° has been observed for cristobalite.

The maximum densification which can be achieved in silica by this mechanism of tetrahedron rotation is about 6 % for a given framework topology.

Structure changes with bond breaking

Higher degrees of densification require changes of framework topology. Fig. 13 and Table 5 suggest that there is a slight tendency of the mean Si-O-Si angles at ambient temperature and pressure of the various silica polymorphs to decrease as the framework density increases. Although the correlation is weak it parallels the strong correlation between Si-O-Si angle and pressure of topologically identical silica polymorphs.

Table 5 indicates an additional interesting structural change with densification. At ambient pressure silica phases are stable which have tetrahedral frameworks whose smallest rings contain six silicon atoms (quartz, tridymite, cristobalite). The minimum ring size of the framework is 5 for keatite which is synthesized under moderate hydrothermal conditions ; it decreases further to 4 in coesite which is stable above about 2.0 GPa. (The pressure/temperature conditions during the formation of natural moganite are unknown the phase has not yet been synthesized).

A similar decrease of the minimum ring size is to be found in the porosil frameworks (Table 5). As the framework density of the clathrasils decreases below that of tridymite, the minimum ring size decreases to 5 in melanophlogite and dodecasils 3C and to 4 in the other clathrasils. This effect can be explained by an increase of internal pressure that is exerted on the silica host framework by the guest molecules which are entrapped in the cages during synthesis. The corresponding effect observed in zeosils seems to confirm this explanation.

Further reduction of the minimum ring size to 3 at even higher external or internal pressure has not been observed for tetrahedral silica frameworks. Instead, at pressures exceeding 8.2 GPa at room temperature, further densification takes place by way of increasing the coordination numbers CN(Si/O) from 4 to 6 and CN(O/Si) from 2 to 3 during the phase transition coesite ⟶ stishovite.

DENSIFICATION OF H_2O FRAMEWORKS

Structure changes below $d_f \simeq 40\ H_2O/1000\ Å^3$

In the range of low framework density d_f there is extensive isotypism between phases in the H_2O and the SiO_2 system (Table 3). This suggests that the mechanism of densification within this range is in principle the same for both systems. In fact, if for the isotypic phases $d_f(H_2O)$ is plotted versus $d_f\,(SiO_2)$, there is a linear correlation between the two

$$d_f(SiO_2) = 0.5246 d_f(H_2O) + 5.0015$$

with a level of significance of $R^2 = 90\ \%$.

Table 5. Some structural data of silica polymorphs:
a. framework density in $n(SiO_2)/1000Å^3$; b. Δ = densification relative to silica-sodalite (d_f = 17.4 $SiO_2/1000Å^3$) ; c. mean value of the Si-O-Si angles as calculated from X-ray diffraction data collected at room temperature and 0.1. MPa. The true values for melanophlogite and the two dodecasils are considerably lower than the tabulated ones, which are falsified by high temperature factors ; d. numbers $n_i \leqslant 8$ of (SiO_4) tetrahedra forming loops (rings) of the framework.

	Phase	d_f a)	Δ b) [%]	∢Si-O-Si c) [°]	Loop sizes d)				
Pyknosils	Coesite	29.3	68	148	4,		6,		8
	Quartz	26.6	53	144			6,		8
	Moganite	26.3	51	ca.151	4,		6,		8
	Keatite	25.1	44	ca.153		5,		7,	8
	Cristobalite	23.2	34	146			6		
	Tridymite	22.9	32	148			6		
Porosils: Clathrasils	Nonasils	19.2	10	163	4,	5,	6		
	Melanophlogite	19.0	9	169c)		5,	6		
	Dodecasils 3C	18.6	7	174c)		5,	6		
	Dodecasils 1H	18.5	6	171c)	4,	5,	6		
	SIGMA-2	17.8	2	?	4,	5,	6		
	Deca-dodecasil 3R	17.6	1	159	4,	5,	6,		8
	Deca-dodecasil 3H	17.6	1	?	4,	5,	6,		8
	Silica-sodalite	17.4	0	?	4,		6		
Porosils: Zeosils	Silica-ZSM-23	20.0	15	?		5,	6		
	Silica-ZSM-48	19.9	14	?		5,	6		
	Silica-ZSM-22	19.7	13	151		5,	6		
	Silica-ferrierite	19.3	11	156		5,	6,		8
	Silica-ZSM-12	18.5	6	?	4,	5,	6		
	Silica-ZSM-50	18.2	5	?	4,	5,	6		
	Silicalite II	17.9	3	154	4,	5,	6		
	Silicalite I	17.8	2	156	4,	5,	6		
	Deca-dodecasil 3R	17.6	1	159	4,	5,	6,		8
	Deca-dodecasil 3H	17.6	1	?	4,	5,	6,		8

Obviously, the influence of external and internal pressure on the minimum ring size is about the same for H_2O frameworks with $d_f(H_2O) \lesssim 40H_2O/1000$ Å^3 as for silica frameworks with $d_f(SiO_2) \lesssim 30$ $SiO_2/1000$ Å^3. Unfortunately, the hydrogen atoms in the H_2O frameworks are not accurately enough known to prove whether a similar correlation as that found between the average Si-O-Si angles and $d_f(SiO_2)$ and/or external pressure exists between the average O-H-O angles and $d_f(H_2O)$.

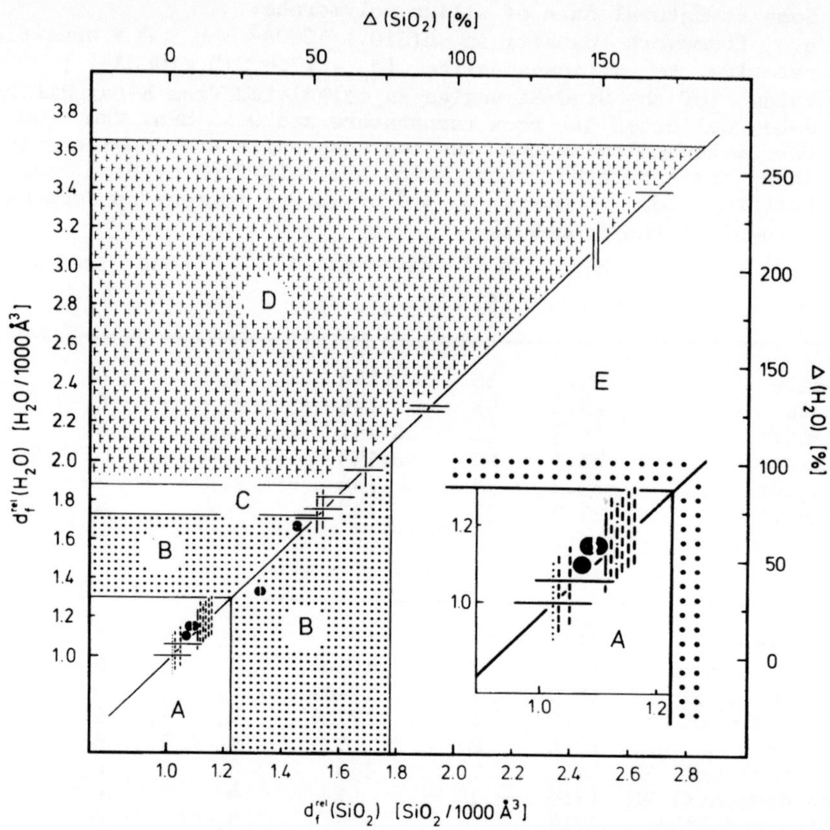

Fig. 14. Correlation between the relative framework density, d_f^{rel}, and densification Δ of silica polymorphs (relative to silica-sodalite) and the corresponding data of phases with 4-connected H_2O frameworks (relative to clathrate hydrate type VI).

A : heteroclathrates (porosils, porohydrates),
B : non-clathrates (pyknosils-[4], pyknohydrates-[4]),
C : autoclathrates (pyknohydrates-[4]),
D : isoclathrates (pydnohydrates-[4]),
E : pyknosils-[6].

Densification by interpenetration

Above $d_f(H_2O)$ = 40 H_2O/1000 $Å^3$ further densification is achieved by interpenetration of O-H-O bonds through rings of the H_2O frameworks. In the autoclathrate ices IV and V the interpenetrating bonds are parts of the same framework as the interpenetrated rings. The smaller the interpenetrated rings the higher the degree of densification.

The portion of rings that are interpenetrated is rather low in autoclathrates. Consequently, only moderate densification is achieved in these two phases. A higher portion of rings is interpenetrated in the isoclathrates, giving rise to much higher framework densities. In fact, in ice VII, VIII, and probably X each six-membered ring of one tetrahedral framework is interpenetrated by an O-H-O bond of the other and vice versa (fig. 11).

Provided the interpretation of the scarce experimental data on ice X[30] are correct, then a further strong increase of framework density is caused by a change from the asymmetric hydrogen bond which has an O...O distance of about 2.85Å, to a symmetrical hydrogen bond of about 2.5Å length.

COMPARISON OF SiO_2 AND H_2O DENSIFICATION

In fig. 14 the densification $\Delta(SiO_2)$ of the silica polymorphs relative to the least dense polymorph, silica-sodalite, is plotted versus the densification $\Delta(H_2O)$ of the H_2O frameworks relative to the least dense one, clathrate hydrate type VI. The phase pairs of the two systems which are isotypic (marked as dots) clearly show the linear correlation between the framework densities mentioned in the previous chapter. Phases which have no istoypic partner in the other system are represented by vertical (SiO_2) and horizontal (H_2O) bars.

From this plot several conclusions can be drawn :
(1) Structural isotypism between SiO_2 and H_2O frameworks is restricted to the low-density ranges $0 \lesssim \Delta(SiO_2) \lesssim 45$ % and $0 \lesssim \Delta(H_2O) \lesssim 70$ %, in other words, it is restricted to the heteroclathrates and the non-clathrates (see also Table 3 to 5).

In this range densification is achieved in both systems by the same mechanism of
(i) reducing the pore size,
(ii) changing the minimum ring size and, at least for silica,
(iii) reducing the ange T-X-T at the two-coordinated atom X.

(2) Above $\Delta H_2O = 70$ % densification by way of interpenetration becomes possible in the relatively open water frameworks (the ratio-large atoms : small atoms is 1:2) because the small hydrogen atom can easily be located in the plane of a ring of six or more oxygen atoms. In silica frameworks, however, which are denser than corresponding water frameworks because of the ratio- large atoms : small atoms = 2:1, the bulky -Si-O-Si- group would require rings of 12 or more tetrahedra for penetration. In fact, only one silicate with interpenetrating frameworks is known ; neptunite $LiNa_2K(Fe, Mg, Mn)_2Ti_2(Si_8O_{22})O$ has interpenetrated rings that are formed by 20 tetrahedra.[51] Due to the low density of such large-ring silicate frameworks no high densification by interpenetration can be expected. Therefore, silica continues to densify by the previous mechanisms (i) to (iii) up to at least $\Delta(SiO_2) = 80$ %.

(3) Above $\Delta(SiO_2) = 80$ % no further densification of 3-dimensional 4-connected silica frameworks seems to be possible. At very high pressures a change of the coordination number of silicon towards oxygen, CN(Si/O), from 4 to 6 leads to a densification value $\Delta(SiO_2) \simeq 150$ % for stishovite and the Fe_2N-type phase of silica (relative to silica-sodalite).

(4) Instead of increasing the coordination number CN(O/H) from 4 to 6, in analogy to the silica phases, water can achieve considerable densification by transformation of the asymmetric hydrogen bonds to symmetric ones. During transformation the O...O distance shortens from about 2.85Å to 2.50Å. The phase transition ice VIII ⟶ ice X is accompanied by a densification change from Δ(ice VIII) = 129 % to Δ(ice X) ≃240 %.

OUTLOOK

As mentioned previously, the large density increase during the transition from pyknosils- [4] to pyknosils- [6] is mainly due to the increase of CN(O/O) from, for example, 6 in quartz to 12 in stishovite and SiO_2 (Fe_2N-type). Despite the high density of ice X, the coordination number CN(O/O) of this H_2O phase is only 8 (fig. 11)*. It can, therefore, be expected that further densification of H_2O is possible by way of increasing CN(O/O) from 8 to 12.

Assuming an O...O distance of 2.50Å, as in ice X, an ultra-dense ice phase having a close-packed oxygen arrangement would have a framework density d_f = 90.5 $H_2O/1000Å^3$ and a densification $\Delta \simeq 290$ % relative to the clathrate hydrate type VI. These figures compare with the corresponding values d_f(ice X) = 80 $H_2O/1000Å^3$ and Δ(ice X) = 240 %.

It may be assumed that such an ultra-dense ice phase is a good proton conductor in which the hydrogen atoms are considerably delocalized.

---

* In ices VII, VIII and X only four of the eight equally distant oxygen atoms are linked by hydrogen bonds to a particular oxygen atom.

REFERENCES

1. P. Bastie and G. Dolino, Phys. Review B31:2857 (1985).
2. M.A. Spackman, R.J. Hill, and G.V. Gibbs, Phys. Chem. Minerals 14:139 (1987).
3. L.Liu, W.A. Bassett, and J. Sharry, J. Geophys. Res. 83:2301 (1978).
4. K.L. Geisinger, M.A. Spackman, and G.V. Gibbs, J. Phys. Chem. 91:3237 (1987).
5. A.F. Wright and M.S. Lehmann, J. Solid State Chem. 36:371 (1981).
6. H. Graetsch, O.W. Floerke, and G. Miehe, Acta Crystallogr. A43:C156 (1987).
7. J. Shropshire, P.P. Keat, and P.A. Vaughan, Z. Kristallogr. 112:409 (1959).
8. D.R. Peacor, Z. Kristallogr. 138:274 (1973).
9. J.H. Konnert and D.E. Appleman, Acta Crystallogr. B34:391 (1978).
10. B. Marler, N. Dehnbostel, H.H. Eulert, H. Gies, and F. Liebau, J. Inclusion Phenom. 4:339 (1986).
11. H. Gies, Z. Kristallogr. 164:247 (1983).
12. H. Gies, Z. Kristallogr. 167:73 (1984).
13. H. Gies, J. Inclusion Phenom. 4:85 (1986).
14. L.B. McCusker, 14th Internat. Congress Crystallogr., Perth, Australia, Poster 12.5-1 (1987).
15. D.M. Bibby and M.P. Dale, Nature 317:157 (1985).
16. H. Gies, Z. Kristallogr. 175:93 (1986).
17. A.C. Rohrman, R.B. Lapierre, J.L. Schlenker, J.D. Wood, E.W. Valyocsik, M.K. Rubin, J.B. Higgins, and W.J. Rohrbaugh, Zeolites 5:352 (1985).
18. J.L. Schlenker, W.J. Rohrbaugh, P. Chu, E.W. Valyocsik, and G.T. Kokotailo, Zeolites 5:355 (1985).
19. B. Marler, Zeolites in press.
20. H. Gies and R.P. Gunawardane, Zeolites in press.
21. R.B. Lapierre, A.C. Rohrman, J.L. Schlenker, M.K. Rubin, and W.J. Rohrbaugh, Zeolites 5:346 (1985).
22. N.A. Briscoe, D.W. Johnson, C.T. Kokotailo, L.B. McCusker, and M.D. Shannon, Zeolites in press.
23. G.T. Kokotailo, P. Chu, and S.L. Lawton, Nature 275:119 (1978) ; D.M. Bibby, N.B. Milestone, and L.P. Aldridge, Nature 280:664 (1979).

24. K.J. Cao, J.C. Lin, Y. Wang, and G.H. Lee, Zeolites 6:35 (1986).
25. A. Weiss and A. Weiss, Z. anogr. allg. Chem. 276:95 (1984).
26. R.P. Gunawardane, H. Gies, and F. Liebau, Z. anorg. allg. Chem. 546:189 (1987).
27. H. Gies, Habilitationsschrift, Universitaet Kiel, (1987).
28. F. Liebau, H. Gies, R.P. Gunawardane, and B. Marler, Zeolites 6:373 (1986).
29. A.F. Wells, "Structural Inorganic Chemistry", 5th edn., Clarendon Press, Oxford, U.K. (1984).
30. K.R. Hirsch and W.B. Holzapfel, Phys. Lett. 101A:142 (1984).
31. W.F. Kuhs, J.L. Finney, C. Vettier, and D.V. Bliss, J. Chem. Phys. 81:3612 (1984).
32. B. Kamb and B.L. Davis, Proc. Nat. Acad. Sci. 52:1433 (1964).
33. K. Yamamoto, Japan. J. Appl. Phys. 21:567 (1982).
34. H. Engelhardt and B. Kamb, J. Chem. Phys. 75:5887 (1981).
35. B. Kamb, A. Prakash, and C. Knobler, Acta Crystallogr. 22:706 (1967).
36. B. Kamb, Acta Crystallogr. 17:1437 (1964).
37. S.J. Laplaca and W.C. Hamilton, J. Chem. Phys. 58:567 (1973).
38. B. Kamb and A.Prakash, Acta Crystallogr. B24:1317 (1968).
39. G.P. Arnold, E.D. Finch, S.W. Rabideau, and R.G. Wenzel, J. Chem. Phys. 49:4365 (1968).
40. W.F. Kuhs and M.S. Lehmann, Nature 294:432 (1981).
41. G.A. Jeffrey, Hydrate Inclusion Compounds, in : "Inclusion Compounds, Vol. 1", J.L. Atwood, J.E.D. Davies, and D.D. MacNicol, eds., Academic Press, London (1984).
42. J.A. Ripmeester, J.S. Tse, C.I. Ratcliffe, and B.M. Powell, Nature 325:135 (1987).
43. F. Liebau, Rend. Soc. Miner. Petrol. Ital., in press.
44. F. Liebau, Fortschr. Mineral. 61, 29 (1983).
45. H.D. Megaw, "Crystal Structures : A Working Approach", p. 453-459, Saunders, Philadelphia (1973).
46. F. Liebau, "Structural Chemistry of Silicates. Structure, Bonding, and Classification", p. 22-30, Springer, Heidelberg, (1985).
47. F. Liebau and H. Boehm, Acta Crystallogr. A38:252 (1982).
48. J. Glinnemann, Dissertation, Rheinisch-Westfaelische Technische Hochschule Aachen (1987).
49. K. Kihara, T. Matsumoto, and M. Imamura, Z. Kristallogr., 177:27 (1986).
50. A. Nukui and H. Nakazawa, Amer. Mineral. 63:1252 (1978).
51. E. Cannillo, F. Mazzi, and G. Rossi, Acta Crystallogr. 21:200 (1966).

GEOMETRICAL METHODS IN THE THEORY OF GLASSSES

Richard Kerner

L.P.T.P.E., Université Pierre et Marie Curie
4, Place Jussieu
75252 Paris Cedex 05

1) In some branches of physics one can observe a surplus of theoretical models coexisting with the dramatic lack of experimental data. The physics of glass is in an opposite situation : the enormous and rich variety of experimental results goes along with an absence of a consistent theory. We rely, in this domain, on some hints or analogies, no more. Somehow the two natural paths which should lead to the theory of glass, namely the theory of liquids and the solid state physics (up to now almost a synonym for the theory of the crystalline state) both come to a dead end or a kind of a no man's land before penetrating into the territory on which the theory of glass should be found.

However, more than fifty years have passed since the fundamental paper by Zachariasen [1] and we have some reasons to be more optimistic than before. There has been considerable progress made in the theory of liquid crystals and more recently quasi-crystals, and in the theory of defects [2,3,4,5,6], all of which bring us certainly closer to the understanding of the amorphous and glassy state. In all these approaches, modern concepts of the differential geometry and topology are widely used, and it is almost sure by now that geometrical methods are likely to produce a coherent theory of glass.

This talk is intended to show some recent applications of such a geometrical point of view, exclusively for the continuous random network mode, adapted for covalent glasses.

2) In order to discuss any model, we would define first the basic notions. We shall take the simplest possible example describing a covalent material, which is a two-dimensional network (random or crystalline) with a fixed coordination number equal to <u>three</u>. The atoms (all of them identical) are placed in the vertices of the network, there are no dangling bonds, so that each atom has three closest neighbors defined by the three covalent bonds of the same length, normalized to 1 in some appropriate units. The distances between the atoms in general must be bigger than 1 ; finally, we suppose that all the polygons (rings) are convex. If the network on the plane is big enough in order to be considered as infinite, the Euler theorem states

$$\frac{1}{\bar{N}_c} + \frac{1}{\bar{N}_p} = \frac{1}{2}$$

where $\bar{N}_c$ is the average coordination number, $\bar{N}_p$ the average number of sides in a polygon forming the network. In our case $\bar{N}_c = N_c = 3$, so that $\bar{N}_p = 6$. The hexagon lattice is the most symmetric and homogeneous solution ; we shall refer to it as "crystal" in this model. If there are other polygons present in the network, and if their relative numbers are given by the probabilities P_k subject to $\sum_k P_k = 1$, then according to Euler's constraint, $\sum_k kP_k = 6$. If we admit only two kinds of deviations from à hexagon, i.e. pentagons and heptagons, then $P_5 = P_7 = (1 - P_6)/2$. The variable P_6 can be chosen as the only independent statistical parameter which determines a class of equivalence of random networks of this kind. If a network is set up and the temperature is so low that the bonds can not be altered, the only remaining degrees of freedom are the bond angles and bond lengths. We shall simplify our assumptions by supposing that the lengths are very close to 1, whereas the angles may vary considerably.

If the forces between the atoms after bond saturation are central and repulsive, the lowest energy configuration for a vertex (tripod) is with all its angles equal to $2\pi/3$; the same repulsive force will tend to maximize the surface of each polygon, which amounts to making it "perfect", with all angles equal to $\pi(k-2)/k$ for a k-sided polygon. Both these tendencies agree for $k = 6$; this is why a hexagon corresponds to the lowest energy state of a polygon. The pentagons and heptagons would represent defects in our network with respect to the "ideal" hexagonal configuration. A pentagon represents positive local curvature, whereas a heptagon represents a negative local curvature of the same order of magnitude. These two deviations from flatness should compensate each other in linear approximation, because the angular excesses (or defaults) are the same :

$$5\left(\frac{2\pi}{3} - \frac{\pi(5-2)}{5}\right) = 7\left(\frac{\pi(7-2)}{7} - \frac{2\pi}{3}\right)$$

Let us denote the absolute value of the energy cost of producing such a defect by ΔE.

Our first aim is to find out what kind of configurations will become dominant during the formation of such a network, i.e. during the cooling of a hot liquid in which all kinds of polygons are about to form.

Let us suppose that when the liquid is cooled enough, relatively stable polygons begin to appear ; let us denote the probabilities of finding a k-sided polygon by $\overset{o}{P}_k$. At this stage only the normalization $\sum \overset{o}{P}_k = 1$ holds, but there is no Euler's restriction. Cooling the liquid a bit more, the polygons begin to cluster into doublets. The simple energy analysis suggests that the probabilities of the corresponding doublets should be altered with the appropriate Boltzmann factor : therefore,

$$\overset{1}{P}_{ij} = \frac{1}{Q} (2 - \delta_{ij}) \overset{o}{P}_i \overset{o}{P}_j \, e^{-|12-i-j|\Delta E/kT}$$

where i,j = 5,6,7, k is Boltzmann's constant, and Q the normalization factor

$$Q = \sum_{i,j} (2-\delta_{ij})\overset{\circ}{P}_i\overset{\circ}{P}_j\ e^{-|12-i-j|\Delta E/kT}$$

Now we can find the relative frequencies of k-sided polygons among all the doublets :

$$\overset{1}{P}_i = \frac{1}{2}\left(2\ \overset{1}{P}_{ii} + \sum_{j\neq i} \overset{1}{P}_{ij}\right)$$

The fixed points of the transformation $P_i \to \overset{1}{P}_k$ are then found as zeros of the functions $\overset{1}{P}_i - \overset{\circ}{P}_i$, which can be given the meaning of a first derivative dP_i/dP_s with respect to some variable s, which itself is a monotonous function of time. All the three variables, P_5, P_6 and P_7 can be represented in an equilateral triangle displayed in fig. 1.

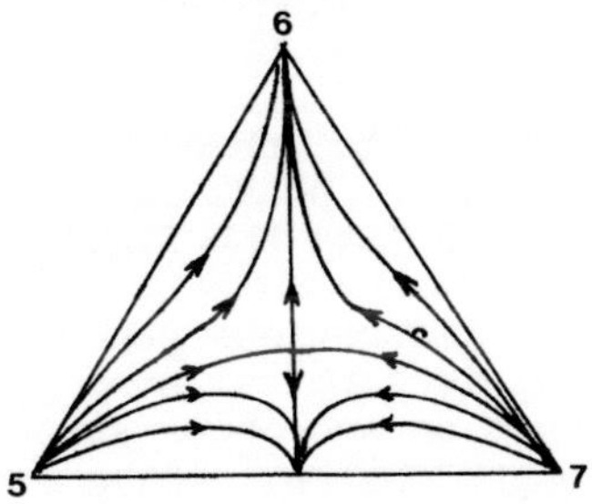

Fig. 1a

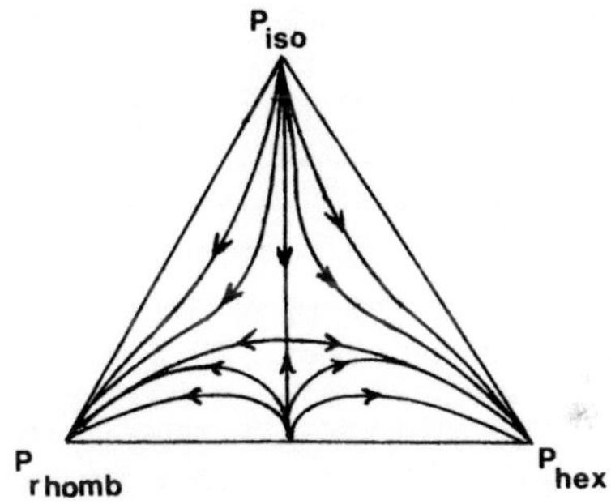

Fig. 1b

The trajectories of the differential equation describing the first step of the clustering of the polygons.

There are five fixed points, out of which two : $P_5 = 1$ and $P_7 = 1$ are repulsive, two : $P_6 = 1$ and $P_5 = P_7 = 1/2$ are attractive, and one : $P_6 = \frac{1 - e^{-\Delta E/kT}}{3 - e^{-\Delta E/kT}}$, $P_5 = P_7 = \frac{1}{3 - e^{-\Delta E/kT}}$ which is a saddle point and whose position depends on T.

The phase trajectories of our system show that two final results are possible : either a crystal (P_6= 1), or a mixture of five-and seven-sided polygons. The saddle point corresponding to an amorphous state can also be reached, but the system can be trapped in it only artificially, if it is quenched by some external influence at an appropriate moment. The same kind of analysis can performed for the second step in clustering process, i.e. the formation of triplets out of doublets and singlets. It yields a second order system, [7,8] which has the same fixed points, but the linearization enables us now to get some more insight into their character : at some temperatures, the eigenvalues of the characteristic equation can become complex, which means that the trajectories in the four-dimensional space of independent variables P_5, P_7 and dP_5/ds, dP_7/ds are converging or diverging spirals, and that it takes much longer to attain the stable regime. Moreover, only above some finite temperature $T_1 \leqslant T$ is the saddle point present ; below it falls on the attractor $P_5 = P_7 = 1/2$.

The energetics of the clustering of elementary configurations we have chosen here can be represented in the symbolic table below, which

	5	6	7
5	2	1	0
6	1	0	1
7	0	1	2

defines the energy as proportional to the deviation from flatness (represented by 6-sided polygons).

In three dimensions, a covalent four-coordinate network presents rather opposite features. It is known[9,10,11], that in such cases a local icosahedral cluster is preferred ; however, the three-dimensional Euclidean space can not be tiled with such clusters without deforming them ; therefore, these two "ideal" local configurations should exclude each other if put closely together ; the deviations or defects represented by an excess or default of some number (say, two) atoms in a cluster should stick together well, because they represent an admissible crystalline symmetry (hexagonal or rhombohedric). The corresponding energetics is represented by a similar table, in which 0's are replaced by 2's and vice-versa :

	hex.	icos.	rhomb.
hex.	0	1	2
icos.	1	2	1
rhomb.	2	1	0

It is easy to check that the corresponding phase-space trajectories look like the fig.1b

Now the pure rhombohedric or hexagonal configurations represent the attractors, whereas their 50 % mixture or a pure icosahedral structure are unstable ; a meta-stable saddle point represents an amorphous state which can be put into correspondence with Franck-Kasper chains[12]. Once again, a system can be trapped in this state if it is cooled rapidly enough during the passage close to the metastable saddle point.

The method of ordinary differential equations can be developed further, but what we have presented here is just a first approximation ; it has to be underlined that in three dimensions the third step is crucial, consisting of addition of a fourth elementary configuration to the three elementary configurations already forming a cell : this becomes a genuine three-dimensional structure by contrast with linear doublets and planar triplets.

3) The forming of glass, or glass transition, is apparently a non-equilibrium thermodynamical process. It seems that one of the most important problems here is to understand what is the nature of the phase space in which the statistical physics of random networks could be formulated. From the mathematical point of view, any network with constant coordination number and constant bond length can be considered as a mapping from $\mathbb{R}^2$ or $\mathbb{R}^3$ into some internal space of configurations which is composed out of all sets of n-pods (vertices). In $\mathbb{R}^2$ for $N_c = 3$, this space is a subset of a 2-sphere given by two independent angles ; α_1, α_2 ($\alpha_3 = 2\pi - \alpha_1 - \alpha_2$) between the bonds ; in $\mathbb{R}^3$ with $N_c = 4$, it is a 5-dimensional space of 5 independent angles between the 4 unit vectors. However, among all such continuous mappings only very special ones can describe real networks : this is because there is an interplay between the external space $\mathbb{R}^2$ and the internal space : if the image of $x \in \mathbb{R}^2$ is $\varphi(x) = (\alpha_1, \alpha_2)$, we know (up to an arbitrary rigid rotation) where the next three vertices have to be. Moreover, if these new three points are x_1', x_2', x_3', then $\varphi(x_i')$ are not

arbitrary : after some number of such steps the polygons must appear, and the bonds can never intersect. These correlations can be interpreted if we introduce a second internal space whose variables describe the polygons surrounding a vertex. On both such internal spaces symmetry groups SO(3) act in a non-linear representation ; the action of SO(3) on the first internal space provokes the action also in the second internal space, but not vice versa : if we deform a vertex, the surrounding polygons must be deformed, too, whereas we can deform the polygons without deforming the central vertex.

Such a situation is also found in field theory, and is described by a fibre bundle space with connection. It leads to the symmetry-breaking potential of the form

$$U(\phi) = \phi^4 = \lambda\phi^2$$

which also appears in Landau-Ginzburg theory.

All these considerations suggest that the phase space of such networks can be represented by the points in our two internal spaces describing fairly enough the local configurations (cells) composed of a vertex and all the adjacent polygons. Then in $\mathbb{R}^2$, a whole class of networks with fixed P_6 can be represented as a collection of average cells composed of three polygons with mean number of sides $n = \Sigma k p_k$ with $p_k = kP_k/6$. The whole phase space is divided into sheaves of given P_6. We assume that it takes much less time for the network to maximize its entropy on one such sheaf then to go from one sheaf to another. This justifies the assumption that the entropy of the network can be separated into two parts, $S = S_{therm} + S_{conf}(P_6)$, thermal and a purely configurational contribution depending on P_6 alone. We shall introduce a mean angular parameter δ defined in the space of 3-polygon cells as follows : let β_{kj} be the angles of k-sided polygon, j = 1,2,...k ; then let

$$\varepsilon_k = \left(\frac{1}{k}\sum\left[\beta_{kj} - \frac{(k-2)\pi}{k}\right]^2\right)^{1/2} \text{; then let } \delta = \left(\frac{1}{3}\sum_{k_1,k_2,k_3} P_{k_1k_2k_3}\sum_{i=1}^{3}\varepsilon^2_{k_1}\right)^{1/2}$$

where $P_{k_1k_2k_3}$ is the relative probability of finding a cell (k_1,k_2,k_2) in the network. We shall imitate the fluctuations around an "ideal" cell by introducing the angular deviations

$$\delta_k = \frac{n(k-2)}{k(n-2)}\,\delta$$

A good approximation of the potential containing the contribution of vertices and polygons is then[13]

$$U = 3\delta^2 + 2\pi^2 \sum\left(P_k/k - \frac{\pi^2}{3} + \lambda\left[\frac{1}{2}\sum P_k S^o_k\ (1 - \cos\delta_k)\right]\right)$$

This expression has been tested on regular lattices, and the error was never bigger than 1 %. U is the internal energy of the network depending on P_6 and δ only, and on the critical parameter λ which depends on the chemical value, the absolute minimum of U is always at $P_6 = 1$, $\delta = 0$, the perfect hexagon lattice. Above the critical value, there is a local minimum at some $P_6 \neq 1$, corresponding to an amorphous state.

In the approximation at which the network can be considered as gas of average cells, we can apply the virial theorem and the equipartition of energy among the cells : as U is the potential energy per one atom, we shall write

$$\overline{N}_d \frac{kT}{2} = \overline{N}_a U(P,\delta)$$

where $\overline{N}_d$ is the average number of degrees of freedom, $\overline{N}_a$ average number of atoms in a cell. Fixing T we obtain the "isotherms" of the system on the (P,δ) - plane. Maybe it is better to speak of curves of constant internal energy at fixed temperature.

Now, if the total entropy of the network can be divided into two parts, one of which depends only on configuration of the network, $S_{conf}(P)$, and if its variations in time are much slower than the variation of the purely thermal part S_{therm}, then the very fact that we have applied the equipartition of energy and the virial to the cells is equivalent with the statement that on the level of cells S_{therm} has attained its maximum. In order to fix the real position of the system on the P-δ plane we have to minimize its free energy (if the volume is kept constant) or its free enthalpy (if the pressure is kept constant). As the thermal part of the entropy is already at its maximum, it means that we have to minimize the expression

$$F = U - TS_{conf}$$

on the "isotherms" given by the virial and equipartition theorems. Before going further, let us show the constant energy curves on P - δ diagram at constant T :

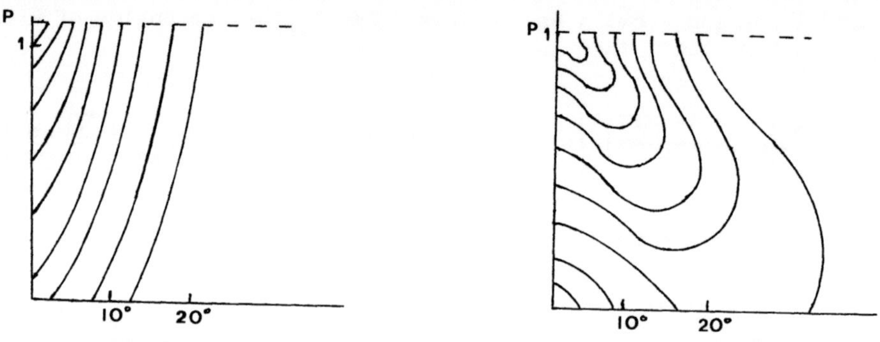

λ small $T_2 > T_1$ Fig. 2 λ big $T_2 > T_1$

As we see, the form of the internal energy surface depends on λ : if λ is small, which corresponds to the bigger energy of bond angles and relatively weak surface contribution of polygons, at any fixed P the minimum is at $\delta = 0$; the absolute minimum is at the undeformed crystalline configuration $P = 1$, $\delta = 0$. If λ grows bigger, which corresponds to the substances for which the surface contribution of rings is important, then, at given P; the local minimum is found at some $\delta \neq 0$.

In order to evaluate the entropy due to the configuration, one is tempted to use the expression

$$S = -k \sum_{i=5}^{7} P_i \mathrm{Log} P_i$$

which can be deduced from a more general expression

$$S = k \text{ Log } \Gamma$$

Γ being the measure of all possible configurations (in a discrete case, just their number) for given value of P_6. If all in all we have N polygons, among which N-2m hexagons and equal number m of pentagons and heptagons, then the number of distinct configurations is

$$\frac{N!}{(N-m)!m!} \cdot \frac{(N-m)!}{(N-2m)!(2m)!}$$

The expression $\sum P_i \text{Log } P_i$ comes out if we take the Log, apply Stirling's formula, and then divide everything by N. The maximum of so obtained entropy is at P = 1/3, but at P = 0 and at P = 1 its derivative with respect to P becomes infinite. This is because Stirling's formula is not applicable when either m or N-m are very small ; the result is the overcount of configurations.

There is another expression for the configurational entropy, which is more suited for our purpose. Let us recall that in this model we make an approximation in which the network is seen as a gas of elementary 3-polygon cells. In such a case, it is more natural to sum up the contributions to the entropy coming from the cells as representatives of small elements of the phase space ; then their statistical weight is the only important information, and after summing up all the contributions, we should obtain

$$S_{conf} = k \sum P_{ijk} \text{Log } W_{ijk}$$

where $P_{ijk} = W_{ijk} P_i P_j P_k$,

$$W_{ijk} = 1 \text{ if } i = j = k, \quad W_{ijk} = 3 \text{ if } i = j \neq k, \quad W_{ijk} = 6 \text{ if } i \neq j \neq k.$$

This gives a polynomial of third order in P :

$$S_{conf} = (1-P)\left[0.267(1-P)^2 + P(1-P) + 1.0986\, P^2\right]$$

It has a maximum at P = 0.34, which is very close to the exact result 1/3, and its derivatives are finite at P = 1 and P = 0.

What is left now is to minimize the expression $F = U - TS_{conf}$ on the curves of constant temperature, on which $\delta = \delta_T(P)$. For each T we can thus find the parameters P(T) and δ(T) describing the state of the network. The result is displayed in fig. 3. We observe the neat phase transition at critical temperature T_c for which the absolute value of the derivative dS_{cont}/dP becomes bigger than $(\partial U/\partial P)|_{P=P(\delta)}$. For the materials for which λ is bigger, there is no phase transition, but a glass transition represented by a continuous curve. In both cases the system comes to $P_6 \approx 1/3$ and corresponding $\delta \neq 0$ at high temperatures ; at low temperatures it falls either to the minimum $P_6 = 1$ corresponding to the crystalline state, or to an amorphous minimum $P_6 < 1$.

In order to obtain this simple description, we must admit a lot of ad hoc hypothesis, which can be recapitulated as follows :

a) Close to the transition temperature the system behaves so that there are two distinct time scales : the time necessary to exchange the thermal energy between the cells (defined by the velocity of phonons) and the time necessary to restructure the bonds ; the second one is supposed to be much bigger than the first :

b) The mean potential for each atom can be decomposed into two contributions, one coming from the bond bending, another from the surface contribution of the polygons (rings). Other contributions have been neglected. The relative strength of these two contributions decides whether the liquid will crystallize or not ;

c) Close to the transition temperature the cells composed of three adjacent polygons are in thermal equilibrium, whereas the atoms inside the cells are not. This is equivalent with the determination of the critical dimension of clustering, which in our case is about 12-13 atoms. In a tetracoordinate 3-dimensional network the elementary cell will contain about 32-33 atoms ;

d) Close to the transition temperature the virial theorem can be applied : this is equivalent to the dominant contribution of harmonic oscillator modes and neglecting the non-linear ones.

Finally, the critical temperature obtained with this model should be close to the value of kT sufficient to break the bonds ; if it is much lower, or much higher, there will be no glass transition possible.

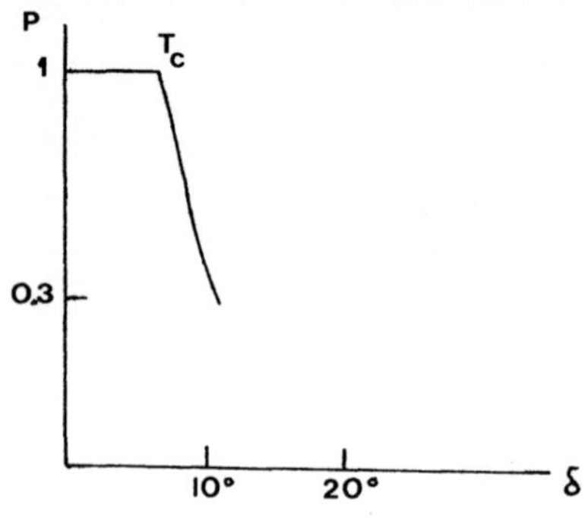

λ small

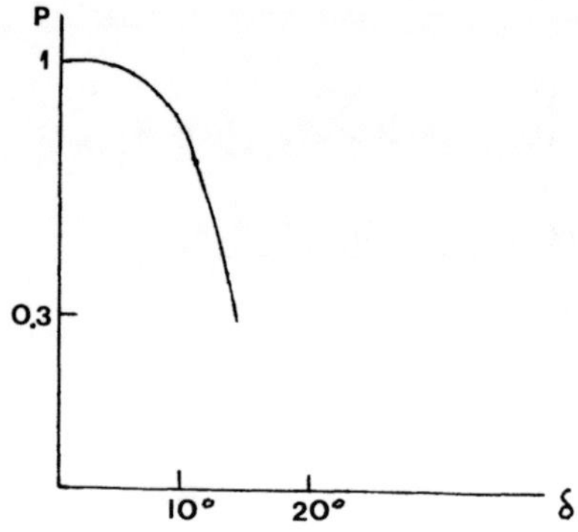

λ big but under λ_c critical

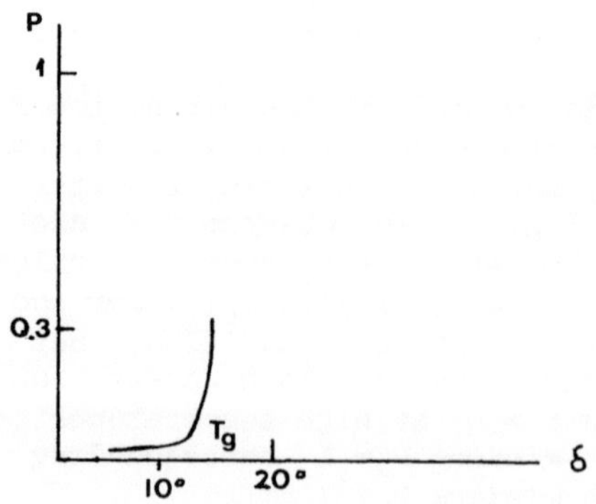

λ above λ_c critical : $\lambda_c \approx 10.2$

Fig. 3

We think that the thermodynamics of glass can be roughly understood along these lines. The three-dimensional generalization of this model can be made quite easily under the assumption that the dihedral angles' contribution to the mean potential is negligible. If such an assumption cannot be made, we have to introduce one variable more, namely the mean square deviation from an ideal dihedral angle ; some tentatives of such a generalization can be found in reference 14.

4) The two quite different approaches presented here have a common source, which is the gauge theory[15,16,17]. The main idea of gauge theories in general is the notion of connection which is a way of prescribing the changes that occur in some internal configuration space of the system when one is following some path in real space. For example, if the internal parameter space is reduced just to the phase factor of the global wave function, as occurs in He-II, the internal space is isomorphic to S^1, a unit circle. If the phase depends on space-time variables, the gauge invariance of the equations imposes the minimal coupling with the compensating field, i.e.

if $\psi \to e^{i\lambda(x)}\psi$, then $\partial_\mu\psi \to \partial_\mu(e^{i\lambda(x)}\psi) = e^{i\lambda(x)}(\partial_\mu\psi + i\partial_\mu\lambda)$; in order to restore the invariance, one introduces the compensating field, or connection $A_\mu(x)$, so that $\mathfrak{D}_\mu\psi = (\partial_\mu\psi + iA_\mu\psi)$; then, this covariant derivative remains invariant under gauge transformations $\psi \to e^{i\lambda(x)}\psi$ if at the same time $A_\mu \to A_\mu + \partial_\mu\lambda$. More general cases are known, when the group of transformations acting on the internal space is non-abelian.

In the case of random networks it has been pointed out that an internal space on which the rotation group acts can be introduced in a natural way. Taking as an example the 3-coordinate network in two dimensions, this internal space can be identified with the space of triplets of unit vectors $(\vec{k}_1, \vec{k}_2, \vec{k}_3)$ divided by the equivalence relation induced by the diagonal action of the rotation group SO(2). This leaves only two independent angles α_1, α_2, because $\alpha_3 = 2\pi - \alpha_1 - \alpha_2$.

To define a network, it is enough to give a mapping from $\mathbb{R}^2$ into this internal space. However, only some very special mappings of this type will correspond to real networks, for which the bonds never cross. In order to describe this extra information, one has to enlarge the internal space so that it would also contain polygons surrounding each triplet of bonds. The connection describes the way of proceeding from one such "point" (elementary cell) to another. The lagrangian modelled as in usual gauge theories gives rise to the quartic potential. The internal energy proposed above is obtained as a mean value of this lagrangian expressed in terms of the internal space parameters averaged over all cells.

On the other hand, the same analogy suggests that the statistical features of the network can be reproduced as a trajectory in the internal space. These trajectories should be similar to the ones proposed in the second section if the the ergodicity is satisfied.

REFERENCES

1. W.H. Zachariasen, Am. Chem. Soc. 54:3481 (1932).
2. R. Zallen, "Physics of Amorphous Solids", (Wiley) (1983).
3. M. Kléman, in "Continuum Models of discrete Systems", Brulin and Hsieh ed., (North-Holland), p. 287 (1981).

4. G. Venkataraman, D. Sahoo, Contem. Phys., 26 579 (1985) and 27:3 (1986) references within
5. G.E. Volovik, N.P. Mineev, Soviet Phys. JETP., 45:1186 (1977).
6. N. Rivier, D. Duffy, J. Physique, 43:293 (1982).
7. R. Kerner, D. Santos, Phys. Rev. B, to appear
8. R. Kerner, C.R. Acad. Sci. Paris, 304:109 (1987).
9. J.F. Sadoc, R. Mosseri, J. Physique, 45:1025 (1984).
10. J.F. Sadoc, R. Mosseri, J. Non-Crystalline Solids, 61:487 (1984).
11. P. Kramer, Acta Crystallogr. A, 38:257 (1982).
12. F.C. Frank, J.S. Kasper, Acta Crystallogr., 12:483 (1959).
13. R. Kerner, J. Non-Crystalline Solids, 71:19 (1985) see also R. Kerner, MRS-Symposia Proceedings 61:39 (1986).
14. D. Santos, J. Physique, 45:1309 (1984).
15. R. Kerner, Phil. Mag. B., 47:151 (1983) ; also ref. 6 above.
16. I.Y. Dzyaloshinskii, G.E. Volovik, J. Physique 39:693 (1978).
17. N. Rivier, Phil. Mag. A, 40:859 (1979).

LOW LYING EXCITATIONS IN SILICA

U. Buchenau

Institut für Festkörperforschung
KFA Jülich, Postfach 1913
5170 Jülich, FRG

ABSTRACT

Neutron scattering measurements in vitreous silica show soft vibrational modes and structural relaxations with the same eignenvector, namely relative rotation of connected SiO_4 tetrahedra. The results imply a common origin of soft vibrations, relaxations and tunneling states.

INTRODUCTION

Glasses exhibit low temperature anomalies in the specific heat, the thermal conductivity, the ultrasonic absorption and other properties connected with low frequency excitations[1]. By low frequencies we denote frequencies below 1 THz, which are low compared to the majority of the vibrational frequencies of the amorphous solid. These low lying excitations dominate the thermal properties below 10 K. In most crystals, one will only find sound waves in this low frequency range. This is fundamentally different in glasses, there one finds a whole variety of different harmonic and anharmonic excitations which strongly influence the mechanical and electrical properties[2].

In this paper we discuss the low-lying excitions for the well-studied case of vitreous silica. One has to distinguish three main groups of excitations which coexist with the sound waves, namely additional harmonic vibrations, relaxations and two-level or tunneling states.

The first group, though by far the biggest in the number of modes, has only recently been identified unambiguously[3,4]. The relaxational and tunneling modes are strongly anharmonic excitations. Both require for their explanation potentials with more than one minimum, in the simplest case a double-minimum potential as postulated in the tunneling model[5,6]. As will be seen, there is experimental evidence that the mode eigenvectors of the harmonic vibrations in the first group are the same as those of the relaxational modes in the second group. This finding supports the hypothesis of a common origin of the low temperature glass anomalies.

HARMONIC LOW FREQUENCY VIBRATIONS

Sound waves are well defined harmonic vibrations in vitreous silica up to about 200 GHz. A linear relation between frequency and wavevector has been measured even farther[7] up to 500 GHz. On the other hand, the heat transport at low temperatures seems to be restricted to phonons below 200 GHz[8,9]. Above this frequency, the mean free path decreases dramatically [10]. This decrease cannot be explained by the scattering from the static disorder in the glass, which at these long wavelengths (10 nm) should be relatively weak[8,9]. Thus one has to look for another explanation for the plateau in the thermal conductivity found at about 10 K. It should be mentioned that this plateau in the thermal conductivity has been generally observed in all known amorphous solids.

One possible explanation for the plateau is the resonant interaction of the sound waves with additional localized vibrational states which by chance happen to have a low frequency. This explanation is supported by the specific heat above and around 5 K, which is markedly higher than the pure Debye T^3 term calculated from the known sound velocities. The evidence is not fully conclusive ; at 20 K also crystalline quartz surpasses the pure Debye heat capacity because of its low-lying TA modes at 2 THz[11]. Cristobalite, another crystalline form of SiO_2, has even a heat capacity which is fairly similar for $T \geqslant 5$ K to that of vitreous silica, probably because in that substance the TA modes are even lower[12]. Nevertheless, though specific heat and thermal conductivity around 10 K do not give an unambiguous picture, they are certainly compatible with a pronounced low-frequency tail in the frequency distribution of short wavelength modes which extends down into the sound wave regime.

A final proof for the validity of this picture has been found in inelastic neutron scattering experiments (for vitreous silica[3,4], for amorphous germanium[13] and for amorphous selenium[14]). Other spectroscopic techniques like Raman scattering[15] and far infrared absorption[16] could only establish the existence of vibrational modes above 500 GHz in a very general way by the Bose temperature dependence of the scattered or absorbed intensities, but could not determine the density of vibrational states in that region because of the uncertainty, in the coupling constants. In the neutron scattering case, the situation is much more favourable. Each harmonic vibration, sound wave or otherwise, is seen with an intensity which oscillates around a common $Q^2.e^{-2W}$ behavior[17,18,4] (Q is the momentum transfer of the scattering process and e^{-2W} is the Debye-Waller factor). For a sound wave, the oscillations occur in phase with the oscillations in S(Q), because in a sound wave the atomic pair correlations remain essentially conserved during the motion. If one has only one atom vibrating in the glass (Einstein mode), there are no oscillations in Q but only the $Q^2.e^{-2W}$ behaviour. For other modes, there can be oscillations in Q which are not in phase with the oscillations in S(Q). Thus one can not only determine the vibrational density of states but one can also get information about the eigenvector of the modes.

In vitreous silica, one finds pronounced oscillations around the $Q^2.e^{-2W}$ behaviour in the whole frequency range from 150 GHz to 2 THz (fig.1). Since the curves at different frequencies can be scaled to a common master curve, there seems to be a predominant eigenvector in this whole low frequency region. This eigenvector is not a sound wave eigenvector because the peak positions are different from those in S(Q)[19]. A reasonable fit can be obtained from a model of coupled rotation of five bonded SiO_4 tetrahedra (see insert of fig.3). The picture is suggested by

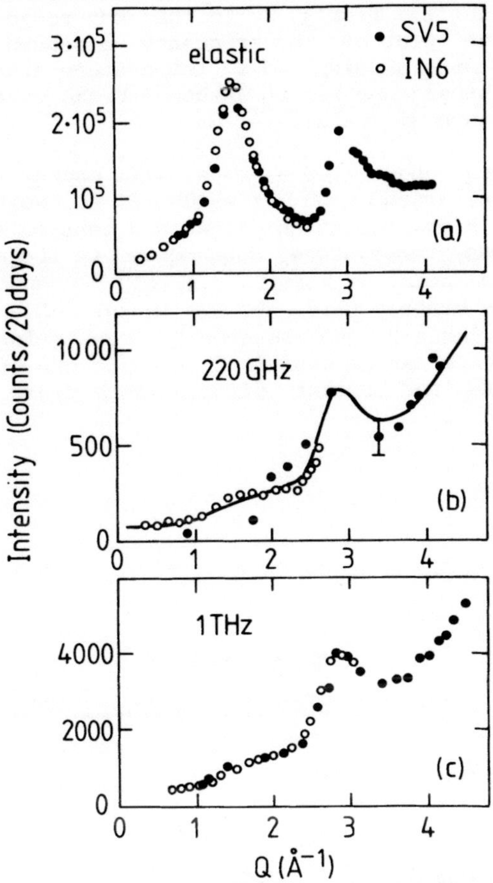

Fig. 1. Momentum transfer dependence of neutron scattering intensities (a) elastic (b) at 220 GHz (from 150 to 290 GHz) (c) at 1 THz (from 0.7 to 1.3 THz). The line in (b) is an eye fit to the points in (c).

the study of the low frequency modes in crystalline quartz[20,21] or cristobalite [22], where below 12 THz the modes are either sound waves or involve the rotation of relatively undistorted tetrahedra. In particular, the soft mode at the α-β- transformation at 846 K[23] is of this type[24].

Using this model one can determine the density of vibrational states on an absolute scale from a comparison of inelastic and elastic intensities in the neutron experiment. The resulting density of states allows calculation of the specific heat expected between 5 and 30 K. The result of this calculation shows excellent agreement with heat capacity data measured on the same sample[4].

RELAXATIONS

Mechanical and dielectric relaxations have been measured in a wide frequency range[25]. One finds a broad relaxation maximum as a function of temperature which shifts from 30 K in the kHz range to about 100 K at 10 GHz. An Arrhenius plot of the frequency dependence yields an average activation energy corresponding to a temperature around 500 K. Since the maximum is fairly broad, one has to assume a broad distribution of barrier heights around this average value.

At temperatures above 100 K, one will expect these relaxations to appear in the range around 100 GHz which is accessible to far infrared absorption and to high resolution Raman and neutron scattering. There is indeed an anomalous temperature dependence in the measured intensities (anomalous from the point of view of a purely harmonic density of states) which begins to be visible below 300 GHz in the infrared absorption[16] and below 500 GHz in Raman[15] and neutron[26] scattering. This anomalous temperature dependence can be explained in a picture of relaxational modes with barrier heights and barrier distributions close to those found at

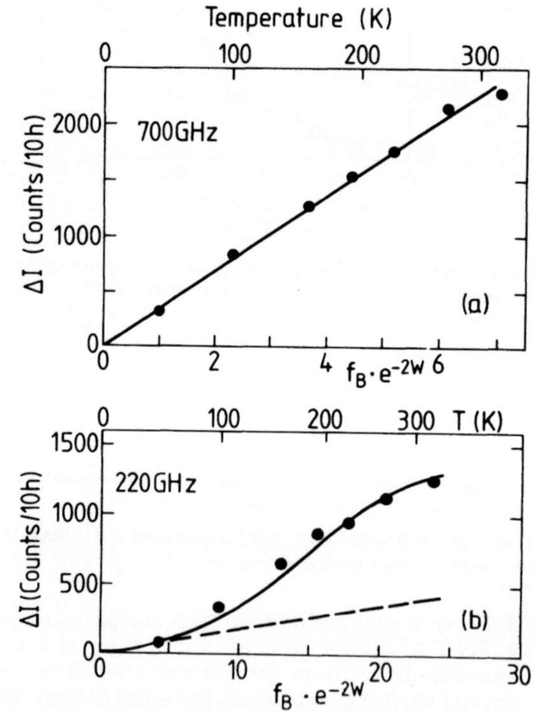

Fig. 2. Temperature dependence of inelastic neutron scattering intensities at (a) 700 GHz (b) 220 GHz (see text). The dashed line in (b) denotes a small harmonic contribution.

lower frequencies. There are problems, however, in accounting for the observed intensities in a picture of relaxational modes alone, especially at the higher frequencies[16,27]. At about 300 GHz, the vibrational modes begin to outweigh the relaxation modes independent of temperature. This is demonstrated by the neutron results[26] at 700 GHz (fig.2 (a)) and at 220 GHz (fig.2 (b)). The intensity is plotted as a function of $f_B.e^{-2W}$ (f_B Bose factor, e^{-2W} Debye-Waller factor), which at these frequencies and temperatures corresponds essentially to a plot vs. temperature. At 700 GHz, the intensity shows the linear increase expected for harmonic excitations. At 220 GHz the behaviour is different and begins to indicate a transition to relaxational behaviour. The intensity rises much higher than one would extrapolate from the point at 50 K. Judging from the low temperature specific heat data[4], most of the signal at room temperature must be due to strongly anharmonic motion. Fig. 1 (b) shows the dynamical structure factor of these relaxations. It coincides within the experimental accuracy with the dynamical structure factor of the harmonic vibrations. This is a very important point because it links the relaxational motion at low frequencies to the harmonic vibrations at higher frequencies. We are forced to conclude

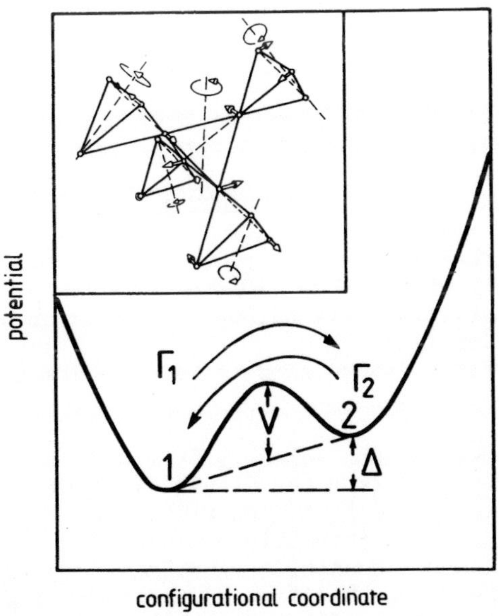

Fig. 3. Asymmetric double minimum potential. Insert : Model of coupled rotational motion of five tetrahedra as a possible configuration coordinate.

that the structural relaxation in vitreous silica from one stable glass configuration to another happens via coupled rotation of essentially undistorted SiO_4 tetrahedra. This implies the existence of double-minimum potentials in that configurational coordinate (fig.3). As we will discuss in the next section, double-minimum potentials are necessary to explain the tunneling states which according to the tunneling model[5,6] are responsible for the low temperature anomalies below 1 K.

TUNNELING STATES

Though the number of modes in this third group of low frequency excitations in glasses is fairly small (1 ppm of the total number of vibrational states) they are by far the most conspicuous. The reason for this is their low frequency. They seem to have a constant density of states below 100 GHz and give consequently rise to a linear term in the specific heat[28] which begins to surpass the T^3 Debye contribution of the sound waves around 1 K. They have been very heavily studied in the last fifteen years. Since there are excellent reviews on this topic[1,2,29], we will only summarize the results. A consistent description of nearly all experiments has been possible in terms of the tunneling model[5,6]. The tunneling model proposes that some atoms or some group of atoms can occupy one of two (or more) potential minima. At low temperatures the quantum-mechanical tunneling of the atoms from one minimum to another gives rise to a very small energy splitting (of the order of 1 K). Assuming slowly varying distribution functions of the energy barriers between the minima and of the energy differences between adjacent minima (the asymmetry of the double minimum potential, see fig. 3) one can explain a wide variety of experimental results.

CONCLUSIONS

The discussion in the preceding sections has shown that there are four main groups of excitations in vitreous silica below 1 THz

(i) Sound waves with a well defined wave vector and a constant velocity
(ii) Harmonic vibrations which predominantly involve coupled rotations of relatively undistorted SiO_4 tetrahedra
(iii) Thermally activated relaxational jumps which seem to have the same eigenvector as (ii)
(iv) Two-level-states which seem to be well described by the tunneling model[5,6].

The first two groups are clearly harmonic in nature, while the last two represent transitions from one stable glass structure to another. The only difference between the last two groups is that the motion in the relaxation case occurs by a thermally activated jump at temperatures above 20 K, while the quantum mechanical tunneling does not need the thermal activation to pass the energy barrier. If the thermally activated motion is indeed a coupled rotation of relatively undistorted tetrahedra, the tunneling motion is probably of the same type. thus, though the evidence presented is not fully conclusive, the experiments give clear indications for a common origin of the low temperature anomalies in vitreous silica.

This conclusion is further supported by theoretical considerations. The way in which frequencies can approach zero in a structure based on linked tetrahedra, and how the structure becomes locally unstable has been demonstrated both by considering the effect of the surrounding solid on a pair of tetrahedra[30] and by model calculations[31]. The resulting two-well potentials are consistent with what is known about tunneling states at low temperatures[29].

REFERENCES

1. Amorphous Solids-Low Temperature Properties, vol. 24 of Topics in Current Physics ed. W.A. Phillips, (Springer-Verlag), (1981).
2. S. Hunklinger, and W. Arnold, Physical Acoustics, vol. XII ed. W.P. Mason and R.N. Thurston, p. 155, (Academic Press), (1976).
3. U. Buchenau, N. Nücker, and A.J. Dianoux, Phys. Rev. Lett. 53:2316 (1984).
4. U. Buchenau, M. Prager, N. Nücker, A.J. Dianoux, N. Ahmad, and W.A. Phillips, Phys. Rev. B34:5665 (1986).
5. W.A. Phillips, J. Low Temp. Phys. 7:351 (1972).
6. P.W. Anderson, B.I. Halperin and C.M. Varma, Phil. Mag. 25:1 (1972).
7. M. Rothenfusser, W. Dietsche, and H. Kinder, Phys. Rev. B27:5196 (1983).
8. M.P. Zaitlin, and A.C. Anderson, Phys. Rev. B12:4475 (1975).
9. J. Jäckle, "The Physics of Non-Crystalline Solids", ed. G.H. Frischat, p. 568, (Transtech), (1977).
10. W. Dietsche, and H. Kinder, Phys. Rev. Lett. 43:1413 (1979).
11. R.O. Pohl. in ref.1, p. 27.
12. N. Ahmad, Thesis, Cambridge, 1987.
13. W.A. Kamitakahara, M. Prager, N. Nücker, and U. Buchenau, (unpublished).
14. U. Buchenau, H.M. Zhou, N. Nücker, and W.A. Phillips, (unpublished).
15. G. Winterling, Phys. Rev. B12:2432 (1975).
16. U.Strom, and P.C. Taylor, Phys. Rev. B16:5512 (1977).
17. J.M. Carpenter, and C.A. Pelizzari, Phys. Rev. B12:2391 (1975).
18. K. Froböse, J. Jäckle, J. Phys. F7:2331 (1977).
19. P.A.V. Johnson, A.C. Wright, and R.N. Sinclair, J. Non-Cryst. Solids 58:109 (1983).
20. N.M. Elcombe. Proc. Phys. Soc. London 91:947 (1967).
21. H.K. Barron, C.C. Huang, and A. Pasternak, J. Phys. C9:3925 (1976).
22. W.A. Phillips, (private communication).
23. J.D. Axe, and G. Shirane, Phys. Rev. B1:342 (1970).
24. H. Grimm, and B. Dorner, J. Phys. Chem. Solids 36:407 (1975).
25. S. Hunklinger, Proc. Ultrasonic Symp. (IEEE), 443 (1974).
26. U. Buchenau, N. Nücker, H. Zhou, W.A Kamitakahara, and W.A. Phillips (unpublished).
27. N. Theodorakopoulos, and J. Jäckle, Phys. Rev. B14:2637 (1976)
28. R.C. Zeller, and R.O. Pohl, Phys. Rev. B4:2029 (1971).
29. W.A. Phillips, to appear in Rep. Prog. Phys. 50 (1987).
30. U. Buchenau, Solid State Commun. 56:889 (1985).
31. L. Guttman, and S.M. Rahman, Phys. Rev. B33:1506 (1986).

NEW METHODS OF IR SPECTROSCOPIC INVESTIGATION OF AMORPHOUS INSULATING FILMS

Klaus Hübner and Ulf Teschner

Section Physics
Wilhelm Pieck University
Rostock, German Democratic Republic

INTRODUCTION

The infrared (IR) spectroscopy of very thin insulating films on semiconductors has some perspective for the non-destructive analysis of corresponding microelectronic systems. Therefore, the derivation of IR optical equations for insulating films (on semiconductors) of decreasing thickness is of increasing importance. On the basis of the FRESNEL theory we derive relations for the determination of the reflection (R), transmission (T), and attenuated total reflection (ATR) of such films from their dielectric functions taking into account oblique incidence of polarized light. In this way we find a theoretical explanation for the detection and identification of transverse <u>and</u> longitudinal optical phonons by means of oblique incidence of IR light, realized within different experimental modes (R, T, ART). Furthermore, this framework leads to a theoretical basis for the ATR spectroscopy of very thin films as well as to a very effective method of determination of the IR optical (dielectric) functions of the film from corresponding experimental spectra, wich is based on the inversion of the optical relations mentioned above. Finally, it will be shown that the theory developed is in quantitative agreement with corresponding measurements performed for SiO_2 and Al_2O_3 on silicon.

THEORY

For a film-substrate system the complex FRESNEL formulas for the reflected and transmitted amplitudes can be written in the form

$$r_j = \frac{r_{1j} + r_{2j} \exp(2\, i\, \delta\, J)}{1 + r_{1j}\, r_{2j} \exp(2\, i\, \delta\, J)}, \tag{1}$$

and

$$t_j = \frac{t_{1j} t_{2j} \exp(i\, \delta\, J)}{1 + r_{1j} r_{2j} \exp(2i\, \delta\, J)}, \tag{2}$$

respectively[1], where (owing to the oblique incidence of light) it is distinguished between perpendicular ($j = \perp$) and parallel ($j = \parallel$) polarization with respect to the plane of incidence. Here, r_{ij} and t_{ij} are the corresponding FRESNEL coefficients of the two interfaces ($i = 1$ - medium of incidence/film, $i = 2$ - film/substrate), $\delta = 2\pi d/\lambda$ is the film thickness d normalized to the wavelength λ, and $J = \sqrt{\varepsilon - \varepsilon_o \sin^2\varphi}$ involves the complex dielectric function $\varepsilon = \varepsilon_1 + i\varepsilon_2$ of the film material and the angle of incidence φ. Then, the corresponding intensities follow from the relations

$$R_j = |r_j|^2, \qquad T_j = \frac{J_s}{J_o} |t_j|^2, \tag{3}$$

where the expressions

$$\left.\begin{aligned} J_s &= \sqrt{\varepsilon_s - \varepsilon_o \sin^2\varphi}\,, \\ J_o &= n_o \cos\varphi_o \end{aligned}\right\} \tag{4}$$

are functions of the refractive indices of the two adjacent media (index o - medium of incidence, index s - substrate, no index - film) and of the angle of incidence.

For very thin films ($\delta \ll 1$) expressions involving the term δJ are approximated by corresponding quadratic functions of this argument. With this approximation the components of the reflection and transmission read

$$R_\perp = \frac{A^2 - 2\delta A\,\varepsilon_2}{B^2 + 2\delta B\,\varepsilon_2}\,, \tag{5}$$

$$R_\parallel = \frac{D^2 - 2\delta D\,(J_o J_s \varepsilon_2 - \varepsilon_o \varepsilon_s G\eta)}{E^2 + 2\delta E\,(J_o J_s \varepsilon_2 + \varepsilon_o \varepsilon_s G\eta)}\,, \tag{6}$$

and

$$T_\perp = \frac{4\,J_o J_s}{B^2 + 2\delta B\varepsilon_2}\,, \tag{7}$$

$$T_\parallel = \frac{4\,J_o J_s \varepsilon_o \varepsilon_s}{E^2 + 2\delta E\,(J_o\, J_s \varepsilon_2 + \varepsilon_o \varepsilon_s G\eta)}\,, \tag{8}$$

respectively, with the abbreviations

$$\left.\begin{aligned} A &= J_o - J_s\,, \qquad & B &= J_o + J_s \\ D &= J_s\varepsilon_o - J_o\varepsilon_s\,, \qquad & E &= J_s\varepsilon_o + J_o\varepsilon_s \\ G &= \varepsilon_o \sin^2\varphi_o & & \end{aligned}\right\} \tag{9}$$

(here, terms of the order δ^2 are not displayed for simplicity). Additionally to the imaginary part of the dielectric function of the film (ε_2), the expressions (5) to (8) involve the so-called energy-loss function $\eta = \mathrm{Im}\,(1/\varepsilon)$ as the second characteristic material parameter. They can be considered as a generalisation of the corresponding special cases derived by Berreman[2], Hübner et al.[3], and Harbecke et al.[4], which is valid also for larger film thickness ($\delta \lesssim 0.1$), if quadratic terms of δ are taken into account[5].

The relation (1) is the theoretical basis for another method for IR spectroscopy of thin absorbing films, namely the attenuated total reflection (ATR). In this case the medium of incidence has a higher value of the refractive index than the substrate ($\varepsilon_o > \varepsilon_s$). Therefore, total reflection appears, if the angle of incidence exceeds the critical value φ_c, which is given by the relations

$$J_s = \theta, \qquad \varepsilon_s = \varepsilon_o \sin^2\varphi_c . \tag{10}$$

If this reflection takes place at an absorbing layer, the light is attenuated in correspondence to its absorption due to the finite penetration depth d_e[6]. It is noteworthy that for the ATR at thin films with $d < d_e$ the condition (10) is independent of the optical properties of the film material. The substitution

$$J_s = i.J_s' = i\sqrt{\varepsilon_o \sin^2\varphi - \varepsilon_s} \tag{11}$$

in the FRESNEL coefficients leads (after representation of expression (1) as a TAYLOR series expansion and taking the amount) the relations

$$\mathrm{ATR}_\perp = \frac{(\varepsilon_o - \varepsilon_s) - 2\delta\left[J_o\varepsilon_2 - J_s'(\varepsilon_o - \varepsilon_1)\right]}{(\varepsilon_o - \varepsilon_s) + 2\delta\left[J_o\varepsilon_2 + J_s'(\varepsilon_o - \varepsilon_1)\right]}, \tag{12}$$

and

$$\mathrm{ATR}_\parallel = \frac{(R^2 + S^2) - 2\delta\left[J_oJ_s'(S\varepsilon_s - R\varepsilon_1) + \varepsilon_o\varepsilon_s G\eta\,(R - S\varepsilon_1/\varepsilon_2) + S\varepsilon_o\varepsilon_s\right]}{(R^2 + S^2) + 2\delta\left[J_oJ_s'(S\varepsilon_2 + R\varepsilon_1) + \varepsilon_o\varepsilon_s G\eta\,(R + S\varepsilon_1/\varepsilon_2) - S\varepsilon_o\varepsilon_s\right]} \tag{13}$$

with the abbreviations

$$R = J_o\varepsilon_s, \; S = J_s'\varepsilon_o . \tag{14}$$

These relations enable quantitative ATR spectroscopy of thin films and lead to an explanation of the similarity of the spectra with corresponding transmission spectra.

It can be seen from the relations (5) to (8), (12), and (13) that the imaginary part ε_2 of the dielectric function and the energy loss function η are the characteristic material parameters, which determine essentially the results of the IR spectroscopy of thin films. The components polarized perpendicular to the plane of incidence represent essentially the function ε_2, whereas the parallel components are influenced by ε_2 as well as η. Since maxima of ε_2 are connected with transverse optical (TO) and maxima of η with longitudinal optical (LO) phonon modes, respectively, under the condition of oblique incidence of light the IR spectroscopy enables the detection and identification of TO- and LO-modes within one and the same experiment. In the special case of the critical angle the parallel component of the ATR couples only with LO-modes.

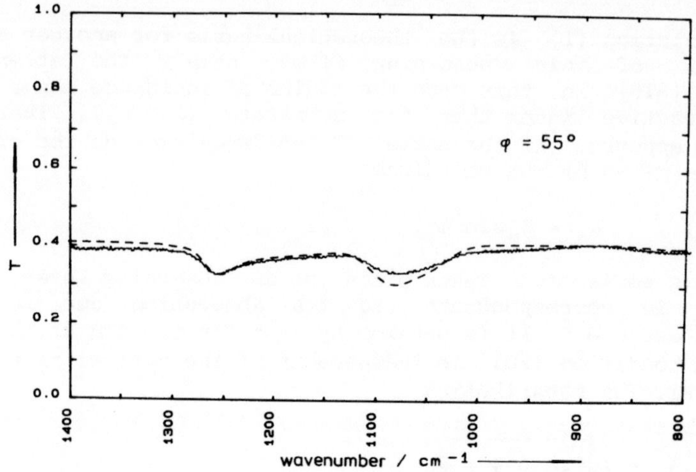

Fig. 1. Transmission T of an SiO_2 film (d = 90 nm) on silicon at oblique incidence of light in the spectral range of the LO- and TO-frequencies of the oxygen-stretching motion (- experimental, --- theoretical).

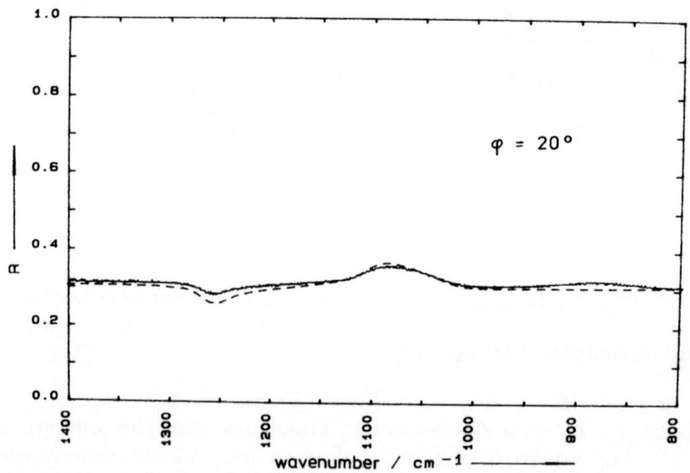

Fig. 2. Reflection R of an SiO_2 film (d = 90 nm) on silicon at oblique incidence of light in the spectral range of the LO- and TO-frequencies of the oxygen-stretching motion (- experimental, --- theoretical).

COMPARISON WITH EXPERIMENT

For the experimental verification of the relations derived, transmission and reflection spectra of a thermally grown 90 nm SiO_2 layer on a silicon substrate were recorded at oblique incidence of light in the spectral range of the oxygen-stretching modes[7] in comparison with a reference Si substrate. Furthermore, corresponding model spectra were calculated from the IR optical functions of quartz glass published by Gaskell and Johnson[8], assuming that the difference between data of film and bulk material can be neglected.

The comparison between experimental and calculated spectra performed in fig. 1 and fig. 2 shows that the relations derived are in qualitative and quantitative agreement with experiment.

Finally, it is shown that the relations derived can be used in a very effective way for the determination of optical functions of thin films from IR spectra. The combination of the expressions (5) and (7) in the form

$$\frac{1 - R_{\perp}}{T_{\perp}} = 1 + \frac{\delta\varepsilon_2}{J_s} \tag{15}$$

leads to a mutual compensation of terms of the order δ^2, which enables an analytical calculation of ε_2. With the help of the expression (6) and (8) for the parallel components one gets the analogous relation

$$\frac{1 - R_{\parallel}}{T_{\parallel}} = 1 + \delta\left[\frac{J_s}{\varepsilon_s}\varepsilon_2 + \frac{\varepsilon_s}{J_s}G\eta\right] + 2\delta^2\varepsilon_2 G\eta, \tag{16}$$

which can be used for the determination of the loss function η.

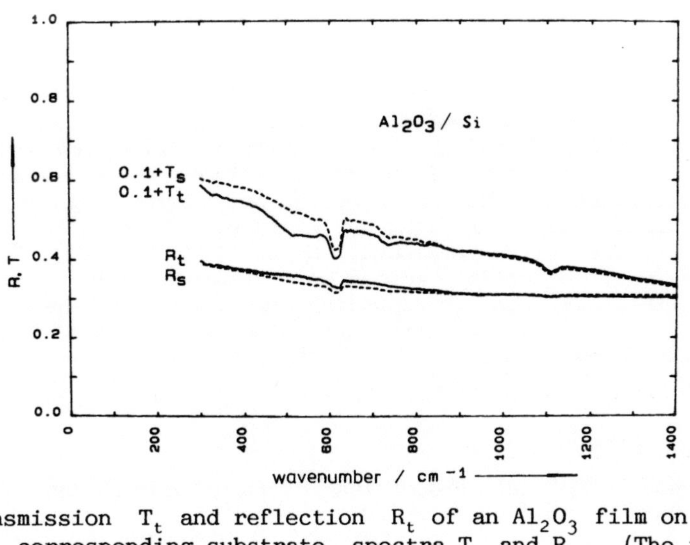

Fig. 3. Transmission T_t and reflection R_t of an Al_2O_3 film on silicon and the corresponding substrate spectra T_s and R_s. (The transmission spectra are shifted by 0.1 units).

In fig. 3 transmission and reflection spectra of an Al_2O_3 film (d = 90 nm) on a polished silicon substrate with rough backside are shown together with the corresponding substrate spectra. From these spectra the pure film spectra were determined by means of a special data transformation[5]. Then, using relation (15) the imaginary part ε_2 of the dielectric function of the film material (fig. 4) was determined from the film spectra. Unfortunately, we did not find an ε_2 spectrum of amorphous Al_2O_3 in the literature for comparison.

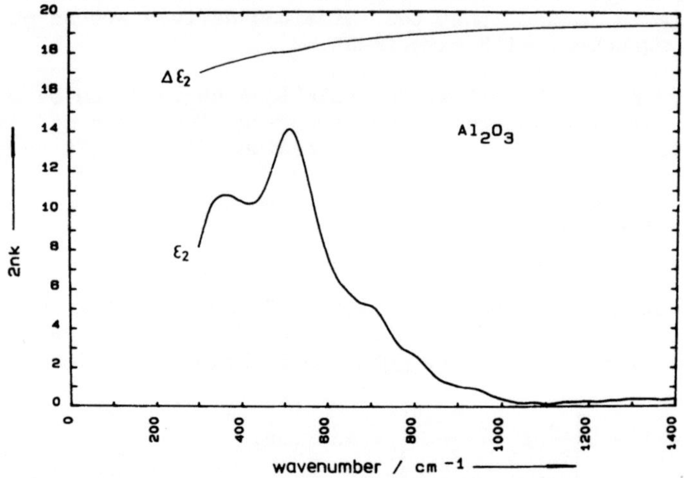

Fig. 4. Imaginary part ϵ_2 of the dielectric function of Al_2O_3 calculated from the spectra shown in fig. 3. The difference between the top abscissa and $\Delta\epsilon_2$ represents the error.

CONCLUSION

In conclusion it was shown that special IR optical equations for the description of the transmission, reflection, and attenuated total reflection of thin films on semiconductors can be derived taking into account oblique incidence of polarized light. These relations give the theoretical explanation for the detection and identification of transverse and longitudinal optical phonons by means of oblique incidence of light realized in different experimental modes (T, R, ATR). Futhermore, we developed a very effective method of determination of the dielectric functions of the film from corresponding experimental spectra. A comparison of calculated model spectra with corresponding experimental spectra shows good quantitative agreement.

REFERENCES

1. O.S. Heavens, "Optical Properties of Thin Solid Films," Butterworths, London, 1955.

2. D.W. Berreman, Phys. Rev. 130 : 2193 (1963).

3. K. Hübner, L. Schumann, A. Lehmann, H.H. Vajen, and G. Zuther, Phys. Stat. Sol. (b) 104 : K1, (1981)

4. B. Harbecke, B. Heinz, and P. Grosse, Appl. Phys. A 38 : 263 (1985).

5. U. Teschner, Thesis, Wilhelm Pieck University, Rostock, 1985.

6. N.J. Harrick, "Internal Reflection Spectroscopy," (Interscience Publishers/ J. Wiley & Sons, New York), (1967).

7. A. Lehmann, L. Schumann, and K. Hübner, Phys. Stat. Sol. (b) 117:689 (1983).

8. P.H. Gaskell and D.W. Johnson, J. Non-Cryst. Solids 20 : 153 (1976).

VIBRATIONAL STUDIES OF AMORPHOUS SiO_2

Paul McMillan

Department of Chemistry
Arizona State University
Tempe
Arizona 85287

INTRODUCTION

There have been many studies of the vibrational structure of SiO_2 glass, using infrared, Raman, neutron and other spectroscopies. Over the past few years, we have used vibrational spectroscopy, especially Raman scattering, to investigate glassy silicas prepared in a variety of ways or subjected to different treatments. Some of our work is summarized here, and the results are discussed in the light of recent models for the structure and dynamics of vitreous SiO_2.

THE RAMAN SPECTRUM OF "NORMAL" SILICA GLASS

The Raman spectrum of "normal" vitreous silica has been described by many authors[1,2,3]. The principal feature is a strong, polarized, asymmetric band near 430 cm^{-1}. Two weak maxima may be distinguished in VV polarized spectra near 270 and 380 cm^{-1}, and a further weak depolarized band may be present near 120 cm^{-1}. These observations suggest that the assymetry in the 430 cm^{-1} band may be due to the presence of unresolved components, and some authors have attempted to decompose this region into Gaussian bands[3]. Two weak, polarized lines are observed at 495 and 606 cm^{-1}, known as the D_1 and D_2 "defect" bands. An asymmetric band occurs near 800 cm^{-1}. This appears to consist of two components : a depolarized band near 790 cm^{-1}, and a component with lower depolarization ratio near 820 cm^{-1}. Dry samples of vitreous silica show[4] a weak polarized feature near 910 cm^{-1}. "Wet" samples show an additional polarized band near 970 cm^{-1} due to vibration of SiOH groups[4-6]. There are two weak depolarized bands near 1060 and 1200 cm^{-1}. The 1200 cm^{-1} band is broad, and some authors have suggested that this contains two component bands[7].

Isotopic substitution experiments[8] and a number of vibrational calculations[2] have allowed some description of the modes responsible for the Raman spectrum of vitreous SiO_2. The major 430 cm^{-1} band is due to predominantly oxygen motion, probably a symmetric oxygen vibration in the plane bisecting the SiOSi linkage. The position of this band is highly sensitive to the SiOSi angle, increasing in frequency with smaller angle. The bands near 800 cm^{-1} have a large component of both silicon and oxygen motion. These may be described as derived from the asymmetric stretching of

SiO_4 groups, or as stretching vibrations of the SiOSi linkage. The bands above 1000 cm^{-1} are described as asymetric stretching combinations of Si-O stretching vibrations within the SiOSi linkage, and contain both Si and O displacement components. These bands shift to lower frequency with smaller SiOSi angle. The shifts in the 430 and 1060/1200 cm^{-1} bands with SiOSi angle may be understood in terms of network dynamics models[9]. Galeener and Lucovsky[10] have suggested that the 1200 cm^{-1} band appears due to TO-LO splitting in the glass, and is a longitudinal mode asociated with the 1060 cm^{-1} band. The hyper-Raman polariton experiments of Denisov and co-workers[11] suggest that the LO component actually lies near 1255 cm^{-1}, and the 1200 cm^{-1} band observed in the normal Raman spectrum is likely due to an independent normal mode of the glass.

The D_1 and D_2 "defect" peaks are much narrower than other features in the Raman spectrum of SiO_2 glass, are highly polarized, and show only oxygen displacement. One interpretation of these observations is that the defect peaks correspond to modes which are vibrationally decoupled from the glass network, and are vibrations of structural units with relatively constant geometry[12]. An alternative view recently suggested by Phillips[13] is that these correspond to highly correlated surface phonon modes associated with internal surfaces within the glass structure. There have been many suggestions for the interpretation of these "defect" peaks[1,13,14] Galeener[14] has assembled a considerable body of evidence to suggest that these modes are due to the symmetric oxygen breathing modes of three-membered (D_2) (boroxol-like) and four membrered (D_1) silicate rings embedded in the glass structure. The strongest arguments for Galeener's model are based on comparison with the oxygen symmetric stretch frequencies of siloxane molecules and crystalline compounds containing three- and four-membered silicate rings, ab initio molecular orbital calculations of the oxygen stretching frequencies expected for three- and four-membered rings[15], and a range of arguments against the likelihood of the various other models[1,14]. The major argument against the surface phonon model suggested by Phillips[13] is that the surface Si=O groups are too unstable to hydration or polymerization reactions to exist in the glass at the concentration needed for the collective surface mode[15]. The three- and four-membered ring model of Galeener provides an economical and physically plausible interpretation of the D_1 and D_2 Raman bands of vitreous SiO_2, and satisfies all the experimental tests available to date. A further "defect" band in vitreous SiO_2 is the weak, polarized Raman mode near 910 cm^{-1}. This is present in both wet and dry samples, and does not scale with the 495 or 606 cm^{-1} defect bands[4,16]. This band occurs in a region commonly assigned to Si-O stretching vibrations, and could well be due to the symmetric Si-O stretching vibration of an $\equiv Si\text{-}O^-$ or $\equiv Si\text{=}O$ group, as suggested in the models of Stolen and Walrafen[4] and Phillips[13].

THE EFFECT OF PRESSURE ON VITREOUS SILICA

A number of authors have studied the effects of densification on the Raman spectra of SiO_2 glass[3]. Walrafen and Krishnan[17] subjected a silica sample to 9 GPa hydrostatic pressure at room temperature and obtained an 8 % densification in the recovered sample. The only changes observed by these authors were a minor narrowing and a slight upward frequency shift of the dominant 430 cm^{-1} band on densification, consistent with a slight decrease in the mean SiOSi angle. There were no observable changes in the D_1 or D_2 bands, but the weak 910 cm^{-1} band was slightly more prominent in the compacted sample. The similarity of the normal and densified glass spectra suggest taht there was little change in the glass network on densification, despite the 8 % decrease in molar volume. This could be rationalized if densification at low temperature were achieved via cooperative rotations of adjacent SiO_4 tetrahedra about SiOSi linkages with no change in the network connectivity. The slight increase in the intensity of the 910 cm^{-1} band could suggest some Si-O bond breaking occured during densification.

McMillan and co-workers[16] studied a sample densified at lower presure (3.95 GPa) but high temperature (530°C) to give a 5.8 % densification (lower than that obtained by Walrafen and Krishnan[17]). In contrast to Walrafen and Krishnan's results, the Raman spectrum of this sample was quite different from that of bulk SiO_2 glass (fig.1). The 430 cm^{-1} band was much narrower (a 35 % reduction in bandwidth, compared to the 6 % reduction reported by Walrafen and Krishnan), and moved to significantly higher frequency (to near 470 cm^{-1}). The weak bands near 1060 and 1200 cm^{-1} decreased in frequency by 40-50 cm^{-1}, the 606 cm^{-1} D_2 band showed an obvious increase in relative intensity, and the weak 910 cm^{-1} band disappeared. The changes in the 430, 1060 and 1200 cm^{-1} bands are consistent with a smaller average SiOSi angle, and a smaller distribution of SiOSi angles, in the densified sample. This is similar to the interpretation of Walrafen and Krishnan's result, but the change is greater in degree for a sample densified at lower pressure to a lower percent densification. The increased intensity of the D_2 band suggests a greater proportion of three-membered siloxane rings in the high temperature densified sample, implying that the higher temperature has allowed bond-breaking and bond-making to occur, to give a densified glass with different network connectivity. Finally, the disappearance of the 910 cm^{-1} band suggests that non-bridging ($Si-O^-$ or Si=O) units present in the undensified glass have been annealed out in the high temperature, 3.95 GPa densification.

Hemley and co-workers[18] have obtained Raman spectra of SiO_2 glass subjected to high pressures up to 27.3 GPa at room temperature, in situ in a diamond anvil cell. Up to 8 GPa, there was a continuous narrowing and shift to higher frequency of the 430 cm^{-1} band (to near 530 cm^{-1} at GPa) and a frequency decrease in the 1060 cm^{-1} band (1052 cm^{-1} at 8 GPa). These effects on the Raman spectrum were reversible up to 8 GPa, consistent with the small changes observed by Walrafen and Krishnan[16] on their quenched

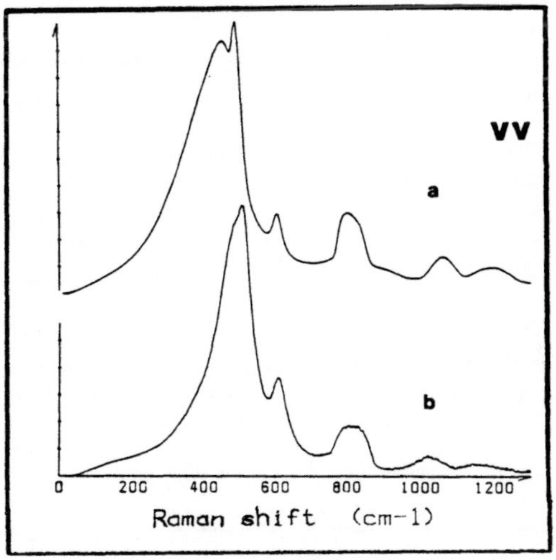

Fig. 1. Raman spectra (VV polarization) of (a) normal and (b) densified glass from ref. 16.

sample, and suggested that no bond breaking took place. At 16.1 GPa and above, the spectra changed irreversibly, with obvious enhancement of the 606 cm^{-1} D_2 band, and perhaps also the D_1 band. This is consistent with the occurence of bond breaking, and reformation of a higher concentration of small siloxane rings in the high pressure glass. These observations are similar to those of McMillan and co-workers[17] at lower pressure (3.95 GPa), but high temperature.

The above experiments suggest that two densification mechanisms operate in vitreous SiO_2. The first involves no bond breaking, but is simply a re-arrangement of the structure to fill available space more efficiently, probably via cooperative rotations about Si-O bonds. The second involves a change in network connectivity, with smaller average SiOSi angles and smaller rings of SiO_4 tetrahedra favoured at high pressure. There is a kinetic barrier to Si-O bond breaking, which may be overcome on a laboratory time scale by application of several hundred degrees Celsius at lower pressure (4 GPa), or static pressures of above 9 GPa at room temperature. The work of Hemley and co-workers[18] suggests a possible third mechanism operating at very high static pressures, above 25 GPa. Their spectra began to broaden and lose Raman scattered intensity at these pressures, which could be associated with the onset of coordination disorder in the local environment around silicon, analogous to changes observed in the spectra of glasses along the SiO_2- Al_2O_3 join[19].

We have recently used Raman spectroscopy to study a suite of natural quartz samples, artificially shocked to peak pressures of near 30 GPa[20]. Below 15 GPa peak shock pressure, there is no change in the quartz

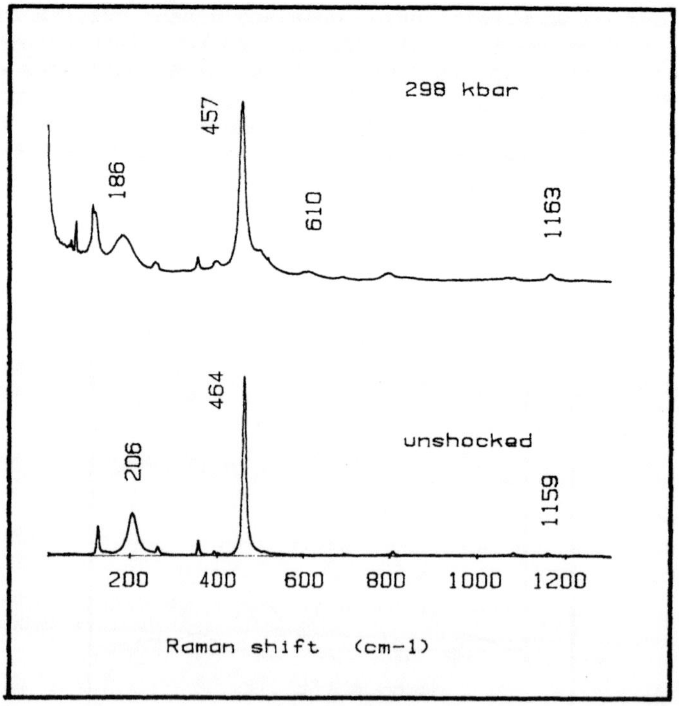

Fig. 2. Effects of 298 kbar shock pressure on quartz[20]. The peak at 610 and the shoulder near 500 cm^{-1} are due to high density diaplectic glass.

spectrum, but above this pressure, the spectra change abruptly (fig.2). All peaks shift in a direction opposite to that observed for quartz under hydrostatic pressure[21], suggesting that the quartz in the shocked samples is under tensile stress. At the same time, a set of weak broad bands appear in the spectrum of the shocked quartz. These correspond to the spectrum of a densified SiO_2 glass, with enhanced intensity of the 600 cm^{-1} D_2 band. We suggest that the effect of the shock wave, above 15 GPa peak pressure, has been to vitrify some of the quartz to form a high density glass with a large concentration of three-membered siloxane rings. On release of the shock pressure, the crystalline part of the sample has attempted to relax to its room pressure volume, but has been constrained by the presence of the permanently densified glass, resulting in tensile stress on the crystal. We are currently analyzing these data[20] to gain a better understanding of the shock process. It is possible that this analysis may also be applicable to the spectra of quartz subjected to intense doses of neutron radiation[22].

THE EFFECT OF WATER AND CHEMICAL VAPOUR DEPOSITION

The spectrum of vitreous SiO_2 containing several hundred to several thousand ppm OH is well known. There is a weak polarized band near 970 cm^{-1} due to the Si-OH stretching vibration of silanol groups, and an asymmetric band with peak maximum near 3690 cm^{-1} due to the O-H stretch[4-6]. Walrafen and Samanta[23] have deconvoluted the O-H stretching region into four component bands associated with four distinct types of OH site. However, McMillan and Remmele[6] concluded that there was no evidence for unresolved component bands, and that the asymmetry in the 3970 cm^{-1} band was due to a continuous range in hydrogen-bonded environments.

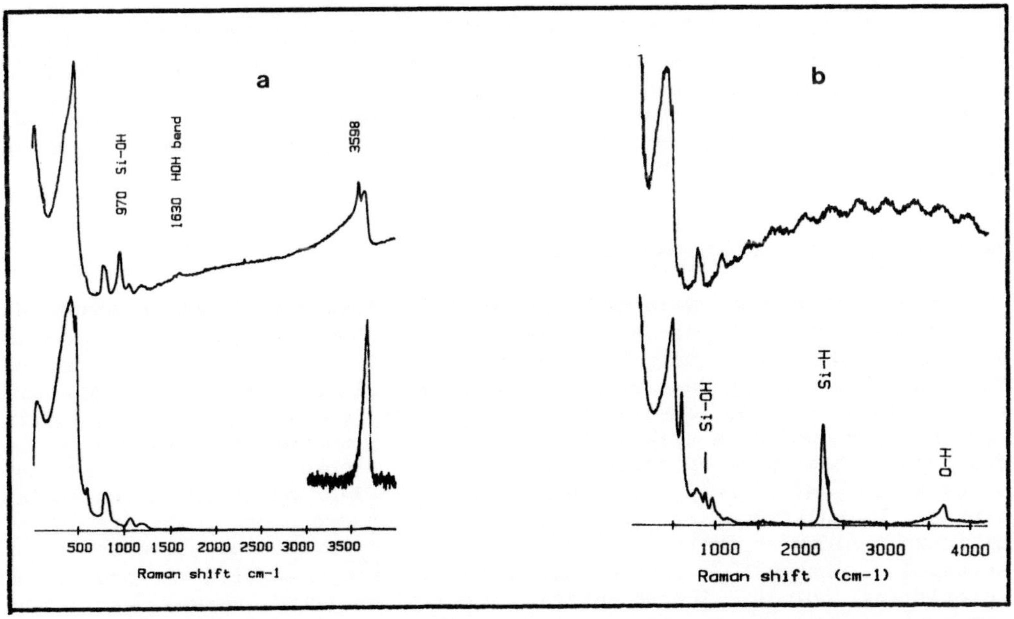

Fig. 3. (a) Raman spectra of SiO_2 glass with 1200 ppm H_2O (Suprasil : bottom) and 6 weight per cent H_2O (top)[6]. (b) Raman spectra of SiO_2 film prepared by chemical vapour deposition at 523 K. Bottom : as-deposited ; top : annealed at 1173 K[25].

McMillan and Remmele[6] and Mysen and co-workers[24] have studied the Raman spectra of hydrous SiO_2 glasses quenched from melts with higher water contents, up to 10 weight per cent H_2O (fig. 3a). The 970 cm^{-1} band increases with increasing water content, while bands due to molecular H_2O dissolved in the glass appear at 1630 cm^{-1} (HOH bending) and in the O-H stretching region. The principal maximum in the O-H stretching region moves to lower wavenumber (3670 cm^{-1}), suggesting increased hydrogen bonding in the silanol groups. The 606 cm^{-1} D_2 peak decreases in intensity with water content, suggesting that three-membered rings are preferentially attacked during the water dissolution process. Finally, a new peak appears in the O-H stretching region at 3598 cm^{-1}, which has not yet been assigned. This peak is narrow and highly polarized, suggesting an O-H species involved in little or no hydrogen bonding.

Huffman and McMillan[25] have investigated the Raman spectra of SiO_2 glasses prepared via chemical vapour deposition from SiH_4-O_2 gas mixtures at 500-700 K. There were considerable changes in the spectra before (as-deposited) and after annealing at 900 K (fig. 3b). The spectra of the as-deposited samples contained bands due to unreacted Si-H groups and O-H species in the glass, which disappeared on annealing. A peak at 970 cm^{-1} was present, associated with SiOH species. A second peak was present at 885 cm^{-1} in the as-deposited samples which has not yet been assigned. This has only been observed in amorphous SiO_2 prepared under certain special conditions[26]. It is possible that this peak might correspond to silanol groups with two OH groups per silicon, $=Si(OH)_2$. This hypothesis has yet to be tested.

The Raman bands due to the SiO_2 framework in the CVD samples also showed considerable changes on annealing. The major band appeared near 470 cm^{-1} and was considerably narrower than the 430 cm^{-1} band of bulk SiO_2. This was consistent with a narrower distribution of SiOSi angles about a smaller mean value in the as-deposited sample. The annealed sample had a spectrum identical with bulk vitreous silica. The 490 and 606 cm^{-1} defect bands were greatly enhanced in the as-deposited sample relative to bulk SiO_2, suggesting either a high concentration of small siloxane rings in the reaction gas phase, or that these units are stabilized at the growing SiO_2 surface. These features also returned to their "normal" relative intensities on annealing[25].

VIBRATIONAL CALCULATIONS ON VITREOUS SIO_2

There have been a large number of calculations designed to model the vibrational behaviour of amorphous SiO_2 [2-9]. Most of these calculations have assumed a local central or valence force field with a small number of Si-O bond stretching and OSiO or SiOSi angle bending terms. Although such calculations can be instructive, the force constants are empirically chosen, and while a given force field may successfully reproduce the observed vibrational spectrum, the calculated atomic displacements are not necessarily realistic[27]. We have begun a series of ab initio molecular orbital calculations in order to obtain non-empirical force fields for silicate molecules, which may be applied to vibrational calculations of condensed phases [28]. These calculations have revealed a number of interesting points. The ab initio calculations give values of around 600 Nm^{-1} for Si-O stretching, 40-50 Nm^{-1} for OSiO bending, and 17 Nm^{-1} for SiOSi bending force constants. These values are similar to those commonly assumed in empirical calculations on silicates, hence vibrations with a large component of these terms alone should be quite reliably modelled.

However, we have also found large terms, on the order of 10-30 Nm^{-1} in absolute magnitude, for stretch-stretch, stretch-bend and bend-bend interaction terms. These terms are commonly ignored in empirical calculations, and will affect the form of the calculated normal modes. Our calculations have also revealed a number of interesting dynamic phenomena which may be important for vitreous SiO_2[28,29]. One low frequency torsional mode of the molecule H_4SiO_4 corresponds to a concerted partial rotation of hydrogens about their Si-O bond axes. This mode is strongly coupled to the OSiO angle bending, and as the hydrogens rotate, the OSiO angle varies from 102° and 116°. This suggests that the SiO_4 groups in silicate structures are not rigid as commonly assumed, but that their OSiO angles are easily deformed, within limits of around 102 to 116°. Buchenau and co-workers[30] have recently suggested the existence of coupled torsional motions of adjacent SiO_4 tetrahedra to rationalize low frequency excitations observed in vitreous SiO_2. Our work suggests that these torsions should also be accompanied by large amplitude OSiO bending modes of the SiO_4 tetrahedra. These experimental and theoretical observations are important for understanding the dynamics of vitreous silica, and may shed light on the nature of displacive phase transitions in crystalline polymorphs of SiO_2.

ACKNOWLEDGEMENTS

Most of the work described here was supported by National Science Foundation grants EAR-8401705 and EAR-8616990.

REFERENCES

1. F.L. Galeener, in "The Structure of Non-Crystalline Materials 1982", ed. P.H. Gaskell, J.M. Parker and E.A. Davis, (Taylor and Francis Inc.) pp. 337-359 (1983).
2. P. McMillan, Am. Mineral., 69:622 (1984)
3. G.E. Walrafen, and M.S. Hokmabadi, in "Structure and Bonding in Non-Crystalline Solids", ed. G.E. Walrafen and A. Revesz, (Plenum Press), pp. 185-202 (1986).
4. R.H. Stolen and G.E. Walrafen, J. Chem. Phys. 64:2623 (1976).
5. C.M. Hartwig, and L.A. Rahn, J. Chem. Phys., 67:4260 (1977).
6. P.F. McMillan and R.L. Remmele, Am. Mineral., 71:772 (1986).
7. B.O. Mysen, D. Virgo, and F.A. Seifert, Rev. Geophys. Space Phys., 20: 353 (1982).
8. F.L. Galeener, and J.C. Mikkelsen, Phys. Rev. B23:5527 (1981), F.L. Galeener and A.E. Geissberger, Phys. Rev. B27:6199 (1983).
9. P.N. Sen, and M.F. Thorpe and F.L. Galeener, Phys. Rev. B22:3078 (1980).
10. F.L. Galeener and G. Lucovsky, Phys. Rev. Letts., 37:1474 (1976), also, in "Structure and Excitations of Amorphous Solids", ed. F.L. Galeener and G. Lucovsky, (American Institute of Physics), pp. 223-228 (1976).
11. V.N. Denisov, B.N. Mavrin, V.B. Podobedov, and Kh.E. Sterin, Sov. Phys. Sol. State., 20:2016 (1978), V.N. Denisov, B.N. Mavrin, V.B. Podobedov, Kh.E. Sterin and B.G. Varshal, J. Non-Cryst. Solids, 64:195 (1984).
12. F.L. Galeener, R.A. Barrio, E. Martinez and R.J. Elliot, Phys. Rev. Letts., 53:2429 (1984).
13. J.C. Phillips, Solid State Phys., 37:93 (1982), J. Non-Cryst. Solids, 63:347 (1984).
14. F.L. Galeener, J. Non-Cryst. Solids, 49:53 (1982), Solid State Comm., 44:1037 (1982).
15. M. O'Keeffe, and G.V. Gibbs, J. Chem. Phys., 81:876 (1984).
16. P. McMillan, B. Piriou and R. Couty, J. Chem. Phys., 81:4234 (1984).
17. G.E. Walrafen and P.N. Krishnan, J. Chem. Phys., 74:5328 (1981).

18. R.J. Hemley, H.K. Mao, P.M. Bell and B.O. Mysen, Phys. Rev. Letts., 57:747 (1986).
19. P. McMillan and B. Piriou, J. Non-Cryst. Solids, 53:270 (1982).
20. G. Wolf, P. McMillan and P. Lambert, in preparation.
21. P.T.T. Wong, D.J. Moffatt and F.L. Baudais, Appl. Spectrosc., 39:734 (1985).
22. J.B. Bates, R.W. Hendricks, and L.B. Shaffer, J. Chem. Phys., 61:4163 (1974).
23. G.E. Walrafen, and S.R.J. Samanta, J. Chem. Phys. 69:493 (1978).
24. B.O. Mysen, and D. Virgo, Chem. Geol. 57:333 (1986).
25. M. Huffman, and P. McMillan, J. Non-Cryst. Solids, 76:369 (1985).
26. W.A. Pliskin, and P.P. Castrucci, J. Electrochem. Soc., 6:85 (1968), A. Bertoluzza, C. Fagnano, M.A. Morelli, V. Gottardi and M. Gugliemi, J. Non-Cryst. Solids, 48:117 (1982).
27. I.M. Mills, in "Infrared Spectroscopy and Molecular Structure", ed. M. Davies, (Elsevier Press), pp. 166-198, R.S. Leigh, B. Szigeti, and V.K. Tewary, Proc. R. Soc., A320:505 (1971).
28. M. O'Keeffe, and P. McMillan, J. Phys. Chem., 90:541 (1986), A.C. Hess, P.F. McMillan and M. O'Keeffe, J. Phys. Chem., 90:5661 (1986).
29. A.C. Hess, P. McMillan and M. O'Keeffe, J. Chem. Phys., submitted.
30. U. Buchenau, N. Nucker and A.J. Dianoux, Phys. Rev. Letts., 53:2316 (1984), U. Buchenau, M. Prageer, N. Nucker, A.J. Dianoux, N. Ahmad, and W.A. Phillips, Phys. Rev. B34:5665 (1986).

RAMAN SPECTRA OF SiO_2 FIBERS AT HIGH TENSILE STRAIN

Hiroshi Kobayashi and Kazuhiro Ema*

National Research Laboratory of Metrology
Skura-Mura, Niihari-Gun
Ibaraki 305, Japan
*Department of Applied Physics
University of Tokyo
Bunkyo-Ku, Tokyo 113, Japan

ABSTRACT

Raman spectra were obtained for SiO_2 fibers of 125 μm diameter at tensile strains ranging form 0 to 3.5. %. The spectrum only for the main peak at 440 cm^{-1} decreased its height with increase in the tensile strain. All the Raman spectra observed were decomposed into their Gaussian components, which are then assigned to normal vibration modes of either SiO_4 or Si_2O molecules. It is concluded that the stress-induced change in the Raman spectrum near 440 cm^{-1} is principally due to a change in the tetrahedral angle of the SiO_4 molecules constituting silica networks and not to either a Si-O bond stretching or a Si-O-Si bond angle broadening.

INTRODUCTION

In the Raman spectrum of SiO_2 glass, there are a main, relatively broad, peak at 440 cm^{-1} and two lower but sharper peaks near 490 and 604 cm^{-1}. The latter two peaks are reduced by annealing and susceptible to neutron irradiation of SiO_2 glass[1].

Using high-strength optical fibers, Walrafen et al.[2] have studied a tensile stress effect on Raman peaks of SiO_2 glass. They found that the peak height at 490 cm^{-1} surpassed that at 440 cm^{-1} by applying a tensile stress of 2.8 Gpa to a fiber and that the peak intensity at 440 cm^{-1} was not affected by tensile stress. However, according to Hibino and al.[3] who carried out a similar experiment, it is the intensity at 440 cm^{-1} that decreases with increasing stress, and the peaks at 490 and 604 cm^{-1} are practically unchanged with applied tensile stress. This is obviously contrary to the conclusion of Walrafen et al.. In this study, by deconvoluting the Raman peaks into Gaussian components, we suggest that the stress induced change at 440 cm^{-1} is caused by the distorsion of the SiO_4 tetrahedron.

EXPERIMENTAL

The samples used were synthetic silica fibers of 125 μm diameter and coated with silicone-resin. They were fabricated by the VAD (vapor-phase axial deposition) method. Figure 1 shows an experimental set up. A terminal portion of the fiber, where the silicone-resin was removed, was cemented to the calipers of a micrometer with an adhesive (Aron Alpha). The fiber was strained only between the calipers with a distance of 16.5. mm. A 514.5. nm argon ion laser beam (NEC) with a 0.85 W output was focused to another end of the fiber. The scattered laser light emanating frome the strained fiber portion was partly introduced into a double monochrometer (JASCO CT-1000D) and led to a photomultiplier. The light intensity, measured with a photon counter (HAMAMATSU), was recorded as a function of wave number.

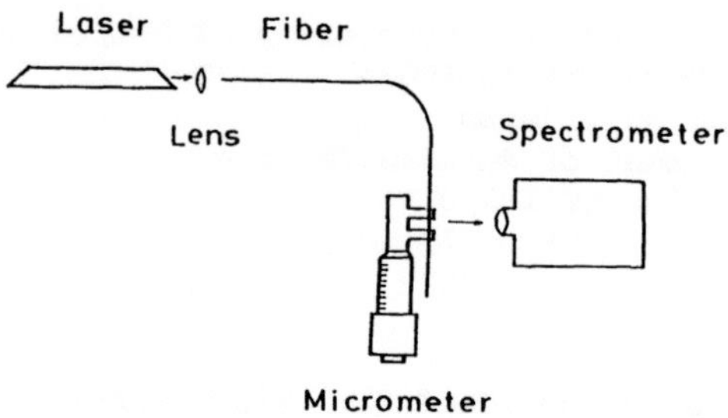

Fig 1. Experimental set up

RESULTS AND DISCUSSION

Figure 2 shows the Raman spectra obtained from the SiO_2 glass at 0 % and 3.5. % tensile strain. The dominant peak at 440 cm^{-1} is seen to be lower at 3.5. % than at 0 % strain. Figure 3 illustrates the results of a deconvolution of these spectra into Gaussain components : they are assigned to normal vibrational modes of either SiO_4 or Si_2O. Hereafter we will principally discuss the change in the V4 mode spectrum due to tensile stress in connection with the glass network structure.

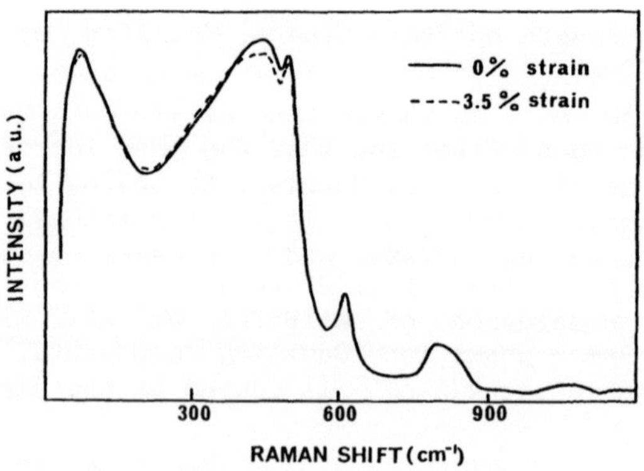

Fig. 2. Raman spectra for a SiO_2 fiber at 0 and 3.5 % tensile strains.

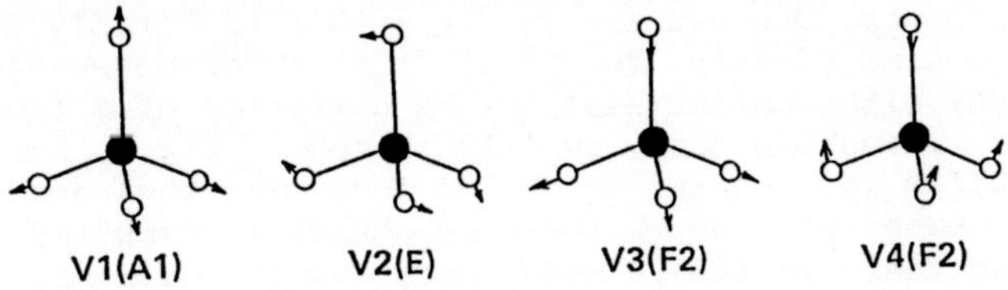

Fig. 3. Normal vibration modes of a tetrahedral molecule.

We can presume that the stress-induced change in bond length must be negligible in comparison with that in bond angles in a stress range such as used in this study, since the force constants for bond-stretching are much greater than those for bond-bending. In other words, stress-induced change in Si-O-Si bond angle θ and O-Si-O bond angle φ must play a significant role in the present analysis of Raman spectrum. In a previous study, however, no tensile stress effects were observed regarding Si-O-Si triangle modes[4]. This indicates that the average of θ is scarcely affected by tensile stress, though θ for each triangle in the glass network may increase or decrease depending on the direction of the Si-Si bond of the triangle concerned, e.g., on whether the direction is parallel or orthogonal to stress axis.

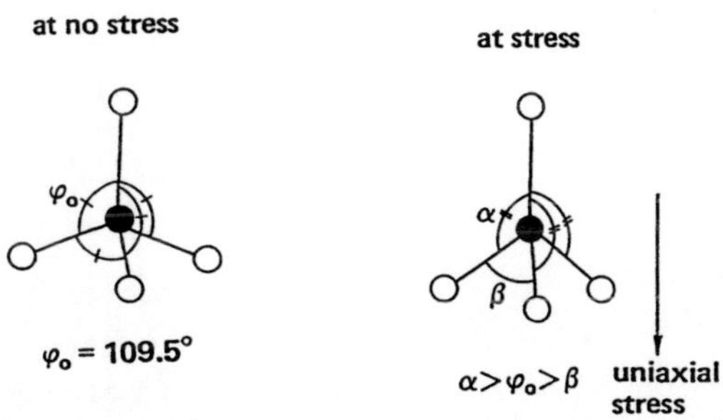

Fig. 4. Schematic illustration for the distortion of a tetrahedral molecule under uniaxial stress

On the other hand, the distortion of a SiO_4 tetrahedron shown in fig.4 must be induced by the uniaxial stress no matter what direction the tetrahedron has (i.e. a tretrahedron has no direction for the sake of its symmetry). We defined the O-Si-O bond angles α and β in the manner shown in fig.4., and calculated the stress dependence of each normal mode frequency of the SiO_4 tetrahedron on α. The results are shown in fig.5, where the dashed line indicates the value at no stress (=109.5°). The tetrahedron has four normal normal modes, i.e., a non-degenerate V1 (808 cm^{-1}), a doubly degenerate V2 (250 cm^{-1}) and two triply degenerate V3 (1065 cm^{-1}),

V4 ($455\ cm^{-1}$) modes. One can see in fig.5 that two triply degenerate modes (V3 and V4) become clearly decoupled into a non-degenerate and a doubly degenerate mode with an increase in the deviation of α from 109.5. Hence, it seems very plausible that, as illustrated in fig.6, the tensile stress will cause shifts in the two decoupled modes and thus change the shape of the V4 mode component. Though the same change is expected for the V3 mode ($1065\ cm^{-1}$) as that in the V4 mode, such change could not be observed in the present study, probably because of the very low intensity at $1065\ cm^{-1}$ equivalent to that of background noise.

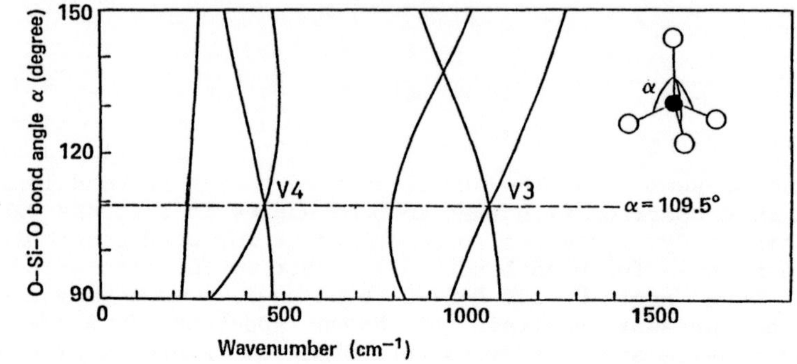

Fig.5. Wavenumber for normal mode vibrations of a SiO_4 tetrahedron as a function of α

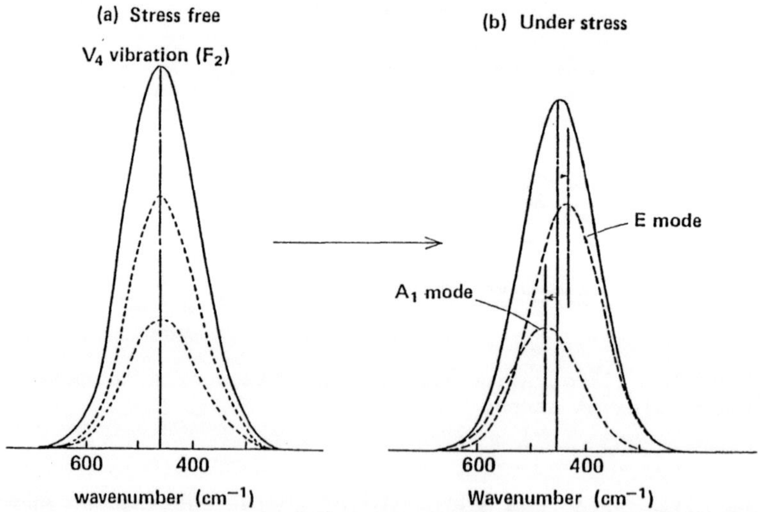

Fig.6. Reduced intensity change associated with the stress-caused decoupling of the V4 mode into a nondegenerate A1 and a doubly degenerate E mode.

We conclude that, under 3.5. % strain, stress-induced effects on the Raman spectra of SiO_2 glass network manifest themselves by way of the distortion of tetrahedrons, and not by way of the bond-to-bond elongation or of the bond breaking. This should be the reason why the elastic hysteresis of SiO_2 glass under 3.5. % strain[5] was not observed so clearly as that for soda glass fibers under 13 % strain[6]. In the glass network, applied tensile stress is considered to induce such change successively with an increase in tensile stress as follows : first the distortion of tetrahedrons, second the elongation of bond lengths and finally bond breakings. A similar study using much higher tensile strain is obviously needed to obtain drastic changes in the Raman spectra, this would be a very difficult experiment.

ACKNOWLEDGMENTS

The authors wish to express great thanks to Professor S. Hyodo for helpful discussions and comments on the manuscript.

REFERENCES

1. R.H. Stolen, and G.E. Walrafen, J. Chem. Phys., 64:2623 (1976).
2. G.E. Walrafen, P.N. Krishnan, and S.W. Freiman, J. Appl. Phys. 52:2832 (1981).
3. Y. Hibino, H. Hanafusa, K. Eam, and S.Hyodo, App. Phys. Lett., 47:812 (1985).
4. K.Ema, Y. Hibino, H. Shigekawa, and S. Hyodo, Jpn. J. Appl. Phys., 26:649 (1978).
5. H. Kobayashi, J. Appl. Phys., 49:4476 (1978).
6. S. Hyodo, and Y. Togami, J. Appl. Phys., 44:2237 (1973).

A COMPARISON OF THE STRUCTURE OF a-SiO_2 PREPARED BY DIFFERENT ROUTES

R. Aujla, R. Dupree, I. Farnan and D. Holland

Physics Department
University of Warwick
Coventry CV4 7AL
United Kingdom

ABSTRACT

Three different silica samples, Heraeus Suprasil I, silica prepared from colloidal silica gel and silica produced by hydrolysis were investigated using Magic Angle Spinning (MAS) ^{29}Si and ^{1}H NMR. Silica produced via the two low temperature routes contained a significant number of Si-OH linkages giving rise to $SiO_3(OH)$ and $SiO_2(OH)_2$ as well as SiO_4 tetrahedra. After heating to 950°C for the hydrolysis prepared sample and 1200°C for the sol-gel material most of the OH was removed and the silicon environment becomes similar to that in melt formed material.

INTRODUCTION

It is well known that vitreous silica is composed of [SiO_4] tetrahedra linked by shared oxygens with the randomness of the structure coming from variations in the intertetrahedral angle [Si-O-Si] . The distribution of these bond angles is a way of describing the short range order present in the glass structure. Neutron and X-ray elastic scattering experiments can give information about the distribution of [Si-O-Si] bond angles but a number of assumptions and many steps are required to deduce the bond angle distribution from the raw data. We have recently demonstrated how it is possible to derive a [Si-O-Si] bond angle distribution from the ^{29}Si MAS NMR lineshape of vitreous SiO_2[1,2]. The ^{29}Si MAS NMR lineshape can thus be used as a measure of the local silicon environment in silica prepared by different routes and at various stages of preparation along those routes.

Silicas which are prepared by gel or hydrolysis routes will contain water or OH groups in amounts which depend on the stage of preparation until the sample temperature is raised to about 1000°C. The presence of protons in the material allows the technique of cross-polarisation[3] to be used. This method involves the transfer of proton magnetisation to ^{29}Si nuclei by applying rotating magnetic fields to both protons and silicons in the sample and arranging for the amplitudes of these fields to be such that a transition in the proton spin system is isoenergetic with a transition in the silicon spin system. This condition brings the two spin systems, which are effectively "hot" (^{29}Si) and "cold" (^{1}H) into thermodynamic contact and order (polarisation) is transferred as they evolve towards a common spin termperature[4]. Those silicons which are closer to protons i.e. those in $SiO_3(OH)$ and $SiO_2(OH)_2$ tetrahedra will be cross-polarised more rapidly and so if the contact time (duration of the two rotating magnetic fields) is short those peaks will be enhanced in the ^{29}Si spectrum. This factor enables the positions and widths of the $SiO_3(OH)$ and $SiO_2(OH)_2$ resonances to be determined with greater accuracy than in the more usual single pulse experiment. However, the cross-polarisation spectrum is not quantitative and quantitative information about the relative amounts of each tetrahedral type is found from the single pulse data where the widths and positions of each component line are determined from the cross-polarisation spectrum.

EXPERIMENTAL

The NMR measurements were carried out using a Bruker MSL 360 spectrometer operating at 71.5 MHz for ^{29}Si. Magic angle spinning at speeds from 3kHz to 4kHz was used together with small angle pulses ($\pi/9$) and relaxation delays of 60 s to obtain the single pulse NMR spectra. Cross-polarisation [^{1}H-^{29}Si] spectra were acquired with a rotating field strength of 50 kHz, contact times from 0.6ms to 20ms and relaxation delays of 5s. In samples where more than one silicon environment was present the relaxation delay in the single pulse experiment was increased to confirm there was no saturation of one resonance relative to another.

Three forms of silica were examined, one produced commercially (Heraeus Suprasil I) and two produced in our laboratory by hydrolysis and by colloidal sol-gel routes.

The gel route requires four stages of preparation, dispersion of fumed silica in alcohol, gelation, drying and consolidation. This sol-gel process is non -aqueous and results in a higher concentration of silica in the gel compared with other gel routes, which reduces both the drying time and shrinkage and drying stresses. The material was produced by dispersing 25 wt% fumed silica (BDH Ltd.) in butanol. Gelation for 4 days was followed by drying at 60°C for 4 days, 120°C for 17 days and finally consolidation at 600°C, 900°C and 1200°C for 1 hour.

The material produced by hydrolysis required the dropwise addition of H_2O to a solution of silicon tetrachloride in ether at 0°C whilst stirring in a nitrogen atmosphere. The solution was then warmed to room temperature and stirred for 2 hours to remove the ether. The resulting white precipitate was washed with distilled water and heated to 225/295°C under vacuum for 24 hours to produce a fine powder. Portions of this sample were then heated to 600°C/2 hours and 950°C/2 hours. A further portion was retained for gravimetric analysis.

RESULTS

Cross-polarisation MAS NMR spectra of the hydrolysis and gel formed silicas at a low temperature stage of preparation are shown in fig. 1. All spectra show features which indicate the presence of SiO_4, $SiO_3(OH)$ and $SiO_2(OH)_2$ units in the samples. The SiO_4 unit in anhydrous vitreous silica gives a peak at -111.5 ppm[2,5] so the peak at -110.2 ppm is consistent with a SiO_4 resonance in the material. The peaks at -91.0 ppm and -101.0 ppm can be compared with shifts obtained for $SiO_2(OH)_2$ (-90.6 ppm) and $SiO_3(OH)$ (-99.8 ppm) units in silica gel[6]. The trend of shifts is also consistent with that observed in other network modified SiO_2 glasses where the number of bridging oxygens bonded to silicon is reduced[7]. A difference can be seen in the hydrolysis formed material between contact times of 5 ms and 10 ms. The shorter contact time enhances the $SiO_2(OH)_2$ peak relative to both $SiO_3(OH)$ and SiO_4 peaks, whereas the longer contact time allows cross-polarisation to silicons in tetrahedra which do not contain an OH group. This is demonstrated by the fact that the SiO_4 peak is resolved for the longer contact time. The relative enhancement of the SiO_4 peak will increase with contact time provided there is not a significant loss of proton magnetisation due to spin lattice relaxation in the rotating frame. The gel formed material appears to show a similar distribution of SiO_4, $SiO_3(OH)$ and $SiO_2(OH)_2$ units to that observed in the hydrolysis formed silica when using a contact time of 5 ms for both samples. However, the single pulse spectrum of the gel formed silica (fig. 2) shows no evidence of $SiO_3(OH)_2$ and the spectrum can be fitted to two gaussians with the $SiO_3(OH)$ concentration being 15 %. The MAS NMR spectrum of the hydrolysis formed silica shows three peaks and these can be fitted using the widths positions of the gaussians established from a fit to the CP spectrum (fig. 3). The relative amounts of the tetrahedral types in the hydrolysis formed silica (56 % Q_4, 34 % Q_3, 10 % Q_2) indicates a distribution which is in agreement with a statistical model for the distribution of OH in the sample. As the gel formed silica has a lower OH concentration very few Q_2 units are expected.

The two forms of silica were subjected to heat treatment at 900°C (gel) and 950°C (hydrolysis) and the MAS spectra are shown in fig. 3 along with the spectra of Suprasil I and the gel formed silica heated to 1200°C. It is evident that the Q_3 and Q_2 type peaks have virtually disappeared in the hydrolysis material when heated to 950°C. This is supported by the proton NMR which gives the OH concentration as 0.1 wt%, which is similar to Suprasil I. In contrast although the gel formed silica retains a relatively

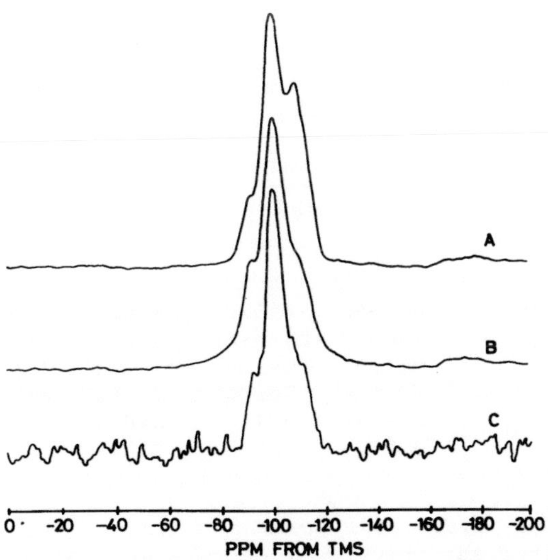

Fig.1. Cross-polarisation Magic Angle Spinning NMR spectra (a) silica formed by hydrolysis and heated to 225°C, 10ms contact (b) silica formed by hydrolysis and heated to 225°C, 5ms contact (c) silica formed from a gel and heated to 225°C, 5ms contact.

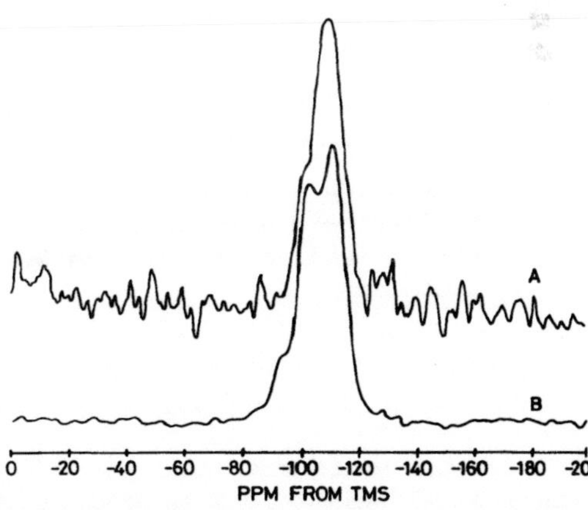

Fig.2. MAS NMR spectra (a) silica formed from a gel and heated to 225°C (b) silica formed by hydrolysis and heated to 225°C.

large concentration of protons as determined by ^{1}H NMR which gives two peaks the ^{29}Si resonance gives no evidence of any $SiO_3(OH)$ units.

Further heating to 1200°C is required to reduce the proton concentration of the gel formed material to the order of Suprasil I (1200 ppm) which then results in a ^{29}Si NMR lineshape similar to Suprasil.

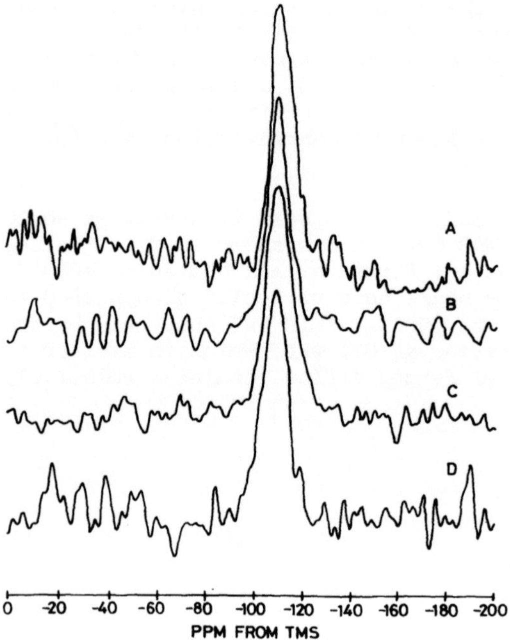

Fig.3. CP MAS NMR spectrum of hydrolysis formed silica heated to 225°C and a fit to three Gaussians.

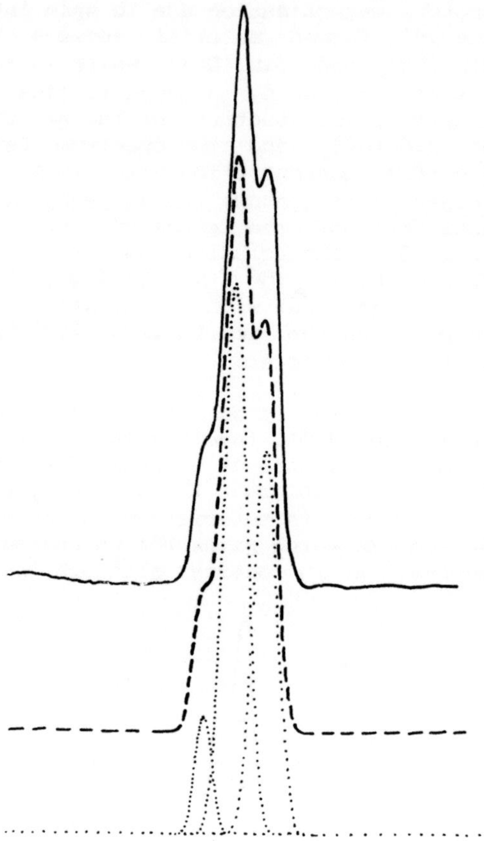

Fig. 4. MAS NMR spectra (a) gel formed silica heated to 1200°C (b) hydrolysis formed silica heated to 950°C (c) Heraeus Suprasil I (d) gel formed silica heated to 900°C.

Thermogravimetric analysis of the hydrolysis formed silica showed high initial weight loss as surface water was removed at temperatures up to 10°C. Further heating resulted in smaller weight losses until at 950°C the sample weight remained constant. By assuming the sample was, by this stage, SiO_2 only the OH content at lower temperatures could be calculated and this is shown in fig. 5.

DISCUSSION

In general the protons retained in these silica materials at the later stage of preparation are, from the NMR spectrum, present as Si-OH. The two

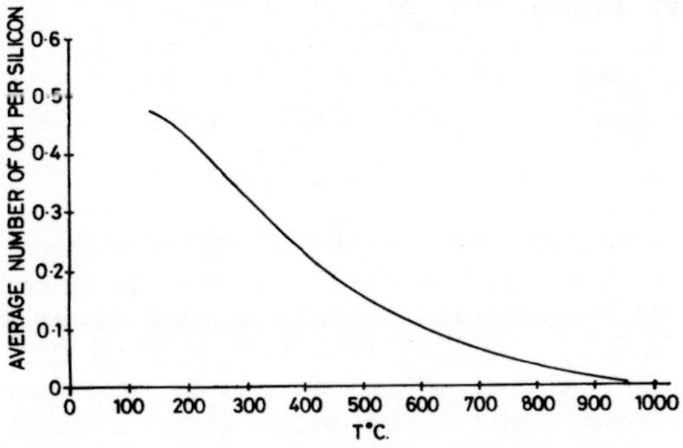

Fig. 5. Thermogravimetric data giving the average number of OH per silicon as a function of temperature for the hydrolysis formed silica.

proton peaks observed in the gel heated to 900°C are shifted from the position in silanol units in agreement with there being no detectable SiO_3 (OH) units in the ^{29}Si spectrum. We speculate that these are due to protons in H_2O which are hydrogen bonded to oxygens either in other H_2O or in Si-O-Si. For crystalline polymorphs of SiO_2 and dealuminised zeolites the ^{29}Si NMR chemical shift varies linearly with the secant of the [Si-O-Si] angle. As shown previously[1,2] the lineshape is determined by a distribution of secants, $\bar{\beta}$, which are an average of the four individual angles, β, of the tetrahedron assuming that these are statistically independent i.e.

$$\bar{\beta} = (\beta_1 + \beta_2 + \beta_3 + \beta_4)/4 \qquad \beta = \mathrm{Sec}\ \theta_i \tag{1}$$

The bond angle distribution can be calculated from

$$F(\bar{\beta}) = X(\beta) * X(\beta) * X(\beta) * X(\beta) \tag{2}$$

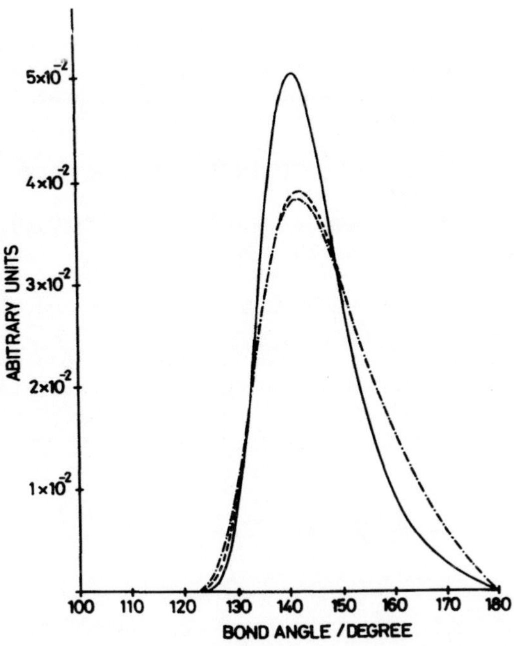

Fig. 6. Si-O-Si bond angle distributions for SiO_4 units in hydrolysis formed silica after heating to 225°C (——) 950°C (---) and Heraeus Suprasil (-.-), normalised to constant area.

where * denotes convolution

and $$n(\theta) = X(\beta) \left|\frac{d\beta}{d\theta}\right| \quad (3)$$

The individual resonances in the CP spectra give an excellent fit to Gaussian lineshapes (Fig. 3) and X(β) can then be calculated analytically. For the MAS spectra a Guassian is still a good first approximation to the linesahpe and the bond angle distributions derived for the Q_4 resonance in the hydrolysis formed silica after low and high temperature treatment are shown in fig. 6 together with the bond angle distribution obtained for Suprasil I. The bond angle distribution of the low temperature hydrolysis silica is narrower and has fewer large angles th Suprasil I, the mean position is also shifted by ~1°. The narrowness of the distribution would indicate a more ordered situation. When the hydrolysis formed silica is heated to 950°C the distribution of bond angles approaches that of Suprasil I, it being only slightly narrower on the low angle side with the same peak angle in the distribution. Similarly the Si-O-Si bond angle distribution in the gel formed material after heat treatment at 1200°C approaches that of Suprasil I.

Thus it would appear that samples of SiO_2 prepared by these different routes converge on a single structure if heated to sufficiently high temperatures but below the softening points of the materials. For the hydrolysed sample a temperature of 950°C is required and for the gel-formed sample a temperature in excess of 900°C.

ACKNOWLEDGEMENTS

We wish to thank the SERC for support of this work.

REFERENCES

1. R. Dupree and R.F. Pettifer, Nature 308:523 (1984).
2. R.A.B. Devine, R. Dupree, I. Farnan and J.J. Capponi, Phys. Rev. 35B:2305 (1987).
3. A. Pines, M.E. Gibby and J.S. Waugh, J. Chem. Phys. 59:569 (1973).
4. S.R. Hartman and E.L. Hahn, Phys. Rev. 128:2042 (1962).
5. L.F. Gladden, T.A. Carpenter and S.R. Elliott, Phil. Mag. B53:L81 (1986).
6. G.E. Maciel and D.W. Sindorf, J. Am. Chem. Soc. 102:7606 (1980).
7. R. Dupree, D. Holland and D.S. Williams, J. Non. Cryst. Sol. 81:185 (1986).
8. R. Dupree, N. Ford and D. Holland, Phys. Chem. Glasses 28:78 (1987).

NMR STUDIES OF NEUTRON-IRRADIATED CRYSTALLINE AND VITREOUS SILICA

S.L. Chan, L.F. Gladden and S.R. Elliott

Department of Physical Chemistry
University of Cambridge
Lensfield Road
Cambridge, U.K.

ABSTRACT

Some results of a ^{29}Si NMR study of neutron-irradiated crystalline and vitreous SiO_2 are reported. The spin-lattice relaxation time is found to change markedly with increasing neutron dose. Pronounced changes are also observed in the shape and position of the magic-angle spinning NMR lineshape. The shift in peak position for the metamict state is associated with a decrease in oxygen bond angle, and this is related to the densificaction of the glass observed upon neutron irradiation.

INTRODUCTION

Silica is an extremely interesting material in view of the large number of crystalline polymorphs which exist and the fact it is often regarded as being the architypal glass-forming material. One of the more intriguing aspects is the behaviour of both crystalline (c-) and amorphous (a-) SiO_2 upon irradiation, particularly by high-energy neutrons. Many properties change with such irradiation, although perhaps the most spectacular is the amorphisation and concomitant density decrease of the c-phase (quartz) and the increase in density of the a-form ; above a certain neutron dose ($\sim 2 \times 10^{20}$ cm^{-2}) a common metamict state is reached[1]. The structural nature of this metamict phase is not known, and its elucidation is the ultimate aim of this work. However, the facility of such neutron-induced structural changes must be related to the relative adaptability of a network of corner-sharing SiO_4 tetrahedra, as evidenced by the existence of a plethora of different c- polymorphs. In addition, it is very interesting to note that a similar (irreversible) densification of a-SiO_2 can be induced by the application of high pressure at high temperatures[2].

Various properties of both c- and a-SiO_2 have been studied as a function of neutron dose. Many of these (e.g. anomalies in specific heat, thermal conductivity and ultrasonic attenuation) are believed to be related to "two-level systems" (TLS), and are present only for "glassy" systems.

Although unirradiated c-SiO_2 exhibits no thermal anomalies associated with TLS states, neutron irradiation induces such anomalies progressively with dose, even though at intermediate doses the material is mostly crystalline with small interspersed damage regions[3]. At the highest neutron doses, the TLS density in the irradiated crystal is approximately 50 % of that found in the unirradiated glass[3]. Conversely, the density of TLS in a-SiO_2 is observed to decrease with neutron irradiation ; for a dose of 2-5 x 10^{19} cm^{-2} the TLS density of the irradiated sample is approximately 70 % of that of the unirradiated glass[3]. Considering the uncertainties associated with the extraction of the TLS density and of comparing doses, this is additional evidence that a common metamict amorphous phase is formed from either the crystalline or the glassy starting material, having very similar structures and (thermal) properties.

The other type of investigation which has been extensively pursued for neutron-irradiated SiO_2 is of point defects in such materials as studied by ESR. An extensive review of ESR studies of intrinsic (and extrinsic) paramagnetic defects in c-SiO_2 has been given by Weil[5] ; similar reviews for the case of a-SiO_2 have been given by Griscom[6,7]. The intrinsic paramagnetic defects can be separated into two categories : those which are Si-related and those which are O-related centers. Unirradiated pure c- or a-SiO_2 are essentially diamagnetic, with no paramagnetic centers being observable by ESR. However, electronic excitation by UV of a-SiO_2 [8] or γ-radiation of c-SiO_2 [5] or of a-SiO_2 [6,7] causes the production of paramagnetic centers from diamagnetic precursors. For the case of both c- and a-SiO_2, E' centers are formed, and for a-SiO_2, in addition oxygen-hole centers (OHC's) associated with both single non-bridging oxygen (NBO) and superoxide (peroxide) radicals are produced [6]. On the other hand, fast neutron irradiation of both c- and a-SiO_2 gives rise to paramagnetic defect centers (in the absence of subsequent further electronic excitation). In c-SiO_2, the magnetic susceptibility measurements [9] indicate that the unpaired spin density increases steadily with neutron dose (at the rate of 3.7 per incident neutron) reaching a maximum of $\cong 4 \times 10^{19}$ cm^{-2}, beyond which N_s decreases and eventually saturates at $\simeq 2 \times 10^{19}$ cm^{-3} for doses in excess of 2×10^{20} cm^{-2}. Comparable data appear not to be available for a-SiO_2 ; however, it is reported[1] that the spin density eventually reaches the same saturation value with neutron dose for both a- and c-SiO_2 starting materials, again indicative of a common metamict state at the highest doses.

In view of the limited amount of structural information available, particularly for the metamict state, it was decided to employ the relatively new technique of magic-angle spinning (MAS) NMR to the problem. In addition to being a chemically-specific probe sensitive to the local atomic structure around a given nucleus, it has recently been demonstrated that ^{29}Si NMR of a-SiO_2 can yield quantitative information about the Si-O-Si bond-angle distribution ($P(\theta_0)$) by analysis of the NMR lineshape[10,11], utilizing a correlation between NMR chemical shift and sec θ_0 established for a variety of c-SiO_2 polymorphs [12]. Thus, the one structural parameter, namely $P(\theta_0)$, which is known from diffraction studies [13] to change the most with neutron irradiation can be monitored directly by NMR without the uncertainties associated with extraction of $P(\theta_0)$ from X-ray diffraction data. In addition, the effect of varying concentrations of paramagnetic centers on nuclear relaxation can be studied for a given sample without the incorporation of foreign paramagnetic species.

EXPERIMENTAL

Samples of ultra-pure c- and a-SiO_2 were used. The quartz single crystals, supplied by S.T.C. Ltd, have very low concentrations of OH and Al impurities ; the a-SiO_2 used was Spectrosil WF supplied by TSL Ltd, which also has a very low OH content. The samples were irradiated in the core of the Herald reactor at AWRE, Aldermaston with fast neutron ($E \geq 1$ MeV) doses varying between 2×10^{17} and 2×10^{20} neutrons cm^{-2}, and in Al cans such that the samples were not in contact with the cooling water. As a result, the temperature rise of the samples during irradiation was estimated to be not more than, ~ 200°C.

The NMR experiments were performed using a Bruker MSL 400 spectrometer operating at 79.5 MHz ; MAS experiments were performed in an Andrew-Beams probe with a spinner rotating at ≃ 2 kHz. The spin-lattice relaxation time (T_1) measurements were made using the progressive saturation method[14] following a 6μs $\pi/2$ pulse on both spinning and static samples. Accurate NMR lineshapes were obtained using $\pi/4$ pulses separated by delay times equal to $2T_1$.

RESULTS AND DISCUSSION

Some preliminary results from our NMR investigations on neutron-irradiated SiO_2 are given below.

Spin-lattice relaxation

T_1 values for the unirradiated glass[11] and crystal are very long ($\geq 10^4$s), and as a result it is difficult to obtain quantitative NMR lineshapes for these materials, especially under MAS conditions where dipolar mediated contributions are removed from the spin-lattice relaxation process. However, we have observed that T_1 (static) for both initially crystalline and glassy SiO_2 changes markedly with neutron irradiation, decreasing and reaching a minimum at a dose of ~ 3×10^{19} cm^{-2}, followed by a further increase (fig. 1). It is very interesting to note that the paramagnetic spin density (at least for neutron-irradiated c-SiO_2) exhibits the opposite behaviour, peaking at approximately the same dose[3] - see figure 1. This behaviour is understandable if an additional spin-lattice relaxation mechanism is operative in the irradiated crystal or glass which involves the paramagnetic centers - in this case, it is expected that $1/T_1$ for both c- and a-SiO_2 plotted versus dose is in qualitative agreement with the behaviour of N_s which, bearing in mind the uncertainties associated with the estimates for T_1, gives some confidence in ascribing the increase in nuclear relaxation rate to interaction with the paramagnetic point defects (E' and OHC) produced by neutron irradiation. There is, however, another possibility which could account for the behaviour of T_1 - this involves TLS states[16]. It is known that the density of TLS, n, in irradiated c-SiO_2 increases with increasing dose and saturates or shows a small peak at a dose of ~ 5×10^{19} cm^{-2} [3]. Thus, since $T_1^{-1} \propto n^2$ [16], this mechanism could also account qualitatively for the behaviour of T_1 for the crystal. However, the TLS density decreases with increasing dose for a-SiO_2[4], and so the behaviour of T_1 for the irradiated glass cannot be understood in this way. We conclude, therefore, that paramagnetic relaxation is the likely mechanism. It should be noted finally that the T_1 values for the highest neutron-dose samples of irradiated c- and a-SiO_2 are the same, further proof of the similarity of the meatmict states.

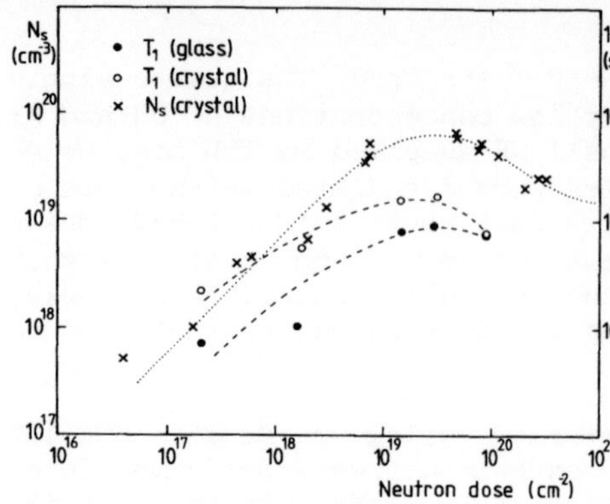

Fig. 1. Spin-lattice relaxation rate $(1/T_1)$ as a function of neutron dose for both crystalline and amorphous SiO_2. Also shown is the dependence of paramagnetic spin density on neutron dose for quartz. The lines are guides for the eye.

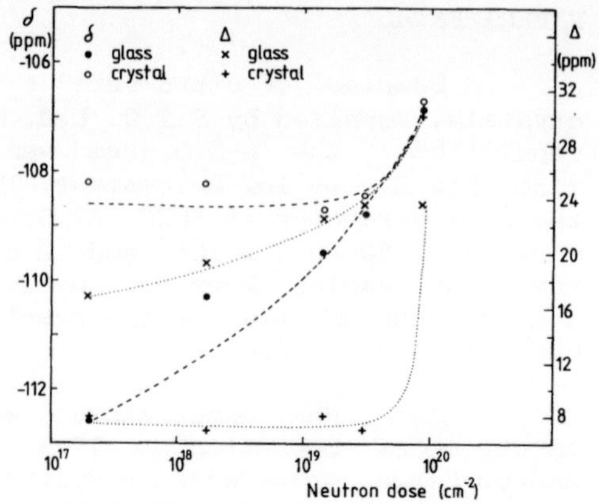

Fig. 2. Peak position δ (chemical shift) and with Δ (FWHM) of the NMR lineshapes of neutron irradiated amorphous and crystalline SiO_2 as a function of neutron dose. The lines are guides for the eye.

NMR LINE SHAPE

The ^{29}Si static lineshape is much narrower in the unirradiated crystal than in the glass (fig. 2), indicative of a greater spread in Si structural environments, i.e. Si-O-Si bond angles, in the latter. On irradiation, the lineshape of c-SiO_2 is invariant in peak position and width until a dose of $\sim 3 \times 10^{19}$ cm^{-2} ; at a dose of $\sim 9 \times 10^{19}$ cm^{-2} the peak shifts by $\sim +4$ppm and its width increases by a factor of four (fig. 2). This behaviour with dose is exactly the same as is observed for the density[1] and in the Raman spectrum[17] - there is very little change with dose up to a value of $\sim 3 \times 10^{19}$ cm^{-2} with a dramatic change in structure-related property thereafter. On the other hand, the (static) NMR peak position and width appears to change continuously with dose for a-SiO_2 (fig. 2), the position shifting to less negative values of chemical shift and the width increasing. At the highest doses, the NMR spectra of irradiated c- and a-SiO_2 are identical (fig. 3) ; this is very strong evidence that a common metamict material is indeed formed.

The MASNMR lineshapes can be deconvoluted to give $P(\theta_0)$ distributions[11] using an empirical linear relationship between chemical shift and sec θ_0 [12]. The results are shown in fig. 4. $P(\theta_0)$ for the glass peaks at $\theta_0 \simeq 143°$ [11], in agreement with that obtained from X-ray diffraction data[18]. The curves for the samples of c- and a-SiO_2 irradiated with the highest doses overlay, indicating an identical structural environment for Si in each case, and the maximum in the distribution is shifted to 134°. It is interesting to note that a sample of Suprasil I a-SiO_2, densified by being subjected to a pressure of 50 kbar at 600°C, has recently been shown to have a $P(\theta_0)$ distribution peaking at 138°[19]. It is

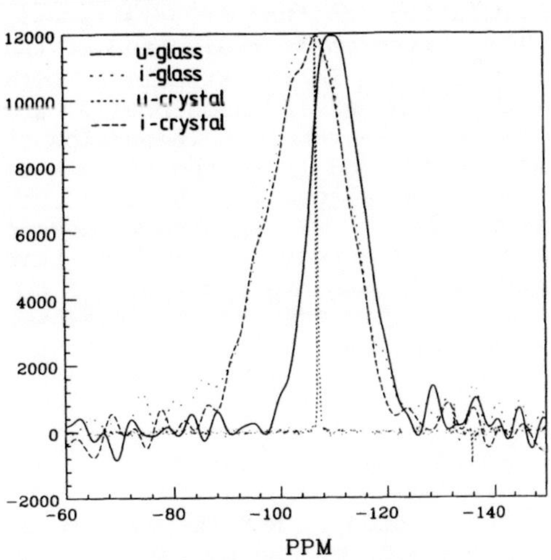

Fig. 3. MASNMR lineshapes for unirradiated (u) and highly irradiated (i) ($\simeq 9 \times 10^{19}$n cm^{-2}) SiO_2 (static for u-c-SiO_2).

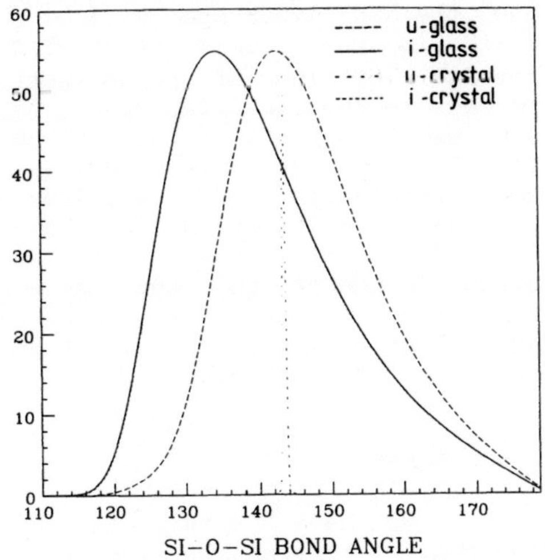

Fig. 4. Si-O-Si angle distributions obtained from NMR lineshapes shown in fig. 3.

apparent, therefore, that densification of the a-SiO_2 structure, by whatever means, is accompanied/caused by a shift in the oxygen bond-angle distribution to smaller angles.

We may speculate as to the cause of the neutron-irradiation (pressure) induced reduction in oxygen bond angle in SiO_2. It is probably associated with a reconstructive transformation resulting in the formation of smaller rings than are found on average in either a-SiO_2 or quartz. It is probably significant that the average oxygen bond angle for three-fold rings is 130.5° for planar rings [20], and ~ 120° for puckered rings as obtained from Monte Carlo studies[21], and we propose, on the basis of the available evidence, that the proportion of 3-rings therefore increases with neutron irradiation. This interpretation is supported by Raman studies[17] on neutron-irradiated a-SiO_2, where the intregrated intensity of the "defect" band at 608 cm^{-1} increases by a factor of seven with respect to the unirradiated glass for a dose of 2×10^{20} cm^{-2}. The significance of this fact is that the 608 cm^{-1} band has been ascribed to breathing motions of the O-atoms in planar 3-fold rings[20], and therefore the increase in Raman intensity of this band with increasing neutron dose is naturally understandable if the concentration of such rings increases with dose, as suggested by the MASNMR results.

However, there is another factor which could have an effect on the NMR lineshape which we are not able to dismiss at this stage. The presence of paramagnetic point defects in the irradiated glass and crystal have a marked effect on T_1 (see fig. 1), but they must also have a concomitant effect on the NMR lineshape. NMR-active nuclei (e.g. ^{29}Si) near such paramagnetic centers will experience a strong additional magnetic field and hence chemical shift - as a result, the line for such centers will be broadened so much as to be unobservable. Such an effect has been observed[22] in a similar system, a-SiO, where again a high concentration of paramagnetic centers exists. The problem is that it is not known exactly how the NMR lineshape will be changed relative to that of diamagnetic a-SiO_2 in the absence of knowledge of the "wipe-out" radius and of the precise location of paramagnetic point defects in the structure (with respect to certain sized rings, for example). Although in these preliminary studies this problem has not been addressed, nevertheless inspection of fig. 4 shows that with neutron irradiation the NMR lineshape appears to move bodily to less negative values of chemical shift ; therefore, the increased contribution to the line shape for less negative values of chemical shift cannot arise from a paramagnetic effect, although at this stage we cannot rule out the possibility that this is the cause of the diminution in intensity on the other side of the peak, at more negative values of chemical shift, nor that there is a paramagnetic-induced bodily shift in the resonance.

CONCLUSIONS

We have investigated for the first time by NMR the structural changes in crystalline (quartz) and amorphous SiO_2 caused by fast neutron irradiation. The spin-lattice relaxation time is found to change dramatically upon neutron irradiation. This behaviour is ascribed to relaxation caused by paramagnetic defects created on irradiation. The magic-angle spinning NMR lineshape of both crystal and glass is found to shift to more positive values of chemical shift with irradiation ; the spectra of both for the highest neutron doses are identical, however. The peak in the oxygen bond-angle distribution of this common metamict state is $\sim 134°$, compared with the value of $\sim 143°$ for unirradiated vitreous SiO_2 and 143.6° for quartz.

REFERENCES

1. E. Lell, N.J. Kreidl and J.R. Hensler, "Progress in Ceramic Science", ed. J.E. Burke (Pergamon), vol. 4, p.3 (1966).
2. See e.g. J. Arndt and D. Stoffler, Phys. Chem. Glasses 10:117 (1969).
3. C. Laermans, "Structure and Bonding in Non-crystalline Solids", eds. G.E. Walrafen and A.K. Revesz (Plenum), p. 329 (1986).
4. T.L. Smith, P.J. Anthony and A.C. Anderson, Phys. Rev. B17:4997 (1978).
5. J.A. Weil, Phys. Chem. Min. 10:149 (1985).
6. D.L. Griscom, J. Non-Cryst. Sol. 73:51 (1985).
7. D.L. Griscom, in "Defects in Glasses", eds. F.L. Galeener et al., Mat. Res. Soc. Symp. Proc. 61:213 (1986).
8. J.H. Stathis and M.A. Kastner, in "Defects in Glasses", eds. F.L. Galeener et al., Mat. Res. Soc. Symp. Proc. 61:161 (1986) ; R.A.B. Devine, C. Fiori and J. Robertson, ibid 61:177 (1986).
9. D.K. Stevens, W.J. Sturm and R.H. Silsbee, J. Appl. Phys. 29:66 (1958).
10. R. Dupree and R.F. Pettifer, Nature 308:523 (1984).
11. L.F. Gladden, T.A. Carpenter and S.R. Elliott, Phil. Mag. B53:L81 (1986)
12. J.V. Smith and C.S. Blackwell, Nature 303:223 (1983).
13. H.D. Bale, R.E. Shepler and G.W. Gibbs, J. Appl. Phys. 41:241 (1970).
14. R. Freeman and H.D.W. Hill, J. Chem. Phys. 54:3367 (1971).
15. W.E. Blumberg, Phys. Rev. 119:79 (1960).
16. M. Rubinstein, H.A. Resing, T.L. Reinecke and K.L. Ngai, Phys. Rev. Lett. 34:1444 (1975).
17. J.B. Bates, R.W. Hendricks and L.B. Shaffer, J. Chem. Phys. 61:4163 (1974).
18. R.L. Mozzi and B.E. Warren, J. Appl. Cryst. 2:164 (1969).

19. R.A.B. Devine, R. Dupree, I. Farnan and J.J. Capponi, Phys. Rev. B35:2560 (1987).
20. F.L. Galeener, Sol. State Comm. 44:1037 (1982).
21. Y.T. Thathachari and W.A. Tiller, J. Appl. Phys. 57:1805 (1985).
22. R. Dupree, D. Holland and D.S. Williams, Phil. Mag. B50:L13 (1984).

ELECTRONIC STRUCTURE OF DEFECTS IN AMORPHOUS SiO_2

J. Robertson

Central Electricity Research Laboratories
Leatherhead, KT22 7SE
United Kingdom

The electronic structure, local bonding and reactions of the principal intrinsic point defects and some impurities (H_2O, H, B, N, Ga, Ge, P, As) in amorphous silicon dioxide are reviewed.

INTRODUCTION

This review describes the electronic structures of various defects in amorphous (a-)SiO_2 and relates them to the experimental properties of bulk fused silica and thermally grown oxide films. Firstly, it is necessary to define a defect and then to review briefly the electronic structure of bulk SiO_2. The structure of a-SiO_2 can be idealised as a topologically disordered lattice containing only Si-O bonds in which all Si sites are four-fold coordinated and all oxygen sites are two-fold coordinated. We define an intrinsic point defect as a site of different coordination or a like-atom bond. Generally, such defects will introduce states into the band gap. This definition excludes the low order rings of bonds discussed by Galeener [1]. These rings introduce sharp Raman features (eg. at 606 cm^{-1}) which have been erroneously attributed to coordination defects in the past.

Fig. 1 shows the density of states (DOS) for crystalline SiO_2 (quartz). The DOS of a-SiO_2 is very similar because it is controlled by the short-range order which is unchanged. The valence band below 0 eV is largely oxygen-like and its features are most easily understood in terms of the single Si-O-Si unit (fig. 1b)[2,3]. The oxygen's p states split into a nonbonding $p\pi$ orbital, a weakly bonding p_x orbital, and a strongly bonding p_y orbital. The nonbonding and weakly bonding states intermix and form the upper valence peak, from 0 to -4 eV. The strongly bonding states form the broad feature from -5 to -13 eV. The conduction band minimum of SiO_2 is surprisingly unaffected by disorder[4,5]; electrons have high mobility in a-SiO_2 and no mobility edge has been detected. In contrast, holes have a high effective mass being π-like and form polarons[6]. However electron spin resonance (ESR) has been unable to detect stationary polarons[7]. The optical gap of 9 eV is dipole-forbidden[8] due to the phase relationships of the π states in the highest valence bands which prevent dipole transitions to Si 3s states. The forbidenes remains in a-SiO_2 as it only depends on the tetrahedral coordination of Si[8]. Transitions from deeper oxygen π states are allowed above ≈12 eV. The first large peak in the optical spectrum at 10.5 eV is the allowed exciton. Its binding energy is ≈ 1.5 e.V. Assuming similar binding, the minimum gap exciton is at ≈ 7.5 eV.

STRUCTURE OF THE INTRINSIC DEFECTS

What type of defects do we expect in a-SiO_2? It is covalent and its network is flexible. We expect the valence electrons, even at defects, to be spin-paired in bonds. We therefore expect the defects to be the like-atom bonds, Si-Si and O-O, and also the diamagnetic configurations of the broken (dangling) bond. This scheme is largely born out by experiment, but not without some complications.

ESR has identified three types of centres in irradiated a-SiO_2 : the nonbridging oxygen, the peroxyl radical and the E' centre (fig. 2)[9]. The nonbridging oxygen is an unpaired electron localized in a pπ orbital of a monovalent oxygen, close to a proton. The peroxyl radical is an unpaired electron localized over two oxygens, one of which is bonded to a Si. Their diamagnetic precursors are believed to the hydroxyl group, -OH, and the peroxide bridge -O-O-, respectively. The oxygen dangling bond is therefore generally present in its hydrated form, -OH. The E' centre is an unpaired electron localized in a single Si sp^3 orbital. It is closely related to the E_1' centre in quartz, the positively charged oxygen vacancy V_o^+. There, the unpaired electron localizes on one of the two Si dangling bonds because of a "pseudo-Jahn-Teller distortion" ; the second Si relaxes away from the vacancy region towards planarity, and is now in the stable geometry for a 3-electron atom (cf. planar boron)[10]. The diamagnetic precursor of V_o^+ is the neutral vacancy V_o^o. Noting that the centre now has two electrons, and that the SiO_2 network is very flexible, O'Reilly and Robertson[11,12] proposed that the two Si sites would draw together to form essentially a full Si-Si bond. Total energy calculations verified that this indeed occurs [13,14]. Thus V_o^o and Si-Si are synonymous in SiO_2. Returning to a-SiO_2, Griscom distinguished four variants of E' centre, one of which E_β' has axial symmetry and presumably involves an isolated trivalent Si precursor[9]. However, one problem is that the photo-induced defects of virgin silica, which presumably should be more "intrinsic" than those of irradiated silica, have not been fully identified[15].

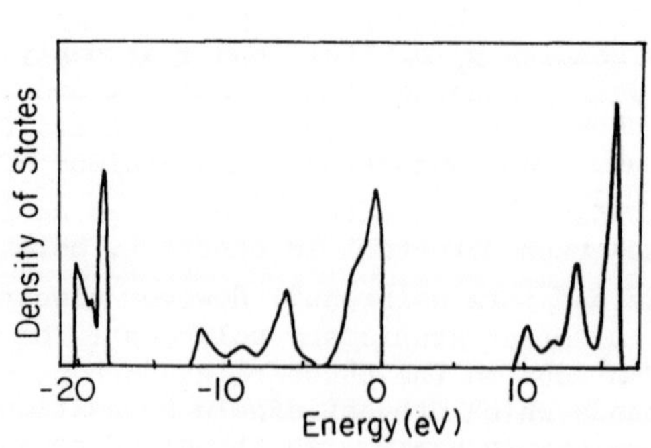

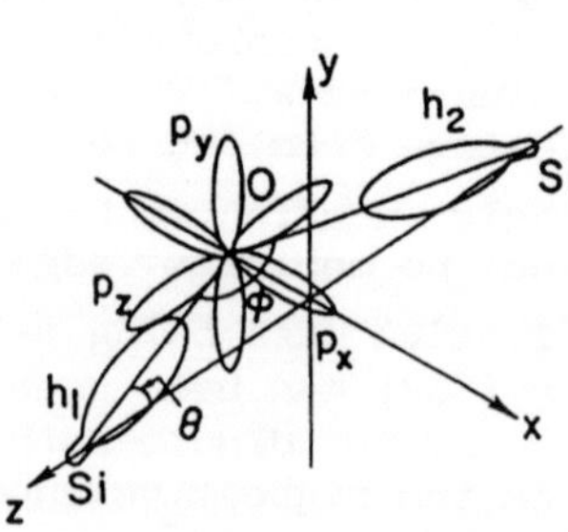

Fig. 1.(a) Density of states SiO_2, (b) bonding in a Si-O-Si unit.

This discussion has confirmed that generally the defects of a-SiO_2 are those expected in a random network. However, a-SiO_2 is also a chalcogenide and so we might also expect a second class of defect, the valence alternation pairs (VAP)[11,16]. These utilise the oxygen pπ electrons to form trivalent oxygen sites. Consider breaking a Si-O bond. This creates either Si_3^+-O_1^-, $Si_3^{\cdot}$-$O_1^{\cdot}$ or Si_3^--O_1^+ according to the sense of charge transfer. Here, the subscript denotes coordination and the superscript denotes the site's formal charge in the covalent limit. We see that for the positive centers the dangling bond orbital is empty. It is therefore able to form a dative bond with a nearby bulk oxygen site ;

$$Si_3^+ + O_2^{\cdot} \rightarrow Si_4^{\cdot} + O_3^+ \qquad (1)$$

giving a trivalent oxygen surrounded by three silicons. The extra bond stabilizes the products. The dangling orbital is partially filled for the neutral site, so that this centre cannot form the dative bond. Thus adding an electron to O_3^+ induces a Si-O bond to rupture, regaining the $Si_3^{\cdot}$ site :

$$O_3^{\cdot} + Si_4^{\cdot} \rightarrow O_2^{\cdot} + Si_3^{\cdot} \qquad (2)$$

THE ELECTRONIC LEVELS OF THE INTRINSIC DEFECTS

The electronic structures of these centres have been calculated by two semi-empirical methods, tight-binding[11] and MINDO/3[13,14,17,19] ; the latter also gives total energies. Figure 3 summarises the calculated positions of gap states of various centres, according to tight-binding[11,20]. All the gap states are strongly localized on the defect site or its nearest neighbour (80-90 % in most cases). The single Si dangling bond is calculated to introduce a state just above midgap. This level is half-filled when neutral, and can therefore trap both electrons and holes. The exact level position depends on its occupancy and the Si bond angle.

The Si-Si bond, or relaxed neutral oxygen vacancy, introduces a filled state of bonding (σ) character at about 1 eV above the valence band edge, E_v, and an empty state of antibonding (σ^*) character at about 1 eV below

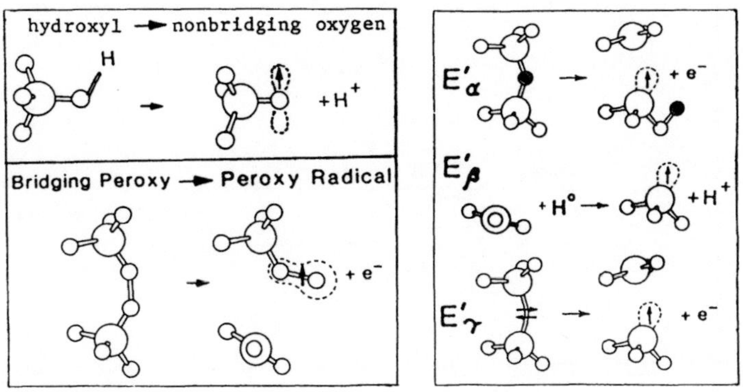

Fig. 2. The paramagnetic centers of silica and their precursors.

the conduction band edge, E_c, according to tight-binding. MINDO/3 finds a similar positioning of these levels but with a slightly smaller splitting of 6.6 eV[14]. A much smaller splitting energy of 1.4 eV was calculated by MINDO by Edwards[19]. However, we believe that this result is not relevant as it corresponds to a triplet excitation rather than an open-shell singlet. The Si-Si bond is also an amphoteric trap. The σ state can trap a hole and the σ^* state can trap an electron. Hole trapping produces an interesting defect reaction. The σ state is highly localized and it is now only singly occupied, i.e. it corresponds to a one-electron bond. Such bonds are unstable, so the silicons move apart, breaking the bond. Any asymmetry causes the single electron to localize on one of the two Si dangling bonds, so we regain the V_o^+, E_1' centre :

$$\equiv Si\text{-}Si\equiv \; + \; e^+ \; \rightarrow \; \equiv Si^o \; + \; \equiv Si^+ \qquad (4)$$

Therefore the Si-Si bond is directly analogous to the oxygen vacancy in its behaviour. Including the effects of relaxation, its thermal +/0 level is found at 3.5 eV[14]. Robertson[12] originally suspected that V_o^- might also break apart. However MINDO/3 finds only a small dilation of the Si-Si bond and the thermal 0/- level is found at ~ 8 eV, close to its optical position[14]. A probable reason for this difference is that the Si_3^+ site in V_o^+ relaxes outwards towards planarity, helping to break the Si-Si bond, whereas Si_3^- in V_o^- prefers to relax pyramidally inwards, opposing the breaking.

The nonbrindging oxygen has two pπ orbitals, one of which gives a gap state at ~ 2 eV and contains the hole of the radical (fig. 3). The superoxide bridge consists of a O-O bond. It introduces a filled O π and O-O σ state below E_v and a filled π and an empty σ^* state above E_v. The superoxide radical has a shorter O-O bond length than the bridge. It introduces three π states, one of which is half-filled and lies in the gap, and also an empty σ^* state. The half-filled π level creates a 0/+ level at 3.5 eV according to MINDO[18].

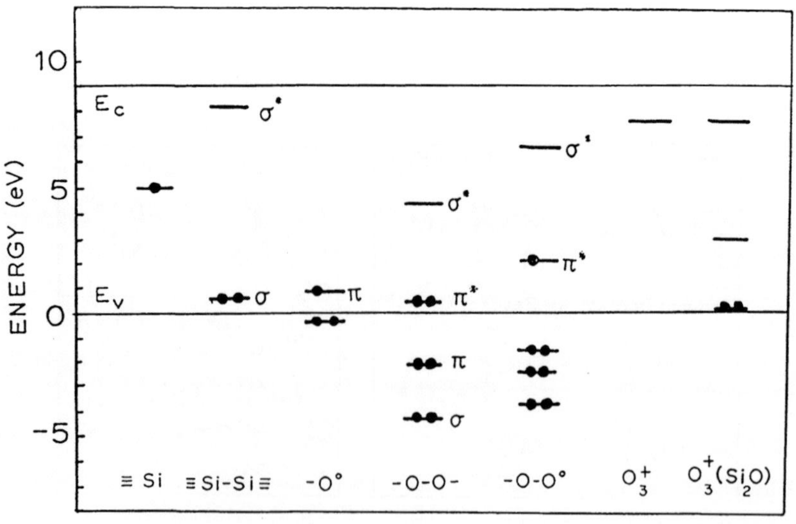

Fig. 3. Gap states associated with various intrinsic centers.

Fig. 3 also shows the gap states of two trivalent oxygen VAP sites. We consider, first the site surrounded by three silicons, O_3. The hypothetical neutral centre O_3^o is calculated not to have a gap state, essentially because the three surrounding Si-O bonds act to expel any possible states from the gap[11]. This centre has an s-like conduction band resonance at ~ 10 eV. The O_3^+, in contrast, has a gap state because of its Coulombic charge. This state is "shallow", bound ~ 1.5 eV below E_c. Fig. 3 also shows the $O_3^+(Si_2O)$ VAP centre formed by overcoordinating a O_1^+ site[11].

Chemical bonding arguments favour the existence of VAPs in silica, particularly the O_3^+. Total energy calculations show that Si_3^+ sites indeed overcoordinate to form O_3^+ [14,16]. However, such diamagnetic centres cannot be detected directly. We must look for evidence of VAP-like properties in the daughter Si_3^o site, formed after electron trapping by reaction (2). Griscom found no evidence of a hyperfine interaction with a fourth oxygen site for Si_3^o and so argued against VAPs[21]. However, this is not necessarily definitive. Firstly, the SiO_2 lattice is very flexible, sufficiently so for the Si^+ site at a E_1' centre to move away and not show much hyperfine interaction. A similar flexibility is possible for O_3^o═ Si_3^o. Secondly, O_3^+ is a logical precursor of the E_β'.

Many optical absorption and luminescence bands have been associated with intrinsic defects in silica, but in general few have been assigned to specific defects, with three exceptions. The 5.8 eV absorption has been linked to the E' centre, and indeed to a single Si sp^3 hybrid[9], but we find it difficult to assign it to a particular transition using our local DOS results[11]. The 7.6 eV absorption has been associated with the peroxyl radical[22]. However, we note that the Si-Si bond, or any positively charged centre with a shallow level and even the indirect exciton may also absorb at this energy[11]. Finally, correlations exist between the 1.8 eV absorption and luminescence and the nonbridging oxygen radical[23,25], although there may be other contributors to these bands. The photo-luminescence data[26,28] cannot yet be fully interpreted.

Defect creation can be classified into three general mechanisms, dispacement (knock-on) damage, electronic rearrangement and radiolysis. To effect displacement damage the incident particle must transfer sufficient momentum to displace an oxygen by breaking two Si-O bonds or to displace a Si by breaking four Si-O bonds. To displace an oxygen, for example, an electron must have a kinetic energy of 70 keV or over[29]. Damage by electron rearrangement implies that carriers are created and trapped separately at pre-existing traps, sometimes inducing a reaction of that trap. An example is the creation of E' centres from the Si-Si bond precursor. The concentration of such centres will saturate when all precursors are transformed. The third process is radiolysis in which defects are generated by the nonradiative recombination of electron-hole pairs created optically or by ionising radiation. It does not saturate until high concentrations are reached, and it is believed to be the principal result of electron and light-ion bombardment[30,31]. The generation of transient E' centres by energetic electron beams is also attracting considerable interest and is believed to occur by radiolysis[29,32,33].

HYDROGEN-RELATED CENTRES

Hydrogen and water are ubiquitous impurities in SiO_2. The energy levels of silanol (≡SiOH) and hydride (≡SiH) groups have been calculated by tight-binding (fig. 4). The Si-H interactions were parameterised against the energy levels of SiH_4 molecule[34] and a self-energy shift was included on the Si sp^3 hybrid. The O-H interactions were scaled with (bond length)$^{-2}$ from those of the C-H bond, which in turn were fitted to CH_4 [34].

The ≡Si-H group is seen to produce a filled σ state just below E_v and an empty σ* state in the gap just below E_c. This group is probably both an electron and hole trap. Such traps associated with water have probably been observed[35]. Thus, hydrogen is unable to passivate Si dangling bonds in silica, as it can in Si itself. It is also probable that hole trapping causes a defect reaction analogous to[3]:

$$\equiv SiH + e^{+} \rightarrow \equiv Si^{+} + H^{o} \qquad (4)$$

Analogously, electron trapping will create negative traps. Support for this is Griscom's suggestion that radiation-induced charge build-up at the Si:SiO_2 interface occurs by the reaction of SiH bonds with radiolytic atomic hydrogen and an electron injected from the silicon substrate[36] :

$$\equiv SiH + H^{o} + e^{-} \rightarrow \equiv Si^{-} + H_2 \qquad (5)$$

Hydride groups are expected in thermally grown silica[37], particularly near the interface due to the reaction of interstitial H_2 molecules with SiOSi bridges. The H_2 is a byproduct of the oxidation of Si by any ambient water.

Our tight-binding calculations find that O-H bonds only produce states well away from the gap (fig. 4). This result is similar to that of Bennett and Roth[38] but contrasts with the empirical calculations of Pantelides[39] and the expectations of Griscom[29]. Our result of an absence of gap states has some important implications, so a confirmation is desirable. Firstly, calculations by the MINDO/3 and CNDO methods produce similar results and indeed the use of hydrogen atoms as terminators of surface bonds in defect cluster calculations is further implicit evidence. Secondly, the water molecule consists of a divalent oxygen and two O-H bonds and so should have a similar electron structure. Its highest occupied molecular orbital is the O pπ state, as in SiO_2, which should therefore lie at a similar energy with respect to the vacuum level E_{vac}, i.e. just below E_v of SiO_2. H_2O contains no molecular orbitals below E_{vac}, which is ~ 1 eV above E_c. It does however possess Rydberg states below E_{vac}, the molecular equivalent of exciton

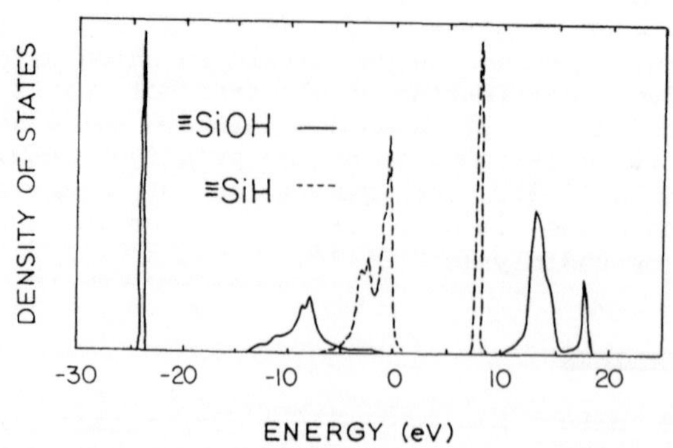

Fig. 4. Hydrogen site DOS associated with Si-H and O-H bonds.

levels. Thus, theoretically we do not expect O-H bonds to produce gap states.

The hydroxyl group is the precursor of the nonbridging oxygen radical. The expected mechanism is hole trapping :

$$-OH + e^{+} \rightarrow -O^{o} + H^{+} \quad (6)$$

but its details are unclear in the absence of a gap level. Radiolysis creates the radicals with great efficiency[29,40]. We believe that an excitonic mechanism is now responsible. The exciton binding energy might be larger locally at an OH bond than in the bulk. The hole is in the O pπ states and weakens the OH bond while the electron will be in a more delocalized H-centred level :

$$-OH + h\nu \rightarrow -O^{o} + H^{o} \quad (7)$$

ELECTRONIC STRUCTURE OF IMPURITIES

Impurity atoms in silica can be classified either as network formers or network modifiers. Network formers enter the covalent network itself and can be either purely substitutional (e.g. tetravalent Ge at Si sites) or they can use the lack of bonding constraints to enter sites of prefered valence (e.g. trivalent As). Network modifiers are interstitial ions, requiring compensation. Some elements are amphoteric : Ga can form substitutional Ga^{-}_{Si} sites where it acts trivalently and requires a fourth electron to form four bonds. These sites can be compensated by Ga^{+} interstitials where Ga is monovalent and its s^2 valence electrons remain bound to the ion.

The electronic structure of network forming impurities has also been calculated using the tight-binding method. For substitutional impurities, the positions of any gap states are easily found using the Green's function method[41]. For impurities substituting at the Si sites, the s-like A_1 states are the principal donor traps. Fig. 5 shows that P_{Si} forms a level just below E_c whereas Ge_{Si} forms a level near E_c [41]. Experimentally, both sites trap electrons and both atoms move off-centre[42]. For impurities

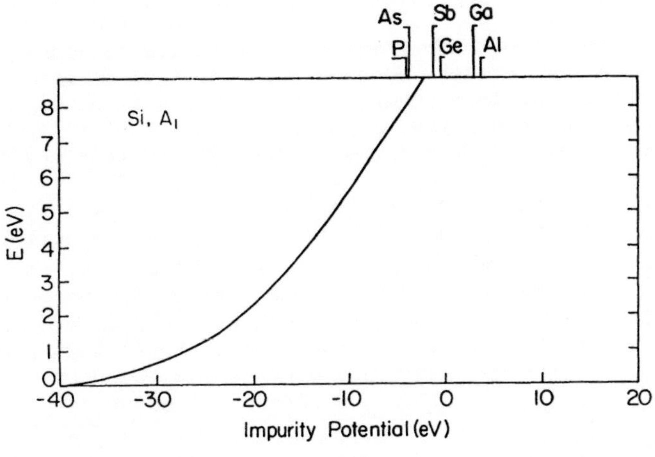

Fig. 5. Gap states of substitutional Si-site impurities.

substituting for oxygen, we conclude that the impurity's nonbonding $p\pi$ orbital, of B_2 symmetry within C_{2v}, always forms the highest level. Fig. 6 shows that the less electronegative chalogens are expected to form gap states and that nitrogen forms a state near E_v.

The electronic structure of nonsubstitutional centres was calculated by embedding the impurity in a large molecular cluster and using the recursion method to calculate the local density of states. Tight-binding two-centre interactions were again scaled as (bond length)$^{-2}$ when the

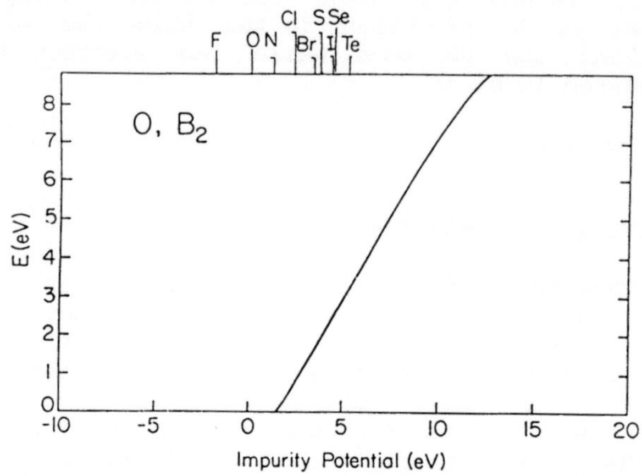

Fig. 6. Gap states of substitutional O-site impurities.

impurity's bond length differs from that of Si-O. Self-energy shifts were applied to any impurity hybrid orbitals not bonded to oxygens, as previously[11].

Halides were studied in their monovalent ≡SiX configurations. The ≡SiF group produces only band states, but the less electronegative halides, even Cl, introduce filled π-like states in the lower gap. These states can act as hole traps and indeed Cl-related hole traps have been observed[43].

Two nonsubstitutional configurations of pnictides were studied, the trivalent ≡X site and the pentavalent ≡X=0 with a terminal oxygen. The trivalent site gives a gap state due to its doubly occupied nonbonding orbital (fig. 7). For P, with bond angles of 100°, this hole trap is calculated to lie at ~ 3 eV (fig. 8). In contrast, the pentavalent site does not produce gap states because the terminal oxygen saturates this nonbonding orbital. It is a general result that when an element exerts its maximum valence it does not give a gap state. The pentavalent states of the heavier pnictides As and Sb are less stable and so these elements are more likely to give traps. Phosphorus is found to produce four paramagnetic species under irradiation or excitation, denoted P_1, P_2, P_4 and POHC (phosphorus oxygen hole centre)[44,45]. P_2 is an electron trapped at substitutional P_{Si}, but which has moved off-centre[44]. P_1 is the phosphorus analogue of the E' centre and is a hole trapped at the ≡P site, which is consistent with fig. 7. Griscom et al.[44] also correlated some absorption bands with the ESR centres. However, the transition energies are not consistent with the energy levels and occupations found by us. Nitrogen can form two types of site in silica, a planar trivalent site, N_3^o, similar to that in Si_3N_4 and a divalent oxygen-like site, N_2^-. Both configurations produce a N π-like state at ~ E_v. The neutral divalent site is

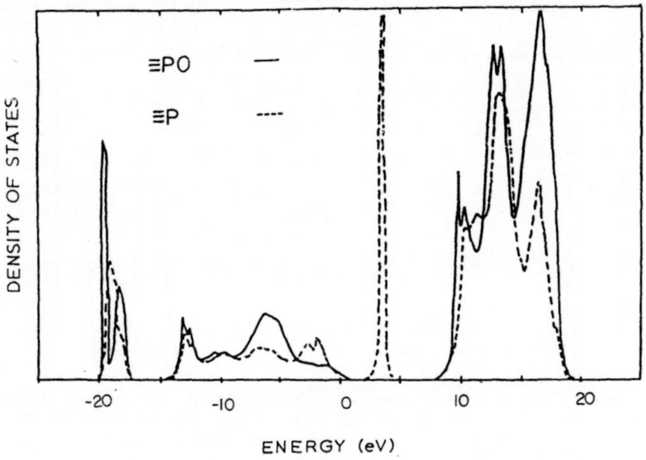

Fig. 7. Local DOS phosphorus at ≡P and ≡P=O sites.

paramagnetic, N_2^o, and has been observed by ESR[46,47].

Boron can form both trivalent and tetravalent sites in silica. The tetrahedral, substitutional site is formally B_4^- and has no deep gap states. However, it is a Coulombic hole trap, with the hole entering the pπ orbital of an adjacent oxygen rather than a B-centred orbital, as in the analogous Al centre in smoky quartz. The trivalent B site can be planar, in which case there is no gap state, only an empty pπ orbital well above E_c (fig. 9). However, when the site distorts pyramidally, this state now drops in energy and traps an electron, becoming the boron E' analogue[48].

Two types of impurity configuration must still be studied, the interstitial ions and the Si-metalloid bonds. To first order, interstitial Na^+ or Ga^+ will introduce a state at the energy of its atomic s orbital below E_{vac}. Configurations with Si-metalloid bonds arise when Si and the

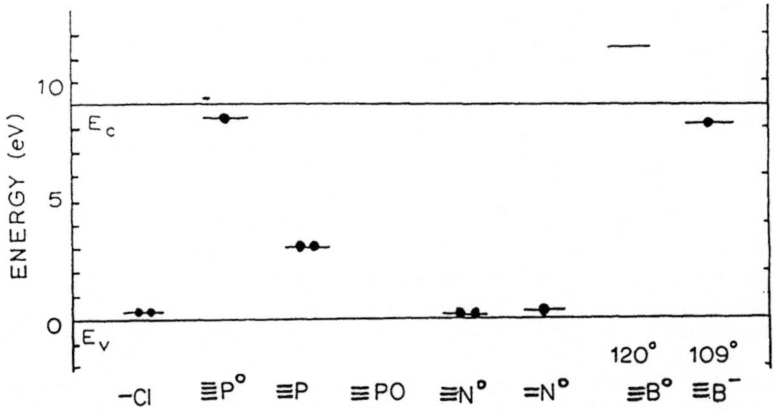

Fig. 8. Gap states associated with various impurity configurations ; monovalent Cl, tetravalent, trivalent and pentavalent P, trivalent and divalent N, and trivalent B with 120° and 109° bond angles.

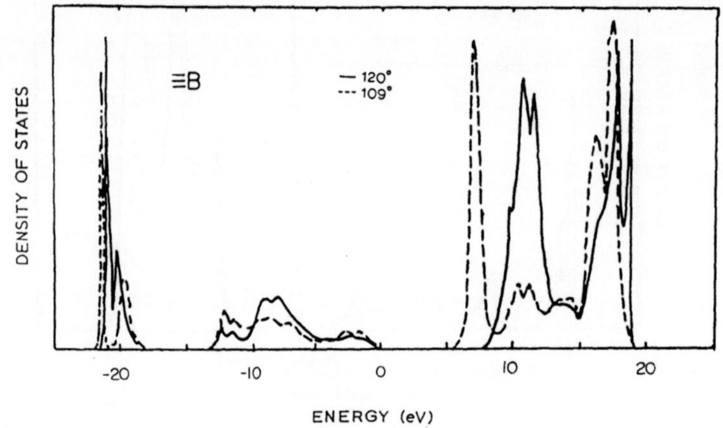

Fig. 9. Local DOS at trivalent B sites, for 120° and 109° bond angles.

impurity must compete for oxygen, for instance if the impurity is implanted. Two configurations of P have been observed by diffusion and photoemission[49,52], and it appears that P and As may even substitute for oxygen and so produce midgap electron traps. However, the simple P_0 site is paramagnetic and it is unlikely that this will survive annealing to 1100 C, at which most intrinsic defects are removed[31]. Hence, the precise configurations are not fully known and require further work such as ESR.

REFERENCES

1. F. Galeener, this volume.
2. J.R. Chelikowsky, M.L. Schluter, Phys. Rev. B 15:4020, (1977).
3. S.T. Pantelides, W.A. Harrison, Phys. Rev. B 13:2667 (1976).
4. N.F. Mott, in "SiO_2 and its Interfaces", ed. S.T. Pantelides, Pergamon, London, (1978).
5. W.Y. Ching, Phys. Rev. Lett. 46:607 (1981).
6. R.C. Hughes, Phys. Rev. B 15:2012 (1977).
7. W. Hayes, T.J.L. Jenkin, J. Phys. C 19:6211 (1986).
8. W.B. Fowler, in "Structure and Bonding in Non-Crystalline Solids", ed. G. Walrafen, A. Revesz, Plenum, New York, (1986).
9. D.L. Griscom, J. Non-cryst. Solids 73:51 (1985).
10. K.L. Yip, W.B. Fowler, Phys. Rev. B 11:2327 (1975).
11. E.P. O'Reilly, J. Robertson, Phys. Rev. B 27:3780 (1983).
12. J. Robertson, J. Phys. C 17L 221 (1984).
13. A.H. Edwards, W.B. Fowler, J. Phys. Chem. Solids 46:841 (1985).
14. J. Rudra, W.B. Fowler, Phys. Rev. B 35:8223 (1987).
15. J. Stathis, Mat. Res. Soc. 61:161 (1986), this volume.
16. J. Robertson, this volume.
17. A.H. Edwards, W.B. Fowler, in ref. (8).
18. A.H. Edwards, W.B. Fowler, Phys. Rev. B 26:6649 (1982).
19. A.H. Edwards, Mat. Res. Soc. Symp. Proc. 61:3 (1986).
20. J. Robertson, Philos Mag. B 52:371 (1985).
21. D.L. Griscom, Phys. Rev. B 22:4192 (1980).
22. M. Stapelbroek, D.L. Griscom, E.J. Freibele, G.H. Sigel, J. Non-cryst Solids 32:313 (1979).
23. E. Friebele, D. Griscom, M. Marrone, J. Non-cryst Solids 71:133 (1985).
24. R.A.B. Devine, C. Fiori, J. Robertson, Mat. Res. Soc. 61:177 (1986).
25. L.N. Skuja, A.R. Silin, Phys. Stat. Solidi A 70:43 (1982).
26. L.N. Skuja, A.N. Streletsky, Solid State Commun 50:1069 (1984).
27. C.M. Gee, M.A. Kastner, Phys. Rev. Lett. 42:1765 (1979).

28. J. Stathis, M.A. Kastner, Phys. Rev. B 35:2972 (1987).
29. D.L. Griscom, SPIE 541:38 (1985).
30. R.A.B. Devine, App. Phys. Lett. 43:1056 (1984).
31. R.A.B. Devine, J. Appl. Phys. 56:563 (1984) ; ibid, 58:3368 (1985).
32. K. Tanimura, T. Tanaka, N. Itoh, Phys. Rev. Lett. 51:423 (1983).
33. W. Hayes, M. Kane, O. Salminen, R. Wood, J. Phys. C 17:2943 (1984).
34. J. Robertson, Phys. Rev. B 28:4658 (1983), Adv. Phys. 35:317 (1986).
35. A. Hartstein, D.Young, App. Phys. Lett. 34:361 (1981).
36. D.L. Griscom, J. Appl. Phys. 58:2524 (1985).
37. A.G. Revesz, J. Electrochem Soc. 126:122 (1979).
38. A.J. Bennett, L.M. Roth, J. Phys. Chem. Solids 32:1251 (1970).
39. S.T. Pantelides, Thin Solid Films 89:103 (1982).
40. N. Azuma, T. Miyazami, K. Fueki, J. Am. Ceram Soc. 69:19 (1986).
41. U.A. Ekenberg, J. Robertson, J.D. Dow, Phys. Rev. B 29:2216 (1984).
42. J. Isoya, J.A. Weil, R.F. Claridge, J. Chem. Phys. 69:4876 (1978).
43. L. Manchanda, J. Vasi, A. Bhattacharyo, App. Phys. Lett. 37:744 (1984).
44. D.L. Griscom, E.J. Friebele, K. Long, J. App. Phys. 54:3743 (1983).
45. H. Kawazoe, J. Non-cryst. Solids 71:279 (1985).
46. J. Mackey, J.W. Boss, M. Kopp, J. Phys. Chem. Glasses 11:205 (1970).
47. E.J. Friebele, D.L. Griscom, T.W. Hickmott, J. Non-cryst Solids 71:351 (1985).
48. D.L. Griscom, G.H. Sigel, R.J. Ginther, J. Appl. Phys. 47:960 (1976).
49. R.F. DeKeersmacher, D.J. DiMaria, S.T. Pantelides, in ref. (4).
50. A.H. Van Ommen, J. App. Phys. 56:2708 (1984).
51. S. Alexandrova, D.R. Young, J. App. Phys. 54:174 (1983).
52. M. Offenberg, P.J. Grunthaner, D.D. Krut, P. Balk, this volume.

ELECTRON AND HOLE TRAPS RELATED TO

A Π-BONDED OXYGEN VACANCY CENTER IN SiO_2

Ireneusz Strzalkowski

Institute of Physics
Warsaw Technical University
Koszykowa 75, 00-662 Warsaw, Poland

INTRODUCTION

An oxygen vacancy is one of the most fundamental defects in SiO_2. It influences many physical properties of silicon dioxide, e.g. charge trapping. Therefore a detailed knowledge of microscopy and chemical bonding of this defect is essential in understanding the physics of many problems in SiO_2.

In an electron lone-pair material like SiO_2 (non-bonding oxygen orbital) many types of defects occur by a bonding rearrangement in which dangling bonds of undercoordinated atoms interact with the lattice by forming chemical bonds with lone pair electrons on neighbouring atoms. In this local reconstruction the π interactions play an important role and should not be neglected in describing properties of material.

A Π-BONDED OXYGEN VACANCY CENTER

Formation Mechanism

It has been shown[1,2] that the lowest energy state of the neutral oxygen vacancy is realized in the following two steps.
1. Transfer of one electron from a non-bonding p oxygen orbital (next to the vacancy) into the opposite Si(1) dangling hybrid, shown in fig. 1.a.
2. Formation of a π bond between the remaining oxygen non-bonding p electron and the nearest Si(2) dangling hybrid, shown in fig. 1.b.

Relaxation and Rehybridization

The presence of a second electron on the dangling hybrid of Si(1) atom causes relaxation and rehybridization of this atom. The largest energy gain needs formation of three p-like hybrids (which form σ-type bonds with the three p orbitals of the surrounding oxygen atoms) and one s-like dangling hybrid occupied by two spin-paired electrons. For energy minimum the Si(1) atom relaxes 0.14 Å from its normal site towards the oxygen vacancy.
Formation of the π bond causes also relaxation and rehybridization of the Si(2) atom. The largest energy gain requires rehybridization from four sp^3 hybrids to three sp^2 (which form σ-type bonds with the three p-orbitals of the surrounding oxygen atoms) and one p-like dangling hybrid which forms a

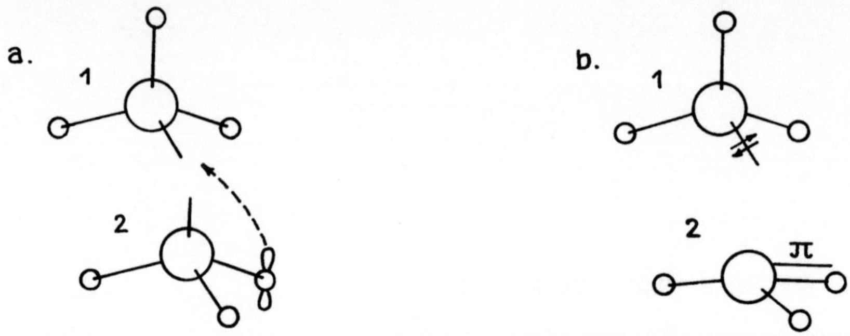

Fig. 1. Formation of a π-bonded oxygen vacancy center.

π bond with the next non-bonding oxygen p-orbital. Minimum energy is obtained when the Si(1) relaxes away from the oxygen vacancy to the plane in which there are three surrounding oxygen atoms.

Formation Energy

The system being considered contains three Si(p) - O(p) σ bonds, three $Si(sp^2)$ - O(p) σ bonds, one Si(p) - O(p) π bond and two electrons on the s-like Si dangling hybrid. Total energy gain can be calculated relative to the system with the same number of electrons and six $Si(sp^3)$ - O(p) bonds, one extended $Si(sp^3)$ - $Si(sp^3)$ and two electrons in the non-bonding O(p) orbital. Calculations can be made within Bond Orbital Approximation of Tight Binding Method using Harrison's matrix elements and Herman-Skillman atomic potentials[1].
Total energy gain in the first step of a π-bonded oxygen vacancy center formation, as shown in the Fig. 2, one can express

$$\Delta E_{tot}^{(1)} = \varepsilon_s(Si) - \varepsilon_p(O) + \Delta E_b^{(1)} + U$$

where $\Delta E_b^{(1)}$ is the rehybridization change of the energy of Si(1) bonds relative to the energy of the sp^3 bonded system (with extended sp^3-sp^3 bond between Si(1) and Si(2), and U is the repulsive Coulomb energy between

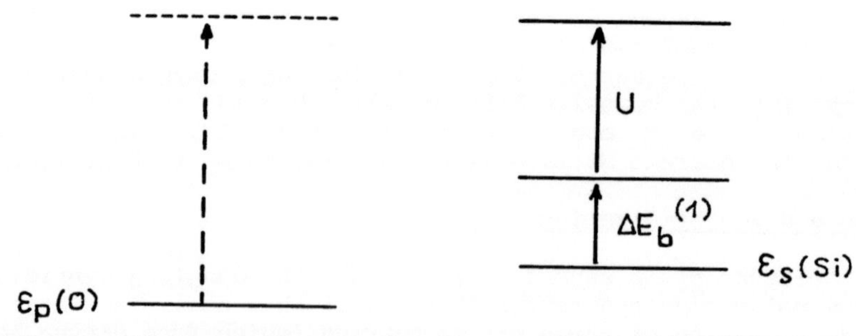

Fig. 2. The energy levels for the first step of a π-bonded oxygen vacancy center formation.

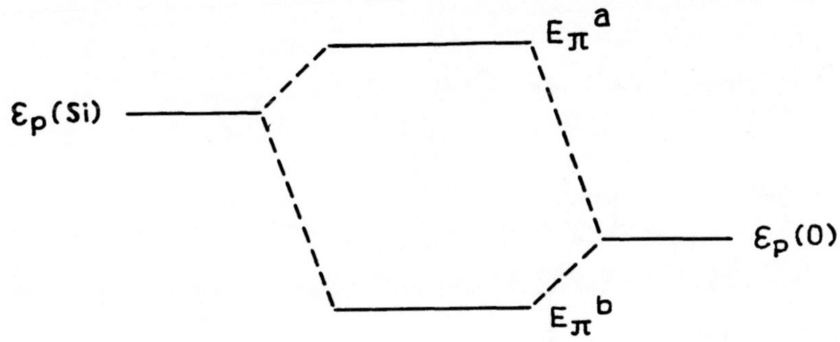

Fig. 3. The energy levels for the second step of a π-bonded oxygen vacancy center formation.

two spin-paired electrons on s-like dangling hybrid of the Si(1) atom. Total energy gain in the second step of a π-bonded oxygen vacancy center formation may be expressed (see Fig.3) :

$$\Delta E^{(2)}_{tot} = 2E^{b}_{\pi} - \varepsilon_p(Si) - \varepsilon_p(0) + \Delta E^{(2)}_{b}$$

where $\Delta E^{(2)}_{b}$ is the rehybridization change of the energy of Si(2) bonds relative to the energy of sp^3 bonded system (with electron pairing in the sp^3-sp^3 extended bond between Si(1) and Si(2),). The gain of the total energy is equal

$$\Delta E_{tot} = \Delta E^{(1)}_{tot} + \Delta E^{(2)}_{tot} = U - 3.9 \text{ eV}$$

The pairing of electrons on Si(1) s-dangling hybrid will be energetically favourable if $U < 3.9$ eV. As was estimated by several authors the effective Hubbard U energy value for sp^3 Si orbital fulfills this inequality. Thus, the pairing of electrons on Si s-dangling hybrid is attractive, principaly due to formation of the π bond.

DISCUSSION

There are three important consequences of including the π interaction in the oxygen vacancy neighbourhood regarding a charge trapping behaviour.

1. Formation of the electron trap (an acceptor level in the upper half of the forbidden energy gap) due to π-antibonding state. The oxygen vacancy centers become negatively charged as the Fermi level is raised towards the bottom of the conduction band.
2. Formation of the hole trap (a donor level close to the valence band) on s-like Si dangling hybrid, where the attractive pairing of two electrons takes place. A hole trapped by neutral oxygen vacancy will recombine with one of the paired electrons on the s-like orbital of the Si(1) atom leading to a positively charged center.
3. Potential of the neutral vacancy center (unfilled traps) in close proximity has a dipolar character. The positive pole is localized at the Si(2) - 0 π-bonded atoms and the negative pole at the Si(1) atom, with the paired electrons on s-like dangling orbitals, as shown in fig. 4.

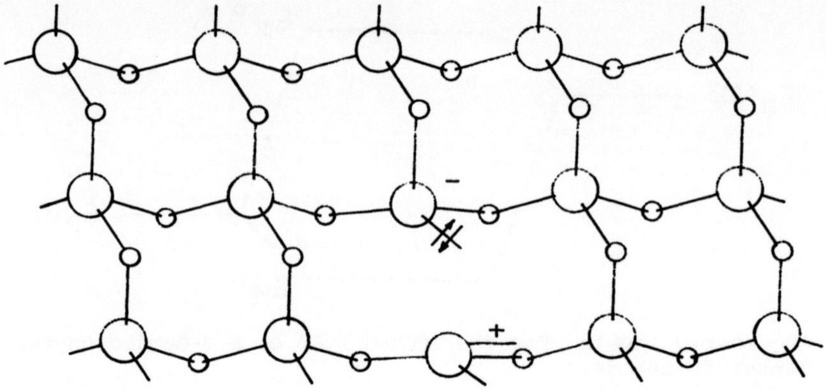

Fig. 4. A dipole type trap related to a π-bonded oxygen vacancy center in the lattice of SiO_2.

Thus, the capture cross-sections for electron and hole traps of this type should have values characteristic for neutral and Coulomb attractive centers, i.e. from 10^{-16} - 10^{-17} cm^2 to 10^{-14} cm^2. These values should also show a weak electric field dependence. The negatively charged oxygen vacancy may capture a second electron on the π-antibonding state. The capture cross-section for electrons should, in this case, be of order 10^{-18} -10^{-20} cm^2.

CONCLUSION

It is demonstrated that the neutral vacancy center in SiO_2 has an amphoteric character and can act as an electron or hole trap center of a dipole type.
There is much experimental evidence for the existence of this type of trap in irradiated and/or ion implanted oxide layers.

ACKNOWLEDGEMENT

The author wishes to thank Professor J.M. Baranowski for many helpful and valuable discussions.

REFERENCES

1. J.M. Baranowski, I. Strzalkowski, M. Marcezewski and M. Kowalski, "6th General Conference of the CMD of the EPS", Stockholm, Europhysics Conference Abstracts, 10a:242 (1986).
2. J.M. Baranowski, I. Strzalkowski, M. Marcezewski and M. Kowalski, in press in J. Appl. Phys. (1987).

THEORY OF OXYGEN-VACANCY DEFECTS IN SILICON DIOXIDE

W. Beall Fowler*, Jayanta K. Rudra*+, Arthur H. Edwards, and Frank J. Feigl***

***Physics Department and Sherman Fairchild Laboratory, Lehigh University, Bethlehem, PA 18015 ; ** U.S. Army E.T. & D. Laboratory, Fort Monmouth, NJ 07703 ; + Present address : Chemistry Department, Tulane University, New Orleans, LA 70118**

ABSTRACT

Large, asymmetric atomic relaxations are important features of oxygen-vacancy-related defects in silicon dioxide. To investigate these defects we have adapted the MINDO/3 and MOPN semiempirical molecular structure methods. In several cases (the E_1', E_2', and E_4' centers) the defect is paramagnetic and its primary characteristic is a single sp^3 electron localized on one Si and oriented towards or away from the O vacancy. Different atomic relaxations and charge states, along with the presence or absence of atomic H, distinguish these defects from one another. One important type of relaxation appears to be the displacement of a Si into a "back-bonds" interstitial position.

INTRODUCTION

In a series of calculations over the past years[1-5] we have established the utility and efficiency of the semiempirical molecular structure method[6] MINDO/3 and its open-shell version [7] MOPN for treating point defects in silicon dioxide. After determining certain parameters we have used these methods in connection with cluster models : a given defect is treated as a large molecule, with dangling bonds terminated by hydrogens, and the structure and properties are then calculated for various charge states and possible atomic relaxations.

Of particular interest has been a class of oxygen-vacancy defects in SiO_2 known generically as E' centers[8]. The common experimental signature of all E' centers[9] is an electron spin resonance (ESR) signal characteristic of an sp^3 electron on a single Si atom. E' centers have been detected in nearly all forms of SiO_2 including alpha quartz[10,11], bulk glass, and thin films[12-14]. While considerable progress has been made in distinguishing the features of different E' centers in bulk glasses[9], the crystalline symmetry of α-quartz has made it possible to resolve and interpret fine details of ESR structure to an extent that detailed models could be introduced and tested. To date, therefore, most of our attention has been paid to such defects in α-quartz, in particular the E_1', E_2', and E_4' centers and their nonparamagnetic charge states. This work has led not only to detailed models but also to the development of considerable insight into the types of atomic relaxations which can take place, and the subtle interplay between the symmetry of the unrelaxed defect and the driving forces leading to relaxation. Such aspects of this recent work are summarized in this paper.

SINGLE OXYGEN VACANCY

Removal of an oxygen atom from otherwise perfect α-quartz leaves a site of low symmetry (C_1) involving two 3-coordinated Si atoms which before relaxation are 3.06 Å apart. As surmised by Robertson[15] and verified by our calculations [5], the system relaxes in such a way that these two Si move together, forming a Si-Si bond with an equilibrium separation of ~ 2.5 Å, only slightly larger than the 2.35 Å found in crystalline Si. This ground state is a singlet ; there will exist a corresponding triplet state, which we compute to lie 1.5 eV higher in energy before further relaxation, and 0.9 eV higher after relaxation. Atomic relaxation in this lowest triplet state leaves the Si atoms 3.0 Å apart. Because we have had little experience using MINDO/3 for triplet systems, these results should be regarded as somewhat preliminary.

There is little evidence for the existence of the neutral O vacancy in SiO_2 ; perhaps the first comes from the tunneling data of Feigl and Zvanut, reported at this conference. One would expect a weak optical absorption from the singlet to the unrelaxed triplet, made allowed by spin-orbit mixing, but no such transition has yet been assigned. One might also expect a long-lived triplet to singlet luminescence band, although our calculations suggest that radiationless transitions may instead occur. It is expected that excited singlet states will also exist, although we do not know at what energy ; optical transitions from the ground state to some of the low-lying singlets are expected to be highly allowed.

Removal of an electron from the neutral O vacancy leaves a paramagnetic, positively charged vacancy. Feigl et al.[16] first suggested that such a defect will undergo an asymmetric relaxation which will localize the unpaired spin on one of the two Si, the other Si(+ charged) moving away from the vacancy into a nearly planar configuration with respect to its three O neighbors. Subsequent detailed molecular orbital calculations by Yip and Fowler[17] were consistent with this model, but they were unable to correctly determine which way the system would relax ; they predicted relaxation along an asymmetric double-well potential-energy surface, but with the electron localized on the "long-bond" side rather than on the "short-bond" side, as observed experimentally[11,18].

Recent calculations by Rudra and Fowler[5] using MOPN and a larger atomic cluster ($Si_2O_6Si_6H_{18}$) yielded results in better agreement with experiment. Relaxation occurs in a way which leaves the unpaired spin localized on the "short-bond" Si. The potential-energy surface involving the relaxation of the "long-bond" silicon, Si(+), and its neighboring oxygens is extremely flat when Si(+) is nearly planar. This flatness along with the presence of a nearby "external" O in the α-quartz lattice led us to expand the cluster by adding the external O along with two H terminators, and allowing Si(+) and its other O neighbors to relax. This leads to an outward relaxation of Si(+) into a puckered configuration (fig 1b), whose location is only 2.1. Å from the external O. In the process Si(+) takes some electron density from the external O, so that in some sense one has a three-fold coordinated, positively-charged oxygen, one of the predicted components of the valence-alternation-pair (VAP) models[19] of defects in SiO_2.

There remains the question of the double-well nature of the E_1' center. Recent calculations by Edwards et al.[20] on a positively-charged oxygen-vacancy cluster constructed to have C_2 symmetry before relaxation have yielded the surprising result that the defect does not relax asymmetrically unless the two "halves" are initially separated by an unrealistically large distance. However, a relatively small symmetry-breaking shift in the positions of the outer cluster atoms leads to a relaxation similar to that calculated for α-quartz.

In view of this result we have further investigated[20] the possible double-well nature of the E_1' center in α-quartz. Even with only the basic cluster (that is, whithout the "external" oxygen) our MOPN calculations indicate that there is no second well. This surprising result, along with the absence of asymmetric relaxation in the C_2 defect, forces us to the conclusion that the "symmetry-breaking" electron-lattice interactions are not as effective as we had expected, and that the inherent asymmetry in α-quartz, even though apparently slight, is in fact crucial in forcing the system to relax off center. We have estimated the relevant parameters in a vibronic model of the E_1' center ; although numerically different, they follow earlier ideas of Schirmer[21], who had included the inherent asymmetry in such a model and had suggested that only one off-centered well exists.

We can speculate on what relevance these recent α-quartz results have with respect to the E' centers in a-SiO_2. To some extent this depends on the structural model for the E' center in a glass. At one extreme is what one might call a "void" model : here it is assumed that some fraction of the Si are three-coordinated, and that the system is neutral when the Si has an unpaired spin. The region towards which the spin is oriented is either perfectly bonded, or far away ; in either case that region is basically irrelevant.

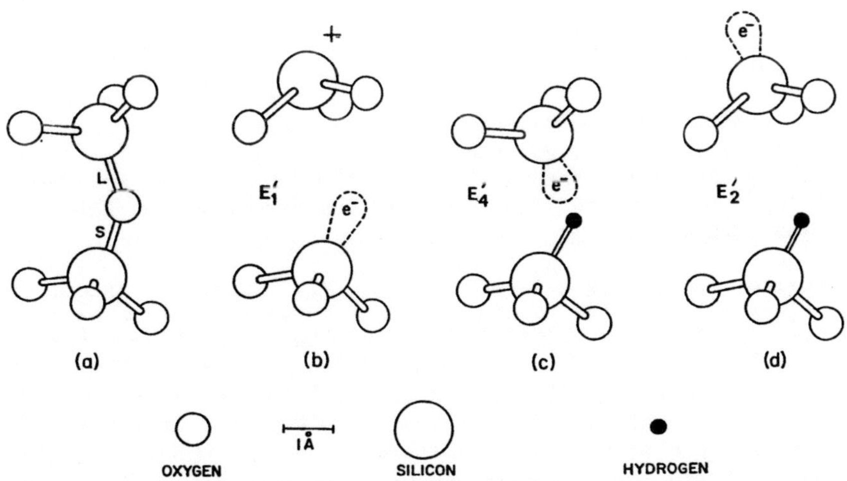

Fig. 1. (a) Fragment of perfect alpha-quartz. "L" stands for long bond and "S" for short bond. (b) Model for the E_1' center (c). Model for the E_4' (d) Model for the E_2' center. In (b)-(d), the unpaired electron is shown schematically as e^-.

The other extreme is an O-vacancy model, similar to that of α-quartz. This E' center would be + charged, as in the E_1' center in α-quartz (unless H were involved ; see the next section). This model is consistent with the identification of positively-charged defects in thin oxide films with E' centers. Even here one might anticipate different vacancy "sizes", depending on preparation conditions. In any case, the amorphous nature of the surroundings would be expected to create the necessary asymmetry to allow off-centered relaxation to occur as in the E_1' center in α-quartz. Again, this is speculative ; we plan to investigate these ideas more fully.

O VACANCY PLUS H

Both the E_2' and the E_4' centers in α-quartz involve a H atom in association with an O vacancy. These defects have been studied by ESR and have been the subject of recent theoretical investigation.

The E_4' center consists of a H atom within an O vacancy in α-quartz. The H moves in an asymmetric double-well potential. This defect is neutral, and paramagnetic ; because there is considerable spin density on both Si, detailed experiments and analysis could be carried out by Isoya et al.[22], who firmly established the model.

Isoya et al. supplemented their experiments with ab initio atomic cluster calculations. Subsequently Edwards and Fowler[2] investigated the E_4' center using MOPN. Both methods yielded qualitatively similar results and were able to show how at the lowest temperature the H atom is bound strongly to one Si (the short-bond side) and the electron spin is localized primarily on the long-bond Si, as shown in fig.1 (c). At high T the H atom is excited to higher vibrational states and spends considerable time near both Si ; the spin density is correspondingly shared by both Si.

It was more difficult to establish a model for the E_2' center. Experimentally one observes an unpaired spin[23,24] on a long-bond Si, and one sees a moderately strong hyperfine interaction associated with a H atom. Rudra's suggestion[4] that one should consider outward relaxation of the Si (that is, a puckered configuration as discussed previously) led to the notion that the E_4' and E_2' centers are identical defects, but in different atomic configurations : in the case of the E_2' center the "long-bond" Si relaxes away from the vacancy and into the interstitial region, while the H atom remains bound to the "short-bond" Si. The unpaired spin is localized on the long-bond Si, as abserved by ESR. Figure 1d illustrates this model.

Detailed calculations by Rudra et al.[4] were consistent with this notion. They obtained the potential-energy curve for motion of the Si from inside to outside the vacancy, finding an energy barrier of ~ 0.6 eV. Furthermore, they found the E_2' configuration to be more stable than E_4' by ~ 0.1. eV. This energy difference is so small that it is not clear that it is meaningful ; what is important, we believe, is that both E_2' and E_4' configurations are stable or metastable, they are nearly degenerate, and they are separated by a barrier which is approximately large enough that either configuration would have a long lifetime against decay into the other.

Hydrogen-related E' centers have been observed[9] in a-SiO_2. None of these appears to have the double -well character of the E_4' center. This may simply mean that the inherent asymmetry of the vacancy site is so great that the H atom remains bound to one Si, even at high temperature. If this were the case, one might expect the major difference between the glassy E_2' and E_4' centers to be the relative strength of the hyperfine splitting associated with the H atom (smaller for E_2' since the H is farther from the unpaired spin). Furthermore, we note again that while E_1' is positively charged, both E_2' and E_4' are neutral ; this point is relevant to the identification of charged defects with E' centers in thin films. We intend to investigate such issues in more detail.

CONCLUSIONS

The calculations on E' defects in α-quartz summarized above have yielded considerable insight into the several aspects of symmetry, bonding, and relaxation necessary to explain their properties. This insight is expected to prove useful in further understanding the corresponding defects in a-SiO_2.

An outstanding feature of all these defects seems to be the integrity of the SiO_3 pyramid on which the unpaired spin is localized. In all three cases we find the locations of the Si above the O basal plane to be identical to within hundredths of an Å and the sp^3 orbital to be within 2 degrees of the basal-plane normal. This defect is in a very real sense the "generic E' center" described by Griscom[9].

ACKNOWLEDGMENTS

The authors thank David Griscom and John Weil for useful discussions. This research was supported in part by the Electronic and Solid State Science Program of the U.S. Office of Naval Research.

REFERENCES

1. A.H. Edwards and W.B. Fowler, Phys. Rev. B 26:6649 (1982).
2. A.H. Edwards and W.B. Fowler, J. Phys. Chem. Solids 46:841 (1985).
3. A.H. Edwards and W.B. Fowler, in "Structure and Bonding in Non-Crystalline Solids", edited by G.E. Walrafen and A.G. Revesz, Plenum, New-York (1986).
4. J.K. Rudra, W.B. Fowler, and F.J. Feigl, Phys. Rev. Lett. 55:2614 (1985).
5. J.K. Rudra and W.B. Fowler, Phys. Rev. B 35:8223 (1987).
6. R.C. Bingham, M.J.S. Dewar, and K.H. Lo, A. Am. Chem. Soc. 97:1285 (1975).
7. P. Bischof, J. Am. Chem. Soc. 98:6844 (1976).
8. R.A. Weeks, and C.M. Nelson, J. Am. Ceram. Soc. 43:399 (1960).
9. D.L. Griscom, Nucl. Instrum. Methods Phys. Res. B 1:481 (1984).
10. R.A. Weeks, J. Appl. Phys. 27:1376 (1956).
11. R.A. Silsbee, J. Appl. Phys. 32:1459 (1961).
12. C.L. Marquardt and G.H. Sigel, IEEE Trans. Nucl. Sci. 22:2234 (1975).
13. P.M. Lenahan and P.V. Dressendorfer, J. Appl. Phys. 55:3495 (1984).
14. W.E. Carlos, Nucl. Instrum. Methods Phys. Res. B 1:383 (1984).
15. J. Robertson, J. Phys. C 17:L221 (1984).
16. F.J. Feigl, W.B. Fowler, and K.L. Yip, Solid State Commun. 14:225 (1974).
17. K.L. Yip and W.B. Fowler, Phys. Rev. B 11:2327 (1975).
18. M.G. Jani, R.B. Bussoli, and L.E. Halliburton, Phys. Rev. B 27:2285 (1983).

19. G. Lucovsky, Phil. Mag. B 41:457 (1980).
20. A.H. Edwards, W.B. Fowler, and F.J. Feigl (to be published).
21. O.F. Schirmer, in "The Physics of MOS Insulators" G. Lucovsky, S.T. Pantelides, and F.L. Galeener, eds, Pergamon, N.Y., (1980) p. 102.
22. J. Isoya, J.A. Weil, and L.E. Halliburton, J. Chem. Phys. 74:5436 (1981).
23. R.A. Weeks, Phys. Rev. 130: (1963).
24. F.J. Feigl and J.H. Anderson, J. Phys. Chem. Solids 31:575 (1970).

TOTAL ENERGY CALCULATIONS FOR INTRINSIC DEFECTS IN AMORPHOUS SiO_2

J. Robertson

Central Electricity Research Laboratories
Leatherhead
Surrey DT22 7SE
United Kingdom

ABSTRACT

The semi-empirical methods MINDO/3 and MOPN are used to calculate the total energies of various intrinsic defect configurations. We find that the positively charged trivalent silicon site gains about 1 eV by overcoordination and that the effective correlation energy of the trivalent silicon defect is positive. The creation energy of valence alternation pairs and like-atom bonds is also calculated.

INTRODUCTION

Considerable progress has been made in the identification of the intrinsic defects in silicon dioxide by using a combination of electron spin resonance (ESR) experiments and total energy calculations. ESR is able to provide information on the wavefunction of the unpaired electron of a paramagnetic defect which allows its symmetry, localization and the chemical identity of key atoms in the defect to be determined, as reviewed by Griscom[1,2]. The total energy calculations are then used to complete the atomic description of the defect by calculating the stable geometry and by rationalising the structure and chemical bonding, as reviewed by Edwards and Fowler[3,4]. These methods were able to identify the three principal defects in irradiated amorphous (a-) SiO_2 (silica) as the Si dangling bond radical or E' center, the nonbridging oxygen radical and the superoxide radical. Each of these paramagnetic defects has a related diamagnetic center, the Si-Si bond, the hydroxyl group and the O-O bond or superoxide bridge, respectively. The Si-Si bond is analogous to the relaxed neutral oxygen vacancy[5], and the hydrogen found in hydroxyl groups is essentially an intrinsic component of silica.

These then are the important defects of irradiated silica. However, it is not apriori clear that these defects, largely created by radiolytic or atomic displacement processes, are necessarily those of lowest total energy. It is this latter type of defect which is likely to be generated thermally, to be important in low energy processes and to be the more important defect in nonequilibrium silicas such as the thermally grown oxide on silicon substrates.

The viscosity of silica and the diffusion of network atoms (Si, O or network-forming impurities) is likely to be mediated by some species of thermally generated intrinsic point defects. In covalently bonded semiconductors like Si, atomic diffusion is mediated by vacancies and interstitials[6]. In a covalently bonded amorphous semiconductor, atomic diffusion can be mediated by dangling (broken) bonds[7]. Therefore, in a covalently bonded amorphous compound such as silica there are a number of possible modes for diffusion, via vacancies, interstitials or broken bonds. Indeed, a model of the viscosity of silica involving valence alternation centers has recently been developed by Mott[3].

Let us first consider the precise configurations of the vacancies and interstitials. We have already noted that the neutral oxygen vacancy is analogous to the Si-Si bond. It is also known that the oxygen interstitial rapidly converts into the superoxide bridge[4,9]. In contrast we may neglect Si site defects as too energetic - we must break four Si-O bonds in order to form a Si vacancy but only break two Si-O bonds in order to form an oxygen vacancy.

The broken bond defects are likely to be the valence alternation pair (VAP) centers found in chalcogenide glasses, as reviewed by O'Reilly and Robertson[10]. Because of ionicity, a Si-O bond will break to form charged rather than neutral defects

$$Si_4^0 + O_2^0 \rightarrow Si_3^+ + O_1^- \qquad (1)$$

Here, the subscripts denote coordination. The Si_3^+ site possesses an empty dangling bond orbital, which is able to form a dative bond with the $p\pi$ electrons of a nearby bulk oxygen site :

$$Si_3^+ + O_2^0 \rightarrow Si_4^0 + O_3^+ \qquad (2)$$

This extra bond might be expected to stabilise the O_3^+ site. Thus, in this paper we present preliminary total energy calculations to study the stabilities of the Si_3^+, O_3^+ and O_1^- sites and of the Si-Si and O-O bonds.

THE CALCULATIONAL METHOD

We have used the semi-empirical molecular orbital method MINDO/3 (and its open shell version MOPN) to calculate the total energies. This scheme, developed by Dewar et al.[11,12], retains only the valence electrons. It uses Slater-type s and p orbitals and retains specific classes of interactions between all sites, which are then parameterised to reproduce known molecular geometries, bond lengths and heats of formation. Most of the parameters are intra-atomic and have been set previously[10], but two are specific to the Si-O bond and must still be determined. One is the two-center interaction which it scales to the wavefunction overlap, using the Wolfberg-Helmholtz approximation, and the other represents core-core repulsions. Both parameters have been recently determined by Edwards and Fowler[13] by fitting the properties of some Si-O bond containing molecules.

The calculations are performed on representative molecular clusters. To study the overcoordination of the Si_3 site, the cluster :

$$(HO)_3\ Si\ ....\ O(Si(OH)_3)_2 \qquad (3)$$

with C_{2v} symmetry is used. To study the relative stabilities of Si-Si, O-O and Si-O bonds, the cluster

$$(H_3SiO)_3Si - X - Si(OSiH_3)_3 \qquad (4)$$

is used, where X represents either no atom, an oxygen atom or an O-O bridge, respectively. Finally, the energy needed to break a Si-O bond is found using the same cluster in the form :

$$(H_3SiO)_3Si\text{-}O \;....\; Si(OSiH_3)_3 \qquad (5)$$

It is seen that hydrogen atoms are used to terminate the surface bonds of the clusters. Generally O-H terminations, as in(3), are preferable to Si-H terminations, as in (4), because the O-H states lie well outside the energy range of possible defect gap states[14], whilst Si-H terminations only repel the states to the band edges. However, computer limitation prevented us from using the larger clusters needed for O-H terminations in cluster[4]. Generally, rather small clusters are adequate for investigation of defects in SiO_2 because of the extreme localization of the defect wavefunctions, although this would not be true for defects in a semiconductor like Si.

In each cluster, the geometry of terminal SiH_3 or OH groups was held fixed, viz bond lengths of 1.47 Å (Si-H) and 0.93 Å (O-H), bond angles of 109.47° (Si) and 165° (O) and the dihedral angles. All other geometrical parameters were allowed to vary to optimise the structure. This optimisation philosophy is appropriate for an "underconstrained" network, and differs from that of Fowler who allows optimisation with fixed positions of terminator atoms, appropriate to a quartz lattice[3,4].

RESULTS

The total energies for cluster (3) were calculated for the two cases of (a) a C_{3h} symmetric cluster, with equal central O-Si bond lengths, and (b) with one of these bonds extrapolated to infinity, for cluster charges of +1, 0 and -1.

It was found firstly from the total energies of the positively charged clusters that the overcoordination of the Si_3^+ was exothermic :

$$\Delta_{over} = E\left(O_3^+\right) - E\left(Si_3^+\right) \approx -1.2 \text{ eV} \qquad (6)$$

Secondly, the effective correlation energy U was calculated. This is defined as the energy penalty of doubly occupying a defect orbital[9]. For the case of $\Delta_{over} < 0$, U is given by

$$U = E\left(O_3^+\right) + E\left(Si_3^-\right) - 2E\left(Si_3^0\right) \qquad (7)$$

The calculated total energies give a value of U of

$$U \approx 4 \text{ eV} \qquad (8)$$

The first result confirms that the general picture of valence alternation defects is valid for SiO_2 but it also shows that the simple bonding models greatly overstimate the magnitude Δ_{over}. We also note that a similar overcoordination was recently found for the Si_3^+ site at the positive oxygen vacancy in quartz[5].

The second result shows that the Si dangling bond is a positive U system, in common with the P_b center at the SiO_2 : Si interface and the Si dangling bond in a-Si:H. Clearly, the energy of overcoordination has been insufficient to produce a negative U, as it is able to in the simplest bonding models of VAPs.

Our calculations do however reveal one fault with the MINDO/3 method. We find that the barrier to overcoordination of the Si_3^- site is only ~ 0.3 eV. This is chemically unrealistic as there should be a strong closed-shell repulsion between the Si_3^- site and the O_2^0 site in this case. This underestimation is a characteristic error of the treatment of MINDO/3 methods of closed-shell interactions[15].

The total energy of the neutral cluster(4) was calculated for two cases, (a) the equilibrium Si-O bond length and (b) for this length extrapolated to infinity. (b) creates a Si dangling bond and an O dangling bond. As expected, these are ionised positive and negative, respectively. The bond breaking energy is calculated to be

$$\Delta_1 \approx 5.8 \text{ eV} \tag{9}$$

This is rather larger than Pauling's estimate of the Si-O bond energy of 4.3 eV, or to the value corresponding to the heat of formation of SiO_2. Combining(6) and(9) gives the VAP creation energy

$$\Delta_{vap} = \Delta_1 + \Delta_{over} \approx 4.6 \text{ eV} \tag{10}$$

Finally, the total energy of the neutral cluster in(4) was calculated for the three cases, to estimate the cost of introducing Si-Si and O-O bonds. This gives the Frenkel pair creation energy :

$$\Delta_{Fr} = E(SiSi) + E(SiOOSi) - 2E(SiOSi) \tag{11}$$

We find

$$\Delta_{Fr} \approx 8.4 \text{ eV} \tag{12}$$

This is also rather high, and we suspect that this is also primarily because of the high estimated strength of the Si-O bond in (9).

The results in (10) and (12) indicate that the VAP is the lowest energy primary defect in silica. We therefore anticipate that network atom diffusion will be mediated by VAPs, if the saddle points between different VAP positions are not too high in energy.

CONCLUSIONS

MINDO/3 calculations for various intrinsic defects in amorphous SiO_2 indicate that trivalent silicon sites will overcoordinate, that trivalent silicon sites have a positive effective correlation energy and that the creation energy of a valence alternation pair is less than that of a Frenkel pair of Si-Si and O-O bonds. We aim to continue these calculations to check the results, using more accurate ab-initio methods where appropriate, and by studying the effect of expanding the basis set to include d orbitals.

ACKNOWLEDGEMENTS

The author is very grateful to Dr A. Edwards for conversations.

REFERENCES

1. D.L. Griscom, in "Physics of SiO_2 and its Interfaces", ed. S.T. Pantelides, Pergamon, London, (1978).
2. D.L. Griscom, J. Non-Cryst Solids 73:51 (1985).
3. A.H. Edwards, W.B. Fowler, in "Structure and Bonding in Non-Crystalline Solids", ed. G. Walrafen and A. Revesz Plenum, New York, (1985).
4. A.H. Edwards, Mat. Res. Soc. Symp. Proc. 61:3 (1986).
5. J.K. Rudra, W.B. Fowler, Phys. Rev. B 35:8223 (1987).
6. R. Car, P.J. Kelly, A. Oshimaya, S.T. Pantelides, Phys. Rev. Lett. 52:1187 (1984).
7. J. Robertson, Philos Mag., to be published.
8. N.F. Mott, Philos Mag. to be published.
9. A.H. Edwards, W.B. Fowler, Phys. Rev. B 26:6649 (1982).
10. E.P. O'Reilly, J. Robertson, Phys. Rev. B 27:3780 (1983).
11. R.C. Bingham, M.J.S. Dewar, D.H. Lo, J. Am. Chem. Soc. 97:1285 (1978).
12. P. Bischoff, J. Am. Chem. Soc. 98:64411 (1976).
13. A.H. Edwards, W.B. Fowler, J. Phys. Chem. Solids 46:841 (1985).
14. J. Robertson, this volume.
15. T. Clark, "A Handbook of Computational Chemistry" Wiley, New York, (1985).

BORON IMPURITY CENTERS IN SiO_2 A TIGHT BINDING CONSIDERATION

Mariusz Marczewski*, Ireneusz Strzalkowski, and
Jacek Baranowski**

*Institute of Physics
Warsaw Technical University
Koszykowa 75, 00-662 Warsaw, Poland
**Institute of Experimental Physics
University of Warsaw
Hoza 69, 00-681 Warsaw
Poland

INTRODUCTION

Boron is a common dopant in silicon technology, its implantation is frequently used in MOS VLSI circuit fabrication. In this way some atoms can be located in the SiO_2. An interesting question is how boron can be incorporated into the SiO_2 network and whether the presence of implanted boron can create electron and/or hole trapping centers.

There are conflicting experimental reports on the trapping properties of boron implanted silica layers. Early papers claimed that proper postimplantation annealing in N_2 removes all trapping centers from boron implanted oxides[1,2]. However, recent work of the Aachen group has given evidence that annealing in pure N_2 cannot reduce the concentration of trapping sites to the level of unimplanted oxides[3,4]. Deep and shallow electron traps and also hole traps have been found in oxide layers after prolonged nitrogen annealing. Similar results have been obtained for oxides implanted with noble gas ions, but the concentration of traps have been lower in this case. It has been shown that short time annealing in oxygen has reduced significantly the concentration of all kinds of post-implantation traps, hence the authors suggest that the investigated traps are oxygen deficiency related, not boron itself. The lack of detailed information about microscopic structure of the considered centers leaves the question open and underlines the need for theoretical consideration.

The aim of this work is to construct a model of the boron containing centers in SiO_2. All calculations presented in this paper are based on Bond Orbital Approximation. Details of this method are widely published[5,6].

---

*This work is partially sponsored within Central Project of Basic Research CPBP 01.08E3.5.

We have used Herman-Skilman atomic potentials[6] and appropriate values of universal matrix elements.

THE INCORPORATION OF BORON INTO THE SiO_2 NETWORK.

There are many kinds of defects in implanted SiO_2 but the oxygen vacancy is the most common. Therefore the probable locations of boron atoms implanted into silicon dioxide layers are either in the lattice or in the vicinity of the oxygen vacancy. We have considered both cases calculating the total energy gain due to replacement of one of the silicon atoms in the network by interstitial boron atom.

Stoichiometric Boron Center

The first case is illustrated in figure 1. It is interstitial boron replacing silicon in the lattice. We named this center "Stoichiometric Boron Center". On the left side of the figure the neutrality of all atoms is preserved but one oxygen bond is broken. On the right side of the figure the number of bonds is the same as in perfect SiO_2 lattice but one electron is transfered from nonbonding p orbital of neighbouring oxygen atom to a boron atom, so an electric dipole is created. The positive pole of the dipole is on oxygen and negative one on boron. Both the number and the nature of bonds is different in these two cases. Three electrons placed

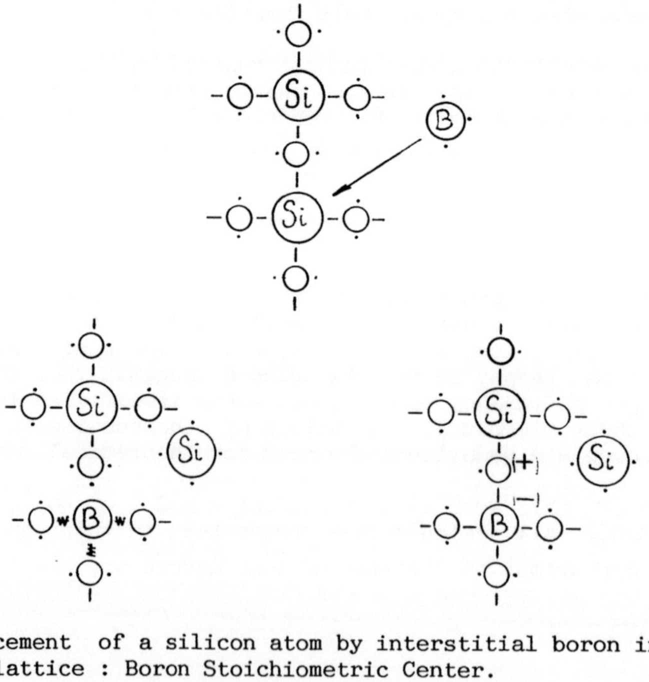

Fig. 1. Replacement of a silicon atom by interstitial boron in the perfect SiO_2 lattice : Boron Stoichiometric Center.
On the left side there is neutral form of BSC, on the right side, dipolar form (energetically favourable).
— (sp^3-p) bonds ; ww (sp^2-p) bonds ;
nonbonding orbitals.

on an atom form sp^2 hybrids laying in one plane so triply coordinated boron lays in a plane of surrounding oxygen atoms. Four electrons on an atom form sp^3 hybrids pointed towards the vertices of a tetrahedron, hence fourfold coordinated boron is in the former silicon position.

Boron Decorated Oxygen Vacancy

The replacement of one of the silicon atoms, forming an oxygen vacancy, is illustrated in figure 2. This center we named "Boron Decorated Oxygen Vacancy". The extended description of pure oxygen vacancy has been given elsewhere[5] and in this proceedings[7]. The most important conclusion from our previous calculations [5] is that the in neutral oxygen vacancy electron must be transfered from one of the oxygen atoms to silicon to enable formation of a π bond between neighbouring silicon and oxygen to obtain a minimum of total energy. The other silicon is negatively charged having two electrons on an s orbital. If boron is placed instead of a silicon atom the whole arrangement is similar but with important differences. The number of electrons is lower by one for this case than for the pure oxygen vacancy, so electron transfer is from oxygen to boron, not to silicon. Then the silicon atom is neutral and forms three sp^2 hybrids to arrange bonds with oxygen. The fourth electron of silicon is located on p orbital. The boron atom is negatively charged having, similar to silicon, four electrons. Three of them are located on sp^2 hybrids forming σ bonds, whereas the fourth is located on a p orbital forming a π bond.

Total Energy Calculations

To check if the above discussed incorporation of boron is possible, one should compare the energy of initial states (with interstitial boron) to the energy of final states (with interstitial silicon). To do so we have calculated the energies of all bonds appearing here.

For BSC two cases have been considered. The total energy of the initial state is the same for both :

$$E_{totI}\ (BSC) = E(B_{int}) + 8E(Si(sp^3)\text{-}O(p)) + E(O(p)).$$

The total energy of final state for the neutral form of BSC (left part of figure 1) is :

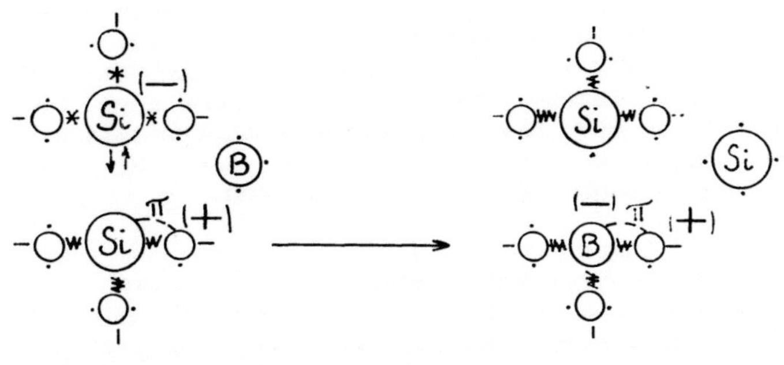

Fig. 2. Replacement of a silicon atom by interstitial boron in an oxygen vacancy : Boron Decorated Oxygen Vacancy.
— (sp^3-p) bonds ; ww (sp^2-p) bonds.
. nonbonding orbitals ; - - (p-p) π bonds.
⇅ pair of electrons on s orbital of silicon.

$E_{totF}(BSC) = E(Si_{int}) + 3E(B(sp^2) - O\text{-}Si(sp^3)) + 2E(O(p)) + E(O\text{-}Si(sp^3))$.

The difference $E_{totF} - E_{totI}$ equals -311.40 - (-312.25) =+ 0.85 eV.

It means that to replace silicon by boron this way one has to provide extra energy of 0.85 eV, which makes this process highly improbable.

The total energy of the dipolar form of BSC is :

$E_{totF}(BSC) = E(Si_{int}) + 4E(B(sp^3) - O\text{-}Si(sp^3)) + U(B\text{-}O) = -317.10 + U(B\text{-}O)$.

For this case the difference of final and initial energy is equal to -4.85 eV + U(B-O).

The subsequent terms in the above equations are the energies of U(B-O) different bonds, atomic orbitals or whole interstitial atoms. $U_{(B-O)}$ is the effective value of repulsive coulombic energy resulting from the electron transfer from oxygen to boron.

The total energy of the initial state of BDOV, calculated in the same manner as above, is equal to -484.86 eV + U(Si-O). The final total energy of BDOV is equal to -486.54 eV + U(B-O). The difference between them is equal to -1.68 eV + U(B-O) - U(Si-O).

There is no detailed calculation of U's appropriate for this case. The Harrison[8] consideration seems to be not relevant here. However some estimations for charged silicon have been made[5] and the value of U(Si-O) considerably smaller than 3.9 eV is expected. For two reasons the value of U(B-O) should be smaller than for silicon : (i) the polarisation of boron-oxygen bonds is greater than silicon-oxygen, so boron, before transfer of electron is more positively charged than silicon ; (ii) the distance between oppositely charged atoms is smaller in the case of boron.

Taking into account considerations about the values of U for boron and silicon, we are convinced that incorporation of interstitial boron into the SiO_2 network is energetically favourable. Hence it probably occurs during a nitrogen high temperature annealing of implanted SiO_2 layers[3,4].

DISCUSSION

Charge Trapping

For a center to become an electron trap must it have an empty level within the energy gap. Both considered centers fulfill this condition. The BSC has an empty level of nonbonding p orbital of positively charged oxygen, very likely located near the valence band edge, so this is a deep trapping center. The BDOV has empty antibonding orbitals of a boron-oxygen π bond. This level is a shallow one. But BDOV can also capture a hole on a nonbonding, singly occupied p orbital of silicon.

Both electron traps are connected with the same electric dipole but the experimentally determined capture cross section of shallow one is expected to be much greater. One can investigate trapping on shallow traps only at low temperature so capturing in the dipole potential well can be effective. We are convinced, that the capture on shallow antibonding level is a two stage process. During the first stage there is trapping in the dipole potential well and in the second stage the electron is carried over to deeper (however still shallow) antibonding orbital of π-type boron-oxygen bond. The capture cross sections for a potential well of a dipole have been estimated in the range of 10^{-15}-10^{-17} cm^2, depending on dipole magnitude, electric field and temperature. However, for a dipole as short as the boron-oxygen distance, the well is so shallow that at room temperature no trapping is possible.

To distinguish the deep traps from the shallow the experiment has to be performed at room or elevated temperature. No two stage capturing, but straightforward transition to the deep state is then only possible, so the capture cross section of trapping connected to the BSC should be expected in the range of 10^{-17} to 10^{-18} cm^2, typical for neutral trapping.

Oxygen Annealing

The trapping properties of both centers considered can be annihilated by incorporation of additional oxygen atoms. In order to annihilate BSC one oxygen atom should bond the charged oxygen p orbital, preventing trapping of an electron on it. For annihilation of BDOV two oxygen atoms are necessary : one to transform BDOV into BSC and the second to saturate it. Both saturated centers have been shown in figure 3. One can expect that oxygen annealing effectively removes boron related traps in SiO_2.

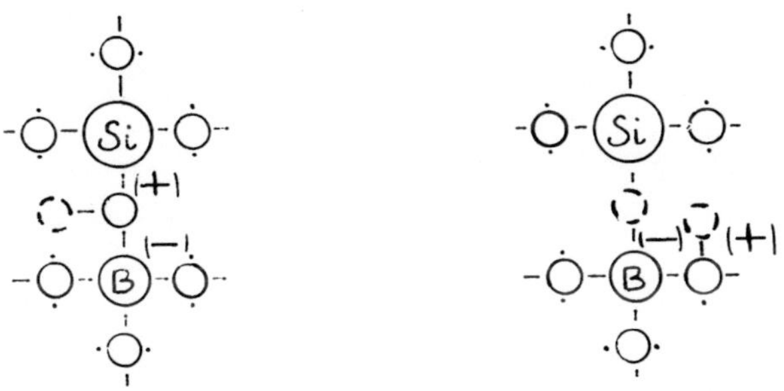

Fig. 3. Saturation of boron trapping centers by incorporation of additional oxygen atoms.
- additional oxygen atoms.

CONCLUSIONS

The major conclusions from our consideration are as follows :
1. Boron implanted into SiO_2 can create two different trapping centers.
2. Boron replacing silicon in the perfect SiO_2 lattice, called by us Boron Stoichiometric Center, has deep electron trap properties.
3. Boron in the place of silicon with an oxygen vacancy creates an amphoteric center called by us Boron Decorated Oxygen Vacancy. This center is shallow as an electron trap but deep as a hole trap.
4. The incorporation of additional oxygen atoms annhilates the trapping properties of both centers.

The proposed models of centers connected with boron atoms incorporated in SiO_2 are in excellent agreement with trapping and annealing experiments[3,4].

REFERENCES

1. M. Marczewski, I. Strzalkowski, Appl. Phys., A 29:233, (1982).
2. D.J. DiMaria, in "The Physics of SiO_2 and its interfaces", S.T. Pantelides, ed. Pergamon Press, New York (1978).

3. M. Offenberg, M. Maier, R. Meyer, P. Balk, J. Vac. Sci. Technol., A 4:1009 (1986).
4. M. Offenberg, T. Johannson, M. Aslam, P. Balk, Physica, 129 B,240, (1985).
5. J.M. Baranowski, I. Strzalkowski, M. Marczewski, M. Kowalski, J. Appl. Phys., in press.
6. W.A. Harrison, "Electronic Structure and the Properties of Solids", Freeman, N.Y., (1980).
7. I. Strzalkowski, this Proceedings.
8. W.A. Harrison, Phys. Rev. B:2121 (1985).

INTRINSIC AND EXTRINSIC POINT DEFECTS IN a-SiO_2

David L. Griscom

Optical Sciences Division
Naval Research Laboratory
Washington, DC 20375
USA

Point defects in amorphous silicon dioxide (a-SiO_2) can be considered as any local deviations from the "perfect" glass structure, which for the sake of discussion will be taken to be a continuous random network of Si $(O_{½})_4$ tetrahedra joined at the corners. Thus, for example, a vacancy-interstitial pair defined on such an otherwise perfect network would be the analogue of a Frenkel pair in crystalline solid. Since in our ideal glass each silicon has four bonds to neighboring atoms while each oxygen has but two, Frenkel pairs involving oxygen atoms (fig. 1a) would seem more probable than those involving silicons. Other possibilities include over or under-coordinated atoms, substitutional or interstitial impurities, or bonds between like atoms. Note in fig. 1a that a neutral oxygen vacancy is tantamount to a Si-Si homo bond.

Point defects in a-SiO_2 may be charged or uncharged, diamagnetic or paramagnetic. Many such defects have associated with them optical absorption bands which may pose problems for long-path-length optical systems (i.e. fiber optics), while the presence of charged defects can perturb the operation of metal-oxide-semiconductor (MOS) devices which employ a-SiO_2 as the gate or field oxide. For these reasons, understanding of the fundamental natures of these point defects is paramount.

Electron spin resonance (ESR) is the one experimental technique which yields sufficiently precise information to identify the atomic arrangements at the defect sites. ESR measures the resonant absorption of microwaves by unpaired electrons as a function of the magnitude of an externally applied magnetic field. The single drawback of this technique is that it is only capable of looking at defects which harbor an unpaired electron "spin". However, this is not a serious impediment, as diamagnetic defects frequently may be activated by photoionization or by the trapping of radiation-induced free carriers. The most important structural information to be gained from ESR derives from the hyperfine (hf) interaction of the unpaired spin with a nearby magnetic nuclide. Here, a magnetic nuclide is defined as one having a non-zero nuclear spin I, and hence a non-zero nuclear magnetic moment μ. If, through an isotopic substitution experiment, a magnetic nuclide should replace a spin-zero nuclide at the defect site, the original one-line spectrum will split into 2I + 1 hf lines, the spacings of which give information pertaining to the wavefunction density of the unpaired electron. More detailed background concerning the theory of ESR and its practice in glasses can be found elsewhere.

The present paper will commence with a brief review of what is known from ESR of intrinsic point defects in pure a-SiO_2 and a-SiO_2:OH. The focus will then shift to certain extrinsic defects associated with (i) a common impurity (Cl) present in nominally pure synthetic silicas and (ii) two common dopants (P and Ge) employed in the fabrication of optical fiber waveguides. In all cases the paramagnetic defects were formed by exposing the samples to ionizing radiations (X rays or γ rays).

INTRINSIC DEFECT IN a-SiO_2

To date, three generic defect types have been identified in a-SiO_2 by ESR. These are the E' center (≡Si·), the nonbridging oxygen hole center (≡Si-O·), and the peroxy radical (≡Si-O-O·). Here, the notation "≡" denotes three back bonds to oxygens in the glass network and "·" signifies an unpaired spin. Shown in fig. 1b is one possible production mode for an E' center, namely, the trapping of a hole at the site of a neutral oxygen vacancy.

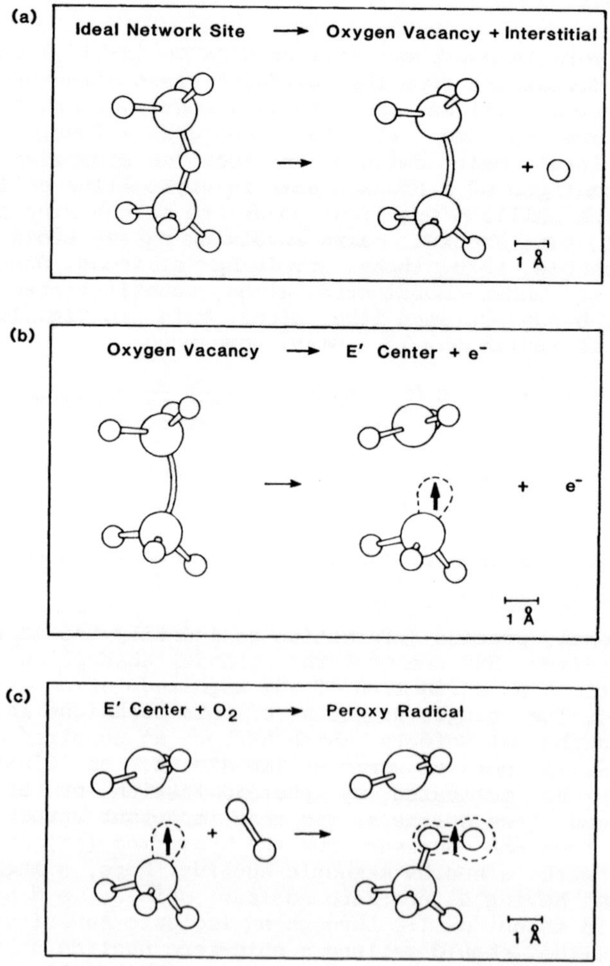

Fig. 1. Models for defects in pure a-SiO_2 : (a) Frenkel pair ; (b) the E'_{γ} center ; (c) the peroxy radical. Vertical arrows denote unpaired spins ; dashed balloons represent their orbitals.

The ESR spectra of E' centers in a-SiO_2 generally exhibit an hf doublet arising from a single ^{29}Si nucleus ($I=\frac{1}{2}$, 4.7% natural abundance), with splittings ($A_{iso} \simeq 42$ mT, $A_{aniso} \simeq 2$ mT) of the magnitudes expected for an unpaired spin localized in a dangling tetrahedral orbital of a single silicon as illustrated in fig. 1b. The asymmetric relaxation of the second (positively charged) silicon at the E'_γ site is strongly supported by theoretical calculation. Although the ESR spectroscopic evidence for the actual presence of the second silicon remains somewhat indirect, a one-for-one correlation of positive charge with E' centers has been demonstrated in SiO_2-on-Si structures.[5] Evidence has been given[6] for the existence of several other E' variants in glassy silica, some of which may not involve simple oxygen vacancies (see below).

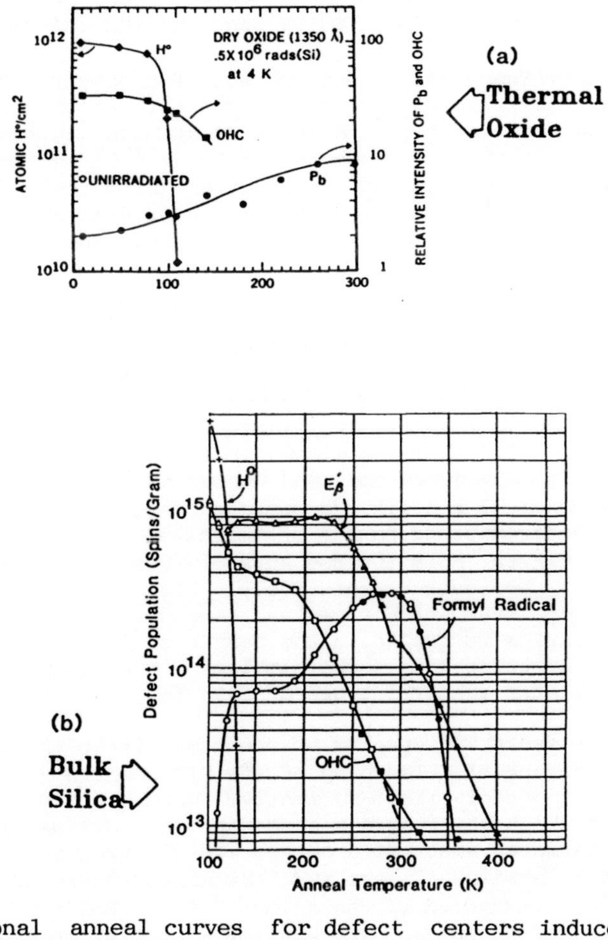

Fig. 2. Isochronal anneal curves for defect centers induced at cryogenic temperatures in (a) γ-irradiated thermally-grown a-SiO_2 on Si (After ref. 8) and (b) x-irradiated bulk synthetic silica containing both 1200 ppm OH and an inadvertent CO impurity (After ref. 7).

The most common production mode for the nonbridging oxygen hole center (NBOHC) is by radiolysis of hydroxyl groups :

$$\equiv Si\text{-}OH \rightarrow \equiv Si\text{-}O\cdot + H\cdot, \quad (1)$$

where the atomic hydrogen H· is also ESR active. As apparent in fig. 2, H· is unstable above 130 K. A variety of evidence strongly indicates that the dominant anneal mechanism for this atomic hydrogen is dimerization :

$$2H\cdot \rightarrow H_2 \quad (2)$$

Above 200 K, the thermal destruction of the NBOHC then continues by a diffusion-limited reaction with the radiolytic molecular hydrogen of eq. (2) :

$$\equiv Si\text{-}O\cdot + H_2 \rightarrow \equiv Si\text{-}OH + H\cdot \quad (3)$$

The highly reactive H· appearing on the righthand side of eq. (3) becomes available for additional reactions, such as combining with dissolved molecules of a CO impurity in silica glass to produce formyl radicals[7] HCO (fig. 2b) or the production of P_b centers $(Si)_3{\equiv}Si\cdot$ in SiO_2-on-Si structures by reacting with $(Si)_3 \equiv Si\text{-}H$ bonds at the interface (fig. 2a)[8,9]

A particularly important production mode for peroxy radicals is by the reaction of an interstitial oxygen molecule with an E' center as depicted in fig. 1c. The fact that the superoxide species is bonded to a single silicon in the network was demonstrated by a ^{29}Si isotopic substitution experiment which showed an hf interaction with only one silicon. The asymmetric spin distribution over the two chemically inequivalent oxygens was shown by substituting ^{17}O (I=5/2) for ^{16}O(I=0) and computer simulating the resulting multi-line hf spectra[12].

DEFECTS ASSOCIATED WITH CHLORIDE IMPURITIES

Chlorine has long been known to be a common impurity in synthetic silicas produced by hydrolysis or plasma oxidation of $SiCl_4$, but only recently[13] have specific Cl-associated point defects been identified in $a\text{-}SiO_2$. Figure 3a reproduces the ESR spectrum of a Cl-containing silica glass X irradiated at 77°K ; fig. 3b shows a computer simulation of the Cl^0 portion of the spectrum based on the hf splittings and g values represented in part c of the figure. Note the two sets of four lines due to ^{35}Cl and ^{37}Cl (75 and 25 % abundant respectively, both with I=3/2). Figure 3d illustrates the spectrum of E' centers expanded from the spectral region defined by the narrow rectangle in the center of fig. 3a. The dashed curve in fig. 3d is a computer line shape simulation comprised of a weighted sum of the three E' center spectral types of fig. 3e.

Figure 4 presents ESR spectra of the same Cl-containing silica sample of fig. 3 after thermal bleaching of the Cl^0 centers[13]. Spectrum (a) was obtained in the normal first-derivative mode, while spectrum (b) was recorded in the high-power second harmonic mode, giving it the appearance of an undifferentiated absorption curve. The outlying pair of lines in (b) indicated by the arrows "1" are the ^{29}Si hf structure of the E'_γ center ; the pair of lines indicated by the arrows "2" are believed the be the ^{29}Si hf structure of the E'_δ center. The smaller hf splitting of E'_δ vis-a-vis E'_γ was interpreted as indicating the unpaired spin to be delocalized over several silicons in the E'_δ case. The bold dashed lines are a computer simulation of the spectrum of a biradical (two unpaired spins in close proximity). The level diagram for such a triplet state is illustrated in (d). The powder patterns for transitions "A" and "B" are drawn in (b) ; the quasi-forbidden half-field transition "C" is shown in (c).

The occurrence of both the E'_{δ} center and the biradical were found to correlate with the yield of Cl^0 in a range of samples, suggesting that all of these centers may be related to the existence of a particular chloride-related precursor structure. It was argued[13] that chloride ions would be most favorably accomodated into the original a-SiO_2 network in

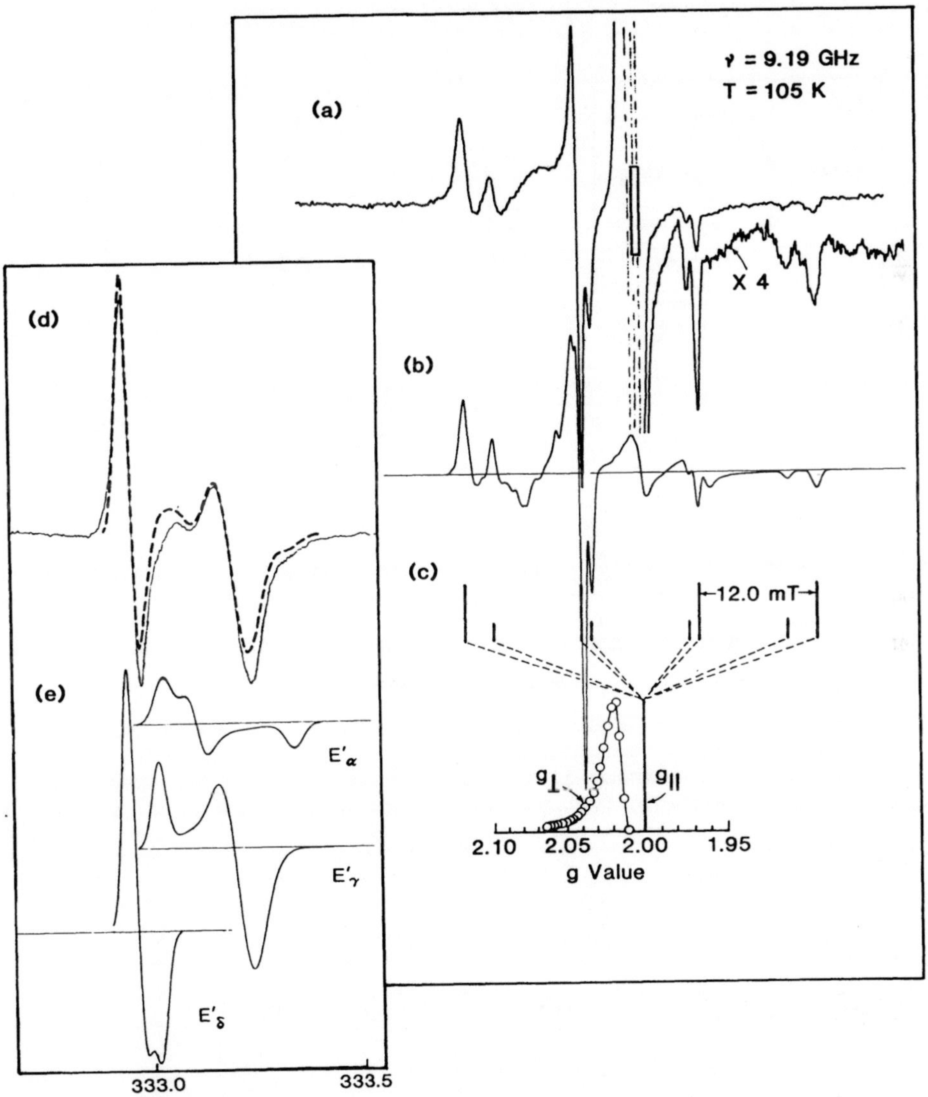

Fig. 3. ESR first-derivative spectra of a Cl-containing silica glass X irradiated at 77 K and measured at 105 K : (a) spectrum of Cl^0 ; (b) computer simulation ; (c) schematic hf splittings and g values (d) spectrum of E' centers from central region of (a) and its computer simulation (dashed curve) ; (e) component spectra for the simulation of (d). (After ref. 13.).

clusters of four (fig. 5a). Assuming such clusters, the models proposed for E'_{δ} (fig. 5b) and the biradical (fig. 5c) represent the removal of one and two atomic chlorines, respectively, from such a precursor site.

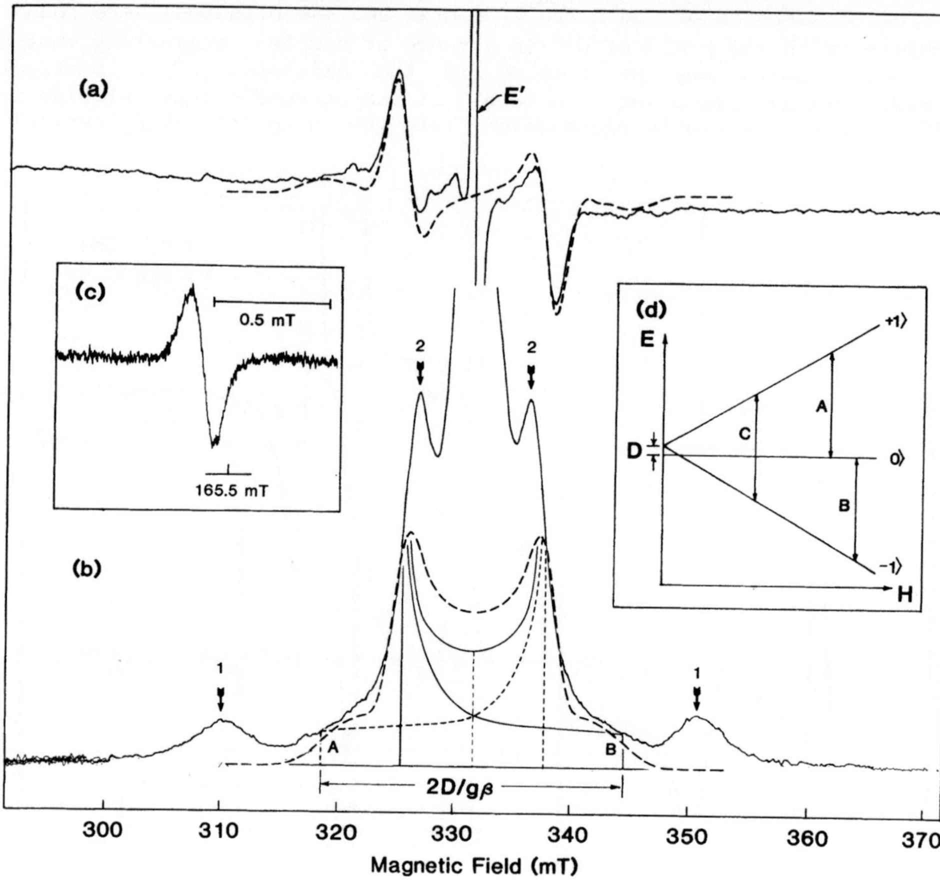

Fig. 4 ESR spectra of a Cl-containing silica glass X irradiated at 77 K and warmed to room temperature before measurement at 225 K : (a) first derivative spectrum ; (b) second-harmonic mode spectrum ; (c) high-gain first-derivative spectrum at half field ; (d) energy level scheme for a biradical with zero-field splitting energy D. (After ref. 13.).

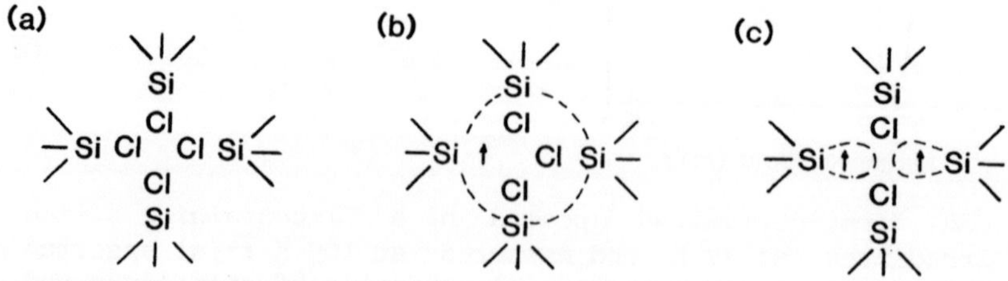

Fig. 5. Proposed models for defect structures in Cl-containing a-SiO_2 : (a) chloride configuration in unirradiated glass ; (b) E'_{δ} center ; (c) biradical. (After ref. 13.).

DEFECTS ASSOCIATED WITH PHOSPHORUS AND GERMANIUM DOPANTS

Figure 5 displays the ESR spectra of a 0.1 P_2O_5-0.9 SiO_2 glass (a) immediately following X irradiation at 77 K and (b) after annealing at 673 K. The dotted curves are computer line shape simulations of doublet spectra due to two separate defect centers undergoing hf interactions with ^{31}P nuclei ($I=\frac{1}{2}$, 100 % abundant). The P_2 simulation of (a) and the P_1 simulation of (b) were optimized by employing the statistical distributions of hf coupling constants illustrated in (c) and (d), respectively.

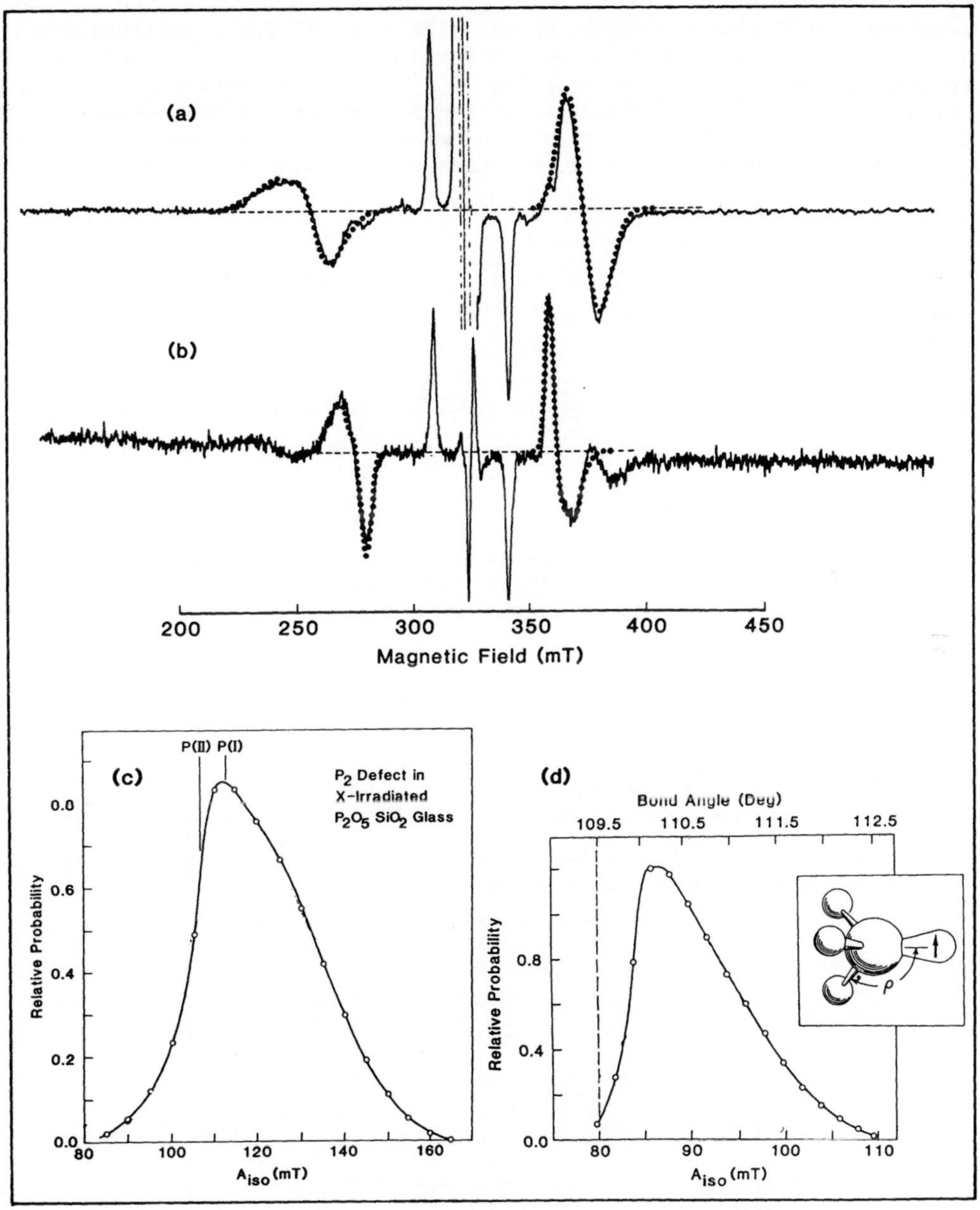

Fig. 6. ESR spectra of an X-irradiated P_2O_5-SiO_2 glass : (a) after irradiation at 77 K ; (b) after warming to 673 K. Dotted curves are computer simulations based on the distributions of hf coupling constants (circles) shown in (c) and (d).

The spectra of two additional defect centers are also manifested in fig. 5. The lines due to the phosphorus-oxygen hole center (POHC) are driven off scale in the center of (a), while the "three-line" spectrum between 300 and 350 mT in (b) arises from the P_4 center. Based on the ESR spectral parameters derived in the computer simulation analyses, the P_1, P_2, P_4, and POHC were identified as the defect structures illustrated in fig. 7. The P_2 center, for example, was shown to be essentially identical with the (crystallographically inter-related) P(I) and P(II) centers[15] in P-doped α quartz, a crystalline polymorph of SiO_2. Note in fig. 6c how the indicated values of A_{iso} for P(I) and P(II) correspond to the peak in the distribution of A_{iso} values for P_2 in glass. In essence, the P_2 center comprises an electron trapped at the site of a P^{5+} ion substitutional for an Si^{4+} ion in the glass network. On the other hand, the P_1, center almost certainly results from the trapping of a hole in the dangling tetrahedral orbital of a P^{3+} ion bonded to just three oxygens in the network. It is possible that the P_4 center, a phosphinyl radical analogue, arises from dissociative electron capture at the same type of site (fig. 7c).

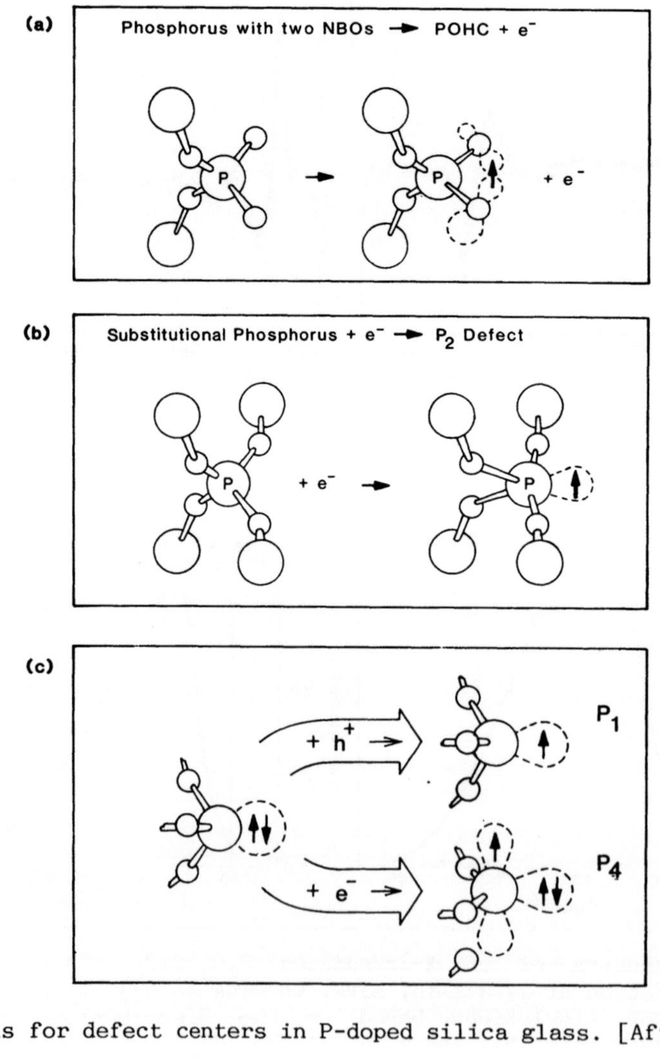

Fig.7. Models for defect centers in P-doped silica glass. [After ref. 14.]

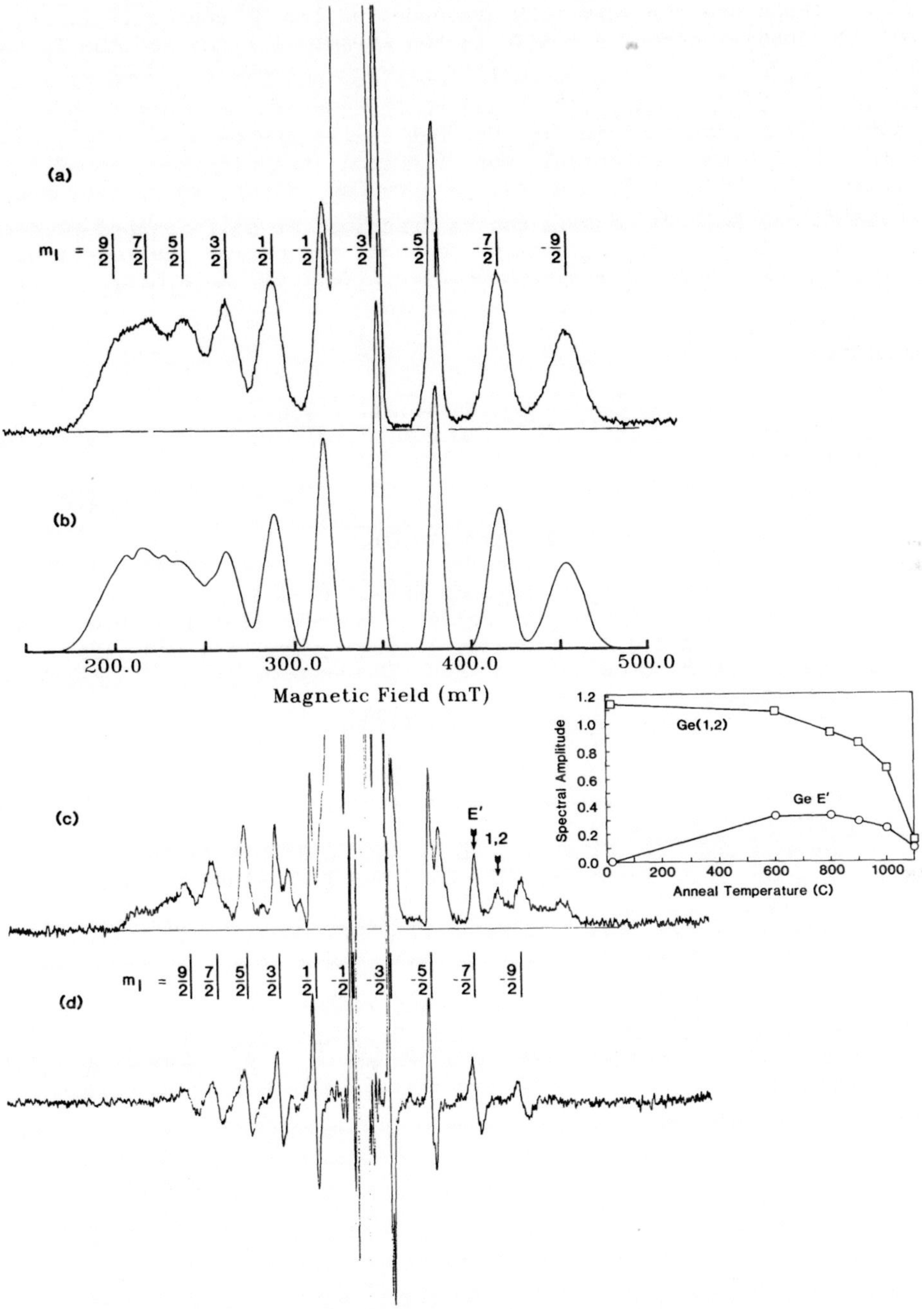

Fig.8. ESR spectra of γ-ray-induced defect centers in GeO_2-SiO_2 glass : (a) as-prepared glass, second harmonic mode ; (b) computer simulation of Ge hf lines of Ge(1,2) centers ; (c) and (d) after vacuum anneal to 1100 C and reirradiation, recorded in second harmonic mode and first derivative mode respectively. ^{73}Ge hf lines in (d) are due solely to Ge E' centers. Inset shows the relative yields of Ge(1,2) and Ge E' centers determined from spectra such as (c) as a function of the temperature of pre-irradiation anneal in H_2.

Only two Ge-related defects have been found in irradiated GeO_2 - SiO_2 glasses. These are the germanium analogues of the E' center (fig. 1b, with a Ge^{3+} ion substitutional for Si^{3+} in the pyramidal site) and the P_2 center (fig. 7b, with a Ge^{3+} ion substitutional for the P^{4+} ion). The latter are known as the Ge(1,2) centers[16,17] and their ^{73}Ge hf structure (I=9/2, 7.6 % abundant) is illustrated in fig. 8a. But unlike the case of the P_2 center, the is no Coulomb potential for electron trapping to account for the stability of the Ge(1,2) centers. In pure GeO_2 glass, where they are also observed[19], the Ge(1,2) centers can be described as self-trapped electrons. This observation leads naturally to the suggestion that electron self trapping may also occur (with much shorter lifetimes) in a-SiO_2.

REFERENCES

1. D.L. Griscom, J. Non-Cryst. Solids 40:211 (1980).
2. D.L. Griscom, E.J. Friebele, and G.H. Sigel Jr., Sol. State Commun. 15:479 (1974).
3. F.J. Feigl, W.B. Fowler, and K.L. Yip, Solid State Comm. 14:225 (1974) ; K.L. Yip and W.B. Fowler, Phys. Rev. B11, 2327 (1975).
4. D.L. Griscom, Phys. Rev. B22:4192 (1980).
5. P.M. Lenahan and P.V. Dressendorfer, J. Appl. Phys. 55:3495 (1984).
6. D.L. Griscom, Nucl. Inst. & Methods B1:481 (1984).
7. D.L. Griscom, J. Non-Cryst. Solids 68:301 (1984) ; D.L. Griscom, M. Stapelbroek, and E.J. Friebele, J. Chem. Phys. 78:1638 (1983).
8. K.L. Brower, P.M. Lenahan and P.V. Dressendorfer, Appl. Phys. Lett. 41:251 (1982).
9. D.L. Griscom, J. Appl. Phys. 58:2524 (1985).
10. A.H. Edwards and W.B. Fowler, Phys. Rev. B26:6649 (1982).
11. D.L. Griscom and E.J. Freibele, Phys. Rev. B24:4896 (1981).
12. E.J. Friebele, D.L. Griscom, M. Stapelbroek, and R.A. Weeks, Phys. Rev. Lett. 42:1346 (1979).
13. D.L. Griscom and E.J. Friebele, Phys. Rev. B34:7524 (1986).
14. D.L. Griscom, E.J. Friebele, K.J. Long and J.W. Fleming, J. Appl. Phys. 54:3743 (1983).
15. Y. Uchida, J. Isoya, and J.A. Weil, J. Phys. Chem. 83:3462 (1979).
16. E.J. Friebele, D.L. Griscom, and G.H. Sigel Jr., J. Appl. Phys. 45:3424 (1974).
17. H. Kawazoe, J. Non-Cryst. Solids 71:231 (1985).
18. E.J. Friebele and D.L. Griscom, Mat. Res. Soc. Proc. 61:319 (1986).
19. T.E. Tsai, D.L. Griscom, and E.J. Friebele, J. Non-Cryst. Solids (submitted).

SELF-TRAPPED EXCITONS IN AMORPHOUS AND CRYSTALLINE SIO_2

Noriaki Itoh, Katsumi Tanimura and Chihiro Itoh

Department of Physics
Faculty of Science, Nagoya University
Furo-cho, Chikusaku, Nagoya 464, Japan

ABSTRACT

Properties of the self-trapped excitons in amorphous and crystalline SiO_2 studied through transient volume and optical absorption change and luminescence are compared. It is emphasized that local lattice relaxation induced upon electronic excitation in crystalline and amorphous materials is similar except that the time decay of the self-trapped excitons in the amorphous is non-exponential. Based on the fact that the self-trapped exciton in crystalline SiO_2 is a close vacancy-interstitial pair, we propose a new mechanism for the luminescence in amorphous SiO_2 exhibiting a non-exponential time decay.

INTRODUCTION

Several amorphous materials such as chalcogenides and silica luminesce in an energy region where no detectable optical absorption is found. The mechanism of the luminescence of amorphous substances has been a topic of general interest and several models have been suggested[1]. In order to elucidate the mechanism it is useful to compare the electron-hole recombination processes in the crystal and in the amorphous phase of the same substance. Such a comparison has been carried out for SiO_2 [2,3] and chalcogenides[4]. In the crystals of both of these substances, photoluminescence with a large Stokes shift and transient optical absorption have been measured and ascribed to the self-trapped excitons, which are annihilated with fixed decay time constants. Further information on the atomic structure of the self-trapped excitons in SiO_2 has been accumulated through the studies of transient volume change[5], of optically detected electron paramagnetic resonance[6] and of polarization of the luminescence[7,8], while information on the atomic structures of the self-trapped excitons in chalcogenides is scanty. Thus it is of interest to examine critically whether the experimental results relevant to recombination luminescence for SiO_2 can be explained in terms of the existing models.

In this paper we compare the major experimental results of luminescence and transient optical absorption and volume change in crystalline and amorphous SiO_2 . Based on the comparison we propose a new

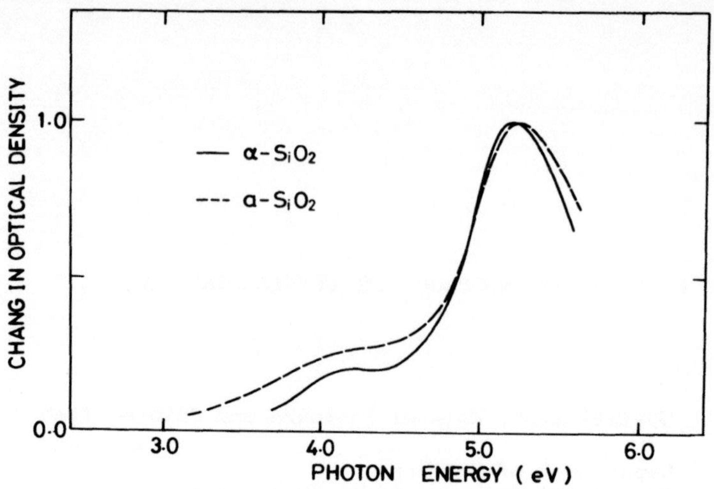

Fig.1. The transient optical absorption spectra induced by irradiation with an electron pulse in crystalline SiO_2 (α-SiO_2) and in amorphous SiO_2 For crystalline SiO_2, optical absorption was measured with light polarized parallel to the c-axis.

mechanism for the non-exponential luminescence decay in amorphous materials.

COMPARISON OF EXPERIMENTAL RESULTS FOR CRYSTALLINE AND AMORPHOUS SiO_2

In previous papers[3,5] we have shown that the transient optical absorption and volume change and luminescence in crystalline SiO_2 induced by an electron pulse decay by the same time constant and all of these property changes have been ascribed to the self-trapped excitons. In fig.1 we plot the transient optical absorption spectrum of the crystal taken through a polarizer with its E vector parallel to the c-axis. The whole spectrum has been shown to be annihilated by the same time constant. Fig.1 includes also the transient optical absorption spectrum for amorphous SiO_2, which show a non-exponential time decay. The peak energy is shifted only by 0.05 eV and the general shape including the satellite at the low energy side, which has been proven to be associated with the self-trapped exciton [8,9], is similar. A comparison of the luminescence spectra induced by an electron pulse for the amorphous and crystalline SiO_2 is given in fig.2. The spectra for the crystal and the amorphous are normalized by the intensity at a delay of 10^{-5}s. For amorphous a blue peak shift is observed with increasing the delay time after bombardment with an electron pulse. The difference in the peak energy for the amorphous and the crystal is larger than for the transient optical absorption. Fig.3 shows a comparison of the decay curves of the luminescence for the amorphous and the crystal, normalized by the intensity at a delay of 10^{-5}s. The decay curve of the optical absorption has been found to be correlated at the later stage but not at the initial stage where non-radiative transition dominates[3,9]. The magnitude of the optical absorption change in the amorphous is smaller only by a factor of 5 than that in the crystal[2], even though the intensity of the luminescence is smaller by two orders of magnitude. The ratio of the volume change per self-trapped exciton in the crystal and the amorphous has been found to be 1.2. Further irradiation with an excimer laser pulse

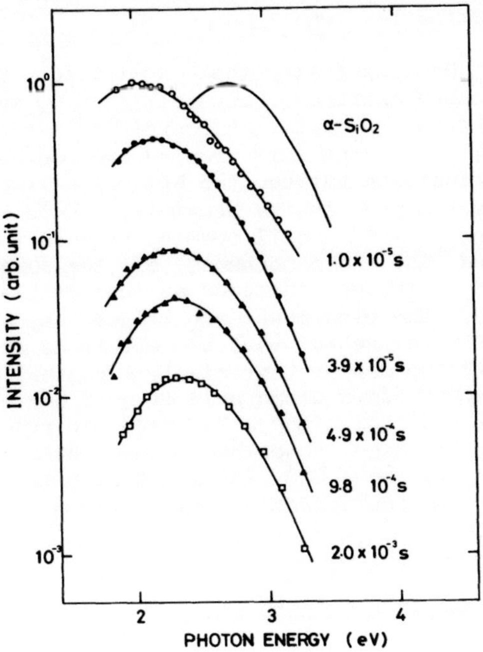

Fig.2. Luminescence spectra of crystalline and amorphous SiO_2 at 77 K. The spectra for the amorphous are normalized by the intensity at a delay of 10^{-5}. The spectrum for the crystal is practically unchanged, while that for the amorphous shows a blue shift.

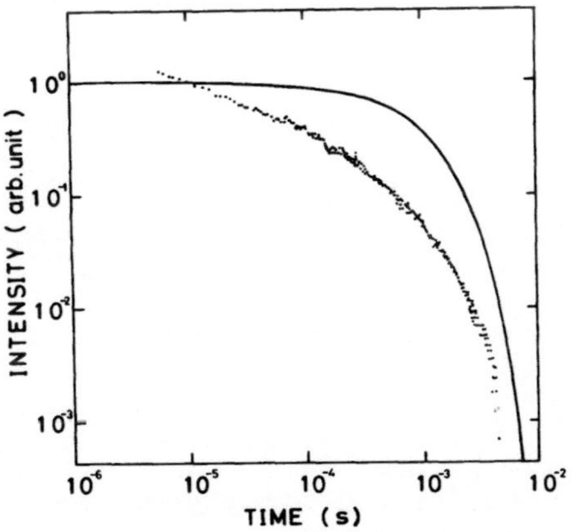

Fig. 3 Decay of the luminescence induced by an electron pulse in amorphous and crystalline SiO_2 at 77 K, normalized by the intensity at a delay of 10^{-5}µs.

(308 nm) of the specimen exhibiting the transient optical absorption has been shown to reduce both the optical absorption and the luminescence.

The results described above indicate that the luminescence in amorphous SiO_2 is originated also from the self-trapped excitons, of which the yield is smaller only by a factor of five than in crystalline SiO_2. The small efficiency for luminescence is not due to the smallness of the yield of the self-trapped excitons but due to the smallness of the radiative transition probability of the self-trapped excitons. Thus we suggest that electronic excitation of the amorphous creates substantially self-trapped excitons which are annihilated either by non-radiative recombination or radiative recombination. In the following we discuss how this model reconciles existing experimental data on SiO_2.

ON THE ATOMIC STRUCTURE OF THE SELF-TRAPPED EXCITONS

An ODMR study by Hayes et al.[6] has indicated that a triplet O_2 structure is included in the self-trapped exciton of crystalline SiO_2 and that the O_2 linkage is spin-decoupled from the electrons comprising a self trapped exciton. Thus the luminescence due to the self-trapped exciton has been ascribed to the charge transfer transition between the holes localized to the O_2 linkage and the electrons trapped in its proximity[8]. Here we consider that the slow decay time arises from the small overlap between the wave functions for the electron and the hole because of the large electron-hole separation. Since the O_2 linkage includes an interstitial oxygen it is natural to consider that the electrons are trapped by an oxygen vacancy. The charge neutrality suggests that two electrons are trapped by an oxygen vacancy (We call the center tentatively F center)*. Thus it is likely that self-trapped exciton is a close pair of an F center and an O_2 linkage. The polarization of the luminescence has been explained in terms of the charge transfer transition described above. Here we consider that a self-trapped exciton in amorphous SiO_2 is also a pair of an F center and an O_2 linkage but with various configurations depending on the local structure of the amorphous material.

RECOMBINATION OF THE SELF-TRAPPED EXCITONS IN AMORPHOUS SiO_2

The blue shift in the recombination luminescence with increasing delay time may be ascribed to the increase in the distance of the charge transfer with increasing the delay time. In the ionic limit the O_2 linkage having two holes and the vacancy having two electrons form a neutral pair. Because of covalency, however, in reality the O_2 molecule is positively charged and the oxygen vacancy is negatively charged. Thus the transition energy $\hbar\omega$ is given as a function of the pair distance r by

$$\hbar\omega = E - e^2/\kappa r, \qquad (1)$$

where E is the energy difference between the F center and the O_2 linkage and κ the dielectric constant. If we assume that the electron is centered on a oxygen lattice point and the hole on the neighboring oxygen lattice point, the value of r for the crystal is 2.6 Å. According to Eq (1), the overall energy shift of 0.02 eV can be explained by assuming that the distance in the amorphous is distributed around 2.6 Å between 3.2 Å and 2.3 Å.

The charge-transfer recombination rate between the F center and the O_2 linkage in the amorphous depends also on the distance between the pair. The wave function of the E_1' center has the character of the silicon sp^3 hybrid orbital[10] hence we expect that it is also the case for that of the F center. The wave function of the O_2 linkage has the character of the oxygen p orbital. Thus unlike the donor-accepter recombination in semiconductors, the relation between the transition rate and the distance of recombination is not straightforward. Here we note some specific features on the distribution of the recombination rate[3]. As shown previously the non-radiative recombination rate dominates for the pairs

---

* The peak position of the electron transition of the self-trapped excitons, probably arising from the F center electrons, is lower than that of the E_1' center by 0.6 eV. The difference in the peak energies in the center with two electrons trapped by a vacancy and that with an electron trapped by a vacancy is not generally very large in oxides.

having high recombination rate. Thus even though the recombination rates are distributed from 10^{-6} to $10^{-2}s^{-1}$ as seen from fig. 3, the radiative recombination rates are distributed within a smaller extent. Since the fastest decay time constant observed for the amorphous is 10^{-6}s, for which the non-radiative recombination dominates, the radiative recombination rate for the shortest pair is even lower. This small radiative transition rate is considered to arise from the strong anisotropic nature of the wave functions and the charge separation rather than the violation of the spin selection rule.

COMPARISON WITH EXISTING MODELS OF THE LUMINESCENCE OF AMORPHOUS SUBSTANCES

In this section we compare the present model for the non-exponential luminescence decay with existing models for the luminescence for amorphous materials in general. Street and Mott[11] have used the concept of the negative U suggested by Anderson[12] to explain the large Stokes shift and the multi-component decay of luminescence in the amorphous. According to these authors free electrons and holes are captured by charged dangling bonds (D^+ and D^- centers) forming D^0 centers. Radiative recombination between trapped carriers occurs when two D^0 centers close to each other exchange an electron to be transformed to a close pair of the D^+- and D^--centers ; this transformation is assumed to be exothermic in the presence of strong electron-lattice coupling. Emin[13], who emphasized the intrinsic nature of the recombination luminescence, suggested the polaron model, which assumes that the recombination luminescence is emitted by recombination of small polaron of opposite signs. The present model is close to the polaron model.

The concept of the self-trapped excitons has been introduced to the recombination luminescence of chalcogenides only recently by Robins and Kastner[4]. Based on the studies of the correlation between the transient optical absorption change and the phololuminescence they concluded that the photoluminescence in the crystalline chalcogenides is due to the triplet self-trapped excitons. The nature of the luminescence in amorphous, particularly its non-exponential time decay, remain unexplained. Self-trapping of excitons in amorphous materials to a configuration in which the electron and the hole are separated has been suggested by Mott and Stoneham[14]. Street[15] has suggested the self-trapped excition is annihilated by non-radiative recombination and ascribed the smallness of the quantum efficiency to the non-radiative recombination through the channel of the self-trapped excitons. According to him the recombination luminescence is emitted only by the defect pair recombination. The model cannot explain the results for SiO_2 where the yields of the self-trapped excitons, as determined by the optical absorption measurement, for the crystal and the amorphous are nearly the same.

CONCLUSION

We compared the experimental results for the transient optical absorption and luminescence of SiO_2 in crystalline and amorphous SiO_2 and suggested a new self-trapped exciton model for non-exponential decay of the luminescence in amorphous materials. The model can account for several important experimental results for SiO_2. It is still an open question whether the model is applicable to the other amorphous materials. The results for SiO_2 and As_2Se_3 are similar ; major difference being that the peak energy of the luminescence band shifts to lower energy in the calcogenide[16] but shifts oppositely in SiO_2. If the self-trapped exciton model is effective for amorphous chalcogenides, this result suggests that the self-trapped excitons in chalcogenide is a neutral pair of a Se vacancy and a Se interstitial.

REFERENCES

1. R. Fisher, "Amorphous Semiconductors", ed. M.H. Brodsky Springer Verlag, Berlin, (1985) p. 159.
2. D.L. Griscom, Proceedings of the Thirty-Third Frequency Control Symposium (Electronic Industries Association, 1979), p. 98.
3. T. Tanaka, T. Eshita, K. Tanimura and N. Itoh, Crystal Lattice Defects and Amorphous Materials 11:221 (1985).
4. L. H. Robins and M.A. Kastner, Phil. Mag. B 50:29 (1984).
5. K. Tanimura, T. Tanaka and N. Itoh, Phys. Rev. Lett. 51:423 (1983).
6. W. Hayes, M.J. Kane, O. Salminen, R.L. Wood and S.P. Doherry, J. Phys. C17:2943 (1984).
7. K. Tanimura and L.E. Haliburton, Phys. Rev. B 34:2933 (1986).
8. C. Itoh, K. Tanimura and N. Itoh, to be published.
9. K. Tanimura, C. Itoh and N. Itoh, to be published.
10. K.L. Yip and W.B. Fowler, Phys. Rev. B 11:2327 (1975).
11. R.A. Street and N.F. Mott, Phys. Rev. Lett. 35:1293 (1975).
12. P.W. Anderson, Phys. Rev. Lett. 34:953 (1975).
13. D. Emin, "Proceedings 7th International Conference on Amorphous and Liquid Semiconductors" ed. W.E. Spear, Edinburg, (1977) p. 261.
14. N. F. Mott and A.M. Stoneham, J. Phys. C 10:3391 (1977).
15. R. A. Street, Phys. Rev. B 17:3984 (1978).
16. G. A. Higashi and M. Kastner, Phys. Rev. B 24:2295 (1981).

IDENTIFICATION OF NATIVE DEFECTS IN a-SiO_2

J.H. Stathis

IBM T.J. Watson Research Center
Yorktown Heigths
N.Y., USA

INTRODUCTION

One of the most powerful experimental tools for the identification of defects in solids is electron paramagnetic resonance (EPR). In a-SiO_2, however, the defects are diamagnetic in their ground state, and are therefore invisible to EPR. Because of this, researchers interested in the structure of defects in SiO_2 have had to rely on the use of ionizing or heavy particle radiation (γ, e^-, n, X,..) in order to generate paramagnetic defects which can then be detected by EPR[1]. While these studies have been very fruitful, and the question of the mechanism by which defects are generated in a glass by the influence of high-energy radiation is an interesting and important one[2], they leave open a fundamental question. Namely, since ionizing radiation may create additional structural defects in the glass, either by direct knock-ons or by a radiolytic process such as a recombination-assisted reaction, one gains little or no information about the native defects present before irradiation. The identity and properties of these native defects in a-SiO_2 are important both for technological reasons, and because they help us to understand the nature of glasses in general.

The existence of defects which are diamagnetic in their ground state is a common feature of many glasses. Any defect which contains an even number of electrons in its neutral state, e.g. an oxygen vacancy in SiO_2, will be diamagnetic, whereas a defect with an odd number of electrons, e.g. a non-bridging oxygen (oxygen dangling bond) will be paramagnetic. However, in semiconducting glasses one finds the additional remarkable result that the Fermi level is strongly pinned. If the pinning is due to defects, then defects near the Fermi level should be singly occupied and paramagnetic. The fact that no paramagnetism is observed is explained[3,4] by negative-U defect pairs, with all defect states being either doubly occupied (negatively charged) or unoccupied (positively charged). The existence of negative-U requires a large electron-phonon coupling. In SiO_2 this is due to the non-bonding oxygen p-electrons, which are available to take part in bond rearrangements accompanying changes in the charge state (occupation) of defects. The lowest energy and hence most numerous defects are therefore believed to be charged over and under-coordinated sites.

In order to study these defects by EPR one therefore requires some way to prepare them in a non-equilibrium or excited paramagnetic state. Various double-modulation techniques, such as optically-detected magnetic resonance or spin-dependent recombination, are useful. This paper describes another method, in which the native defects are put in a metastable paramagnetic state by optical excitation, and are then measured by standard EPR. We have found[5-7] that paramagnetic states may be produced by sub-band-gap optical excitation of a-SiO_2. The paramagnetic state is metastable at room temperature.

Since the absorption of sub-band-gap light can only occur at defects with states in the gap, the spins that are observed must result from direct photoionization of native defects, plus subsequent re-trapping of the free carrier at other sites. The use of sub-band-gap light therefore has the significant advantage that, in contrast to ionizing radiation, the energy is inadequate to create new defects. One may therefore be reasonably sure that the resulting EPR signal indeed reflects the structure of the native defects in the as-quenched material.

Of course, whether the paramagnectic states are generated by ionizing radiation or optical excitation, the fact remains that EPR only probes the excited state of the defect, so that one can only speculate about the diamagnetic ground state from which the paramagnetic state was created. It is here that photoexcitation has another important advantage, since the excitation energy can be tuned. Since different defects have different ionization energies, varying the photon energy of the light allows their selective excitation. Finally, it is possibly to exploit the fact that the excitation can be polarized to obtain completely new information about the properties of the defects.

Large electron-phonon coupling is associated with a number of other interesting defect metastabilities as well. Perhaps best known is the DX center in AlGaAs, whose behavior is most successfully explained by a model involving large lattice-relaxation as the defect occupation changes[8]. Other examples include the M center in InP[9] and several extrinsic centers in silicon[10]. At least one such center, the As_{Ga} antisite defect in GaAs, exhibits a metastable photoinduced EPR effect as well[11].

In this paper we describe our observations of optically-induced metastable spin-resonance signals in SiO_2. We find that the dominant photo-induced spin center is unusual in that it is created by capture of either an electron or a hole. This behavior is explained by a negative-U defect pair model. Several other types of defects are observed as well.

PHOTOINDUCED EPR IN SiO_2

To generate paramagnetic centers in a-SiO_2 we have used three different photon energies, available as laser transitions from various halogen/rare-gas mixtures. These energies (5.0, 6.4, and 7.9 eV) are all well below the band gap, and they sample several orders of magnitude in the absorption coefficient (~ $10^{-3}cm^{-1}$, $10^{-2}cm^{-1}$, and 10 cm^{-1}, respectively). The typical excitation density was ~ 10^{20} absorbed photons per cm^{-3}.

Samples of Suprasil-W1 were obtained from Heraeus-Amersil[12]. After exposure, the samples were annealed in air for 15 minutes at successively higher temperatures, in steps of 100 C, and the EPR spectra were re-measured at room temperature after each annealing step. Samples irradiated with the three different photon energies were annealed simultaneously. EPR measurements were carried out at ~ 10 Ghz using a Varian E-9 spectrometer equipped with a TE011 cylindrical cavity.

The EPR spectra resulting from exposing samples of Suprasil-W1 to 7.9 eV, 6.4 eV, and 5.0 eV photons and subsequent annealing are shown in figures 1, 2, and 3. The uppermost curve in each of these figures is the initial spectrum measured several days after irradiation at room temperature. It is clear, from the shape of the spectra and from the variation with photon energy, that each of these spectra consists of several superimposed components arising from a number of different defects.

As the samples are annealed the spectra change shape, reflecting the fact that the different centers have different activation energies for returning to their diamagnetic ground state. As seen in figures 1-3, the sharp features anneal first. We note that part of the shoulder originally at $g \simeq 2.025$ results from a resonance at $g = 2.019$ whose amplitude actually increases during annealing at 400 C. This resonance disappears completely after annealing at 500 C. The overall amplitude of the remaining resonances decreases upon further annealing, and after annealing at 700 C the signal is no longer measurable for any of the excitation energies.

By subtracting the spectra shown in figure 1-3 from one another we have been able to extract the major components of the photoinduced EPR. Such a procedure, in which one hopes the subtraction will cancel out parts of the curves to leave isolated components, is of course susceptible to error because of the necessity to normalize the spectra prior to subtraction. In performing the analysis we were able to take advantage of significant end points, for example at the highest annealing temperature where all but one or two centers have annealed, or the lowest excitation energy where fewer defects can be ionized, or by the use of high microwave power where some resonances are saturated. Similar analysis of such isochronal annealing data can of course be applied to radiation-induced defects as well. In the present case, however, the additional parameter of variable photon energy makes the procedure significantly less ambiguous.

The result of this analysis is shown in figure 4. These are the components, designated A, B, and C, which constitute the major portion of the 7.9 eV photo-induced EPR and which describe the EPR induced by the lower photon energies after partial annealing. These components account for more than 80 % of the spins in the case of 7.9 eV excitation. The remainder, and those which dominate in the case of 6.4 eV or 5.0 eV excitation, comprises the 'low energy' centers, so called because 1) they have a low threshold energy for photogeneration ; 2) they have a low activation energy for thermal annealing ; and 3) they saturate at low microwave power, i.e. they have a long spin-lattice relaxation times T_1. These centers may be separated from the other components by their microwave saturation properties. Whereas the components A, B, and C did not saturate at even the highest microwave powers available (200 mW waveguide power), the low-energy centers exhibited saturation unless the power was kept at 0.2 mW or lower. When the low-energy centers are not present in too high a density (as in the case of 7.9 eV excitation) it is therefore possible to completely or almost completely eliminate them from the measured spectrum by performing the measurement at high microwave power. Conversely, the spectrum of the low-energy components present in the 7.9 eV-induced EPR can be isolated by subtracting spectra taken at low and high microwave powers.

The result[6] is a spectrum essentially identical to that obtained at lower photon energies.

Having extracted the spectra of the various components contributing to the photoinduced EPR, it is now possible to fit each of the original spectra with a sum of the components. The fitting gives the amount of each component present at each annealing temperature. The isochronal annealing

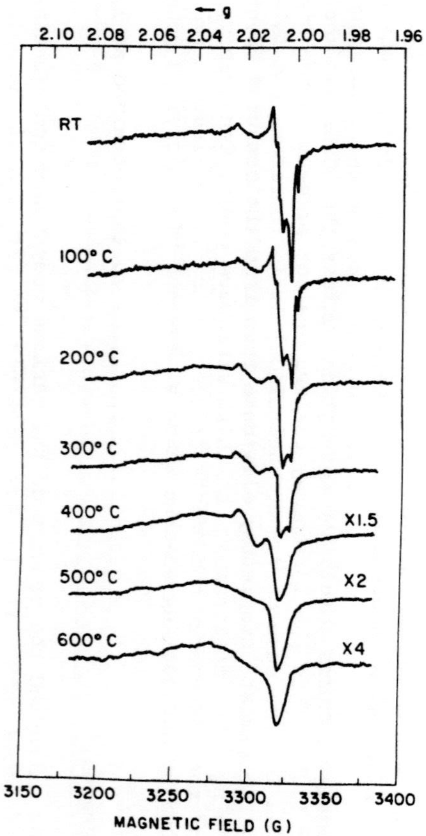

Figure 1. 7.9 eV excitation.

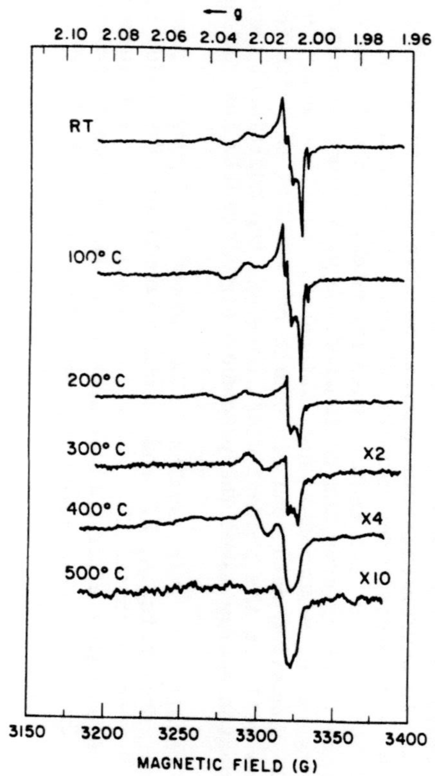

Figure 2. 6.4 eV excitation.

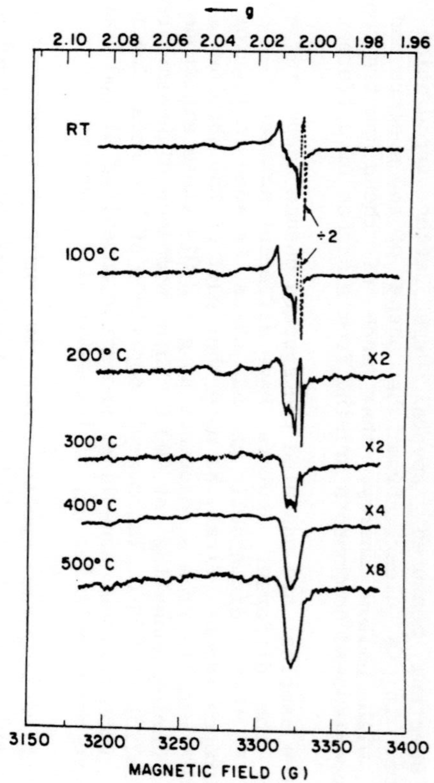

Figure 3. 5.0 eV excitation.

Figures 1-3. Evolution of photoinduced EPR in a-SiO_2 as a function of annealing temperature.

curves for the case of 7.9 eV excitation are shown in figure 5. The top two curves are the total spin density, measured by double numerical integration of the experimental spectra. The solid line, labelled 0dB, was measured at high microwave power (200 mW) so that the low energy centers are saturated. This curve therefore measures the total of only the broader lines which have been resolved into the components A, B, and C. On this figure the total density of low energy centers is therefore given by the difference between the measured spin densities at high (0dB attenuation) and low (30 dB attenuation) microwave powers.

There are several interesting features in figure 5. The growth of the C center (here resolved into two gaussian components) is seen clearly. This growth appears to correlate with the beginning of the A-center annealing ;

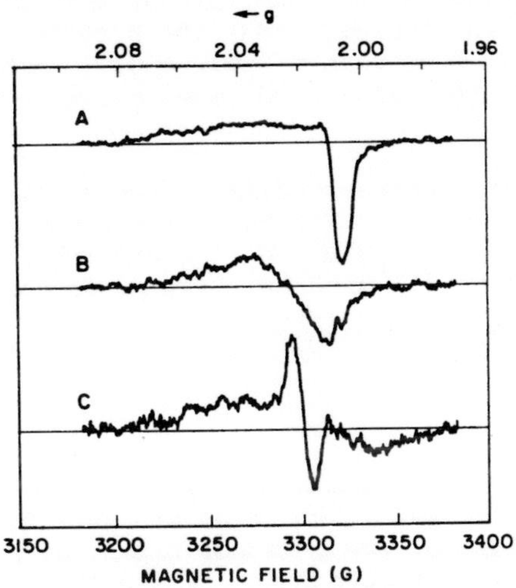

Fig. 4. Spectra of the major components of the photoinduced EPR.

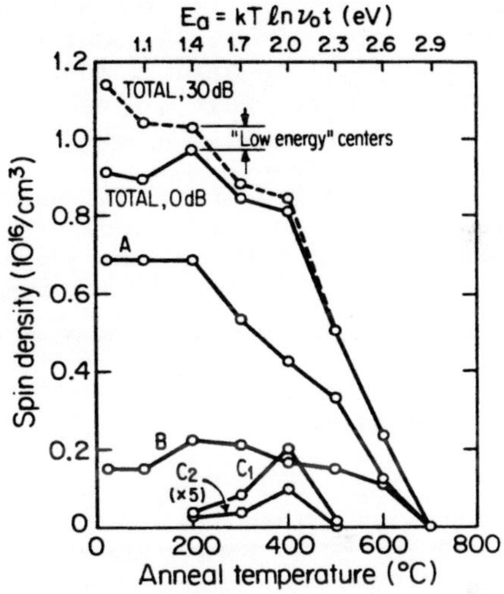

Fig. 5. Isochronal annealing curves of the photoinduced paramagnetic centers in 7.9 eV irradiated Suprasil-W1. The spectrum of the C center was fitted to two gaussian components whereas the other components were fitted using the experimentally derived spectra shown in figure 4.

some fraction of the carriers released from A centers are apparently retrapped at C precursor sites (i.e., C centers in a diamagnetic configuration). Notice that the annealing of the A center is very gradual, implying a broad distribution of trap depths, and that there are A centers both shallower and deeper than the C centers.

Second, the number of B centers also increases, and this increase correlates with the annealing of the majority of the low energy centers, which can again be understood by assuming that some of the carriers released from low-energy sites are retrapped at diamagnetic precursors to the B centers. Finally, figure 5 shows the striking result that more than half (~ 70 %) of the spins are in the A center.

For both 6.4 eV and 5.0 eV the A center also accounts for more than 50 % of the spins after annealing of the low energy centers at 300 C. The most surprising feature of the annealing studies for these lower excitation energies is the absence of B centers. The significance of these results will be discussed in the next section.

Note that the density of A centers is well above that of any impurities except for hydrogen and chlorine. For the other centers an impurity role cannot be ruled out from these number densities alone. It is interesting to note that the density of E' centers (the narrow resonance, 2.4 gauss wide, at $g \simeq 2.001$), is only $\sim 10^{13} cm^{-3}$, even for 5.0 eV excitation where this center is most strongly excited. We believe this is strong evidence that the E' center is not a dominant native defect in a-SiO_2, and that the large density of E' centers typically observed in radiation-damage experiments is a result of the radiolytic formation of oxygen vacancies. In thermal oxides, on the other hand, 5.0 eV excitation results in $\sim 10^{17} cm^{-3}$ E' centers[13]. This may indicate that the structure of thermal SiO_2 is significantly different from that of bulk SiO_2 ; another possibility is that oxygen vacancies are generated when carriers are injected from a silicon substrate.

Trap depths, i.e. activation energies corresponding to the annealing temperatures are also indicated for reference in figure 5. These energies correspond to setting the thermal activation rate, $\nu = \nu_0 \exp(-E_a/kT)$, equal to the inverse of the annealing time (15 minutes). The prefactor ν_0 was taken to be a typical phonon frequency, $\nu_0 = 10^{-12} sec^{-1}$.

INTERPRETATION

The EPR lines shapes have been discussed in some detail previously[6]. Since in general it is not possible to ascertain the atomic structure of a defect based on a knowledge of the g-tensor alone, we will not attempt in this paper any direct interpretation of the these EPR spectra. Rather, we focus on certain striking aspects of the annealing data of fig. 5.

In fig. 5 we saw that in the 7.9 eV-irradiated sample more than half (roughly 70 %) of the spins are in the A center. This is a very surprising result. Because the excitation photon energy is smaller than the band gap, the spins we see must result from direct photoionization of native defects, leaving behind a trapped hole (electron), with subsequent retrapping of the photoexcited electron (hole) at a different site. While lattice relaxation can, and probably does occur following changes in the defect occupancy, this cannot change the fact that equal numbers of trapped electrons and holes are generated since the sample must remain overall neutral. Thus even if there are, for example, many different types of electron trap but only one variety of hole trap, there seems to be no way, barring multiple excitations of the same defect, to have a majority of the paramagnetic centers be of one type.

It turns out, however, that this behavior is predicted by the valence-alternation model of Kastner et al.[4]. Mott[14] applied the valence-alternation idea to SiO_2, proposing the defect pair $C_1^-/C_3^+(Si_2O)$ as the most important in SiO_2. In this notation, C stands for chalcogen, the coordination number is indicated by the subscript, and the nearest neighbor atoms of the overcoordinated species are given in parenthesis. These defects are illustrated on the left side of fig. 6.

Consider the effect of removing an electron from the C_1^- and a hole from the $C_3^+(Si_2O)$, as shown in fig. 6. Assuming steric constraints allow the oxygen from the C_1^- to find a nearby bridging oxygen, as shown, these two defects will both relax to the <u>same</u> neutral paramagnetic $C_3^0(Si_2O)$ structure, since this involves no large atomic motions for either center. Thus trapped electrons and trapped holes give the same EPR signal.

Other explanations for the data cannot be entirely ruled out. There might be additional resonances too broad to detect, and if a single defect is doubly-ionized, or traps two carriers, then some electrons or holes would go undetected. With these caveats, the data presented here may be the most direct evidence to date for the applicability of the valence-alternation model to a-SiO_2.

A second interesting feature in fig. 4 is the 70 % increase in the number of B centers which occurs at 200 C. It was pointed out that this increase appears to be correlated with the disappearance of the low energy centers, which can be understood by assuming that some of the carriers released from the latter sites are retrapped at diamagnetic precursors of the B centers. However, the B center is never observed after 6.4 eV or 5.0 eV excitation, even though the same number of low-energy centers are produced and are observed to anneal at the same temperature. If our interpretation of the growth of the B center is correct, then it must be that its diamagnetic precursor is, in fact, created by the 7.9 eV light and is not present in thermal equilibrium. Otherwise, the disappearance of the low energy centers in the 6.4 eV or 5.0 eV-irradiated samples should result in the appareance of the B center in these samples.

This behavior can be explained by another defect model. Street and Lucovsky[15] pointed out that there are two possible types of three-fold coordinated oxygens a-SiO_2. Starting from a broken Si-O bond, and following the approach of Kastner et al., they proposed that the neutral dangling bonds would be unstable toward the formation of a $C_1^-/C_3^+(Si_3)$ pair. The overcoordinated center is illustrated on the left side of fig. 7.

As illustrated in this figure, the defect pair $C_1^-/C_3^+(Si_3)$ does not interconvert, so that removing two holes from the $C_3^+(Si_3)$ generates a T_3^- (T is for tetrahedron, i.e. group IV element), which was not present in thermal equilibrium. If one assumes, for the sake of argument, that 7.9 eV photons, are energetic enough to remove two holes, in succession, from the positively charged defect, then some spins will be observed in the form of $C_3^0(Si_3)$ but there will also be some new diamagnetic T_3^- centers created. These new negative centers are then available for retrapping of holes released from the low-energy centers, accounting for the observed growth of the B center. On the other hand, if the photon energy is insufficient to produce any paramagnetic B centers initially, as is the case for 6.4 eV or 5.0 eV excitation, then it will certainly be incapable of removing <u>two</u> carriers. The negative center will then not be available for retrapping of the released holes and thus no B-center formation will accompany the disappearance of the low-energy centers, exactly as observed in the case of 6.4 eV or 5.0 eV excitation. Note that it is very likely that the creation of doubly ionized state is a two-step process, because even a 7.9 eV photon probably has insufficient energy to remove two carriers simultaneously.

It should be emphasized that there is no strong supporting evidence, such as hyperfine structure, for the identification of the observed EPR spectra with any specific structural models such as those depicted in figs. 6 and 7. Nonetheless, these simple models have properties which are consistent with our measurements of the annealing behavior and photon energy dependence of the photoinduced paramagnetic centers in a-SiO_2. For the B center, the model in fig. 7 is not unique ; any charged defect that can be doubly ionized would suffice. For the A center, however, the only existing model which has the required amphoteric behavior is that of valence-alternation pairs.

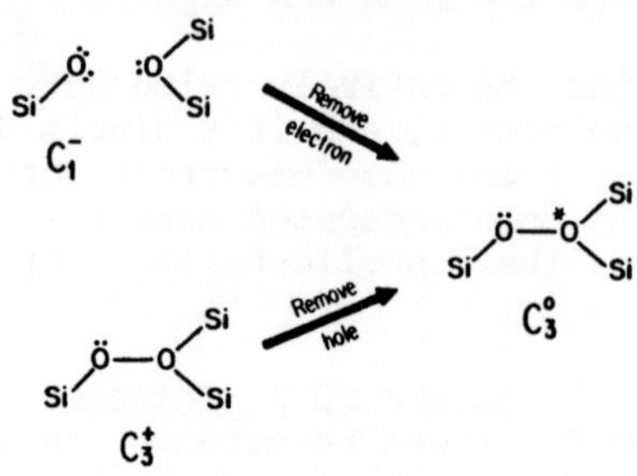

Fig. 6. Capture of appropriate carrier by C_1^- and $C_3^+(Si_2O)$ resulting in the same paramagnetic defect.

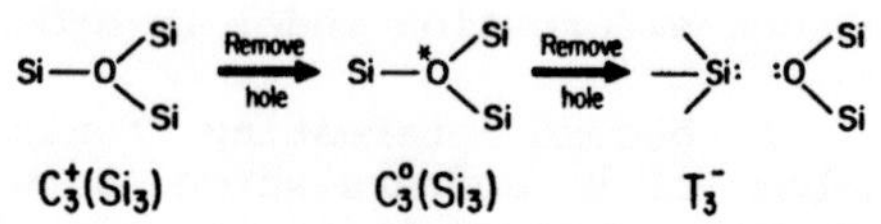

Fig. 7. Removal of two holes in in succession to form first $C_3^0(Si_3)$ and then its negative precursor. This can explain behavior of the B center.

ORIENTED DEFECTS IN GLASS

In addition to being able to selectively excite certain defects by tuning the excitation photon energy, it is possible to exploit the fact the excitation can be polarized. In the usual sort of experiment in which defects are generated by high-energy irradiation, the defects assume a random orientation because of the inherently random structure of the glass. Instead of observing discrete, angle-dependent EPR lines corresponding to defects with specific orientations, one instead observes a powder pattern, corresponding to an angular average over all orientations.

We recently[16] described a novel effect in which the paramagnetic centers are created with a preferential orientation along an externally-defined axis, by using polarized light to ionize native defects in a-SiO_2. If the formation mechanism of the paramagnetic center involves the ionization of a localized defect state, with for example an electron being excited from a localized p-orbital into an s-like band state, then the transition matrix element will be proportional to the cosine of the angle between the polarization direction and the axis of the p-orbital.

Figure 8 shows the EPR spectrum induced in Suprasil W1 by polarized 6.4 eV light. Only the low-energy centers are present in these data. The spectra were obtained at two different orientations of the magnetic field relative to the polarization direction. It can be seen that there is a distinct, although small, orientation dependence.

The data in fig. 8 are complicated by the fact that some components of the EPR spectra are unpolarized. Since the excitation is below the band gap, the mechanism for producing the paramagnetic centers is almost certainly one involving the ionization of native defects, with subsequent trapping of the ionized electron or hole at a second site[6,7]. This trapping process should not depend on the polarization of the exciting light, so at least half of the centers are expected to be unpolarized. Close inspection shows that the spectra are composed of at least two distinct centers, and

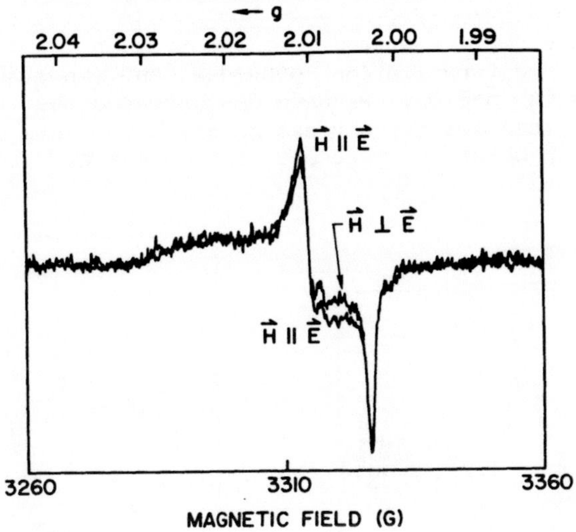

Figure 8. Orientation dependence of the EPR induced in a-SiO_2 by polarized 6.4 eV light.

it is in fact possible to separate these components by their annealing behavior[7]. Furthermore the centers, even if initially created with maximum polarization, may have partially relaxed and randomized before the EPR was measured. Indeed, the EPR was re-measured after allowing the sample to anneal at room temperature for a period of one week and it was found that the anisotropy had relaxed somewhat. For these reasons it is more meaningful to look at the polarization-induced anisotropy, i.e. the difference between the EPR spectra measured with the magnetic field perpendicular and parallel to the polarization direction. This difference is shown in fig. 9.

If one assumes that the underlying symmetry of the defect is axial, then it is possible to obtain closed-form expressions for the polarized-light-induced powder patterns[16], $P(g)=\sin\theta D(\theta)|\delta\theta/\delta g|$ where $D(\theta)$ is the angular distribution of the defect symmetry axis with respect to the magnetic field direction. The theoretical predictions are shown in fig. 10, for two orientations of the magnetic field and for the two simplest cases for $D(\theta)$, depending on whether the resulting paramagnetic centers are produced with their unique axes along the $\vec{E}$ vector of the light (case I), or perpendicular to the light (case II). The experimental difference spectrum resembles quite closely the prediction for case (II). Note in particular the agreement in the relative magnitudes of the two minima (indicated by arrows). Of course the difference spectrum is rather noisy, and is not in precise agreement with the theoretical prediction, e.g. in the region between the two minima. Such discrepancies in the exact shape of the spectra may result from the assumption of perfect axial symmetry in the theoretical spectrum. This assumption could be relaxed in a numerical calculation.

SUMMARY

Sub-band-gap photoexcitation provides an important new tool for studying defects in a-SiO_2. We have demonstrated that by using different photon energies we can selectively excite different sets of native defects. The additional parameter of variable photon energy provides invaluable

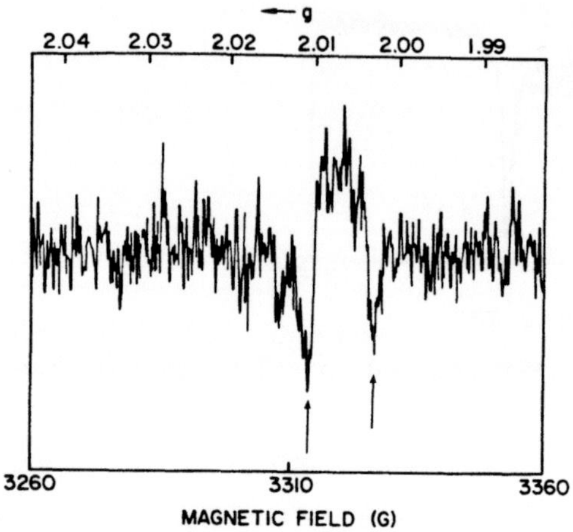

Fig. 9. Polarization-induced anisotropy, i.e. the difference $(\vec{H}\perp\vec{E}) - (\vec{H}\parallel\vec{E})$ of the two spectra in fig. 8. This figure is plotted with a gain of 5 relative to fig.8

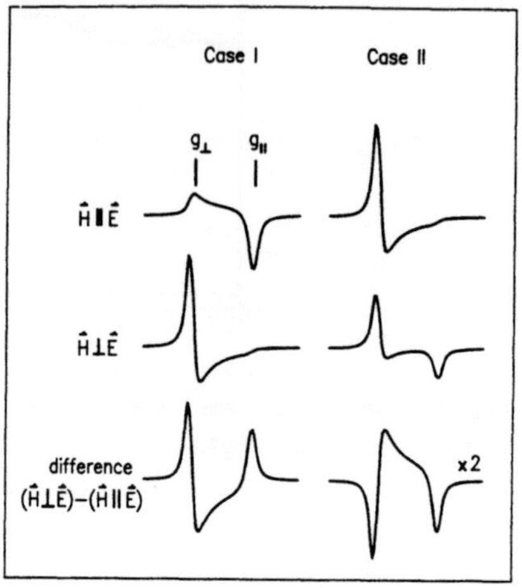

Fig. 10. Theoretical predictions for powder patterns produced by polarized light, for centers of axial symmetry. The bottom two curves are the predicted polarization-induced anisotropy.

assistance in isolating the EPR spectra of individual defects. This may also make the goal of correlating the EPR signals with various optical bands much easier[17].

Most importantly, we are now able to investigate the nature of native defects, rather than being limited to defects produced by radiation damage as has been the case previously. This is being used, for example, to look at drawing-induced defects in optical fibers[18].

The extension of this technique by the use of polarized light leads to the exciting ability to produce partially oriented paramagnetic centers in amorphous materials, opening up a host of possibilities for new experiments. This was demonstrated with the paramagnetic centers produced by 6.4 eV light. Application to the A, B, and C centers, which predominate at higher excitation energy, could provide much more insight into their identity. The evolution of such anisotropic EPR spectra during annealing could tell us much about the mechanism by which the defects transform.

REFERENCES

1. D.L. Griscom, Mat. Res. Soc. Symp. Proc. 61:213 (1986).
2. D.L. Griscom, SPIE J. 541:38 (1985).
3. P.W. Anderson, Phys. Rev. Lett. 34:953 (1975).

4. M. Kastner, D. Adler, and H. Fritzsche, Phys. Rev. Lett. 37:1504 (1976).
5. J.H. Stathis and M.A. Kastner, Phys. Rev. B. 29:7079 (1984).
6. J.H. Stathis and M.A. Kastner, Mat. Res. Soc. Symp. Proc. 61:161 (1986).
7. J.H. Stathis, Ph.D. Thesis, Massachusetts Institute of Technology, (unpublished), 1986.
8. D.V. Lang, R.A. Logan, and M. Jaros, Phys. Rev. B 19:1015 (1979).
9. M. Stavola, M. Levinson, J.L. Benton, and L.C. Kimmerling, Phys. Rev. B 30:832 (1984).
10. A. Chantre and D. Bois, Phys. Rev. B 31:7979 (1985) ; A. Chantre and L.C. Kimmerling, Appl. Phys. Lett. 48:1000 (1986).
11. E.R. Weber, Mat. Res. Soc. Symp. Proc. 46:169 (1985) ; M. Baeumler, U. Kaufmann, and J. Windscheif, Appl. Phys. Lett. 46:781 (1985).
12. Suprasil-W1 is a trade name of Heraeus-Amersil, Sayerville, NJ 08872.
13. C. Fiori and R.A.B. Devine, Phys. Rev. Lett. 52:2081 (1984).
14. N.F. Mott, Adv. Phys. 26:363 (1977) ; J. Non-Cryst. Solids 40:1 (1980).
15. R.A. Street and G. Lucovsky, Solid State Commun. 31:289 (1979) ; G. Lucovsky, Philos. Mag. B 41:457 (1980).
16. J.H. Stathis, Phys. Rev. Lett. 58:1448 (1987).
17. R.A.B. Devine, C. Fiori, and J. Robertson, Mat. Res. Soc. Symp. 61:177 (1986).
18. J. Simpson, J. Ritger, and F. Di Marcello, Mat. Res. Soc. Symp. 61:333 (1986).

UV AND VUV OPTICAL ABSORPTION DUE TO INTRINSIC AND LASER INDUCED DEFECTS IN SYNTHETIC SILICA GLASSES

H. Imai, K. Arai, T. Saito, S. Ichimura, H. Nonaka,
J.P. Vigouroux*, H. Imagawa**, H. Hosono*** and Y. Abe***

Electrotechnical Laboratory
1-1-4 Umezono, Sakura-mura, Niihari-gun, Ibaraki 305 Japan
**Departement of Applied Chemistry, Tokyo University
2100 Kujirai Nakanodai, Kawagoe-shi, Saitama 350, Japan
***Inorganic Departement, Nagoya Institute of Technology
Gokisyo, Showa-ku, Nagoya 466, Japan

ABSTRACT

We examined twelve kinds of synthetic silica glasses prepared by various methods, such as flame hydrolysis plasma-CVD, CVD-soot remelting and sol-processes. Each synthetic method and preparation process could berelated to particular intrinsic defects in the glasses such as hydroxyl groups, peroxy-linkages and oxygen-deficient defects. It was found that excess absorption near the band edge was caused in each case by the dominant intrinsic defects. The E' center induced by ArF laser irradiation in the glasses containing oxygen-deficient defects was found quite stable at room temperature, whilst for glasses containing other dominant defects, it was unstable. The irradiation effects on the 7.6 and 5.0 eV absorption bands strongly suggest that these bands originate from oxygen-deficient defect.

INTRODUCTION

The optical properties of synthetic silica glass in UV and VUV (vacuum-ultraviolet) regions and the durability of such glass against excimer lasers have become increasingly important with the rapid progress in excimer lasers and their expanding scope of application to fields such as VLSI lithography. Although these properties are known to be sensitive to intrinsic defects and impurities in glass, systematic studies on them have not been carried out.

Kaminow et al.[1] reported that absorption near the band edge of Spectrosil and Suprasil varies markedly with hydroxyl content and that a peak or shoulder is observed at 7.6 eV in hydroxyl-free samples.

---

* On leave from Centre d'Etudes Nucleaires de Saclay, France.

Table 1. Preparation methods and impurity contents

Sample	Preparation method	Content (ppm) OH	Cl	F
P1	Ar plasma-CVD	N.D.	200-500	N.D.
P2	O_2 plasma-CVD	N.D.	200-500	N.D.
Suprasil W	plasma-CVD	1-2	-------	-------
H1	flame hydrolysis	500-600	100-200	N.D.
H2	flame hydrolysis	800-900	N.D.	N.D.
S1	CVD-soot remelting	N.D.	N.D.	200-500
S2	CVD-soot remelting	N.D.	50-100	200-500
S3	CVD-soot remelting	100-200	50-100	N.D.
S4	CVD-soot remelting	-------	-------	-------
SG1	sol-gel	90	-------	-------
SG2	sol-gel	300	-------	-------
SG3	sol-gel	20	-------	-------

Historically, the origin of the 7.6 eV band observed both in as-prepared and irradiated silica glasses is controversial. Griscom et al. assume that the oxygen-deficient defect (≡Si-Si≡) causes the band and suggest that the peroxy-radical center is also related to absorption around 7.6 eV.[3] Moreover, little information is available on absorption of other defects possibly present, such as the peroxy-linkage (≡Si-O-O-Si≡) and the hydroxyl group (- OH). Recently a few studies on the defects produced by excimer lasers have been reported[4,5]. Induced paramagnetic defects in Suprasil were found sensitive both to excitation energy and OH content[4]. Devine et al.[5] studied the laser induced oxygen-related center in OH free silica glass (Suprasil W1) in relation to accumulated UV dose, irradiation temperature and annealing behavior. Defects caused by UV and VUV irradiation may originate by direct photo-ionization of intrinsic defects or trapping of photo-ionized charge carriers at other pre-existing defects, while γ-rays and X-rays with much higher energy possibly create new defects by the knock-on processes. Thus, the study of laser irradiation effects should facilitate understanding of the relationship between intrinsic and induced defects.

EXPERIMENTAL

The glasses were classfied into four types according to the synthetic method of production, such as flame hydrolysis(H), plasma-CVD(P), CVD-soot remelting(S) and sol-gel(SG) processes as shown in Table 1 along with impurity content. VUV absorption measurements at 6.2-8.3 eV were made using synchroton radiation from a storage ring at our laboratory, through a 1m monochromator (Nikon/McPherson 225). UV absorption below 6.2 eV, IR absorption, luminescence and EPR measurements were made using commercial spectrometers. Some samples were treated with O_2, H_2, N_2 and ordinary air atsmospheres at 800 or 900°C for 3-240 hours. The ArF excimer laser (Lambda Physik model EMG 102 MSC) for the irradiation experiment was operated at 30 Hz with a 20 ns pulse width. The maximum energy per pulse of a rectangular beam (8×28 mm^2) was 100 mJ unless otherwise noted.

RESULTS AND DISCUSSION

Absorption Properties in the VUV Region

The absorption behavior of the twelve silica glasses near the band edge was found to be classifiable into the three types shown in fig. 1. Hydroxyl-free glasses prepared by the CVD-soot remelting process (S5 and S6) had distinct peaks at 7.6 eV. The samples prepared by plasma-CVD (P1, P2 and Suprasil W), which were almost hydroxyl-free, showed tail absorptions prominent down to 7.0 eV. The other samples (H1, H2, S3, S4 and SG1-SG3), containing varying amounts of water, showed no prominent absorption below 7.5 eV but shifted band edges to the lower energy side with increasing OH content, regardless of the preparation method. Each excess absorption was attributable to an intrinsic defect, an oxygen-deficient defect, peroxy-linkage and OH group, respectively, as will be discussed below.

The 7.6 eV band was about two times larger than that observed by Kaminow[1] in Spectrosil WF but attribution of the origin of this band to the peroxy-radical[3] was ruled out since no paramagnetic center was observed in any as-prepared sample. This absorption band measured for a 0.2 mm thick sample, decreased to ~ one tenth the initial intensity by heat-treatment in H_2 at 800°C for 3 hours and to ~ one third after 30 hours in O_2 at 900°C. It did not change at all in an N_2 atmosphere. Thus, the defect responsible for the 7.6 eV band reacts with H_2 and O_2 but is not simply thermally reduced. The reaction is explained by the tarnishing reaction between the defect and H_2 or O_2, the defect density was estimated to be $10^{18}cm^{-3}$. No OH group was detected by IR measurement in the samples treated with H_2, indicating little existence of the peroxy-linkage. Thus, the origin of 7.6 eV is likely to be oxygen-deficient defect. The atomic structure of the defect has not been identified yet. The defect, ≡Si-Si≡, as suggested by Griscom[2] is a possible origin. But ≡Si-H, possibly produced following H_2 treatment, was not observed by IR since the estimated content was below our detection limit[6]. Griscom[2] assumes that the oxygen-deficient defect, ≡Si-Si≡, accompanies the 5.0 eV absorption band with 4.3 eV luminescence as well. We confirmed the presence of a 4.3 eV luminescence and a 5.0 eV excitation peak in these samples having the 7.6 eV band. It decreased with heat-treatment in O_2 or H_2 as did the 7.6 eV band. This fact supports Griscom's assumption. However, the results of the ArF laser irradiation indicated two different states for the oxygen-deficient defect.

An absorption tail down to 7.0 eV was removed by heat-treatment in H_2, resulting in the production of a certain amount of OH groups observed by IR. Shelby[7] suggests that the chemical reaction may be described as a type of tarnishing reaction.

The reaction, ≡Si-O-O-Si≡ + H_2 = 2 x (≡Si-OH), occurs in excess oxygen samples having peroxy-linkages. The findings for plasma-CVD samples correspond to this reaction. The intensity of the infrared OH band at 2.7 μm was completely satured following heat-treatment in H_2 at 900°C for 30 hours, so that the density of the peroxy-linkages in these excess oxygen samples was of the order of $10^{18}cm^{-3}$. The absorption coefficient from 7.0 to 8.0 eV was proportional to the thus determined peroxy-linkage content. Thus, the peroxy-linkage causes an absorption tail down to 7.0 eV.

OH containing samples had no prominent absorption below 7.5 eV but the absorption edge shifted to the lower energy side with increase in OH content, regardless of the preparation method. Thus, excess absorption above 7.5 eV is dominated by OH content in the samples.

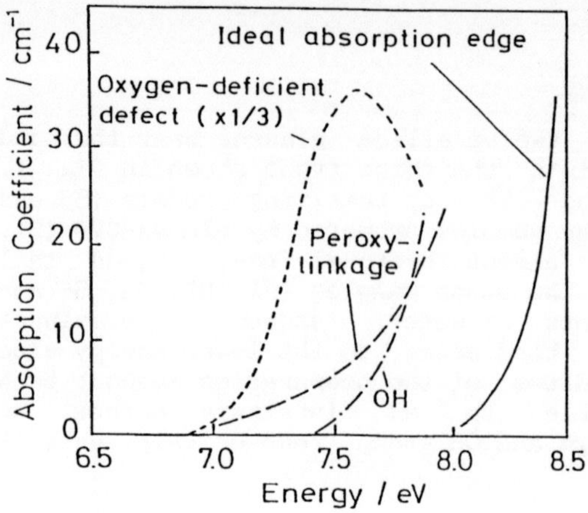

Fig. 1 Typical absorption features of twelve samples near the band edge and classified into three types.

In conclusion, absorption in the VUV region of synthetic silica glasses is characterized by individual dominant defects, such as oxygen-deficient defects, the peroxy-linkage and OH group. Other impurities such as chlorine and fluorine may not influence absorption properties. The defects may be reasonably considered related to the synthetic methods and the preparation processes. The ideal absorption edge appearing after subtracting excess absorption should be above 7.8 eV.

E' Center Induced by ArF Excimer-Laser Irradiation

Although the photon energy of the ArF excimer laser, 6.4 eV, was well below the band gap, ArF laser irradiation at room temperature created an E' center in all samples except Suprasil 1 and also produced an oxygen-related

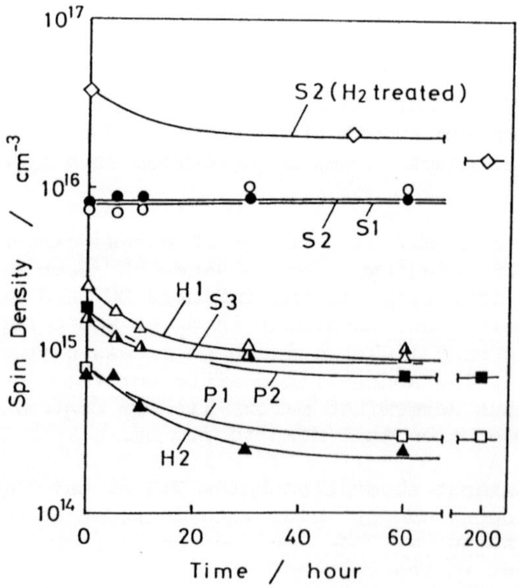

Fig. 2. Decay of induced spin density after laser irradiation at room temperature.

center in glasses having peroxy-linkages as reported by Stathis et al.[4]. Both centers were accompanied by absorption bands at 5.7 eV and 4.8 eV, respectively, resulting in reduction of the UV transparency. The E' center is the well-known paramagnetic defect ascribed to a silicon dangling bond. How defect production is related to the intrinsic defects warrants attention. We found that the manner of formation and stability of the E' center strongly depends on the existence of intrinsic defects.

Fig. 2 shows the decay of spin density induced after ArF laser irradiation at room temperature. The E' centers induced in the samples containing oxygen-deficient defects did not decrease. In the other samples with OH groups or peroxy-linkages, the number of E' centers produced was about one order less that in the former samples and gradually decreased to about one third its initial value after 30 hours and remained almost constant. Thus, the E' center induced in the former samples is quite stable, while that in latter ones is unstable. In samples treated in H_2, the induced E' center increased by about one order and showed a decreasing tendency was similar to that of the latter. In the case of O_2 treatment, the number of E' centers produced was reduced by more than one order. Regarding the growth curve of the E' center as a function of accumulated photon doses, the E' center induced in the samples having oxygen-deficient defects increased with laser photon dose. On the other hand, the defect density induced in samples containing OH groups rapidly reached the saturated value. Although detailed information on formation and the extinction mechanism of the E' centers are not yet clear, OH groups, peroxy-linkages and Si-H bonds should be essential to these processes through structural relaxation and successive defect reactions.

Oxygen-deficient defects

The ≡Si-Si≡ bond, an oxygen-deficient defect, is the likely precursor of the E' center[2], since ≡Si-Si≡ = $\equiv Si^{\bullet} + (\equiv Si + e^-)$. This process should be accompanied by large structural relaxation because the reverse process does not occur at room temperature. Nagasawa et al.[8] reported the 5 eV absorption band observed in oxygen-deficient samples is decreased by KrF laser irradiation. Thus, the E' center may be created at the expense of an ≡Si-Si≡ bond. However, they observed rather rapid recovery of band intensity at room temperature. In fig. 3, the 4.3 eV luminescence intensity, originating from the 5.0 eV absorption band, can be seen to be reduced by ArF irradiation. It may be reasonably assumed that the luminescence intensity change is directly related to that of the absorption band. The reduced luminescence intensity, however, did not recover at room temperature, corresponding to the case of the induced spin density of the E' center. Thus, as far as our experiment is concerned, the defect causing the 5.0 eV absorption band appears to be the precursor of the stable E' center.

The 7.6 eV absorption band, assumed to have the same origin as the 5.0 eV band by Griscom, was not changed by ArF irradiation within our detection limits, in contrast to the case of the E' center and 5.0 eV band ; this strongly suggests that the origin of these absorption bands should be attributed to individual structural defects. But these defects are the same with respect to oxygen-deficiency. The number of defects concerned with the 7.6 eV band was $1 \times 10^{18} cm^{-3}$ as stated before. Thus, it would be difficult to detect a decrease in band intensity if, for example, some of the defects causing the 7.6 eV band changed into the E' centers, (say as many as $10^{16} cm^{-3}$). The density of the defects related to the 5.0 eV band should be of the same order as the induced E' center density. Both bands diminished qualitatively in parallel as the result of chemical treatment as mentioned above. Several models have been proposed for

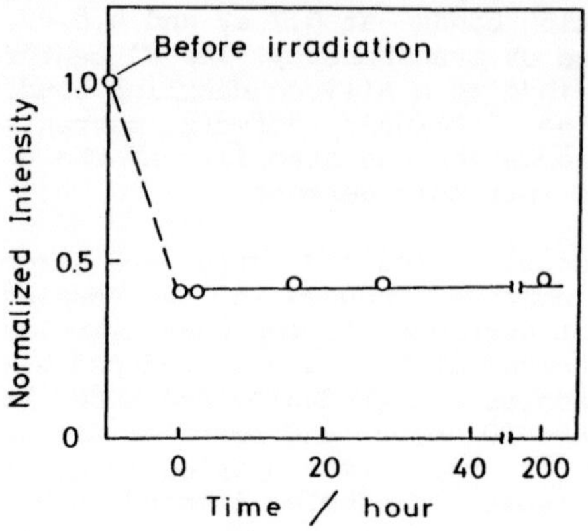

Fig. 3. Change in the 4.3 eV luminescence intensity by ArF laser irradiation.

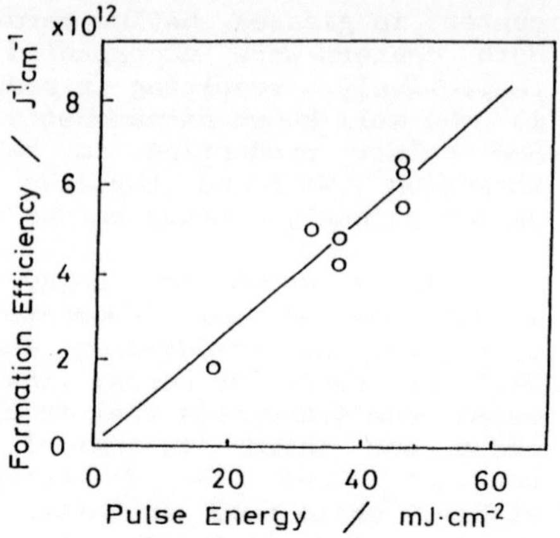

Fig. 4. Formation efficiency of the E' center as a function of ArF pulse energy (mJ/cm^2) in oxygen-deficient glasses.

oxygen-deficient defects. In addition to the previously mentioned ≡Si-Si≡ bond model, Skuja et al.[9] state that the 5.0 eV absorption band and 4.3 eV luminescence are due to twofold coordinated silicon defects. It should be noted that these defect structures are easily interchangable[6], elucidation of the exact nature of oxygen-deficient defect will require additional reseach.

The mechanism of excimer laser excitation in synthetic silica glass is another problem related to laser induced damage. Silica glass essentially has no absorption at an ArF laser photon energy of 6.3 eV. Soileau et al.[10] suggest that two-photon absorption of the second harmonic of Nd:YAG laser exists in silica glass. We observed the formation efficiency of the E' center in the samples having oxygen-deficient defects to be linearily proportional to laser pulse energy, as shown in fig. 4. This means the spin density produced is proportional to the square of laser pulse energy. Thus, the two-photon process is suggested to be responsible for defect formation in the case of ArF excitation.

ACKNOWLEDGMENT

The authors express their sincere gratitude to the Shin-Etsu Chemical Co. Ltd., Nippon Quartz Products Co. Ltd., Asahi glass Co. Ltd., and Seiko Epson Co. Ltd. for supplying the samples and to K. Endo, Y. Ohawatano and K. Ando for allowing use of the excimer lasers and to S. Yamasaki and M. Aizawa for conducting the EPR experiment.

REFERENCES

1. I.P. Kaminow, B.G. Bagley, and C.G. Olson, Appl. Phys. Lett., 32:98 (1978).
2. D.J. Griscom, J. Non-Cryst. Solids, 73:51 (1985).
3. M. Stapelbroek, D.J. Griscom, E.J. Friebele, and G.H. Sigel Jr., J. Non-Cryst. Solids, 32:313 (1979).

4. J.H. Stathis and M.A. Kastner, Phys. Rev. B, 29:7079 (1984).
5. R.A.B. Devine, C. Fiori, and J. Robertson, in : "Defects in glasses", F.L. Galeener, D.L. Griscom, and M.J. Weber, ed. Material research Society, Pittsburgh, USA (1986).
6. Detailed discussions will be presented elsewhere by H. Imagawa et al.
7. J.E. Shelby, J. Appl. Phys., 51:2589 (1980).
8. Private communication, to be presented in this conference.
9. L.N. Skuja, A.N. Streletsky, and A.B. Pakovich, Solid State Comm., 50:1069 (1984).
10. M.J. Soileau, N. Mansour, E. Canto, and D.L. Griscom, in : "Defects in glasses", F.L. Galeener, D.L. Griscom, and M.J. Weber, ed. Material Research Society, Pittsburgh, USA (1986).

6. J.H. [illegible] and B.A. Kaufman, Phys. Rev. B, [illegible] (1979).
7. [illegible] effects in glasses"
[illegible] Materials Research [illegible]
[illegible] (1980).
8. [illegible] diffusion [illegible] by [illegible] et al.
[illegible] (1980).
9. [illegible] Smith, [illegible] and [illegible], [illegible] (1981).
10. [illegible], [illegible], [illegible] and [illegible], [illegible], Material [illegible]
Research Society, Pittsburgh, PA (1981).

INCOMMENSURATE PHASE OF QUARTZ: MICROSCOPIC ORIGIN AND INTERACTION WITH DEFECTS

G. Dolino

Laboratoire de Spectrométrie Physique
Laboratoire associé au C.N.R.S. L.A. 08
B.P. 87 - 38402 Saint Martin d'Hères, Cédex, France

Silicon dioxide is well known for its extensive polymorphism[1] : in addition to about 20 crystalline phases, it is also easily obtained in an amorphous state which is a prototype of glass structure. With the exception of Stishovite, these phases consist of three dimensional frameworks of corner sharing SiO_4 tetrahedra, giving structures with different topological connections. Furthermore the low pressure phases (quartz, cristobalite, tridymite) present displacive transitions produced by small displacements of the SiO_4 tetrahedra, without breaking any atomic bond. In this way quartz at 846 K transforms from the low temperature α phase of trigonal symmetry to the high temperature β phase of hexagonal symmetry. Although this transition has been studied for nearly a century, it was only in 1980 that Bachheimer discovered that the α-β transition was not direct[2] but occured through a new intermediate phase, later characterized as an incommensurate (inc) phase[3]. In an inc structure some property (atomic position, electronic or spin density ...) is modulated with a period λ which is not commensurate with the lattice period a. In a diffraction experiment satellite peaks are observed in addition to the usual lattice reflections. Indeed in the inc phase of quartz satellites have been observed by diffraction experiments with neutrons[3], X rays[4] and electrons[5]: satellites are observed along the 6 equivalent <100> directions of the hexagonal reciprocal lattice at a small distance $q \simeq 0.03\ a^*$ from the Bragg peaks. Furthermore this is one of the few inc structures which has been observed directly by electron microscopy : a periodic structure of equilateral triangles, produced by the local superposition of 3 waves at 120° has been observed[5,6]. (Recently it has been shown that the application of a uniaxial compressive stress along [100] produces an inc structure with a single modulation wave along [100][7]).

The late discovery of this new phase is due to the fact that it is observed only in a small temperature range of 1.4 K, in between the classical α and β phases. However in this small range large variations of most physical properties are observed[8] : the wavevector modulation q presents a 30 % decrease from .033 to .022 a^* while the thermal expansion coefficient α_x increases 10 times from 10^{-4} to $10^{-3}\ K^{-1}$.

Origin of the inc phase

The first suggestion for the existence of an inc phase in quartz is found in the theoretical work of Aslanyan and Levanyuk[9] who pointed out that the hexagonal symmetry of β quartz allows the existence of a coupling term between strains U_{ij} and the spatial derivatives of the order parameter η, which can be written as :

$$A\,(U_{xx} - U_{yy})\,\frac{\delta\eta}{\delta x} - 2\,U_{xy}\,\frac{\delta\eta}{\delta y} \qquad (1)$$

(in the α phase, the order parameter η corresponds to a rotation of SiO_4 tetrahedra around the OX-type axis, associated with the appearence of the d_{xxx} coefficient of the piezoelectric tensor).

If the coupling constant A is large enough, an intermediate inc phase is expected in between the classical α and β phases. In a lattice dynamic picture, the gradient interaction of eq.(1) corresponds to an anisotropic coupling between acoustic modes and the soft mode associated with the order parameter. The inc modulation along the OY axis is due to a strong coupling between the transverse acoustic shear mode (associated with U_{xy}) and the soft mode.

Experimental results obtained from inelastic neutron scattering measurements are in agreement with this model : a strong anisotropy of phonon branches has been observed in the XY plane, with a large premonitory diffuse scattering along Y axis[10]. Furthermore clear evidence of the coupling between the transverse acoustic U_{xy} mode and the soft mode has been obtained in a recent experiment[11] performed above 1000 K. This high resolution experiment has also shown that the diffuse scattering observed along OY has a dynamical origin which can be attributed to a low frequency branch, resulting from the interaction of the acoustic modes with the soft mode, with an overdamped harmonic oscillator behaviour.

Further insight is gained by considering a more microscopic model. The first step was to consider the possible vibration modes of rigid SiO_4 tetrahedra[12] : this geometrical approach leads to a strong coupling of the soft mode with the acoustic U_{xy} mode, along Y axis. Further improvements with deformable tetrahedra, where only the SiO distance is constant, lead to a good description of the 3 lowest frequency modes along OY. The atomic displacements in the inc phase result from a mixing of a shear wave with the tetrahedra rotations of the α-β soft mode. It is possible that these low frequency excitations may be related to those recently observed in silica glass[13].

Interaction with defects

Progressively it has been realized that inc phases are very sensitive to the presence of defects : in a perfect lattice, the energy of an inc phase is not changed by a translation of the modulation wave. This translation invariance is broken by the presence of point defects which in general introduce a pinning of the modulation, leading to irreversible effects. In this respect, quartz is an interesting case, as one can obtain very pure samples, with well characterized impurities, although very sensitive experiments are needed to detect these irreversible effects in the small temperature range of 1.4 K of existence of the inc phase. The first evidence for irreversible behaviour was obtained in a measurement of the modulation wave vector q by γ-ray diffraction on a natural sample[14]: during heating and cooling cycles, a thermal hysteresis was observed in the temperature variation of q, reaching 0.2 K near T_c the transition temperature between the inc phase and the α phase.

A systematic study of this hysteresis behaviour has been performed using birefringence measurements. Fig. 1a shows the variation of the birefringence observed upon cooling and heating inside the inc phase : different variations are observed upon heating and cooling. By performing a temperature cycle of .15 K, one obtains the small hysteresis cycle shown in Fig. 1b. In the high quality synthetic samples investigated, with Al concentrations < 5 ppm, the maximum hysteresis varies from 15 to 60 mK. The hysteresis effect is very sensitive to small variations in the defect density.

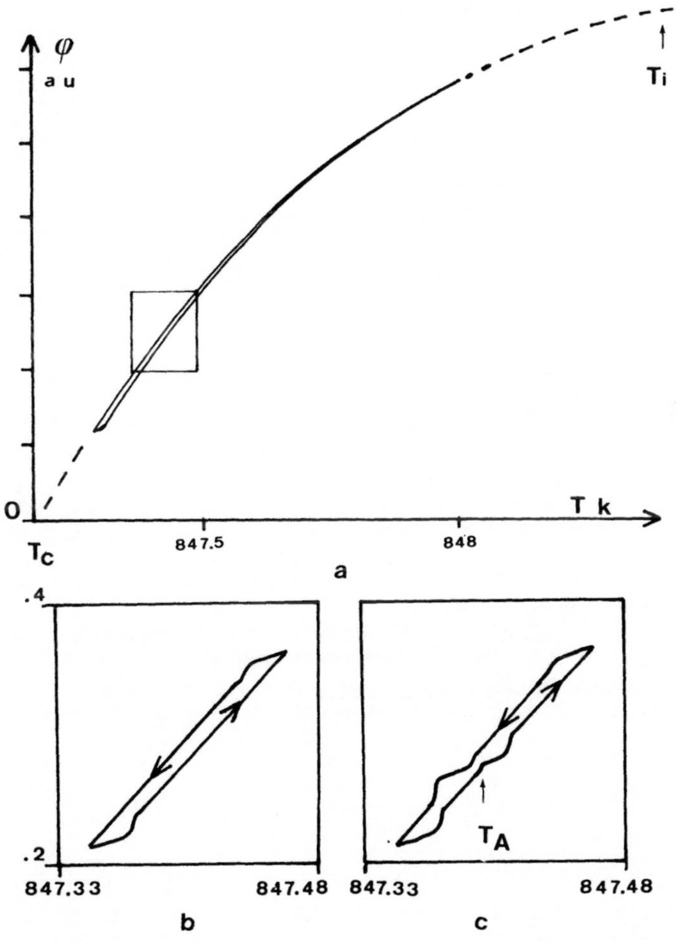

Fig. 1. a) irreversible behaviour of the temperature variation of the birefringence φ (arbitrary units) in the incommensurate phase.
b) Hysteresis cycle, corresponding to the rectangle of fig. 1a.
c) Memory effects obtained after waiting 10 h at T_A.

In most samples with narrow hysteresis a remarkable "memory effect" is observed as shown in fig. 1c. Before starting the temperature cycle, the samples was stabilized for 10 hours at a constant temperature T_A. The sample remembers that it has spent a long time at T_A. This effect is explained by a slow migration of mobile defects during the long waiting time at T_A. The interaction with the modulation wave leads to the growth of a "defect density wave" which keeps its period in subsequent temperature variations[15]. When the temperature comes in the vicinity of T_A, a lock-in occurs between the modulation wave and the defect density wave producing the small steps observed in the birefringence curve.

The inc phase of quartz reveals very interesting irreversible behaviour such as temperature hysteresis and memory effects. One has to realize that an inc modulation is a distortion from a perfect crystalline order which produces small changes in atomic distances. One can hope that the understanding of the microscopic interactions at the origin of these irreversible effects can be used to understand the behaviour of the more disorded structure of amorphous materials.

REFERENCES

1. R.B Sosman, "The Phases of Silica", Rutgers University Press, New Brunswick (1965).
2. J.P. Bachheimer, J. Physique Lett. 41:L561 (1980).
3. G. Dolino, J.P. Bachheimer, B. Berge and C. Zeyen, J. Physique 45:361 (1984).
4. K. Gouhara and N. Kato, J. Phys. Soc. Japan 1868 (1985).
5. E. Snoeck, C. Roucau and P. Saint Grégoire, J. Physique 47:2041 (1986).
6. J. Van Landuyt, G. Van Tendeloo, S. Amelinckx and M.B. Walker, Phys. Rev. B 31:2986 (1985).
7. G. Dolino, P. Bastie, B. Berge, M. Vallade, J. Bethke, L.P. Regnault and C. Zeyen, Europhys. Lett. 3:601 (1987).
8. G. Dolino, in "Incommensurate Phases in Dielectrics 2", R. Blinc and A.P. Levanyuk eds, North Holland, Amsterdam (1986).
9. T.A. Aslanyan and A.P. Levanyuk, Solid. State Comm. 31:547 (1979).
10. J. Bethke, G. Dolino; G. Eckold, B. Berge, M. Vallade, C. Zeyen, T. Hahn, H. Arnold and F. Moussa, Europhys. Lett. 3:207 (1987).
11. B. Berge, G. Dolino, M. Vallade and F. Moussa (Unpublished).
12. B. Berge, J.P. Bachheimer, G. Dolino, M. Valade and C. Zeyen, Ferroelectrics 66:73 (1986).
13. U. Buchenau, M. Prager, N. Nucker, A.J. Dianoux, N. Ahmad and W.A. Phillips, Phys. Rev. B 34:5665 (1986).
14. P. Bastie and G. Dolino, Phys. Rev. B 31:2857 (1985).
15. P. Lederer, J.P. Jamet and G. Montambaux, Ferroelectrics, 66:25 (1986).

GAMMA-RAY INDUCED 2 eV OPTICAL ABSORPTION BAND IN PURE-SILICA CORE FIBERS

Kaya Nagasawa*, Yoshimichi Ohki and Yoshimasa Hama

***Sagami Institute of Technology
Fujisawa, Kanagawa 251
Waseda University
Shinjuku-ku, Tokyo
Japan**

INTRODUCTION

The mechanism of optical absorption near 2 eV induced by γ-irradiation in pure silica glass has been studied by many workers[1-6]. It appears that non-bridging oxygen hole centers (NBOHC), which were revealed by Friebele et al. through the study of electron spin resonance (ESR) spectra[1], are the most probable factor causing the optical absorption[2-5]. However, it is also clear that the 2 eV band cannot be ascribed to only one kind of defect center, because, as the OH-group content of the sample increases, the peak wavelength shifts from 630 nm to 600 nm[3,5] and the oscillator strength decreases[3]. Friebele et al. thus assumed that the 2 eV band is caused mainly by NBOHC in high-OH silica (i.e., silica which contains a high amount of OH groups) and by a non-paramagnetic center such as $\equiv Si:^-$ in low-OH silica[3]. Contrary to the above assumption, the present authors thought that the difference in the 2 eV band induced in high-OH silica and in low-OH silica is caused by hydrogen bonds between the NBOHC and an OH group near the NBOHC, and that the 2 eV band in both low- and high-OH silica is caused only by NBOHC[2]. In this paper, experimental results are presented followed by discussion of the model derived from these results.

EXPERIMENTAL

The sample fibers used are listed in Table I. All the fibers have pure silica core and silicone cladding. The manufacturing methods were :

1/ the direct glass deposition method, using oxyhydrogen gas flame,
2/ the plasma method, with $SiCl_4$ oxydized directly in an RF plasma without using oxyhydrogen gas flame, and
3/ the soot method.

The "pre-treated" samples were prepared by the following four consecutive procedures : hydrogen gas treatment (3 ata) at room temperature for 100 hours → γ-irradiation at room temperature with a dose rate of 200 Gy/h up to a dose of 16 kGy → a second hydrogen gas treatment (3 ata) at room temperature for 100 hours → heat treatment at 90°C for 100 hours to remove the hydrogen molecules from the inside of the fibers.

Table I. Sample list

Sample Name	Manufacturing Method	OH-group content [ppm]	Cl Content [ppm]
P6f	$Ar+O_2$ plasma	6	600
P3f	$Ar+O_2$ plasma	4	300
Prf	Ar plasma	3	1700
P2f	O_2 plasma	5	240
Sff	soot	1	0
Srf	soot	1	1050
Df8	direct	800	0
D17	direct	700	100

Pre-treated samples and non-treated samples were irradiated up to 40 kGy using ^{60}Co γ-rays at a dose rate of 200 Gy/h at room temperature in air. Their optical loss spectra were measured seven days after the irradiation by the cut-back method. The irradiated part of the sample was 10 m long, and the leader-fiber lengths of the light-injecting side and of the other side were 5 to 15 m and 4 to 5 m, respectively.

Pre-treated Srf, P6f and D17 fibers and non-treated Srf, P6f, D17 and Df8 fibers (see Table I) were irradiated by γ-rays at a dose rate of 200 Gy/h for 2 to 7 days at room temperature in air, and their optical loss spectra were measured 8 to 13 days after the irradiation. ESR measurement was also carried out on these types of fiber after they had been irradiated and kept in air in the same manner as for the optical measurement. The sample for the ESR measurement was cut into pieces, each 8 ~ 10 cm long and these were then placed in a silica tube. The intensity of the ESR signal was normalized to the intensity of Mn^{2+}. Measurements were made using X-band with 100 kHz field modulation. The microwave power was set to 5 mW.

The effect of raising or dropping the temperature on the spectrum of the 2 eV band was measured by the following procedure : the samples were P6f and D17 fibers that were irradiated with γ-rays (80 kGy) and kept in air for about 50 days after the irradiation. First, the optical loss spectrum of the samples was measured at room temperature. After this measurement, the irradiated part of the samples was immersed in water at 90°C or in ethylene glycol at -35°C, and then the optical loss spectrum was measured. Next, the spectrum at room temperature was measured again by taking out the sample into the air. Such optical measurements were repeated four to five times.

RESULTS AND DISCUSSION

Figure 1 shows the wavelength dependence of the absorption loss induced in non-treated and pre-treated P6f and D17 fibers irradiated with γ-rays to 40 kGy. Non-treated samples of both P6f and D17 fibers clearly show the 2 eV band. If the contribution from the shorter-wavelength absorption is subtracted, the peak appears at 630 nm for P6f fiber, which has low-OH content silica core. The peak shifts to 600 nm in case of D17 fiber (high-OH content core)[2,3,5]. In both fibers, the pre-treatment suppresses the 2 eV band[6]. It is thought that precursors of the 2 eV band are changed by the pre-treatment to NBOHC which in turn vanishes by forming a covalent bond with hydrogen ; precursors $\xrightarrow{\gamma}$ NBOHC $\xrightarrow{H_2}$ ≡Si-OH. The suppression of the 2 eV band by the pre-treatment is larger in the P6f fiber than in the D17 fiber. Similar experiments revealed that Df8 fiber, a

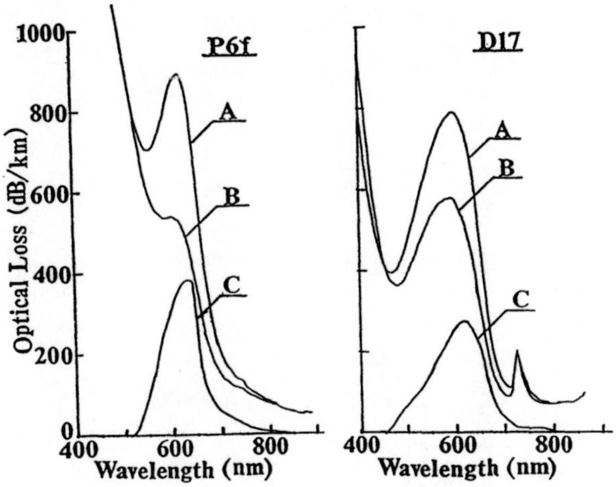

Fig.1. Effect of the pre-treatment on the growth of the α-ray induced 2 eV band.
A : Untreated sample, B : Pre-treated sample,
C : Decrease in the loss spectrum due to the "pre-treatment".

a high-OH fiber, has similar characteristics to the D17 fiber, and that P3f, Prf, P2f and Srf fibers, all containing only a few ppm of OH groups, have similar characteristics to the P6f fiber. The facts concerning the difference in the effect of the pre-treatment suggest that there are least two kinds of precursors responsible for the 2 eV band ; one exists mainly in high-OH fibers and the other in low-OH fibers. Further, the fact that the peak wavelength depends on the OH content suggests that there are two different states of the NBOHC, assuming the two different peaks are caused by NBOHC. This will be corroborated later.

Curve C in fig. 1 shows the decrease of the absorption loss due to the pre-treatment. The decrease is largest around 630 nm for both fibers. This indicates that the precusors responsible for the 630 nm absorption are eliminated effectively by the pre-treatment and that such precursors exist in high-OH fibers as well as low-OH fibers.

Figure 2 shows the relation between the absorption intensity of the γ-ray induced 2 eV band and the ESR intensity of the NBOHC. Measurements were made 8 to 13 days after the γ-irradiation had ceased. The ESR and the optical measurements of each sample were made on the same day. Similar experiments were done using pre-treated Srf and P6f fibers, but no quantitative analysis was obtained because of weak ESR intensity. The absorption intensity varies linearily with the amount of NBOHC in the case of non-treated P6f and Srf fibers (low-OH content silica core, see line A in fig. 2). Pre-treated Df8 fiber, a high-OH fiber, also shows linear relation with a different slope (see line B). These facts indicate that the 2 eV band is caused by NBOHC in both fibers and that the quantum efficiency of NBOHC is different between non-treated low-OH and pre-treated high-OH fibers. This in turn suggests that NBOHC must have two different states. Data of non-treated high OH fibers (i.e., D17 and Df8) lie between lines A and B. This must be related to the fact that there are two kinds of precursors in non-treated high-OH fibers as discussed above.

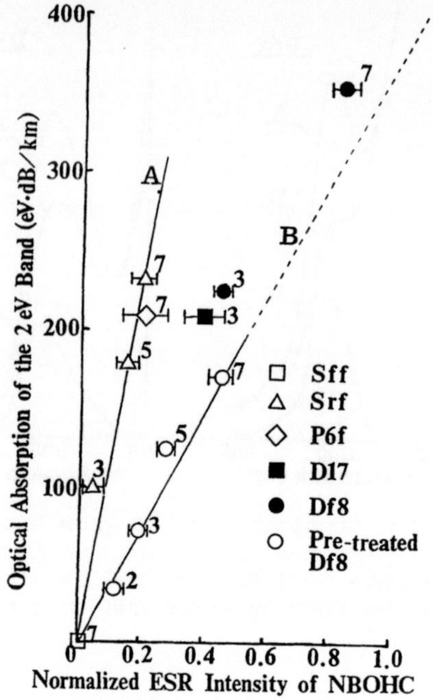

Fig. 2. Relation between the absorption intensity of the γ-ray induced 2 eV band and the ESR intensity of NBOHC. Numerical figures next to the data points show the number of days during which the sample was irradiated at 200 Gy/h (≃ 4.8 kGy/day).

Walrafen[8] reported that some of the OH groups in pure silica form hydrogen bonds in such forms as those shown on the left side of the following equation[7,8].

$$\left.\begin{array}{l} \equiv \mathrm{Si{-}O}\;..\mathrm{H}..\;\mathrm{H{-}O{-}Si} \equiv \\ \\ \equiv \mathrm{Si{-}O} \overset{\mathrm{H}}{\underset{\mathrm{H}}{\quad}} \mathrm{O{-}Si} \equiv \end{array}\right. \xrightarrow{\gamma\text{-RAY}} \equiv \mathrm{Si{-}O}^{\bullet}.....\mathrm{H{-}O{-}Si} \equiv + \mathrm{H}^{\bullet}$$

The authors also obtained experimental results confirming the existence of OH group which form hydrogen bonds from the temperature dependence of optical absorption around 1.39 μm [9]. When these OH groups are changed to NBOHC by γ-irradiation, NBOHC will have hydrogen bonds as seen in the above equation. It follows that such NBOHC can be formed easily in high-OH fibers but hardly at all in low-OH fibers. In other words, NBOHC in low-OH fibers seldom forms hydrogen bonds. It is hypothesized that NBOHC with and without hydrogen bonds corresponds to the above-proposed two different states of NBOHC. The following experiments were done to verify the hypothesis.

Hydrogen bonds are generally weak and sensitive to temperature. Therefore, change in the absorption loss spectra of P6f and D17 fibers, each γ-irradiated to 80 kGy, was examined by increasing the temperature to 90°C or by decreasing it to -35°C. When the fiber is heated to 90°C for the first time, the 2 eV absorption increases. But when it is cooled down to room temperature, the absorption decreases more than that gained during the preceding heating. This is called the 'temperature annealing effect'. After several repetitions of such temperature treatment cycles, the increase and the decrease of absorption become nearly equal. Under this condition, the change of the absorption spectra was measured. The results are shown in fig. 3.

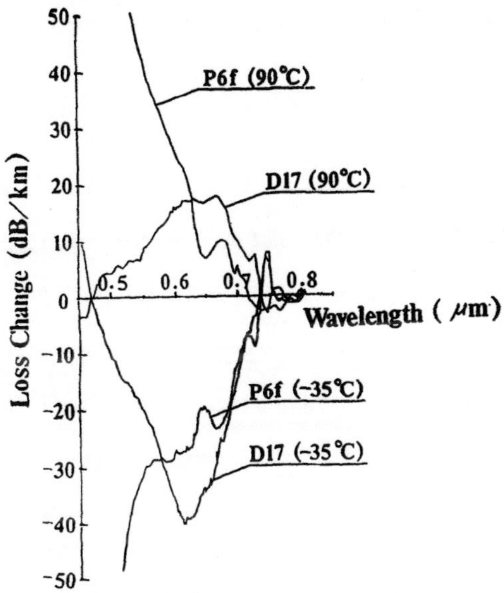

Fig. 3. Change in the loss spectra of the 2 eV band when the samples were heated from room temperature to 90°C or cooled from room temperature to -35°C.

The curves have a peak around 670 nm, after being heated up to 90°C. This is probably related to the absorption caused by the silicone cladding. The absorption around 620 nm increases markedly in the D17 fiber, a high OH fiber, by heating. On the other hand, it is very small in the low-OH P6f fiber although the absorption tail from the shorter wavelength region increases.

In general, hydrogen bonds are weakened by heating. This will increase the quantum efficiency of the 2 eV band and shift the position of the absorption peak to longer wavelength in the case of the high-OH fibers. This may be the reason for the increase in absorption around 620 nm, which is the longer-wavelength component of the 2 eV band in D17 fiber. A negligible amount of hydrogen bonds in the P6f fiber must be responsible for the small change of the absorption spectra. The change of the absorption spectra by cooling to -35°C is almost the opposite of what is seen by heating to 90°C. The cooling is supposed to make the hydrogen bonds stronger, and to make the quantum efficiency smaller.

As described above, it is concluded that the 2 eV absorption is caused by the NBOHC, which absorbs light around 600 nm when it forms hydrogen bonds in the form of $\equiv$Si-O$^{\bullet}$ .... H-O-Si$\equiv$ and around 630 nm when it does not ($\equiv$Si-O$^{\bullet}$).

The authors would like to express their thanks to Drs. Toshiaki Kuroha and Akira Iiono of the Furukawa Electric Co., Ltd., and to the Shin-Etsu Quartz Products Co., Ltd., and Shin-Etsu Chemical Co., Ltd., for providing the sample fibers. This work was partly supported by a Grant-in-Aid for Scientific Research from the Ministry of Education, Science and Culture of Japan (No. 60460123).

REFERENCES

1. M. Stapelbroek, D.L. Griscom, E.J. Freibele and G.H. Sigel Jr, J. Non-cryst. Solids 32:313 (1979).
2. K. Nagasawa, Y. Hoshi, Y. Ohki and K. Yahagi, Jpn. J. Appl. Phys. 25:464 (1986).
3. E.J. Friebele, D.L. Griscom and M.J. Marrone, J. Non-cryst. Solids 71:113 (1985).
4. L.N. Skuja and A.R. Silin, J. Non-cryst. Solids 63:431 (1984).
5. K. Nagasawa, Y. Hoshi and K. Yahagi, Annu. Rep. IEEE Conf. Electrical Insulation and Dielectric Phenomena (IEEE Service Center, New Jersey, 1984) 114.
6. K. Nagasawa, R. Tohmon and Y. Ohki, Jpn. J. Appl. 26:148 (1987).
7. J. Stone and G.E. Walrafen, J. Chem. Phys. 76:1712 (1982).
8. G.E. Walrafen and S.R. Samanta, J. Chem. Phys. 69:493 (1978).
9. K. Nagasawa, U. Yokomachi, R. Tohmon and Y. Ohki, Proceedings of the 6th International Conference on the Physics of Non-Crystalline Solids, Paper No. C.9 (to be published in a special issue of J. Non-Cryst. Solids).

ON THE DECAY OF X-RAY INDUCED LUMINESCENCE OF SiO_2

I. Godmanis[1] and W. Hohenau[2]

1: Peter Stucka University Riga, USSR
2: Karl-Franzens-Universität Graz, Austria

INTRODUCTION

A radiation induced luminescence in crystalline and glassy SiO_2 has been widely studied[1,2] but, only few investigations of its afterglow behaviour have been reported[3,4,5]. This paper is dedicated to the study of the influence of usual and different added impurities on the x-ray induced luminescence and especially on it's afterglow in crystalline and glassy SiO_2.

EXPERIMENTAL RESULTS

X-ray flashes (1 us) induce a visible luminescence both in crystalline and glassy SiO_2. The luminescence spectra of synthetic and different types of natural quartz (Brasil, smoky, morion) crystals consist of broad overlapping bands, lying between 2.5 and 3.3 eV. The relative intensity of these bands changes with each sample. But the afterglow of luminescence has the same temperature dependent decay in all investigated quartz crystals. Its decay time of about 1 ms at 100 K decreases exponentially with the rise of temperature above 170 K and is characterized by an activation energy of 370 meV (fig. 1). The afterglow with the same activation energy occurs in all investigated quartz crystals, containing very different concentrations of usual impurities such as Al, Li+, Na+, K+, and H+.

The same activation energy (370 meV) for the decay of afterglow was measured for x-ray induced luminescence in Ge-doped quartz and even in crystalline GeO_2 (fig. 2). However, the luminescence spectrum (maximum at 2.3 eV) in SiO_2, GeO_2 and the decay time of 660 us at 100 K in GeO_2 differ from those in Ge-free crystalline quartz.

This characteristic temperature dependent afterglow totally diminishes after electrolysis of quartz by Ag+ and Cu+ ions. A fundamentaly different luminescence at 3.4 eV with a temperature independent afterglow up to 500 K and a decay time of 46 us occurs in x-irradiated crystalline SiO_2-Cu+. Also a temperature independent afterglow for the silver center luminescence at 2.2 and 4.75 eV was observed in SiO_2-Ag+ (fig. 3).

The afterglow of integral luminescence in different x-irradiated quartz glasses (Suprasil, Corning, Herasil) does not depend on temperature till 600 K. Generally the intensity of luminescence in quartz glasses is somewhat weaker as in crystalline quartz and is characterized approximately by two decay times ; 50 us and 450 us. Similarly, as in crystalline quartz, an electrolysis with Cu+ and Ag+ ions in quartz glasses drastically changes the luminescence spectrum and its afterglow behaviour. In copper containing glass the decay time of luminescence at 2.4 eV slightly rises from 139 us to 153 us with decrease of temperature from ambient to 80 K. There is no change in the decay time of afterglow detectable in SiO_2-Cu+, between 300-600 K. Spectrally the same luminescence band at 2.4 eV is measured in glassy SiO_2-Ag+. But in contrast to Cu containing glass, the decay time in SiO_2-Ag+ is shorter than lus. Contrary to SiO_2-Cu+, in glassy SiO_2-Ag+ the intensity of luminescence drastically decreases at temperatures above 500 K.

DISCUSSION

The same activation energy of 370 meV observed for the decay of afterglow, indicates the existence of one type of trap, lying 370 meV deep in the gap and common in all types of crystalline quartz. The constant value of this activation energy is independent of type and concentration of usual impurities in quartz like Al, Li+, Na+, K+, H+, demonstrating that these ions are not included in the defects, which form traps with a depth of 370 meV. There is also no connection between these traps and defects in the silicon "sublattice" of quartz, given by replacing Si with Ge in the center of the $(SiO_4)4-$ tetrahedron in germanium doped synthetic quartz. Moreover, the defects forming traps with 370 meV depth, exist even in crystalline GeO_2.

Consequently, we have to conclude, that only fundamental defects in the "opposite", oxygen sublattice are responsible for forming such traps with 370 meV depth in crystalline quartz. These fundamental defects, common in crystalline SiO_2 and GeO_2, have a local structure, insensitive to the neighbourhood of different "cations" like Si or Ge. We suggest oxygen vacancies as a model for these oxygen-associated defects in crystalline SiO_2, SiO_2-Ge and GeO_2. During irradiation the oxygen vacancies may trap holes. This follows from the absence of a concurence in trapping process between them and germanium impurities, which are effective electron traps in quartz. In opposition to Ge and Me+ (Li+, Na+, K+), the Cu+ and Ag+ ions in quartz may trap not only electrons, producing Ag_o and Cu_o, but also holes (Ag++, Cu++). Therefore a more effective hole trapping at Cu+ (Ag+) in comparison to oxygen vacancies prevents the formation of hole traps with 370 meV depth in crystalline SiO_2-Ag+ and SiO_2-Cu+. This determines a quenching of the characteristic afterglow after an electrolysis of quartz by Ag+ and Cu+ ions. Such an effective concurrence in hole trapping process is only possible, if Cu+, Ag+ ions are located near to oxygen vacancy defects in the quartz lattice, which are connected with the 370 meV traps.

During electrolysis Ag+ and Cu+ replace monovalent interstitials like Me+ (Li+, Na+, K+). So we conclude, that prior to electrolysis a part of Me+ is located nearby oxygen vacancy defects, forming isolated defect complexes in crystalline quartz. These complexes may trap holes during irradiation, which are thermally activable by 370 meV, and produce furthermore by recombination an afterglow of luminescence in crystalline quartz. The traps with a characteristic 370 meV depth for crystalline quartz are not found in quartz glasses. We explain this a consequence of the drastically different structure of oxygen-associated defects in

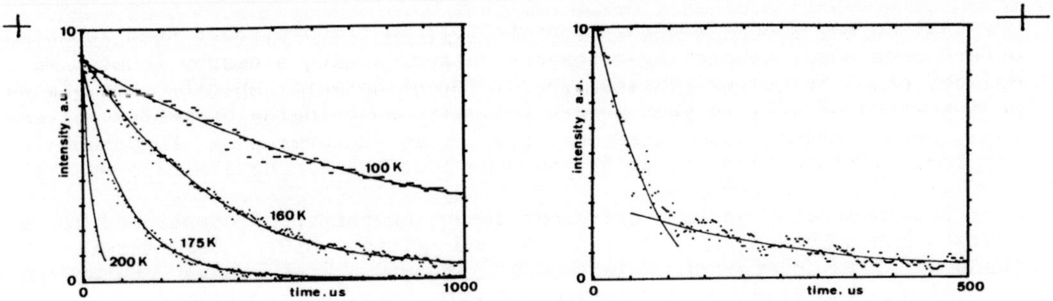

Fig.1. Afterglow of luminescence : crystalline SiO_2 (quartz)

glassy SiO_2

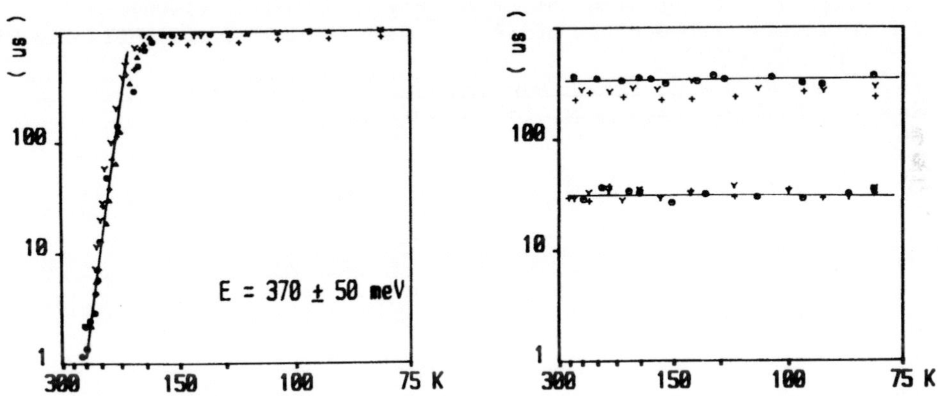

Fig.2. Temperature dependence of the afterglow (Arrhenius-plot)
of crystalline SiO_2
O Morion ; Δ Brazil
y synthetic ; + Ge-doped

of glassy SiO_2
O Suprasil ; y Corning
+ Plasmenij

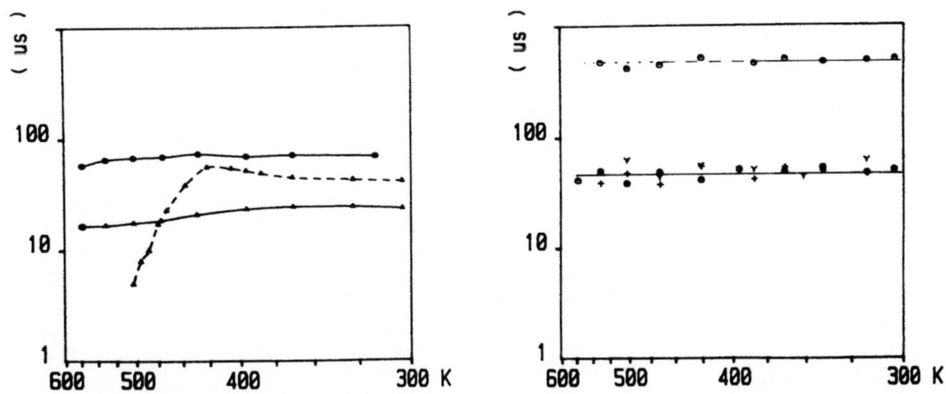

Fig.3. Temperature dependence of afterglow at high temperatures
O ——— Ag-quartz 4.75 eV
Δ ——— Ag-quartz 2.2 eV
Δ ---- Cu-quartz 3.4 eV

O Suprasil
y Corning
+ Plasmenij

crystalline and glassy modification of SiO_2. We propose, that such isolated defect complexes, consisting of oxygen vacancies with a nearby located Me+, do not exist in quartz glasses. The different defects, specific for glassy modification of SiO_2 (oxygen peroxy radicals, nonbridging oxygens a.o.) are the main oxygen associated defects. They determine a fundamentally different luminescence and afterglow behaviour than in crystalline quartz.

The new data on the different decay character of copper and silver center luminescence in crystalline and glassy SiO_2 demonstrate the fundamentally different electronic structure of these centers in crystalline and glassy modifications of SiO_2.

REFERENCES

1. G.H. Sigel, Jr., J. Non-Crystalline Solides, 13:372, (1974).
2. K. Tanimura, L.E. Halliburton, Phys. Rev. B 34:2933, (1986).
3. D.L. Griscom, Proceedings of 32 Frequency Control Symp., Atlantic City, N.J., p. 98, (1979).
4. W. Hohenau et al, J. Phys. Chem. Solids, 45:61, (1984).
5. W. Hohenau, Phys. Stat. Sol. (a) 88:267, (1985).

ON THE ROLE OF O_2^- IN THE LUMINESCENCE OF AMORPHOUS AND CRYSTALLINE SiO_2

M. Guzzi, M. Martini, F. Pio, G. Spinolo and A. Vedda

Dispartimento di Fisica
Università di Milano
Via Celoria 16
20133 Milano, Italy

INTRODUCTION

The study of defects in the various forms of SiO_2 has been greatly assisted by the use of optical methods. However, few conclusive assignments have been made of the known absorption or emission bands to particular defects. This is partly due to the difficulties in correlating the data obtained using different techniques.

In our research program on defects in SiO_2 we have been studying photolumincescence (PL), Thermally Stimulated Luminescence (TSL) and Phosphorescence (PP) as a function of different irradiations and thermal treatments : of the results obtained so far ; the identification of the O_2^- molecular ion as one of the luminescent centers[1] in silica appears the most interesting up to now.

In the present paper we report new data which confirm the role of O_2^- in the luminescence of both silica and quartz. Such a contribution was proposed in the 100°C TSL emission as well[2] ; this assignment is supported by the analysis of PP spectra at room temperature, which has now been carried out.

Of great interest appears to be the growth of the emission band at ~ 2.7 eV following UV irradiation that was suggested to correspond to the creation of a new center[3] ; our preliminary results indicate the possibility of an activation of pre-existing centers via electronic capture.

EXPERIMENTAL PROCEDURES

The material we used was natural and synthetic quartz (Premium quality bars by Sawyer Res. Products inc., Eastlake, Ohio) and various types of silica (mainly Herasil by Heraeus Amersil).

The TSL glow curves were obtained using a photon counting apparatus with a heating rate of 10°C/s starting from room temperature (RT). The presence of one or more peaks at about 100°C in the TSL glow curves of all the types of SiO_2 allows measurement of their spectral distribution through the analysis of the phosphorescence (PP) which is the RT decay of the 100°C TSL peaks. PP spectra were taken after irradiations using an exposure of ~ $7x10^5$ Roentgen by an X-ray tube operated at 35 KV.

PL spectra were recorded at temperatures from 300 down to 60 K, the excitation source being a 50 W deuterium lamp. Both PL and PP spectra were corrected for the spectral response of the optical system.

Further details of the experimental techniques and instrumentation are reported elsewhere[1,4].

RESULTS

a) Photoluminescence

Our crystalline samples, either synthetic or natural, showed no measurable luminescence under UV excitation, within the detection limits of our apparatus. On the other hand, in type I and II amorphous silica (which present the B_2 band in absorption[5]) we observed emission bands at ~ 4.2 eV, ~ 3.1 eV and ~ 2.7 eV ; following ref.1 we will refer to them as α, β and γ respectively.

In this section we report recent PL results regarding type II silica (Herasil) under 4.96 eV excitation : similar behaviour has been observed in type I silicas as well. In the RT emission spectrum only the α and β bands are present.

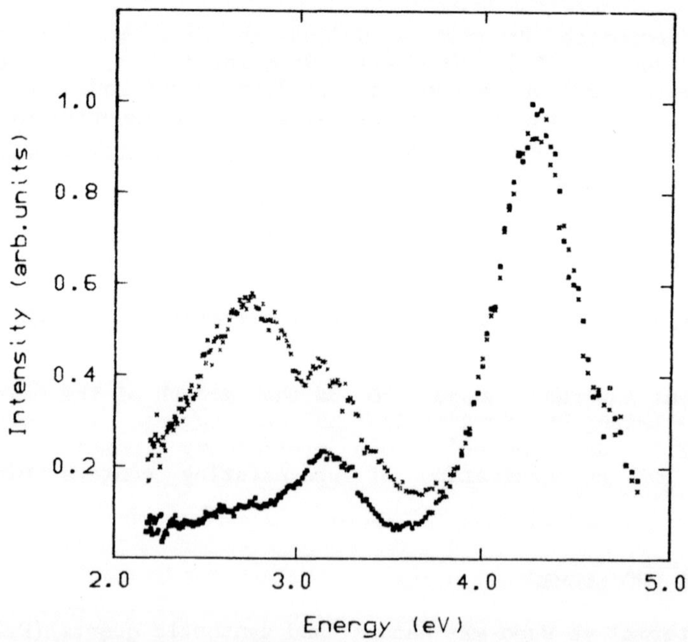

Fig. 1. LNT photoluminescence spectra of Type II Silica (Herasil) measured under 4.96 eV excitation. The two spectra are measured immediately after the sample cooling (squares) and after 5 hours exposure to the exciting photons (crosses).

The LNT photoluminescence is shown in fig.1, the α, β and γ emission bands are all present.

We observed a strong enhancement of the γ component after UV irradiation (4-6 eV) at LNT, while the intensity of the α band did not change. Fig.1 displays a 5-fold increase of the γ band after a 4-hours exposure to 4.96 eV photons (no saturation effect was observed) ;the γ band keeps increasing even after the UV excitation has been interrupted. In any case the structured shape of the PL emission in the 2-4 eV region [1] is confirmed, the vibrational subbands being spaced ~ 1250 cm^{-1}.

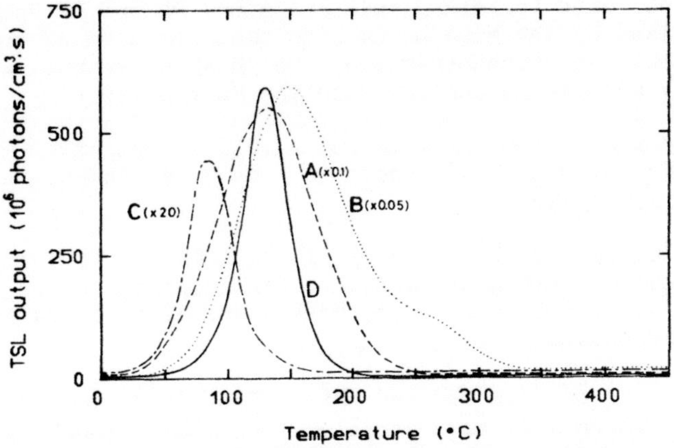

Fig.2. TSL glow curves of type I silica (A), type II silica (B) and Sawyer PQ quartz (C) 0.5 mm thick. Substraction of the black body emission has been performed. Heating rate = 10°C/s ; dose = 1.0 Gy from the ^{90}Sr-^{90}Y source. TSL glow curve of thermal SiO_2 film with thickness of 1.28 μm (D) is also shown for comparison.

However the data reported here add new elements to this analysis. Looking at fig.1 it appears as if the structures were present only when the γ band is more intense, possibly indicating that the structured band is the γ rather than the β one. A deconvolution of the spectrum results is rather

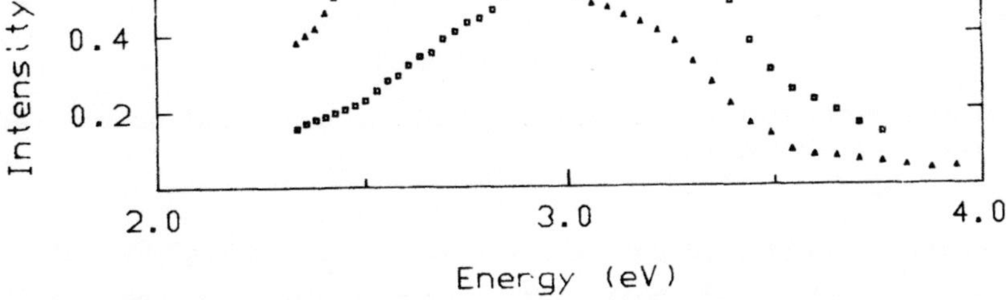

Fig. 3. Phosphorescence spectra of two different type I silica samples recorded at room temperature.

complicated ; nevertheless the present results are a further support to the proposal that the O_2^- molecular ion is one of the emitting centers in SiO_2.

b) Phosphorescence and thermally stimulated luminescence

When considering the TSL glow curves of the various kinds of SiO_2 a common feature is evident : a peak at 100-150°C is always present (see fig.2). This fact has been discussed elsewhere [4] leading to the conclusion that the traps and recombination centers related to the 100°C TSL peak are of intrinsic origin. The analysis of the room temperature phosphorescence spectrum can be used to study this emission. PL and PP spectra in silica are characterized by the same bands with the exception of the α band which is always absent in phosphorescence. No phosphorescence was detected in type III and IV silica, in analogy with the PL results.

Wide differences from sample to sample are observed in the relative intensity of the γ band with respect to β band, as shown in fig.3 for two different type I samples.

The 100°C TSL peak exhibits in quartz a typical increase of sensitivity known as "pre-dose" effect : this consists of an enhancement in its intensity as a response to a unit dose, after repeated irradiations followed by appropriate heating.

In a recent paper [2] we studied this effect in Sawyer synthetic quartz, in samples in which the concentrations of alkali ions (Li^+ and Na^+) had been increased or decreased by electrodiffusion. We gave evidence that only unswept or partially swept quartzes show the pre-dose effect and proposed a relation between the formation of $Al\text{-}OH^-$ centers and the 100°C TSL intensity. The enhancement of the luminescence is thought to be due to the following process : oxygen hole centers containing hydrogen are modified

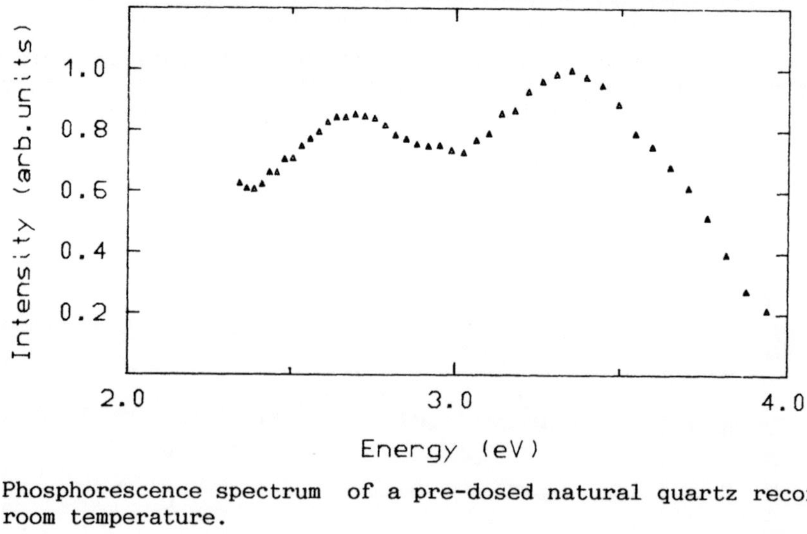

Fig. 4. Phosphorescence spectrum of a pre-dosed natural quartz recorded at room temperature.

into O_2^- molecular ions, responsible for the luminescence. The H^+ ions, made available in such a process, are responsible for the $Al\text{-}OH^-$ center formation. Such a model receives confirmation from the study of the PP emission. Up to now no PP spectra of quartz had been recorded, also due to the low defect concentration. As a consequence of the possible enhancement of luminescence following irradiations and heating (pre-dose effect), we have been able to observe the RT phosphorescence in a "pre-dosed" natural quartz, as shown in fig.4. The spectrum is very similar to the PP spectra of silica, even if much less intense, confirming the presence of the same luminescent centers. Due to the decreasing sensitivity of the PP optical system, only an indication can be given of an additional peak below 2.3 eV.

CONCLUSIONS

The analysis of PL, PP and TSL in quartz and silica gives analogous results for both of them.

The evidence for O_2^- structure in the PL spectra of silica, together with the similarity between the various luminescences and emissions in quartz and silica, is in good agreement with the proposal that O_2^- participates to both PL and PP in SiO_2.

Of great relevance is the growth effect on the 2.7 eV band after UV irradiation (see fig. 1). Such an effect has recently been discussed by Stathis and Kastner [3] who reveal a 2.8 eV band in the PL spectra at low temperature and not at RT in agreement with our recent results. [1] They propose that UV excitation creates new defects as does electron beam[6] and X-ray [7]excitation. Our results agree with such a proposal. However, the growth of the 2.7 eV band even in the absence of activation, suggests that thermal release of carriers from very shallow traps may transform suitable precursors into emitting centers. The role of O_2^- in this mechanism is in agreement with the interpretation of both Tanimura et al.[6] and Hayes et al. [7] who consider the 2.8 eV emission as related to oxygen vacancy-interstitial pair formation.

ACKNOWLEDGEMENTS

We are grateful to Miss C. Melosi for the help in the recent PL measurements. This work has been supported by the GNSM of the Italian National Council of Research (CNR) and by the CISM of the Italian Ministry of Education (MPI).

REFERENCES

1. M. Guzzi, M. Martini, M. Mattaini, F. Pio and G. Spinolo, Phys. Rev. B 36: (1987).
2. M. Martini, G. Spinolo and A. Vedda, J. Appl. Phys. 61:2486 (1987).
3. J.H. Stathis and M.A. Kastner, Phys. Rev. B 35:2972 (1987).
4. M. Martini, G. Spinolo and A. Vedda, J. Appl. Phys. 60:1705 (1986).
5. R. Bruckner, J. Non-Cryst. Solids 5:123 (1970).
6. K. Tanimura, T. Tanaka and N. Itoh, Phys. Rev. Lett. 51:423 (1983).
7. W. Hayes, M.J. Kane, O. Salminen, R.L. Wood and S.P. Doherty, J. Phys. C 17:2943 (1984).

TRANSFORMATION OF RADIATION INDUCED DEFECT CENTERS AS A PROBE OF MOLECULAR DIFFUSION IN a-SiO_2

R.L. Pfeffer

US Army Electronic Technology and Devices Laboratory (LABCOM)
Fort Monmouth, NJ 07703-5000
USA

INTRODUCTION

An important cause of ionizing radiation induced degradation in MOS integrated circuit devices is the buildup of trapped positive charge in gate and field oxides[1]. Recent electron paramagnetic resonance (EPR) studies[2] have identified virtually all trapping sites in the oxides as E' centers (paramagnetic bridging oxygen vacancies). The formation of these vacancies during irradiation by energetic photons or electrons is predominantly by ionization processes[3], i.e., bond rearrangements after electron-electron collisions, rather than displacements from electron-atom collisions.

The annealing of E' centers in dry a-SiO_2 occurs through two distinct routes. At temperatures over 700°C the annealing process, which is irreversible, involves thermally activated local bond rearrangement[4]. However an alternate annealing process occurs at lower temperatures (about 300°C). This process, which is reversible by re-exposure[5], appears to be a nonlocal one involving the thermally activated diffusion of some species to the E' site and its subsequent reaction to form a metastable entity, rather than merely reflecting a change in the E' electrical state such as hole detrapping. Edwards and Fowler[6] have suggested that low-temperature annealing in dry SiO_2 is mediated by the diffusion of dissolved O_2 and its subsequent capture at E' centers to form superoxide radicals (E' + O_2 → SiOO·). Various annealing studies have been performed to test this idea[7] ; these have been either EPR observations of E' concentrations in MOS oxides or bulk SiO_2 or of oxide trapped charge densities in MOS structures. Results have been consistent with it in general but have failed to agree in detail with the predictions of a diffusion-limited bimolecular reaction model[8]. A distribution of activation energies[9] which purportedly characterize O_2 diffusion in SiO_2 has frequently been invoked to ameliorate the discrepancy. We have sought a physical basis for the seemingly ad hoc assumption that a distribution of activation energies exists through a computer simulation of the diffusion process based on the Anderson-Stuart model[10]. We found[11] that even a wide distribution of doorway and interstice radii in the SiO_2 network yielded diffusion characterized by a single effective activation energy.

To improve our modeling of the diffusion and annealing process we have conducted annealing experiments involving bulk samples of dry a-SiO_2 which were specially prepared with widely differing amounts of dissolved O_2. Our intention was to modify the annealing kinetics by controlling the average path length of a diffusing molecule. In addition to tracking the evolution of E' center concentrations we also monitored that of SiOO· centers. After conducting an initial experiment[12], we are engaged in several followup studies using samples pre-loaded with different gas concentrations, irradiated to different doses, and isochronally and isothermally annealed. We here review the results of the initial experiment and report on the first of these followups.

INITIAL EXPERIMENT

Procedure

The samples were 4 x 23 x 0.5 mm slabs of Suprasil W1 (Heraeus-Amersil, Somerville, NJ), a very low-OH fused silica. One sample was degassed by a 900°C vacuum bake at 7 mTorr for 5 days, one was oxygen-stuffed at 900°C for 5 days at 8000 psi, and one was left untreated. All then received room-temperature irradiations of 10^8 rad(Si) by gamma rays of 1.17 MeV and 1.22 MeV from a ^{60}Co source at a dose rate of 0.7 Mrad/hr. After a 6-month period at room temperature, a sequence of cumulative 10-min isochronal anneals was conducted in air on all samples, with spin concentrations measured after each by EPR. The spectrometer was a standard Varian E-9 type operating near 9.1 GHz in the absorption mode at a modulation frequency of 100 kHz, with the samples at -170°C in a TE-102 cavity. The E' signals were recorded using a 0.4 G modulation amplitude and a 0.2 μW power level (switching to 1 mW out of phase for the weakest signals) ; for the SiOO· signals a 1 G modulation amplitude a 40 mW power level were used.

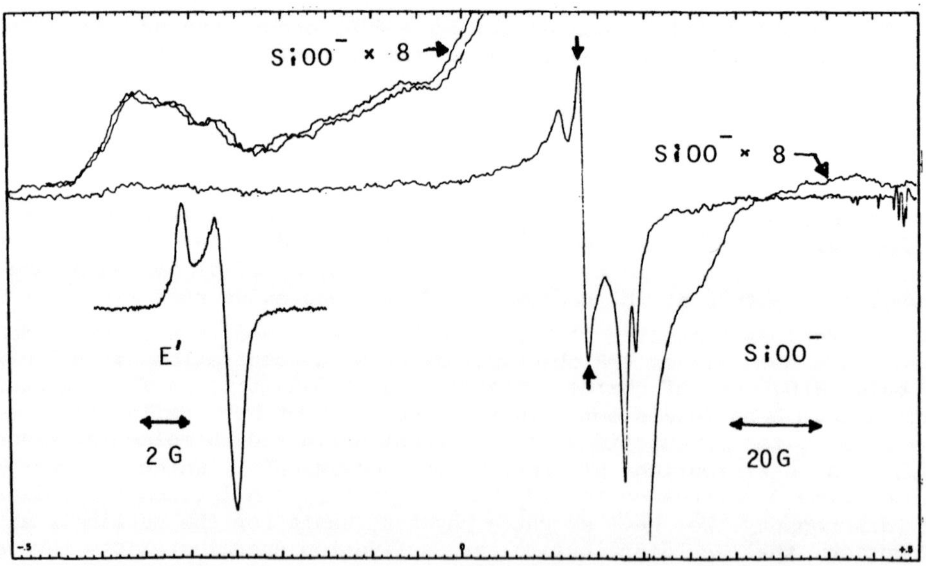

Fig. 1. Representative EPR traces (here from the untreated sample after 10 min at 70°C). The spin densities were gotten from the peak-to-peak resonance amplitudes, e.g. the arrows shown for SiOO·. (C.f. refs. 13, 14 for detailed discussions of these resonances).

The spin concentrations (total [OHC] and [E']) were determined in one sample by numerical integration and comparison with a strong pitch standard, with microwave saturation effects taken into account. The [SiOO·] for that sample was estimated as a fraction of the total [OHC] by comparison with previous simulations. The relative [E'] and [SiOO·] in all other cases were obtained by comparing the peak-to-peak amplitudes of each type of resonance to those in that sample.

RESULTS

The fraction of E' centers remaining in each sample after each anneal is shown in fig. 2. It is evident that the E' centers persist to a much higher temperature in the degassed sample than in the others ; however their annealing kinetics in the oxygen-stuffed and untreated samples are almost indistinguishable. It might be thought that the stuffing has little effect on the properties of the SiO_2. However this is not the case : as can be seen from the absolute concentrations shown in fig.3, the stuffed sample has five times the defect density of the untreated one.

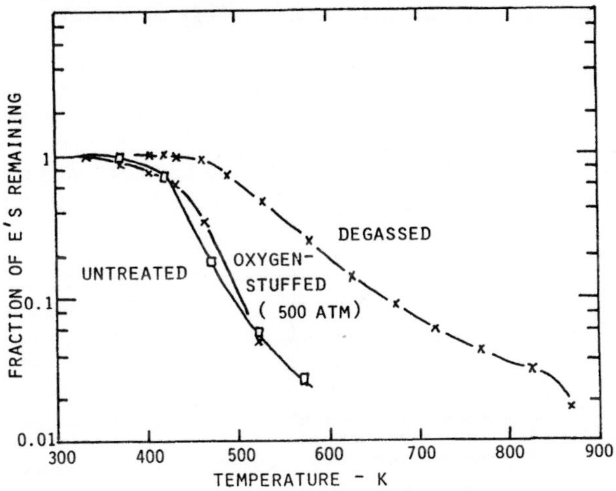

Fig. 2. Fraction of E' centers remaining in Suprasil W1 samples after 10 min isochronal air anneals at indicated temperatures. The untreated sample is a close match to Griscom's E'_γ curve for Suprasil W1 (ref.15).

The results of this initial experiment qualitatively support the Edwards-Fowler model : as is evident from fig. 3, (1) all SiOO· densities show increases which exactly match corresponding decreases in E' densities, and (2) there is a pronounced persistence of E' centers at higher temperatures in the degassed sample. Quantitatively, however, this data provided a poor fit to Waite's diffusion-limited bimolecular reaction model.

FOLLOWUP

Procedure

As before, the samples were slabs of Suprasil W1. This time, however, all samples were degassed by a 900°C vacuum bake at 10^{-6} Torr for 8.5 days.

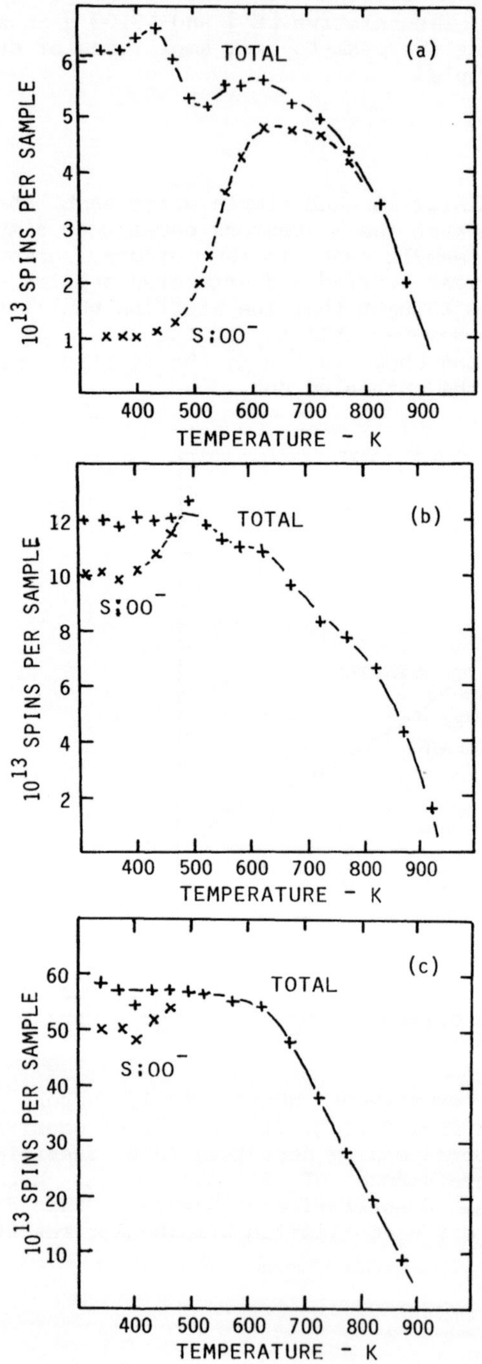

Fig. 3. Absolute spin concentrations in Suprasil W1 samples after 10 min isochronal air anneals at indicated temperatures. Curves labelled TOTAL are the sum of contributions from SiOO· and E' centers. Note that these curves are relatively constant for temperatures up to 600 K suggesting transformation of E' into SiOO .
(a) degassed sample,
(b) untreated sample,
(c) oxygen-stuffed sample (500 atm).

Then one (sample O_1) was oxygen-stuffed at 875°C for 8 days at 14.7 psi, another (sample O_{22}) at 850°C for 7 days at 370 psi, a third (sample O_{500}) at 850°C for 8 days at 7580 psi, and a fourth received no further treatment. All were then ^{60}Co-irradiated as before. After a 3 week period at room temperature, a sequence of cumulative 10-min isochronal anneals was conducted in air on all samples, and EPR spectra were recorded as before except that a (011) cylindrical cavity was used at a temperature of -135°C.

The E' concentration was determined in one sample by numerical integration and comparison with a strong pitch standard, with microwave saturation effects taken into account. This, time, however, the absolute [SiOO·] was not measured, athough the relative [E'] and [SiOO·] vs. anneal temperature were obtained by comparing the peak-to-peak amplitudes of each type of resonance.

Results

The fraction of E' centers remaining in each sample after each anneal is shown in fig.4. The difference from fig.2 is apparent : these samples all show a remarkable exponential fall off in the E' fraction with increasing anneal temperature, implying an eventual exponential decrease in [E'] with time whose rate constant depends only on the initial ratio of [E'] to $[O_2]$ dissolved, and <u>not</u> on temperature.

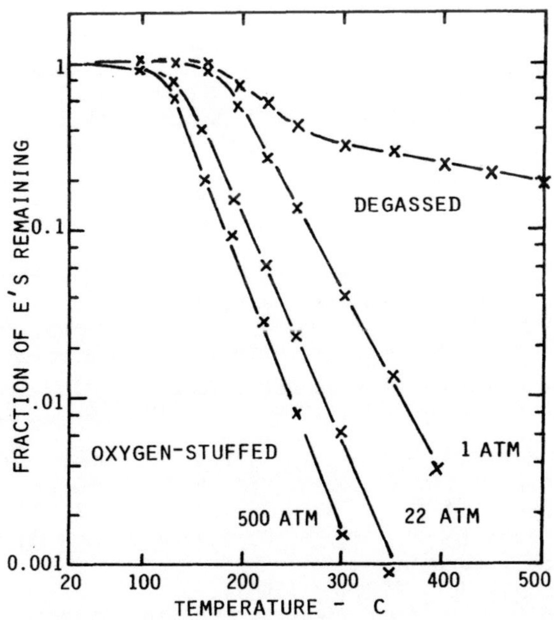

Fig.4. Fraction of E' centers remaining in previously degassed Suprasil W1 samples after 10 min isochronal air anneals at indicated temperatures. The actual initial concentrations were $4.7(10)^{15}$ cm^{-3} (degassed sample), $4.7(10)^{15}$ cm^{-3} (1 atm sample), $17(10)^{15}$ cm^{-3} (22 atm sample), and $34(10)^{15}$ cm^{-3} (500 atm sample). Note the exponential falloffs.

As before, all increases in [SiOO·] tracked corresponding decreases in [E'] ; although a one-to-one replacement could not be verified (the absolute [SiOO·] was not measured) a similar trend was observed. All the SiOO· centers annealed at the same temperature (much higher than the E's), suggesting a local rather than an oxygen diffusion mediated process. And finally each of the pre-anneal defect concentrations, i.e., both the initial [E'] and the initial [SiOO·], varied directly with dissolved oxygen content.

CONCLUSIONS

1. The annealing kinetics of E' centers in dry a-SiO_2 are seen to be generally consistent with the Edwards-Fowler model of an E' to SiOO transformation mediated by the diffusion of dissolved molecular oxygen.

2. The data suggest that Suprasil W1 as manufactured appears to contain a significant amount of dissolved gas whose identity and concentration are undetermined. Previous difficulties in reconciling anneal data with a diffusivity characterized by a single activation energy are thus probably attributable to use of a poorly characterized material, and are likely to be resolved by a thorough purge of this gas by appropriate preparation.

3. High defect densities seen in high-oxygen-content samples imply that MOS oxides grown under high pressure (HIPOX) are likely to be radiation-soft unless properly annealed.

ACKNOWLEDGMENTS

The author is grateful to M. Harmatz, P. Caplan, D. Brocking, E. Potenziani, and J. Conrad for sample preparations ; to J. Rodgers and the staff of ATT Bell Laboratories (Murray Hill) for irradiations ; to G. Gerardi for spin concentration determination, and above all to D.R. Griscom for his invaluable guidance.

REFERENCES

1. E.H. Nicollian and J.R. Brews, "MOS Physics and Technology" Wiley, New York, (1982), chap. 11 and refs. cited.
2. P.M. Lenahan and P.V. Dressendorfer, J. Appl. Phys. 55:3495 (1984).
3. R.L. Pfeffer, J. Appl. Phys. 57:5176 (1985) and refs. cited.
4. R.A.B. Devine and C. Fiori, J. Appl. Phys. 58:3368 (1985).
5. A. Golanski, R.A.B. Devine, and J.C. Oberlin, J. Appl. Phys. 56:1572 (1984).
6. A.H. Edwards and W.B. Fowler, Phys. Rev. B56:6649 (1982).
7. D.B. Brown, D.I. Ma, C.M. Dozier, and M.C. Peckerar, IEEE, Trans. Nucl. Sci. NS-30:4059 (1983) ; A. Golanski and T. Nicolle, in "Induced Defects in Insulators", ed. by P. Mazzoldi, Les Editions de Physique, Les Ulis (1984), p. 255.
8. T.R. Waite, Phys. Rev. 107:463 (1957).
9. R.A.B. Devine, J. Appl. Phys. 58:716 (1984), A. Golanski and T. Nicolle, op cit.
10. O.L. Anderson and D.A. Stuart, J. Am. Ceram. Soc. 37:573 (1954).
11. R.L. Pfeffer, unpublished.
12. R.L. Pfeffer and D.L. Griscom, unpublished.
13. D.L. Griscom, Phys. Rev. B 20:1873 (1979).
14. M. Stapelbroek, D.L. Griscom, E.J. Friebele, and G.H. Sigel J. Noncryst. Solids 32:313 (1979).
15. D.L. Griscom, Nucl. Instrum. Methods Phys. Res. B 1:481 (1984).

OBSERVATION OF THE NEUTRAL OXYGEN VACANCY IN SILICON DIOXIDE

M.E. Zvanut, F.J. Feigl, W.B. Fowler, and J.K. Rudra

Department of Physics and Sherman Fairchild Laboratory
Lehigh University
Bethlehem, PA, 18015
USA

ABSTRACT

We have studied electron-trapping defects in SiO_2 films incorporated into Metal-Oxide-Silicon capacitors. The film consisted of SiO_2 sputtered onto the native oxide formed on silicon. A process of direct band-to-trap tunneling of electrons into defects within the oxide film was isolated. This trap filling process, and the reverse emptying process, were analyzed using the assumptions that the trapping defects had a fixed spatial location and an extended energy level distribution. For a trap located at the sputtered SiO_2/native SiO_2 interface, we found a trap depth of 4 eV and a trap relaxation of 1 eV upon capture of an electron. These experimental results are consistent with theoretical calculations for the formation of a neutral oxygen vacancy in quartz by the addition of an electron to a positively-charged oxygen vacancy (that is, to an E_1' center). We therefore may have obtained the first direct experimental evidence for a neutral oxygen vacancy in SiO_2.

INTRODUCTION

We have used the techniques and analyses appropriate to the metal-oxide-semiconductor (MOS) microelectronics technology to study the transfer of electrons between band states within a silicon crystal and localized band-gap states within an overlying silicon dioxide film. We have also determined the trap depth and the relaxation energy, upon capture of an electron, for the defect responsible for the localized states. These results compare favorably with the results of theoretical computations on oxygen vacancies in crystal silicon dioxide (alpha-quartz). This indicates that the oxide film defect may be some form of an oxygen vacancy, such as a dangling silicon defect associated with the interface between two distinct silicon dioxide films. If so, we have finally experimentally isolated the neutral state of a vacancy and therefore the "precursor" for an E' center[1].

EXPERIMENTAL DETAILS

The SiO_2 films were ion beam sputtered from a bulk glassy SiO_2 target in high vacuum and deposited onto polished (100) or (111) silicon substrates at room temperature. Both p-type and n-type substrates with resistivities of 0.2, 3, or 70 ohm-cm were used. They were not acid cleaned prior to deposition ; therefore, a 0.5-1.5 nm thick native oxide layer existed between the Si surface and the sputtered SiO_2 film[2,3]. The sputtered SiO_2 was deposited to a measured total oxide thickness of 40 to 60 nm[4]. A matrix of MOS research capacitors was fabricated on each substrate by sputtering onto the surface of the deposited SiO_2 an array of circular Al electrodes, each of area 0.01 cm^2. In order to make electrical contact, a thick coating of Al was also sputtered onto the back of each substrate. The entire MOS fabrication sequence, including the SiO_2 sputter deposition and the Al metallization steps, was executed at the Honeywell Physical Science Center in Minneapolis, Minnesota, USA under the direction of Dr. J. David Zook.

Standard instrumentation and computer interfacing were used to perform constant-current, constant-bias, and isochronal electron injection on individual research capacitors. These processes provide means of introducing electrons into the oxide from the aluminum electrode or the silicon substrate[5]. All the results to be discussed below concern injection of electrons from the substrate. Periodically during the experiment, the injection is interrupted and the capacitance (C) and/or the current (I) is measured as a function of applied voltage (V). Any shift of the I-V or C-V characteristic along the voltage axis reflects the presence of charge trapped in the oxide film or at the oxide-silicon interface. Quantitatively, the centroid-weighted amount of oxide trapped charge Q_{ot} is related to the voltage shift ΔV through the equation[5] $Q_{ot} = x_1 Q_0 = -C_{ox}\Delta V$. In this equation, Q_0 is the net amount (zeroth moment)of the charge distribution within the oxide film, x_1 is the charge distribution centroid (first moment) normalized to the oxide thickness, and C_{ox} is the parallel-plate capacitance of the oxide dielectric film. A comparison of the voltage shift from the C-V data with that from the I-V data may provide an estimate of the position of the trapped charge within the oxide film[5,6].

EXPERIMENTAL RESULTS

Experimental dark current I-V characteristics and 1 MHz C-V characteristics of an as-prepared research capacitor were generally similar to those of MOS capacitors incorporating conventional thermal oxide dielectrics. The I-V and C-V data indicated that positive charge exists within the as-prepared oxides and that negative oxide charge is generated under current or voltage stress.

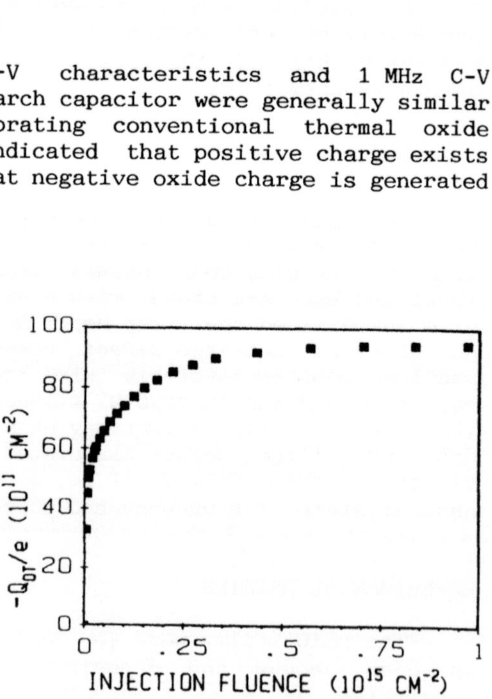

Fig. 1. Oxide charging data from a constant current injection sequence on an MOS capacitor fabricated on p-type Si. Data is presented as net oxide charge $-Q_{ot}/e$ vs. electron fluence Jt/e, both in units of electronic charges/cm^2.

Figure 1 presents data from a constant-current injection sequence. The density of oxide trapped charge, Q_{ot}/e, is plotted as a function of the injected electron fluence. This experiment is done at gate voltages sufficiently high to produce band-to-band tunneling of electrons from the Si into the SiO_2, and is done under constant current conditions. Therefore the fluence, the number of electrons/cm^2 passed through the oxide film, was calculated as the product of the electron current particle flux, J/e, and the time, t.

Figure 2 illustrates the results of an isochronal bias-stress sequence executed on a single capacitor. The as-fabricated capacitor was subjected to an initial one-second stress at systematically reduced biases, to a minimum of -55 V. This is the sequence labelled "empty" in fig. 2. Subsequently, the capacitor was subjected to a sequence of one-second stresses at systematically increased biases, to a maximum of +48 V. This is the sequence labelled "fill" in fig. 2. Variations of this experiment were executed on MOS capacitors with varying substrates and oxides, and for a number of different starting conditions. The detailed shape of the fill and empty curves was similar for all of these cases. The fill and empty sequences were also studied by using a separate capacitor for each bias in the sequence. To account for the variation among capacitors, each measurement at a selected stress voltage was repeated on several capacitors. Within significant statistical error, the fill and empty curves obtained using data from several capacitors agreed with those in fig. 2.

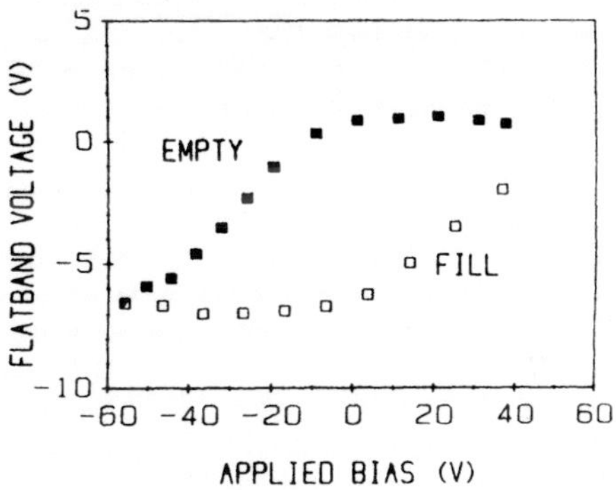

Fig. 2. Data from an isochronal bias-stress sequence on an MOS research capacitor fabricated on n-type Si. The applied bias is increased in the FILL sequence (open points) and decreased in the EMPTY sequence (solid points).

The voltage shifts determined from C-V and I-V data measured during the fill and empty sequences were approximately equal, indicating that the charge is more than 0.5 nm from the Si-SiO_2 interface[6]. Attempts to study the time dependence of C-V voltage shifts under constant-bias conditions indicated that the fill and empty transient times were less than the time resolution of our experiments (of order one second). In terms of the band-to-trap tunneling model to be presented below, this result indicates that the positive oxide charge is less than 3 nm from the Si-SiO_2 interface[7].

ANALYSIS OF EXPERIMENTAL DATA

The constant current injection data shown in fig. 1 consists of two distinct regions. One is an increase of negative trapped charge from 6.0 to 8.5×10^{12} electrons/cm^2. This increase scales with the fluence. It is therefore the result of electron capture from mobile states in the conduction band of the SiO_2 into localized states in the band-gap. The mobile electrons were generated by band-to-band tunneling from the conduction band of the Si substrate (dark current injection). The trapping process was analyzed as a first order kinetic process, as is routinely done

Fig. 3. Schematic diagram of the process of direct tunneling of electrons from the surface inversion or accumula- layer in Si into traps in the SiO_2 (A-C), and the reverse process of trap-to-band tunneling after relaxation (D-F).

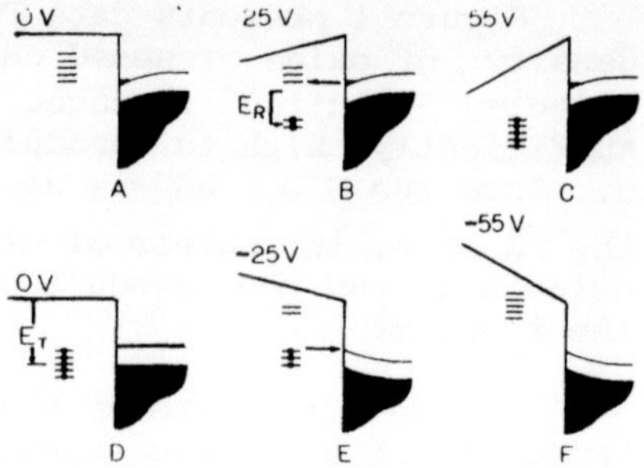

for trapping during constant current injection[8]. This will not be discussed further.

The other region of the constant current injection curve is an increase of negative trapped charge from 0 to 6 x 10^{12} electrons/cm^2. This increase scaled with time, and not with injected electron fluence. Furthermore, the amount of oxide trapped charge Q_{ot}/e produced by this process exceeded the injected electron fluence Jt/e. This process has been analyzed as a direct tunneling of electrons from the silicon substrate to defects within the SiO_2 film. For defects located spatially very close to the SiO_2/Si interface such a band-to-trap tunneling process can occur with Q_{ot}/e greater than Jt/e.

We have developed a model for the band-to-trap tunneling process, based on the two assumptions that the electron trapping defects have a fixed spatial location within the MOS oxide film and an extended energy level distribution within the SiO_2 band gap. The energy level distribution is presumed to be associated with electron trapping defects which are spatially localized at the interface between the native oxide and the sputtered SiO_2. The "fill" and "empty" sequences of isochronal bias-stress data in fig. 2 can be interpreted in terms of two distinct distributions of trapping state energies. In these experiments, a trap level distribution is determined from the variation of the amount of trapped charge with the applied bias voltage. The model, illustrated schematically in figure 3, may be described as follows.

In the "fill" sequence, the effect of the applied bias is to tilt the bands of the SiO_2 resulting in a change of the trap level with respect to the Si conduction band edge. This, band, located 3.25 eV below the SiO_2 conduction band edge, contains either inversion layer electrons (p-type substrates) or accumulation layer electrons (n-type substrates). Therefore, it provides the source of trapped electrons in both resistivity types. When the traps are empty, as at negative bias, the entire energy level distribution must be above the Si conduction band (fig. 3A). As the applied bias is increased to 25 V, for example the oxide bands tilt such that a portion of the distribution may be filled by electrons (fig. 3B). Finally, after sufficiently high bias the energy level distribution is completely filled. A further increase in the field has no effect. At this point, the entire distribution must be located below the Si conduction band edge (fig. 3C).

After capturing an electron the energy of a trap does not remain fixed at its original level below the SiO_2 conduction band edge. Rather, the defect structure relaxes and its energy level drops by an amount E_r to the thermodynamic equilibrium value, E_t, for the occupied trap. Thus, it takes a relatively large change in bias to empty the trap level distribution. This is seen at positive bias values of the "empty" sequence data in fig. 2 and schematically in fig. 3D. As the applied bias is decreased to negative values, the oxide bands tilt such that a portion of the distribution may be emptied as electrons tunnel to the Si conduction band (fig. 3E). Finally, after sufficiently low bias the energy level distribution is completely emptied. A further decrease in the field has no effect. At this point, the entire distribution must again be located above the Si conduction band edge (fig. 3F). Note that the bias is of opposite polarity during the fill and empty processes. Therefore, the Si conduction band may be, in the former case, the initial state and, in the latter case, the final state of the tunneling process.

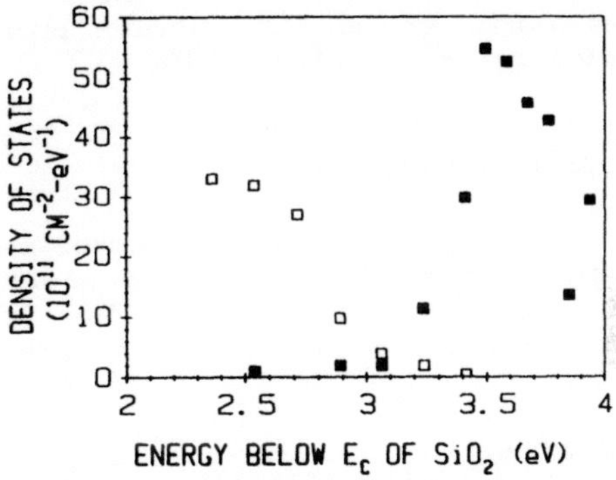

Fig. 4. Results of calculation of the trap state distributions for the unrelaxed and relaxed electron trapping defect. E_t - E_r (open points) is obtained from the FILL data and E_t (solid points) is from the EMPTY data on fig. 2.

In summary, the model states that an electron tunnels <u>into</u> an <u>unrelaxed</u> defect configuration located at a energy E_t - E_r below the SiO_2 conduction band. However, the electron tunnels <u>out of</u> the <u>relaxed</u> state of the defect, located at trap depth E_t. Using appropriate equations and numbers generated from the data in fig. 2, it is possible to determine both the unrelaxed and relaxed energy distributions. The results are shown in figure 4, assuming that all traps are located 1 nm from the native oxide/silicon interface.

THEORETICAL MODEL

Theoretical calculations indicate that the results in fig.4 are reasonable for oxygen vacancy defects in SiO_2 [9]. These results consist of semiempirical molecular structure calculations of the large, asymmetric relaxation known to be characteristic of these defects[10]. In particular, we have investigated singlet and triplet states of the neutral oxygen vacancy in crystalline alpha-quartz. The calculations extended over a range of molecular relaxations spanning the equilibrium configurations of both the singlet and triplet neutral state and the E_1' paramagnetic positive state of this defect. These calculations predict that the energies associated with the capture of an electron on the E_1' center and its subsequent relaxation are E_t = 5 eV and E_r = 1.5 eV. These theoretical results are to be compared with our experimental results for E_t and E_t-E_r in fig. 4.

DISCUSSION AND CONCLUSIONS

In analyzing the isochronal bias-stress data, we have assumed that electron trapping defects were located at the native oxide/sputtered oxide interface. As mentioned previously, the tunneling model requires that the location of the traps is 0.5-3 nm from the native oxide/Si interface. Furthermore, the tunneling instability reported herein is unique in our experience with MOS SiO_2 oxides. What is unique about these oxides is the room temperature native oxide/sputtered oxide interface, located approximately 1 nm from the Si surface[3]. We have concluded therefore, that this region is the most probable location of the trapping defect.

We would like to suggest here that the trapping defect is an E' center[11,12,13]. This center has been observed in sputtered SiO_2 [14] and in irradiated thermal oxide films[15,16] via electron paramagnetic resonance. They have been similarly observed, in concentrations appropriate to our data, in films which we have studied ; this was done by P.J. Caplan and E.H. Poindexter. In conclusion, we suggest that the trapping center located at or near the native SiO_2/sputtered SiO_2 interface is an E' defect. The results reported herein then constitute the first direct experimental observation of an E' center precursor and possibly the neutral oxygen vacancy defect in glassy SiO_2.

ACKNOWLEDGEMENTS

This research was supported in part by the US Navy Office of Naval Research. Miss Zvanut gratefully acknowledges support as a Graduate Fellow from the IBM Corporation. The full cooperation of Drs. J.D. Zook and E.H. Poindexter were critical to these studies.

REFERENCES

1. M.G. Jani, R.B. Bossoli, and L.E. Halliburton, Phys. Rev. B27:2285 (1983).
2. S.I. Raider, R. Flitsch, and M.J. Palmer, J. Electrochem. Soc. 122:413 (1975).
3. A. Ourmazd, D.W. Taylor, J.A. Rentschler, and J. Bevk, Bull. Am. Phys. Soc. 32: 774 (1987).
4. A. Kalb in "Laser Topics" (Summer, 1986), page 11.
5. F.J. Feigl in N. Einspruch and G. Larrabee, eds. "VLSI Electronics Microstructure Science", Vol. 6 Academic, New York, 1983, page 119.
6. P. Solomon, J. Appl. Phys. 48:3843 (1977).
7. E.C. Ross and J.T. Wallmark, RCA Rev. 30:366 (1966).
8. F.J. Feigl, D.R. Young, D.J. DiMaria, S. Lai, and J. Calise, J. Appl. Phys. 52:5665 (1981).
9. J. Rudra (PhD dissertation, Lehigh University, 1986), unpublished.
10. W.B. Fowler, J. Rudra, A. Edwards, and F. Feigl, these proceedings.
11. F. Feigl, W.B. Fowler, and K. Yip, Sol. State Comm. 14:225 (1974).
12. J. Rudra, W. Fowler, and F. Feigl, Phys. Rev. Lett. 55:2614 (1985).
13. M.H. Woods and R. Williams, J. Appl. Phys. 47:1082 (1976).
14. T.W. Hickmott, J. Appl. Phys. 42:2543 (1971).
15. D. Griscom, J. Appl. Phys. 58:2524 (1985).
16. P.M. Lenahan and P. Dressendorfer, J. Appl. Phys. 55:3495 (1984).

NEW INSIGHT INTO THE STRUCTURE OF SiO_2 GLASS FROM A POINT DEFECT STUDY

K. Nagasawa*, H. Mizuno, Y. Yamasaka, R. Tohmon,
Y. Ohki and Y. Hama

*Sagami Institute of Technology
Fujisawa, Kanagawa
Waseda University
Shinjuku-ku, Tokyo
Japan

INTRODUCTION

Formation of defect centers in pure silica glass (a-SiO_2) depends greatly on the manufacturing process of the glass. Non-bridging oxygen hole centers (NBOHC : ≡Si-O•) are dominantly created by γ-ray irradiation in high OH-group content (≥ 700 ppm) silica made by the direct glass deposition process[1]. In low OH-group content silica made by the plasma or soot method, oxygen content in the glass strongly affects the defect formation[2,3] ; drawing induced peroxy radicals, drawing induced NBOHC, γ-ray induced peroxy radicals and the 1.52 μm band induced by hydrogen treatment are created in oxygen-rich silica, but the 5 eV (245 nm) band and drawing induced E' center (≡Si•) are created in oxygen-deficient silica. Based on these experimental results, the authors proposed[2] the defect in the form of ≡Si-Si≡ as a model of the oxygen deficiency. In the present paper, further experimental evidence which supports the authors' model and its proof by numerical calculation are presented. The structure of glass is also discussed.

EXPERIMENTAL

Samples used are listed in Table 1. They were rod shaped with a diameter of 10 to 15 mm and a length of 40 to 45 mm. The Df8, P3f and Prf-A samples were irradiated at room temperature using a photon beam with a wavelength of 248 nm (KrF excimer laser). The laser was flashed in a series of pulses with a frequency of 2 Hz up to the total number of 200 to 600 pulses. The energy of one shot was 40 mJ. The optical absorption and electron-spin-resonance (ESR) signals were measured before and after laser irradiation. The optical measurement was performed in the wavelength region from 200 to 800 nm using a double-beam spectrometer. Further, using samples with a disk shape (10 to 15 mm Φ, 0.6 to 1 mm thick), vacuum-ultraviolet spectroscopy was measured before the laser irradiation.

METHOD OF NUMERICAL CALCULATION

To examine the energy diagram of the structure, ≡Si-Si≡, numerical calculation was done using the Ab Initio method of molecular orbitals. The structure, $(HO)_3Si\text{-}Si(OH)_3$ was used as a model for calculation.

Table I. Sample list

Name	Manufacturing method	Cl content [ppm]	OH-group content [ppm]	7.6eV band	5.0eV band⁺	γ-ray or drawing induced peroxy radicals
Sff-A	soot	free	free	o	o	x
Sff-B	soot	free	free	o	o	x
Sf1-A	soot	free	60	x	x	o
Sf1-B	soot	free	(60)	x	o	x
S1f-A	soot	(100)	free	o	o	x
S1f-B	soot	(100)	free	o	o	x
S11	soot	(100)	(100)	x	o	x
Df8	direct	free	800	x	x	x
D17	direct	100	700	x	x	x
P3f	O_2 plasma	300	free	x	x	o
P6f-A	O_2 plasma	600	free	x	x	o
P6f-B	Ar plasma	600	free	o	o	x
Prf-A	Ar plasma	4000	free	o	o	x
Prf-B	Ar plasma	3900	free	o	o	x

free : less than a few ppm, ( ) : estimated value,
o : observed,
x : not observed,
⁺ : judged from the occurence of photoluminescence by excitation at 5eV

First, optimization of the structure in the ground state was made with the GAUSSIAN 80 : HITAC Version. The energy in the excited state was then calculated for the structure obtained by using an improved program, based on the GAUSSIAN 76 : IBM Version[4], which is capable of calculation based on the configuration interaction method with single-electron excitation, or the SE-CI method[5]. In the calculation of the Hartreee-Fock orbitals, the highest 19 occupied orbitals and the lowest 13 unoccupied orbital were considered.

Figure 1 shows the model structure for calculation, where X denotes the oxygen vacancy. Two X-Si$(OH)_3$ parts form two regular tetrahedrons, twisted with respect to each other at 60°. The angle Si-X-Si was set to 144°, which is the average bonding angle of Si-O-Si in amorphous glass[6]. The distance between an oxygen atom and the center of gravity of each tetrahedron, G, was kept constant at 1.6 Å, the regular distance between silicon and oxygen atoms in glass. Hydrogen atoms were put in the X-G-O plane, at the point 1.0 Å away from the oxygen atom.

It was hypothesized that two silicon atoms can move fairly freely, maintaining symmetry with respect to X, as they are the nearest neighbors of X, and that oxygen atoms are fixed and immobile in the glass network. Based on the above hypothesis, the distance between X and Si and the angle ∠HOG were optimized, changing the distance between X and G in the range from 1.0 to 1.8 Å. The most stabilized structure and the corresponding energy, namely the energy at the ground state, was thus obtained. In the calculation, the Restricted Hartree-Fock method was used with the STO-3G basis set. Further, the excitation energy was calculated with the improved GAUSSIAN 76 for the optimized-structure obtained.

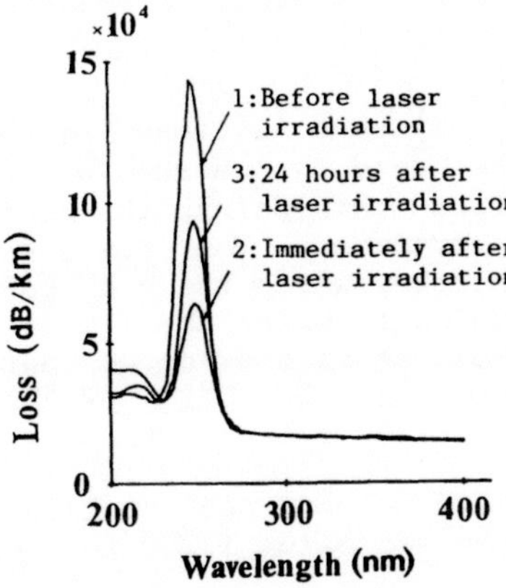

Fig. 1. Model structure for numerical calculation.

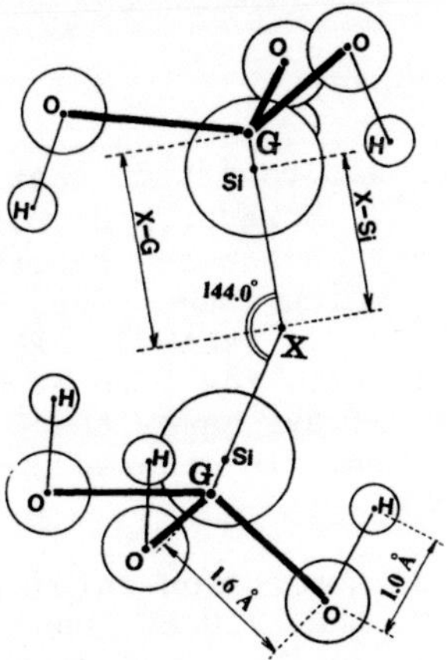

Fig. 2. Change in loss spectra resulting from laser irradiation.

EXPERIMENTAL, RESULTS AND DISCUSSION

(5 eV band)

Figure 2 shows the loss-spectra change of the Prf-A sample caused by the irradiation. Peak absorbance in the 5 eV (= 245 nm) band is reduced from 12×10^4 dB/km to 4×10^4 dB/km by the laser irradiation (see Curves 1 and 2). Contrary to this reduction, absorption at 5.8 eV (= 215 nm), which is caused by the E' center, is increased. The once-reduced 5 eV-band absorption increases back to 7×10^4 dB/km by leaving the sample at room temperature for 24 hours as shown by the Curve 3. In this 24 hours, the absorption at 5.8 eV decreases to about half of what was induced by the laser irradiation. Detectable change in the absorption was observed neither at 5 eV nor at 5.8 eV by laser irradiation of the Df8 and P3f samples which do not have the 5 eV band.

Silin[7] and Arnold[8] proposed -O-$\overset{\bullet}{\underset{\bullet}{\text{Si}}}$-O- and ≡Si-Si≡, respectively, as models for the oxygen deficiencies responsible for the 5 eV band. As mentioned in the Introduction, the authors reported that the E' centers created by fiber drawing are observed only in oxygen-deficient fibers, and that based on this result ≡Si-Si≡ is considered to be the most probable form of the oxygen deficiency [2]. This hypothesis is strongly supported by the present results. That is to say, it is considered that as the structure of ≡Si-Si≡ changes to ≡Si• •Si≡ by the laser irradiation, the absorption decreases at 5 eV and increases at 5.8 eV. After the laser irradiation was ceased, the bi-radical structure, ≡Si• •Si≡, is thought to change back to ≡Si-Si≡ as time proceeds, making the absorption at both wavelengths return to the original values.

On the other hand, the -$\overset{\bullet}{\underset{\bullet}{\text{Si}}}$- model for the oxygen deficiency gives no clear explanation to the present results. Therefore, it is concluded that the absorption at 5 eV is caused by the oxygen vacancy in the form of ≡Si-Si≡ (*).

(7.6 eV band)

It was reported that the 7.6 eV (= 163 nm) band is caused by peroxy radicals[1]. However, a possibility that the band is caused by oxygen deficiency has been reported also[9]. To confirm this point, the kinds of samples which have the 7.6 eV band were examined. The results are summarized in Table I. Although not all the samples where the 5 eV band is observed show the 7.6 eV band (**), the samples in which the 7.6 eV band appears always have the 5 eV band. Therefore, it is considered that the 7.6 eV band is caused by oxygen deficiency in the form of ≡Si-Si≡ structure.

The absorption intensity of the 5 eV band is very much smaller than that of the 7.6 eV band, the ratio of the two being 1:800 to 1:4000, depending on the samples examined.

RESULTS OF AB INITIO CALCULATION

The optimized values of the distance between X and Si and of the angle ∠HOG are listed in Table II as a function of the distance between X and G. The optimized distance between X and Si becomes smaller than the distance between X and G except for the cases where the X-G distance is 1.0 and 1.1 Å. This indicates that two silicon atoms approach to each other a little to get stabilized. However, when the distance between X and G is 1.0 or 1.1 Å, two silicon atoms move apart as the distance between two silicon atoms is smaller than the regular Si-Si distance in silicon single crystals, 2.35 Å.

The energy values of the ground state and of the excited states, calculated based on the data listed in Table II, are shown in fig. 3. The excitation energy, i.e. the difference between the energies of two states, is shown in fig. 4. From fig. 3, it is found that the energy of the ground state is minimum when the distance of X and G is 1.3 Å. The structure with the X-G distance of 1.3 Å, however, gives the excitation energy of about 10 eV, as shown in fig. 4, far higher than the value obtained from the experiments. On the other hand, the excitation energies to the excited singlet-state and to the excited triplet-state are 7.6 and 5 eV, respectively in the case where the X-G distance is 1.65 Å. These energy values coincide with the experimentally-obtained absorption energies.

---

* It has been recently found by the authors that there are probably two types of defects causing the absorption at 5 eV. The 5 eV absorption accompanies photoluminescence only at 270 nm in some glasses, and at 280 and 380 nm in other glasses. The oxygen vacancy ≡Si-Si≡ is thought to cause the absorption which accompanies one luminescence band. The details will be presented at the 6th International Conference on The Physics of Non-Crystalline Solids (C.10.) in Kyoto, July 1987.

** Glasses with the 7.6 eV band have a luminescence band at 270 nm when excited by light with an energy of 5.0 eV. In the case of the glasses which have the absorption band at 5.0 eV but not at 7.6 eV, photoluminescence appears at 280 and 380 nm. From the experimental results described in the footnote (*), it is considered that the 7.6 eV band is related to the structure in the shape of ≡Si-Si≡.

Table II. Results of structure optimization

X-G distance(Å)	optimized X-Si distance(Å)	Si-Si distance(Å)	optimized ∠H-O-G(°)
1.0	1.09	2.07	118.8
1.1	1.12	2.13	114.0
1.2	1.15	2.19	109.6
1.3	1.19	2.26	105.8
1.4	1.23	2.35	102.6
1.5	1.29	2.45	100.0
1.6	1.35	2.57	98.0
1.65	1.38	2.63	97.2
1.7	1.42	2.71	96.6
1.8	1.51	2.87	95.9

As is well known, direct excitation from the singlet ground-state to an excited triplet-state is a forbidden transition. This means that the transition probability is very small. As described previously, the absorption at 5 eV is in fact far weaker than the absorption at 7.6 eV. Consequently, it is concluded that the cause of the 5 eV band is the excitation to the triplet state and that of the 7.6 eV band the excitation to the singlet state.

The distance, 1.65 Å, between X and G is nearly the same as the regular distance, 1.6 Å, between Si and O in glasses. This indicates that the oxygen vacancy has little effect on the position of the second nearest-neighbor oxygen atoms. This, in turn, means that such an oxygen vacancy as is observed in crystalline SiO_2 exists in SiO_2 glasses. Concerning the short-range order, the SiO_2 glass synthesized for as a preform for optical fibers has good periodicity. It is not clear whether the whole glass or a part of it keeps this periodicity.

CONCLUSIONS

The absorptions at 5 eV and at 7.6 eV are caused by the oxygen deficiency with the structure, ≡Si-Si≡. The former absorption is due to excitation to the triplet state and the latter to the singlet state. The absorption at 5 eV is about 3 orders in magnitude smaller than the 7.6 eV.

ACKNOWLEDGMENTS

The authors are deeply indebted to Mr. K. Sasagane of Waseda University for his valuable advice and helpful discussions on ab initio calculation. We are also grateful to Drs. S. Arai and A. Takematsu of the Institute of Physical and Chemical Research for their help in laser irradiation, and Prof. T. Takada and Mr. A. Yoshida of Musashi Institute of Technology for their help in vacuum-ultraviolet spectroscopy measurement. Glass samples were provided by courtesy of The Furukawa Electric Co., Ltd., Shin-Etsu Quartz Products Co., Ltd., and Shin-Etsu Chemical Co., Ltd. This work was partly supported by a Grant-in-Aid for General Scientific Research from the Ministry of Education, Science and Culture (60450123).

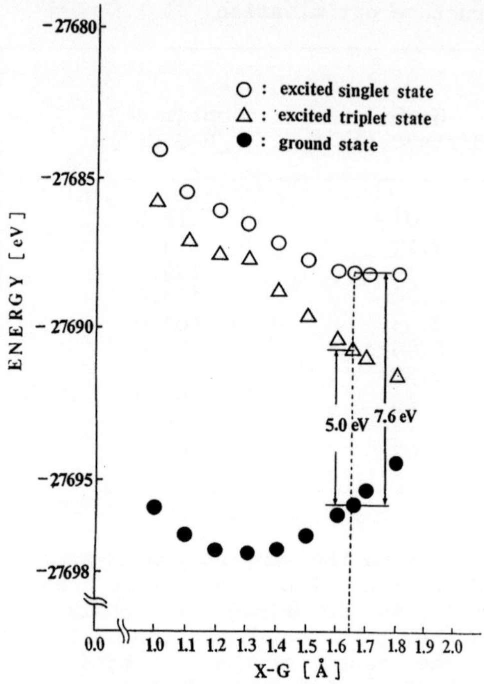

Fig. 3. Energy values of the ground state and the excited states.

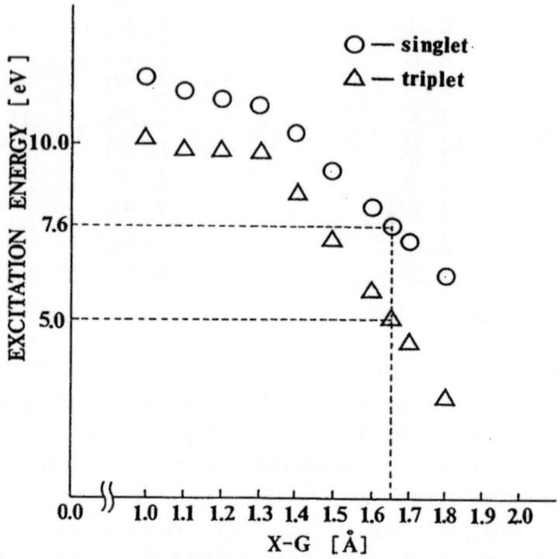

Fig. 4. Excitation energy for sin-singlet to singlet (○) and singlet to triplet (△) transitions.

REFERENCES

1. M. Stapelbroek, D.L. Griscom, E.J. Frieble and G.H. Sigel Jr, J. Non-Cryst. Solids 32:313 (1979).
2. K. Nagasawa, Y. Hoshi and Y. Ohki, Jpn., J. Appl. Phys., 26:L554 (1987).
3. K. Nagasawa, T. Todoriki, T. Fujii, Y. Ohki and Y. Hama, Jpn., J. Appl. Phys. 25:L853 (1986).
4. C.M. Cook, QCPE, 12:391 (1980).
5. K. Sasagane and R. Itoh, "Contribution from the Research Group on Atoms and Molecules, No 17, Progress Report, XIV, October, 1984" by Research Group on Atoms and Molecules, Department of Physics, Faculty of Science, Ochanomizu University, 2-1-1, Otsuka, Bunkyo-ku, Tokyo 112, Japan.
6. A.H. Edwards and W.B. Fowler, J. Phys. Chem. Solids 46:841 (1985).
7. A.R. Silin and L.N Skuja, J. Non-Cryst. Solids 71:443 (1985).
8. G.W. Arnold, IEEE Trans. Nucl. Sci. NS-20:220 (1973).
9. D.L. Griscom, J. Non-Cryst. Solids 73:51 (1985).

HYDROGEN BONDS BETWEEN PEROXY RADICALS AND HYDROGEN MOLECULES IN SiO_2 GLASS

K. Nagasawa*, Y. Yokomachi, R. Tohmon, Y. Ohki and Y. Hama

*Sagami Institute of Technology
Fujisawa, Kanagawa, Japan
Waseda University
Shinjuku-ku, Tokyo, Japan

INTRODUCTION

Optical fibers are one of the most important components supporting communication systems in these modern days. Many studies have been devoted to clarifying the mechanism of loss increment caused by hydrogen-diffusion which affects adversely the long-term reliability of optical communication systems. In addition to the absorption band at 1.24 μm caused by hydrogen molecules[1] and the one spreading from 1.39 to 1.41 μm caused by OH groups[2], the absorption bands at 1.42, 1.45 and 1.52 μm are known to be related to hydrogen[3,6]. Among them, the absorption at 1.52 μm is important because it influences the transparency at 1.55 μm where single-mode fibers are operated. In this paper, the cause of this absorption is discussed.

EXPERIMENTAL

Table I shows the samples used in this investigation along with the characteristics of each sample reported earlier[4]. The sample fibers are

Table I. Sample list

Sample name	Manufact-turing method	Cl cont. [ppm]	OH cont. [ppm]	5eV [x10^4 dB/dm]	peroxy radical D	peroxy radical γ	2eV band D	2eV band γ	E' D	E' γ	oxygen cont.*
A	soot	U	2	3.0	x	x	x	x	o	o	deficient
B	direct	U	800	7	x	x	x	o	x	o	(unknown)
C	O_2+ Ar	300	4	0.2	o	o	o	o	x	o	rich

U : undetected, O_2+Ar: O_2+Ar Plasma, E': E' center, o: observed,
x : not observed.
'D' and 'γ' represent whether the defect is induced by drawing or by γ-ray irradiation, respectively.
* Low OH-group content pure-silica can be classified into two types ; one is oxygen rich and the other is oxygen deficient[4].

silicon clad fibers with a pure-silica core of diameter 125 μm. For Fourier Transform Infrared (FTIR) measurements, fiber preforms with a thickness from 0.6 to 0.8 mm, mirror-polished on both sides, were used.

Figure 1 shows the experimental apparatus for measuring optical loss in fibers. Optical absorption spectra in the range from 0.6 μm to 1.6 μm were measured in samples A and C before and after the treatment mentioned below ; hydrogen-treatment was performed by keeping the samples in a chamber filled with hydrogen gas. The pressure and the duration for each sample will be specified later. The hydrogen treatments were all performed at room temperature. Gamma-ray irradiation was performed using a ^{60}Co source at room temperature at a dose rate of 120 Gy/h to the dose specified. Time dependence of the optical absorption during γ-ray irradiation was measured in situ.

Prior to optical measurements, sample A was treated in hydrogen (3 ata, 41 h) and then irradiated by γ-rays (20 kGy). Sample C was treated in hydrogen (3 ata, 150 h) and then the time dependence during γ-ray irradiation (up to 10 kGy) was measured.

FTIR measurements were made on each sample which had been 1) untreated, 2) irradiated by γ-rays (20 kGy) followed by hydrogen treatment (1.7 ata, 130 h), and 3) treated in hydrogen (3 ata, 120 h) followed by γ-ray irradiation (20 kGy).

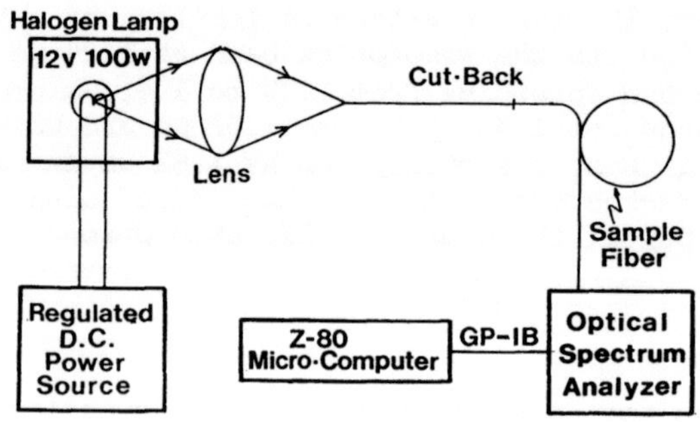

Fig. 1. Experimental Apparatus for measuring fiber losses.

RESULTS AND DISCUSSION

There have been reports which attribute the cause of the 1.52 μm band to the structure in the form of ≡Si-H, because the second overtone of the Si-H stretching vibration appears at 1.52 μm[5,7]. The present study was carried out in order to examine the validity of this model.

Figure 2 shows the IR spectra of various samples which were not treated (Curve 1) and treated (Curves 2 and 3) by a combination of γ-ray irradiation and hydrogen treatment. The absorption at 2250 cm^{-1} is the fundamental Si-H stretching mode. It is clear that this absorption appears

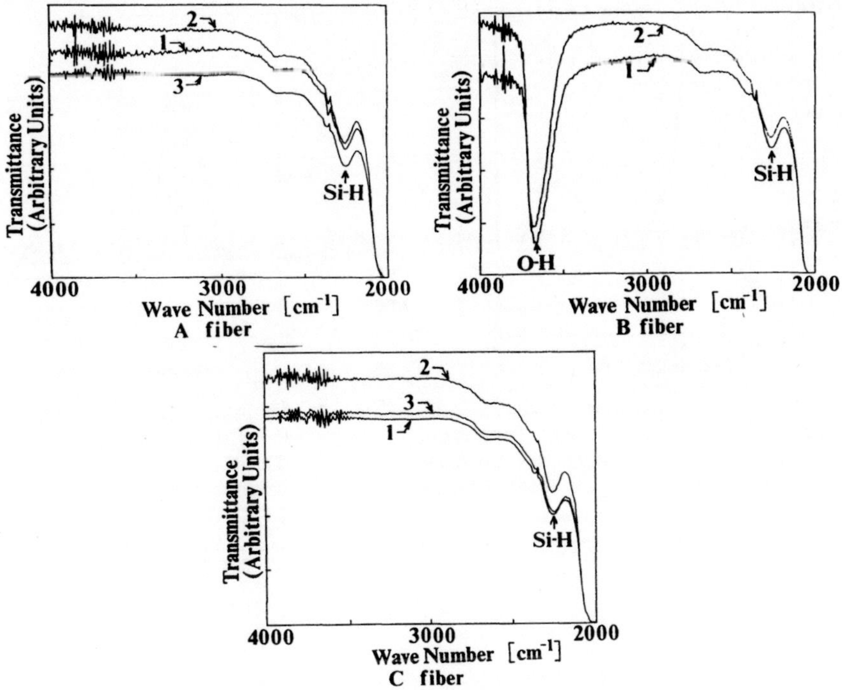

Fig. 2. IR spectra of each sample, curve 1: untreated, curve 2: γ-ray irradiation followed by hydrogen treatment, curve 3 : hydrogen treatment followed by γ-ray irradiation.

with a similar intensity in all the samples and that the intensity hardly changes by any treatment examined.

Figure 3 shows the effect of pre-treatment in hydrogen on the spectra of γ-ray irradiated A fiber. Although the broad absorption from 1.50 to 1.55 μm caused by the silicone cladding is seen, the 1.52 μm band is not induced in the fiber by γ-ray irradiation following hydrogen pre-treatment. The B fiber, too, has no induced absorption at 1.52 μm. On the other hand, combined treatment of hydrogen and γ-rays, irrespective of their order, induces the 1.52 μm band in C fiber. (No peak is seen at 1.52 μm in C fiber in the untreated state). As an example, change in loss spectra by γ-ray irradiation in C fiber pre-treated in hydrogen, is shown in fig. 4. As a summary of the above-mentioned results, the appearance and the intensity of the 1.52 μm band depend on the sample and the treatment. This is not the case in the results of IR absorption concerning the fundamental Si-H stretching mode shown in fig. 2. If the 1.52 μm band is the second overtone of Si-H stretching, the optical absorption at 1.52 μm and the IR spectrum at 2250 cm^{-1} must have the same behavior. It is thus concluded that the Si-H structure is inadequate as a model for the 1.52 μm band.

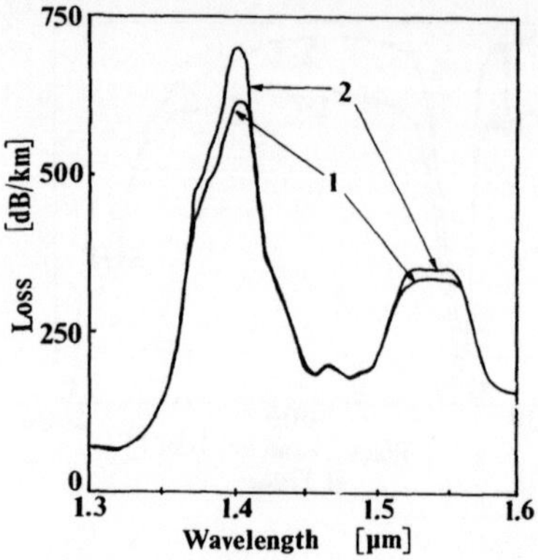

Fig. 3. Loss spectra of γ-ray irradiated A fiber, not treated (curve 1) and treated (curve 2) in hydrogen gas beforehand

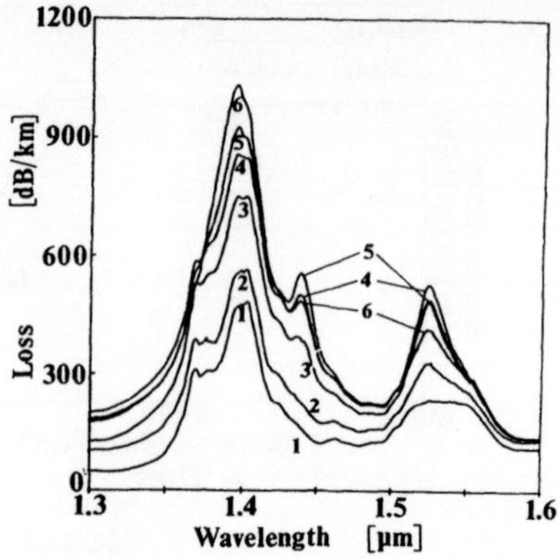

Fig. 4. Change in loss spectra of C fiber during γ-ray irradiation. The fiber was pretreated in hydrogen.
1: 0 Gy, 2: 480 Gy, 3: 1.4 kGy, 4: 2.9 kGy, 5: 5.8 kGy 6: 10 kGy.

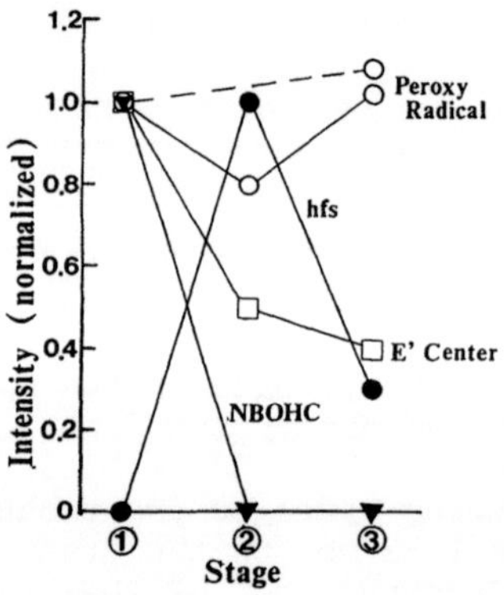

Fig. 5. Change in intensity of various ESR signals in C fiber at the following stages : 1: γ-ray irradiation, 2: 1 + hydrogen treatment (100 h), 3: 2 + heat treatment (6 h) [after reference 4].

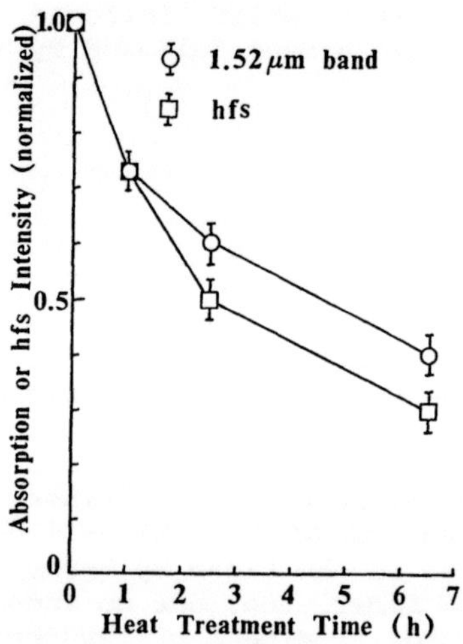

Fig. 6. Decay of 1.52 µm band and of hfs during heat treatment at 90°C [after reference 4].

In a previous paper[4], the authors reported on a hyperfine structure (hfs) in the ESR spectrum which had not been previously reported, and assigned this hfs to be related to peroxy radicals. The authors also reported in the same paper[4] that the 1.52 μm band is thought to be caused by a hydrogen molecule which forms a hydrogen bond with a peroxy radical (≡Si-O-O ...H-H), based on the similarity of the change properties between the hfs and the 1.52 μm band induced by various treatments (See figs. 5 and 6).

ACKNOWLEDGEMENTS

Samples were provided by courtesy of the Furukawa Electric Co., Ltd., Shin-Etsu Quartz Products Co., Ltd., and Shin-Etsu Chemical Co., Ltd. This work was partly supported by a Grant-in-Aid for General Scientific Research from the Ministry of Education, Science and Culture (60460123).

REFERENCES

1. K. Mochizuki, Y. Namihira and H. Yamamoto, Elect. Lett. 19:743 (1983).
2. N. Uchida, N. Uesugi, Y. Murakami, M. Nakahara, T. Tanifuji and N. Inagaki, 9th European Conf. on Optical Communication, post deadline paper, Oct. 1983.
3. M. Ogai, A. Iino, K. Matsubara and K. Okubo, Proceedings 12th European Conference on Optical Communication, September, 1986, (TELEFONICA ECOC, Beatriz de Bobadilla, 3, 28040, Madrid, Spain) MA2.
4. K. Nagasawa, T. Todoriki, T. Fujii, Y. Ohki and Y. Hama, Jpn. J. Appl. Phys. 25:L853 (1986).
5. K.T. Nelson, P.J. Lemaire, K.L. Walker, K.J. Wesser, H.T. Shang, "The Sixth International Conference on Integrated Optics and Optical Fiber Communication", (1987) WA2.
6. M.G. Blankenship, A.J. Morrown, D.R. Powers, "The Sixth International conference on Integrated Optics and Optical Fiber Communication", (1987) WA3.
7. S.Sakaguchi, Appl. Phys. Lett. 47:15 (1985).

In a previous paper[?] the authors reported on a hyperfine structure (HFS) in the [illegible] spectrum which [illegible] [illegible] the [illegible] [illegible] in [illegible] [illegible] the [illegible] [illegible] [illegible] [illegible] [illegible] [illegible] [illegible] [illegible] [illegible] [illegible] [illegible] [illegible] [illegible] the [illegible] [illegible] [illegible] [illegible] [illegible] [illegible] [illegible] [illegible] [illegible] [illegible] [illegible] [illegible] [illegible] [illegible] [illegible] of [illegible] [illegible] between the HFS and [illegible] [illegible] [illegible] [illegible] [illegible] [illegible] [illegible] [illegible].

ACKNOWLEDGEMENTS

Samples were provided by courtesy of the Furukawa Electric Co., Ltd., Sumitomo Electric Industries Co., Ltd., and [illegible] Co., Ltd. [illegible] [illegible] [illegible] [illegible] [illegible] [illegible] from the Ministry of Education, Science and Culture (Japan).

REFERENCES

1. [illegible], [illegible] and [illegible], [illegible] Lett. [illegible] (19[illegible]).
2. [illegible], [illegible], [illegible], [illegible], [illegible] and [illegible], [illegible] Communication, [illegible] [illegible], 19[illegible].
3. [illegible], [illegible], [illegible] and [illegible], [illegible] [illegible] [illegible], [illegible] [illegible] [illegible] [illegible] [illegible].
4. [illegible], [illegible], [illegible] and [illegible], [illegible] [illegible] (19[illegible]).
5. [illegible], [illegible], [illegible], [illegible] and [illegible], [illegible] [illegible] [illegible] [illegible] [illegible] (19[illegible]).
6. [illegible], [illegible], [illegible], [illegible] [illegible] [illegible] [illegible] [illegible] (19[illegible]).
7. [illegible], Appl. [illegible] [illegible] [illegible] (19[illegible]).

ESR STUDIES OF PARAMAGNETIC DEFECTS FORMED IN AMORPHOUS SiO_2 BY HIGH ENERGY HEAVY IONS

E. Dooryhee[1], Y. Langevin[1], J. Borg[1],
J.P. Duraud[2], and E. Balanzat[3]

[1]Laboratoire René-Bernas
91406 Orsay
[2]Dept de Physico-chimie
C.E.N. Saclay
91191 Gif
[3]C.I.R.I.L.
14040 Caen
France

INTRODUCTION

High energy heavy ions in insulators induce the formation of defects which have been studied by track etching methods and small angle X-ray scattering[1-3]. These previous studies have shown that the formation of defects is linked to electronic energy losses. However, the processes which result in lattice defects following such interactions are not well understood. In order to characterize the defects formed and their lattice environment, we studied amorphous SiO_2 (dry Tetrasil SE) irradiated by heavy ions using Electron Spin Resonance (ESR). The paramagnetic defects formed in this material by γ-ray, X-ray and electron irradiation have already been extensively studied[4-7]. After such irradiation, two major types of defects have been observed : the E_1' center[8] (hole trapped by an oxygen vacancy) and the oxygen hole center [9] (or OHC, associated with a peroxy radical). The density of defects observed was closely related to the total energy deposited in the sample. We previously showed [10] that high energy heavy ions also induce the formation of E_1' centers and OHC's. However, the ion irradiated samples present specific characteristics, which are linked to the very high density of energy deposited near the path of heavy ions[11]. We present here a study of the dependence of the defects on the residual energy, the atomic number and the fluence of the incident ions. We show that, in contrast to γ-ray irradiations, the total energy deposited is not the single parameter controlling the formation of paramagnetic defects by high energy heavy ions.

EXPERIMENTALS

Irradiations have been performed at GANIL (Oxygen Argon, Krypton and Xenon) and UNILAC Darmstadt (Uranium). The total energy available was always larger than the energy corresponding to the maximum of electronic energy losses. The fluence varied from 10^{11} ions.cm^{-2} (Uranium) to 7.10^{13} ions.cm^{-2} (Oxygen). The flux was kept lower than 10^9 ions.cm^{-2}.s^{-1} so as to minimize temperature increases. We irradiated stacks of silica slabs with thicknesses ranging from 200 µm to 1 mm. Slabs corresponding to a given energy range were then placed in a 3 mm x 3 mm x 30 mm ESR sample so as to determine the evolution of the defect populations with the residual range while avoiding cavity fill-up effects. We performed annealing of several samples for 2 hours at temperatures from 0 to 700°C.

ESR measurements were obtained on a Varian E-9 spectrometer in the first derivative microwave absorption mode at X-band, either at room temperature or at 170°K. Optimal observation conditions for both E_1' centers and OHC were determined from studies of modulation and microwave power saturation. We selected as a reference a sample irradiated with 10 Mrad γ-rays, and compared ion irradiated samples to this reference using double numerical integration or the product of the peak to peak amplitude and the linewidth (which is quite consistent with the numerical integration for the E_1'). The accuracy of these relative evaluations can be estimated to be ~ 20 %.

Absolute concentrations are less reliable, resulting from comparisons with a weak pitch standard, the E_1' signature and the OHC spectra being obtained under different experimental conditions. We will therefore discuss mainly the results in terms of relative concentrations. Room temperature annealing of paramagnetic defects resulted in decreases of up to 50 % in spin densities during the first few weeks, after which they decreased at a much slower rate. We therefore compared spin densities determined during this second phase.

RESULTS AND DISCUSSION

The observed E_1' signature is characterized by a double peak near g_3, which has already been observed[12] after γ-ray irradiation of chlorine rich silica. The linewidth is observed to increase with the atomic number of the incident ion (fig. 1). This effect can be interpreted in terms of spin-spin interactions in regions where spin densities become very large.

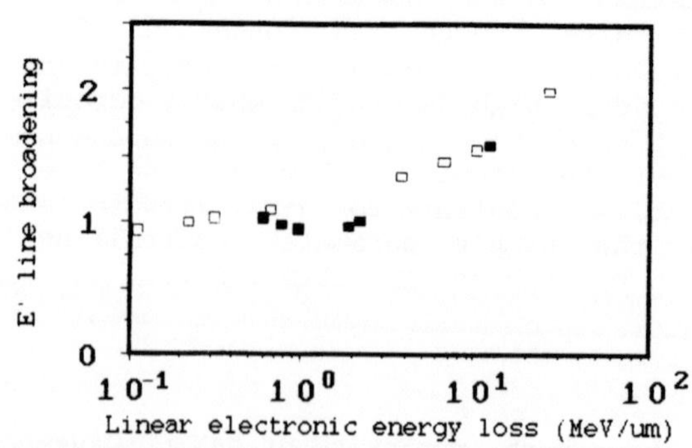

Fig. 1. Linewidth of E_1' negative peak as a function of the atomic number of incident ion.

It can therefore be linked to the very high density of energy deposited which is characteristic of heavy ion irradition. It is therefore interesting to examine the effects of increasing the level of energy deposited. Such an increase can be obtained by moving closer to the maximum of energy loss, by increasing the atomic number, or by increasing the fluence above the overlapping threshold.

We consider first only samples irradiated at low fluences of oxygen ions, so as to minimize effects induced by track overlap. Defect densities plotted as a function of the residual range show that the density of E_1' centers increases slower than the energy loss when approaching the maximum of energy loss (fig. 2). On the other hand, OHC are not observed at high energies, and their density rapidly increases over the last mm of the oxygen ion range (fig. 3). In this region, the fraction of OHC in the population of paramagnetic defects increases while the production efficiency of E_1' centers decreases. This shows that defect densities are not simply proportional to the energy deposited. In samples irradiated with Argon ions, the production efficiency of E_1' centers is approximately constant. The production efficiency of OHC increases with decreasing range,

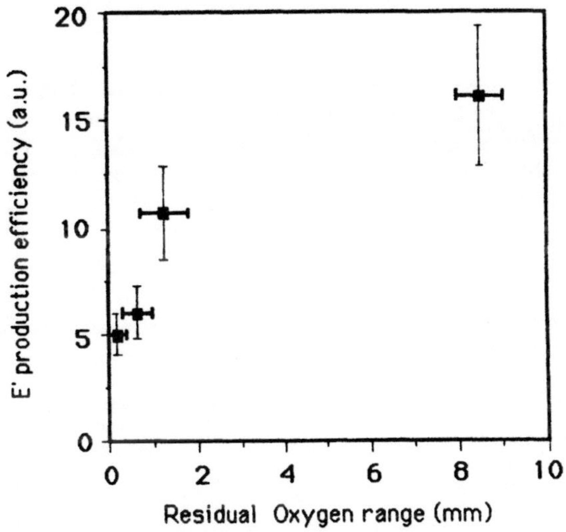

Fig. 2. Defect concentrations of E_1' centers as a function of the residual range of Oxygen ions.

but OHC are now observed at the highest energy available (76 MeV/amu). For Krypton ions, both the E_1' and OHC production efficiencies increase by a factor of ~ 2 from the maximum energy (42.7 MeV/amu) to the end of their range. Finally, the available energies for Xenon and Uranium (27 and 16.5 MeV/amu) did not allow us to determine profiles, but confirmed the trend towards high OHC/E_1' ratios for high energy deposits.

When comparing the defect production efficiencies for different ions, it is interesting to select energy ranges for which the energy losses are comparable : for instance, Argon ions between 21 MeV/amu and 0 lose between 5 and 1.4 MeV/μm, while Krypton ions between 42.7 and 33.9 MeV/amu lose ~ 3.8 MeV/μm. The observed defect concentrations show that at similar levels of deposited energy, Krypton ions are 6 to 10 times more efficient than Argon ions in producing paramagnetic defects. This shows that the level of energy deposited is not the only parameter controlling the formation of defects by high energy ions. Similarly, we show that Xenon

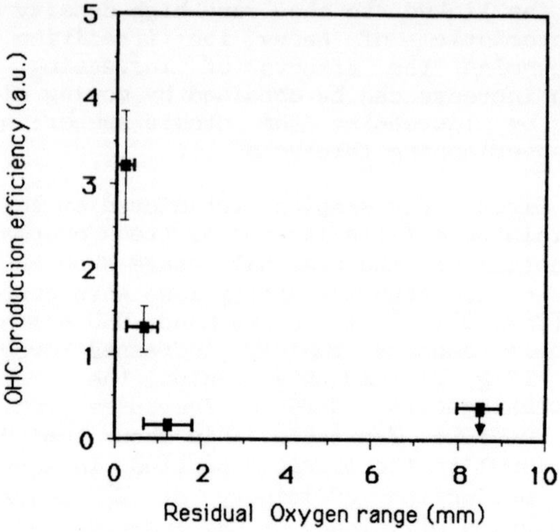

Fig. 3. Defect concentrations of OHC as a function of the residual range of Oxygen ions.

ions are ~ 2 times more efficient that Krypton ions, while low energy Oxygen ions are more efficient than Argon ions in producing E_1' centers and OHC (see fig. 4). There are therefore three ranges in atomic number where the behavior of the production efficiencies is quite different : from Oxygen to Argon, where the E_1' efficiency decreases, from Argon to Xenon, in which both the E_1' and the OHC production efficiencies increase monotonically, and from Xenon to Uranium, where both production

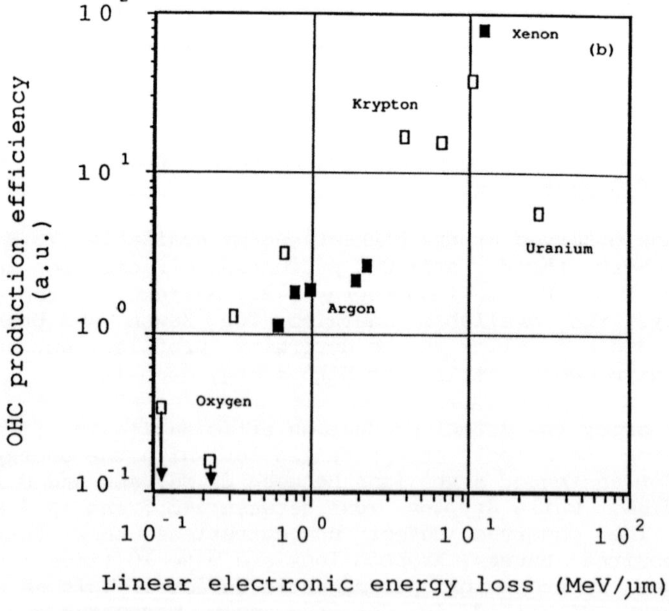

Fig. 4. OHC production efficiency as a function of the mean energy deposited.

decrease. Such transitions should be linked to changes in the structure of the track.

Oxygen ions induce very few OHC in their high energy range (93 to 88 MeV/amu). The OHC production efficiency drastically increases with increasing fluence between 3.10^{12} and 10^{13} ions.cm^{-2}, whereas the E_1' production efficiency decreases by a factor of ~ 10. At very high fluences (7.10^{13} ions.cm^{-2}), the E_1' production efficiency in the medium energy range (36 MeV/amu to 22 MeV/amu) decreases by a factor of 4, and that in

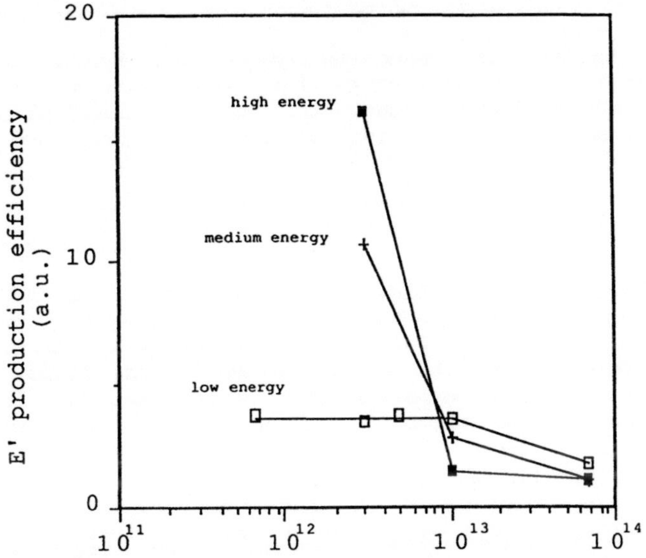

Fig. 5. E_1' production efficiency as a function of the fluence of Oxygen ions for 3 energy ranges (93 to 88, 36 to 22, and < 22 MeV/amu)

the low energy range (22 MeV/amu to 0) by a factor of 2.5. The production efficiency therefore becomes similar in all three energy ranges, which suggests that a steady state is reached between E_1' formation and ionization annealing (see fig. 5).

These features can be interpreted if the energy deposited by high energy oxygen ions is lower than a threshold of local energy deposited below which OHC do not form. For large fluences, the total energy deposited rises above this threshold due to track overlap, and OHC are formed at the expense of the E_1' population. It is interesting to note that a similar transfer from E_1' to OHC is observed upon low temperature annealing, and is interpreted as temperature enhanced atomic diffusion[6]. This observation is consistent with the concept of a "thermal spike" during which lattice defects would reorganize in the wake of the incident ion.

It is interesting to compare these results with those obtained on the structure of ion tracks using track etching methods and small angle X-ray scattering[1,2]. These studies showed that ion tracks are constituted of two types of defects : the point defects, which are observed even with light ions, and the clusters of defects, which appear above a critical atomic number depending on the studied mineral (Z ~ 20 for feldspar). The size and linear density of these clusters increase with the atomic number of the incident ion, and the linear density decreases much faster than the energy loss above the maximum of energy loss. For very heavy ions (Z > 70), these clusters join to form a continuous damaged region around the maximum of energy loss[3].

We note that our ESR results, although dealing only with the paramagnetic part of the defect population, confirm the non linearity of the damaging process with the deposited energy, and the important role of the atomic number of the incident ion. The large increase in defect production efficiency between low Z ions (O and Ar), and high Z ions (Kr and Xe) may be related to the appearence of clusters of defects. Conversely, the decrease observed between Xenon and Uranium may correspond to the formation of a continuous amorphous region along in the ion path.

However, the absolute concentrations of paramagnetic defects (up to a few thousands per μm of ion range for Xenon ions) are lower than those required to account for the characteristics of chemical etching of the tracks. Complementary studies using optical spectroscopy are in progress to study the diamagnetic defects.

CONCLUSION

Using Electron Spin Resonance, we studied silica samples irradiated with high energy ions with atomic numbers ranging form 8 to 92. These observations showed that the formation of paramagnetic defects could not be simply interpreted as due to the deposited energy alone, in contrast to γ-ray irradiations : the atomic number of the incident ions is a major factor influencing the formation of defects. A broadening of the line shapes and an extra peak in the E_1' signature are linked to the very high local density of energy deposited which is specific to heavy ion irradiation. As a function of the atomic number, three different regimes have been observed : from Oxygen to Argon, from Argon to Xenon, and from Xenon to Uranium. This may correspond to the structural transitions observed in ion tracks using small angle X-ray scattering and track etching techniques. ESR observations yield information on the paramagnetic part of the defect population. Complementary studies using optical spectroscopy are in progress to study the diamagnetic defects produced by heavy ions in amorphous silica.

REFERENCES

1. E. Dartyge, J.P. Duraud, Y. Langevin, M. Maurette, Phys. Rev. B23:5213 (1981).
2. J.P. Duraud, PhD thesis, Université Paris XI (1978).
3. J.P. Duraud, Y. Langevin, Nucl. Inst. Meth., Phys. Res. B1:398 (1984).
4. R.A. Weeks and C.M. Nelson, J. Am. Soc. 43:399 (1960).
5. D.L. Griscom, E.J. Friebele, Rad. Eff. 65:63 (1982).
6. D.L. Griscom, SPIE 541:38 (1985).
7. R.L. Pfeffer, J. Appl. Phys. 57:5176 (1985).
8. D.L. Griscom, Nucl. Inst. Meth., Phys. Res. B1:481 (1984).
9. M. Stapelbroek, D.L. Griscom, E.J. Friebele, G.H. Sigel Jr., J. Non-Cryst. Solids 32:313 (1979).
10. Y. Langevin, E. Dooryhee, J. Borg, J.P. Duraud, C. Lecomte, E. Balanzat, Appl. Phys. Lett. 49:1699 (1986).
11. T.A. Tombrello, C.R. Wie, N. Itoh, T. Nakayama, Phys. Lett. 100A:42 (1984).
12. D.L. Griscom, E.J. Friebele, Phys. Rev. B34:7524 (1986).

ON THE RADIATION INDUCED COLORATION OF SIO_2

W. Hohenau[1], I. Godmanis[2]

1: Karl-Franzens-Universität Graz, Austria
2: Peter Stucka University, Riga, USSR

INTRODUCTION

The radiation induced absorption spectra of crystalline and glassy SiO_2 generally consist of broad overlapping A (450-650 nm), B (290-350 nm) and C (200-250 nm) bands[1]. Their relative intensities and distinct spectral positions slightly change in different types of crystalline and glassy SiO_2 [2]. The absorption in the A band is attributed to the aluminium hole (Al/e+) and the C band region to E' type defect centers[3,4]. Contrary to this is, the nature of the color centers forming the B band, is not determined. A more effective detection of this band in short time experiments as during stationary absorption measurements could indicate an unstable character of corresponding color centers in SiO_2 [5].

In this paper we have studied the stability of color centers, by measuring a bleaching effect in the radiation induced absorption of crystalline SiO_2, SiO_2-Ge and glassy SiO_2.

EXPERIMENTAL PROCEDURE

We have investigated different types of crystalline SiO_2 (Morion, Brasil, Smoky, synthetic and Ge-doped quartz) as well as glassy SiO_2 (Suprasil, Herasil, Corning). The samples were irradiated at ambiant temperature by X-rays (80 kV, 30 mA) with different doses (20 krad). Immediately after irradiation the absorption spectra were measured using a Hewlett-Packard (8451 A) diode array spectrometer. The alteration of the complete absorption spectrum between 190 and 820 nm was detectable in every 0.2 seconds with a resolution in optical density of about 0.001.

RESULTS

The new results consist mainly in an observation of the alteration of optical absorption in SiO_2 during and after X-irradiation. Weak but clear resolved bands in B and C regions become detectable at low doses of irradiation. Furthermore their intensity saturates at doses of about 20 krad, where a generation also of A bands occurs, followed by a strong overlapping between all mentioned bands. Immediately after irradiation a bleaching process in the induced coloration in SiO_2 is detectable. This bleaching is dependent on wavelength and is proportional to the previously induced coloration both in crystalline and in glassy SiO_2.

In natural quartz (Brasil, Smoky, Morion) during irradiation the most intense absorption band was detected at 460 nm (A region), followed by broad bands at 650 nm (A) and 230 nm (C region)(fig. 1a). Formation of B band (300 nm) is detectable only in alteration spectra of absorption (fig. b). However also in this spectrum, an overlapping effect does not allow one to distinguish the exact maximum positions of all registered bands. In contrast to that, a clear resolved B band at 300 nm occurs in the bleaching spectrum of crystalline quartz (fig. 1). Furthermore a bleaching in the C band (230 nm), overlapped by other bands in the vacuum UV is detectable. But no bleaching effect is detectable in the A band region.

In Ge-doped quartz besides an intensive B band at 290 nm, dominating in the radiation induced absorption spectrum, a C band at 220 nm is detectable. The coloration in A region is sometimes weaker as in natural quartz. Generally similar as in Ge-free quartz, the structure of the bleaching spectrum consists of a well defined B (300 nm) and C band (200 nm). A spectral shift was detected between B band positions in bleaching (300 nm) and coloring (290 nm) spectra of SiO_2-Ge.

In quartz glasses the radiation induced absorption spectrum consists of only two defined bands at 300 nm (B) and 215 nm (C). As in crystalline quartz a bleaching in quartz glasses occurs in defined C (220 nm) and B (31 nm) bands. Similarly as in crystalline SiO_2-Ge, a red shift of the bleaching B band (315 nm) compared with the coloring B band (300 nm) takes place (fig. 2). In Ge-doped quartz as well as in quartz glasses no bleaching effect in A band region was detectable.

DISCUSSION

Considering the experimental results there is no doubt, that the coloration of SiO_2 by X-irradiation is a complex, many-stage process. A comparison of usual SiO_2 absorption spectra with those measured during irradiation, indicates an alteration of induced color centers. Moreover, the investigated bleaching effect demonstrates a decay of certain unstable color centers without additional thermal excitation.

The correlated coloring and bleaching of each band in the C and B region of the absorption spectrum indicates a direct connection between the generation and decay of the corresponding two types of color centers. One of the E' type centers are trapped hole centers. The observed E' decay simultaneously with B color centers could be explained as a result of electron-hole recombination between these two centers, emphasising the trapped electron character of B color centers in quartz.

An additional information about the character of B color centers is given by an investigation of Ge-doped quartz. The radiation induced B band in Ge-doped quartz is definitively connected with trapped electron (Ge e-/Me+) centers[6]. However, the observed spectral shift between coloring (290 nm) and bleaching (300 nm) bands indicates that the absorption in B region is composed of several bands. The different color centers, forming these bands, are envisaged to be different positions or types of monovalent interstitials Me+ (Li+, Na+) in the (Ge e-/Me+) centers. One type of these centers is similar to B color centers in natural quartz and unstable at ambient temperature.

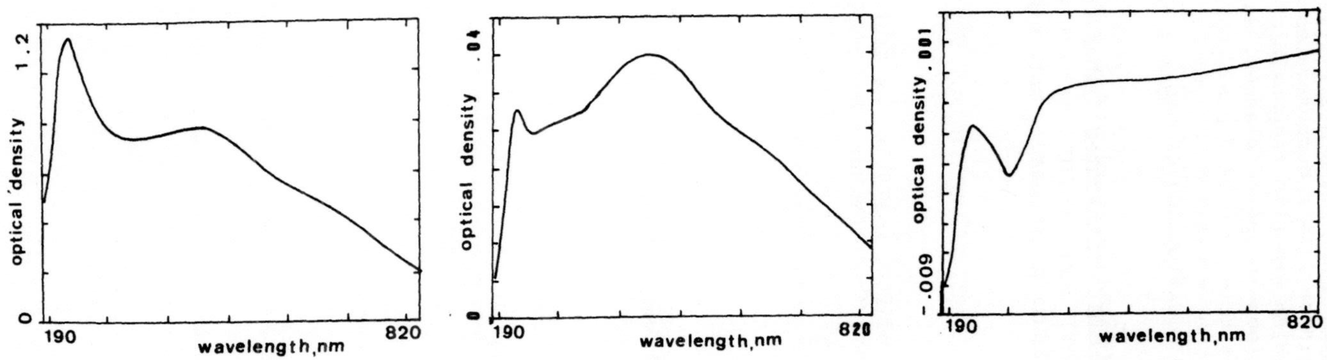

Fig.1: crystalline SiO_2: absorption spectrum X-ray induced change of absorption bleaching spectrum

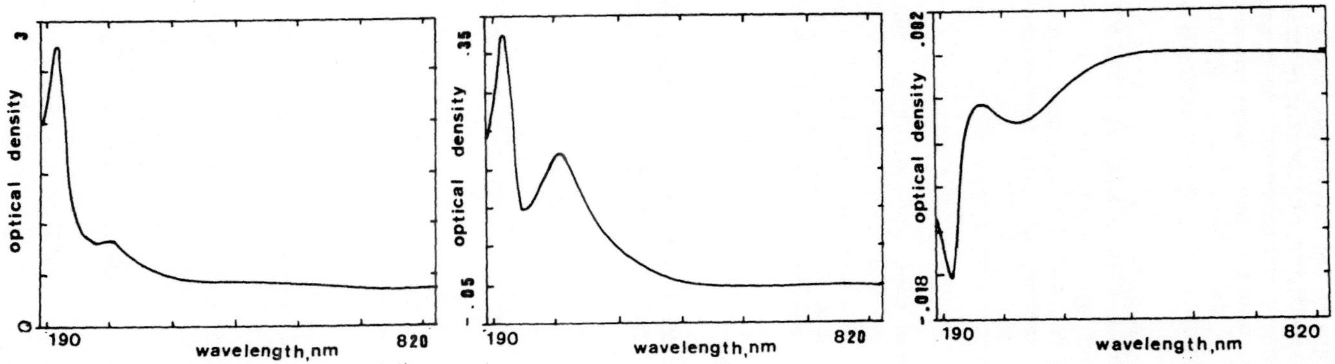

Fig.2: glassy SiO_2: absorption spectrum X-ray induced change of absorption bleaching spectrum

Generally an equivalent behavior of coloring and bleaching processes was observed in different types of quartz glasses. This demonstrates the same composition of unstable color centers which are also responsible for absorption and bleaching in B and C band regions in quartz glasses. As in Ge-doped quartz, the observed spectral shift between B band positions in coloring and bleaching spectra could be explained by different configurations of B color centers in the fused modification of SiO_2.

The absence of the bleaching effect in the A band region in crystalline SiO_2, SiO_2-Ge and in glassy SiO_2 indicates that the corresponding aluminium hole centers (Al/e+) are stable at ambient temperature and do not take part in recombination processes with other unstable centers at this temperature.

SUMMARY

A new type of radiation induced color center, unstable at ambient temperature and absorbing at 300 nm is observed. These centers are trapped electron centers in crystalline quartz and take part in the recombination process with E' centers. Structurally modified such centers exist also in Ge-doped crystalline quartz and quartz glasses.

REFERENCES

1. E.W.J. Mitchell, E.G.S. Paige, Phil. Mag., 1:1985 (1956).
2. G.W. Arnold Jr, J. Phys. Chem. Solids, 13:pp. 306 (1960).
3. B.K. Meyer et al, J. Phys. C17:L31 (1984).
4. K.L. Yip, W.B. Fowler, Phys. Rev. B11:2237 (1975).
5. V.I. Sptisyn, Isvestiya Akademii Nauk SSSR, Neorganicheskie Materialy, 18:74-78 (1980).
6. A. Halperin, J.E. Ralph, J. Chem. Phys.,39:63 (1963).

EFFECTS OF PROCESSING ON RADIATION DAMAGE THRESHOLDS IN SILICA GLASSES

R.F. Haglund, Jr., D.L. Kinser, H. Mogul, N.H. Tolk, P.W. Wang and R.A. Weeks

Center for Atomic and Molecular Physics at Surfaces
Vanderbilt University, Nashville, TN 37235 USA

ABSTRACT

We report experimental evidence linking near-surface radiation damage mechanisms to identifiable variations in the manufacturing process of silica glasses. We have bombarded a well-characterized set of Suprasil and Spectrosil glasses with ions and electrons under ultrahigh vacuum conditions. Relative yields of atoms and molecules desorbed from the sample glasses were obtained from the characteristic optical spectra ; the time dependence of the bulk luminescence arising from electronic excitations in the near-surface bulk was also monitored. Comparison of these data with other work suggest a correlation between radiation damage susceptibility and the OH content of the fused silica.

INTRODUCTION

Surface damage is a significant problem in fabrication, analysis, and application of amorphous silica (a-SiO_2) in optical and electronics technology. Low-energy radiation sources used in many manufacturing processes -- ion and electron beams, ultraviolet, and x-ray photons and pulsed lasers -- are especially likely to produce desorption (i.e., surface bond-breaking) by electronic mechanisms, leading ultimately to erosion, etching, and other changes to surface morphology. Because the susceptibility of dielectric materials to desorption induced by electronic transitions (DIET) is determined by the available pathways through which incident electronic energy is absorbed, localized, and transformed, it is plausible that DIET can be influenced by variations in materials processing, particularly because of processing-related differences in impurity content. Indeed, DIET is generally more sensitive to processing-related changes, particularly in dielectrics, than secondary-ion mass spectroscopy[1].

In previous studies of electron- and photon-irradiated alkali halides, it has been shown that the electronic energy deposited in the near-surface bulk is rechanneled primarily into luminescence from the near-surface bulk and desorption of ground-state and excited-state neutral atoms[2]. However, it is also becoming increasingly clear that this pattern of near-surface radiation damage is a generic property, not only of such simple crystalline materials as alkali halides, but indeed of all materials which can support the creation of self-trapped excitons which relax to form permanent mobile defects[3]. Since a-SiO_2 also fits this pattern of behavior when electronic energy is absorbed[4], it is to be expected that electronic transitions will play a role in surface radiation damage.

In recent exploratory experiments on surface radiation hardness in a-SiO_2, we have observed significant differences in bulk luminescence and excited-state emission from desorbed Si^* atoms for various silica glasses under low-energy ion irradiation, and characteristic differences in the time dependence of bulk luminescence decay from electron-irradiated fused silica glasses. The data suggest that the spectroscopic techniques employed here can provide clues to process control of radiation damage thresholds.

EXPERIMENTAL PROTOCOL

The general features of DIET experiments with optical detection techniques have been described previously[5], and are shown schematically in figure 1. The fused silica samples were mounted on a precision micromanipulator in ultrahigh vacuum (nominal base pressure 10^{-10} torr). The measurements presented here were made at room temperature ; however, the sample can be either cooled or heated if desired. The radiation source, shown as incident from the left, was a Colutron ion gun or a high-current electron gun. Optical emissions from the target region were observed via a 0.3-m grating spectrometer coupled to a photomultiplier tube ; the sample volume was imaged onto the spectrometer slits by an f/5.6 lens system. The slits on the monochromator were set to 1 mm, giving 25 Å resolution over the wavelength range 2000 Å - 8000 Å. For time-resolved measurements, the output of the photomultiplier was fed into a multichannel analyzer with a minimum dwell time of 100 ns/channel. The laser shown at the right of the target is used for laser-induced fluorescence monitoring of ground-state desorbing atoms, but was not employed in the experiments described below.

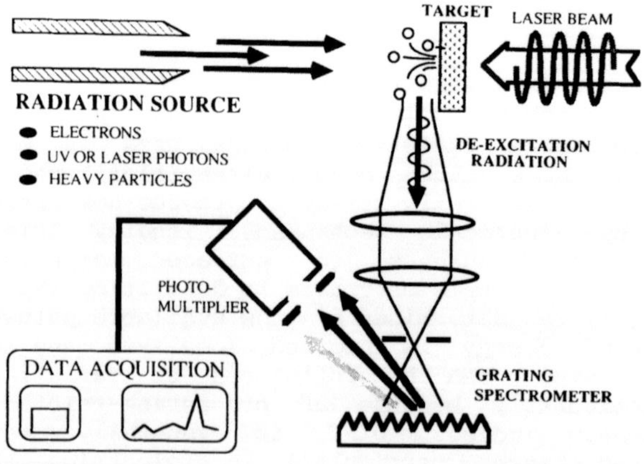

Fig. 1. Experimental geometry for surface damage experiments with ions and electrons. The irradiating beam is incident from the left ; desorbing particles and the bulk luminescence are observed by a spectrometer- photomultiplier combination linked to a Macintosh personal computer through a CAMAC interface.

Because the low-energy ions and electrons used in these studies stop in a thin near-surface layer, there is less difficulty separating primary from cascading secondary effects, such as those which occur in MeV ion irradiation of dielectrics. At the same time, one avoids some of the difficulties of thin films, since the surface studies use bulk material for samples.

Four different high-purity fused silica samples were used in the experiments reported here : Suprasil 1 and Spectrosil are type III glasses, produced using an oxygen-hydrogen flame ; Suprasil W-1 and Spectrosil WF, on the other hand, are type IV glasses, processed using a water-vapor-free, plasma flame. The principal impurity species in the glasses are OH and Cl ; the differing impurity concentrations are listed in Table 1, with the four glasses grouped in pairs as "high-OH" and "low-OH" samples.

Table 1. Impurity Content of Ion-and Electron-Irradiated Fused Silica Glasses

Silica tradename	OH content (ppm)	Cl content (ppm)
Spectrosil	200	50
Suprasil-1	1200	130-180
Spectrosil-WF	<2	180
Suprasil-W1	<5	230-280

Wafers were cut to size (1x2 cm^2), chemically cleaned, then inserted into the UHV chamber and sputter cleaned by Ar^+ bombardment for about half an hour. For comparisons between different samples, the front surface of each sample was aligned to a preselected line of sight in the spectrometer imaging system, with an estimated tolerance of 0.2 mm.

ION BOMBARDMENT STUDIES

We have compared the near-surface damage susceptibility of the four candidate glasses by bombarding them with argon, nitrogen and hydrogen ions. Figures 2 and 3 show spectra from Suprasil sample glasses bombarded by 9 keV Ar and H_2 positive ions over a wide range of current densities (10^{-8} - 10^{-6} mA/cm^2). The optical emissions typically consist of a broad maxima, which we identify with the bulk luminescence, punctuated by sharp peaks which can be identified with desorbed, excited atoms and ions radiating out in front of the sample surface. Thus the line radiation is indicative of damage via a surface erosion process, while the bulk luminescence is the characteristic signature of a non-desorptive, radiative electron-hole recombination which can lead to desorption (generally of ground-state atoms) if the defect produced through recombination can diffuse to a nearby surface.

In addition to Ar^+ and N^+, the samples were irradiated with 9 keV H_2 ions to ascertain the effects associated with varying momentum transfer ; the Ar and N ions cause damage primarily through the collision cascade mechanism[6], whereas the H_2 ions lose energy primarily through electronic processes[7].

The spectra of fig. 2 are particularly noteworthy in that the classical collision cascade picture of heavy-particle sputtering -- which is generally believed to be valid in this case -- offers no explanation for the dramatic differences in the ways the incident ion energy is being transformed in the near-surface bulk of samples which differ only in the impurity content at the ppm level. Since the bulk fluorescence band centered at 450 nm is believed to be associated with the creation of E' centers and other radiation defects[8], this seems to indicate that the low-OH Suprasil W1 is less prone to ion-induced damage than its high-OH cousin.

Figure 3 shows a similar pair of spectra, this time for H_2 positive ions irradiating Spectrosil. Here again we see characteristic differences which do not appear to be attributable simply to "knock-on" damage characteristic of ordinary physical sputtering. Instead, we see a marked change in the relative fraction of the bulk fluorescence yield in the 450 nm band, and a significant difference in the yield of excited hydrogen atoms between the two samples. This lack of agreement with a simple momentum-transfer collisional damage process agrees with Griscom's observation[9] that even damage done by much more energetic ions appears to be primarily "radiolytic" in character.

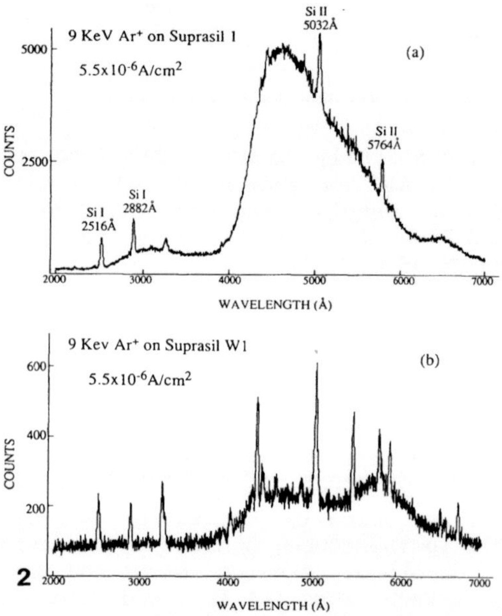

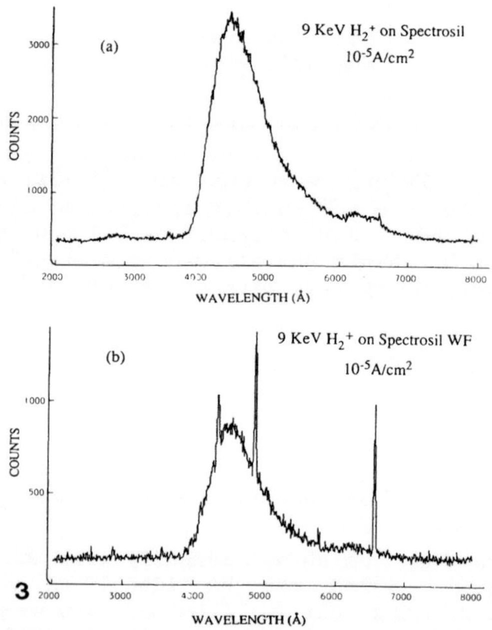

Fig. 2. (at left) Optical fluorescence yields (in counts/s) as a function of wavelength for Suprasil bombarded at normal incidence by 9 keV Ar ions at a beam current density of 5.5 $\mu A/cm^2$. Line radiation is primarily from desorbing silicon atoms in excited states (both first-and second-order peaks) ; the broad peaks are bulk luminescence.

Fig. 3. (at right) Spectra for Spectrosil samples bombarded by 9 keV H_2 ions at normal incidence and at a current density of 10.0 $\mu A/cm^2$. The line radiation in the lower spectrum is the Balmer β-series of atomic hydrogen lines.

In assessing these experiments, it is critical to understand that the excited silicon atoms are probably not the dominant desorbing species. Experiments in many materials exhibiting the same generic damage mechanisms have shown repeatedly that the dominant particle loss channels involve ground-state neutral atoms and molecules. Nevertheless, the excited state neutrals are usually associated with the desorption of ground-state neutrals, and their appearance in the spectrum often signals the onset of surface erosion and damage.

ELECTRON BOMBARDMENT STUDIES

We attempted to assess the relative importance of electronic vs. momentum-changing processes in the creation of radiation damage by bombarding Spectrosil and Spectrosil WF with 320 eV electrons at normal incidence, following the same experimental protocol described for the ion irradiation experiments. Lifetimes of the luminescence bands were obtained by chopping the electron beam and then monitoring the decay of the luminescence during the "beam off" intervals. The chopper speed was adjusted for long decay times to be four to five e-folding times for the given decay. The electron current was monitored on the sample holder and maintained at approximately 30 μA on both the Spectrosil and Spectrosil WF samples.

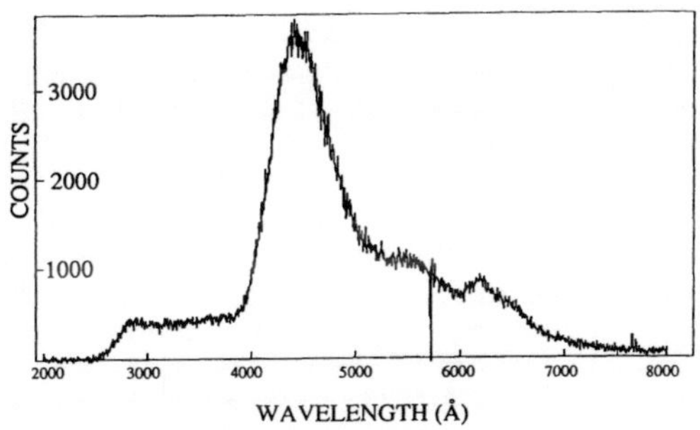

Fig. 4. Bulk luminescence measured at the surface of Spectrosil WF bombarded by 320 eV electrons at normal incidence. The four identifiable luminescence peaks are located at 2880 Å, 4500 Å, 5590 Å and 6370 Å, respectively.

Four distinct luminescence bands appear during electron bombardment of Spectrosil WF and Spectrosil at room temperature ; the spectrum for Spectrosil WF, shown below in fig. 4, is typical. The total bulk luminescence intensity from Spectrosil WF was invariably much larger than from Spectrosil, and no visible line radiation from desorbing excited atoms was observed for either of these two fused silicas. This does not necessarily mean that there were no excited state atoms desorbing, of course, since some of the most prominent Si and O lines occur in the ultraviolet and would not have been detected in the present experiment. Moreover, as noted in the section on the experimental protocol, the present experiment included no provisions for detecting ground-state silicon or oxygen atoms ; these would have to be identified using multiphoton resonance ionization techniques not presently available in our experiments.

A coarse search to find approximate decay times was made by measuring the relative intensities of the four luminescence bands while changing the chopper speed by factors of ten. The intensities of the bands at 2880 Å, 5590 Å and 6370 Å were increased, but that of the 4500 Å band was decreased, indicating that the 4500 Å band has a longer lifetime than the other luminescence bands. Measurements of the luminescence yields centered at 4500 Å and 2880 Å carried out with the chopped electron beam are shown in figures 5 and 6. The measured lifetime of the 4500 Å band is 2.1 ms, while the lifetime of 2880 Å band is less than 10 μs. The bands at 5590 Å and 6370 Å show the same short lifetime.

5

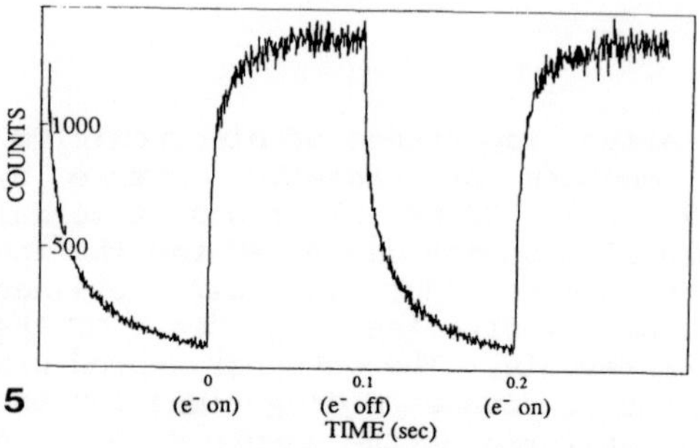

6

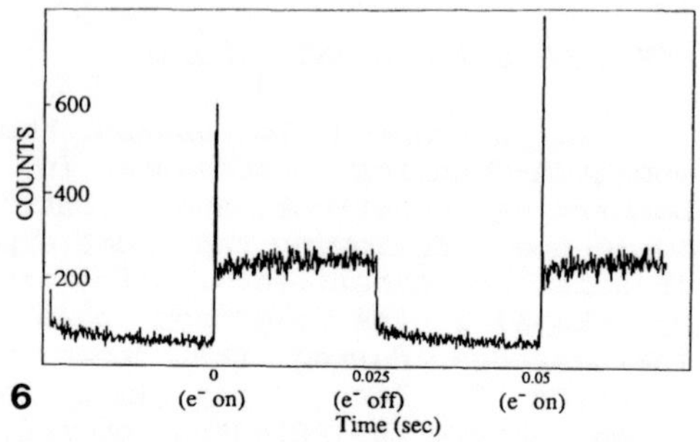

Fig. 5 (at left) Chopped-beam measurement of the bulk fluorescence yield at 4500 Å, showing a decay time of 2.1 ms after the electron beam pulse is turned off.

Fig. 6. (at right) Chopped-beam measurement of the bulk fluorescence yield at 2880 Å, showing a decay time of less than 10 μs. The spike is an electronic artifact.

The long lifetime of the luminescence band at 4500 Å suggests that a metastable state exists between the excited state and ground state of the luminescence center, preventing the direct transition to the ground state (which EPR measurements suggest may be the E' center). This slow light-emission process may result from the diffusion of a defect to some specific position in the glass before relaxation and desorption, similar to the way F-center diffusion mediates the desorption of Li° from LiF surface under electron bombardment[10,11].

The variation in the total bulk luminescence intensity between Spectrosil WF and Spectrosil appears to be correlated with the differing OH content in the two glasses[12] and the same result has been seen for Suprasil 1 and Suprasil W1[13]. Since the bulk luminescence represents a non-desorptive channel for the incident energy, the higher luminescence yield may indicate lessened surface damage. However, a definitive conclusion on this score will require measurement of the ground-state neutral atom desorption.

CONCLUSION

The experiments show that the OH content of fused silica glasses is correlated with distinctive differences in the mechanisms and onsets of ion- and electron-induced radiation damage at the surface and in the near-surface bulk. Because optical spectroscopy can monitor the major particle-emission channels and the bulk fluorescence from irradiated surfaces, DIET experiments of the type described here appear to have great promise for studies of processes-related variations in damage susceptibility in glasses, and for identifying specific mechanisms leading to desorption, erosion and surface damage. Moreover, comparison of the optical emission data and the data produced in our ongoing electron paramagnetic resonance, ellipsometry and thermally stimulated current measurements are expected to permit detailed correlation of the observed microscopic radiation damage events with those macroscopic observables (such as impurity densities and conductivity) which depend on processing parameters.

ACKNOWLEDGEMENT

We gratefully acknowledge the support of this work by the U.S. Naval Research Laboratory under contract N00014-86-C2546.

REFERENCES

1. N.H. Tolk, L.C. Feldman, J.S. Kraus, R.J. Morris, M.M. Traum and J.C. Tully, Phys. Rev. Lett. 46:134 (1981). N.H. Tolk et al., Phys. Rev. Lett. 49:812 (1982);
2. N.G. Stoffel, R. Riedel, E. Colavita, G. Margaritondo, R.F. Haglund, E. Taglauer and N.H. Tolk, Phys. Rev. B 32:6805 (1985).
3. R.F. Haglund Jr. et al., Nucl. Instrum. Meth. in Phys. Res. B 13:525 (1986).
4. N. Itoh and T. Nakayama, Nucl. Instrum. Meth. in Phys. Res. B 13:550 (1986).
5. N.H. Tolk, R.F. Haglund Jr., M.H. Mendenhall and E. Taglauer, in "Desorption Induced by Electronic Transitions (DIET-II)", W. Brenig and D. Menzel, eds., Springer-Verlag, Heidelberg (1985)
6. P. Sigmund, Phys. Rev. 184:363 (1969).
7. See, for example, R. Behrisch, in "Sputtering by Particle Bombardment II", ed. R. Behrisch, ed., Springer-Verlag, New York (1983).
8. A. Manara, M. Antononini, P. Camagni and P.N. Gibson, Nucl. Instrum. Meth. in Phys. Res. B1:475 (1984).
9. D.L. Griscom, in "Radiation Effects in Optical Materials", Society of Photo-optical Instrumentation Engineers Conference Proceedings 541:38 (1985).
10. G.M. Loubriel et al., Phys. Rev. Lett. 14:1781 (1986).
11. T.A. Green, G.M. Loubriel, P.M. Richards, N.H. Tolk, and R.F. Haglund Jr., Phys. Rev. B 35:1781 (1986).
12. J.P. Mitchell and D.G. Denure, Solid-State Electronics 16:825 (1973) ; C.E. Jones and D. Embree, in : "The Physics of SiO_2 and its Interfaces", S.T. Pantelides, ed., Pergamon Press, New York (1978).
13. P.W. Wang et al., to be published in "Desorption Induced by Electronic Transitions (DIET-III) ed. M.L. Knotek.

THE STRUCTURE OF DEFECTS IN DOPED SiO_2

M. Offenberg, P.J. Grunthaner*, D.D. Krut* and P. Balk

Institute of Semiconductor Electronics
Technical University Aachen, D-5100 Aachen, FRG
*Jet Propulsion Laboratory, California Institute of
Technology, Pasadena, Calif. 9110G, USA

INTRODUCTION

The common silicon dopants B, P and As tend to be incorporated in SiO_2 as network formers, i.e. they will attempt to occupy Si sites in the glassy oxide structure. The present study deals with dopants introduced by ion implantation. Since the amount of oxygen is not sufficient to fulfill the bonding requirements of silicon and the dopant the implantation process leaves an oxygen deficient system. Annealing in an O_2-free ambiant will restore the bonding in the network but will not affect this deficiency. The latter is only possible by an O_2 treatment. Consequently, the oxide will have a different defect structure in the two cases. In the present paper we will review data on diffusion, XPS, IR absorption and electron trapping studies supporting this model.

EXPERIMENTAL

The diffusion of dopants was studied in 1 μm thick thermal oxide films grown on (100) Si substrates implanted with 2×10^{15} to 3.5×10^{16} cm^{-2} B, P or As atoms. Profiles were measured as implanted or after a high temperature N_2, O_2 or N_2/H_2 anneal using an Atomika SIMS instrument.

To determine the bonding state of the dopants IR absorption and XPS measurements were performed on similar samples. In the latter case also thinner oxide films (100 nm) were used. The XPS study was carried out using a HP 5950 A ESCA spectrometer with an Al Kα 1486.8 keV source. By measuring after stepwise thinning of the oxide in a dilute HF solution a profile of the bonding state of the dopant at various depths in the oxide was obtained.

The samples for studying the carrier trapping behavior of the film were 0.1 Ωcm (100) p-Si with relatively thin (50 or 100 nm) oxide films implanted with lower doses (5×10^{12} to 1×10^{14} cm^{-2}). Again high temperature annealing steps in N_2, sometimes followed by O_2, were used. The characterization of the trapping properties (trap densities and cross-section) was performed by avalanche injection of electrons using evaporated Al electrodes.

RESULTS

Figure 1 shows data on the P diffusion in SiO_2 at 1100°C. In a N_2 ambient a distinct redistribution takes place ; however, it cannot be described by a simple diffusion mechanism. Particularly near the peak of the concentration profile very little movement is observed. The profiles are best explained by assuming that only 20 % of the P is mobile, with a diffusion coefficient of 5×10^{-15} cm^2/s.[1] This is close to the value published in the literature for the diffusion of P through SiO_2 using P_2O_5[2]. Our experimental findings suggest that P is incorporated into SiO_2 in two different configurations. A similar behavior was observed for the diffusion of As in O_2-free ambients[3].

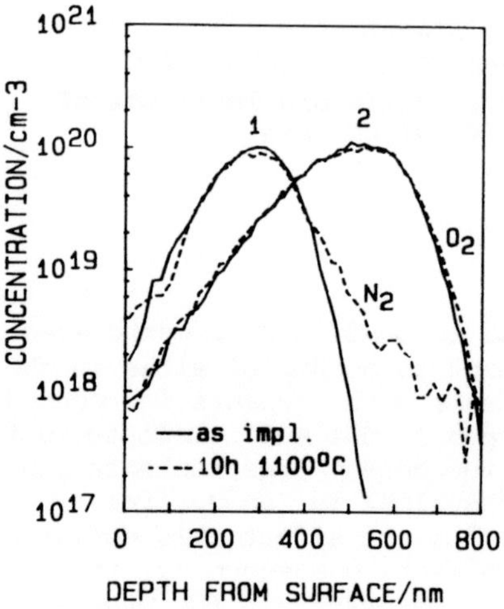

Fig. 1. P diffusion in thermally grown SiO_2 during N_2 and O_2 anneal ; implants :
1 : $2 \times 10^{15}/cm^2$, 200keV ;
2 : $3 \times 10^{15}/cm^2$, 340keV

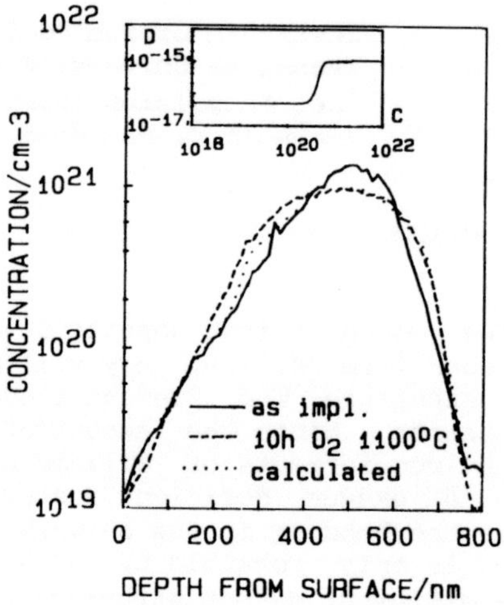

Fig. 2. Redistribution of high dose P implant in SiO_2 during an O_2 anneal ($3.5 \times 10^{16}/cm^2$, 340keV)

In an O_2 ambient hardly any change in the P profile is observable for the same annealing conditions as used with N_2. The diffusion coefficient is below $5 \times 10^{-17} cm^2/s$ in the O_2 case. However, for high concentration implants (peak concentrations around $10^{21} cm^{-3}$) the situation is rather different. A broadening and flattening of the profile near the peak is observed (fig. 2). This profile may be fitted using a concentration dependence of the diffusion coefficient as proposed by van Ommen[4] and indicated in the insert of fig. 2. The concentration at the flat top appears to coincide with the liquidus concentration at this temperature[5]. The appearance of a liquid phase offers a simple explanation for our observations. A very similar behavior was found for As implanted oxides ; the explanation is most likely the same as given for the P data.

In contrast to P and As, implanted B shows extremely slow diffusion not only in O_2 but also in N_2 (fig. 3). However, at very high concentrations (near $10^{21} cm^{-3}$) again a broadening of the distribution near

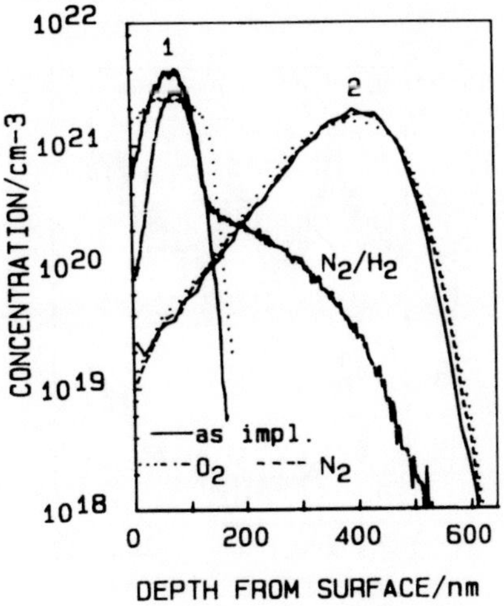

Fig. 3. Diffusion of implanted B in N_2, O_2, and N_2/H_2 ambient (10h, 1100°C) 1: $2x10^{16}/cm^2$ 20keV, 2: $3x10^{16}cm^2$, 100keV

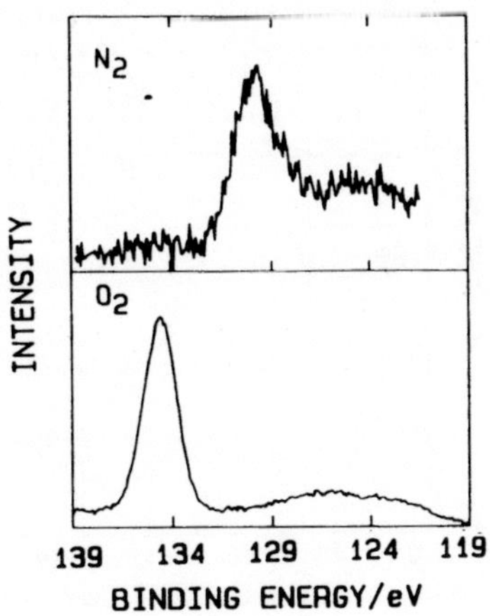

Fig. 4. P 2p core level spectra of P-implanted SiO_2 annealed for 10h at 1100°C in N_2 or O_2.

the peak is observed. We presume that also in this case a liquid phase is obtained. Only in H_2 containing ambients, like forming gas (N_2/H_2), B diffuses relatively fast at low concentrations.

Fitting yields a concentration dependent diffusion coefficient with values up to $5 \times 10^{-15} cm^2/s$ at 1100°C. Previous annealing in O_2 strongly reduces the diffusivity obtainable by a forming gas anneal.

To clarify the bonding conditions in the doped SiO_2 films XPS[6] and IR-absorption [7] measurements were performed. By XPS profiling of the films it was established that a significant shift accross the oxide of the core level binding energy does not occur. Thus, the spectra obtained after removal of half of the oxide were taken as representative and are presented here. There is a clear distinction between in the P signal in P-implanted SiO_2 found for N_2 and O_2 annealed samples, the 2p level being at 129.7 eV in the former and at 134 eV in the latter case (fig. 4). A similar difference was found for these two annealing ambients for As-implanted films (41.4 resp. 45.2 eV, fig. 5). This finding is indicative that the bonding environments of the dopant after treatment in the two annealing ambients are not the same.

B-implanted films show a different behavior (fig. 6). Samples subjected to a high temperature N_2 treatment yield two separate peaks (at 187.4 and 192.6 eV). On the other hand, an O_2 treatment gives rise to only one signal at 192.6 eV.

Interesting details regarding the bonding in P- and B-doped oxides were obtained from an IR absorption study. O_2 annealing led to an increase in the densities of P-O and B-O bonds compared to those in N_2 annealed films. A forming gas anneal did not affect the density of B-O bonds

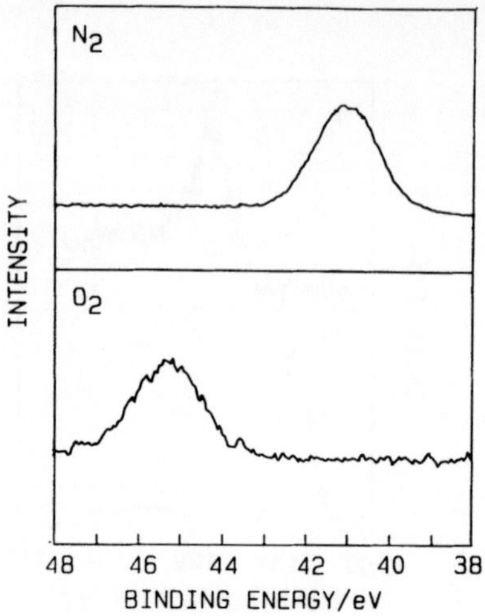

Fig. 5. As 3d core level spectra of As-implanted SiO_2 annealed in N_2 or O_2 (10min 1100°C)

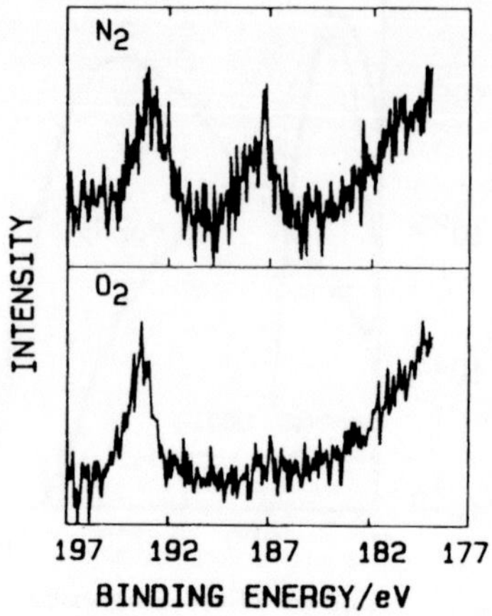

Fig. 6. B 1s core level spectra of B-implanted SiO_2 annealed for 10min in N_2 or O_2 at 1100°C.

compared to films which had been treated in N_2. However, the density of Si-O-B configurations decreased upon treatment in H_2 containing ambients.

Studies on electron[7,8,9] and hole[10] trapping behavior yield further insight into the defect structure of B-, P- and As-doped SiO_2 films. Here some results obtained from investigations of electron injection will be

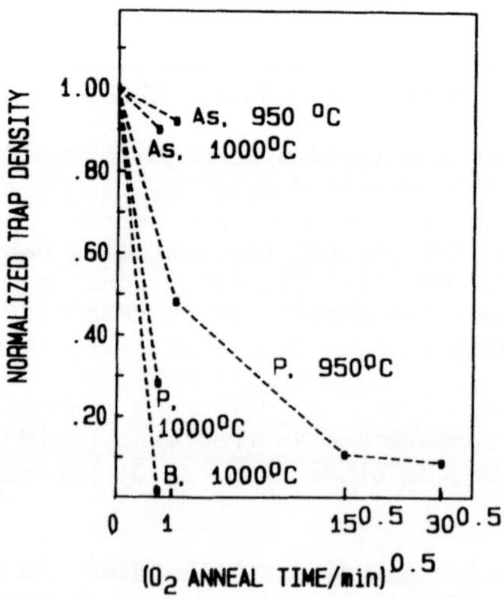

Fig. 7. Normalized density of deep electron traps generated by implantation of dopants vs O_2 annealing time$^{0.5}$.

summarized. The number of traps generated per implanted ion and determined after a N_2 anneal is highest for As, lowest for B. It is somewhat dependent on the dose and amounts to 1.5-2.0 traps per ion for As, 1.0-1.5 per ion for P and 0.05 per ion for B. A subsequent brief O_2 anneal leaves the trap density in As-implanted films nearly unaffected, but causes a considerable decrease in P-implanted and particularly in B-implanted layers (fig. 7). For a given annealing time this decrease is faster at higher temperatures.

Implantation of noble gas ions with mass comparable to those of the dopant ions also gives rise to electrically active defects[11]. However, the defect concentration is lower in this case. These electron traps are not removed by high temperature N_2 annealing, only by O_2 annealing. It appears likely that this type of "radiation damage" also occurs in dopant implanted SiO_2.

DISCUSSION

The data presented in the foregoing indicates that annealing of implanted oxides in an O_2-free ambient tends to result in a defective oxide structure. This is the cause of the trapping behavior of the doped oxides and in the case of P- and As-implanted films of the relatively fast diffusion of these impurities. The distribution of the oxygen deficiency over the network formers will be determined by their relative affinities for oxygen. The trend of this parameter may be estimated from the free enthalpies of oxide formation ΔG_F per mole O_2 for these elements. The values of interest are in kJ/mol [12] :

$1/2\lvert\Delta G_F\ (SiO_2)\rvert$	$2/3\lvert\Delta G_F\ (B_2O_3)\rvert$	$1/5\lvert\Delta G_F\ (P_2O_5)\rvert$	$2/3\lvert\Delta G_F\ (As_2O_3)\rvert$
1020	964	718	563

In the competition for oxygen it is considerably more likely for B to have its valencies saturated by bonding to O than for P or As. Apparently, this is the reason for the strong bonding environment of B and consequently the very slow diffusion of this element also during a N_2 anneal. The oxidation of B proceeds at the expense of that of Si. Thus, in this case the oxygen deficiency and consequently the electron traps are located at Si centers (trivalent Si atoms). Apparently, hydrogen is capable of mobilizing particularly those B atoms which are incompletely bonded to O. This fraction is represented by the low energy XPS signal in the B 1s region. The IR results suggest that a BOH group is involved in the diffusion process. It fits quite well with this concept that an O_2 anneal transforms all B atoms into the fully oxidized state, reduces the possibility of their mobilization by hydrogen and eliminates the Si-centered trapping centers.

Because of the lower oxygen affinity of As and P atoms the oxygen deficiency in N_2 annealed films tends to be concentrated at these impurities. The binding energies derived for these films from XPS measurements are close to those for elements As and P. This suggests that As and P clusters are formed in addition to incompletely bonded dopant atoms. Indeed, spherical As-rich inclusions were directly observed by transmission electron microscopy[13]. Such clusters were only formed in films containing more than 1 % As and after extended high temperature annealing. They appear to be quite immobile. The question if, in analogy to P and As clusters in P and As implanted films, Si clusters may form in B-implanted and N_2 annealed SiO_2 remains open.

The formation of dopant clusters may also explain the observation of immobile P in N_2 annealed P-doped SiO_2. After the long annealing time the segregation is probably fairly complete. Only a relatively small fraction of the dopant remains dispersed through the film, and is incompletely bonded to oxygen. For this reason it is fairly mobile. In the XPS spectra there is no evidence of P-O bonding, which should show up as a higher energy core level signal. However, such signals were reported for P-implanted SiO_2 which had been annealed in O_2-free ambients[14,15]. A possible explanation for this apparent discrepancy may be that in our case long annealing times (10 h) were used compared to those for the quoted literature data (20 to 30 min.). Shorter annealing times may lead to incomplete cluster formation and leave a larger P-O signal.

In the case of the samples used for trapping studies cluster formation is also less probable. The implanted concentrations were considerably lower than those in the earlier discussed samples. Also, the annealing times were shorter. In all likelihood the electrical data therefore are related to point defects rather than the extended defects. The noble gas implantation experiments indicate that independent of the oxygen affinity of the dopant displacement damage remains after high temperature treatment in O_2-free gases. We suggest that these defects are incompletely bonded Si atoms. These add to the complexity of the trap structure of the oxide.

Annealing in O_2 tends to restore the oxygen bonding of dopant and silicon atoms. Due to the higher free enthalpy of oxide formation P is more likely to become fully oxidized than As. The high bonding energy measured in XPS for P is close to the value reported for P_2O_5. These fully oxidized P centers do not trap electrons. Along with the defects centered at the P atoms those at the Si atoms (trivalent Si) will also become fully oxidized. In contrast, As centers are not completely oxidized in an O_2 ambient due to the lower affinity of As for oxygen. The 3d binding energy for this element is observed halfway between the values for As_2O_3 and As_2O_5, suggesting that not fully oxygen saturated $O=As=O_2$ configurations may be the dominant electrically active defects in this case.

CONCLUSIONS

We have shown that the oxide stoichiometry is disturbed by the implantation of dopant atoms into a silicon dioxide layer. At low concentrations the oxygen deficiency results in an increased trap density and at high concentrations, additionally, in the formation of new dopant-rich or possibly Si-rich phases. Annealing in oxygen restores the stoichiometry, reduces the diffusivity at low concentration and removes the dopant related traps except for As. These observations may be explained by simple models involving partially oxidized dopant and Si atoms or also clusters of these species. At high concentration a liquid phase of silicate glass is formed in the fully oxidized systems.

REFERENCES

1. M. Offenberg, M. Maier, F. Schulte and P. Balk, Insulating films on Semiconductors, J.J. Simone and J. Buxo, eds. Elsevier, Amsterdam, (1986) p. 195.
2. M. Ghezzo and D.M. Brown, J. Electrochem. Soc. 120:146 (1973).
3. R. Singh, M. Maier, H. Kräutle, D.R. Young and P. Balk, J. Electrochem. Soc. 131:2645 (1983).
4. A.H. Van Ommen, to be published in "Applied Surface Science", Proceedings INFOS 87.

5. J.M. Eldridge and P. Balk, Trans. Met. Soc. of AIME 242:539 (1968).
6. P.J. Grunthaner, D.D. Krut, M. Offenburg and P. Balk, to be published.
7. M. Offenberg, M. Maier, R. Meyer and P. Balk, J. Vac. Sci. Technol. A 4:1009 (1986).
8. M. Offenberg, T. Johannson, M. Aslam and P. Balk, Physica 129B:240 (1985).
9. S. Alexandrowa and D.R. Young, J. Appl. Phys. 54:174 (1983).
10. H. Neitzert, M. Offenberg and P. Balk, to be published in "Applied Surface Science", Proceedings INFOS 87.
11. M. Offenberg and P. Balk, to be published in "Applied Surface Science", Proceedings INFOS 87.
12. I. Barin and O. Knacke, Thermochemical Properties of Inorganic Substances, Springer, Berlin,(1973).
13. C.K. Celler, L.E. Trimble, K.W. West, L. Pfeiffer and T.T. Sheng, Appl. Phys. Lett. 50:664 (1987).
14. K.T. Wu and A.W. Saxena, J. Electrochem. Soc. 132:932 (1985).
15. K. Hoh, M. Saitoh and Y. Miura, J. Electrochem. Soc. 128:1613 (1981).

ELECTRICAL AND OPTICAL CHARACTERISTICS OF VANADIUM DOPED AMORPHOUS SILICON DIOXIDE FILMS PREPARED BY CVD.

K.V. Krishna, J.J. Delima, A.J. Snell and A.E. Owen

Department of Electrical Engineering
University of Edinburgh
The King's Buildings
Edinburgh EH9 3JL, Scotland

INTRODUCTION

Previous experiments on pure and vanadium (V) doped amorphous silicon dioxide (a-SiO_2) films prepared by r.f. sputtering have shown that both the V-doped and undoped films can be switched reproducibly between two stable states of widely differing conductivity by field stressing the films[1,2]. Similar experiments using chromium, colbalt or titanium doping have also resulted in significant changes in the electrical properties of the oxide, but in these cases the material was less stable, even when annealed, with instabilities present at low to moderate fields. The behaviour of such materials is of great interest from both pure and applied aspects, but greater control over preparation is desirable. In the present paper we report the electrical and optical properties of V-doped a-SiO_2 films prepared by CVD and compare them with the previous results on sputtered films.

MATERIAL PREPARATION

An "Advanced Semiconductor Materials" atmospheric pressure CVD silicon oxide reactor (Model No. 10-2) was used for the deposition of the pure and V-doped a-SiO_2 films. Oxygen and 5 % silane in nitrogen were used as reactive gases, with the oxygen to silane ratio set at 8:1. The oxides were generally deposited on n-type 1 ohm.cm resistivity silicon wafers although some samples were grown on tungsten sheets to provide metal-insulator-metal (MIM) structures. Undoped a-SiO_2 films were deposited initially, and the quality of the films was characterised by deposition rate (~ 500 A/min.), uniformity, refractive index and IR absorption.

Vanadium was introduced into the oxide by passing nitrogen through a bubbler filled with $VOCl_3$ at room temperature, and hence into the reactor through a dopant line. It was observed that for low dopant gas flow rates there was no appreciable change in either the deposition rates, the refractive indices or other properties of the resulting films. For flow rates above a critical value, the deposition rate was reduced to about half the original rate whilst the refractive indices of the films increased. Initially it was assumed that the reduction in growth rate was due to insufficient oxygen, as $VOCl_3$ is known to react vigorously with oxygen.

However, further increase in either oxygen or dopant gas flow rates always resulted in a further drastic reduction of the deposition rate, and the films obtained had an opalescent, milky appearance. The refractive indices of these films, as determined by a combination of taly-surf and ellipsometer measurements, were in excess of 2.5, although it is doubtful if such measurements are valid because of the hazy nature of the films. It was also noted that the adhesion of such films to silicon and metal surfaces was very poor.

Decreasing the oxygen to silane ratio from the original 8:1 to 5.5:1 did not seem to have any significant effect on the deposition rate or refractive indices of the doped films, in contrast to the undoped case where the reduction of this ratio acts to reduce the deposition rate. A range of films with oxygen to silane ratio from 5.5:1 to 8:1 were made and the experimental results on these films are reported below.

IR RESULTS

Infra-red measurements were carried out using a Perkin-Elmer 598 spectrophotometer. Typical transmission measurements for both undoped and V-doped films are shown in figure 1, from which it can be seen that the undoped films absorb at 1100 and 800 cm^{-1} while for V-doped films an absorption around 900 cm^{-1} is also observed. These results are consistent with previous observations on films prepared by co-sputtering[3]. No changes in the IR results were observed in films grown with a range of oxygen flow rates as described earlier. The values of the wave numbers at which peaks were observed before and after annealing together with other parameters are given in Table 1.

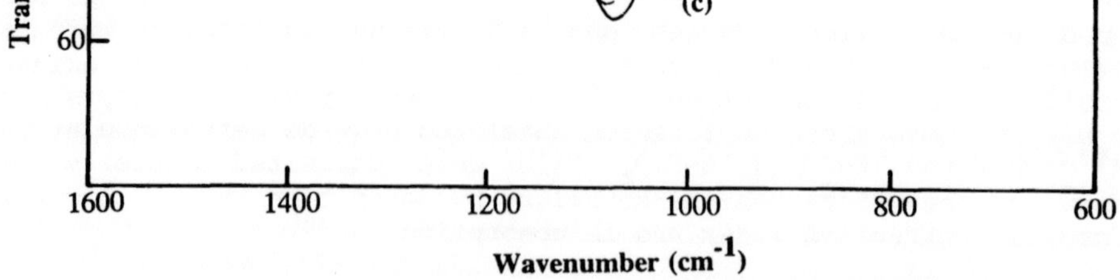

Fig. 1. IR spectra for pure and V-doped SiO_2 films grown by CVD. The total nitrogen flow rate was 45 l/min and the doping line flow rate was 270 cc/min. (a) Undoped oxide, O_2 flow rate 450 cc/min. (b) Doped oxide, O_2 flow rate 450 cc/min. (c) Doped oxide, O_2 flow rate 300 cc/min.

Table 1. Details of IR and refractive index observations for pure and V-doped CVD oxide films.

	Undoped	Doped O_2 flow 300cc/min	Doped O_2 flow 350cc/min	Doped O_2 flow 400cc/min	Doped O_2 flow 450cc/min
Growth rate (A/min)	500	245	210	210	260
Thickness	1000 A	980 A	840 A	840 A	1040 A
Ref. Ind.	1.45	1.55	1.60	1.55	1.55
Position of IR peaks (cm^{-1}) before annealing	1062 --	1067 915	1065 905	1065 905	1070 910
After annealing	1073 --	1080 920	1080 910	1080 910	1080 915

From the table it is clear that the as-deposited undoped films are slightly oxygen deficient as deduced from the value of the peak wavenumber, and annealing brings these values close to the values reported for thermal oxides. In the case of the doped films the 1100 cm^{-1} peak occurs at a slightly higher wavenumber than in the undoped films. This is contrary to previous observations on co-sputtered films, in which the value of the wavenumber of the peak is always slightly less for V doped films than for undoped oxides[4]. This may be attributed to differences in the film preparation and thickness. The 900 cm^{-1} peak was also observed to move to higher wavenumbers after annealing, in line with the previously reported results[4].

X-ray energy dispersion analysis (EDAX) measurements carried out on these samples do show the presence of small percentages of vanadium. While this is a qualitative result rather than a quantitative one, it does appear that in these oxides the quantity of vanadium is less than in a reference doped film prepared by co-sputtering where the vanadium content was estimated to be about 4 % from Auger and RBS measurements[4].

MEASUREMENTS

After annealing the samples at 900 C in nitrogen, sintered nichrome contacts were made to the back of the wafer. Aluminium contacts were evaporated on to the oxide through a shadow mask to define simple MOS structures with a range of areas.

Current-voltage measurements were conducted in a nitrogen ambient after baking the device at 150 C to remove any absorbed moisture and hence minimise surface conduction. The results obtained can be divided into two main categories. V-doped oxides deposited with the highest concentrations

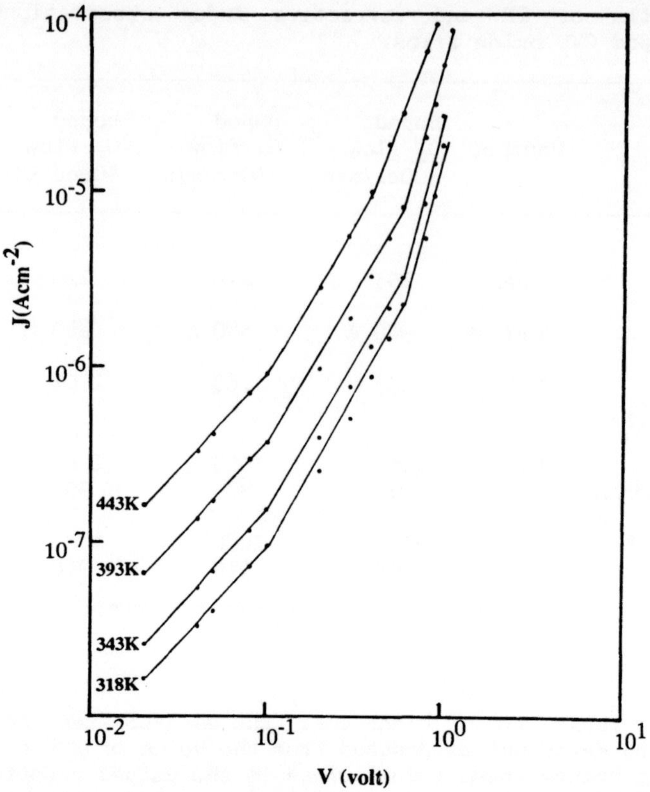

Fig. 2. J-V measurements for a doped oxide prepared with O_2 flow rate of 300 cc/min.

of oxygen showed resistance instabilities at all temperatures, even at the lowest applied fields. Under these conditions it was not possible to make any steady state measurements on the devices. The current instabilities observed were more pronounced than those previously observed in chromium and titanium co-sputtered films. Oxides deposited with lower oxygen flow rates showed ohmic behaviour at low fields and non-ohmic behaviour at higher field strengths. The activiation energy obtained is 0.22 eV, and this compares well with values obtained for co-sputtered V-doped films in their high conducting state[5].

The high field conductivity behaviour is shown in fig. 2 ; space charge limited current conduction accounts well for the observed behaviour at moderate and high fields[5]. One major difference noted in these observations compared with previous results was that in the CVD grown films with low flow rates of oxygen it was not possible to cycle between the low- and high- conducting states by the application of a field. In other words the bistable switching seen in co-sputtered material is absent in CVD films deposited with low oxygen flow but is present in CVD films deposited using higher oxygen flow rates, resulting in a family of unstable stages as a consequence of the resistivity fluctuation.

Resistance fluctuations have been recently observed in small dimensional (submicron) MOS devices by various authors[6,8]. These effects, known as "two level fluctuation" (TFL), metastable states or "random telegraph signals" (RTS), are believed to be a fundamental property of the amorphous material ; namely the availability of multiple charge equivalent states via lattice relaxation and lattice reconstruction. Most of these observations are on thermally grown tunnelling oxides where the two level and multi level transitions in resistance are thought to be a result of ensembles of strongly interacting traps emptying and filling simultaneously. Our observations may be related, and if so it appears that the trap state distribution in the CVD deposited films is affected by the oxygen flow rate. Moreover, the evidence of earlier work[5] suggests that r.f. sputtered oxides (doped or undoped) are also different in this respect. At low oxygen flow rates no switching is observed while at higher rates the switching is more complex than previously observed in co-sputtered V-doped films. A careful study of electrical noise as a function of temperature may be helpful in understanding such behaviour in these doped films.

In the light of the above observations, it would be of interest to investigate the behaviour of small-area tunnelling structures such as MISS devices or floating gate memory devices incorporating these oxides. If the multi-level fluctuations are an intrinsic property of these doped oxides, it is expected that the switching delay time observed in MISS structures as a function of voltage will be discontinuous. Also, the threshold voltages at which these devices operate is likely to be modified compared to conventional devices.

REFERENCES

1. J.J. Delima, K.V. Krishna and A.E. Owen, Phil. Mag. B53:115 (1986).
2. K.V. Krishna, J.J. Delima and A.E. Owen, J. Non Cryst. Sol. 77 & 78:1321 (1985).
3. J.J. Delima, A.J. Snell, K.V. Krishna and A.E. Owen, J. Non Cryst. Sol. 90:291 (1987).
4. J.J. Delima, Ph.D. Thesis, "The Electronic Properties of Pure and Transition Metal Doped a-SiO_2 films" (1987), University of Edinburgh.
5. K.V. Krishna, J.J. Delima, F.C. Eze and A.E. Owen, Physica B129:245 (1985).
6. K.R. Farmer, C.T. Rogers and R.A. Buhram, Phys. Rev. Lett., 58:2255 (1987).
7. C.T. Rogers and R.A Burham, Phys. Rev. Lett., 55:859 (1985).
8. M.J. Kirton, M.J. Uren and S. Collins, INFOS 87 Proceedings, to be published.

ACKNOWLEDGEMENTS

The authors would like to thank Gerry McDade for his technical help. Financial assistance received from Hughes Microelectronics Ltd., the Science and Engineering Research Council and the conference organizers is also gratefully acknowledged.

THEORY OF DANGLING ORBITAL DEFECTS AT THE <111> Si/SiO_2 INTERFACE

Arthur H. Edwards

U.S. Army LABCOM
Ft. Monmouth
N.J. 07703-5000
U.S.A.

ABSTRACT

We review the experimental and theoretical pictures of the dangling orbital defects at the <111> Si/SiO_2 interface. We show that recent theory agrees in detail with the observed hyperfine and super hyperfine. Also, the estimated level positions obtained from cluster calculations imply that the dangling orbital on the <111> interface is a normal, positive U defect. This is at variance with recent local-density Green's-function calculations on the dangling orbital in crystalline silicon, that predict that the dangling orbital is a negative U defect. We discuss possible sources of the discrepancy. Finally, calculations on the recently proposed model for the P_b defect involving a three-fold coordinated oxygen atom yield general disagreement with spin resonance data, as well as with electrical measurements. Hence, the original, simple model of a trivalent silicon atom continues to be strongly preferred.

INTRODUCTION

The interface between bulk crystalline silicon, and amorphous SiO_2 continues to be great scientific and technological interest. Technological interest arises from the deleterious effects of charge traps localized at this interface. The first metal-oxide-semiconductor (MOS) devices exhibited large densities of these interface states, as measured implicity by soft inversion thresholds, and explicity by voltage-dependent capacitance and conductance methods (C-V, G-V)[1], as well as the related deep level transient spectroscopy (DLTS)[2]. Subsequently, it was found that a low temperature (450° C) hydrogen anneal reduced the interface state density[3,4]. However, the effects of this anneal can be eliminated if the interface is subjected to ionizing radiation[5] . Hence, the need for MOS devices that are tolerant to a variety of radiation environments ensures technological interest in the fundamental structure of the Si/SiO_2 interface.

Scientific interest arises, in part, from our apparent ability to study a certain class of defects, dangling orbitals of well-characterized crystallographic orientation, at this interface. These defects have been shown to contribute to the spectrum of technologically interesting, electrically active defects, so that their study is more than idle curiosity[2,5]. In fact, some investigators believe that all the electrically active interface states are dangling bond defects[6].

In this paper, we endeavor to review the current understanding of dangling orbital defects on the <111> Si/SiO_2 interface. This could encompass a large area of research, including electronic structure, creation and annihilation kinetics, interaction with radiation and injected charge, and interaction with extrinsic species, i. e. H. F. Cl. We will limit this discussion to electronic structure and microscopic models for defects observed in electron spin resonance, and in the various electrical measurements listed above. Specifically, we will review the experimental and theoretical studies of the P_b defect at the <111> Si/SiO_2 interface.

EXPERIMENTAL BACKGROUND

The first spectroscopic data on charge traps at the <111> Si/SiO_2 interface were from electrical measurements[1]. These measurements yield electrical level positions of interface states in the silicon band gap. In principal, these techniques are also sensitive to deep levels within the depletion layer. However, these have been ruled out, except for a narrow dead layer, ~ 100-300 A thick, by DLTS performed on Schottky diodes. Early measurements showed that the spectrum of interface states, D_{it}, was a rather featureless, U-shaped continuum[1]. Later, Johnson et al.[2] found that for very large interface state densities, $D_{it}(E)$ exhibited peaks, as shown in fig. 1. We note that Poindexter et al. have virtually reproduced this data using the low frequency C-V technique[7]. As we will discuss below, these peaks arise from two inhomogeneously broadened discrete levels associated with the silicon dangling bond at the <111> Si/SiO_2 interface.

Over the past 18 years several groups have studies the <111> Si/SiO_2 system using spin resonance[5,8-11]. Early studies discuss multiple (up to three) paramagnetic centers. Of these centers, Poindexter showed that one is associated with the conduction electron in phosphorus doped silicon, and another with the neutral Fe center.

The remaining signal, labeled P_b by Nishi, has been studied intensely over the past ten years. Poindexter et al.[11] was the first group to argue convincingly that the P_b center was a trivalent silicon atom at the Si/SiO_2 interface. The experimental features that supported this model are

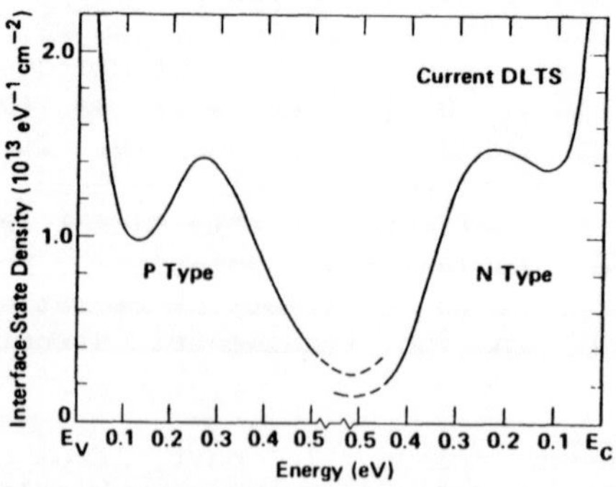

Fig. 1. Interface state distribution. E_v and E_c correspond to the valence -and conduction- band edges respectively. From ref. 2.

these :(1). The g-dyadic is axial- consistent with a C_{3v} defect attached to the silicon lattice. Furthermore, the g-dyadic anisotropy is of similar magnitude to other dangling orbital defects in crystalline silicon[12] (2). Only one of four equivalent <111> orientations was observed and this one was perpendicular to the Si/SiO_2 interface. (3). The signal dissappeared after an HF etch.

While the work of Poindexter et al. lent very strong support to the trivalent silicon model of the P_b center[13] unambiguous identification awaited Brower's ^{29}Si hyperfine data[14] . Because this data is of great importance to the calculations we discuss below, we include it as fig. 2. This data firmly established that the P_b resonance arises from an unpaired electron localized strongly on a single silicon atom in an s-p hybrid orbital that is perpendicular to the Si/SiO_2 interface. The current picture of this defect is shown in fig. 3.

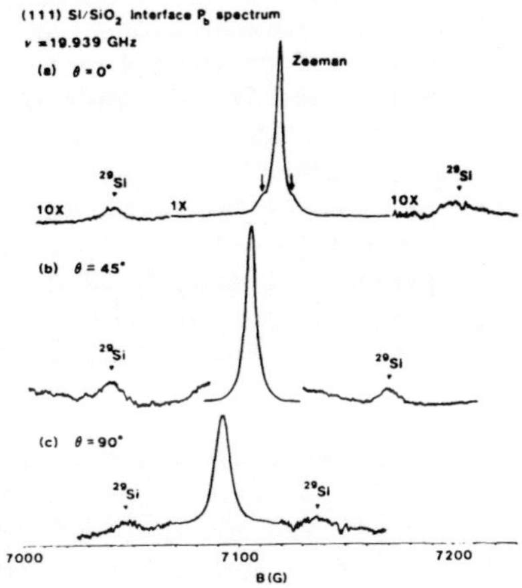

Fig. 2. Experimental hyperfine spectra for P_b center at the <111> Si/SiO_2 interface, θ measures the angle between the magnetic field and the <111> direction. After Ref. 14.

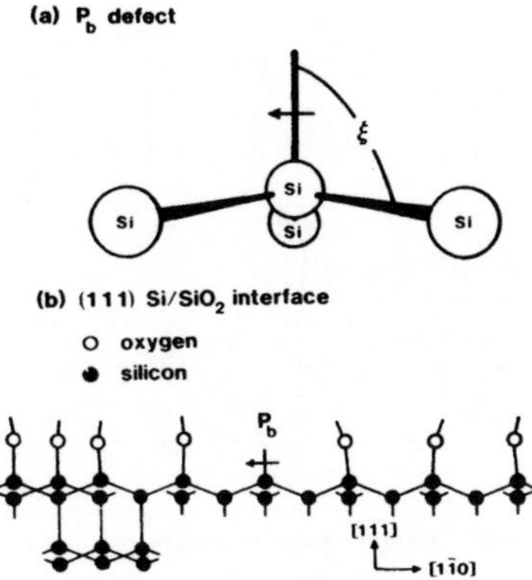

Fig. 3. Current model for P_b center. After Ref. 21.

Recently, Brower has performed several very detailed studies on the line shape of the P_b resonance[15-17]. In these he considered line broadening due to magnetic dipole-dipole interactions, and strain. These studies are important because they estimate the extent to which the simple model shown in fig. 3 is valid.

This study of strain-broadening addressed the question of atomic flatness of the interface, and the uniformity of the defect environment [15,17]. Brower found the strain manifest in a distribution of values for both $g_{\perp}$ and A_{iso}. From the variation in $\delta g_{\perp}$ as a function of magnet angle[18], and using the simplest relationship between the bond angle (ξ) in fig. 3, and the 3p component of the defect wave-function[18], Brower estimated that the maxium standard deviation of ξ would be approximately 5°. This is not to say that the orbital will 5° away from the <111> direction, but that the defect atom will move along the <111> direction ~0.2 A to achieve this 5° variation.

The linewidth of the hyperfine lines is a much more direct measure of the strain, as the hyperfine interactions are proportional to matrix elements of the ground state defect wave-function, while, in the second order perturbation theory expansion, $\delta g_{\perp}$ depends on ground-to-excited state energy differences as well (changes in these were ignored in Brower's simple treatment)[20]. From estimated changes in the s and p wave-function components as a function of atomic position[19], Brower estimates that the standard deviation in ξ is approxiamtely 0.5°, i. e. the vertical position of the defect atom is uncertain to only 0.02 Å. We will argue below that this estimate is roughly a factor of five too small. However, even uncertainly of 0.1 Å reflects a surprisingly flat interface in the neighborhood of the P_b center.

The study of magnetic dipole-dipole broadening is much less conclusive [15,16]. No resolved dipole-dipole structure is observed, indicating that neighboring P_b centers are separated by at least 13 Å. Furthermore, of the total linewidth, 2.1 Gauss, the component due to dipole effects is only 0.35 G, which is too small to detect. This implies that P_b centers are either spin paired (diamagnetic) at reasonable distances, or that they are anticorrelated, i.e. that they avoid each other. This might be true if they were associated with strain relief.

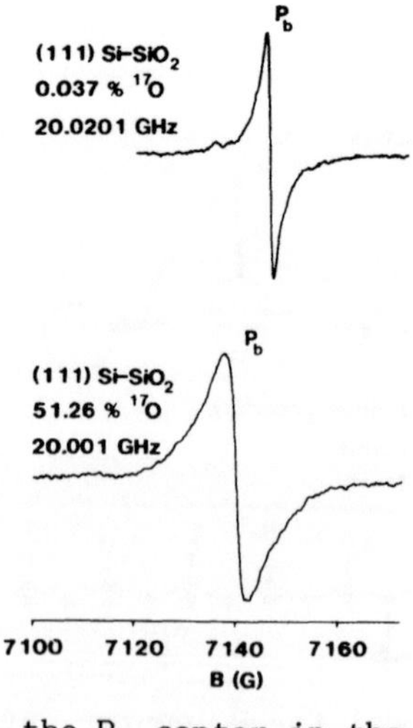

Fig. 4. ESR Spectra of the P_b center in the presence of two levels of ^{17}O oxide enrichment after Ref. 21.

Brower has also performed ESR studies on ^{17}O enriched oxides grown on <111> silicon surfaces. The results of these studies are shown in fig. 4. Here, we note that there is only broadening, indicating that oxygen is probably not a nearest-neighbor.

Several groups have performed electric field-modulated ESR measurements[2,5,7,10]. In these studies an electric field is applied to the surface of the SiO_2, either by corona discharge[5,10] or by use of a metal gate electrode[2,7]. This field modulates the surface potential at the Si/SiO_2 interface Psi, thereby modifying the occupation of the interface states. It was found that for Psi < 0.3 eV and for Psi > 0.85 eV (measured relative to the valence-band edge), there was no P_b spin signal. However, for 0.3 V < Psi < 0.85 eV, the P_b spin signal was observed. This is consistent with the two peaks in fig. 1 arising from the +/0 and 0/- transitions for the P_b center. Hence, this defect has an effective correlation energy, U, of approximately 0.55 eV.

Before we turn to the theoretical studies of this defect, we should point out the striking similarities between this defect and the D^o defect in a Si[22]. First, a powder pattern spectrum for the P_b center, wherein the P_b signal is averaged over all angles, bears striking resemblance to the ESR data for the D° center[23]. It was on this basis that the D° center was assumed to be a dangling orbital. Second, the hyperfine interaction for D° is also primarily on a single silicon atom, although the splitting is somewhat reduced[24]. Cook has discussed this difference at length, and shown that this difference does not imply a different structure, but rather reflects subtle but important differences in the defect environment[25]. Finally, the level structure for the defect is quite similar to that of the D° center[23] That is to say, both defects have a level in the lower and upper halves of the band-gap, and both are paramegnetic with the Fermi-level between these two levels. The U for D°~0.4 eV, is slightly lower than the value for the P_b center, 0.55 eV. These similarities lead us to assume that the theory for these two defects is approximately transferable. Calculations performed on the P_b center will probably hold at least qualitatively for D°, and vice-versa. There is notable exception to the strong qualitative similarity between these two defects. Johnson and Jackson have reported optical data for the P_b center[26]. Using an electric field modulation technique similar to those described above (in their case, they used a gate electrode), they have switched the P_b occupation between zero and one electron (between the + and 0 charge states) and, using phase-sensitive detection, they have measured optical absorption. From this study, they conclude that the dominant absorption arises from electron capture from the valence-band into the neutral, singly-occupied state. For the D° center, the principal absorption is between the singly occupied defect and the conduction band[27].

THEORETICAL STUDIES

Over the past eleven years, many groups have studied the dangling orbital at the <111> Si/SiO_2 interface[28-41]. We have previously discussed these at length[39], so that here we will only consider those calculations that can be compared directly with experiment.

We will first discuss the estimation of hyperfine parameters. Because these are sensitive to lattice relaxations, we consider only those calculations that include atomic motion explicity. The first such set was performed by Redondo et al[28,29]. Using an ab-initio, generalized valence-bond technique, they calculated equilibrium geometries of a Si_4H_9

cluster for three charge states (+, 0, -). In these calculations, the three backbonded silicon atoms were held in crystalline silicon positions, the Si-H bond directions were along appropriate crystalline silicon direction, and R_{Si-H} = 1.48 Å, the equilibrium bond length in SiH_4. The equilibrium position of the defect atom in the positive, neutral, and negative charge states were -0.38 Å, -0.08 Å, and +0.17 Å respectively, where - (+) indicate that the defect atom moves toward (away from) the plane of the three backbonded silicon atoms, always along the <111> direction. These relaxations have been qualitatively reproduced in other ab-initio and semiempirical calculations[39]. Furthermore, we expect it on intuitive grounds. That is to say, in the neutral state we expect the defect atom to relax toward its three nearest neighbors to strengthen the three bonds at the expense of the dangling orbital. While Redondo et al didn't calculate the hyperfine interactions, they do quote the localization and decomposition of the defect wave function (93 % of the wave function on the central silicon atom with 93 % p-character). Using these estimates, we obtain hyperfine values that are strikingly too low. This result arises from an apparent over-estimation of p-character, rather than underestimation of localization. It may also reflect neglect of spin polarization of the valence-bands (i. e. discussion of only the wave-function rather than the full spin density matrix).

Recently Edwards[33,40] and Cook[25,40] have performed calculations on large atomic clusters. Although their techniques were quite different (semiempirical molecular orbital technique[33,39] versus scattered-wave X-α [25,40]), their results were strikingly similar. Their calculated hyperfine parameters are given in Table I, along with the experimental values.

Table I. Calculated and Experimental hyperfine and super-hyperfine parameters for the P_b center at the <111> Si/SiO_2 interface. All values in Gauss.

$A_{\parallel}$	$A_{\perp}$	
187.7	99.0	Ref. 40
179.2	123.9	Ref. 39
156.0	91.0	Ref. 14 (exp.)

We note that these were calculated with the defect atom in the same position relative to its three nearest-neighbors, but with different surface termination (OH groups[39] versus H atoms[40]). Cook's values were calculated from first principles within the X-α approximation, while Edwards combined his calculated spin density matrix with ab-initio Hartree-Fock atomic values for $|\psi_s(0)|^2$ and $\langle \psi_p | r^{-3} | \psi_p \rangle$[42].

To appreciate the quality of these results, we show in fig. 5 the dependence of the calculated hyperfine parameters on the position of the defect atom above the plane of its three nearest neighbors. We have included results from Edwards[39] (solid lines) and Cook[40] (dashed lines). Several features bear comment. We note the remarkably close correspondence between these two calculations. Of particular interest is the virtual equivalence in slopes. These slopes speak directly to the range of vertical positions one would predict from the hyperfine line width. Recall that, based on elementary orthogonality considerations, Brower estimated that ~9 Gauss of excessive linewidth broadening implies -0.02 Å variation in defect atom position. Using the same assumed broadening, fig. 5 predicts about

0.1 Å variation (larger by a factor of 5). Because the slopes are virtually identical, this estimate is independent of theoretical technique. Finally, considering the sensitivity of the hyperfine interactions to small changes in atomic position, we consider the agreement with experiment (within 10 % at the MOPN equilibrium) to be excellent.

It is important to note that both of these calculations predict the existence of resolvable super hyperfine interactions on three second-nearest-neighbors, labeled B in fig. 6. The predicted magnitudes (21 G[39] and 13 G[40]) are in good-agreement with the observed value (~ 18 G) denoted by the arrows in fig. 2. Also the intensity of the super hyperfine signal is much larger than the principal hyperfine signal (note the vertical scale of the principal hyperfine is ten times larger than that of the central line). We would expect a factor of three increase, because there are three possible second-nearest-neighbors that could contribute to this signal, and hence three times the probability of a magnetic nucleus compared to the central atom.

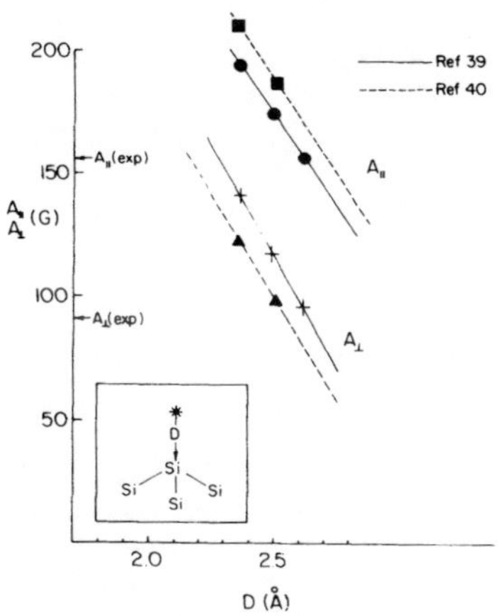

Fig. 5 Defect atom hyperfine interactions as a function of atomic position. Motion is along <111> direction. D = 2.35 Å is ideal tetrahedral position. D_{eq} = 2.44 Å.

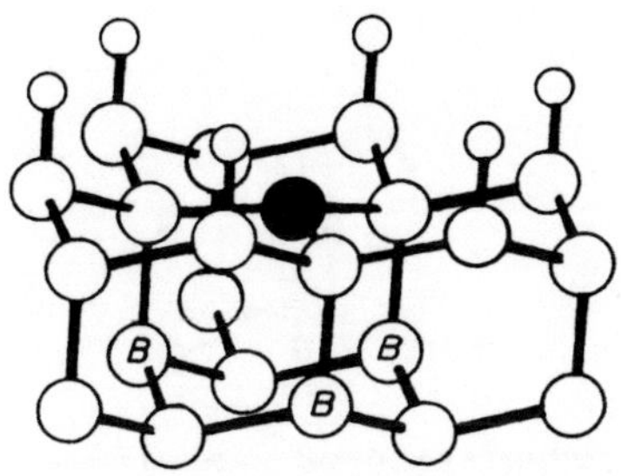

Fig. 6. Largest cluster used for P_b calculations (see ref. 39). B denotes atoms with large super hyperfine interactions.

The degree to which the theory agrees with experimental hyperfine argues strongly that the simple model shown in fig. 4 is a very good approximation for the real defect. We note that the understanding of both the principle and super hyperfine interactions could not be obtained from the most exact calculation on clusters without the correct second-nearest-neighbor coordination. Finally, we note that in both calculations spin polarization of the valence-band contributed significantly to the defect atom hyperfine values.

We now discuss the calculated electrical levels for this defect. Various groups have made estimates of these. Several groups performed self-consistent tight-binding calculations of the dangling bond in either crystalline or a-Si. These calculations do not include lattice relaxations, so that they could be inaccurate by - 0.25 eV. The estimates from these techniques range between 0.5[35-37] and 0.75 eV.[41]

There are only two total energy calculations in the literature where-in lattice relaxations have been explicity included. In his cluster calculations, Edwards estimates the effective correlation energy (U) to be between, 0.32 and 0.6 eV, depending on the method of estimation. On the other hand, in a calculation on the dangling orbital in crystalline silicon, Bar-Yam and Joannopoulous[38] obtained a value of -0.2 eV. In their calculation, they used a self-consistent super-cell Green's-function local-density technique. It is important to note that a negative U implies that in thermal equilibrium the neutral paramagnetic species would not be observed[43]. To see this we point out that U is defined by

$$U = E^{+} + E^{-} - 2\,E^{o} \tag{1}$$

where the superscripts refer to charge states of the defect, and E refers to total energy. If this is negative it implies that two neutral P_b centers would spontaneously transform into one positive center and one negative center.

There are three possible conclusions to be drawn. (1) The Bar-Yam-Joannopoulos calculation does not include some important piece(s) of physics, and this (these) omission(s) lead to an erroneous value for U (2). The calculation is correct, but it only applies to the specific defect considered, and not to the P_b center. (3) The calculation is correct and the model for P_b must be abandoned. It is our belief that (1) is probably the case. Recall that local density calculations rely on an ansatz that the exchange energy has the same form in an atom, or a solid that it does for

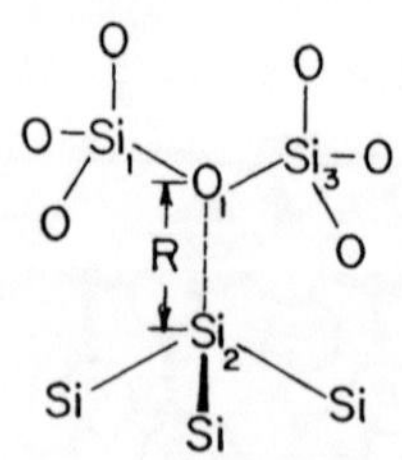

Fig. 7. Schematic representation of Pantelides model of the P_b defect.

the free electron gas. While calculations using this approximation have yielded good agreement with a variety of experimental properties, it underestimates the band-gap in silicon by 0.5 eV[44]. We also point out that the Bar-Yam-Joannopoulos calculation is not spin polarized.
If the neutral state were spin polarized, the extra variational freedom would probably lower the energy of this state. The factor of 2 in Eq. 1 increases the possible importance of spin polarization. We point out that there are several well-understood defects in crystalline silicon that involve dangling orbitals. Calculations of these defects using local-density formalism would go far to establish the credibility of the Bar-Yam-Joannopoulos calculation.

Recently, on the basis of the Bar-Yam-Joannopoulos calculation, Pantelides has proposed a new model for the P_b center[45]. This is shown schematically in fig. 7. Pantelides argued that the oxygen atom would influence the U of the defect, but not the spin resonance parameters.

We have performed a set of calculations to understand the physics of this model defect[46]. Using the cluster shown in fig. 6, to which we added a $Si_2O_7H_6$ as shown in fig. 7, we performed spin unrestricted molecular orbital calculations using the semiempirical program MOPN[47]. From these we obtained level positions and hyperfine parameters at equilibrium geometries. We investigated the variations in our results as a function of the position of the $Si_2O_7H_6$ unit respect to the interface. Our method of choosing this position deserves discussion. We start each calculation with the $Si_2O_7H_6$ at some prescribed distance, denoted R in fig. 6, from the defect. The Si-O-Si bond angle is set at 144°, and all R_{Si-O} are set to 1.63 Å, the MOPN average equilibrium Si-O bond length. During the calculation, the outer O_3H_3 units are fixed, while the central oxygen and silicon atoms are allowed to search for the minimum energy positions. Our results our summarized in Table II.

Table II. Hyperfine interactions (^{29}Si and ^{17}O) and defect energy levels as a function of R for the overcoordinated model of the P_b center. R is shown in fig. 6. Hyperfine interactions are in Gauss, R is in Angstroms, and the defect energy level position with repect to the valence band is in eV.

R (Å)	$A_\perp$ ^{29}Si	$A_\parallel$ ^{29}Si	A ^{17}O	$E_d - E_{v.b.}$
2.0	308.0	300.0	30.0	1.2
3.0	136.0	188.0	5.0	0.25
∞	123.0	179.0	0.0	0.05

Our estimated effective correlation energy (calculated from one-electron energy levels) is 1.0 eV at R = 2.0 Å, compared with 0.6 eV for R = ∞.

Several features of these calculations bear comment. First, contrary to Pantelides assertions, the hyperfine parameters on the central silicon atom change dramatically if the oxygen atom is part of this defect. The reason for the dramatic change is the strong silicon s admixture. This results from motion of the defect atom toward the oxygen atom, i. e. out of the plane of its three nearest-neighbors. Also, the oxygen atom has enough spin density to cause measureable hyperfine interaction. We recall that Brower observed no ^{17}O interaction that would agree with our prediction. Second, the energy levels also change dramatically, with the neutral level traversing the entire band-gap. This was also seen, by Sakurai and Sugano in their tight-binding Green's -Function treatment of the <111> interface[32] The increase can be understood intuitively. As we move the stoichiometric $Si_2O_7H_6$ unit toward the defect we are admixing states of this moeity that are usually unoccupied hence raising the energy. Because the band gap for SiO_2 is ~8.5 eV, admixture of these unoccupied states would cause large changes, as we have observed. Our last point is not shown in Table II, though we should have expected it. It turns out that the $Si_2O_7H_6$ unit breaks the C_{3v} symmetry of the P_b center so that the defect atom actually goes off center. This leads to significant spin density on one of the nearest neighbors (180 G !).

These results indicate that for each observable we could calculate, the Pantelides model gives results at variance with experiment. Also, we could trace the degradation of agreement to rather simple physical origins. We conclude from this that the simple model of a trivalent silicon atom in C_{3v} symmetry is the best model to date for the P_b center. Furthermore, calculations on this model have yielded results that not only agree with experimental hyperfine, but that add new insight into the origin of the observed super hyperfine interactions.

REFERENCES

1. See fig. 3 in M.H. White and J.R. Cricchi, IEEE Trans. Electron Dev. ED-19:1280 (1972).
2. N.M. Johnson, D.K. Biegelsen, M.D. Moyer, S.T. Chang, E.H. Poindexter, and P.J. Caplan, Appl. Phys. Lett. 43:563 (1983).
3. B. Deal, McKenna, and Jastro, J. Electrochemical Soc. 116:997 (1969).
4. P. Balk, Electrochem. Soc. Meeting, Buffalo, Oct. 1965, Abstract 111.
5. P. Lenahan and P. Dressendorfer, J. Appl. Phys. 55:3495 (1984).
6. S. Lyon, (private communication).
7. E.H. Poindexter, G.J. Gerardi, M.E. Rueckel, P.J. Caplan, N.M. Johnson, and D.K. Biegelsen, J. Appl. Phys. 56:2844 (1984).
8. Y. Nishi, Jap. J. Appl. Phys. 10:52 (1971) ; Y. Nishi, K. Tanaka and A. Ohwada, Jap. J. Appl. Phys. 11:85 (1972).
9. A.G. Revesz and B. Goldstein, Surf. Sci. 14:361 (1969).
10. C. Brunstrom, and C. Svensson, Solid State Communication, 37:399 (1981).
11. P.J. Caplan, E.H. Poindexter, B.E. Deal, and R.R. Razouk, J. Appl. Phys. 50:5847 (1979).
12. See, for example, G.D. Watkins and J.W. Corbett, Phys. Rev. 134:A1359 (1964).
13. A review of their most important spectroscopic results is found in E.H. Poindexter and P.J. Caplan, Prog. in Surf. Sci. 14:201 (1983).
14. K.L. Brower, Appl. Phys. Lett. 43:111 (1983).
15. K.L. Brower, J. Elect. Mat. 14:485 (1985).
16. K.L. Brower and T.J. Headley, Phys. Rev. B34, 3610 (1986).
17. K.L. Brower, Phys. Rev. B33:4471 (1986).
18. To first order, for an axially symmetric defect, Watkins and Corbett have shown that $\delta g_{||} \sim 0$, so that the angular variation of the g-dyadic is due only to variation in $g_\perp$. Furthermore, any distribution in g-values as a function of strain will arise from a distribution in $\delta g_\perp$. See ref. 12.
19. C.A. Coulson, Valence Oxford University Press, London, (1961), p. 203.
20. A. Abragam and B. Bleaney, Electron Paramagnetic Resonance of Transition Metals, Clarendon Press, Oxford, (1970).
21. K.L. Brower, Z. Phy. Chem. 151:165 (1987).
22. The D° defect has often been identified as the dangling orbital although recently Pantelides has raised questions about this identification (See S.T. Pantelides, Phys. Rev. Lett. 57:2979 (1986)).

23. D.K. Beigelsen, N.M. Johnson, M. Stutzmann, E.H. Poindexter, and P.J. Caplan, Applications of Surface Science. 22/23:879 (1985)
24. D.K. Biegelsen and M. Stutzmann, Phys. Rev. B33:3006 (1986).
25. M. Cook, This conference proceedings.
26. N.M. Johnson and W.B. Jackson, Jour. Elect. Mat. 14:499 (1985).
27. W.B. Jackson and Nabil M. Amer, Phys. Rev. B25, 5559 (1982).
28. A. Redondo, W.A. Goddard, and T.C. McGill, J. Vac. Sci. Technol. 21:649 (1982).
29. A. Redondo, W.A. Goddard, T.C. McGill, and T.C. Surratt, Solid State Comm. 20:733 (1976).
30. R.B. Laughlin, J.D. Joannapoulos, and D.J. Chadi, Phys. Rev. B21:5733 (1980).
31. K.L. Ngai and C.T. White, J. Appl. Phys. 52:320 (1981).
32. T. Sakurai and T. Sugano, J. Appl. Phys. 52:2889 (1981).
33. A.H. Edwards, J. Elec. Mat. 14:491 (1984).
34. J. Bernholc, J. Electronic Mat. 14:781 (1984).
35. A.S. Carrico, R.J. Elliott, and R.A. Barrio, Phys. Rev. B34:872 (1986).
36. W.B. Fowler and R.J. Elliott, Phys. Rev. B34, 5525 (1986).
37. R.A. Barrio, R.J. Elliott, and A.S. Carrico, Phys. Rev. B34:879 (1986).
38. Y. Bar-Yam and J.D. Joannopoulos, Phys. Rev. Lett. 56:2203 (1986).
39. A.H. Edwards, Phys. Rev. B (in press).
40. M. Cook and C.T. White, Phys. Rev. Lett. (in press).
41. J. Petit, M. Lannoo and G. Allan, Solid State Comm. 60:851 (1986).
42. K.L. Brower, Phys. Rev. B26:6040 (1982).
43. P.W. Anderson, Phys. Rev. Lett. 34:953 (1975).
44. M.T. Yin and M.L. Cohen, Phys. Rev. B26:5668 (1982).
45. S.T. Pantelides, Phys. Rev. Lett. 57:2979 (1986).
46. A.H. Edwards, unpublished.
47. R.C. Bingham, M.J.S. Dewar, and D.H. Lo, J. Am. Chem. Soc. 97:1285 (1975) ; P. Bischof, J. Am. Chem. Soc. 98:6844 (1976).

STRUCTURE AND HYPERFINE INTERACTION OF $Si_3\equiv Si\bullet$ DEFECT CLUSTERS

Michael Cook and C.T. White

Code 6129
U.S. Naval Research Laboratory, Washington, D.C. 20375

The P_b center located at the Si(111)/SiO_2 interface[1,2] and the D-center in a-Si and a-Si:H[3-5] each have g-tensors whose isotropic part $\bar{g}$ ~ 2.0055. A value of $\bar{g}$ = 2.0055 ± 0.0005 is considered to be diagnostic of radicals with the schematic structure $Si_3\equiv Si\bullet$, in which the trivalent defect atom is bonded into the bulk material through three Si nearest neighbors. When the nearest neighbors are not all Si atoms (e.g., the E' center in silica, $O_3\equiv Si\bullet$) characteristically different ESR signatures are obtained. The similarity of the P_b and D-center g-factors suggests that the two defects might resemble each other closely not only in their schematic structures, but also in their detailed physical conformation and electronic structure.

However, this is not the case. There have recently been a number of experimental studies of hyperfine interactions in the P_b center[6-8] and the a-Si:H D-center[9-11] which show that the hyperfine couplings of these two defects are very different. In the P_b center the isotropic parts of the observed hyperfine and superhyperfine tensors are ~ 105-115 G and 13 G respectively[6-8] the corresponding values for the D-center are 60-74 G and 20-26 G[9-11]. As compared to P_b, the hyperfine interaction in the D-center is greatly reduced and the superhyperfine is increased. This striking contrast has led to speculation that one (or perhaps even both) of these defects has a structure different from the standard $Si_3\equiv Si\bullet$ dangling-bond model[12].

In the present article we suggest that no radically different structure is called for in order to explain the difference in hyperfine interactions between the P_b and the D-center : the observed values are consistent with a model in which the threefold-coordinated atom relaxes nearly into the plane of its nearest neighbors in the amorphous material, but is prevented from doing so in c-Si by the rigidity of the crystalline lattice. We support this picture by semiempirical (MOPN) calculations for the geometry of cluster models and by spin-polarized multiple-scattering Xα (MS-Xα) calculations for the hyperfine interactions in the model clusters.

The first clue that the $Si_3\equiv Si\bullet$ unit is nearly flat in the D-center is offered by the example of small molecular Si radicals. The conformation of AB_3 radicals is known to be a sensitive function of the relative electronegativity of A and B ; the observed trends are well-understood in terms of simple Walsh diagrams[13,14]. When the ligands are more electronegative than the central atom (e.g., SiH_3) the radical is pyramidal. The singly-occupied orbital can have a significant amount of s-character on the central atom, leading to a large direct Fermi-contact interaction. When A and B are of comparable electronegativity, however, (e.g., CH_3) the radical is planar. In this case the A s-character of the orbital containing the unpaired electron goes to zero by symmetry, and the spin density at the nucleus of A is much smaller, being caused entirely by spin-polarization of the doubly-occupied levels. An excellent example of this trend is given by the sequence of radicals $Si(CH_3)_{3-n}(Si(CH_3)_3)_n$, in which methyl ligands are progressively substituted by the less electronegative trimethylsilyl group as n increases from 0 to 3. Experimentally, for n = 0 the isotropic hyperfine coupling constant at ^{29}Si is 183 G ; for n = 1, 137 G ; n = 2, 71 G ; and for n = 3, 65 G.[15] The isotropic coupling constant decreases by a factor of three as the substitution of less electronegative ligands flattens the defect atom into the plane of its nearest neighbors. Significantly, the coupling constants for n = 2 and n = 3 (71 and 65 G) are effectively identical to that observed in the D-center (60-74 G).

In this picture the question to ask is not why the D-center becomes planar, but why the P_b center does not. The difference between the a-Si and c-Si lattices is regularity, and presumably one consequence of the greater regularity of the crystalline material is a greater rigidity. It is plausible that a-Si:H is sufficiently flexible to allow the $Si_3{\equiv}Si\bullet$ unit to relax into its planar conformation by simply bending the nearest-neighbor bonds ; the motion of the nearest neighbors could be accommodated by low-energy torsional distortions of the network. In the crystalline material the nearest neighbors are locked into place by the lattice, so that as the defect atom relaxes from its regular tetrahedral position towards the nearest-neighbor plane it must compress bonds rather than bend them. Since bond compression is a high-energy process, the defect atom reaches an equilibrium position while it is still in a markedly pyramidal conformation.

This qualitative picture is supported by the results of explicit calculations. A variety of ab initio and semiempirical studies have been made for the relaxation of the defect atom in the P_b center, using cluster models with the nearest-neighbor atoms held fixed [16-19]. The results are moderately sensitive to the size of the cluster mode, to the theoretical method, and to such computational details as the size of the basis set, but the computed vertical relaxations tend to lie in the range 0.05-0.15 Å. In a previous article[20] we reported the results of MS-Xα calculations on a P_b cluster model in which the defect atom (Si' in fig. 1) was relaxed by a representative value, 0.09877 Å ;[18] this corresponds to an Si_{s1}-Si'-Si_{s1} angle of 111.7° and an out-of-plane angle of 17.2° between the Si'-Si_{s1} vector and the plane of the three nearest neighbors (the (111) plane, in a P_b cluster model). This is a comparatively small variation from the ideal tetrahedral values of 109.47° and 19.47°. Using the relaxed geometry, very good agreement was found between the MS-Xα hyperfine tensors and the experimental P_b center values.

In the present study we model the D center by an $Si_{13}H_{27}$ cluster. The semiempirical MOPN method (open-shell MINDO/3[21]) was used to help determine the cluster geometry (fig. 1). A full geometry optimization was carried out on an Si_4H_9 cluster[22], and a partial optimization on $Si_{13}H_{27}$.[23] For both cluster sizes the equilibrium out-of-plane angle is ~ 7°, corresponding to $\angle Si_{s1}$-Si'-Si_{s1} ~ 118.5°. As expected, this is much closer to a planar $Si_3\equiv Si\bullet$ conformation than is found in the P_b cluster optimizations, where Si_{s1} were not allowed to move.

During the MOPN optimizations the D-center clusters retained approximate C_{3v} symmetry, with three of the second-nearest neighbors (Si_{b2} in fig. 1) lying nearly in vertical reflection planes. For the geometry to use in the MS-Xα calculations this C_{3v} symmetry was enforced rigorously : the Si-Si bond lengths were fixed at the c-Si value of 2.352 Å [24] and Si-H bonds were given the SiH_4 distance of 1.48 Å [25]. All bond angles were kept tetrahedral with the exception of the Si_{s1}-Si'-Si_{s1} angle, which was given the MOPN-optimized value of 118.5 (Si'-Si_{s1} vector 7° out-of-plane).

The SCF spin-polarized MS-Xα[26-28] calculations were carried out using the same α values and atomic sphere radii as in our study of the P_b center [20]. A polarization basis set was employed, including angular momentum functions up to $\ell = 2$ on Si, $\ell = 1$ on H, $\ell = 4$ on the outer sphere. The origin of the outer sphere was located at the average of nuclear coordinates and its radius was chosen to overlap the outlying H atom spheres by 0.20 a_o.

Our theoretical MS-Xα hyperfine tensors are reported in Table I and are compared to the experimental results for the tris (trimethylsilyl) silyl radical (a molecular D-center analogue) and for the D-center itself. The isotropic Si' coupling constant a(Si') is in good agreement with the experimental values. The principal component of the anisotropic Si' tensor

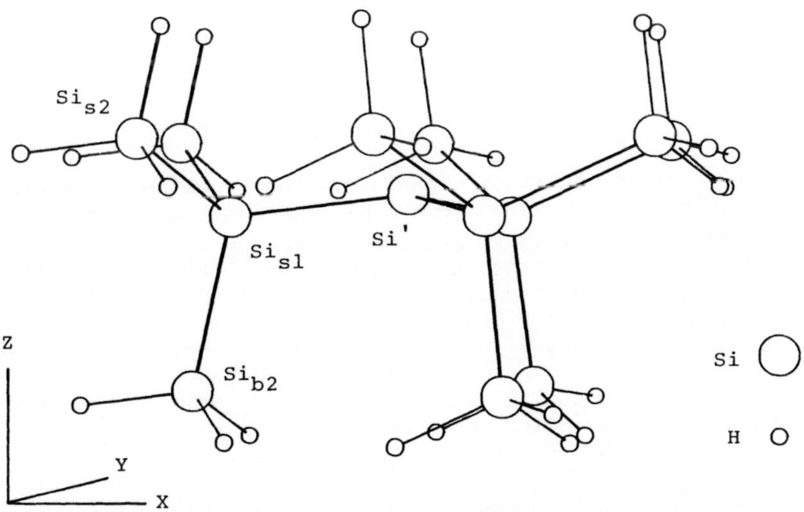

Fig. 1. Ball-and-stick model of the C_{3v}-symmetry $Si_{13}H_{27}$ defect cluster. The Z-axis is the C_3 axis and the XZ plane is one of the three equivalent Si_{s1} is the 1st neighbor on the (111) surface ; Si_{b2} is the 2nd neighbor in the bulk.

is found to be large (-58.63 G), reflecting the large Si'_p character of the singly-occupied molecular orbital.

The nearest-neighbor atoms Si_{s1} have a = + 8.40 G, corresponding to negative spin density (net minority spin) at the Si_{s1} nuclei. In the P_b center the nearest-neighbor hyperfine is found to be very small (< 1 G) because the majority-spin contribution from direct spin delocalization is largely cancelled by the minority-spin contribution due to spin-polarization of the valence orbitals[20]. In the D-center, the direct delocalization contribution decreases almost to zero as the $Si_3\equiv Si\bullet$ unit becomes more nearly planar, while the spin-polarization contribution is not greatly affected. The net result is a sizeable minority-spin coupling at the nearest neighbors.

The major question in analysing the D-center results in Table I is to identify the origin of the large superhyperfine interaction observed in

Table I. Model D-center Hyperfine Tensors (G)[a]

		Theoretical (MS-Xα)	$Si(Si(CH_3)_3)_3$ Exp't	D-Center Exp't		
Si'	a	-56.50	65[d]	74,[e]	71,[f]	60,[g]
	A_{zz}	-58.63				
Si_{s1}	a	8.40	6.2[d]			
	$A_{x'x'}$	0.96				
	η[b]	0.30		25 ± 5,[e]	26[f]	
	θ[c]	-11.0°				
Si_{b2}	a	-11.56				
	$A_{z'z'}$	-2.87				
	η[b]	0.01				
	θ[c]	-17.0°				
Si_{s2}	a	0.26				
	$A_{y'y'}$	-0.31				
	η[b]	0.49				

a) a is the scalar isotropic part of the hyperfine tensor. $\underline{A}$ is the traceless dipolar part. The principal axes of $\underline{A}$ coincide with local coordinate axes only for Si'. For other atoms the principal component of $\underline{A}$ is labeled by the coordinate axis with which it makes the smallest angle (primed subscripts). The experimental results do not include determination of the signs of the coupling constants.
b) The asymmetry parameter $\eta = (A_{11}-A_{22})/A_{33}$, where A_{ii} denotes the principal components in order of increasing magnitude.
c) Counterclockwise rotation angle about the axis out of the paper in coincidence with the principal axes of $\underline{A}$.
d) Ref. 15 e) Ref. 9 f) Ref. 11 g) Ref. 10.

the a-Si:H D-center. Biegelsen and Stutzmann[9] have estimated a superhyperfine interaction of 25 ± 5 G from the line broadening observed in ESR experiments on a-Si:H enriched to 93 % ^{29}Si. Yokomichi et al.[11] carried out ENDOR-detected ESR experiments on natural-abundance material, and observed a 26 G doublet that presumably represents the same superhyperfine interaction. In our theoretical calculations we do not find any single a(Si) to be this large in the $Si_{13}H_{27}$ cluster. The most plausible explanation for this is the highly symmetric structure away from Si' are assigned their equilibrium crystal values. Although the high-symmetry structure of fig. 1 is a local minimum in the potential surface, the energy for rotation about an Si'-Si_{s1} bond is presumably small, and the torsional disorder about this bond induced by the a-Si network might modify the second-neighbor hyperfine interactions sufficiently to produce a superhyperfine coupling as large as the observed value. The superhyperfine coupling at the second-neighbor positions is highly anisotropic ($a(Si_{b2})$ = -11.56 G, $a(Si_{s2})$ = 0.26 G) and dependent on the interaction between the second neighbor and the p_z orbital on Si'. Torsional disorder and the variation in Si_{b2}-Si'-Si_{s2} bond angles when the cluster is embedded in the amorphous material could modulate the second-neighbor superhyperfine coupling to a significant degree. Alternate cluster geometries are presently being studied in order to identify the source of the large observed superhyperfine interaction.

In summary, cluster models for the D-center in a-Si:H have been studied using the MOPN and MS-Xα theoretical methods. The calculated geometries and the good agreement of the theoretical Si' hyperfine interaction with experiment suggest that the D-center is a $Si_3 \equiv Si\bullet$ defect in which the amorphous network allows the defect atom to relax into the planar or nearly-planar conformation that would be expected from electronegativity trends in small molecular Si radicals.

ACKNOWLEDGEMENTS

We thank Art Edwards and Dave Griscom for valuable discussions. MC is grateful to the National Research Council for support as an NRC/NRL Research Associate, and to the Office of Naval Technology for present support as an ONT/ASEE Research Associate. This work was supported by ONR.

REFERENCES

1. Y. Nishi, Japan, J. Appl. Phys. 10:52 (1971).
2. P.J. Caplan, E.H. Poindexter, B.E. Deal and R.R. Razouk, J. Appl. Phys. 50:5847 (1979.
3. M.H. Brodsky and R.S. Title, Phys. Rev. Lett. 23:581 (1969).
4. R.A. Street and D.K. Biegelsen, Solid State Comm. 33/1159 (1980).
5. T.R. Askew, H.J. Stapleton and K.L. Brower, Phys. Rev. B33:4455 (1986).
6. K.L. Brower, Appl. Phys. Lett. 43:1111 (1983).
7. K.L. Brower and T.J. Headley, Phys. Rev. B34:3610 (1986).
8. W.E. Carlos, Appl. Phys. Lett. 50:1450 (1987).
9. D.K. Biegelsen and M. Stutzmann, Phys. Rev. B33:3006 (1986).
10. S. Yamasaki, M. Kaneiwa, S. Kuroda, H. Okushi and K. Tanaka, Phys. Rev. B35:6471 (1987).
11. H. Yokomichi, I. Hirabayashi and K. Morigaki, Solid. State. Comm. 61:697 (1987).
12. S.T. Pantelides, Phys. Rev. Lett. 57:2979 (1986).
13. L. Pauling, J. Chem. Phys. 51:2767 (1969).
14. T.A. Albright, J.K. Burdett and M.H. Whangbo, "Orbital Interactions in Chemistry", Wiley, New York, (1985) pp. 133-136, 140-145.
15. J. Cooper, A. Hudson and R.A. Jackson, Mol. Phys. 23:209 (1972).
16. A. Redondo, W.A. Goddard III, T.C. McGill and G.T. Surratt, Solid. State Comm. 20:733 (1976).
17. L.C. Snyder and Z. Wassermann, Surf. Sci. 77:52 (1978).
18. A.H. Edwards, J. Elect. Mater. 14a:491 (1985).
19. D. Deaven, M. Cook and C.T. White, unpublished MNDO, MOPN and GAUSSIAN 82 calculations.

20. M. Cook and C.T. White, submitted for publication.
21. P. Bischof, J. Am. Chem. Soc. 98:6844 (1976).
22. A.H. Edwards, personal communication.
23. M. Cook, unpublished.
24. W. Parrish, Acta Cryst. 13:838 (1060).
25. D.R.J. Boyd, J. Chem. Phys. 23:922 (1955).
26. J.C. Slater, "The Self-Consistent Field for Molecules and Solids : Quantum Theory of Molecules and Solids v.4" Mc Graw-Hill, New York, (1974).
27. D.A. Case, Ann. Revs. Phys. Chem. 33:151 (1982).
28. M. Cook and D.A. Case, "XASW : A FORTRAN Programm Package for Atomic Xα and Molecular Multiple-Scattering Xα Electronic Structure Calculations", QCPE Program # 465, Quantum Chemistry Program Exchange, Bloomington, Indiana.

ON THE RELATIONSHIP BETWEEN THERMAL GROWTH AND THICKNESS INHOMOGENEITIES IN VERY THIN SiO_2 FILMS

X. Aymerich, J. Suñé, F. Campabadal and Y. Placencia

Centro Nacional de Microelectronica (CSIC-UAB)
Universidad Autonoma de Barcelona. Facultad de Ciencias
08193-Bellaterra, Barcelona, Spain

INTRODUCTION

Roughness on the atomic scale at the Si-SiO_2 interface caused by thickness inhomogeneities of the oxide film in MOS structures has been found to have strong influence on the electrical characteristics of IC devices when dimensions are scaled down[1-4]. Thus, there is need for a deeper physical and mathematical understanding of local thickness inhomogeneities in thin oxide layers.

In this paper we present a model which, taking into account the mechanisms of thermal growth of SiO_2, allows us to build up the statistical distribution of thicknesses that shows the presence of heterogeneities. Our results do not overestimate the dispersion of oxide thicknesses as in the case of previously reported models[5,6]. In this approach, the theoretical oxidation models that exist in the literature are applied at a local scale. They are not valid for the mean growth of the layer since no reference to heterogeneity is made in them. The area that defines this scale is also estimated in this work.

ROUGHNESS MODEL

In order to obtain a model able to describe the thickness inhomogeneities, one has to consider the growth mechanisms on a microscopic scale. The already published oxidation models have to be applied to a scale where the SiO_2 thickness is constant in the sense that one is not affected by roughness at a smaller scale. First, one needs to know the size of the area that defines this local scale.

As our interest is in the electrical properties of the gate oxide of microelectronic devices, we will consider the roughness at an electronic scale. Two phenomenological approaches have been performed. In a steady-state framework one considers the silicon protrusions at the Si-SiO_2 interface as two dimensional potential wells. When the protrusion area is very small, electrons cannot be retained in the well, but, when the area becomes larger, the localized fundamental state has an energy low enough to be occupied by one electron. Assuming that the energy of the fundamental state has to be of the order of kT, the minimal area that defines the roughness scale, S_o, can be estimated by the expression :

$$S_o = \Pi^2 h^2 / m^* \, kT \sim 25 \text{ nm}^2 \tag{1}$$

m^* being the electron effective mass in the silicon.

From a dynamical point of view one can estimate the minimal area by considering the Heinsenberg uncertainty principle. The smaller the area of the silicon protrusion, the larger the number of Bloch states needed to "localize" the electronic wave packet. Since only the states with energy of the order of kT are available, the minimum roughness scale is fixed. This approach yields an area $S_o \sim 3$ nm^2. Note that in both cases the roughness scale is at least one order of magnitude larger than the atomic one (typically $S_{at} \sim 0.1$ nm^2). Nevertheless, the scale is small enough to allow statistical fluctuations in the oxidation process to be important. These fluctuations are responsible for the intrinsic roughness of the SiO_2 layer. Furthermore, inhomogeneities of the oxidation parameters (temperature, substrate asperities, substrate defects, doping, etc..) would produce an additional roughness at a larger scale. From now on we will only consider the intrinsic roughness.

To obtain the probability distribution of local inhomogeneities we divide the SiO_2 layer in columns with equal base area, S_o, and discrete local thickness ha, where "a" is the thickness of a unity step and "h" the number of the already grown unity steps. Then, we define the local rate of growth, V_h in such a way that $(V_h)^{-1}$ is the mean time needed to increase the local thickness in a unity step, from ha to (h+1)a. The dependence of V_h is due to the fact that oxygen must spend a larger time diffusing through the SiO_2 layer when it is thicker, so that it limits the rate of growth. So, V_h is a decreasing function of h. As a first approximation, one can expect that the local growth function, V_h is given by the rate of growth when the mean thickness is ha. This approximation is only valid if the local rates depend linearly on h, but this behaviour is not experimentally observed. In second order, the relationship between the two rates depends on the second derivative of the local rate with respect to h :

$$\langle V_h \rangle = V_{\langle h \rangle} + (\sigma^2/2a)\left(d^2 V_h/dh^2\right)\Big|_{h = \langle h \rangle} \tag{2}$$

σ being the standard deviation of local thicknesses.

The published models on oxidation are well understood in such a local framework since they deal with flat oxides, which is just the case when these models are applied to one column. Thus, V_h is, in fact, the mean growth rate derived from these models. In order to determine the statistical distribution of thicknesses we can build up the continuity equation for the percentage of columns with thickness ha at instant, t, f(h,t) :

$$df(h,t)/dt = -V_h f(h,t) + V_{h-1} f(h-1,t) \tag{3}$$

On the right hand side, the first term represents the decreasing rate of f(h,t) due to the columns that grow from ha to (h+1)a, and the second, the increasing of f(h,t) due to the growth from (h-1)a to ha. Solving the above equation and after tedious algebraic transformation we get a non-recurrent expression for f(h,t) :

$$f(h,t) = \sum_{k=0}^{h} \left(\prod_{i=0}^{h-1} V_i\right) \Big/ \left(\prod_{i=0}^{h} (V_i - V_k)\right) \exp(-V_k t) \tag{4}$$

Note that as initial condition the silicon surface is assumed to be flat. Thus, $f(0,0) = 1$ and $f(h,0) = 0$ $(h>0)$.

If one considers the limit case where the local rate of growth does not depend on the local thickness (V_h=const.) expression (4) becomes a Poisson distribution. This situation corresponds to the unrealistic linear growth of SiO_2, since it assumes that the oxide diffusion time does not depend on oxide thickness. Moreover, as it is easy to show, the Poisson distribution overestimates the interface roughness.

DISCUSSION

Considering the SiO_2 growth from a local of view and from the above results, one is able to see how the entire SiO_2 layer grows. In this framework one needs to know the set of local growth rates (V_h) to get a full description of the statistical distribution of thickness inhomogeneities.

The mean growth law, $d = d(t)$, can be obtained from expression (2) after some mathematical transformations :

$$d(t) = a\langle h\rangle = a\sum_{j=0}^{\infty}(t^j/j!)\sum_{h=0}^{j} h\left(\prod_{i=0}^{h-1} V_i\right)(-1)^{j+h}\langle V_o,\ldots,V_h\rangle^{j-h} \qquad (5)$$

where $\langle V_o,\ldots,V_h\rangle^{j-h}$ is the polynomial of order $(j-h)$ on each rate V_o, ..., V_h, with unitary coefficients (i.e.$\langle V_o,V_1\rangle^3 = V_0^3 + V_1^3 + V_0^2V_1 + V_0V_1^2$).

If one uses the oxidation models available in the literature to obtain the local growth rates (i.e. the Deal Grove model), the proposed model can be applied to calculate the mean thickness grown as a function of oxidation time. Furthermore, it is also possible to obtain information about the SiO_2 layer roughness calculating, for instance, the standard deviation of thicknesses, σ, or the effective thickness, d_{eff}, that the oxide presents when analyzing capacity or current-voltage characteristics. Note that σ or the effective thicknesses can be obtained directly from the thickness distribution, $f(h,t)$. These roughness parameters obviously lay in the range defined by a flat oxide ($\sigma = 0, d_{eff} = d.$), and the ones given by the linear growth model ($\sigma^2 = (da), d_{eff} = d$)[7].

In another way, the local growth model is able to relate the mean thickness dependence on the oxidation time with the intrinsic roughness produced. Thus, the influence of a change in the oxidation conditions (temperature, pressure, etc..) on the SiO_2 roughness is accurately evaluated from the changes produced in the local growth rate. This evaluation is straightforwardly performed comparing expression (5) with a Taylor's series expansion of $d(t)$:

$$d^j d/dt^j\big|_{t=0} = \sum_{h=0}^{j} h\left(\prod_{i=0}^{h-1} V_i\right)(-1)^{j+h}\langle V_o,\ldots,V_h\rangle^{j-h} \qquad (6)$$

Note that from the jth derivative one can evaluate the local rate V_{j-1} since the previous V_1 are already computed. So, from these derivatives one can obtain the distribution function of thickness inhomogeneities, $f(h,t)$, and every roughness parameter. If the experimental oxidation conditions are modified, a different set of V_h would be obtained and obviously, the roughness parameters would not be the same anymore. In this way one can choose the oxidation conditions to optimize the degree of roughness and the set of process parameters in order to improve the oxide quality.

A deeper analysis of the relationship between the oxidation conditions and the quality of the obtained SiO_2 layer would be very interesting to be developed in future work.

As a conclusion, a model of SiO_2 roughness compatible with a non-linear growth (i.e. parabolic) has been presented for the first time. It can be seen that from a local interpretation of the theoretical oxidation models the mean growth law together with the description of the SiO_2 roughness can be obtained.

REFERENCES

1. P.O. Hahn and M. Henzler, J. Appl. Phys. 54:6492 (1983).
2. Z.A. Weinberg and A. Hartstein, J. Appl. Phys. 54:2517 (1983).
3. Y.C. Cheng and H.L. Chau, IEEE Elect. Dev. Lett., 10:191 (1980).
4. J. Suñdé, Y. Placencia, F. Campabadal and X. Aymerich, in : Abstracts of the 9th European Conference on Surface Science, Luzern, (1987).
5. Z. Hurych, Solid. State Elect., 9:967 (1966).
6. Y.P. Song, R.L. Meirhaeghe, W.H. Laflere and F. Cardon, Solid. State Elect., 29:633 (1986).
7. X. Aymerich, P. Galindo Tomas, F. Campabadal and F. Serra-Mestres, in Proceeding of the International Conference "Dielectric Layer in Semiconductors : Novel technologies and devices", Strasbourg (1986), p. 329.

THE P_b CENTER AT THE Si-SiO_2 PRECIPITATE INTERFACES IN BURIED OXIDE MATERIALS : ^{29}Si HYPERFINE INTERACTIONS AND LINEWIDTHS

W.E. Carlos

Naval Research Laboratory
Washington, DC 20375
USA

INTRODUCTION

Electron Spin Resonance (ESR) has recently been very successfully applied to the problem of thermal oxide - Si interfaces[1-3]. That work resulted in the identification of the P_b center as the primary fast interface trap and the determination that its basic structure is a trivalent Si atom at the Si-SiO_2 interface. In addition to the technological motivations, there is also fundamental interest in this defect, which is at the interface between a crystalline solid and an amorphous one, and, therefore, might be expected to have features characteristic of defects in both types of materials. The detailed ESR study of the structure of this defect is inhibited by the small fraction of interfacial atoms in a typical Si-SiO_2 structure. It has recently been shown that the principal paramagnetic defect observed in silicon on insulator materials formed by oxygen implantation is a P_b center at the interface between Si and SiO_2 precipitates in the Si film over the buried oxide layer[4-6]. The total precipitate surface area can be much greater than the simple surface area of an Si-SiO_2 structure. This increased number of "interfacial" atoms affords the opportunity to conduct more detailed ESR studies of the P_b center than readily possible with Si-thermal oxide structures. In addition these interfaces are formed in a significantly different manner than the thermal oxide interfaces and a comparison of the P_b centers formed in the two manners may offer insights into their formation. In this work we observe that the ^{29}Si hyperfine structure for the P_b center at the Si-SiO_2 precipitate interface is very close to that for the P_b center at the Si-thermal oxide interface. The linewidths indicate that the P_b centers in these materials are remarkably similar to those observed at the interfaces between Si and thermal oxides.

One of the most promising silicon on insulator techniques is the SIMOX (separation by implanted oxygen) process[7,8] in which isolation is achieved by the implantation of large doses ($\sim 2\times10^{18}/cm^2$) of oxygen ions into Si. During implantation and subsequent annealing, a large number of small (typical dimensions ~20 to 1000 Å) SiO_2 precipitates form, in addition to a buried oxide layer (~0.2 μm below the top surface). During the annealing procedure most of these precipitates are dissolved and incorporated into the oxide layer leaving a high quality Si film as the top layer. ESR has recently been applied to SIMOX materials by Barklie and coworkers[4], by Makino and Takahashi[5] and by Carlos[6]. All found two ESR centers in the Si film : an amorphous Si-like line and the P_b center at interfaces between the Si and SiO_2 precipitates in the top Si layer. The P_b center for thermal oxides on Si has been widely studied[1-3] and it has been shown that this defect is a trivalent Si atom backbonded to three other Si atoms at the Si-SiO_2 interface. The current work focuses on the P_b centers in the SIMOX materials, in particular on the ^{29}Si hyperfine interactions of this defect and its linewidths and spin lattice relaxation processes.

Samples used in this work were implanted with 0.5 to 2.6×10^{18} 160 keV or 180 keV oxygen ions per cm^2 at 450°C-575°C. Little spectral variation, other than defect density, was observed for samples implanted under the different conditions. Samples were annealed in dry N_2 or Ar at temperatures between 500°C and 1200°C. The ESR measurements were performed on a Varian E9 X-band spectrometer at T = 4.2 K to 300 K. The neutral phosphorous donors (P°) in one n-type sample served as standards both for calibrating the intensity of the defect lines and for determining their positions precisely.

HYPERFINE INTERACTIONS

The central ^{29}Si hyperfine (hf) structure of the P_b center was first resolved by Brower[3] using a stack of very thin oxidized Si pieces having a total surface area of 30.5 cm^2. The central hf structure for a SIMOX (area = 5.4. cm^2) sample which had been annealed for 1 hour in dry Ar at 800°C is illustrated in Fig. 1. The central lines were first observed by Barklie et al.[4] who deduced that these were due to P_b-like centers at the interfaces of Si and the SiO_2 precipitates in similar samples. The resolution of the ^{29}Si hyperfine structure confirms that identification.

Table I. The spin Hamiltonian parameters of P_b centers in SIMOX and Si-thermal SiO_2 interfaces and those calculated by Cook and White.

	SIMOX	(111)Si-SiO_2^a	Theory[b]
$g_{\parallel}$ (111)	2.0011	2.0016	----------
$g_{\parallel}$	2.0082	2.0090	----------
$\delta g_{\perp}$	0.0010	0.00075	----------
Central Atom			
Orientation	(111)	(111)	(111)
Aiso	106G	113G	129G
Aaniso	23G	22G	30G
δAiso	9G	10G	----------
Second Nearest Neighbor			
Orientation	(111)	(111)	(111)
Aiso	13G	----------	10G
Aaniso	1G	----------	1.5G

[a] ref. 3
[b] ref.10

The g-tensor and the hyperfine interaction have near axial symmetry about the (111) axis and their values closely match those of the P_b center at the Si (111)-thermal oxide interface as seen in Table 1.

Following Watkins and Corbett[9] treatment of similar defects in Si, a simple linear combination of atomic orbitals (LCAO) is used to write the wave function of the P_b center as

$$| \psi \rangle = \Sigma n_i \left(\alpha_i \ | s\rangle + \beta_i \ | p\rangle \right), \tag{1}$$

where the summation is over all atomic sites and n_i^2 is the amplitude of the wave function at site i and α_i^2 and β_i^2 give its relative s and p character. These parameters are readily derived from the hyperfine tensor using tabulated atomic orbital parameters. For the central atom of the P_b center in SIMOX (values in parenthesis are taken from ref. 3 for the Si(111)-thermal oxide interface) n^2 = 0.84 (0.80), α^2 = 0.11 (0.12) and β^2 = 0.89 (0.88). The P_b centers at the $Si\text{-}SiO_2$ interfaces formed by implantation and annealing are clearly quite similar to those formed by thermal oxidation.

In addition to the ^{29}Si hyperfine interaction with the central Si atom, the superhyperfine (shf) interaction with three neighboring Si atoms

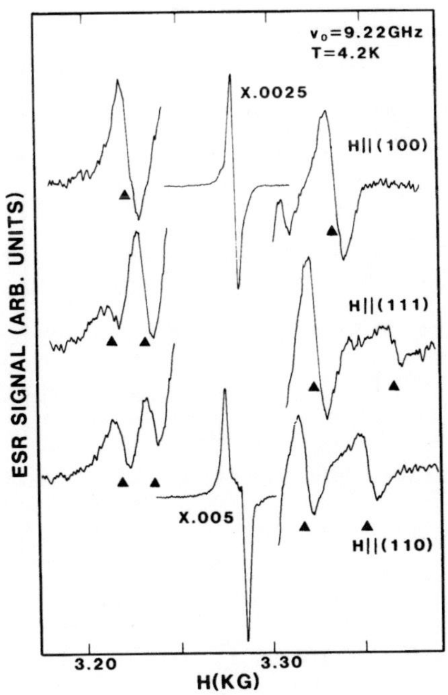

Fig. 1. ESR spectra of a SIMOX sample. The triangles indicate the ^{29}Si hyperfine structure.

is also resolved, as shown in fig. 2. (the sample is identical to that used for fig. 1 except that the annealing temperature is 1000°C). Those are second derivative spectra, obtained by detecting the second harmonic of the modulation frequency. For the (100) orientation shf structure is resolvable in a first derivative spectra as well (note the shoulder on the low field side of the central line in fig. 1.). The measured shf interaction, given in Table I, is in good agreement with the weak structure observed by Brower and Headley[10] for the (111) Si-thermal oxide interface. The intensity of the these lines is about 12 % of the central line, consistent with an assignment to 3 neighboring atoms. From Eq. (1) we find $n^2 = 0.04$, $\alpha^2 = 0.26$ and $\beta^2 = 0.74$ as expected for an interaction with an sp^3 orbital. Cook and White[11] have recently calculated the hyperfine interactions of the P_b center with neighboring atoms and found the largest hyperfine interaction, aside from the central interaction, to be with the three next nearest neighbors in the bulk of the Si. The interactions with the 3 nearest neighbor and 6 next nearest neighbor surface atoms are considerably weaker. As seen in Table I their results are in excellent agreement with the experimental results.

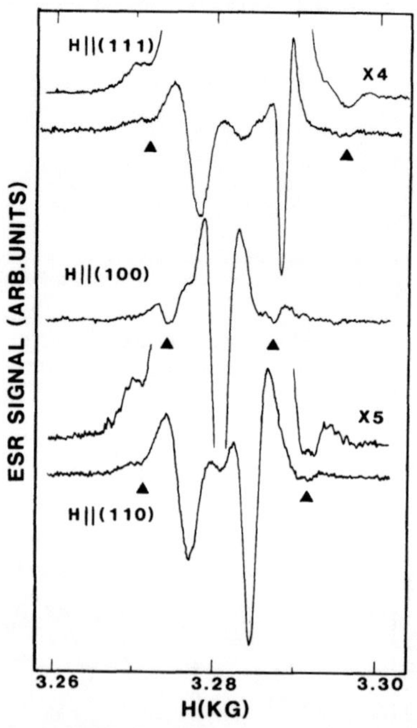

Fig. 2. Second derivative ESR spectra illustrating the superhyperfine structure.

RELAXATION PROCESSES AND LINEWIDTHS

Brower[12] has shown that the linewidths of the Zeeman and hyperfine lines are primarily due to distributions in the g-tensor elements and hyperfine tensor elements respectively. His work indicates that the inhomogeneous broadening of the central line is entirely due to a distribution in g and that the line is homogeneously broadened for an applied magnetic parallel to the principal axis. The degree of homogeneity in an ESR line is reflected in the dependence of the intensity on microwave power[13]. The power dependence of the intensity of a homogeneously broadened line is given by

$$I = A\sqrt{P}/(1+BP), \tag{2}$$

where A and B are parameters related to the number of centers and their relaxation times, respectively, and P is the applied microwave power. For completely inhomogeneous line the relationship becomes

$$I' = A'\sqrt{P}/(1+B'P)^{1/2} \tag{3}$$

Note that Eq. (2) reaches a maximum for $P = B^{-1}$ and the intensity is

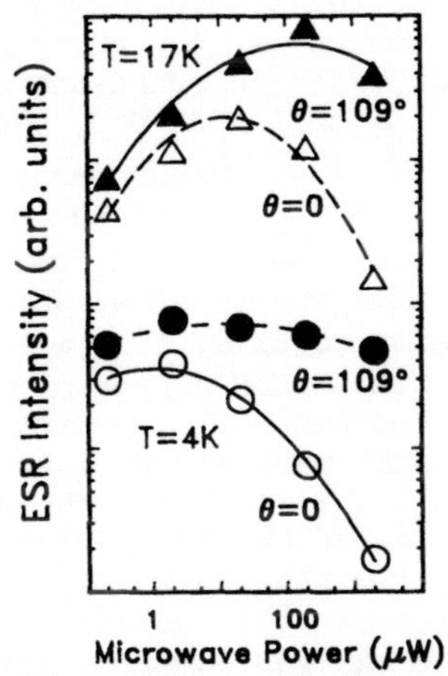

Fig. 3. The ESR signal intensity as a function of microwave power for two temperatures. All data were taken with the magnetic field along a (111) axis. The angle, θ is the angle between the principal axis of the center and the magnetic field.

proportional or inversely proportional to $\sqrt{P}$ for $PB \ll 1$, respectively, while Eq. (3) approaches a constant value at high power levels. In fig. 3 the intensity of the P_b centers are plotted as a function of microwave power for the case of H parallel to the (111) direction. For this orientation the bond direction of the P_b center is either parallel to the magnetic field or at an angle of 109.5° to it. As seen the power dependence for those parallel to the bond direction follows Eq. (2), the dependence for those at an angle of 109.5° is much closer to Eq. (3). From θ = 0 linewidth (1.3G) a spin-spin relaxation time, T_2 = 50nsec. is determined. With the saturation data this yields spin-lattice relaxation times, T_1 = 23msec. (4 K) and T_1 = 2.4msec. (17 K).

Again referring to Table I, it is seen that the distributions in spin Hamiltonian parameters, $\delta g_\perp$ and δA_{iso} are very similar for P_b centers in the SIMOX material and at the Si-thermal oxide interface. Furthermore, both the Zeeman and hyperfine linewidths are nearly constant with annealing temperature. Brower argues that these variations in spin Hamiltonian parameters are due to variations in lattice distortions (i.e., strain broadening). Ion scattering measurements do indicate that the Si-SiO_2 interface is strained[14]. It can then be argued that the origin of these variations in lattice distortions are fluctuations in the interfacial strain and that the P_b centers act as local "strain gauges" to measure the distribution of strains. Considering the current data from this perspective, it is then surprising that the distributions in $g_\perp$ and A_{iso} observed in the current work and those for thermal oxide interfaces are nearly identical and that these distributions are nearly invariant with annealing. It would certainly seem that the distributions of interfacial strain for interfaces between the precipitates and Si would differ from that observed for thermal oxide interfaces and that these strain distributions might be modified with annealing.

An alternative source of the disorder needed to give a distribution in the spin Hamiltonian parameters is the amorphous character of the oxide itself. From this perspective, the interactions of the defect wave function with near neighbor O and Si atoms in the SiO_2 yield small variations in the character of the wave function. Measurements on ^{17}O-enriched Si-SiO_2 interfaces reveal a broadening due to weak unresolved hyperfine interactions,[15] indicating that the P_b wave function does interact with the oxygen in the SiO_2. Given that the local structure of the oxide around the P_b center will vary from site to site, this will lead to a distribution of wave function parameters such as the degree of localization and relative s and p character. It is emphasized that the distributions in parameters are quite sharp ; Brower calculated variations of the localization and relative s and p character on the order of 1 %[3]. These variations lead to slight variations in the angle between the dangling bond and the three backbonds of the trivalent Si atom (~0.5°). This is very close to the half width (~0.7°) of the distribution of the comparable angle for the E' center in amorphous SiO_2 [16] which also has the unpaired spin localized on a trivalent Si atom. The distinction between these two perspectives on the Si-SiO_2 interface is subtle but it is important. If variations in interfacial strain were responsible for the line-widths, it would be natural to expect differences in line-widths from interface to interface, and in particular, between the interfaces with thermal oxides and those with oxide precipitates. Furthermore, one would expect to see changes in the lineshape as the interfaces are annealed. Neither of these effects is observed. However, if the linewidths are due primarily to the amorphous nature of the oxide, the distributions in spin Hamiltonian parameters are seen as an inherent part of the P_b center's ESR signature.

In summary it has been shown that the dominant paramagnetic defect in buried oxide SOI materials is a P_b center at the many interfaces between SiO_2 precipitates and the single crystal Si film. The detailed structure of these centers is remarkably similar to that of the P_b center at the interfaces between thermal oxides and Si. This indicates that the amorphous SiO_2 layer is important in determining the details of the ESR spectrum of this center.

REFERENCES

1. E.H. Poindexter, P.J. Caplan, B.E. Deal and R.R. Razouk, J. Appl. Phys. 52:879 (1981). E.H. Poindexter, G.J. Geradi, P.J. Caplan; N.M. Johnson and D.K. Biegelsen, J. Appl. Phys. 56:2844 (1984).
2. P.M. Lenahan and P.V. Dressendorfer, J. Appl. Phys. 55:3495 (1984).
3. K.L. Brower, Appl. Phys. Lett. 43:1111 (1983).
4. R.C. Barklie, A. Hobbs, P.L.F Hemment and K. Reeson, J. Phys. C 19:6417 (1986).
5. T. Makino and J. Takanashi, Appl. Phys. Lett. 50:267 (1987).
6. W.E. Carlos, Appl. Phys. Lett. 50:1450 (1987).
7. P.L.F. Hemment, Mat. Res. Soc. Symp. Proc. 53:207 (1986) ; H.L. Lam and R.P. Pinizzotto, J. Crystal Growth, 63, 554 (1983).
8. G.D. Watkins and J.W. Corbett, Phys. Rev. 134:A1359 (1964).
9. K.L. Brower and T.J. Headley, Phys. Rev. B34:3610 (1986).
10. M. Cook and C.T. White, Phys. Rev. Lett. (in press).
11. K.L. Brower, Phys. Rev. B33:4471 (1986).
12. A. Abragam, "The Principles of Nuclear Magnetism", Oxford University Press, London, (1961), chapter 3.
13. R. Haight and L.C. Feldman, J. Appl. Phys. 53:4484 (1982).
14. K.L. Brower, Z. Phys. Chem. 151:177 (1987).
15. D.L. Griscom, E.J. Friebele, G.H. Sigel, Jr., Solid State Commun. 15: 479 (1974).

METASTABLE AND MULTIPLY-CHARGED INDIVIDUAL DEFECTS AT THE $Si:SiO_2$ INTERFACE

M.J. Kirton, M.J. Uren and S. Collins

Royale Signals and Radar Establishment
St. Andrew's Road
Great Malvern Worcestershire WR14 3PS
United Kingdom

ABSTRACT

In small-area silicon metal-oxide-semiconductor field-effect transistors (MOSFETs), the fluctuating occupancy of individual $Si:SiO_2$ interface states generates Random Telegraph Signals (RTSs) in the drain current. We have observed a new class of RTS which exhibits anomalous behaviour. We demonstrate that these signals are due to individual interface states which can exist in two or more charge-equivalent metastable states. We present results on a single defect which exhibits two-electron capture, metastability and negative-U-like behaviour.

INTRODUCTION

Discrete switching behaviour, commonly referred to as burst noise, has been observed in a wide variety of semiconductor devices[1]. Ralls et al[2] found these so-called random telegraph signals (RTSs) in the drain current of small-area (< 1 μm^2) MOSFETs and showed they were due to the fluctuations in occupancy of individual defects at the $Si:SiO_2$ interface. The fluctuations give rise to localised changes in channel resistance and thus discrete switching in the current. Uren et al[3] demonstrated that 1/f noise in large-area MOSFETs is the result of the superposition of RTSs from many interface states. Through the study of single RTSs as a function of temperature, Kirton and Uren[4] developed a novel model for the capture and emission kinetics of individual $Si:SiO_2$ interface states residing in the bandgap of silicon. In particular, they showed that capture takes place via a thermally activated multi-phonon process and they were able to evaluate capture cross-sections and entropies of ionisation of single interface states. In this paper, we shall report on our recent finding of a new class of RTS which exhibits anomalous and sometimes very complex behaviour. We will show that this behaviour is a direct demonstration of defect metastability : i.e. the ability of a given charge state of a defect to exist in two, or more, different reconstruction modes. In addition, we will present results on an anomalous three-level signal which exhibits two-electron capture, metastability and negative-U-like behaviour.

In figure 1a, we display a conventional RTS generated by a single $Si:SiO_2$ interface trap in an n-channel MOSFET with electrical channel dimensions of 0.5 by 0.75 μm. The times in the up and down states correspond to single-electron capture and emission, respectively[3,4]. Kirton

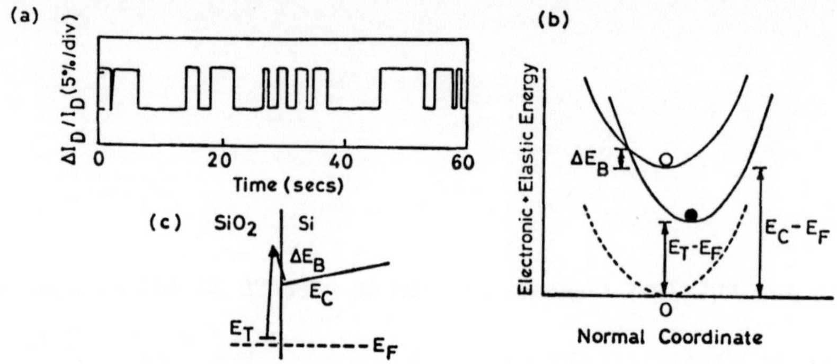

Fig. 1. (a) Random Telegraph Signal. Change in current versus time. Active device area is 0.4 μm², V_d = 10 mV, V_g = 0.94 V, I_d = 6.4 nA, T = 295 K.
(b) Configuration-coordinate diagram for capture of inversion-layer electron into interface state. The dashed curve shows the the empty trap before creation of a free electron in the conduction band ; 0 labels the empty trap + free electron ; ● marks the filled state.
(c) The relation of (b) to the usual band diagram.

and Uren[4] showed that capture proceeds with a cross-section of the form $\sigma = \sigma_o \exp[-\Delta E_B/kT]$; an appropriate configuration-coordinate (CC) diagram for the process is shown in figure 1b.

Figure 2 depicts the anomalous switching behaviour found in the drain current of an n-channel MOSFET. This complex behaviour was relatively common : out of a total of 320 RTSs that we observed, 13 were found to be anomalous. Within the time window t_1 to t_2, the rapidly switching RTS shows the conventional behaviour of figure 1a. However, during periods such as t_2 to t_3 the rapid switching completely disappears and the RTS maintains its low level. It thus appears that the fast-switching RTS is modulated in time, with the envelope of modulation itself being an RTS of the same amplitude.

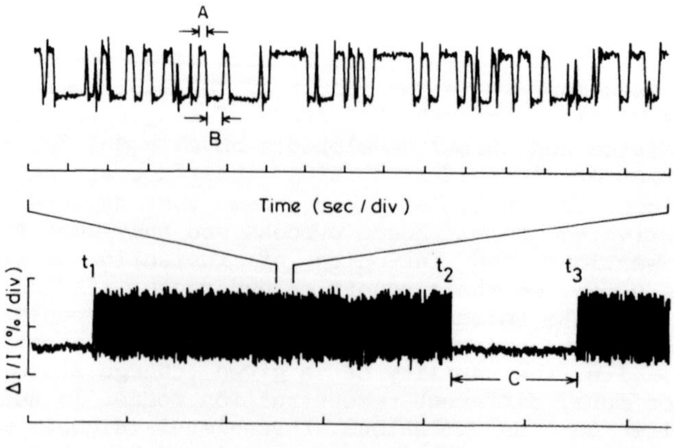

Fig. 2. Fluctuations in current versus time showing a rapidly-switching RTS modulated by an envelope of the same amplitude. The upper trace is an expansion of part of the lower trace. V_g = 1V, V_d = 10 mV, I_d = 10.3 nA, T = 293 K.

DISCUSSION

We have previously shown in detail that neither Coulombic interactions between randomly distributed defects that happen to lie in close proximity, nor current flow along a percolation channel provide a satisfactory explanation for this phenomenon[5]. Percolation models can be ruled out since the phenomenon occurs in strong inversion where percolation is not a major current-carrying mechanism. For Coulombic interactions to be the cause, then it is required that the defects be clustered around inactive point (not line or planar) defects and that their separation never be greater than ~ 2 nm. At such small separations any pair of defects would show significant hybridisation of their electronic wave functions and should really be considered to be a single composite defect. The only consistent explanation involves a single defect with two reconstruction modes (metastable states) available for the filled trap, with the trapping probability being different for the two configurations. This hypothesis immediately accounts for the fact that the amplitudes of the underlying RTS and its envelope are equal and the total absence of a third level which would result from Coulombic interactions. Moreover, it is particularly appropriate in view of the established evidence of metastability in glassy systems[6,7,8] and our own observations[4] of significant electron-lattice coupling at individual $Si:SiO_2$ interface states.

Figures 3a,b show schematic CC diagrams of two models which exhibit metastability ; the models differ only in the way the various states inter-communicate. Rapid electron capture and emission proceeds via total-energy minimum $\alpha 1$ and accounts for section t_1 to t_2 of figure 2. In the model shown in figure 3a, after electron capture into $\beta 1$ it will take a time, $\overline{C}(\overline{C} \gg \overline{A},\overline{B})$ for the filled trap to re-emit the electron. During this time interval no switching of the RTS will occur, and the drain current will remain fixed at its low level as in period t_2 to t_3. For the model of figure 3b, the transformation from $\alpha 1$ into state $\beta 1$ should be thermally

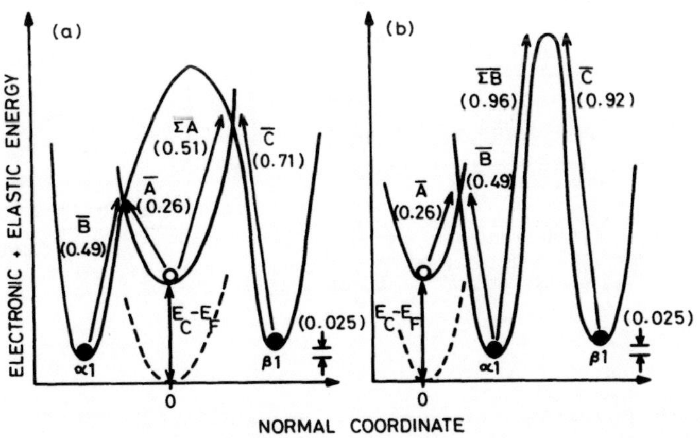

Fig. 3. Two possible CC diagrams for the defect whose RTS is shown in figure 2. The observed transitions are labelled with the corresponding average times from figure 2. (a) both metastable states are in communication with the inversion layer. The activation energies in eV are given in brackets (ref. 3). (b) Only $\alpha 1$ is able to capture an electron directly. Activation energies for the transformations $\alpha 1 \leftrightarrow \beta 1$ were evaluated assuming $\tau = \tau_o \exp(E/kT)$.

activated and gate-voltage independent, with $\overline{C}$ representing the time taken for the emission of the trapped electron from β1 into α1. Referring to figure 2, we see that the time for electron capture into α1 is given by the average of time A, $\overline{A}$; the emission time from α1 is $\overline{B}$. The capture time from α1 is $\overline{B}$. The capture time into β1, $\overline{\Sigma A}$, can be obtained by summing the time A over an average time window t_1 to t_2 ; alternatively, the transformation time from α1 to β1 is given by $\overline{\Sigma B}$.

Using the gate-voltage dependence[2] of $\overline{A}$ and $\overline{\Sigma A}$, we found that the metastable states α1 and β1 were located the same distance into the oxide. Using the Grand Partition Function[9] to evaluate equilibrium occupancies, we found that β1 lay above α1 by 0.025 eV. In fact, both models shown in figure 3 were consistent with our data and we were unable to distinguish between them. The important point to note, however, is that the concept of metastability provides a simple yet elegant explanation for the anomalous RTS of figure 2, independent of transformation mechanism.

So far, we have discussed the behaviour characteristic of the majority of anomalous RTSs that we observed. We would now like to consider one particular RTS which showed complex and interesting behaviour : see figure 4a. The first point to note is that it is a three-level signal with the separation between levels 0 and 1 equal to the separation between levels 1 and 2. In addition, the rapid switching represented by times U and V occurs only between levels 1 and 2 showing that the signal is not just a straightforward superposition of two distinct and independent RTSs. These facts taken together imply that the signal represents a sequential two-electron capture process : transition 0 → 1 being capture of the first electron ; transition 1 → 2 capture of the second electron.

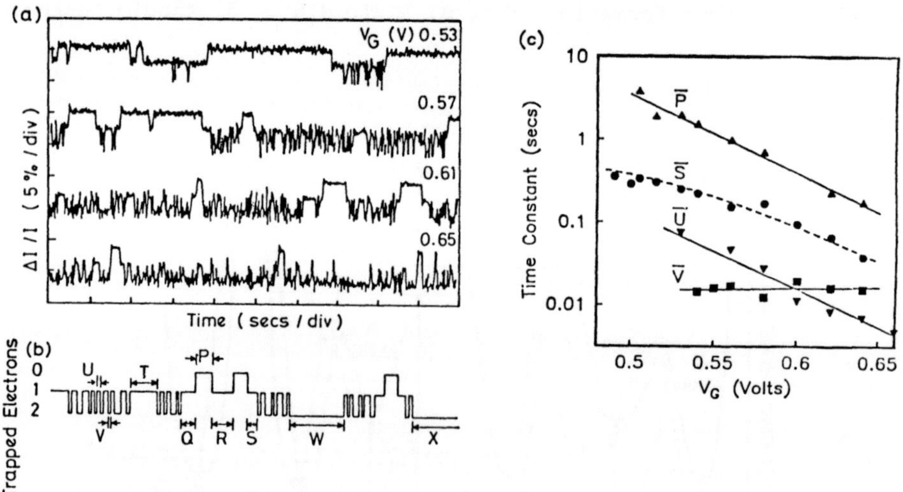

Fig. 4. (a) Gate voltage dependence of a complex RTS generated by a single defect which shows one- and two-electron capture. T = 290 K, V_d = 20 mV, V_g, I_d = 0.53 V, 47 pA ; 0.57 V, 97 pA ; 0.61 V, 229 pA ; 0.65 V, 519 pA. (b) Schematic RTS showing all the features which were observed for this defect. (c) Gate-voltage dependence of representative times $\overline{P}$, $\overline{S}$, $\overline{U}$, $\overline{V}$. The full lines are regression lines and the dashed line is a fit assuming simple competition between emission and capture in the θ1 state.

Within both the one-electron and two-electron levels, several different time constants are visible, reflecting the presence of metastable states. The interesting question is, how do these various states inter-communicate ? To address this, we investigated the gate-voltage dependence of the various time constants : see figure 4c. $\overline{P}$ and $\overline{U}$ are both strongly gate-voltage dependent with the same slope. This demonstrates that they represent straightforward single-electron capture : $\overline{P}$ of the first electron ; $\overline{U}$ of the second electron.
However, immediately after the capture of the first electron and before a transition to the charge-2 level takes place, the RTS spends a period of time, $\overline{S}$, in the charge-1 level. At high gate voltages, $\overline{S}$ has the same gate-voltage dependence as both $\overline{P}$ and $\overline{U}$, whereas at low gate voltages it becomes independent of gate voltage. These two facts imply that $\overline{S}$ represents the capture of a second electron in competition with emission of the first electron. $\overline{V}$ was found to be a simple emission process. It was not possible to study W and X in any detail : W was quite a rare event lasting around 0.5 seconds ; and X turned off all fluctuations for several hours.

On the basis of the above information, the principal routes for metastable-state inter-communication were identified and are shown in figure 5. The charge-1 level consists of two metastable states, θ1 and λ1. Capture of the first electron from the inversion layer, with time constant

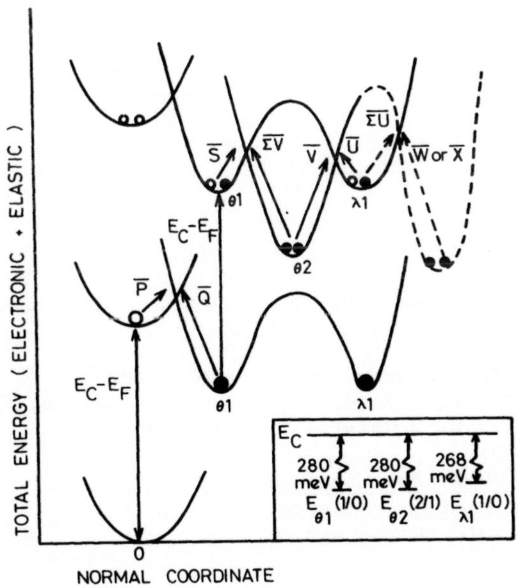

Fig. 5. Schematic CC diagram for the defect whose RTS is shown in figure 4. The diagram shows the total energy of the defect for the three cases of 0, 1 and 2 electrons removed from the reservoir (at energy E_F). The arrows show the various transitions we have identified ; the dashed section is speculative and incomplete. Varying the gate voltage changes $E_C - E_F$, the relative positions of the three sets of curves, and hence, the occupancy of the defect. The inset gives the trap energy levels we evaluated.

$\overline{P}$, takes place into the θ1 state and not into the λ1 state. Capture of the second electron initially takes place from θ1 into θ2 with time constant $\overline{S}$. Thereafter, fluctuations in occupancy of θ2 take place : via θ2 ↔ λ1 with relatively fast time constants $\overline{U}$ and $\overline{V}$; and via θ2 ↔ θ1 with slower times $\overline{T}$ (= $\overline{S}$) and $\overline{\Sigma V}$. Two important observations are :
(i) Direct, thermally activated (and gate-voltage independent) transformation between the metastable states, θ1, and λ1, of the charge-1 level was not observed.
(ii) The occupancy levels $E_{\theta_1}(1/0)$ and $E_{\theta_2}(2/1)$ were degenerate to within a few millivolts. Although not an unequivocal demonstration of a negative-U system, this clearly shows that the strong electron-lattice coupling is able to offset the electron-electron interaction of two trapped electrons[10]. In addition, it is worth noting that the complexity of this defect is not so surprising in view of recent numerical simulations which showed metastability arising from the rotations of up to seven SiO_4 tetrahedra[8].

Finally, we would like to point out that Knott[11] and Kandiah[12] have observed so-called intermittent burst noise in bipolar transistors in which the current fluctuations bear a strong resemblence to the RTSs of figures 2 and 4. It is thus quite likely that the intermittent burst noise is due to defects exhibiting metastable behaviour either residing in the bulk of the device or at a $Si:SiO_2$ interface.

ACKNOWLEDGMENT

We would like to thank R. Lambert of the GEC Hirst Research Centre for the devices.

REFERENCES

1. M.J. Buckingham, "Noise in Electronic Devices and Systems", Ch. 7, Ellis Horwood, Chichester, (1983).
2. K.S. Ralls, W.J. Skocpol, L.D. Jackel, R.E. Howard, L.A. Fetter, R.W. Epworth and D.M. Tennant, Phys. Rev. Lett. 52:228 (1984).
3. M.J. Uren, D.J. Day and M.J. Kirton, Appl. Phys. Lett. 47:1195 (1985).
4. M.J. Kirton and M.J. Uren, Appl. Phys. Lett. 48:1270 (1986).
5. M.J. Uren, M.J. Kirton and S. Collins, submitted for publication (1987).
6. P.W. Anderson, B.I. Halperin and C.M. Varma, Phil. Mag. 25:1 (1972).
7. W.A. Phillips, J. Low Temp. Phys. 7:351 (1972).
8. L. Guttmann and S.M. Rahman, Phys. Rev. B. 33:1506 (1986).
9. C. Kittel and H. Kroemer, "Thermal Physics", W.H. Freeman, San Francisco, (1980).
10. P.W. Anderson, Phys. Rev. Lett. 34:953 (1975).
11. K.F. Knott, Solid State Electron. 21:1039 (1978).
12. K. Kandiah, private communication.

THE INFLUENCE OF DISORDER ON THE Si_{2p} XPS LINESHAPE AT THE Si - SiO_2 INTERFACE

R.A.B.. Devine and J. Arndt*

CNET, BP98, 38243 Meylan, France
*Petrographisch-Mineralogisches Institut
Universität Tübingen, D-7400 Tübingen, FRG

ABSTRACT

The origins of X ray photoelectron spectroscopy Si_{2p} core level lineshapes and lineshifts in thermally grown SiO_2 on Si are discussed. It is demonstrated that photoelectron lineshapes are related to bridging bond angle distributions through convolutions which result in linewidths being substantially narrower than initially expected. We conclude, by comparison with experiment, that broadening effects are present in SiO_2 Si_{2p} experimental spectra which mask the intrinsic linewidth expected from bond angle distribution arguments.

INTRODUCTION

The nature of the physical bonding in bulk, amorphous silica and at the Si - SiO_2 interface remains a question open to some conjecture. A detailed picture requires data on the Si-O-Si bridging bond angles between adjacent SiO_4 tetrahedra and on the Si-O bondlengths within a given tetrahedron. The region in the immediate vicinity of the Si-SiO_2 interface is particularily interesting because of the very large mismatch between the lattices of the Si substrate and the overgrown oxide. Indeed, there is some debate as to how the amorphous SiO_2 network bonds itself to the substrate and whate role, if any, is played by residual strain. Various authors[1-7] have suggested that apart from dangling and strained bonds, the interface involves Si-Si_4/O_4 tetrahedra in different oxidation states (e.g. Si-Si_2O_2, Si-SiO_3, etc.) and that the immediately overlaying, stoichiometric oxide is somehow different[1] from bulk SiO_2. Some authors, however, have questioned both the need to invoke different oxidation states[6] and the suggestion[1] that the superficial, stoichiometric oxide at the interface involves a limited number of SiO_2 ring structures[4,8].

The most commonly used technique[1-7] for the study of the Si-SiO_2 interface is X ray photoelectron spectroscopy (XPS) of the Si_{2p} or O_{1s} core levels. The line shift observed, with respect to the Si_{2p} level in pure Si, is directly related to the Si → O electron transfer due to bonding. Most of the work to date has addressed the problem of the origin of lineshifts whereas none appears to have questioned the origin of the observed XPS linewidths. It should be mentioned that electron spin resonance has been used to investigate dangling bonds[9] and oxygen vacancy centers near the interface[10] but both of these are defects and neither form a very high percentage of the total number of interfacial species. In the following we shall discuss various problems resulting from the interpretation of the XPS spectra and in particular, we shall address the question of the origin of the linewidths. The discussion will be made in the light of calculations of the electron transfer (Si → O) in Si-O bonds as a function of the Si-O-Si bridging bond angle and of the distribution of these angles. The relevance of interfacial stress is discussed.

A study of the Si_{2p} core level spectra in ultra-thin (<20 Å) films of SiO_2 grown on Si at low temperatures and O_2 pressures has revealed[3] that five oxidation states of Si may be simultaneously present in the interfacial region corresponding to tetrahedra having the configurations :

$Si\text{-}Si_4$ (Si^0) ; $Si\text{-}Si_3O$ (Si^+) ; $Si\text{-}Si_2O_2$ (Si^{2+}) ; $Si\text{-}SiO_3$ (Si^{3+}) and $Si\text{-}O_4$ (Si^{4+})

the latter state being the regular SiO_4 tetrahedron, it should presumably be representative of amorphous SiO_2. The measured energy shifts of the Si_{2p} level with respect to the value for pure Si are -1 eV (Si^+), -1.8 eV (Si^{2+}), -2.6 eV (Si^{3+}) and -3.5 eV (Si^{4+}), these values have been confirmed by independent authors [3,5]. The results suggest an average energy shift of -0.9 eV per SiO bond (replacing an Si-Si bond). In fig. 1 we show the results of calculations [1] of the charge transfer per bond and shift of the Si_{2p} level versus Si-O-Si bridging bond angle. The value of -0.9 eV is seen to correspond to an electron transfer from the Si towards the O of approximately 0.38 electrons, smaller than the value expected in amorphous SiO_2 ~ 0.5 electrons per bond. However, as the oxide thickens, the Si_{2p} level shift increases rapidly to -4.4 eV, the thickness at which this happens varying from >11Å[3] to >50Å[1]. The shift to -4.4 eV corresponds to the value expected for bulk SiO_2 yielding an Si → O charge transfer per bond of the order of 0.49 electrons. We draw attention to the fact that as the oxide thickens, the energy separation between Si_{2p} and O_{1s} core level spectra does not appear to vary but remains essentially constant[4] except in

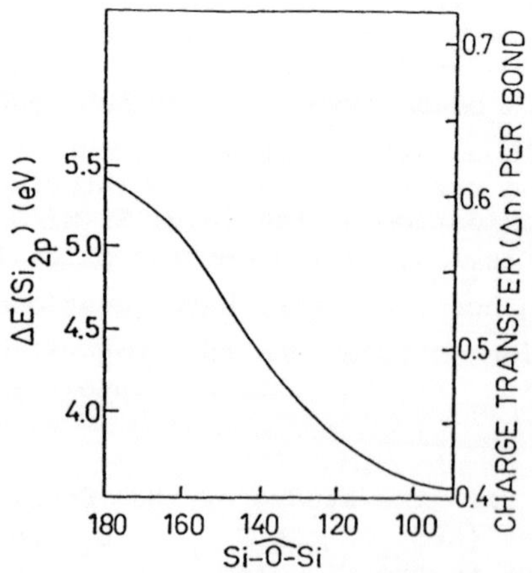

Figure 1

The calculated charge transfer, Si→O, per SiO bond (in numbers of electrons) and Si_{2p} core level shift (ΔE (Si_{2p})) as a function of Si-O-Si brindging bond angle taken from reference 1.

one case[3] of oxides grown at low O_2 pressure and low temperature. The result of an increase in charge transfer towards the O from the Si should result in an increase in the energy separation of the O_{1s} and Si_{2p} peaks (ΔE (Si_{2p} - O_{1s})). We know that a 0.1 electron transfer shifts the Si_{2p} peak to lower energy by ~ -0.9 eV but we do not know what the corresponding shift of the O_{1s} level should be, only that the splitting ΔE (Si_{2p} - O_{1s}) should decrease. Although this is born out by some measurements it is not clear that the shift ΔE (Si_{2p} - O_{1s}) would necessarily be large enough to be resolved given the combined resolution of the two peaks (> 1eV). Crucial to this type of argument, however, is the result of calculation shown in fig. 1. If we assume that the mean bridging bond angle of amorphous SiO_2 is 144° then ΔE (Si_{2p}) is -4.5 eV. From this curve we see that a value of -3.5 eV corresponds to a mean bond angle ~ 90° which is energetically extremely unfavourable[11] and consequently, unlikely. Consequently, the suggestion that the Si_{2p} line observed at -3.5 eV in thin oxides is due to Si^{4+} tetrahedra in amorphous SiO_2 is unlikely. In particular, this conclusion eliminates the possibility that such large shifts are produced by residual interfacial strain which has been cited[6] but is much too small[12] to provoke such enormous bond angle variations[13]. The exact state of the SiO_4 tetrahedron in thin oxides therefore leaves some questions open.

We now turn to the question of the influence of the bridging bond angle distribution on the width of the Si_{2p} peak originating from SiO_4 tetrahedra in thick oxide. The narrowest observed peak (from synchroton radiation studies[3]) is shown in fig. 2 after removal of an estimated 0.2 eV width due to instrumental resolution. The peak is shown in terms of energy shift which is directly related to electron transfer in the Si - O bonds. It is important, now, to realize that the peak observed results from an <u>average</u> charge transfer produced by the sum of the transfers of the four Si - O bonds attached to a given tetrahedron. The average transfer, Δn, is written :

$$\overline{\Delta n} = \frac{\Delta n_1 + \Delta n_2 + \Delta n_3 + \Delta n_4}{4}$$

The joint probability function, $P(\Delta n_1, \Delta n_2, \Delta n_3, \Delta n_4)$, which is linked to the distribution function for each transfer, Δn_i (and consequently Si-$\widehat{O}$-Si through fig. 1) given by $X(\Delta n_i)$ is taken as

$$P(\Delta n_1, \Delta n_2, \Delta n_3, \Delta n_4) = X(\Delta n_1)\, X(\Delta n_2)\, X(\Delta n_3)\, X(\Delta n_4)$$

The experimentally measured lineshape function, $F(\overline{\Delta n})$, is given by

$$F(\overline{\Delta n}) = X(\Delta n)*X(\Delta n)*X(\Delta n)*X(\Delta n) = X_4(4\Delta n)$$

where * denotes convolution and we have assumed the Δn_1 are independent variables. From Fourier transformation of the experimental lineshape we obtain :

$$FT\{F(\Delta n)\} = (FT\{X(\Delta n)\})^4$$

from which we may deduce the real $X(\Delta n)$. The ease of obtaining the fourier transform depends upon the form of the experimental lineshape. If, for simplicity, it is assumed Gaussian :

$$F(\Delta n) = \exp - \left[\frac{\left(E_{Si_{2p}} - E^{o}_{Si_{2p}}\right)^2}{a}\right]$$

$$\text{then } X(\Delta n) = \exp - \left[\frac{\left(E_{Si_{2p}} - E^{o}_{Si_{2p}}\right)^2}{4a}\right]$$

In general the lineshape is the more complex Voigt shape[5], however, if we assume that given in fig. 2 to be approximately Gaussian then we obtain

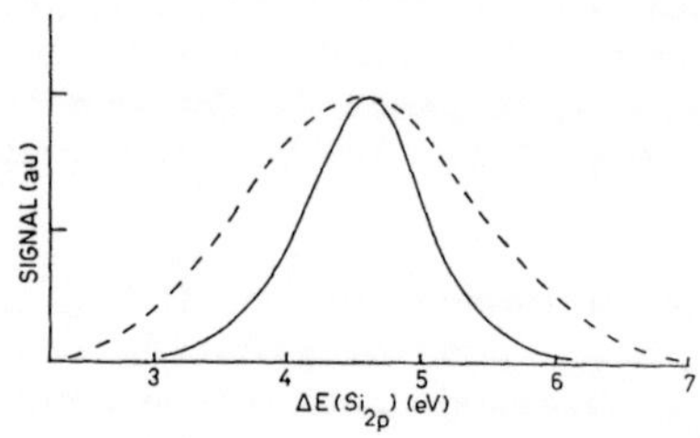

Figure 2

The experimental XPS Si_{2p} core level spectrum (solid line) in amorphous SiO_2 after removal of instrumental broadening. The dashed line is the deconvolution of the experimental line taking into account the mutual independence of the four bridging bond angles associated with each SiO_4 tetrahedron.

the deconvoluted form $X(E(Si_{2p}))$ shown in fig. 2. The breadth at half maximum height is twice the experimental value. If we use the deduced distribution $X(E(Si_{2p})$ together with fig. 1 we note immediately that the width of the distribution considerably exceeds the range of shift values anticipated on the basis of the bond angle variation model. From the figure these limits are -5.4 eV $< \Delta E\ (Si_{2p}) < -3.6$ eV whereas the distribution predicted spreads over the range -6 eV to -2.5 eV. This result strongly suggests the presence of some supplementary broadening process artificially enlarging the experimental curve. We can ask whether or not we can obtain some idea of the anticipated experimental width based on bond angle

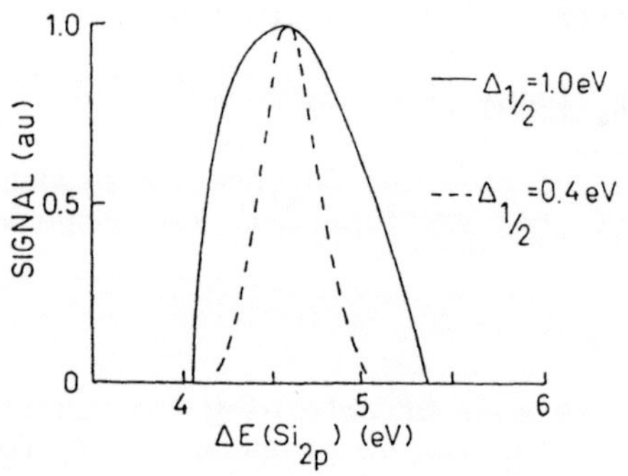

Figure 3

The XPS Si_{2p} line expected in amorphous SiO_2 from known Si-O-Si bridging bond angle distributions before (solid line) and after (dashed line), convolution to allow for the four mutually independent bond angles on each SiO_4 tetrahedra.

on bond angle distribution widths ? In fig. 3 we show the distribution of E (Si_{2p}) expected on the basis of the known bond angle distribution[14,15], its full width at half peak amplitude is ~ 1eV. We have performed numerical convolution of this distribution to simulate the experimental lineshape, the result is also shown in fig. 3. The simulated full width at half peak height is 0.39 eV, considerably less than the measured value (fig. 2) of 1 eV. This again emphasizes the presence of an unidentified broadening mechanism. It must be underlined that this broadening mechanism cannot be due to phonons for the following reason : nuclear magnetic resonance measurements, also sensitive to charge transfer due to bond angle distributions, reproduce[13] the distribution width at half peak height as determined by X ray scattering[14,15] to within < 30 %. Consequently, in a similar measurement such as XPS it would be unreasonable to imagine phonons increased the width by a factor of 2.5.

We have endeavoured to demonstrate how bridging bond angle distributions intervene in the interpretation of XPS Si_{2p} spectra. The results suggest that experimental XPS lines are broadening by a factor 2.5 times the width expected on the basis of known bond angle distributions. In very thin oxides a shift in energy of the Si_{2p} peak associated with the SiO_4 tetrahedron cannot be readily interpreted in terms of a modification in mean bond angle or interfacial stress. The origin of the shift must lay with the local environment around the SiO_4 which does not result in the Si → O charge transfer found is SiO_2. Instructive data might be obtained if measurements could be performed on samples at low temperatures (say liquid nitrogen).

ACKNOWLEDGEMENTS

This work was supported in part by a NATO International Collaborative research grant n° 87-18.

REFERENCES

1. F.J. Grunthaner, P.J. Grunthaner, R.P. Vasquez, B.F. Lewis, J. Maserjian and A. Madhukar, Phys. Rev. Letts. 43:1683 (1979).
2. A. Ishizaka, S. Iwata and Y. Kamigaki, Surf. Science 84:355 (1979).
3. G. Hollinger and F.J. Himpsel, Appl. Phys. Lett. 44:93 (1984).
4. A. Iqbal, C.W. Bates Jr. and J.W. Allen, Appl. Phys. Lett. 47:1064 (1985).
5. P.J. Grunthaner, M.H. Hecht, F.J. Grunthaner and N.M. Johnson, J. Appl. Phys. 61:629 (1987).
6. A. Darlinski and J. Halbritter, J. Vac. Science Techn. 5A:(in press).
7. M. Grundner and H. Jacob, Appl. Phys. A39:73 (1986).
8. R.A.B. Devine, J. Appl. Phys. 60:468 (1986).
9. N.M. Johnson, D.K. Biegelsen, M.D. Moyer, S.T. Chang, E.H. Pointdexter and P.J. Caplan, Appl. Phys. Lett. 43:563 (1983).
10. P.M. Lenahan and P.V. Dressendorfer, J. Appl. Phys. 55:3495 (1984).
11. A.G. Revesz and G.V. Gibbs, in "The Physics of MOS Insulators" edited by G. Lucovsky, S.T. Pantelides and F.L. Galeener Pergamon, New York (1980) p. 92.
12. C. Camelin, G. Demazeau, A. Straboni and J.L. Buevoz, Appl. Phys. Lett. 48:121 (1986).
13. R.A.B. Devine, R. Dupree, I. Farnan and J.J. Capponi, Phys. Rev. B35:2560 (1987).
14. R. Couty, Thesis, Université Pierre et Marie Curie, 1977 (unpublished).
15. R.L. Mozzi and B.E. Warren, J. Appl. Crystallog. 2:164 (1969).

$Si - SiO_2$ INTERFACES - A HRTEM STUDY*

N.M. Ravindra, D. Fathy*, O.W. Holland* and J. Narayan**

Vanderbilt University, Nashville, TN 37235
*Oak Ridge National Laboratory, Oak Ridge, TN 37381
**North Carolina State University, Raleign, NC 27695-7907

ABSTRACT

A High Resolution Transmission Electron Microscopy (HRTEM) study of the Si - SiO_2 interfaces is reported here. The study has been carried out on the thermal oxidation of (a). Si and (b). Ge implanted Si. Evolution of the Si-SiO_2 interface with oxide thickness and the presence of small amounts of Ge, i.e., one mono-layer at the interface and its influence on the oxidation kinetics is discussed.

INTRODUCTION

The Si-SiO_2 interface has been the most studied solid-solid interface. Despite the tremendous efforts that have been invested to understand the precise nature of the interface and its correlation with the oxidation kinetics, the exact role of the interface structure in determining the electrical behavior has not always been explicit.

Various characterization techniques such as High-Resolution X-ray Photoelectron Spectroscopy[1], Spectroscopic Ellipsometry[2], Low Energy Electron Diffraction[3], Pulsed Laser Atom Probe[4], Transmission Electron Microscopy[5], and Electron Paramagnetic Resonance[6] have been employed to study the nature of the Si-SiO_2 interface. All these techniques seem to lead to one conclusion - the Si-SiO_2 interface consists of a transition region of a silicon rich oxide. Recent HRTEM studies[7] show that the Si-SiO_2 interface exhibits roughnesses of varying magnitude depending on the (i) oxidation temperature, (ii) the ambient during oxidation, (iii) cleaning procedure involved and (iv) oxide thickness. Apart from all these aspects, the starting silicon wafer is known to have influence on the interface integrity.

This roughness region which is a part of the transition region is known to be the result of build-up of stress at the interface. Although the SiO_2 film is amorphous, the difference in thermal expansion between the Si and SiO_2 accounts for most of the accumulation of stress at the interface. Once the viscous flow sets in (the temperature of which is 975°C) stress is expected to be relieved during thermal oxidation. Thus, a high temperature anneal as a post-oxidation step is always favoured to result in improved oxide quality.

It is imperative that when a comparison of $Si-SiO_2$ interfaces are made, the experimental conditions under which the oxides were grown need to be considered. With a view to understanding the differences in the structure of these interfaces, we present here a HRTEM study of thermally grown SiO_2 on Si.

EXPERIMENTAL DETAILS

In the first case, we consider the conventional and the simplest case of thermal oxidation of silicon. Single crystal silicon wafers (p-type Czochralski) of (100) orientation and 2 ohm cm. resistivity were used for growing the oxides. The cleaning procedure consisted of the conventional RCA technique followed by a HF dip and a thorough rinsing with deionized water. The oxides were thermally grown at 800°C for time durations of 1 to 300 minutes. Ultra high pure oxygen (H_2O less than 0.5 ppm) was used in the oxidation process.

In the second part of this work, single crystals of n-type silicon with resistivities in the range of 1-15 ohm cm. and (100) orientation were used. Germanium was implanted at 77 K using singly charged $^{74}Ge^{+}$ ions with an average current density of 10 microamps over the implanted area. An area of each sample was masked during implantation. Oxidation was done in a tube furnace at 900°C and 1000°C in a steam ambient. Oxide thicknesses were measured on implanted and non-implanted regions on the same sample in order to permit a valid comparison.

Ion beam thinning procedure was used to prepare cross sections for HRTEM studies. High-resolution phase contrast images of the interface were taken at Scherzer defocus value of 65 nm and (110) orientation using a JEOL (200CX) TEM at 200 kV with a 0.26 nm point-to-point resolution.

RESULTS & DISCUSSION

In fig. 1, we present a low magnification image of the Si/SiO_2 including the interface. Such images have been utilized to determine the oxide thickness. As can be seen in the figure, the oxide thickness is quite uniform. In fig. 2, a high magnification HRTEM of the Si/SiO_2 for oxide thicknesses ranging from 3 to 22 nm is presented. From this figure, it appears that the thickest oxide has the roughest interface with protrusions of crystalline silicon of the order of 1 nm. It should be noted that our results of HRTEM do not show any evidence of the presence of crystalline SiO_2 at the $Si-SiO_2$ interface as has been claimed by some investigators[8].

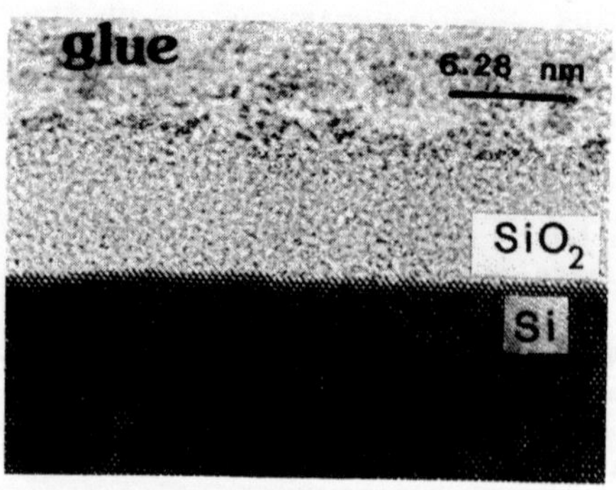

Fig. 1. Low magnification image of the Si/SiO_2 interface.

These investigators[8] have actually studied the oxidation of silicon films grown on crystalline silicon by molecular beam epitaxy.

We are of the opinion that the micrographs presented in fig. 2 should represent the most typical feature of thin films of SiO_2 on Si by virtue of the fact that the conditions of oxidation and the cleaning procedure employed represents the situation in a semiconductor device processing laboratory. Further, the oxide thickness of almost 20 nm is compatible with the gate oxide thickness for a one micron Complimentary Metal Oxide Semiconductor process.

The interface roughness is essentially determined by stress relaxation, strain distribution and diffusion limited processes (in thick oxides). At 800°C, the stress relaxation time is known to be about 5100 hours[9]. Thus, the stress is expected to be accumulated at the interface irrespective of the oxide thickness. Further, it is anticipated that at least in the oxide thickness range of 3 to 22 nm, the stress at the interface would increase with oxide thickness. This should result in increased roughness with increasing oxide thickness. However, for large oxide thickness (greater than 350 nm), the interface roughness is also determined by contributions from diffusion limited processes.

It has earlier been proposed that there exists a transition layer consisting of a silicon rich oxide at the $Si-SiO_2$ interface with a thickness of 0.6 nm for an oxide thickness of 100 nm[10]. These evaluations, which have been based on single wavelength ellipsometry, yield a refractive index of 2.8 at a wavelength of 5461 Å for this transition layer. Our on-going studies of the spectroscopic ellipsometry of thin SiO_2 films on Si

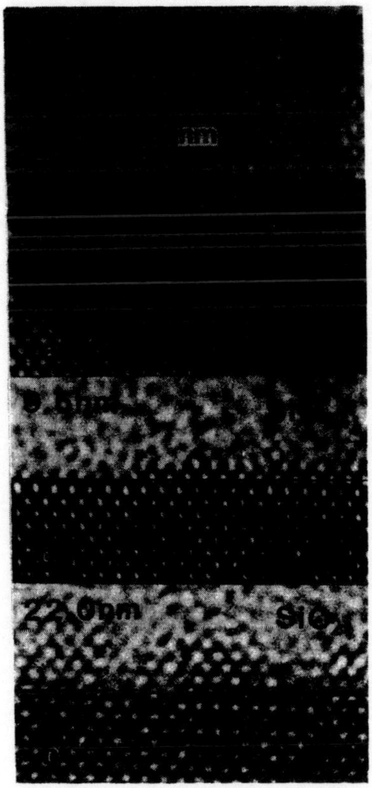

Fig. 2. High-resolution transmission electron micrographs of the Si/SiO_2 interface for samples of different oxide thicknesses. The oxidation temperature is 800°C.

(oxide thickness = 3 to 22 nm)[11] have yielded thicknesses for the transition layer in agreement with our HRTEM studies. Thus, within the experimental limitations of our SE & HRTEM studies, at least for these thin oxides, it seems that the transition layer is really a mixture of protrusions of crystalline silicon and amorphous SiO_2.

Fig. 3. High-resolution image of the $SiO_2/Ge/Si$ structure of 2×10^{15} cm^{-2} Ge^+ implant.

In fig. 3, a low magnification image of the SiO_2/Ge is presented. We have carried out studies on the oxidation of Ge implanted silicon at various doses, oxidation temperatures and for different oxidation times. These studies reveal interesting structural differences from that of SiO_2/Si. Further, the Ge implantation into silicon is seen[12] to enhance the oxidation rate and seems to act as an intermediary layer for stress release. Thus, the SiO_2/Ge interface appears flat as is seen in fig. 3. The interface structure does not exhibit any significant roughness. The reasons for the interface flatness are attributed to :

(a). The degree of freedom for SiO_2 to release its stress is enhanced by virtue of the distribution of stress across the Ge and the underlying silicon substrate,

(b). The reduced lattice mismatch of SiO_2/Ge compared to SiO_2/Si,

(c). The thickness of SiO_2 considered here is 340 nm at an oxidation temperature of 900°C. The stress relaxation time at 900°C is about 21 hours (much less than that at 800°C).

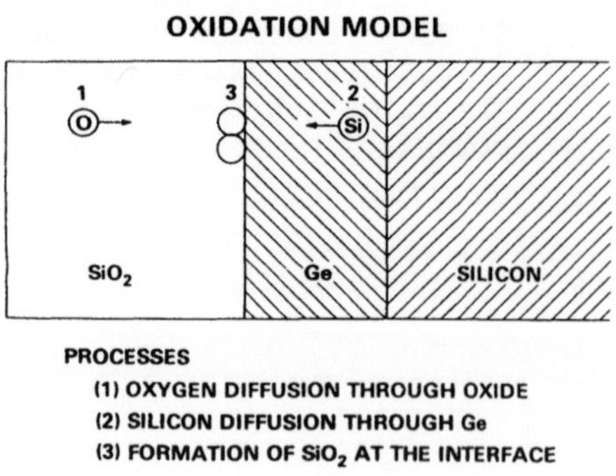

Fig. 4. Illustration of the model to explain the oxidation of Ge implanted Si.

The model to explain these results is illustrated in fig. 4. Process 1 is typical of the thermal oxidation of silicon - oxygen diffuses through the SiO_2 into the underlying substrate. Process 2 involves the diffusion of Si through the Ge layer. This is made possible because of very thin layer of Ge and the diffusivity of Si in Ge being high. Process 3 is the formation of SiO_2 at the interface. Thus, with increasing oxide thickness, the Ge layer is pushed down. Studies have shown that the presence of one mono-layer of Ge is sufficient to complete processes 1, 2 & 3.

CONCLUSIONS

HRTEM techniques have been employed in the above study to characterize the structure of Si-SiO_2 interfaces. Protrusions of crystalline silicon with a thickness of almost 1 nm have been observed in 22 nm thick SiO_2 films thermally grown on Si. The oxidation of Ge implanted Si leads to an enhanced oxidation rate with the Ge layer segregating towards the interface. The SiO_2/Ge interface is seen to be smooth.

REFERENCES

1. F.J. Grunthaner, P.J. Grunthaner, R.P. Vasques, B.F. Lewis, J. Maserjian and A. Madhukar, Phys. Rev. Lett. 43:1683 (1979).
2. D.E. Aspnes and J.B. Theeten, Phys. Rev. Lett. 43:1046 (1979).
3. M. Henzler and P. Marienhoff, J. Vac. Sci. Technol. B 2(3):346 (1984).
4. C.R.M. Grovesnor, A. Cerezo and G.D.W. Smith, in "Materials Research Society Symp. Proc.", MRS Pittsburg, vol. 37, 199 (1985).
5. A.H. Carim and R. Sinclair, Materials Letters 5:94 (1987).
6. E.H. Poindexter, G.J. Gerardi, M.E. Rueckel, P.J. Kaplan, N.M. Johnson and D.K. Biegelsen, J. Appl. Phys. 56:2844 (1984).
7. N.M. Ravindra, D. Fathy, J. Narayan, J.K. Srivastava and E.A. Irene, Materials Letters, 4:337 (1986).
8. A. Ourmazd, Private Communication.
9. E.A. Irene, E. Tierny and J. Angilello, J. Electrochem. Soc., 129:2594 (1982).
10. E. Taft and L. Cordes, J. Electrochem. Soc., 126:131 (1979).
11. N.M. Ravindra, D. Fathy, J. Narayan, A.R. Heyd and K. Vedam, Communicated (1987).
12. D. Fathy, O.W. Holland and C.W. White, Materials Research Society meeting, Anaheim, CA, April (1987).

*Work at Oak Ridge sponsored by the Division of Materials Sciences, U.S. Department of Energy under contract DE-AC05-840521400 with Martin Marietta Energy Systems, Inc.

ELECTRICAL AND INTERFACE PROPERTIES OF MOS STRUCTURES OF GETTER TREATED SILICON

N.M. Ravindra, Patrick Smith* and J. Narayan**

Vanderbilt University, Nashville, TN 37235
***Microelectronics Center of North Carolina, Research Triangle**
****Park, P.O. Box 12889, NC 27709**
North Carolina State University, Raleigh, NC 27695-7907

ABSTRACT

Electrical and micro-structural properties of Metal-Oxide-Semiconductor (MOS) capacitors fabricated utilizing silicon wafers, treated differently for gettering of impurities, is reported in this study. The treatment of silicon wafers for gettering purposes, considered here, essentially involves (i) back side argon ion implantation at doses of 2.5×10^{15} cm^{-2} and 5.0×10^{15} cm^{-2}, at energies of 195 and 350 keV respectively, and (ii) back side polysilicon deposition. The results of these measurements have been compared with those of ungettered silicon. Results indicate that some improvement has been attained in minority carrier lifetimes and maximum breakdown voltages in MOS capacitors fabricated particularly using back side argon ion implanted silicon.

INTRODUCTION

With the onset of sub-micron technology, the inter-relationships between the device processing, device performance and the starting material has become very important. In particular, device properties such as minority carrier lifetime and breakdown voltage are known to depend on the quality of the starting silicon wafers. During the processing of silicon, defects and impurities are generated in addition to those already present in the as-grown material.

Results on the electrical and Electron Beam Induced Currents (EBIC) studies[1] of thin oxide MOS capacitors have shown that the defects responsible for poor dielectric properties are impurity decorated stacking faults located in silicon near the Si-SiO_2 interface. These impurities have been identified to be metals such as Cu, Ni, Fe, Zn, Sn. Upon heat treating the silicon wafers to high temperatures, these metallic impurities tend to diffuse towards and aggregate at dislocations in the host silicon lattice. This phenomena is known as gettering. The presence of impurities in the starting silicon results in energy levels in the band gap of the material. This leads to an enhanced generation of electron-hole pairs.

Gettering in silicon will help remove these impurities away from the vicinity of the electrically active regions. This is known to result in improved oxide integrity[2] and helps in controlling the minority carrier lifetimes[3]. Various methods of gettering in silicon have been proposed. These include extrinsic gettering by misfit dislocations generated at epitaxial silicon interfaces by incorporation of 0.1-1 % Ge[4], back surface argon ion implantation (BSII)[5], laser damage gettering[6,7] and back surface mechanical abrasion[8].

In the present study, we report some experimental studies of the electrical and structural properties of MOS capacitors fabricated utilizing getter treated silicon.

EXPERIMENTAL DETAILS

Single crystal silicon wafers (p-type Czochralski) of (100) orientation and 2 ohm cm resistivity were used in these studies. The cleaning procedure consisted of the conventional RCA technique followed by a HF dip and a thorough rinsing with deionized water. In Table I, the rest of the process steps are outlined.

The oxide thicknesses were measured using a carefully aligned Gaertner ellipsometer. Several nine-point measurements were made on each wafer. The measurements yielded thicknesses within ± 0.5 nm. Breakdown voltage measurements were made using a Rucker and Kolls automatic prober stepper model 682. The steady-state ramp method[9] was employed to determine the lifetime.

Ion beam thinning procedures were used to prepare cross sections for Transmission Electron Microscopy (TEM) studies. High-resolution phase contrast images of the interface were taken at Scherzer defocus value of 65 nm and (110) orientation using a JEOL (200 CX) TEM at 200 kV with a 0.26 nm point to point resolution.

RESULTS & DISCUSSION

In Table II, we present a summary of the electrical measurements on the four sets of MOS devices. As can be seen in the table, the maximum breakdown voltage (mbv) has a maximum value of 10.5-11.0 MV/cm for set C. These have been fabricated using BSII silicon with an argon dose of $2.5 \times 10^{15} cm^2$ at 195 keV. All the breakdown voltage measurements have been made for a threshold current of 10 microamps. The defect density DD has been determined using the relation,

$$DD = - \frac{\ln (P)}{A}$$

where, P is greater than $0.8E_{max}$, P is the probability of a breakdown event, E_{max} is the maximum breakdown field and A is the area of the MOS capacitor. MOS capacitors fabricated with BSII silicon exhibit the maximum defect density (Table II - set C & D). However, the improvement in the percentage of the number of devices exhibiting maximum lifetimes for these devices should more than off-set the defect density.

Our results of the breakdown voltage of BSII Si are in complete accord with those of Yun[2]. However, E_{max} obtained in this study is higher than those obtained by Yun.

TABLE I

Process steps employed to fabricate MOS Capacitors of the following four sets.

Set	A	B	C	D
	20 nm SiO_2 800°C dry	back-side polysilicon LPCVD.500 nm	back-side Ar^+ $2.5x10^{15} cm^{-2}$; 195 keV ; * through 40 nm screen oxide	Implantation* $5.0x10^{15} cm^{-2}$ 350 keV
	Al evap. from side 32 mil dots	20 nm SiO_2 800°C dry	CVD Si_3N_4	125 nm
	backside etch	Al evap. front side 32 mil dots	CVD SiO_2	50 nm
	Al evap. backside	backside oxide etch	1050°C anneal for 90 minutes in argon	
	Post-metal anneal forming gas 400°C-10 min.	backside Al evap. 500 nm	450 nm SiO_2 at 950°C	
		Post-metal anneal forming gas 400°C-10 min.	etching steps for nitride/oxide stack	
			20 nm SiO_2 at 800°C	
			Al evaporation front side 32 mil dots	
			back side oxide etch	
			back side metallization Al 500 nm	
			post-metal anneal in forming gas 400°C - 10 min.	

In fig. 1, a low magnification image of the SiO_2/Si is presented. As can be seen in the figure, the oxide appears to be uniformly thick. In fig. 2, a cross-sectional TEM image of the polysilicon/Si is shown. Grains characteristic of the polycrystalline structure are noticeable in the figure.

CONCLUSIONS

Back side argon ion implantation into silicon is seen to favour improved device characteristic. This has been shown for the particular case

TABLE II

Summary of the Breakdown voltage and lifetime measurements.

Set	A	B	C	D
t_{ox} (nm)	21.0	20.8	20.7	20.7
E_{max} (MV/cm)	9.5 - 10.0	10 - 10.5	10.5 - 11.0	10 - 10.5
No. of Devices (%) at E_{max}.	96	72	70	72
Defect Density (cm^{-2})	6	10	25	27
No. of devices at Maximum lifetime (%)	66.8	61.1	70.0	82.0

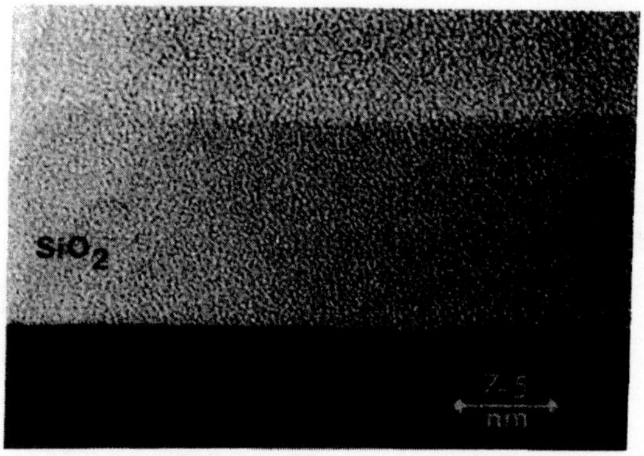

Fig. 1. Low magnification image of the SiO_2/Si interface. Oxide thickness is about 20 nm.

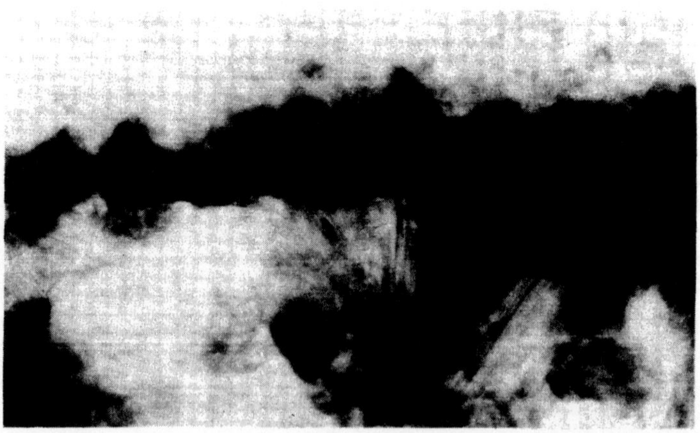

Fig. 2. Cross-sectional TEM image of the polysilicon/silicon interface.

of MOS capacitors. The device performance has been seen to be better than that of MOS structures fabricated using silicon wafers with polysilicon deposited at the back.

REFERENCES

1. P.S.D. Lin, R.B. Marcus and T.T. Shen, J. Electrochem. Soc. 130:1878 (1983).
2. B.H. Yun, Appl. Phys. Lett. 39:330 (1981).
3. H.F. Kappert, G. Sixt and G.H. Schwuttke, Phys. Stat. Sol. (a), 52:463 (1979).
4. A.S. Salih, H.J. Kim, R.F. Davis and G.A. Rozgonyi, Appl. Phys. Lett. 46:419 (1985).
5. A.G. Nassibian and B. Golja, IEE Proc. 127:29 (1980).
6. G.E.J. Eggermont, D.F. Allison, S.A. Gee, K.N. Ritz, R.J. Falster and J.F. Gibbons, in "Laser and Electron-beam Interaction with Solids", B.R. Appleton and G.K. Celler, editors, Elsevier, New York (1982).
7. Y. Hayafuji, T. Yanada and Y. Aaki, J. Electrochem. Soc. 128:1975 (1981).
8. T.J. Magee, C. Leung, H. Kawayoshi, B.K. Furman and C.A. Evans Jr., Appl. Phys. Lett. 38:891 (1981).
9. E.H. Nicollian and J.R. Brews, MOS (Metal Oxide Semiconductor) Physics and Technology, Wiley, New York (1982).

INFLUENCE OF DIFFERENT PREPARATION METHODS ON INTERFACIAL (SiO_2/Si) PARAMETERS OF VERY THIN SiO_2 LAYERS

Georges Pananakakis, Panagiota Morfouli, and Georges Kamarinos

Laboratoire de Physique des Composants à Semiconducteur
ENSERG/INP-G, 23 Avenue des Martyrs, 38031 - Grenoble cédex
France

INTRODUCTION

Very thin Silicon dioxide layers (thickness less than 100 Å) have a growing importance for the VLSI integrated circuits. Indeed they are used in EEPROM (memories) and in a variety of bistable devices (as the MISS for example). Besides they could be used as intermediate dielectric tunnel layers serving to enhance injection in Schottky contacts for bipolar circuit applications.

However, their fabrication is not yet "standardized" and different techniques are in competition. In this paper we compare the electrical properties of thin silicon dioxide films obtained by four different methods :

1. Thermal dry oxidation under low oxygen pressure (LPO_2)
2. LPCVD
3. Anodic oxidation
4. Dry thermal oxidation under chlorine ambient.

The interfacial state density (and their energy distribution and cross sections) as well as the semiconductor surface potential and the voltage drop across the oxide, have been measured by a variety of experimental methods exploiting the static I-V and dynamic C-V and G-V measurements at various temperatures ($77\ K \leq T \leq 300\ K$), and levels of illumination.

Internal photoemission in the range from UV up to I-R radiations has also been used, in order to obtain the metal-oxide and oxide-semiconductor barrier heights.

DESCRIPTION OF THE SAMPLES

The very thin (thickness $25\ Å \leq \delta \leq 100\ Å$) oxide layers have been obtained on Si by four different methods :

a) Thermal oxidation under low oxygen pressure (LPO_2). The silicon substrate (<100> oriented n-type silicon ; $N_d = 10^{15} cm^{-3}$) is oxidized at $T = 950°C$ at a pressure of 1.0 Torr, for 25 minutes ($\delta \simeq 25\ Å$).

b) Low Pressure Chemical Vapor Deposition of SiO_2 (LPCVD)[1]. The SiO_2 film is obtained by chemical reaction in vapor phase in $SiH_2\ Cl_2$ and N_2O at T = 880°C under a pressure of 0.2 Torr during 9 minutes ($\delta \simeq 25$ Å) ; it was deposited on <100> oriented n-type silicon substrate with $N_d = 10^{15}cm^{-3}$. The 25 Å tunnel devices of both technologies were fabricated by Al deposition on the oxide layer.

c) In order to obtain thin SiO_2 layers free of any contamination, anodic oxidation of silicon in ultra pure water (P = 18 MΩ.cm at 20°C under a constant current density of 10 $\mu A/cm^2$) was used. $Cr\text{-}SiO_2\text{-}Si$ or $Al\text{-}SiO_2\text{-}Si$ capacitors of two different thicknesses (50 Å and 100 Å) were fabricated on 5 Ωcm p-type <100> oriented silicon substrate[2].

d) Thermal oxidation under chlorine ambient[3] (0,2 % HCl 10% O_2 and N_2). $Cr\text{-}SiO_2\text{-}Si$ capacitors of several thicknesses (53 Å and 74 Å) were fabricated on p-type silicon <100> oriented with $N_a = 10^{15}cm^{-3}$.

I-V MEASUREMENTS

I-V characteristics allow us to test electrically the quality of the devices studied. In the case of very thin (30 Å) devices, the comparison between the experimental and theoretical characteristics leads to the determination of the interface state density using a detailed and original model of the working of MIS tunnel structures[1]. For the thicker devices (50 Å up to 100 Å) the Fowler Nordheim barrier height has been determined from these characteristics and has been found in good agreement with the barrier values obtained using optical measurements.

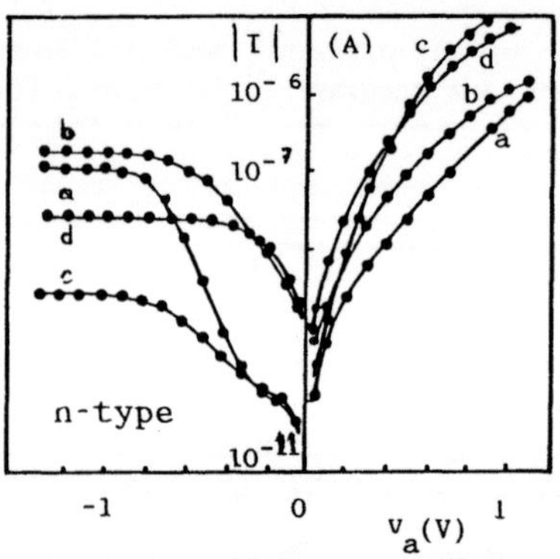

Fig. 1a) Experimental I-V characteristics of MIS samples
a) LPO_2 non annealed samples.
b) LPCVD non annealed samples.
c) LPO_2 annealed samples.
d) LPCVD annealed samples.

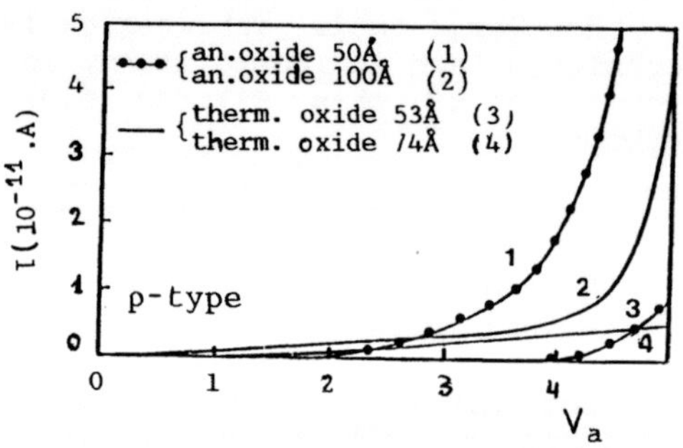

Fig. 1b) Experimental I-V characteristics of anodic and thermal samples.

OPTICAL MEASUREMENTS OF BARRIER HEIGHTS

The measurement of metal-oxide and oxide-semiconductor barrier height performed using internal photoemission, allows a complementary characterization of the quality of the insulating layers. In the case of tunnel devices, the effective metal-oxide barrier Φ_m, as well as the Schottky barrier Φ_{bn}, could be determined using chopped infrared radiation of photons energy $h\nu<E_g$ using a monchromator and a silicon filter (fig. 2). Φ_{bn} and Φ_m values of the order of 0.1 eV and 0.8 eV have been so deduced in agreement with other results obtained using the conductance technique and in with values used for the computer simulation of the I-V characteristics of the tunnel devices[4] studied.

For the other devices, the barrier heights were measured using UV radiation sources with 230 nm $\leqq \lambda \leqq$ 300 nm.

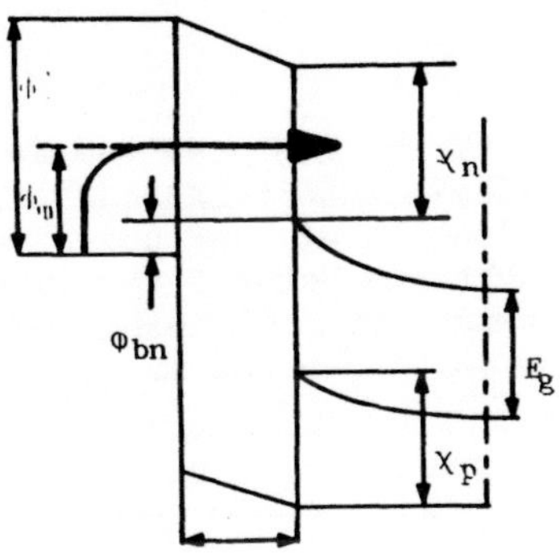

Fig. 2a) Energy band diagram of a MISS tunnel structure. Φ'_m : metal-oxide barrier height ; Φ_m : effective tunneling barrier height ; $\chi_n + E_g$: semiconductor-oxide barrier.

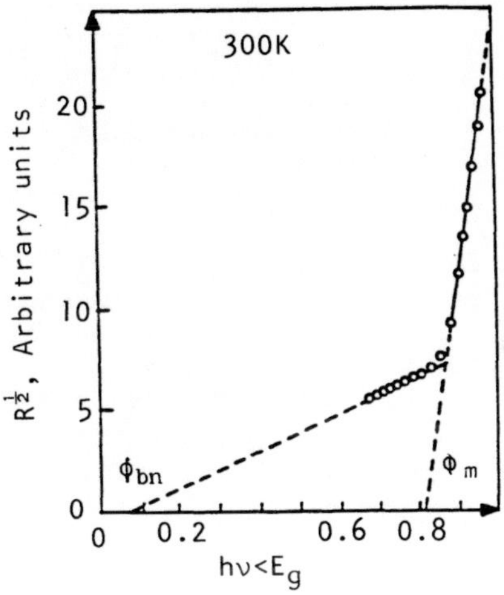

Fig. 2b) Plot of the square root of the photoresponse $R^{1/2}$ vs photon energy $h\nu$.

For the thermal oxide devices (74 Å) the metal-oxide barrier height has been found in the range of 3 eV (fig. 3). The semiconductor-oxide barrier height has been determined to be in the range of 4 eV. These values depend upon the applied voltage owing to the variation of the oxide layer trapezoïdal barrier depth.

In the case of thinner thermal oxides (53 Å) the semiconductor-oxide barrier was found to be practically the same as for 74 Å thick oxides.

INTERFACE STATE DENSITY MEASUREMENTS

The measurement of the interface state density N_{ss} near the silicon mid gap can be performed using the well known conductance technique[5].

In order to explore surface states located near the majority carrier energy band, we have applied this method in a range of temperatures from 77 K to 300 K. The lowering of T provokes an increase of the surface states time-constant and it is possible to measure the N_{ss} using frequencies from 50 Hz to 1 MHz[6].

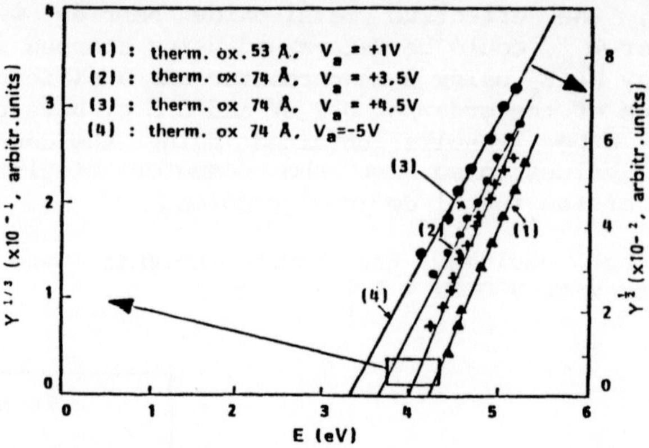

Fig. 3. Optical measurements of the metal-oxide and semiconductor-oxide barriers.

The characterization of the other part of the gap near the minority carrier energy band can be done using the illumination technique[7]. In the presence of a sufficiently high illumination level and under reverse samples bias, it is possible to enhance exchanges between the interface states and the minority carrier energy band. By combining these two methods, a complete exploration of the silicon forbidden gap is possible[6].

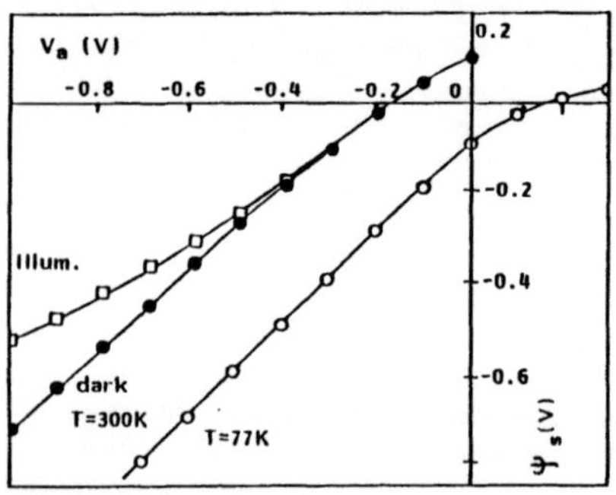

(a)

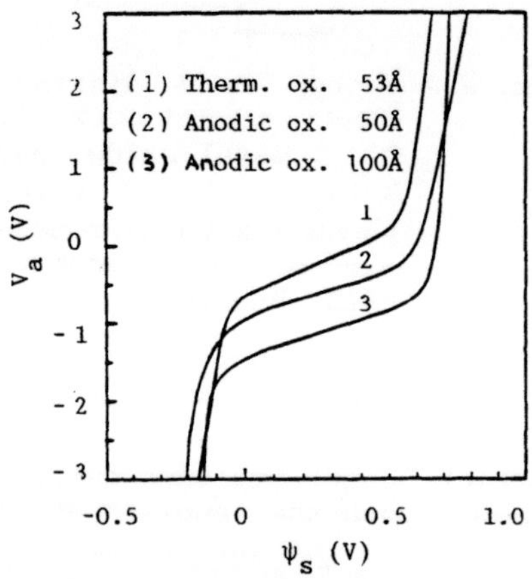

(b)

Fig. 4. Surface potential vs V_a for (a) LPCVD devices and (b) for anodic and thermal oxide devices.

In order to obtain the density of states N_{ss} vs E, the surface potential ψ_s (V_a) has been determined using the Berglund integral method[8]. Fig. 4a shows typical $\psi_s(V_a)$ plots for a very thin (25 Å) LPCVD oxide device in darkness and under illumination at T = 300 K. A ψ_s (V_a) characteristic at T = 77 K is also shown. All these characteristics have been obtained by performing capacitance measurement at f = 50 Hz. In the case of thermal or anodic thicker oxide devices, the $\psi_s(V_a)$ plots have been obtained using quasistatic capacitance measurements. It is worth noticing that these devices reach the inversion regime. By comparison the very thin oxide (25 Å) devices do not present any inversion layer.

Fig. 5a and 5b show $N_{ss}(E)$ plots for tunnel oxide (25 Å) devices and fig. 6a and 6b for thicker oxide devices.

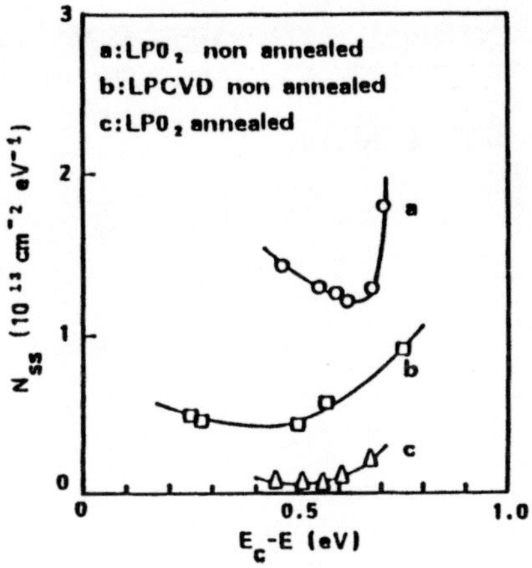

Fig. 5a) N_{ss} vs E plots for LPCVD and LPO_2 devices.

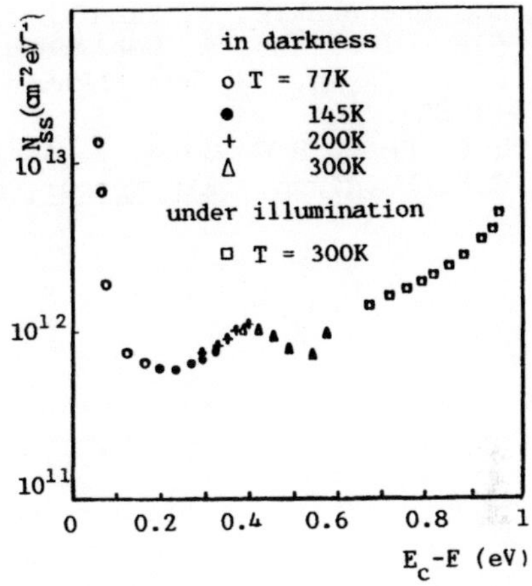

Fig. 5b) N_{ss} vs E plots for LPCVD device.

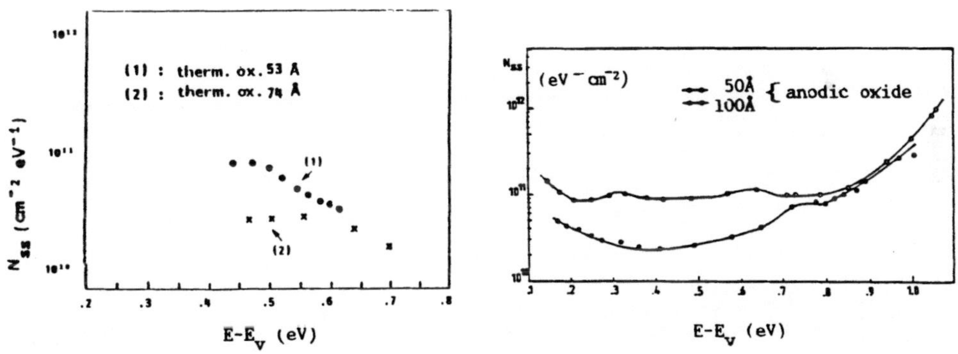

Fig. 6 N_{ss} vs E plots for thermal (a) and anodic (b) devices.

CONCLUSION

Our results show that in the case of very thin SiO_2 layers ($\delta \leqq 30$ Å) the deposited (LPCVD) oxides exhibit lower densities of states and are more homogeneous than the thermally grown oxides (LPO_2).

In the case of thicker oxides ($50 \leqq \delta \leqq 100$ Å) we found that the interfacial parameters of anodic and thermal oxide are comparable with respect to the interface state densities. However, it seems that the thermal oxides have more defects (i.e. pinholes) than the anodic insulating layers.

REFERENCES

1. G. Pananakakis, G. Kamarinos, M. El-Sayed, Solid. St. Electron, 26:415 (1983).
2. F. Gaspard, A. Halimaoui, G. Sarrabayrouse, Revue Phys. Appl., 22:65 (1987).
3. J.L. Prom, F. Rossel, C. Solano, T. Doconto, G. Sarrabayrouse, Journées GCIS, 18-19 Mai 1987, Toulouse.
4. G. Pananakakis, G. Kamarinos, Surface Science, 168:657 (1980).
5. E.H. Nicollian, A. Goetzberger, The Bell System J., XLVI:1055 (1967).
6. M. El-Sayed, G. Pananakakis, G. Kamarinos, Solid. St. Electron, 28:345 (1985).
7. T.C. Poom, H.C. Card, J. Appl. Phys., 51:6273 (1980).
8. C.N. Berglund, IEEE Trans., ED-13:701 (1966).

THERMAL OXIDATION OF SILICON

A.G. Revesz, B.J. Mrstik, and H.L. Hughes

Naval Research Laboratory
Washington, D.C. 20375

ABSTRACT

Thermal oxidation of silicon results in the formation of a non-crystalline (nc) SiO_2 film through which the oxidizing species, oxygen or OH, diffuses. A key feature of nc-SiO_2 is the wide range of Si-O-Si bond angles (120°-180°) associated with a very small variation (≈0.1 eV) in the Si-O bond energy. The Si substrate influences the structure of the oxide by affecting the short-range-order (SRO) of the (newly formed) oxide at the interface. Thus the structure of the as-grown oxide resembles that of some high density polymorph of SiO_2 ; this effect is labelled quasi-expitaxy. As further growth occurs at the interface, pushing away the previously grown oxide, the structure then relaxes to more closely resemble that of vitreous SiO_2. This relaxation process is similar to that which occurs after pressure compaction of vitreous SiO_2, as indicated by measurements of the changes in the index of refraction and density of the oxide. This process is essentially a pseudo-polymorphic transformation of the oxide structure in the course of which the SRO of the oxide mimics that of various crystalline SiO_2 polymorphs and results in substantial accomodation of the large volume expansion (≈126 %) associated with the transfer of Si atoms from the substrate to the oxide. This transformation differs from viscous flow, which has often been used to account for the relaxation of stress in the oxide, primarily in that it does not require the breaking of bonds, as does the viscous flow model. We point out several difficulties of the viscous flow model in explaining experimental observations.

Because the transformation does not involve breaking bonds, in dry oxides some remnants of the originial quasi-epitaxial layer remain in the oxide in the form of intrinsic structural channels which are oriented preferentially normal to the interface. Because transport of O_2 molecules along these channels is easier than through the bulk of the oxide, they affect the oxidation kinetics of the Si. Hence, the oxidation kinetics in dry O_2 is described in terms of two transport processes ; one which describes the normal random-walk interactive diffusion through the oxide and depends on the oxide thickness, and one which describes a non-interactive oxygen transport through the structural channels and which is relatively independent of the thickness of the oxide. The random-walk diffusion is associated with a limited exchange process between O_2

molecules in solubility sites and network O atoms. This interaction has been observed by ^{18}O tracer experiments as affecting mostly the outer surface region of the $Si^{16}O_2$ films. Such a model for the oxidation kinetics and mechanism produces very good agreement with experimental growth rate data over a wide range of oxide thicknesses without assuming any special mechanism for the initial "fast growth" regime and/or the oxygen exchange process.

INTRODUCTION

The thermal growth of oxide on silicon occurs at the Si/SiO_2 interface utilizing oxygen molecules which have diffused through the previously formed oxide. If it is assumed that the newly formed oxide is confined to the volume occupied by those silicon atoms which have become part of the oxide, then the oxide must be highly compressed, since the molar volume of the oxide is about 2.26 times greater than that of silicon. This molar volume difference would be expected to produce a stress at the interface of about 10^{11} dyne/cm^2, as calculated from the elastic constants of silica. From measurements[1] of the warpage of the substrate during oxide growth, however, the actual stress at the interface has been found to be on the order of only 10^9 dyne/cm^2. The question then arises as to how this enormous reduction of the interfacial stress occurs.

The mechanism which has been used most often to explain this stress relief involves viscous flow of the oxide. This viscous flow model was originally used to explain the observation[2] that the warpage is independent of growth temperature for temperatures above 925 C, implying that above this temperature the oxide is able to flow so that the substrate doesn't undergo any additional warpage. This model was later extended[3] to explain why the index of refraction depends on the growth temperature, as shown in fig. 1. Again, there seems to be a temperature above which the index of refraction is constant, suggesting that above this temperature the oxide is able to flow, thereby relieving the stress, whereas at lower temperatures the stress results in a compacted oxide with a higher index of refraction. The temperature at which the oxide flows is apparently dependent on its water content, as shown in fig. 1. The viscous flow model has also been used[4,5] to explain the initial fast oxidation regime. In this model the diffusivity of oxygen is assumed to be dependent on the stress in the oxide. Since in the viscous flow model the stress anneals out with time, the oxygen diffusivity varies with the distance from the interface, rather than being thickness independent as assumed in the derivation of the linear-parabolic growth rate law.

There are a number of serious problems with the viscous flow model, however :

A) Oxidation in dry oxygen is found[2] to produce much less stress than oxidation in wet oxygen, which implies that the viscosity of dry oxides is less than the viscosity of wet oxides. This is contrary to measurements[6] which indicate that the viscosity of dry oxides is about two orders of magnitude greater than the viscosity of oxides containing OH groups.

B) The measured stresses are much too small to explain the large changes, observed in the refractive index. For an oxide grown at 930 C, for example, a stress of about 4×10^9 dyne/cm^2 has been measured. Data of Brueckner[6] indicates that this amount of stress should increase the index of refraction by about 0.002, whereas the actual index of refraction is about 0.015 larger than for the stress free high temperature oxide.

C) The predicted relaxation times do not agree with the experimental observations. The relaxation time τ is given by

$$\tau = \eta/B \tag{1}$$

where η is the bulk viscosity and B is the bulk modulus. Measurements of the bulk viscosity are not available, but if we assume that it is close to

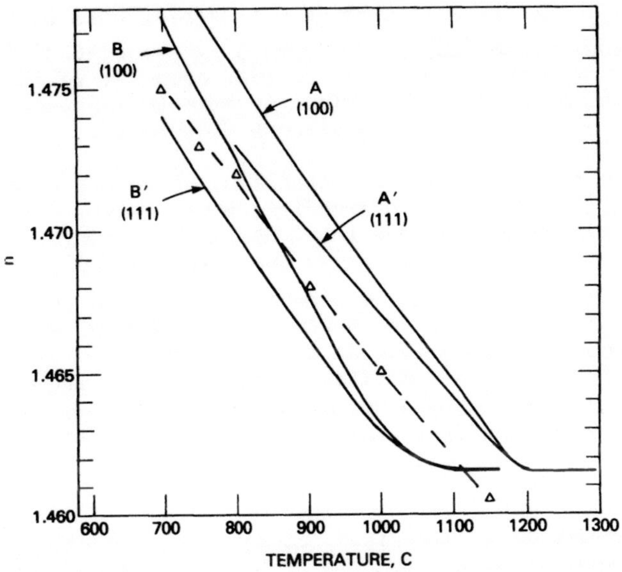

Fig. 1. Refractive index, n, of SiO_2 films as a function of growth temperature. Curves A and A' refer to oxidation in dry O_2 (ref. 3) and curves B and B' to oxidation in steam (ref. 7). Results for growth on (100) and (111) silicon substrates are shown. The triangles are from ref. 8 ; the dashed line represents their trend. All n values were determined at a wavelength of 546.1 nm.

the shear viscosity, then the calculated relaxation times are about 100 times larger than the observed relaxation times. For example, at 1000 C the calculated relaxation time is about 10^5 seconds, whereas an anneal of only about 1000 s was found[8] to reduce the index of refraction close to its unstressed value. At 800 C the calculated relaxation time is about 5100 hours, so that the stress of about 10^{11} dyne/cm^2 expected on the basis of molar volume considerations should not be reduced to the observed values in the 10^9 range.

D) The activation energy of oxygen transport during oxide growth ($\approx$1.2 to $\approx$2.6 eV, depending on the model used in the analysis) is much less than that of viscous flow, which is $\approx$7 eV and $\approx$5eV for dry and OH-containing SiO_2, respectively[7].

Therefore, although viscous flow may play some role, particularly in OH-containing SiO_2 films, there must be additional mechanisms responsible for stress relaxation at the interface. We will next suggest a new stress relief mechanism which is based more on a chemical and structural description of the growth process.

PROPOSED GROWTH MECHANISM

Our proposed growth mechanism is based on three major observations :

A) The SiO_2 structure is extremely flexible, that is, a change in the Si-O-Si bond angle from 120 to 180 degrees is associated[9] with a change in bond energy of only about 0.1 eV, so that even at relatively low growth temperatures the oxide structure might be able to rearrange itself without breaking bonds, as required in the viscous flow model.

B) Although the oxide is noncrystalline, there are several indications that the oxide may have some intermediate range order. XPS measurements[10], for example, have shown evidence for 4 member Si-O rings at the interface, and RHEED measurements[11] have revealed the existence of crystalline SiO_2 at the interface. It has also been reported[12] that coesite precipitates form in oxygen-containing silicon at temperatures below 800 C, which also suggests that crystalline SiO_2 is stable at the interface. Additionaly, a numer of properties of thermally grown films, including the diffusivity of oxygen and the index of refraction[3], are dependent on the substrate orientation. All of these observations indicate that structural effects may be important in understanding the nc-oxide. We'll refer to this as a quasi-epitaxial behavior.

C) For crystalline materials atomic rearrangement can occur by polymorphic transformations. Because of the quasi-epitaxial nature of SiO_2 we suggest that a similar pseudo-polymorphic rearrangement of atoms may also occur in SiO_2. This type of transformation is a cooperative restructuring of the material which involves all the atoms rather than just the nearest neighbor atoms.

It is instructive to compare the index of refraction of thermal oxide films and pressure compacted SiO_2. An oxide film grown in dry oxygen at 700 C has an index of refraction of about 1.475. Vitreous silica can attain this same index of refraction if it is compacted under a pressure[6] of 3×10^{10} dyne/cm^2 at 700 C. Although this stress is much larger than the 10^9 which is observed during thermal growth, it is less than the 10^{11} dyne/cm^2 expected on the basis of molar volume considerations alone, and it suggests the possibility that there is some similarity in the structure of the two materials.

MacKenzie[13] studied the decompaction of pressure compacted v-SiO_2 during annealing, and found that the process is definitely <u>not</u> viscous flow. His results show that the annealing process is characterized by a very rapid additional decompaction. The extent of the decompaction depends both on the annealing temperature and on the initial density of the material. For example, when annealed at 700 C, a sample which had initially been compacted to a density of 2.24/cm^3 decompacts to a density of 2.21 g/cm^3, whereas a sample originally compacted to 2.44 g/cm^3 decompacts to a density of 2.27 g/cm^3. Of course, the final density also depends on the anneal temperature. When annealed at 700 C, for example, the oxide decompacts considerably more than when it is annealed at 400 C. This annealing behavior is described by the following equation[13].

$$V_1 = V_\infty - 1.0053(V_\infty - V_o)/[1+5.3\times10^{-3}\exp(0.006T)] \quad (2)$$

which gives the molar volume after a 1 hour anneal, V_1, as a function of the anneal temperature, T. V_o is the molar volume immediately after compaction, and V_∞ is the molar volume of uncompacted vitreous silica ($27.25\ cm^{-3}$).

If we assume that the thermally grown SiO_2 films are compacted when initially formed at the interface, and that they decompact during subsequent growth following this same relationship, we can determine what the density of oxide was when it was initially formed. The results are shown in Table I for growth at several temperatures on the Si (111) surface. At each temperature we used the value of the index of refraction as determined by Taft[3], then used the data of Arndt and Stoeffler[14] to find the molar volume of compacted silica which corresponds to this value of the refractive index. We took this as V_1, and calculated the molar volume

TABLE I

CALCULATION OF DENSITY OF NEWLY GROWN OXIDE AT THE INTERFACE

Growth Temperature (C)	Refractive index (3)	V_1 (cm^{-3})	V_o (cm^{-3})	Density (g/cm^3)
700	1.4767	26.965	23.64	2.54
800	1.4733	26.215	22.81	2.63
900	1.4700	26.680	23.26	2.58
1000	1.4670	26.811	22.01	2.72
1100	1.4635	27.004	22.10	2.71
		Average density		$=2.64 \pm 0.07\ g/cm^3$

immediately after compaction using MacKensie's equation[13]. The values determined for the density of the newly grown material at the interface are shown in the last coluum of Table I. These calculations indicate that the density of the oxide originally formed was about $2.64\ g/cm^3$, considerably denser than uncompacted SiO_2, and about the same as quartz (Table II). This result is in agreement with the XPS[10] and RHEED[11] reports mentioned earlier of a crystalline type structure at the interface.

TABLE II

DENSITY OF SEVERAL SiO_2 POLYMORPHS

Uncompacted v-SiO_2	2.20 g/cm^3
Keatite	2.50
Quartz	2.65
Coesite	2.90

The decompaction of the newly formed oxide occurs as additional oxide is growing at the interface. This decompaction involves only modifications in the way in which the SiO_2 tetrahedra are arranged, and does not involve breaking bonds, like the viscous flow model does. Because bonds are not broken, the intermediate-range-order of the final oxide might be expected to resemble the initially grown oxide. Some indication that this is in fact the case is suggested by comparison of the polarizabilities of various oxides, calculated from their density and index of refraction using the Lorenz-Lorentz relationship. As shown in fig. 2, when plotted as a function of density the polarizabilities of the polymorphs, shown by the solid points, lie below the trend of those for compacted vitreous silica, shown by the line. Accurate measurements of the density of thermal oxide films are very difficult, but a few measurements have been made. The results are shown in fig. 2 for oxides grown at 900 or 1000 C. Although the experimental uncertainty in these points is large, they all lie below the trend line of compacted vitreous silica, suggesting that they may in fact have an intermediate-range-order which more closely resembles a crystalline polymorph than vitreous silica.

Additional support for this mechanism of oxide growth and stress relief at the interface comes from other evidence that remnants of the original compacted silica polymorph remain in the oxide film far from the growing interface. These remnants may be responsible for the surprising orientation dependence in the growth rate and index of refraction of thermal oxides. In the next section we show that detailed studies of the oxidation kinetics of dry thermal oxides also provide evidence of the existence of such remnants. In particular, we show that they are responsible for the well known initial "fast growth" oxidation of thermal oxides grown in dry oxygen.

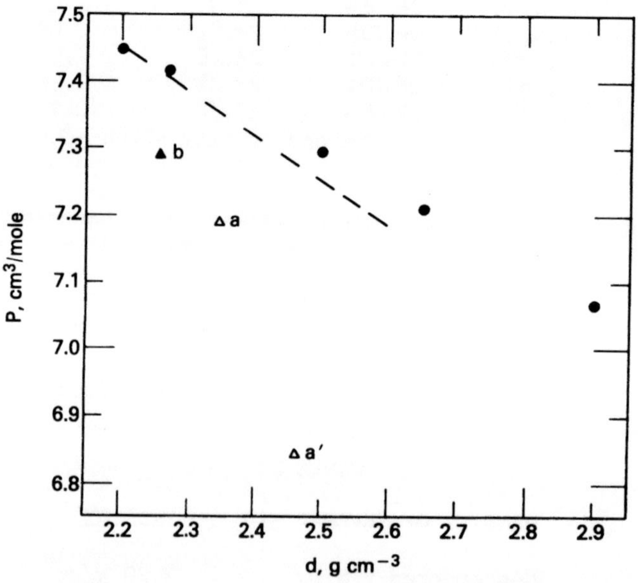

Fig. 2. Mole polarizability, P, as a function of density of SiO_2, the full circles represent v-SiO_2, α-tridymite, keatite, α-quartz, and coesite in order of increasing density[15]. Points a and a' refer to SiO_2 films with n=1.457 grown on silicon substrates at 800 C using the extreme d-values (2.35 and 2.47 g/cm^3) given in reference 16. Point b refers to an SiO_2 film with n=1.461 and d=2.26 g/cm^3 grown at 1000 C[16]. The dashed line showns the trend for pressure compacted v-SiO_2.

EFFECT OF GROWTH MECHANISM ON OXIDATION KINETICS

The rate, dx/dt, at which the oxide thickness, x, increases can be described in general by

$$dx/dt = F/C_o \tag{3}$$

where $C_o = 2x10^{22} O_2/cm^3$ is the concentration of oxygen in v-SiO_2. Since the oxide growth occurs at the Si/SiO_2 interface by the transport of O_2 molecules through the oxide, the total oxygen flux, F, is limited by the rate at which oxygen molecules can be transported through the growing oxide, and by the rate at which they can react with silicon atoms at the interface. F can therefore be written as

$$1/F = 1/F_S^O + 1/F_T^O \tag{4}$$

where F_S^O is the flux consumed by the surface reaction $Si+O_2 \rightarrow SiO_2$ when $F_T^O \rightarrow \infty$, and F_T^O is the flux controlled by the transport for $F_S^O \rightarrow \infty$. If the surface reaction rate is constant and the transport occurs by a random walk type of diffusion process, then

$$F_S^O = K_S C_O \tag{5}$$

and

$$F_T^O = C_O KT/x, \tag{6}$$

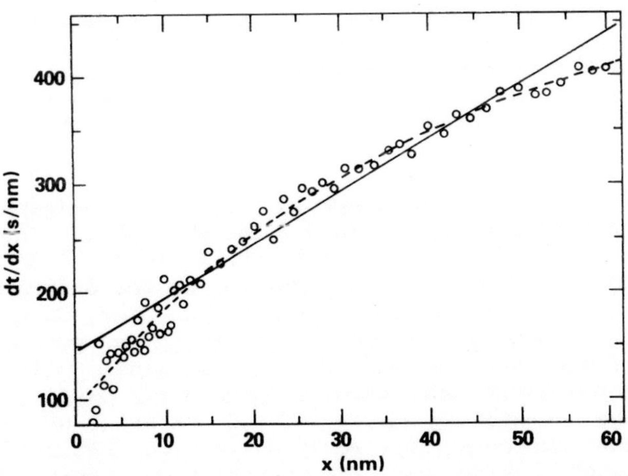

Fig. 3. Inverse growth rate as a function of oxide thickness. Incremental inverse growth rate between adjacent experimental data points is shown by circles. Inverse growth rate calculated from parameters determined by fit to linear-parabolic model (Eq. 7) and "channel" model (Eq. 9) are shown by solid and dashed lines, respectively. Data is for an oxide grown on a (111) silicon substrate at 870 C in an atmosphere of dry oxygen[17].

which define the surface and transport parameters K_S and K_T in terms of oxide thickness. Eq. (3) then becomes

$$dt/dx = 1/K_S + x/K_T \tag{7}$$

Eq. (7) is the well-known linear-parabolic rate law[18], which has been found to accurately describe the growth of oxides in an atmosphere containing water vapor. For oxides grown in dry oxygen however, significant departures from this rate law have been observed. In fig. 3, for example, is shown the growth rate as a function of oxide thickness for an oxide grown in an atmosphere of dry oxygen at 870 C. The experimental data points, determined by in situ ellipsometry measurements during oxide growth[17] are shown by the open circles, and the best least-squares fit of Eq. (7) to the data is shown by the solid line. The failure of Eq. (7) to adequately describe the experimental results is well-known, and is generally considered to reflect the presence of an additional growth mechanism which results in an initial "fast growth" regime for thin oxides grown in dry oxygen. Thus Eq. (7) is often considered to be accurate only for oxides thicker than some critical thickness, typically taken to be about 30 nm. As fig. 3 shows, however, the dt/dx versus x relationship has curvature everywhere, not just in the thin oxide regime, so that the inclusion of an additional growth mechanism just for the thin oxide regime does not totally remedy the problems of the linear-parabolic rate law.

We suggest that the discrepancies shown in fig. 3 are not due to a separate growth mechanism which operates only in thin oxides, but rather, are due to the "structure" of the non-crystalline oxide, as discussed above. In particular, we suggest that remnants of the pseudo-polymorph which formed at the growing interface are important in modifying the flux of oxygen which is transported through the oxide. This possibility arises from the fact that the remnants may result in regions of the oxide having, for example, six member Si-O rings with an internal diameter of ≈0.3 nm, comparable to the diameter of an O_2 molecule[19], through which the oxygen could move with little resistance. Such paths are somewhat similar to the "structural channels" along the C-axis of tridymite and quartz crystals. The remnants of the originally grown polymorph may thus present regions in the oxide which have a similar intermediate-range-order, and therefore enable a similar "fast" transport of oxygen molecules through these regions of the oxide.

We point out that the remnants from the original growth may be responsible for the differences in dry and wet oxides. Whereas the growth of dry oxides proceeds by the relatively inert transport of oxygen molecules through the oxide, the growth of wet oxides proceeds by the transport of OH groups which move through the oxide by continual interchange with the network structure, breaking and reforming Si-O-Si and Si-OH bonds. Thus the remnants of the original oxide structure are negligibly affected during subsequent growth in dry oxygen, but are rapidly destroyed by the scrambling action of the exchange of OH groups during oxidation in wet ambients. This would explain why wet oxides have neither the "fast growth" regime nor the orientation dependent growth rate characteristic of dry oxides.

To account for the increased oxygen flux which can be transported through the oxide as a result of these structural "channels", we can modify Eq. (6) for the transport flux as follows :

$$F_T^o = F_{T1}^o + F_{T2}^o = C_O K_T/x + K_O \tag{8}$$

where F_{T1}^0 is the (random-walk) transport flux without channels and F_{T2}^0 is the transport flux through the channels, which, to first approximation, is taken to be independent of oxide thickness because it is assumed to be much less dependent on thickness than F^0_{T1}. The growth rate law then becomes

$$dt/dx = 1/K_S + x/(K_T + xK_0) \tag{9}$$

The least-squares fit of the experimental data to Eq. (9) is shown by the dashed line in fig. 3, and clearly provides an extremely good fit over the entire range of oxide thickness. We have used Eq. (9) to fit growth rate data[17] obtained at other temperatures and oxygen pressures, and have obtained similar close fits over the entire data range[20] of those data sets as well.

From their temperature dependences, the activation energies of the parameters K_0 and K_T are found to be about 0.4 eV and 2.6 eV, respectively. The low activation energy of K_0 suggests that the transport of O_2 molecules in the structural channels proceeds without significant interaction with the SiO_2 network. The value of 0.4 eV is similar to the 0.2 to 0.45 eV range characteristic of the inert mode of H_2 diffusion in v-SiO_2 [21]. The higher activation energy observed for K_T suggests that the thickness-dependent transport, $F_{T1}x$, involves interaction between O_2 molecules and SiO_2. This interaction has been observed by ^{18}O tracer experiments[22] as affecting mostly the outer surface region of Si $^{16}O_2$ films. These activation energies are therefore in qualitative agreement with our physical interpretation of the two transport processes, i.e., one which is non-interactive and one which is interactive.

SUMMARY

We have pointed out several problems that the widely used viscous flow model has in explaining stress relief at the interface during growth, and we have suggested a new model of the oxide growth based on the following key features :

A) Oxide growth at the interface is quasi-epitaxial, that is, its structure depends on the structure of the substrate. The structure at the interface apparently resembles that of quartz.

B) Stress in the newly formed oxide is relieved by a pseudo-polymorphic transformation involving bond angle rotations rather than the breaking of bonds, as in the case of the viscous flow model.

C) Remnants of the originally grown oxide structure remain in dry oxides after this transformation making the intermediate range order of the thermal oxide films different from that of vitreous silica (as indicated by their calculated polarizabilities), and leading to orientation dependent properties of the nc-oxide.

D) Remnants of the originally grown oxide are quickly destroyed during oxide growth in wet ambients by the scrambling effect of the transport of OH groups through the oxide. Thus wet oxides have neither a "fast growth" oxidation regime or an orientation dependent growth rate.

E) Deviations of the observed oxide growth rate from the expected linear-parabolic rate law are the result of structural properties (i.e., the remnants of the originally grown oxide) of the nc-oxides.

REFERENCES

1. E.P. EerNisse, Appl. Phys. Letters 30:290 (1977).
2. E.P. EerNisse, Appl. Phys. Letters 35:8 (1979).
3. E.A. Taft, J. Electrochem. Soc. 125:968 (1978).
4. A. Fargeix and G. Ghibaudo, J. Appl. Phys. 54:7153 (1983).
5. G. Camera Roda, F. Santarelli and G.C. Saiti, J. Electrochem. Soc. 132:1909 (1985).
6. R. Brueckner, J. Non-Cryst. Solids 5:123 and 176 (1970).
7. E.A. Taft, J. Electrochem. Soc. 127:923 (1980).
8. E.A. Irene, E. Tierney and J. Angilello, J. Electrochem. Soc. 129:2594, (1982).
9. B. Borie, C.J. Sparks and J.V. Cathcart, Acta Met. 10:691 (1982).
10. F.J. Grunthaner, P.J. Grunthaner, R.P. Vasquez, B.F. Lewis, J. Masserjian and A. Madhukar, Phys. Rev. Letters 43:48 (1984).
11. B. Agius, S. Rigo, F. Roche, M. Froment, C. Maillot, H. Roulet and G. Dufour, Appl. Phys. Letters 44:48 (1984).
12. A. Bourret, J. Thibault-Dessaux and D.N. Seidman, J. Appl. Phys. 55:825 (1984).
13. J.A. Mackenzie, J. Am. Ceram. Soc. 46:470 (1963).
14. J. Arndt and D. Stoeffler, Phys. Chem. Glass 10:117 (1969).
15. A.G. Revesz, Phys. Rev. Letters 27:1576 (1971).
16. E.A. Irene, D.W. Dong and R.J. Zeto, J. Electrochem. Soc. 122:1216 (1975).
17. M.A. Hopper, private communication.
18. B.E. Deal and A.S. Grove, J. Appl. Phys. 36:3770 (1965).
19. A.G. Revesz and H.L. Schaeffer, J. Electrochem. Soc. 129:137 (1982).
20. A.G. Revesz, B.J. Mrstik, H.L. Hughes and D. McCarthy, J. Electrochem. Soc. 133:586 (1986).
21. D.L. Griscom, J. Non-Cryst. Solids 68:301 (1984).
22. F. Rochet, B. Agius and S.J. Rigo, J. Electrochem. Soc. 131:914 (1984).

A FRAMEWORK FOR INCORPORATING MEMORY EFFECTS OF STRUCTURAL RELAXATION IN MODELS FOR THERMAL OXIDATION OF SILICON

R.W. Rendell and K.L. Ngai

Naval Research Laboratory
Washington, D.C. 20375-5000

INTRODUCTION

Thermal oxidation of silicon occurs as a result of the transport of oxidizing species through the existing oxide to react with silicon at the Si-SiO_2 interface and form SiO_2. The resulting SiO_2 films are similar to silica glass in their index of refraction, density and other properties, although there are subtle but important differences in their structures. In the present work, we summarize and discuss the physical ingredients necessary for a quantitative understanding of time-dependent structural relaxations of the oxide. For glasses formed by quenching from the melt, the time-dependent recovery of a nonequilibrium structure to its equilibrium state has been widely documented by measurement of quantities such as specific volume, index of refraction, enthalpy, and others[1,2] These exhibit nonlinearity with respect to the magnitude of departure from equilibrium, asymmetry with respect to the sign of the departure and memory effects which are sensitive to the thermal and annealing history. Such phenomena are also to be expected in thermally grown silicon dioxide layers. Indeed, measurements have indicated that the density of SiO_2 can be a time-dependent quantity which eventually relaxes toward a final value[3]. Such structural changes are expected to modify the oxide growth process. For instance, it has been shown in vitreous silica and other glasses that the diffusion of inert gasses is dependent upon the thermal history of the specimen[4]. Dilatometry measurements of volume relaxation as a function of annealing temperature and time have in fact revealed a direct correspondence between the diffusion coefficient and the deviations δ of volume from its equilibrium state[4]. Thermal oxidation of silicon ought to reveal the same nonlinearity, assymetry and memory effects of structural relaxation through the O_2 diffusion coefficient. Thermal and annealing history dependences for index of refraction and density of thermal oxides have already been demonstrated[5,6] and possible modification of the oxygen diffusion coefficient from these effects have been discussed[7]. During sequential oxidations, there has also been evidence for memory effects in which the initial growth rates for the oxidation steps depend strongly on the time-temperature history of the sample[8,9]. Interpretations in terms of variations of the diffusion coefficient of oxygen in SiO_2 caused by structural changes in the oxide during annealing and subsequent oxidation have been suggested[10].

A framework for quantitative model descriptions of structural relaxation as a function of sample history in glass-forming systems has been developed over the past decade by several groups[1,2]. It has been found that necessary ingredients for such a description are : (i) a slow nonexponential relaxation function and (ii) relaxation time(s) which are a function of the structural state, as well as temperature, $\tau(\delta,T)$. More recently, a third ingredient has been found necessary[11] : (iii) the origin of the slow nonexponential relaxation function is in large part due to a recently discovered "coupling mechanism" and it cannot be treated as simply a distribution of relaxation times.

FRAMEWORK FOR STRUCTURAL RELAXATION

Let δ denote the deviation from equilibrium of a structural dependent quantity such as volume, index of refraction, enthalpy, etc. For instance, with volume, $\delta(t) = (v(t) - v_\infty)/v_\infty$ where $v(t)$ is the instantaneous specific volume and v_∞ is its equilibrium value at the same temperature and pressure. A nonequilibrium state can be attained by, for example, a sudden change in temperature or pressure. The volume must then approach its new equilibrium value v_∞ at sufficiently long times. Fig. 1 shows typical behavior for $\delta(t)$ from the initial states $\delta(o) = \delta_o, \pm 2\delta_o$. The asymmetry between $\delta>0$ and $\delta<0$ as well as the nonlinearity is evident. Behavior of this type has been observed in materials such as silica glass, soda-lime glass, B_2O_3, $ZnCl_2$, and several glassy polymers. Volume, index of refraction, enthalpy, and other quantities often show similar, though not necessarily identical, behavior. A single constant relaxation time is inadequate to describe this behavior. The next simplest possibility is a single δ-dependent relaxation time,

$$d\delta/dt = -\delta/\tau(\delta), \text{ which has the solution } \delta = \delta_o \exp\left[-\int_0^t dt'/\tau[\delta(t')]\right].$$

For a given function $\tau = \tau(\delta)$, the solution for δ must be obtained self-consistently. Physically, we would expect $\tau(\delta)$ to decrease as δ increases (i.e. as the volume expands). Such a dependence now introduces both asymmetry and nonlinearity into the problem. However, a description of the detailed shapes of data such as in fig. 1 requires many different relaxation times or some deviation from the concept of a single exponential decay. This is understandable since even near structural equilibrium relaxations in glasses, amorphous materials and polymers exhibit slow nonexponential decay. These relaxation data have often been interpreted using a distribution of relaxation times, $\delta(t) = \delta_o \sum g_i \exp[-t/\tau_i]$. Alternatively, they often conform to the empirical Kohlrausch function[11]

$$\delta(t) = \delta_o \exp\left[-(t/\tau^*)^{1-n}\right] \quad , 0<n<1. \tag{1}$$

For the nonequilibrium situation, Kovacs et al.[2] have used a distribution of δ-dependent relaxation times $\delta(t) = \delta_o \sum g_i \exp\left[-\int_0^t dt'/\tau_i \ (T,\delta(t'))\right]$.

Alternatively, Moynihan et al.[1] have used a nonexponential function $\delta(t) = \delta_o \exp\left[-\left(\int_0^t dt'/\tau[T,\delta(t')]\right)^\beta\right]$ where $0<\beta<1$. These are the simplest type of relaxation functions containing ingredients (i) and (ii). The Kovacs and Moynihan expressions produce similar results and have been demonstrated to be successful in reproducing many features observed in structural data[1,2]. However, these phenomenological models are difficult to interpret physically and therefore little insight is gained into the fundamental physical processes involved. In addition, some important features of experiments cannot be reproduced for any choice of the parameters and functions $\tau(\delta,T)$. A prominent feature which has remained

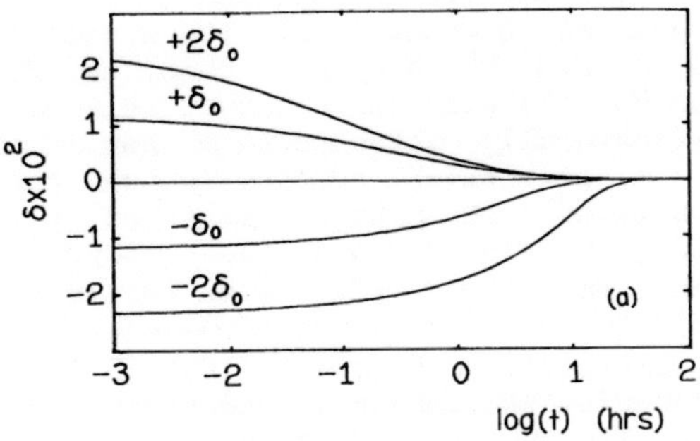

Fig. 1. Illustration of typical time-dependences observed in structural relaxation from initial states $\pm\delta_o$, $\pm 2\delta_o$.

unexplained for many years is the anomalously large difference observed in the rate of change of volume deviation as equilibrium is approached during volume expansion ($\delta<0$) for small differences in the initial states[11]. Although identification of the ingredients (i) and (ii) was an important advance in developing descriptions of structural relaxation, it is clear that further progress requires a more fundamental approach.

COUPLING MECHANISM FOR RELAXATION

A traditional physical interpretation for the origin of slow nonexponential relaxation is that material randomness produces a distribution of different relaxation times τ_i throughout the material. It has recently been found that slow nonexponential relaxation can be produced by yet another mechanism which can be experimentally distinguished from the effects of randomness. Even if there is little material randomness so that the relaxation times have nearly the same value τ_o from site to site, within some domain of short-range order, the relaxation process can still become slowed-down due to cooperative interactions with its surroundings. For example, relaxation of a structural unit in a glass requires cooperative movements of other structural units. It has been found[11] that such cooperativity modifies the constant relaxation rate, $d\delta/dt = -\tau_o^{-1}\delta$, so that it becomes explicitly time-dependent beyond a time ω_c^{-1} at which cooperative adjustments are important : $d\delta/dt = -\tau_o^{-1}(\omega_c t)^{-n}\delta$. This slowing down of rates due to cooperativity is referred to as the "coupling mechanism". The coupling parameter n ($0<n<1$) is a measure of the strength of the coupling to cooperativity. The solution for δ is a special type of nonexponential function characteristic of this coupling mechanism[11].

$$\delta(t) = \delta_o \exp\left[-\int_0^t dt'(1-n)t'^{-n}/\tau^{*1-n}\right] \tag{2}$$

$$\tau^* = \left[(1-n)\omega_c^n \tau_o\right]^{\frac{1}{1-n}} \tag{3}$$

Both $n = n(\delta)$ and the effective relaxation time $\tau^* = \tau^*(T,\delta)$, defined in Eq. (3), are in general dependent on the structure δ. Near structural equilibrium, where n and τ^* are not changing, Eq. (2) reduces to the Kohlrausch function Eq. (1) with the observed relaxation time τ^*. Thus the coupling mechanism naturally introduces a new concept involving two relaxation times : a "primitive" process (which can often be identified with molecular quantities) attempts to relax as a single exponential on time-scale τ_o but is slowed down due cooperativity resulting in nonexponential decay, Eq. (1) or (2), with the longer observed timescale τ^*. In equilibrium, the coupled predictions Eqs. (1) and (3) have been verified for many relaxation processes[11]. Eq. (3) in particular is unique to the coupling mechanism and cannot be understood in terms of randomness. For structural nonequilibrium, Eq. (2) retains ingredients (i) and (ii) but contains the additional ingredient (iii) of the coupling mechanism. For independently measured n and $\tau^*(T,\delta)$, it quantitatively reproduces volume relaxation data including the anomalies during expansion[11].

STRUCTURAL RELAXATION AND THERMAL OXIDATION

Similarities between thermally grown SiO_2 films and pressure-compacted vitreous SiO_2 have been alluded to by Irene et al.[3] and explored in detail by Mrstik et al.[12]. It has been suggested that the as-grown oxide resembles some high-density polymorph of SiO_2 due to the influence of the substrate. With further growth, the previously grown oxide moves away from the interface and its structure relaxes to resemble that of vitreous SiO_2 although there may be differences in these structures due to the presence of heterogeneities. The time dependent volume expansion of densified silica and boron oxide glasses was investigated by Mackenzie[13] for different thermal and annealing histories. Nonlinearity and asymmetry were observed in the deviations from equilibrium and Mackenzie stressed that the annealing kinetics could not be described by a single relaxation time. In the absence of a self-consistent description of this process, he arbitrarily defined a time τ' at which δ decreased to 1/e of its original value. For silica glass, this was found to have apparent energy of activation of $E_a' = 55$ kcal/mole = 2.4 ev. However, the physical interpretation of such an activation energy is certainly problematical. A similar situation occurs in B_2O_3 where more experimental information is available. Bucaro et al.[14] have measured both pressure-jump volume relaxation and dynamic light-scattering spectroscopy (which is a measure of the longitudinal stress modulus) for B_2O_3. Their results at low temperature are accurately described by the Kohlrausch function Eq. (1) with $n = 0.4$ and $\tau^* = (8.5\times10^{-31})\exp(3.6\text{eV}/kT)$ s. The deviations from equilibrium were very small, $\delta \sim 10^{-3}$, and the δ dependences of τ^* are possibly not significant here. In this case, the decay function Eq. (2) would reduce to Eq. (1) as observed. Shelby [4] also measured volume relaxation in B_2O_3 but with δ values an order of magnitude larger than Bucaro. He noted that the relaxation is nonexponential and also defined the 1/e time τ'. From his volume data, it is found that $\tau' = (1.64 \times 10^{-5}) \exp(0.86\ \text{eV}/kT)$. This differs greatly from the result of Bucaro and in particular $E_a' = 0.86 \ll E_a^* = 3.6$. In addition to possible differences in samples, a large part of this difference might be attributed to the importance of structural evolution in Shelby's case. More precisely, τ' as defined is the time at which the integral in Eq.(2) equals one. If the structural relaxation time τ^* has activation energy E_a^*, then it can be shown directly from Eq.(2) that τ' is activated with energy $E_a' = E_a^*[\tau^*(T,\delta(\tau'))/\tau'(T)]^{1-n}$. For simplicity, n is assumed constant. From self-consistent numerical solutions of Eq.(2), it is found that E_a' can be both independent of T and significantly smaller than E_a^* if $\tau^*(\delta)$ has sufficient dependence on δ. A decrease of τ^* which is roughly exponential with δ is compatible with known dependences from other systems[1,2,11]. Although insufficient information is available to test this particular case in detail, it provides a plausible explanation of the difference between the Bucaro and Shelby data and illustrates the importance of a proper description for structural relaxation. It has been previously shown that Eqs. (2) and (3) can quantitatively describe structural changes in terms of physical quantities measured near equilibrium. Bucaro et al.[15] have also used dynamic light scattering spectroscopy on Suprasil W silica glass in structural equilibrium for T>1305C. It conforms to the Kohlrausch function Eq. (1) with n = 0.3. Shear modulus measurements by Mills[16] on Infrasil at lower temperatures over the range 1000 - 1200 C also show Kohlrausch decay but n in this range has increased to 0.37. This suggests that Eq. (2) is relevant for the nonequilibrium silica glasses of Mackenzie. As with the B_2O_3 case, it is possible that $E_a^* > E_a' = 2.4$ eV due to structural changes. Though information is not available for the values, we show an illustrative example choosing $E_a^* = 4$ eV, n = 0.4 and $\log \tau^*(\delta)/\tau^*(0) = 2.4$ at $\delta = 0.12$ and decreasing linearly to 0 at $\delta = 0$. The results, calculated self-consistently for T = 500C and also including a jump to T = 700C at 0.7 hr, are typical of Mackenzie's situation and are shown in fig. 2. For these curves, the E_a' is 2.4 eV. With more experimental information, this type of analysis can be made more quantitative and can give insight into the structural processes involved. For instance, one could attempt to physically identify the activation energy for the underlying primitive relaxation τ_o which, from Eq. (3), is $E_a = (1-n)E_a^*$.

As mentioned earlier, these nonlinear structural changes may be reflected in the SiO_2 growth. One possibility is through the O_2 diffusion coefficient. Shelby measured helium diffusion through B_2O_3 in conjunction with the volume relaxation measurements. The diffusion coefficient $D(\delta)$ was found to change nearly linearly with δ. The larger O_2 molecule may be expected to be even more strongly influenced by compactness of the structutre. The structural relaxations addressed in this paper have referred to the homogenous structure of a glass. However, a thermally grown oxide film may be pictured as a series of slices, each of which begins to relax as it moves away from the interface. From Eq. (2), we can consider the quantity $\delta(t,\bar{t}) \equiv \delta_o \exp\left[-\int_{t-\bar{t}}^{t} dt'(1-n)(t'-t+\bar{t})^{-n}/\tau^{*1-n}\right]$ which can be interpreted as the structure of a slice of oxide at time t which was originally grown at time $\bar{t}$, where δ_o is the structure of the as-grown oxide. Fargeix et al.[17] have attempted to incorporate a stress dependent diffusivity in a model of the growth process by modifying the Deal-Grove expression for the growth rate $r(t) = dx/dt$ to allow a diffusivity profile across the layer. Within the specific context of stress relaxation by viscous flow, this model used a single relaxation time with a self-consistent evolution of stress. For the present discussion of structural changes, an analogous modified Deal-Grove expression for the growth rate could be examined :

$r^{-1}(t) = k_L^{-1} + 2D(o)k_p^{-1}\int_0^t dt' r(t')/D(\delta(t,t'))$. Other possibilities such as the effects of structure on the reaction kinetics at the oxide-silicon interface, cooperativity between the diffusing species and the oxide, and changes in $n(\delta)$ and $\tau^*(T,\delta)$ due to water content could also be considered. For example, if the diffusion of the oxygen (or other species) requires cooperativity with the oxide or with the other diffusing species (e.g. near the interface) then the diffusion process itself might be subject to the coupling mechanism. Ionic conductivity in glasses which involves ion-ion correlations has given strong evidence of the coupling mechanism[11]. Water content in glasses is known to alter the value of the coupling strength n and significantly effect the annealing of structure[11].

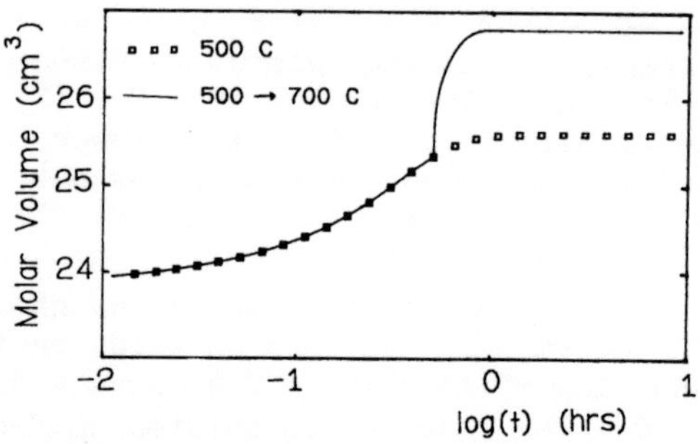

Fig. 2. Illustrative simulation from Eq. (2) representative of expansion of densified vitreous silica

Simulations for these and other effects must be developed in light of experimental information. However, it is known from the current understanding of time-dependent structural changes that the ingredients (i) - (iii) discussed in this paper are required and these should be incorporated into descriptions of thermal oxidation.

Supported in part by ONR Contract No. N0001487WX24039

REFERENCES

1. C.T. Moynihan, Ann. N.Y. Acad. Sci. 279:15 (1976).
2. A.J. Kovacs, Ann. N.Y. Acad. Sci. 371:38 (1981).
3. E.A. Irene, E. Tierney and J. Angilello, J. Electrochem. Soc. 129:2594 (1982);
4. J.E. Shelby, J. Non-Cryst. Solids 14:288 (1974).
5. E.A. Taft, J. Electrochem. Soc. 125:968 (1978).
6. E.A. Irene, D.W. Dong and R.J. Zeto, J. Electrochem. Soc. 127:396 (1980).
7. E.A. Taft, J. Electrochem, Soc. 132:2486 (1985).

8. C.R. Helms, J. Electrochem. Soc. 129:2883 (1982).
9. M. Hamasaki, Solid-State Electron. 25:479 (1982).
10. C.J. Han and D.R. Helms, J. Electrochem. Soc. 132:402 (1985).
11. K.L. Ngai, R.W. Rendell, A.K. Rajagopal and S. Teitler, Ann. N.Y. Acad. Sci. 484:150 (1987) and references therein.
12. B.J. Mrstik, A.G. Revesz, M. Ancona, and H.L. Hughes, J. Electrochem. Soc., in press (1987).
13. J.A. Mackenzie, J. Am. Ceram. Soc. 46:470 (1973).
14. J.A. Bucaro, H.D. Dardy and R.D. Corsaro, J. Appl. Phys. 46:741 (1975).
15. J.A. Bucaro and H.D. Dardy, J. Non-Cryst. Solids 24:121 (1977).
16. J.J. Mills, J. Non-Cryst. Solids 14:255 (1974).
17. A. Fargeix and G. Ghibaudo, J. Appl. Phys. 83:589 (1984).

ANALYSIS OF STRESS RELAXATION AND GROWTH KINETICS FOR TWO-STEP THERMAL OXIDATION OF SILICON

Gérard Ghibaudo

Laboratoire de Physique des Composants à Semiconducteurs
UA-CNRS 840, Enserg
23 rue des Martyrs, 38031 Grenoble, France

INTRODUCTION

Thermal oxidation of silicon has been the subject of much research. The anomalous initial regime of dry oxidation is still a key point for the understanding of the oxidation process. This phenomenon has been studied by many workers[1-5] and, in particular, has been attributed to the existence of stress during growth[6,7].

It has been considered throughout all stress state models of oxidation that the stress developed during growth induces a decrease in the oxygen diffusivity across the oxide[6,7]. To test the validity of such an hypothesis, the role of stress on the oxygen diffusivity has been studied with the help of two-step thermal oxidations[8-12]. In these experiments, the extra oxide thickness of the second oxidation is correlated, via the oxygen diffusivity, with the stress state of the initial oxide layers[10-12].

The aim of this work is, first, to re-analyse the two-step oxidation data and then to generalize the stress state model of oxidation[6,13] for the interpretation of the kinetics and of the diffusivity activation energy observed during the second oxidation.

THEORY

X_i being the thickness of the initial oxide layer and ΔX_{ox} the extra thickness after the second oxidation, the inverse of the oxidation rate, dt/dX_{ox}, is readily expressed as : [6,12]

$$\frac{dt}{dX_{ox}} = \frac{1}{k_l} + \frac{2\,X_i}{k_{p1}} + \frac{2\,\Delta X_{ox}}{k_{p2}} = \frac{1}{k_{leff}} + \frac{2\,\Delta X_{ox}}{k_{p2}}, \tag{1}$$

where k_{p1} (k_{p2}) is the parabolic constant in the first (second) oxide, k_l is the linear kinetic constant and $k_{leff} = (1/k_l + 2X_i/k_{p1})^{-1}$ is the effective linear constant for the second oxidation.

According to the stress state model, k_{p1} and k_{p2} are related approximately to the stress levels in region 1 and 2 by :[13]

$$k_{p1,2} = k_{po} \exp(- \sigma_{1,2}\Delta V/kT) \; ; \; \sigma_1 = \sigma_{10} \exp(-t/\tau) \qquad (2)$$

where σ_{10} is the average stress due to the initial oxidation and τ is the stress relaxation time.

Similarly, the stress level in region 2 is given by :

$$\sigma_2 = \sigma_m \exp(- t/\tau), \qquad (3)$$

where σ_m is the maximum stress generated at the interface[14].

RESULTS

We consider that the second oxidation temperature T_2 is fixed, say 800°C, and vary arbitrarily the stress relaxation time. Three different situations can then be predicted from Eq. 1 for the second oxidation kinetics (see fig. 1). The first concerns an initial oxide grown (or annealed) at high temperature in order to be in a free state ($\sigma_{10} \approx 0$). In this case, the effective linear constant is equal to its maximum value during all the second oxidation (see curve (a) of fig. 1).The second possible situation is an initial oxide grown at low temperature in order to be highly stressed ($\sigma_{10} \neq \sigma_m$). Moreover, if the second oxidation time is smaller than the stress relaxation time, then the stress amount in the initial oxide is almost constant and maximum during the second growth. So, the effective linear constant in Eq. 1 remains constant during the second oxidation and the inverse of the oxidation rate is a linear function of ΔX_{ox} (see curve (b) of fig. 1). The third, intermediate, situation takes place when the initial oxide is highly stressed and when the oxidation duration is greater than the stress relaxation time. The effective linear constant increases as the second oxidation proceeds inducing a net decrease in the inverse of the oxidation rate that lies between the two preceding limits (see curve (c) of fig. 1).

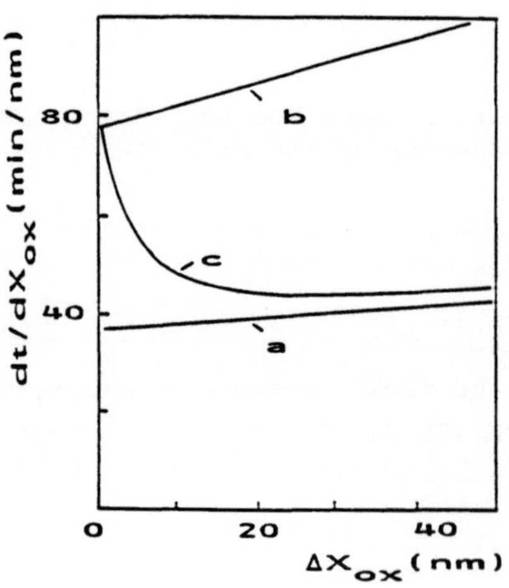

Fig. 1 : Theoretical variations of dt/dX_{ox} as a function of the extra thickness ΔX_{ox} as obtained from our analysis for three different situations
a) fully relaxed initial oxide,
b) highly stressed initial oxide and stress relaxation time greater than second oxidation time and
c) highly stressed initial oxide and stress relaxation time slightly below second oxidation time.

The two previous cases have been experimentally studied by Landsberger and Tiller[15] for two-step oxidations conducted on thermally (or corona) relaxed or unrelaxed initial oxides (see fig. 2a). The third intermediate case has been observed by Srivastava and Irene[11] on thermally annealed or unannealed initial layers (see fig. 2b). It is worth noting that, although the second oxidation temperatures were said to be the same in both references 11 and 15, the results are not identical. This suggests that some discrepancies in the second oxidation temperatures and/or in the oxidation conditions must exist between these two experiments.

Nevertheless, consideration of these results as representative data for second oxidation kinetics allows a comparison of fig. 1 and fig. 2 to show that a qualitative description of the second oxidation kinetics behaviour can be obtained with our analysis (cf. fig. 1). Particularily, it points out the strong influence of stress on the oxygen diffusivity in the initial oxide throughout the modification of the effective linear constant k_{1eff} with oxidation time.

These kinds of experiments demonstrate that, for dry thermal oxidation of silicon, a correlation between the oxidation rate and the intrinsic stress does exist. The stress can be indirectly determined from the values of the refractive index[10,15,16]. In particular, in such two-step experiments, the enhancement of the oxidation rate during growth is clearly correlated with a decrease in the refractive index of the initial oxide [10,15]. All these results strongly support the conclusion that the oxygen diffusivity for thermal oxidation of silicon depends on the intrinsic oxidation stress and on its relief during growth.

Furthermore, one can extract from the data by Srivastava and Irene[11] the value of the relaxation time associated with the decrease in the inverse of the oxidation rate with the oxide thickness (fig. 2b). According to this data, this relaxation time is estimated to be about 15-20 hours. The corresponding viscosity η of the oxide can then be deduced ($\eta = \tau\mu$; μ is the shear modulus[6]). This value compares favorably to those previously obtained either from oxidation kinetics fitting[6] or from direct in situ refractive index measurements[16].

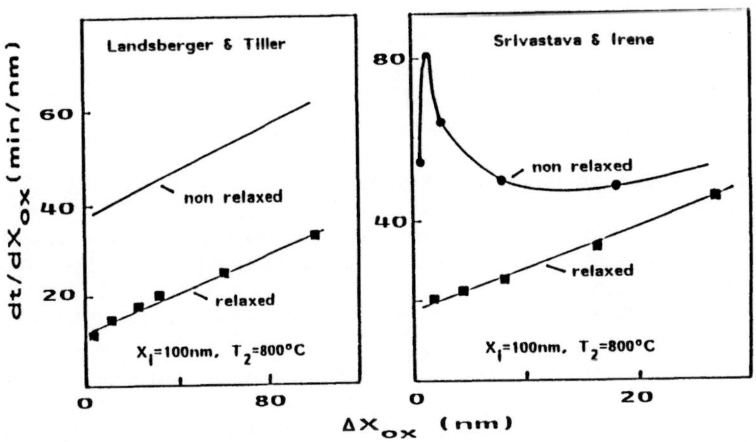

Fig. 2. Variations of dt/dX_{0X} as a function of the extra thickness ΔX_{0X} as obtained in two-step oxidation experiments : a) after Lansberger and Tiller[15] and b) after Srivastava and Irene[11].

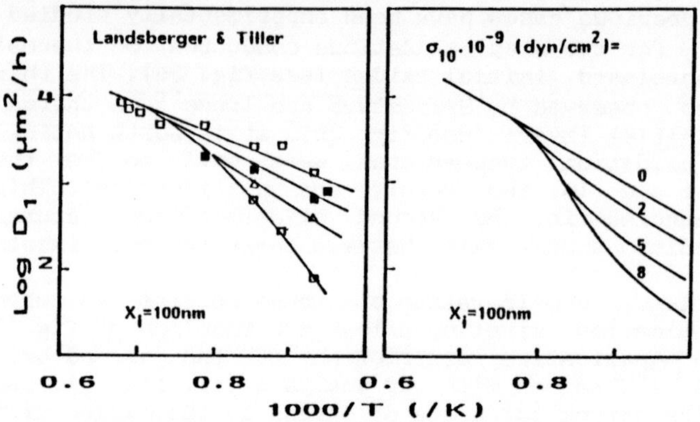

Fig. 3 Variations of the oxygen diffusity D_1 as a function of the inverse of the second oxidation temperature for different relaxation states of the initial oxide : a) experimental results from Landsberger and Tiller[12,25] and b) theoretical results obtained by our model using various values of the stress in the initial oxide σ_{10}.

Another point to examine is the oxidation temperature dependence of the effective linear constant k_{1eff} for different levels of the stress of the initial oxide. For pre-formed oxides of thickness around 100 nm, k_{1eff} is dominated by the oxygen diffusivity term since $2X_i/k_{p1} \gg 1/k_1$ [12,15]. The temperature dependence of k_{1eff} is thus directly indicative of the temperature dependence of the parabolic constant k_{p1} and, so, of the oxygen diffusivity D_1 in the initial oxide ($k_{p1} \approx 2vCD_1$; v : molecular oxide volume, C : oxygen solubility)[13].

Experimental data of oxygen diffusity D_1 as a function of the second oxidation temperature have been obtained by Landberger and Tiller[12,15] (fig. 3a). The corresponding theoretical plots obtained by our model are reported in fig. 3b. These curves exhibit the variations of the effective diffusity $D_1 = k_{1eff}/(2vC)$ with the inverse of the second oxidation temperature and are parametrized by the value of the stress σ_{10} in the initial oxide. They were determined numerically as in ref. 13 but using slightly different values of the parameters (intrinsic activation energy of diffusivity : E_a =1.0 eV, oxygen diffusion volume : $\Delta V = v$,[16] ; activation energy of viscosity : E_a = 4.5 eV).

The comparison between experiment and theory of fig. 3a and b shows that the residual amount the stress σ_{10} in the initial oxidation follows the behaviour of these Arrhenius plots. Moreover, it shows that the reduction of diffusivity at low oxidation temperature can be suppressed for fully relaxed initial oxides ($\sigma_{10} = 0$). In contrast, at high temperature (T_2>950°C), the influence of the residual stress σ_{10} is attenuated because of the rapidity of stress relaxation. Therefore, the effective activation energy of D_1 in these plots depends strongly on the residual stress level σ_{10}. The activation energy of fully relaxed oxides has a nearly constant value associated with the intrinsic activation energy of the oxygen diffusivity in the free state. On the other hand, for not fully relaxed initial oxides, the effective activation energy is, in the low temperature range, a function of stress[13].

CONCLUSION

We have shown that two-step oxidation experiments give evidence for stress dependence of the oxygen diffusivity during dry thermal oxidation of silicon. Moreover, it has been pointed out that the stress state model,[6,13] proposed earlier, is also well adapted for the analysis and the interpretation of two-step oxidations. Especially, it gives a reasonably good description of the second growth kinetics and of the peculiarity observed in the Arrhenius plots of the oxygen diffusivity after relaxation of the initial oxide.

REFERENCES

1. H.Z. Massoud, J.D. Plummer and E.A. Irene, J. Electrochem. Soc, 132:2693 (1985).
2. V. Samalam, Appl. Phys. Lett., 47:736 (1985).
3. S.A. Shafer and S.A. Lyon, J. Appl. Phys., 47:154 (1985).
4. A.G. Revesz, B.J. Mrstik, H.L. Hughes and D. McCarthy, J. Electrochem. Soc., 133:586 (1986).
5. V. Murali and S.P. Murarka, J. Appl. Phys., 60:2106 (1986).
6. A. Fargeix and G. Ghibaudo, J. Appl. Phys., 54:7153 (1983).
7. G. Camera Roda, F. Santarelli and G.C. Sarti, J. Electrochem. Soc, 132:1909 (1985).
8. M. Hamasaki, Sol. Stat. Electron., 24:479 (1982).
9. E. A. Irene, J. Electrochem. Soc., 129:413 (1982).
10. E.A. Taft, J. Electrochem. Soc., 132:2486 (1985).
11. J.K. Strivastava and E.A. Irene, J. Electrochem. Soc., 132:2815 (1985).
12. L.M. Landsberger and W.A. Tiller, Appl. Phys. Lett., 49:143 (1986).
13. A. Fargeix and G. Ghibaudo, J. Appl. Phys., 56:589 (1984).
14. A. Fargeix and G. Ghibaudo, J. Phys. D, 17:2331 (1984).
15. L.M. Landsberger and W.A. Tiller, Electrochemical Society Fall Meeting, San Diego, oct. 19-24, 1986 (Abstract 353).
16. G. Ghibaudo, Phil. Mag. B., 55:147 (1987).

PHOTO-INDUCED OXIDATION PROCESSES IN SILICON

E. Fogarassy

Centre de Recherches Nucléaires - Laboratoire Phase

23, Rue du Loess - F-67037 Strasbourg Cedex France

ABSTRACT

We review in this paper the different ways which have been explored to photo-induce oxidation of silicon using CW and pulsed lasers of different wavelengths working both in solid and liquid phase regimes. The specific influence of visible and ultraviolet intense light sources on the oxidation processes will be detailed with special attention to the non thermal effects suggested to be present when using UV photons of high energy (> 3.5 eV).

INTRODUCTION

The study of oxidation processes is of primary importance in silicon device technology. It has become increasingly desirable, in recent years, to form good quality SiO_2 films at lower substrate temperature or with a significant reduction in high temperature processing times. By contrast to conventional thermal treatment, lasers offer the possibility to confine high temperature processing to only near surface regions with good spatial resolution during very short periods of time.

Different ways have been explored in this field using both CW and pulsed lasers of different wavelengths working in the solid or in liquid (pulsed) phase regime. Pulsed laser annealing of oxygen implanted Si layers was one of the first experiments[1] involving laser beams to produce silicon dioxide. Experiments[2-5] have been performed using CW lasers of different wavelengths (Ar and Co_2) to grow directly SiO_2 layers on Si surfaces heated by the laser beam in an oxygen rich environment. Several studies demonstrated the possibility of using pulsed lasers working either in solid[6-8] or in liquid[9-15] phase regimes to induced direct oxidation of silicon and silicon monoxide[16-17]. In this area, excimer lasers, which emit pulses of short duration (10-100 nsec) at different UV wavelengths (193, 248, 308 nm) with a good uniformity and high repetition rate (up to 500 Hz), present several advantages compared to other types of visible and IR pulsed lasers (Ruby, Yag, CO_2) : UV wavelengths are strongly absorbed in silicon ($\alpha < 10^6$ cm^{-1} for $\lambda < 300$ nm / fig. 1) resulting in absorption depths less than 100 Å. Consequently the effects of irradiation are concentrated very near the surface. A second advantage of a short wavelength source comes from the high energy of the photons (6.4 eV at

193 nm) which offers the capability for inducing photochemical reactions both in the gas phase and in the solid by breaking chemical bonds in Si or at the Si/SiO_2 interface. Finally, in the field of future development of photo beam assisted technologies in microelectronics, the spatial resolution of a process can be improved when the wavelength of the light source is reduced.

The major difficulty in the interpretation of photonic oxidation processes is to evaluate accurately the thermal contribution resulting from the interaction between the incoming light and the solid. Surface heating during laser irradiation is difficult or impossible to measure experimentally, especially for pulsed lasers. Consequently as shown in the next part of this paper the thermal contribution has to be evaluated numerically by solving the heat flow equation. These calculations show that, following the conditions of irradiation, we have two different regimes : at low intensity the oxidation process takes place in the solid phase with specific contribution from photochemical reactions when using UV light sources. At high intensity, with the pulsed lasers, the oxidation of Si results from the fast diffusion of oxygen from air (or from O_2 controlled atmosphere) into liquid Si, which is trapped at the liquid-solid interface during solidification. These different points are now considered in more detail.

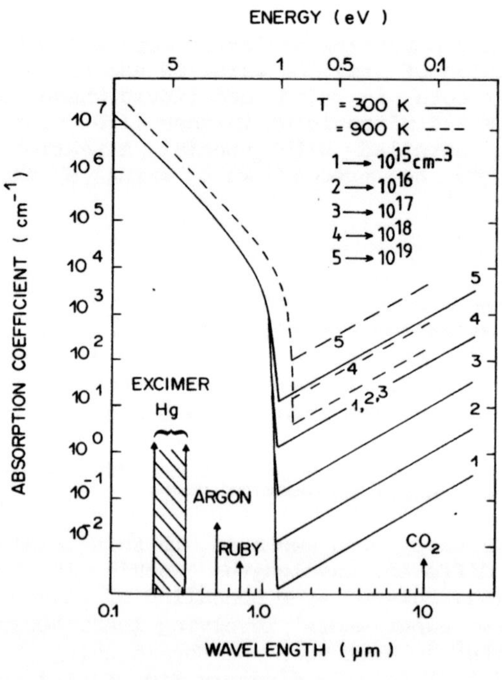

Fig. 1. Absorption coefficient of Si as a function of wavelength.

MODELLING OF THE THERMAL CONTRIBUTION

With a knowledge of the optical and thermal properties of the sample and the intensity I_o of the light source, the temperature rise induced in the material by the laser beam may be estimated from the numerical resolution of the heat flow equation[18] . The accuracy in the calculated temperature is therefore strongly related to how precisely we know the experimental values of these different parameters : Optical absorption (α)

and surface reflectivity (R) of silicon which depend strongly upon wavelength of the light source as shown in figs. 1 and 2 respectively, are specially important since they control the amount of incident light participating in the heating process through the general relation :

$$G(x,t) = I_o(t) \cdot (1 - R) \cdot \alpha \cdot \exp(-\alpha, x)$$

where $G(x,t)$ represents the energy absorbed per unit volume per second at the depth x.

We can see in fig. 1 that the physical process responsible for the coupling of the laser radiation to the Si lattice may be very different depending upon the wavelength : in the visible and UV range, above Si band gap, band-to-band transitions are responsible for the absorption process which may be very strong ($> 10^6 cm^{-1}$) below 300 nm. In contrast, in the IR range, optical coupling is only possible through free carrier absorption. This explains why CO_2 radiation can be effectively coupled to the Si lattice only by raising the substrate temperature or by working at high doping levels (fig. 1). It is important to notice that the coupling of CO_2 radiation (9.6-10.6 μm) with the surface becomes higher during the oxidation of silicon because of strong absorption ($\alpha \sim 3.5 \times 10^4 cm^{-1}$)[19] by the Si-O stretching vibrational mode located around 1000-1100 cm^{-1}.

Another major difficulty is related to the value of the optical reflectivity to be used in the thermal calculations because this parameter may be strongly modified during the growing of the oxide as shown in fig. 2 for the visible and UV range. In particular, at UV wavelengths we have a complicated situation resulting from interference effects for ~ 1000 Å thick SiO_2 layers.

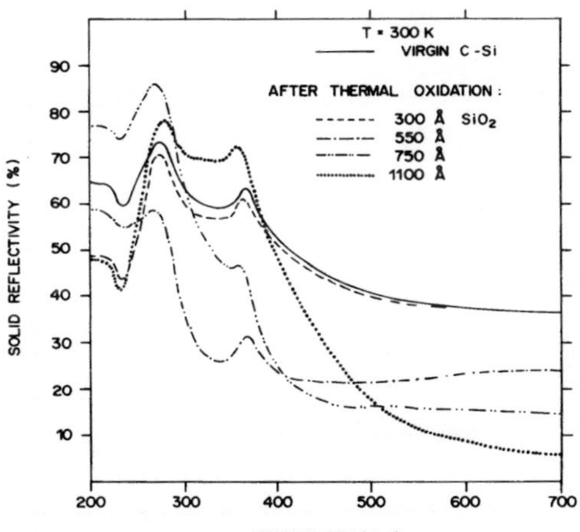

Fig. 2. Reflectivity of Si and SiO_2 on Si as a function of wavelength.

Dependent upon the duration of the heating cycle responsible for the temperature change, we have to consider two different modes of laser beam processing. We have compared in fig. 3 and 4 the time dependent temperature increase calculated for single crystal silicon (c-Si) irradiated respectively with a CW Ar (fig. 3) and pulsed excimer laser (fig. 4). With the CW laser, the irradiation is sufficiently long for heat to flow completely through the thickness of the semiconductor (thermal flux mode). In these conditions the temperature of the sample increases rapidly during the first microseconds before reaching an equilibrium state characterized by a constant value of the temperature as long as the laser beam is on. By contrast, with the excimer laser, the pulse is sufficiently short (~ 23 nsec) that little heat is lost by diffusion from the absorption layer (~ 100 Å) during the irradiation, and so nearly all the energy absorbed is utilized in raising the surface layer temperature (adiabatic mode) for a very short time (a few tens of nanoseconds). Above an energy density of 600 mJ/cm^2, the laser irradiation leads to the melting of the c-Si surface to a depth not exceeding a few thousand angstroms and rapid solidification (100 to 200 nsec) by liquid phase expitaxy from the underlying single crystal substrate. For amorphous silicon (a-Si) the melting threshold is considerably lower (100-200 mJ/cm^2) mainly due to the reduced thermal conductivity of a-Si compared to c-Si.

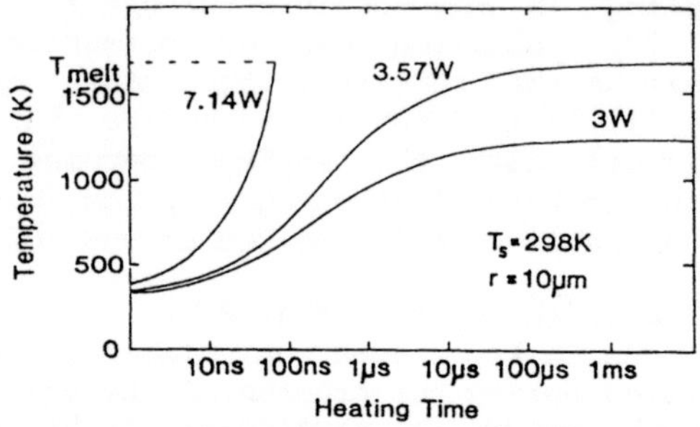

Fig. 3. The time dependent temperature increase of c-Si during cw-Ar laser irradiation (Ref. 7).

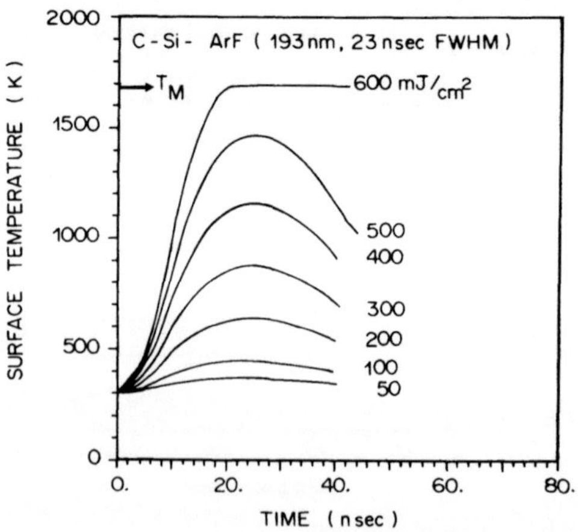

Fig. 4. The time dependent temperature at the surface of c-Si irradiated with a pulsed ArF laser. Influence of the pulse energy density.

NON THERMAL CONTRIBUTION TO THE SOLID PHASE OXIDATION

In addition to the purely thermal contribution resulting from the optical absorption of the incident light by the sample, several authors invoked recently a small non thermal component to explain observed enhancements in the Si oxidation rate, especially when the photon energy is greater than the silicon band gap. Young and Tiller[20] demonstrated that at lower laser power density the enhanced oxidation rate of Si is linearly proportional to the photon flux density for photon energies in the range 2.4 to 2.7 eV. In addition, the enhancement effect appears to be greater for <100> Si than for <111> Si. Very recently, an interesting experiment has been proposed by Boyd and Micheli[21] confirming the photonic enhacement in the same range of photon energy (2.5 eV). All these studies emphazise the important role played by the photon generated excess carrier population relative to the thermal level in the vicinity of Si/SiO_2 interface. Oren and Ghandhi[22] were the first to demonstrate that UV irradiation provided by a high pressure Hg lamp emmiting at 253 nm photons of 4.8 eV, was able to increase the growth rate of a thermally grown oxide. More recently, Shafer and Lyon[23] have observed a spectrally dependent enhancement during dry oxidation with an additional contribution for UV photons ($h\nu \geq 3.5$ eV). The UV photons are able to induce internal photoemission of electrons, at the $Si\text{-}SiO_2$ interface, from the conduction band of Si which needs photons of 3.3 and 4.2 eV respectively[24]. These photoinjected electrons are able to combine with oxygen dissolved in SiO_2 to form negatively charged species which may play an important role in the oxidation process. Fiori[6] and Fiori and Devine[25] have suggested that the UV photons provided by a pulsed KrF excimer laser working at 248 nm (5 eV) may interact in a non thermal fashion on chemical bonds in the solid at an $Si\text{-}SiO_{x\leq 2}$ interface inducing not only bond rearrangements between oxygen and silicon atoms for $h\nu \leq 4.3$ eV, but also stable structural defects such as oxygen vacancies in SiO_2. These processes could be helped by the creation of a high density of electron-hole pairs ($> 10^{21} cm^{-3}$) in the Si lattice by the intense pulsed light provided by the excimer laser, which may weaken Si-Si bonds. It is also important to notice that the diffusion of oxygen through the SiO_2 layer toward the $Si\text{-}SiO_2$ interface may be promoted by the UV photodissociation, in the gas phase, of O_2 molecules into atomic oxygen ($h\nu \geq 5.1$ eV) when using a pulsed ArF excimer laser ($h\nu = 6.4$ eV). In this case, the formation of ozone in the gas phase resulting from O_2 dissociation could play a role which is not yet clarified. The last question is related to a possible influence of water vapor on the photo-oxidation process.

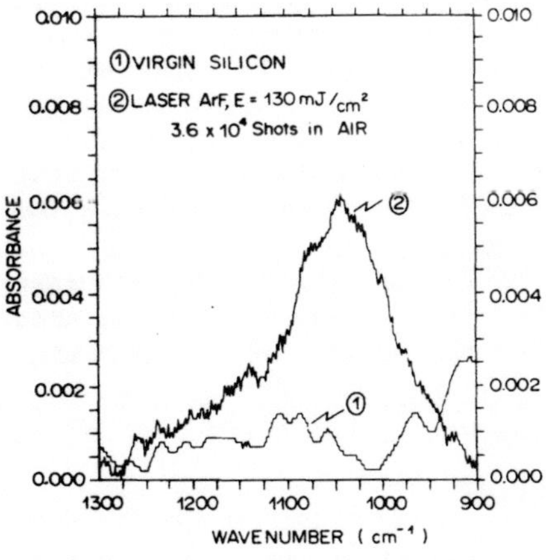

Fig. 5. IR absorption spectrum (Si-O stretching band) of a 30 Å thick oxide layer grown on c-Si irradiated with a pulsed ArF laser.

The specific influence of UV light on the oxidation process has found an interesting application in direct growth of thin films of SiO_2 (≤ 100 Å) on Si at low temperature. As deduced from the differential IR absorption spectrum of fig. 5, corresponding to the stretching vibration of the Si-O bond, a 30 Å thick oxide layer was grown at the Si surface following a 10 ns pulsed ArF laser irradiation in O_2 at 60 Hz repetition rate with an energy density per pulse of 130 mJ/cm^2. After 30 minutes irradiation, the grown oxide layers were 50 Å thick on virgin Si and 120 Å on doped Si. In all these experiments the surface temperature was estimated to be less than 400°C from the calculations of fig. 4. The peak position of the Si-O stretching vibrational mode is located at 1050 cm^{-1} close to the value found both for very thin (≤ 50 Å) CO_2 laser and furnace grown oxides[26]. Boyd and Wilson[26] suggested that this behaviour could be related to the presence of compressive stresses in the oxides arising from the structural transition from crystalline Si to amorphous SiO_2 for very thin films (≤ 100 Å). The width of the IR band (~ 100 cm^{-1}) at FWHM is however significantly larger than for CO_2 laser and furnace grown oxides (~ 70 cm^{-1}) indicating a higher degree of disorder in the oxide grown at low temperature using a pulsed excimer laser.

LASER INDUCED OXIDATION IN LIQUID PHASE

In the liquid phase regime the oxidation process includes incorporation and diffusion of oxygen into the molten phase, trapping at the liquid-solid interface during resolidification and rapid reactions to form chemical bonds with atoms of silicon and possibly with impurities which may be present at the interface.

i) Single-pulse radiation

Liu et al.[11] were the first to observe the strong incorporation of oxygen and the formation of oxide (up to 350 Å thick) in the amorphous Si layer produced on the surface of crystalline Si that was exposed, in air, to an intense 266 nm UV pulsed laser (Nd : Yag) with energy densities ranging between 1 and 4 J/cm^2. The formation of an amorphous Si layer and surface degradation of the sample appear to be well correlated with the strong incorporation of oxygen resulting in the formation of SiO_2 but also of other kinds of Si-O complex as revealed by differential fourier-transform IR spectroscopy. Bentini et al.[15] reported oxygen incorporation into Si submitted to a single pulse radiation performed in O_2 atmosphere (4 atm) with a Q-switched Ruby laser (λ = 694 nm). In these experiments, a massive incorporation of oxygen resulting in the formation of a 500 Å thick SiO_2 layer is only observed when the energy density of the pulse (~ 3 J/cm^2) is about three times higher than that required for Si melting. A tentative model for the oxidation mechanism assumes that the temperature reached by the sample surface is above the melting point of SiO_2. It is important to notice that the energy threshold for oxygen incorporation (~ 3 J/cm^2) corresponds also to the energy where very high surface roughness is observed at the surface of irradiated Si.

ii) Multi-pulse irradiation

In order to avoid or to limit the structural damage which develops at surface of Si samples during single pulse oxidation for energy densities in the range 3 to 4 J/cm^2 and which may be unacceptable for the performance of electronic devices, oxidation has been recently performed at lower energy densities using multipulse irradiation provided by excimer lasers[8,13]. These present, as already mentioned, in addition to their high repetition rate, the advantage of being considerably superior to solid lasers (Yag and Ruby) in terms of spatial beam homogeneity. Orlowski and Richter[13] formed high quality, patterned SiO_2 layers (> 3000 Å thick after 4×10^3 shots) on Si substrates irradiated in an O_2 atmosphere with a pulsed XeCl (308 nm) excimer laser working close to the Si melting threshold (0.9 J/cm^2). By using a pulsed ArF (193 nm) excimer laser, Fogarassy et al.[8] demonstrated the possibility of oxidizing in air or in O_2 heavily ion implanted Si at much lower energy density (0.1 to 0.2 J/cm^2). RBS spectra in fig. 6 recorded in random and channeling conditions for As (40 KeV, $3 \times 10^{16} cm^{-2}$) implanted Si, show clearly the influence of the number of laser pulses (E = 150 mJ/cm^2) on oxygen incorporation and the formation of a stoichiometric thick SiO_2 layer (900 Å) after 10^4 shots. In the same figure, we observe a strong redistribution of arsenic, which extends both into the grown oxide layer and silicon at a depth which corresponds to the existence of a molten phase regime.

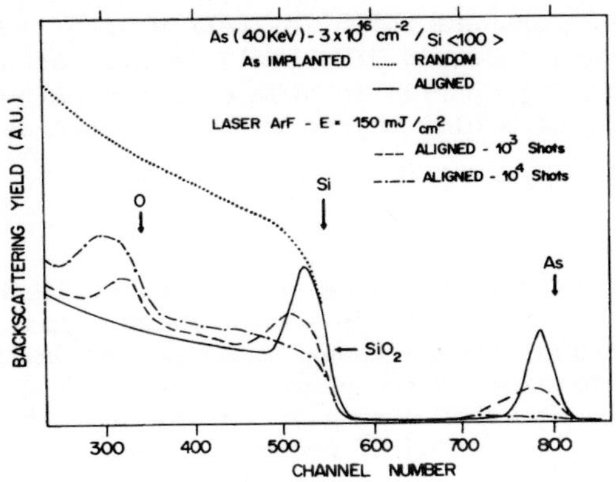

Fig. 6. RBS spectra in random and channeling conditions for As (40 KeV) implanted Si before and after multi-pulsed ArF laser irradiation (10^3, 10^4 shots).

Evidence for the formation of a stoichiometric and low disordered SiO_2 layer is also provided by the position (~ 1075 cm^{-1}) and the width (~ 70 cm^{-1}) of the IR band as shown in fig. 7.

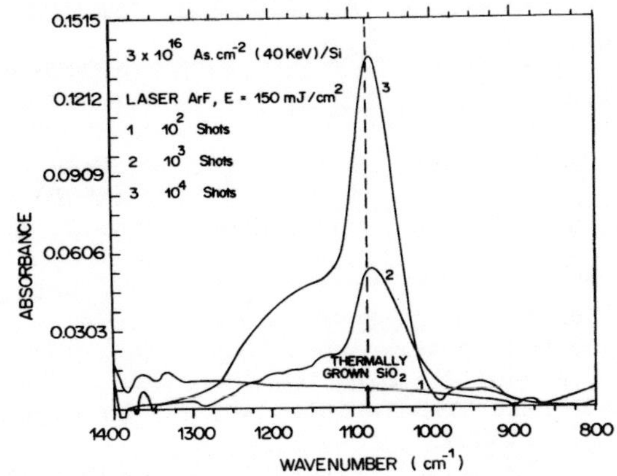

Fig. 7. IR absorption spectra for As (40 KeV) implantation after multipulsed ArF laser irradiation (10^2, 10^3, 10^4 shots).

Very similar behaviour has been found for nitrogen and fluorine implanted Si. By contrast we do not detect any significant oxygen incorporation into Si implanted silicon samples for the same range of laser energy densities[8]. These results indicate a strong influence of impurities on the oxidation of silicon following multi-pulse laser irradiation close to the melting threshold. The oxidation process is not yet clearly understood. Due to the large differences in the thermal and optical properties of SiO_2 when compared to Si, the behaviour of the grown oxide surface layer when submitted to pulsed UV laser close to the melting threshold for Si is questionable. The energy provided by the laser beam could be sufficient to melt silicon by keeping the surface grown oxide layer in the solid phase. In these conditions, the incorporation of oxygen would be only possible by solid diffusion through the SiO_2 layer which could stay solid at a temperature close to the melting point of Si (1410°C). This is realistic if we consider the diffusion coefficient of O_2 in SiO_2 at this temperature which is fairly high (10^{-7} cm^2/sec)[27]. The diffusion process through solid SiO_2 could be also enhanced if we speculate that the main diffusing species is atomic oxygen which results from the UV photodissociation of molecular oxygen by the excimer laser.

CONCLUSION

Laser induced oxidation, using both CW and pulsed lasers of different wavelengths working in the solid or liquid phase regimes appears to be very attractive to replace conventional thermal oxidation, in the field of localized beam processing for microelectronics. The development of these new technologies is however strongly related to the performance of new types of lasers, such as pulsed UV excimer lasers.

REFERENCES

1. S.W. Chiang, Y.S. Liu and R.F. Reihl, Appl. Phys. Lett. 39:752 (1981).
2. J.F. Gibbons, Jpn. J. Appl. Phys. Suppl. 19:121 (1980).
3. I.W. Boyd, J.I.B. Wilson, J.L. West, Thin Solid Filsm 83:L173 (1981).
4. I.W. Boyd and J.I.B. Wilson, Appl. Phys. Lett. 41:162 (1982).
5. I.W. Boyd, Appl. Phys. Lett. 42:728 (9183).
6. C. Fiori, Phys. Rev. Lett., 52:23, 2077 (1984).
7. J. Siejka, R. Srinivasan, J. Perrière "Dielectric Layers in Semiconductors" : Novel Technologies and Devices" E. MRS, ed. by G.G. Bentini, E. Fogarassy, A. Golanski, Les Editions de Physique, Les Ulis, (1986) p. 213.
8. E. Fogarassy, A. Slaoui and C. Fuchs, to be published in the NATO Proc. Workshop on "Emerging Technologies for in situ Processing", (Cargesse, May 1987).
9. G. Battaglin, G. Della Mea, A.V. Drigo, C. Foti, G.G. Bentini and M. Servidori, Phys. Status Solidi (a) 49:347 (1978).
10. K. Hoh, H. Koyama, K. Uda and Y. Miura, Jpn. J. Appl. Phys., 19:L375 (1980).
11. Y.S. Liu, S.W. Chiang and F. Bacon, Appl. Phys. Lett. 38:1005 (1981).
12. A. Cros and F. Salvan, Appl. Phys. A28:241 (1982).
13. T.E. Orlowski and H. Richter, Appl. Phys. Lett. 45:241 (1984).
14. E. Fogarassy, C.W. White, D.H. Lowndes and J. Narayan, "Beam Solid Interactions and Phase Transformations", ed. by H. Kurz, G.L. Olson and J.M. Poate, 51:173 (1985).
15. G.G. Bentini, M. Berti, A.V. Drigo, E. Jannitti, C. Cohen and J. Siejka, "Laser Chemical Processing of Semiconductor Devices" extended Abstracts p. 126 ed. by F.A. Houle, T.F. Deutsch, R.M. Osgood Jr. (1984).

16. S.E. Blum, K.H. Brown and R. Srinivassan, Appl. Phys. Lett. 43:1026 (1983).
17. E. Fogarassy, S. Unamuno, J.L. Regolini and C. Fuchs, Phil. Mag. B55:253 (1987).
18. H.S. Carslaw and J.C. Jaeger, "Conduction of Heat in Solids", 2nd edit. Clarendon Press, Oxford (1959).
19. J.L. Regolini, see Ref. 7, p. 297.
20. E.M. Young and W.A. Tiller, Appl. Phys. Lett. 42:63 (A983).
21. I.W. Boyd and F. Michelli, see Ref. 8.
22. R. Oren and S.K. Ghandhi, J. Appl. Phys. 42:752 (1971).
23. S.A. Schafer and S.A. Lyon, J. Vac. Sci. Technol. 21:422 (1982).
24. A.M. Goodman, Phys. Rev. 152:785 (1966).
25. C. Fiori and R.A.B. Devine, Phys. Rev. Lett. 52:2081 (1984).
26. I.W. Boyd and J.I.B Wilson, Appl. Phys. Lett. 50:320 (1987).
27. S.M. Sze, "Physics of Semiconductor Devices", J. Wiley and Sons, N.Y., (1969).

GROWTH AND STRUCTURE OF ARGON LASER GROWN SiO_2

Francesca Micheli and Ian W. Boyd

Electronic and Electrical Engineering
University College London
London WCIE 7JE
United Kingdom

INTRODUCTION

The use of intense beams of photons from 193 nm to 10.6 μ to oxidise single crystal silicon (c-Si) in dry oxygen (O_2) has been widely studied using lasers and incoherent light sources[1]. Depending upon the precise wavelength of the radiation used, thermal and/or non-thermal reactions can be initiated. Here we review the results of a recent experiment with visible radiation, which indicates that the induced reaction is thermally dominated, and that there is an additional wavelength dependent component to the overall mechanism[2]. We also present new results of a study by infrared spectrometry of the bonding nature of the films grown by this method.

EXPERIMENTAL RESULTS

Radiation at 488 or 514 nm from a Coherent Innova 100-20 argon ion laser was directed a few degrees off normal incidence onto 400 μm thick samples of electronic grade c-Si of 9 mm^2 in area. The samples were cleaned using standard production line techniques, and held by a quartz support in near thermal isolation in a stainless steel chamber containing a closed atmosphere of pure dry O_2 (BOC Electra II grade). The incident beam power was adjusted to compensate for differences in c-Si reflectivity at the two wavelengths so that a constant power of 4W was absorbed. The initial temperature induced by the absorbed radiation was determined by an optical pyrometer to be approximately 850°C at the centre of the irradiated zone. As will be shown later, it was not necessary to know the precise processing temperature. The nature of the Gaussian profile of the laser beam, together with the irradiation geometry, was such that the radial temperature profile initially induced across each sample decreased by roughly 10 % between the center and the edge.

Different film thicknesses were grown by varying the irradiation times. After exposure, a three dimensional contour mapping of the oxide was obtained by an automatic scanning ellipsometer with a sampling beam size of 26 by 50 μm. The peak thickness at the centre of each sample varied from 100 to 850 Å.

Although the precise mechanisms of Si oxidation are still somewhat controversial and have yet to be satisfactorily formulated, the temperature dependance of the rate constants appropriate to the different reaction regimes observed suggests that they can be fitted by Arrhenius expressions of the form[3] :

$$C \exp [-E_a/kT] \qquad (1)$$

where E_a is the effective activation energy, k is Boltzmann's constant, and T is the induced temperature[3], which is regulated by the amount of absorbed radiation and the energy lost by the sample to the surroundings.

As we have previously shown, the absorption of the laser light is affected during irradiation by the multiple reflections within the growing transparent film[2]. As the oxide grows to about 850 Å, the local reflectivity, R, of the film/substrate system will decrease, and more radiation will be coupled into the silicon, thereby causing the temperature to rise. As indicated by Eq. 1, this will encourage the reaction to proceed faster. However, as the oxide thickness increases, it will induce, through the multiple interference behaviour, even more energy absorption leading to higher temperatures and faster oxidation. Thus, a positive feedback is introduced into the reaction rate.

This characteristic may be used to compare nearly identical reaction rates induced by slightly different processing conditions. Any small differences between the initial rates induced by exposure to the 488 and 514 nm radiation will therefore be amplified, and since we have ensured that the thermal components of the reactions are essentially identical, differences due to non-thermal effects will be amplified by this positive feedback for films up to approximately 850 Å. Thus, as a function of time, much thicker films will be grown by whichever process starts at the fastest reaction rate.

Small differences in the thermally controlled growth rates induced by the two wavelengths may of course be expected. The absorption coefficients for the two wavelengths are about 10 and 7.5×10^4 cm^{-1} near 800°C[4]. Therefore, it may be argued that the shorter wavelength radiation, being absorbed more strongly, would induce higher temperatures, and therefore faster reaction rates. Further, if one examines the predicted behaviour of R, for the system, as a function of oxide thickness, the shorter wavelength effects a mildly faster decrease. This behaviour would further tend to favour faster reaction rates for the shorter wavelength.

The thickness of the oxide films grown by identical absorbed powers at the two different wavelengths for various irradiation times is shown in figure 1. It shows that while at short irradiation times the films grown were of similar thicknesses, at longer exposures there is a clear difference in the film thicknesses ; those grown by the longer wavelength are significantly thicker. The 514 nm radiation evidently induces an observable increase in the reaction rate over that initiated by the 488 nm photons, which is contrary to the expectations provided by purely thermal arguments. We have shown that the difference between the rates induced by these two wavelengths is equivalent to an increase in the thermal reaction rate of about 20 %[5].

It is well known that either Si, O_2, or SiO_2 absorb preferentially 514 nm to 488 nm radiation. However, for equal power levels, a beam of longer wavelength radiation will contain a larger photon flux F than a shorter wavelength (a 5 % difference in our case). Therefore, 514 nm

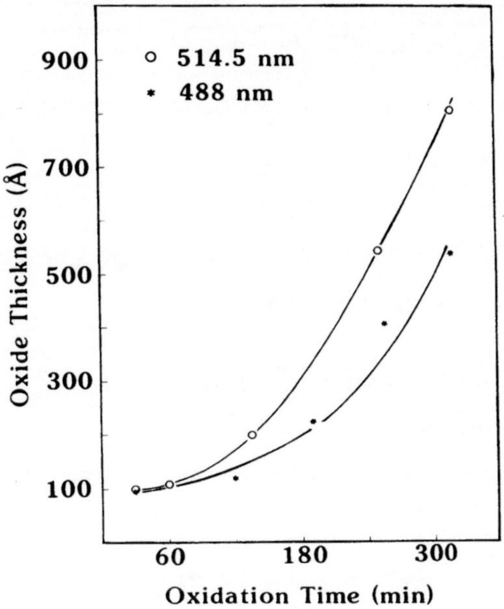

Fig. 1. Oxide thickness versus growth time for incident radiation at 488 and 514 nm^2.

radiation will create around 5 % more electron-hole pairs per unit area on the Si surface than the 488 nm light at the same power level, assuming a generation efficiency of 1. As to why this should affect the oxidation reaction, several models have highlighted the importance of

(a) available silicon bonds for the reaction with oxygen,
(b) conduction-band electrons in the Si which may be further excited into the conduction band of SiO_2,
(c) the formation of negatively charged oxygen species.

Each of these criteria may be affected by a change in the total population of conduction band electrons in the c-Si. Further experimentation is in progress to investigate the further effects of red and infrared radiation, as well as UV, on the reaction of oxygen with silicon in the solid phase.

INFRARED CHARACTERIZATION

A preliminary characterisation of our laser grown oxides has been performed with FT infrared (IR) spectroscopy. The data were taken at room temperature using a Brucker IFS 45 spectrometer fitted with a microscope attachment resulting in a sampling spot diameter of 80 μm, a compromise between ultimate resolution and signal to noise ratio. Each spectrum was recorded as the summation of 800 scans with a resolution of 2 cm^{-1}. A smoothing routine was used to reduce the remaining noise and interference fringes. All the IR data were collected from float zone, p-type, lightly doped Si, with an identical, unoxidized, sample used as a reference to eliminate background spectral features from the a crystal silicon substrate.

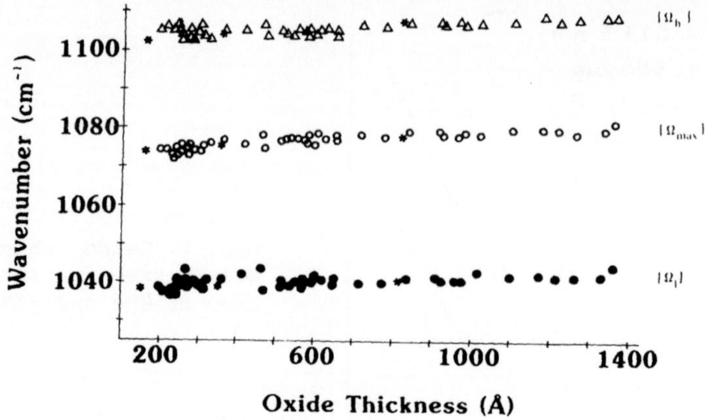

Fig. 2. Peak position of the Si-O stretching mode for various laser grown oxides. The * represents furnace oxides.

The three main vibrations in the spectral response of SiO_2, associated with the various vibrational modes of atoms in the Si-O-Si group can be found at 450 cm^{-1} (rocking mode), 810 cm^{-1} (bending mode), and 1075 cm^{-1} (stretching mode). We have limited our interest to the strongest absorption feature, the 1075 cm^{-1} stretching band, principally for reasons of sensitivity. No SiOH, H_2O or SiH peaks (around 3600, 1640 and 2250 cm^{-1} respectively) were observed within the sensitivity of the instrumentation.

The infrared transmission minima near 1075 cm^{-1} (9.2 µm) of thermally grown SiO_2 films have previously been shown[6, 7] to obey Lambert-Bouguer's law, within experimental error, for film thicknesses from 3.5 µm to approximately 28 Å. This law states :

$$A = \log I_o/I = 0.434\, \alpha_{app}(x-xo) \quad (2)$$

where A is the absorbance, I_o and I the incident and transmitted intensities, respectively, α_{app} the apparent absorption coefficient and x_o the oxide thickness in the reference sample, i.e the usual native oxide (≈15 Å) resident on the surface of microelectronic grade silicon. We have verified the law for our Ar laser grown oxides, in the range 150 to 1400 Å, finding $\alpha_{app} \approx 3.32 \times 10^4 cm^{-1}$, in good agreement with previous results.

Fig 2 shows the thickness dependence of the Si-O stretching mode absorption peak Ω_{max} and of the two frequencies Ω_1 and Ω_h wich define the full width at the half-maximum (FWHM). A similar degree of asymmetry as was recently reported for furnace and CO_2 laser grown oxides was found[8]. Figure 3 shows the behaviour of the FWHM on oxide thickness. Again, as with furnace prepared films, the data show only a mild thickness dependence[8]. This suggests that the oxides are very similar to conventional thermally grown layers, and not CVD deposited films, which exhibit very different IR characteristics. Some points obtained from furnace grown and annealed oxides are also plotted on the graphs. It has been suggested that the

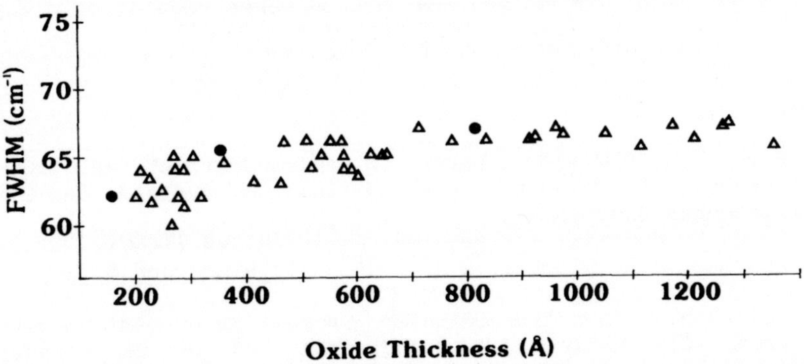

Fig. 3. FWHM of the Si-O stretching band for various thicknesses of laser oxides. The dots represent furnace oxides.

stretching band at 9.2 μm is strongly infuenced by the chemical bonding character, stoichiometry, porosity, density and level of compressive strain in the film[7-9]. Recently it has also been shown[10] that for thermal oxides, the growth temperature can affect the density, and produce changes in the Si-O-Si angle and Si-Si distances which would shift the position of the stretching vibration peak. It should be stressed that our oxides are not grown at a uniform temperature, as discussed above. In fact, we estimate that for the thickest samples, the temperature varied by as much as 200°C during the growth period. Together with the results obtained previously with furnace and CO_2 laser grown oxides[8], this would indicate that temperature is not the only controlling influence on the position and FWHM of this absorption feature. Further studies in this area will focus on the IR properties of thinner films grown using this laser method. Correlation of the spectral properties with the electrical characteristics as derived from C-V measurements is also under way.

SUMMARY

We have confirmed that a photonic effect exists in the reaction of oxygen and silicon induced by visible radiation by minimizing the dominant thermal contributions to the reaction and amplifying the small differences arising from non-thermal contributions. IR absorption studies indicate that films thicker than about 200 Å grown by Ar laser are very similar in structure to layers prepared by conventional furnace techniques.

ACKNOWLEDGEMENTS

We acknowledge the invaluable contribution of C Brown of GTM Ltd and Dr J.I.B. Wilson of Heriot-Watt University in providing many of the ellipsometry measurements. We are also indebted to Dr R. Thompson, of Hughes Micro-electronics Ltd, and Dr A Hodge of RSRE for supplying Si samples and confirming several ellipsometric data points. This work was partly funded by a UGC capital grant, SERC contract no. GR/E 16090 and an award from Nuffield foundation. FM acknowledges receipt of a 2 year SERC grant.

REFERENCES

1. I.W. Boyd, in "Dielectric Layers in Semiconductors", ed. G.G. Bentini, E. Fogarassy, A. Golanski, Les Editions de Physique, Les Ulis (1986) and references therein.
2. I.W. Boyd, F. Micheli, Electronics Lett., 23:298 (1987).
3. H.Z. Massoud, J.D. Plummer, E.A. Irene, J. Electrochem. Soc. 132:1745 (1985).
4. G.E. Jellison, in "Pulsed Laser Processing of Semiconductors", ed. R.F. Wood, C.W. White, R.T. Young, Vol. 23 of Semiconductors and Semimetals, Academic, New York, (1984).
5. F. Micheli, I.W. Boyd, Appl. Phys. Lett., 51:to be published, (1987).
6. J.E. Dial, R.E. Wong, J.N. Fordemwalt, J. Electrochem. Soc. 115:327 (1968).
7. I.W. Boyd, J.I.B. Wilson, J. Appl. Phys. 53:4166 (1982).
8. I.W. Boyd, J.I.B. Wilson, Appl. Phys. Lett. 50:320 (1987).
9. W.A. Pliskin, H.S. Lehman, J. Electrochem. Soc. 112:1013 (1965).
10. G. Lucovsky, M. Manitini, J.K. Srivastava, E.A. Irene, J. Vac. Sci. Techn. B5:530 (1987).

TRANSPORT PROPERTIES OF PLASMA ENHANCED CVD SILICON OXYNITRIDE FILMS

Yves Cros, Jean Christophe Rostaing

LEPES*, C.N.R.S.
BP 166
38042 Grenoble Cedex, France

INTRODUCTION

Up to now, few transport property measurements have been made on low temperature deposited Silicon Oxide or Silicon Oxynitride Films. In SiO_2 prepared by RF sputtering and annealed at 570 K, the current was not measurable at 300 K for low electrical field, but subsequently presented linear Ln J versus $E^{1/2}$. For electric field E between $6x10^5$ V cm^{-1} and 1.2 10^6 V cm^{-1} [1,2], ionic Na^+ transport with an activation energy of 1.1 eV has been observed above 470 K[2]. For plasma Enhanced Chemical Vapor Deposited (PECVD) low temperature Silicon Oxide/Oxynitride, in spite of considerable work on composition and mechanical properties[3,4,5], the electrical properties at low field remain unknown. Silicon rich-Silicon dioxide has been studied with the dynamic ramp I-V technique for fields above 5 10^6 V cm^{-1}i.e. currents above 10^{-10} Amp corresponding to Fowler Nordheim characteristics before breakdown occurs[6]. Recently [7,8,9] electrical quality improvement was obtained when a large dilution with of He and low deposition rate were used in the PECVD technique, but transport properties at low and intermediate electric field were not published.

We present here transport measurements on $SiO_xN_yH_z$ films prepared by PECVD at 430°C and 320°C with SiH_4 and N_2O reactive gases. These samples were previously studied by Elastic Recoil Detection Analysis and IR spectroscopy to determine their composition and local chemical ordering, [10,11] by ESR to measure the spin concentration [11] and by optical absorption up to 9 eV to follow the variation of the optical gap and Urbach tail with composition [12].

EXPERIMENTAL AND RESULTS

Computer-controlled current voltage measurements were performed on MIS ($Al/SiO_xN_yH_z$/n c Si) structures under high vacuum using a Keithley 617 electrometer. Currents lower than $5x10^{-15}$ A may be detected. Our samples

---

*Laboratory associated with Université Scientifique, Technologique et Médicale de Grenoble.

exhibit current time decay in response to a bias step, as generally observed in insulators [13,14]. The measuring procedure must therefore be chosen with care. Apart from the few seconds long initial capacitance response of the measuring circuit, long lasting time decreasing currents may be explained by transient ionic drift or/and electronic charge trapping. Temperatures higher than 500 K must generally be achieved to activate ionic currents. Na^+ are the most likely ions to be present in SiO_2. Extrapolating high temperature (400 K - 900 K) ionic mobility variations [15,16] to room temperature and calculating assuming very high Na^+ concentrations leads to conductivity values at least by two orders of magnitude lower than those we actually observed. Moreover, observed capacitance-voltage hysteresis effects such as those shown in fig. 1 (up to 450 K) do not agree with ionic drift results as shown in figure 2.

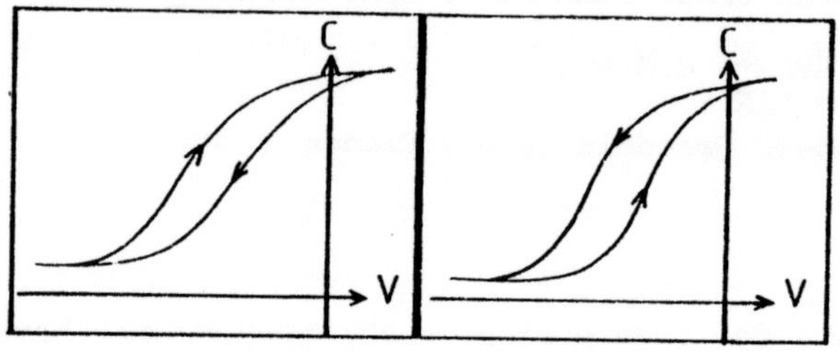

Fig. 1 C(V) hysteresis from electronic charge trapping.

Fig. 2 C(V) hysteresis from ionic drift.

We therefore conclude that the time decay of the current is not related to ionic motion. Two decay regimes occur in sequence [19,3]. The first not lasting more than a few minutes, is ascribed to the electric field lowering induced by electronic space charge accumulation at a partially blocking electrode. The second is much slower and results from the increase of the trapped space charge, thus altering the potential profile, and is governed by generation - recombination processes[17]. These effects are observed for either low or high applied voltages (fig. 3).

From these considerations we will assume that the current value measured just a few seconds after the voltage step is a proper approximation of the actual initial electronic current, i.e. the internal field has not been

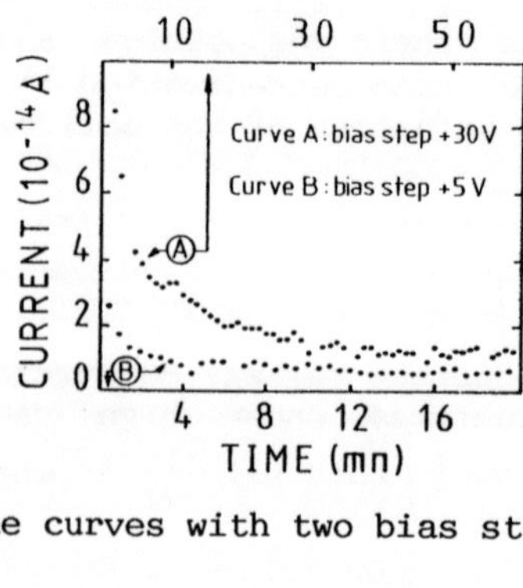

Fig. 3 Current time curves with two bias steps for sample 16.

noticeably disturbed yet. After the current has been established, bias is suppressed, resulting in a reverse current which decays to zero within few seconds to a few minutes, although it may be superposed on some instability. The next biasing will take place afterwards. Hence to test

Table 1

Chemical composition, preparation temperature T, thickness d, optical gap E_{op}, spin density and the associated g factor, Dc conductivity σ_{DC} and maximum electric field E_m for the linear zone.

Sample N°	$SiO_xN_yH_z$ x	y	z	T °C	d Å	E_{op} eV	ESR results Spin, cm^{-3}	g factor	Low Field results σ_{DC} $\Omega^{-1}cm^{-1}$	Em $V\ cm^{-1}$
2	1.94	0.17	0.12	430	3300	7.8	$4x10^{17}$	2.0006	$1.8x10^{-17}$	$6.1x10^{5}$
10	1.91	0.29	0.10	320	1100	-	$2x10^{16}$	2.0008	$1.6x10^{-16}$	very lim.
16	0.99	0.28	0.31	320	2500	3	10^{18}	2.0043	$1.4x10^{-16}$	$6.x\ 10^{5}$
5	0.7	0.24	0.25	430	2450	2.54	10^{18}	2.0040	$4.5x10^{-16}$	$2.5x10^{5}$

the low field J (E) reproducibility using point by point measurement, the biasing sequence consisted of zero volt centered cycles of increasing excursion. This in turn allowed us to monitor irreversible effects as well. These effects associated with departure from linearity of the J (E) curves will not be considered further in this paper in which we focus on the low field regime.

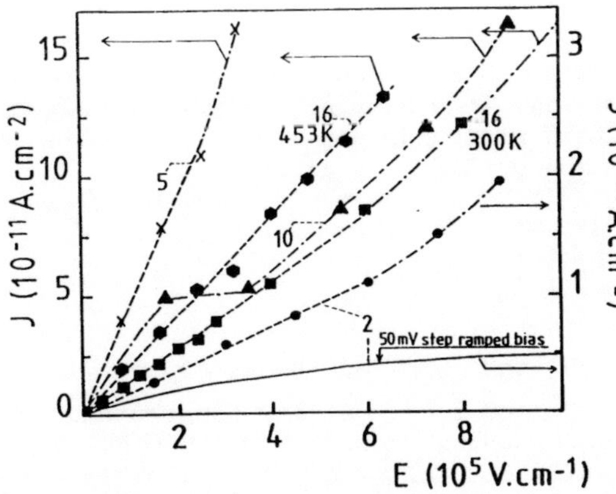

Fig. 4. J (E) characteristics at 300 K showing the linear range. J (E) non linear curve obtained for sample 2 with a 50 mV step ramped bias. Linear J (E) curve at 453 K for sample 16.

Figure 4 displays the J (E) characteristics obtained at 300 K under a positive gate biasing of oxynitride samples of table 1. Experimental points are fully reproducible on cycling. Samples 2, 16 and 5 exhibit a linear characteristic up to E_m while sample 10 shows a non linear J (E) curve. We define the low field DC conductivity σ_{DC} as the J/E slope of the linear curve (table 1) compared to the point by point linear curve of sample 2. We also present in figure 4 the non linear J (E) curve obtained with the same sample using a monotonic increasing bias in 50 mV steps. This demonstrates the effect of electron trapped charge growth resulting in a strong decrease of the current density flowing through the sample during the measurements time. Due to the high noise ($\sim 10^{-13}$A) generated by the temperature regulator, it was not possible to measure σ_{DC} during heating for the most resistive sample n° 2. Only careful manual operation allowed us to obtain the J (E) curve for sample 16 at 453 K (fig. 4). The linear field range is not reduced by raising the temperature.

The σ_{DC} variation with T between 300 K and 453 K is given fig. 5. It appears that we cannot attribute the weak variation to an activated regime following an exp (-A/T) law.

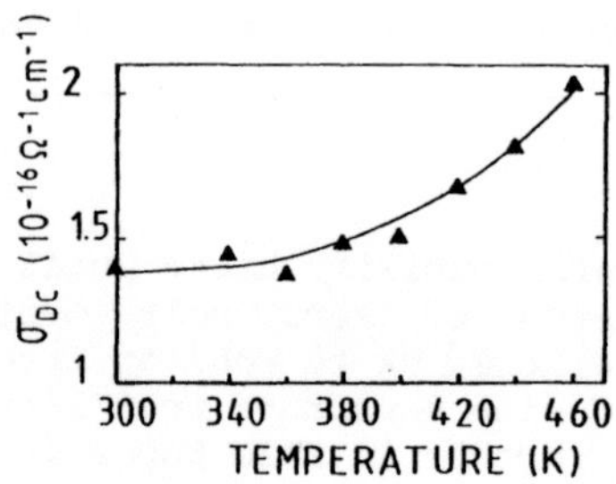

Fig. 5 - Variation of the DC conductivity with temperature for sample 16.

DISCUSSION

Deep donor levels in the band gap of insulators induce electronic conduction by tunneling. This hopping conduction induces an ohmic linear J (E) regime at low field preceding non linear conduction at intermediate and high fields. The DC conductivity is given by :

$$J/E = \sigma_0 \exp\left[-\left(\frac{T_0}{T}\right)^{\alpha}\right] \quad (1)$$

where σ_o depends on the applied bias, T_o on the material and α is a coefficient between 0 and 1.[18]. At low temperature, $\alpha = 1/4$ is generally observed and at temperatures around 300 - 500 K, the so-called activated law with $\alpha = 1$ is still not followed. We certainly are in an intermediate situation for sample 16 (fig. 5) with only a 1.4 multiplying factor between 300 K and 460 K. In RF sputtered SiO_2, none of the numerous mechanisms currently proposed has been able to explain the observed experimental J (E) curves between $6x10^{-5}$V cm^{-1} and $1.3x10^{6}$V cm^{-1} which are fitted by[1,2] :

$$LnJ \propto -\left(\frac{T_0}{T}\right)^{n} + E^{1/2}/F(T) \quad (2)$$

with n = 1/4 or 2/5. Authors have proposed a model of hole like polaron hopping controlled by non bridging oxygen defects. This model is unlikely to explain our linear J (E) low field curves. Up to now, the theoretical aspect of the hopping regime in insulating materials is not clear and remains an open question, so we shall not attempt to use any formula to fit our σ_{DC} conductivity values. The discussion will be limited to the possible nature of the deep centers involved in the hopping process in our $SiO_xN_yH_z$ for which the band gap varies from 8 eV to 2.5 eV (table 1).

Samples 2 and 10 present low incorporation of Nitrogen (respectively 5 and 9 atomic %) and Hydrogen (respectively 4 and 3 atomic %) with an oxygen incorporation approaching 2. For these, we suppose that the band gap is the same as in thermal SiO_2 although the presence of a few percent of N an H may introduce new states in this band gap. Our ESR measurements have revealed only Si dangling bonds with the same g value (g = 2.0008) as the E' centers in thermal SiO_2. No signal associated with non bridging oxygen hole centers (g = 2.009, ΔH = 13 G) has been detected. So we limit the discussion to states in the thermal SiO_2 band gap associated with Si dangling bonds. In figure 6 we indicate the energy levels of the three charge states Si_3°, Si_3^{+}, and Si_3^{-} as proposed by Griscom[19] and in figure 7 the Si_3° and Si_3^{+} levels calculated by O'Reilly and Robertson[20].

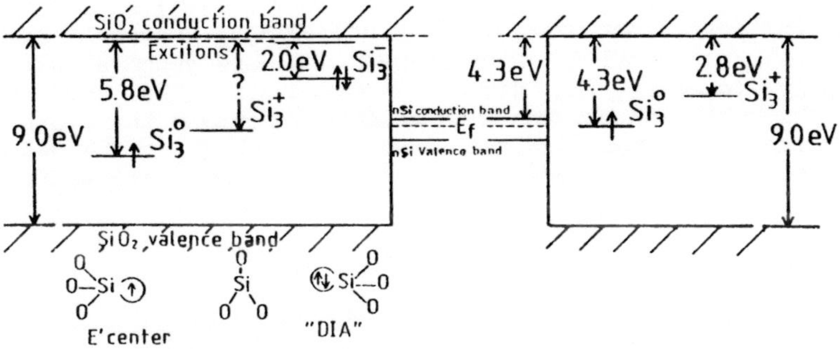

Fig. 6 Optical transition in thermal SiO_2 from ref. 19.

Fig. 7 Optical transitions in thermal SiO_2 from ref. 20

Compared to sample 2, the conductivity of sample 10 is increased by a decade while the E_1' center concentration is divided by 20. Sample 2 undergoes many bias cycles up to $\pm 3\times10^6$ V cm^{-1} without breakdown. Sample 10 presents a very limited linear range, a high charge trapping at intermediate field and breaks around 1.5×10^6 V cm^{-1}. Sample 10 may be compared to sample (1) of refs. 10,11,12, with $SiO_{2.06}N_{0.09}H_{0.18}$ for which are simultaneously observed the absence of a linear J (E) range, a very large gap states tail in the optical absorption between 3 eV and 8 eV, a very low breakdown field and the absence of ESR signals. Obviously in sample 10, it is likely that charge trapping has transformed Si_3° neutral paramagnetic centers into Si_3^{-} or Si_3^{+} diamagnetic centers explaining the lower ESR spin concentrations. Owing to the 4.3 eV barrier height between the $SiO_xN_yH_z$ sample and n.cSi substrate assumed identical to that at the SiO_2/n.c Si interface, the hopping level most likely to connect with the Fermi level in c-Si is Si_3^{+} in figure 6 and Si_3° in figure 7. In contrast to the case of amorphous Si_3N_4 where Si_3° centers, found in high concentrations, are responsible for hopping conduction at low electric field[21] it is difficult to explain our low field hopping linear regime owing to the low concentration ($4x10^{17}$ cm^{-3}) of $Si_3°$ ESR centers found in sample 2. If Si_3^+ centers are the hopping states (by a mechanism to be clarified) it would explain why the reduction of ESR signal, maybe from increasing Si_3^+ concentration, is correlated with increasing conductivity ; the loss of a linear regime in sample 10 would not be explained. Using only these facts one cannot propose a firm model for the hopping process, complementary studies are needed.

In samples 16 and 5 with off stoichiometric composition and optical gaps around 3 eV (table 1), the barrier height with n c-Si and the location in the band gap of the three charged states of Si dangling bonds are unknown. The neutral Si_3^o ESR centers correspond to $°SiOSi_2$ (g = 2.0037), $°SiSi_3$ (g = 2.0055) or some mixed °SiONSi configurations. The DC conductivity is inversely proportional to the mean hopping distance, but depends strongly on the energy barrier height between these hopping centers. Assuming the hypothesis of $Si_3°$ hopping centers and despite the fact that their concentrations are the same in sample 16 and 5, we can understand the higher σ_{DC} of sample 5 as due to the lowering of the barrier height between these hopping centers because of the narrower band gap in sample 5 compared to sample 16. The reduction of the linear J (E) range in sample 5 may also be understood in a same manner.

To go further in our understanding of $SiO_xN_yH_z$ transport properties we need complementary results. Photoinduced ESR measurements must be made to unambiguously identify the states involved in the hopping process. Internal photoemission studies must be done to obtain the evolution of the barrier height on n-cSi with the composition, which is the starting point in understanding the conduction regimes in the intermediate and high electric field ranges.

REFERENCES

1. M. Meaudre, R. Meaudre, J.J. Hauser, J. Non Cryst.Sol. 58:145 (1983)
2. M. Meaudre, R. Méaudre, J. Non Cryst. Sol., 68:281 (1984)
3. J.R. Hollahan, J. Electrochem. Soc. 126:930 (1979)
4. E.P.G.T. Van de Ven, Sol. Stat. Technol. 24:167 (1981)
5. A.C. Adams, F.B. Alexander, C.D. Capio, T.E. Smith, J. Electrochem.Soc. 128:1545 (1981)
6. S. Yokoyama, D.W. Dong, D.J. Di Maria, S.K. Lai, J. Appl. Phys. 54:7058 (1983)
7. P.G. Pai, S.S. Chao, Y. Tagaki, G. Lucovsky, J.Vac. Sci. Technol. A4: 689 (1986)
8. S.S. Chao, G. Lucovsky, S.Y. Lin, C.K. Wong, J.E. Tyler, Y. Tagaki, P.G. Pai, M.J. Mantini, J.Vac. Sci. Technol, A4:1574 (1986).
9. J. Batey, E. Tierney, J. Appl. Phys. 60:3136 (1986).
10. Y Cros, J.C. Rostaing, MRS Strasbourg, Ed. G.G. Bentini - Les Editions de Physique, Les Ulis, 77 (1986).
11. Y. Cros, D. Jousse, J. Liu, J.C. Rostaing, J. Non Cryst. Sol. 90 (1987).
12. Y. Cros, J.C. Rostaing, J. Peisner, G. Lévèque, C. Ance J.A.P. June 1987.
13. R.H. Walden J. Appl. Phys. 43:1178 (1972).
14. C.M. Osburn, E.J. Weitzman, J. Electrochem. Soc., 119:603 (1972).
15. G.H. Frischat, "Ionic diffusion in Oxide glasses", Trans Tech. Publications (1975).
16. R. Meaudre, M. Meaudre A. Deguin, J. Non Cryst. Sol. 33:391 (1979)
17. R. Meaudre, M. Meaudre, Phys. Rev. B 19:3252 (1979)
18. P. Hesto, "Nature of electronic conduction in Instabilities in Silicon Devices, Vol. 1" G. Barbottin, A. Vapaille editors, North Holland, N.Y. (1986) p. 305.

19. D.L. Griscom, J. Non Cryst. Sol. 73:51 (1985)
20. E.P. O'Reilly and J. Robertson, Phys. Rev. B 27:3780 (1983)
21. J. Robertson, M.J. Powell, Appl. Phys. Lett. 44:415 (1984).

CHARACTERISTICS OF SiO_2 AND SiO_XN_Y, OBTAINED BY RAPID THERMAL PROCESSES

N. Chan Tung, Y. Cartini*, R. Pantel, and J.L. Buevoz

Centre National d'Etudes des Télécommunications
BP. 98 - 38243 Meylan, France
* Atelier Electro Thermie
Zirst - 38243 Meylan, France

ABSTRACT

Thin gate oxides of 30-150 Å are currently grown in a rapid thermal processing machine. Oxidation kinetics have been studied in the temperature range of 1000°-1250°C for an oxidation time of 5 to 60 s. An activation energy E_A of 1.4 eV has been obtained. Rapid thermal nitridation of a 96 Å SiO_2 film has been performed at a temperature of 1150°C for a nitridation time up to 150 s. The electrical characteristics of MOS capacitors were compared to those of MIS and an average breakdown field of 14.6 MV/cm has been obtained. The evolution of the stoichiometry of Nitrogen profiles has been studied for various nitridation times and will be discussed.

INTRODUCTION

Thin gate dielectrics have been investigated for several device processing technologies. As MOSFET dimensions are scaled down, it is necessary to reduce the gate oxide thickness and it has been established that submicron channel length VLSI MOSFETS will use oxide layers thinner than 150 Å [1]. Thin tunnel oxides are also used in memory devices such as EEPROMS. Conventionally, thin SiO_2 films for device applications have been grown using standard furnaces with oxidation times usually long (~ 20 min) at low partial pressures of oxidant. Nitrided thermal SiO_2 films have shown improved chemical and electrical characteristics over pure thermal oxides [2-4]. Previous nitridation was accomplished using excessively long furnace times or plasma-enhanced processes[5], however, these techniques suffered from the drawback of precise control of nitrogen spatial distribution in the oxides. The aim of this paper is three fold : firstly, to report fairly extensive silicon kinetics data in the short time regime of 5 to 60 s, secondly, to compare the electrical characteristics of Al-gate MOS capacitors with MIS capacitors for a rapid thermal nitrided oxide, with particular emphasis on breakdown field statistics and thirdly to investigate the stoichiometry of Nitrogen profiles obtained by Auger electron spectroscopy.

EXPERIMENTAL PROCEDURE

The rapid thermal machine (Addax R 1000 manufactured by AET) used for our experiments has been described elsewhere[6-8]. The temperature measurement was made with an optical pyrometer used in a closed loop temperature control. Prior to each set of experiments, the pyrometer was calibrated against a thermocouple attached to dummy wafers.

Both n and p type <100> 4 inch silicon wafers with resistivities ~5-8 and 6.5-9Ω.cm respectively were chemically cleaned with H_2SO_4 : H_2O_2, HF and then rinsed with DI water. These cleaned wafers with a native oxide of about 10 Å were immediately processed in the rapid thermal machine. The SiO_2 film thickness was characterized by means of a monochromatic ellipsometer (HeNe 6328 Å) with the refractive index fixed at 1.46. The thickness was cross-checked by spectroscopic ellipsometry with the wavelength varying between 2500 and 10000 Å. MOS and MIS capacitors were delineated by conventional lithography processes and annealed in a forming gas atmosphere at 450°C for 30 min post metallization.

Two sets of experiments were carried out. In the first set, rapid thermal oxidation (RTO) kinetics was studied. In the second, a rapid thermal annealing (RTA) of the oxidized wafers was carried out at a temperature of 1100°C and 1200°C for 60 s in an argon atmosphere, followed by nitridation at 1150°C in an ammonia ambient.

RESULTS AND DISCUSSIONS

Oxide

Fig. 1 shows the oxide thickness vs time for oxidation temperatures ranging from 1000 to 1250°C. The experimental data represent the mean value of 81 thickness measurements made at different locations on the wafer. The solid curves represent straight lines fitted to the experimental data by the least square method. From these results, linear growth kinetics are obtained over the investigated time (5-60 s) and temperature range. An activation energy of 1.4 eV [8] is deduced. Although the linear-parabolic model of Deal-Grove[9] is not applicable to the very short time regime, it is worthwhile nevertheless to compare the linear growth rate B/A with our results. B/A has an activation energy of 1.76 eV for a temperature T lower than 1000°C and 3.2 eV for T>1000°C [10]. The activation energy, 3.2 eV, is higher than our results. However, recently published results compare the activation energy value for three different machines. For the Eaton machine, E_A = 1.3 eV [11] whilst that of AG Associates, E_A = 1.44 eV [11]. With a Peak machine, an E_A value of 1.3 eV [12] was obtained. Our oxide homogneneity results show a σ value of 1.9 Å [13] for a mean oxide thickness of 102 Å measured at 149 points across a 4 inch wafer.

A histogram of the percentage of occurance of break down as a function of the electrical field is shown in fig. 2 for an RTO of 1150°C-20 s. The statistical measurements were performed on 196 MOS capacitors. A mean electrical breakdown field of 10.1 MV/cm and σ value of 18 % have been obtained. Such high value of breakdown field compares very favourably with thermally grown oxide at low pressure[14].

A comparison of the breakdown histogram of an RTO 1150°C-20 s (fig. 2) with that of an RTO 1150°C-20 s + RTA 1100°C-60 s shows that in the latter case, the mean electrical breakdown field increases slightly to $\overline{E}$ = 10.3 MV/cm with a σ ~ 19 %. An increase of the RTA temperature, i.e.,

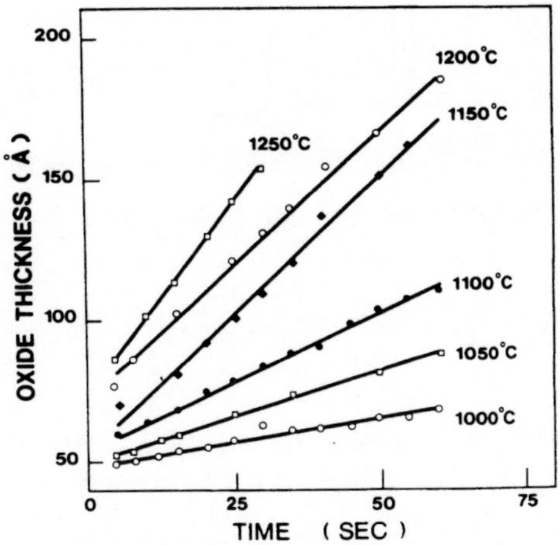

Fig.1 Oxide thickness vs time

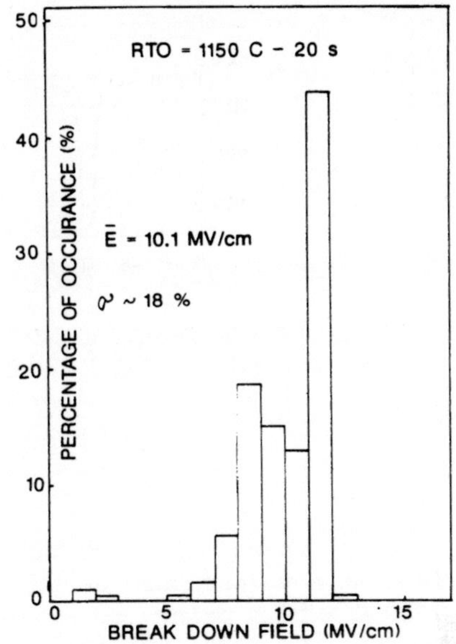

Fig.2 Histogram of breakdown field for MOS capacitors RTO 1150°C-20 s

RTO 1150°C-20 s + RTA 1200°C-60 s causes a net increase of $\bar{E}$ which is equal to 12.3 MV/cm with a $\sigma \sim 25$ % and a maximal electrical field of 17 MV/cm. Thus, RTA enhances the mean electrical breakdown field, the maximal breakdown field and at the same time broadens the histogram statistics.

Although Nulman[15] has reported the enhancement of the electrical breakdown field with RTA, the issue of the value for histogram statistics has not been addressed nor, by any other author to our knowledge. This will be discussed in the second part of this paper. High resolution transmission electron microscopy (HREM) of RTO 1150°C - 20 s +RTA 1200°C - 60 s shows a very_ sharp Si/SiO_2 interface with practically no protusions. The increase of $\bar{E}$ value with increasing RTA temperature may be due to the reduction of the protrusions which may be present following the RTO cycle by oxidizing the silicon atoms closest to, or within the interface, thus ensuring a uniform oxide thickness and hence improving the break down characteristics. RTA would also permit a reduction in the stress.

Nitridation of oxide

The interface trap density of a 96 Å SiO_2 (a) and that of an RTN oxide at 1250°C-60 s (b) is shown in fig. 3. The high frequency C-V curves are continuous lines whilst the quasi-static are dashed ones. A D_{it} value of 1.10^{10} cm^{-2} eV^{-1} has been obtained in both cases. The area of the capacitors was 0.23 mm^2. Nitridation shows an increase of C_{ox} value, from 902 pf (a) to 1096 pf (b), suggesting that the dielectric constant of the nitrided oxide has at least increased by 21 % if we assume the same thickness value for the pure oxide and for the nitrided one.

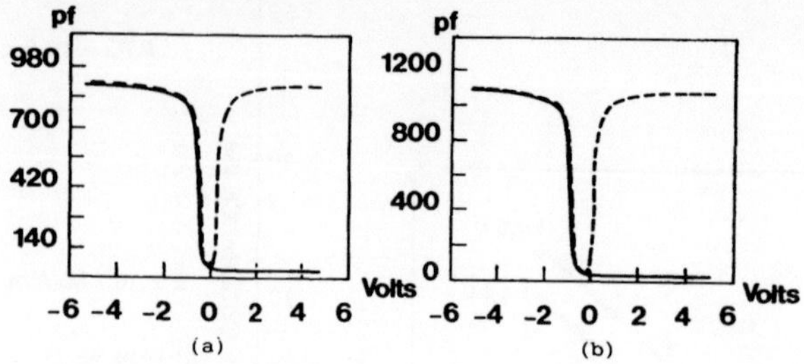

Fig. 3 High and low frequency C-V measurements of (a) 96 Å SiO_2 and (b) RTA 1200°C - 60 s + 1250°C - 60 s.

Fig. 4 shows the breakdown field histogram of a 96 Å SiO_2 film nitrided at a temperature of 1150°C for (a) 120 s and (b) 150 s with an RTA of 1200°C - 60 s prior to nitridation. For the 120 s RTN, an $\bar{E}$ value of 12.6 MV/cm has been obtained with a σ value of 14 %. For the 150 s RTN, $\bar{E}$ = 14.6 MV/cm with σ = 12 %. These results show that by increasing the RTN time, the mean breakdown field is enhanced. Comparing the breakdown fields in nitrided oxide with non nitrided, that is, ($\bar{E}$ = 10.1 MV/cm for RTO only

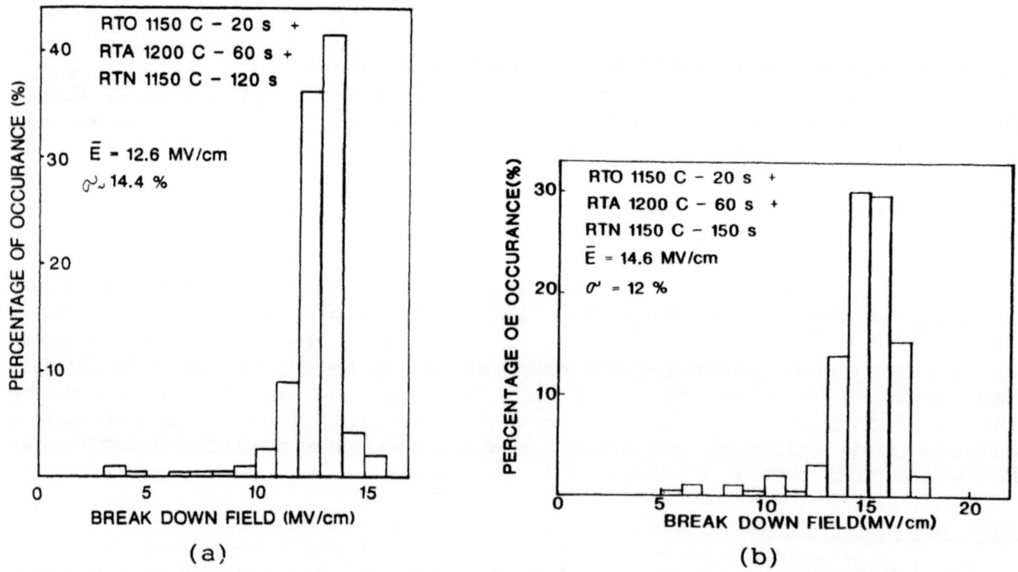

Fig. 4. Histogram of breakdown field for MIS capacitors with an RTN of 1150°C for (a) 120 s and (b) 150 s with an RTA of 1200°C - 60 s.

and $\overline{E}$ = 12.3 MV/cm for RTO + RTA 1200°C - 60s) shows that nitridation enhances the breakdown field. If we compare the results of fig. 4 with those of nitrided oxide that have not undergone any RTA, that is, ($\overline{E}$ = 11.8 MV/cm with σ = 12.6 % for RTO + RTN 1150°C - 120 s and E = 12.6 MV/cm with σ = 6.8 % for RTO + RTN 1150°C - 150 s), we find that the RTA process at 1200°C - 60 s enhances the mean breakdown electrical field and broadens the histogram statistics for the oxynitride. A similar result has been obtained for RTN at 1200° and 1250°C.

The oxygen and nitrogen stoichiometry of the oxynitride (SiO_xN_y) have been investigated by Auger spectroscopy. The depth is normalized to the position where the Si (LVV) signal is equal to 50 % of that of the Si substrate. For the calculation of the X value, it is assumed that the maximal oxygen signal for the oxide obtained from RTO + RTA corresponds to x = 2. From the original Auger profiles, the N stochiometry was deduced taking into consideration the respective sensitivity of the O_{KLL} and N_{KLL} Auger peaks. In our case, the sensitivity was equal to 1.6. The concentration profiles of O and N are shown in fig. 5(a) for various nitridation times ranging from 5 to 150 s with RTN = 1150°C (RTO = 1150°C -20 s + RTA 1100°C -60 s).

For RTO + RTA only fig. 5(a), the oxygen stoichiometry is very similar to that of the 5 s RTN and is not shown for clearness of the figure. All the profiles of (a) and (b) show N peaks at the surface region and corresponding depletion of O indicating the presence of a highly nitrided SiO_2 at the surface and at the interface. The mean X value decreases from 2 to 1.6 whilst $\overline{Y}$ increases from 0.1 to 0.6 (fig. 5(a)) for increasing RTN time (5-150 s).

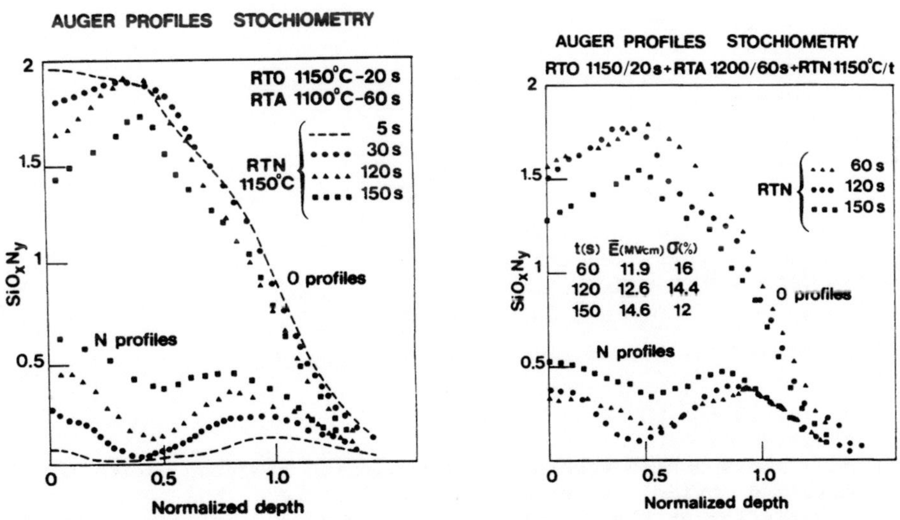

(a) (b)

Fig. 5. Stoichiometry profiles of SiO_xN_y for (a) RTO 1150°C-20s + RTA 1100°C-60s + RTN 1150°C (t) (b) RTO 1150°C-20s + RTA 1200°C-60s + RTN 1150°C (t).

In order to facilitate the comparison between different Auger profiles, an equivalent Si_3N_4 thickness is defined relative to area concentrations of N profiles. The nitrogen profiles at the surface are equivalent to 1 Å Si_3N_4 for 5 s and to 16 Å for 150 s. At the interface, an equivalent thickness of 3.7 to 17 Å Si_3N_4 is obtained for increasing nitridation time from 5 to 150 s.

For the RTO 1150°C - 20 s + RTA 1200°C - 60 s + RTN 1150°C (fig. 5(b)), the equivalent surface thickness increases from 6.6 Å to 14.7 Å Si_3N_4 for increasing RTN time (60 - 150 s). Comparing figs. 5(a) and (b) for the 150 s RTN, (i.e. increasing RTA temperature from 1100° to 1200°C) does not cause any significant variation in the equivalent Si_3N_4 thickness at the surface or at the interface.

Fig. 6 shows the effect of increasing RTN time from high frequency C-V measurements. The C-V curves shift to negative voltages with a maximal shift occuring for the 30 s RTN. The shift can be attribued to the presence of fixed charges at the interface with a maximal value of $1.10^{10}cm^{-2}$.

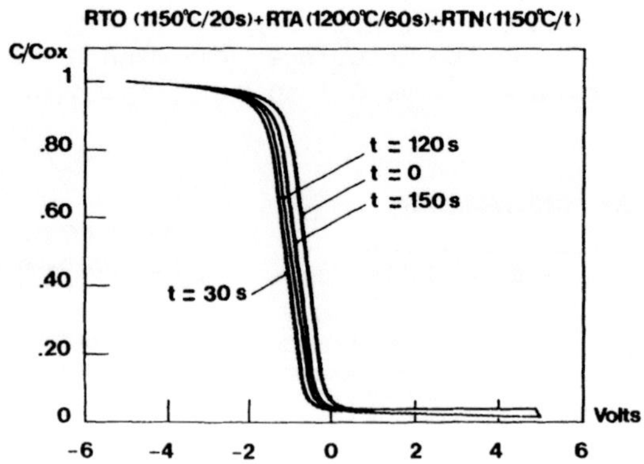

Fig. 6. High frequency C-V measurements for RTO 1150°C - 20 s + RTA 1200°C - 60 s + RTN 1150°C.

Fig. 7 shows a high resolution transmission electron microscopy (HREM) of a nitrided oxide. A high quality interface has been obtained. No contrast variation at the interface of the oxide and silicon is noted. This means that the structure at the interface is not silicon nitride but SiO_xN_y as shown by the previous Auger profiles.

CONCLUSION

Thin oxides of 30-150 Å are currently grown by rapid thermal processing with good homogeneity (σ = 1.9 Å) for a mean thickness value of

Fig. 7 High resolution TEM of an oxynitride for an RTO+RTA 1200°C-60s + RTN 1150°C-120s.

102 Å. A mean electrical breakdown field of 10.1 MV/cm has been obtained. Nitridation of SiO_2 shows an enhancement of the electrical breakdown field. A mean value of 14.6 MV/cm has been obtained for an RTN of 1150°C-150 s. RTA increases the mean electrical field both for the oxide and oxynitride, whilst causing a broadening of the histogram statistics. An increase of RTA temperature from 1100° to 1200°C does not cause any significant variation in the equivalent Si_3N_4 thickness at the surface nor at the interface. The interface trap density is $10^{10} cm^{-2}\ eV^{-1}$ both for the oxide and the oxynitrided.

REFERENCES

1. R.H. Dennard, F.H. Gaensslen, H.N. Yu, V.L. Rideout, E. Bassous and A.R. Le Blanc, IEEE J. Solid State Circuits 9:256 (1974).
2. S.S. Wong, C.G. Sodini, T.W. Ekstedt, H.R. Grinolds, K.H. Jackson and S.H. Kwan, J. Electrochem. Soc. 130:1139 (1983).
3. F.L. Terry Jr, R.J. Aucoin, M.L. Naiman and S.D. Senturia, IEEE Electron. Device Lett. EDL4:191 (1983).
4. T. Ito, T. Nakamura and H. Ishikawa, IEEE Trans. Electron Devices ED29:498 (1982).
5. F.H.P.M. Habrakev, A.E.T. Kuiper, Y. Tamminga and J.B. Theeten, J. Appl. Phys. 53:6996 (1982).
6. N. Chan Tung, J. Electrochem. Soc. 132:914 (1985).
7. N. Chan Tung, D.P. Vu and C. Le Pen, 4th Europ. Mat. res. Conf. Proc., Edited by V.T. Nguyen and A.G. Cullis, Editions de Phys., Les Ulis, (1985) 255.
8. N. Chan Tung and Y. Caratini, Electronics Lett. 22:694 (1986).
9. B. Deal and A. Grove, J. Appl. Phys. 36:3770 (1965).
10. H. Massoud, J. Plummer and E. Irene, J. Electrochem. Soc. 132:1745 (1985).

11. C.A. Paz de Araujo, Y.P. Huang, J.C. Gelpey and R. Kwor, to be published in Materials Research Society meeting, Anaheim, U.S.A. (April 1987).
12. S.A. Lassig, T. Debolske and J. Crowley, to be published in Materials Research Society meeting, Anaheim, U.S.A. (April 1987).
13. N. Chan Tung, Y. Caratini and J.L. Buevoz, to be published in Materials Research Society meeting, Anaheim, U.S.A. (April 1987).
14. A.C. Adams, T.E. Smith and C.C. Chang, J. Electrochem. Soc. 127:1787 (1980).
15. J. Nulman, Solid State Technol. 4:189 (1986).

EVIDENCE FOR OXYGEN BUBBLES IN FLUORINE DOPED AMORPHOUS SILICON DIOXIDE THIN FILMS

A.G. Dias**, E. Bustarret* and R.C. da Silva
**Centro de Fisica Molecular
Universidades de Lisboa
1000 Lisboa, Portugal
*LEPES, CNRS
B.P. 166, 38042 Grenoble Cédex
France
Laboratorio Nacional de Engenharia
e Tecnologia Industrial
E.N. 10, 2685 Sacavém
Portugal

ABSTRACT

The superstoichiometry (x from 2.0 to 2.5) observed by R.B.S. in P.E.C.V.D. fluorine doped silicon dioxide thin films deposited at 300°C was correlated with the density of bonded oxygen detected by IR absorption spectroscopy. These results were interpreted in terms of structurally inhomogeneous $Si\ O_x\ H_y\ F_z$ films consisting in an a-Si:O:H:F phase and oxygen bubbles accomodated by a SiO_2 tissue. This description is further supported by SEM, which shows sample blistering and bubble explosion upon annealing at 800°C.

INTRODUCTION

The lower refractive index of fluorine doped amorphous silicon dioxide, as compared with conventional silicate glasses, has aroused much interest in fluorine doping of vitreous silica for optical communications [1,2]. Further, as low deposition temperatures of thin amorphous suboxydes of silicon were developed, the study of their oxygen bonding has received increased attention[3,4]. This communication deals with structural and compositional studies of fluorine doped silicon dioxide thin films deposited at low temperatures by plasma enhanced chemical vapour deposition (P.E.C.V.D.). We propose a three phase structural model for these materials involving a Si:O:H:F region and oxygen bubbles accomodated by a silicon dioxide tissue.

EXPERIMENTAL DETAILS

The fluorine doped amorphous silicon dioxide thin films were deposited at substrate temperatures of 300°C in $SiH_4/F_2/Ar$ mixtures, with increasing silane-fluorine gas ratios ($r=F_2/SiH_4$) using a capacitively coupled apparatus described elsewhere[5]. The relevant deposition parameters are listed in Table I. The high oxygen content of the resulting $Si\,O_x\,H_y\,F_z$ films, as derived from infrared absorption measurements (I.R.) and Rutherford backscattering spectroscopy (R.B.S.) was traced back to the fluorine container and to the etching of the Pyrex wall of the reaction chamber by impinging fluorine active species[6].

TABLE I - Deposition conditions

Total flowrate :	20 s.c.c.m.	substrate temperature :	300° C
Total pressure :	3.8 torr	base pressure :	10^{-5} torr
r.f. frequency :	12.9 M Hz	power density :	0.35 W/cm^3
$r=F_2/SiH_4$:	0 to 0.8	SiH_4/Ar :	1:10

The R.B.S. spectra were obtained for samples deposited onto graphite and crystalline silicon substrates. The energy of the helium analyzing beam was 1.6 MeV, the backscattered particles being detected at 140° with a solid state detector with resolution of 15 KeV. The refractive index of films deposited during the same run onto 7059 Corning glass substrates, were derived from the interference fringes of their visible absorption spectra, recorded using a Cary 17E visible-UV spectrophotometer. I.R. absorption spectra of samples deposited onto crystalline silicon substrates, were recorded using a Perkin Elmer 683 double beam spectrophotometer in the transmittance mode between 4000 cm^{-1} and 400 cm^{-1}. The absorption coefficients were calculated from the transmittance data with film thicknesses derived from I.R. interference patterns. Scanning electron microscopy (S.E.M.) observations were performed with a Jeol JSM-35CF apparatus.

RESULTS

Films produced without the presence of fluorine in the gaseous mixture (r=0), in conditions that should yield microcrystalline silicon films, were oxygen free[6,7]. As previously reported by Pai et al.[8] only hundreds of p.p.m. of oxygen in the gas phase of a silane glow discharge are needed to produce silicon bonded oxygen in the percent range. So, taking into account the strong O_2^+ signal detected in a plasma etching helium-fluorine radio-frequency (r. f.) discharge by Vasile[9] we think that our particular set of deposition conditions (namely the high r. f. power density ~0.35 W/cm^3) favours strong oxygen incorporation observed in films produced in the presence of fluorine[10].

The experimental values of refractive index (n_{exp}), sample composition and densities of (Si-O), (Si-H) and (O-H) bonds, derived respectively from visible absorption measurements, R.B.S. and I.R. spectroscopy for various Si:O:H:F films are summarized in Table II.

The density of (Si-O) bonds was calculated using the expression :

$$(Si\text{-}O) = 4.2 \times 10^{18} \alpha_{max} \qquad (1)$$

where α_{max} is the absorption coefficient maximum in the frequency range of the Si-O-Si stretching modes [11]. The (Si-H) bond density is calculated

TABLE II - Refractive indices, sample composition and bond densities (experimental values)

Sample r	n_{exp}	Si_x x	O_x	H_y z	F_z	x10^{21} (Si-O)	bonds (Si-H)	cm^{-3} (O-H)
0.1	1.88	2.1		- . -		1.6	0.4	- . -
0.2	1.81	1.9		0.06		4.5	1.4	- . -
0.3	1.73	2.3		0.30		6.8	2.0	- . -
0.4	1.69	2.5		0.25		10.5	2.6	4.4
0.5	1.78	2.2		0.30		17.1	4.8	5.1
0.6	1.79	2.2		0.26		19.6	5.6	5.7
0.7	2.08	2.0		0.30		24.3	7.3	6.4
0.8	1.93	2.1		0.16		46.9	8.4	7.1

assuming :

$$(\text{Si-H}) = 16 \times 10^{19} \int \frac{\alpha(\omega)}{\omega}\, d\omega \tag{2}$$

where $\alpha(\omega)$ is the absorption coefficient of the film at wave number ω in the region of the Si-H wagging mode[12]. To calculate the density of (O-H) bonds we derive the following expression :

$$(\text{O-H}) = 2.3 \times 10^{19}\, \alpha_{3650} \tag{3}$$

where α_{3650} is the absorption coefficient maximum ascribed to the O-H stretching mode from silenol groups, located around 3650 cm^{-1}. This expression was derived from the corresponding mass percentage determination of Pliskin[13], using the atomic weights of the species involved and assuming for the atomic and mass density of silica respectively, 7×10^{22} atoms/cm^3 and 2.2 g/cm^3.

FORMULATION

The superstoichiometric nature of the great majority of our Si:O:H:F compounds (see Table II), when correlated with previous I.R. data[6] suggested an interpretation of their structure as inhomogeneous, in respect to the presence of oxygen molecules. Assuming that, in a continuous random network, each atom satisfies its valency with simple bonds (with the exception of molecular oxygen), we have deduced the following relations between the atomic concentration (denoted by $|\ |$) and the corresponding bond densities :

$$\begin{aligned} 4\,|Si| &= 2\,(\text{Si-Si}) + (\text{Si-O}) + (\text{Si-H}) + (\text{Si-F}) \\ 2\,|O| &= (\text{Si-O}) + (\text{O-H}) + 4\,(O_2) \\ |H| &= (\text{Si-H}) + (\text{O-H}) \\ |F| &= (\text{Si-F}) \end{aligned} \tag{4}$$

where (O_2) represents the density of molecular oxygen. Due to the low hydrogen and fluorine content of the Si:O:H:F compounds the presence of hydrogen, fluorine or hydrogen fluoride molecules was not considered, nor were oxygen-fluorine bonds, whose vibrational modes were not detected in the I.R. spectra. The (Si-O), (Si-H) and (O-H) bond densities were

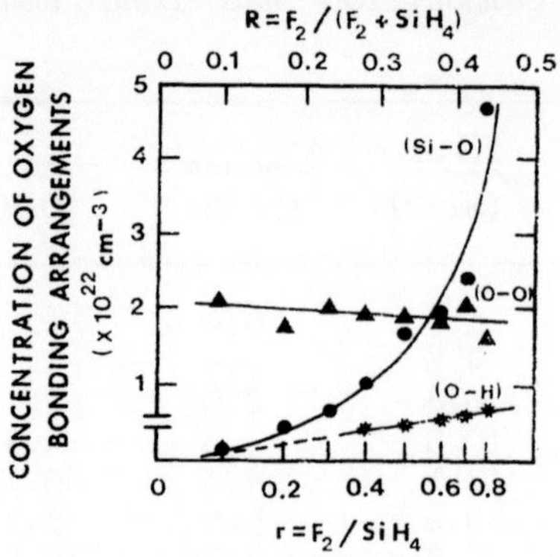

Fig. 1 Concentration of various oxygen bonding arrangements in the Si O_x H_y F_z films as a function of fluorine-silane gas ratio (r).

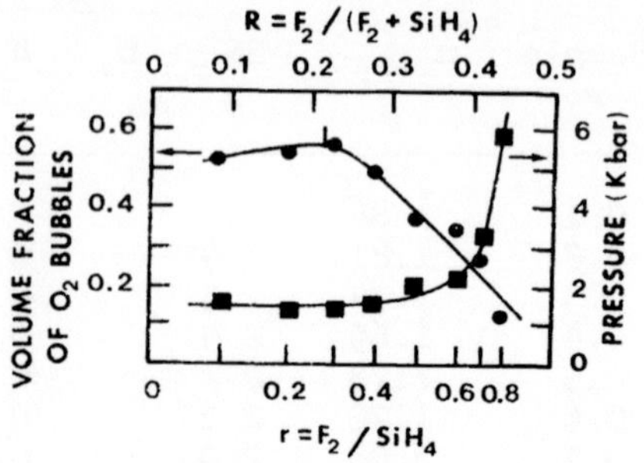

Fig. 2 Volume fraction of oxygen bubbles (f) and pressure as a function of the fluorine -silane gas ratio (r).

calculated as described above, their concentration is shown in Table II. Fig. 1 displays the dependence of the different oxygen bonding arrangement densities in the Si O_x H_y F_z films as a function of the fluorine-silane gas ratio.

The experimental refractive index values were fitted using a bonding density weighted linear combination of the indices associated with each phase[14] :

$$n_{exp} = 1 + \frac{n_{Si} - 1}{2|Si|_{a-Si}} (Si\text{-}Si) + \frac{n_{SiO_2} - 1}{2|O|_{SiO_2}} (Si\text{-}O) + \frac{n_W - 1}{2|O|_W} (O\text{-}H) \qquad (5)$$

where $n_{Si}=2.9$, $n_{SiO_2}=1.44$ and $n_w=1.33$ are respectively the refractive indexes of a hydrogenated amorphous silicon reference sample (a-Si:H), thermal silica (th-SiO_2) and water. The corresponding values of the atomic densities of silicon in a-Si:H, oxygen in th-SiO_2 and in water are respectively $|Si|_{a-Si} = 4.5\times10^{22}/cm^3$; $|O|_{SiO_2} = 4.6\times10^{22}/cm^3$ and $|O|_W = 3.3\times10^{22}/cm^3$. The atomic concentration of silicon, hydrogen, fluorine and oxygen and the (Si-Si) bond density, calculated from expressions (4) and (5) for samples under investigation are shown in Table III.

The volume fraction of oxygen bubbles (f) was calculated from these results, using the expression :

$$f = 1 - \frac{(Si\text{-}Si)}{2|Si|_{a-Si}} - \frac{(Si\text{-}O)}{2|O|_{SiO_2}} - \frac{(O\text{-}H)}{2|O|_w} \qquad (6)$$

their pressure (p) was estimated using the ideal gas equation. The dependences of both p and f on the fluorine to silane gas ratio (r) are shown in fig. 2.

TABLE III - Atomic concentration and density of (Si-Si) bonds (calculated values).

Sample r	$\times 10^{21}$ |Si|	atoms |O|	cm^{-3} |H|	|F|	$\times 10^{21}$ bonds cm^{-3} (Si-Si)
0.1	20.2	42.4	0.4	-	39.4
0.2	19.6	37.3	1.4	1.2	35.6
0.3	9.4	44.6	2.0	5.8	31.5
0.4	18.3	45.8	7.0	4.6	27.8
0.5	22.2	48.8	9.9	6.7	30.2
0.6	22.8	50.2	11.3	5.9	29.9
0.7	28.7	57.4	13.7	9.4	37.3
0.8	28.4	59.6	15.5	4.8	26.8

DISCUSSION

Our results, when correlated with previous I.R. data[6] , can be explained in terms of structuraly inhomogeneous $Si\,O_x\,H_y\,F_z$ films, consisting of a silicon rich Si:O:H:F amorphous phase and oxygen bubbles enveloped in a SiO_2 tissue.

From fig. 1 we concluded that the increasing values of the (Si-O) bond density with r are associated with a decrease in the concentration of (O-O) bonds, being the oxygen atoms mainly incorporated in the growing SiO_2 phase. As previously reported[14] the observed increase of (O-H) groups is also associated with the formation of this SiO_2 tissue. As seen in fig. 2 the decrease in the volume fraction of oxygen bubbles, observed in films produced with $r>0.4$, is related with the growth of the SiO_2 phase leading to an increase in their pressure.The corresponding increase in values of the (Si-Si) bond density is related to a segregation of the oxygen, fluorine and hydrogen atoms taking place in the Si:O:H:F phase, promoting the formation of silicon clusters, as detected in the Raman spectra of these samples[7].

Further support for the proposed structural description comes from S.E.M. observations of the surface of films deposited onto crystalline silicon substrates then high vaccum annealed at 800°C. Fig. 3a. shows a S.E.M. micrograph of an annealed film, produced with r = 0.4, with bubbles in the process of formation and exploded. The medium bubble diameter is about 60 μm. The inner surface of the exploded central bubble is shown in fig. 3-b) where the bubbles (diameter ~ 1 μm), which originated the violent mechanical destruction of the film can also be seen.

CONCLUSION

We have produced P.E.C.V.D. $Si\,O_x\,H_y\,F_z$ films, deposited at 300°C. A three phase model is proposed for these structurally inhomogeneous films, consisting of a Si:O:H:F zone and oxygen bubbles enveloped in a SiO_2 film.

Depending on the fluorine-silane gas ratio, oxygen is preferentially incorporated in the Si:O:H:F phase and in molecular form ($r<0.4$), or as a reduced volume fraction of oxygen bubbles leading to a more extended SiO_2 phase ($r>0.4$).

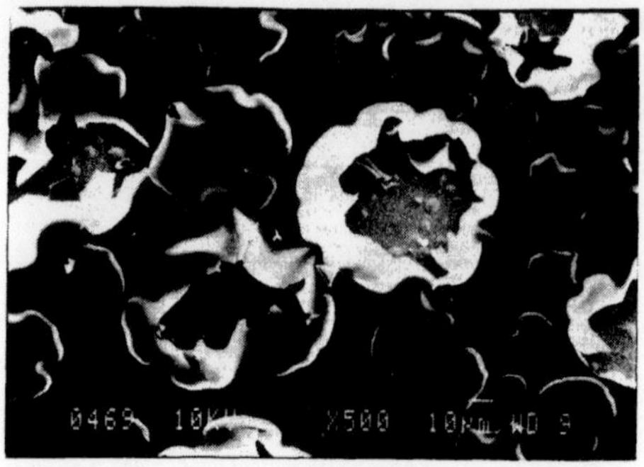

Fig. 3a

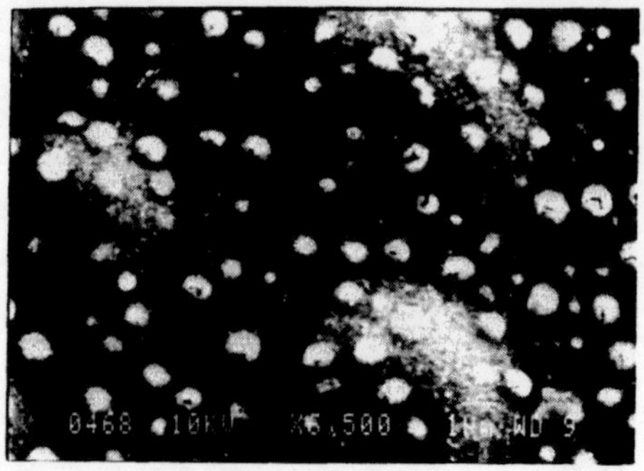

Fig. 3b

Fig. 3. S.E.M. micrographs of blistering and exploded bubbles in 800°C annealed $SiO_xH_yF_z$ samples produced with r = 0.4 ; b) Inner surface of a bubble.

ACKNOWLEDGEMENTS

We are indebted to Prof. Marat Mendes from "Universidade Nova de Lisboa", Prof. Rom ão Dias from "Centro de Quimica Estrutural", and Prof. Amaral Fortes from the "Universidade Técnica de Lisboa" respectively for the use of the visible spectrophotometer, I.R. spectrophotometer and S.E.M. facilities. Fruitful discussions with Dr. Yves Cros from the "Laboratoire d'Etudes des Propriétés Electroniques des Solides" are also acknowledged. This work was made possible by the attribution of scholarships to A.G. Dias and E. Bustarret by the French M.R.E. and Embassy in Portugal.

REFERENCES

1. E.M. Rabinovch, Phys. and Chem. of Glass. 24:54 (1983).
2. P. Dumas, J. Corset, W. Carvalho, Y. Levy and Y. Neuman, J. Non-Cryst. Sol. 47:239 (1982).
3. G. Luckovsky, P.D. Richard, D.V. Tsu, S.Y. Lin and R.J. Markunas, J. Vac. Sc. Techn. A4:681 (1986).
4. S.S. Chao, G. Lucovsky, D.V. Tsu, S.Y. Lin, P.D. Richard, Y. Takagi, P. Pai, J.E. Keem and J.E. Taylor, J. Non-Cryst. Sol. 77/78:929 (1985).
5. R. Martins, A.G. Dias and L. Guimarães. J. Non-Cryst. Sol. 57:9 (1983).
6. A.G. Dias, Proc. "Colloque Int. sur la Science des Matériaux pour l'énergie", Trieste (1986).
7. A.G. Dias and E. Bustarret, Proc. "2º Simposio Ibérico de Fisica da Matéria Condensada", Sevilla (1986).
8. P.G. Pai, S.S. Chao, Y. Takagi and G. Lucovsky, J. Vac. Sc. Techn. A4:689 (1986).
9. M.J. Vasile, J. Appl. Phys. 51:2503 (1980).
10. R.C. da Silva, A.G. Dias, L. Guimarães and M.F. da Silva, Proc., "2º. Simposio Ibérico de Fisica da Matéria Condensada", Sevilla (1986).
11. Y. Cros and J.C. Rostaing (private communication).
12. H. Shanks, C.J. Fang., L. Ley, M. Cardona, F.J. Demond and S. Kalbitzer, Phys. Status Sol. (b) 100:43 (1980).
13. W.A. Pliskin, J. Vac. Sc. Techn. 14:1064 (1977).
14. Y. Cros and J.C. Rostaing, Proc. "MRS - Symposium D", Les Editions de Physique, Les Ulis (1986) 77.

LOW TEMPERATURE P.E.C.V.D. SILICON RICH SILICON DIOXIDE FILMS DOPED WITH FLUORINE

A.G. Dias*, L. Guimarães* and M. Brunel**

*Centro de Fisica Molecular
Universidades de Lisboa
1000 Lisboa, Portugal
**Laboratoire de Cristallographie
CNRS
B.P. 166, 38042 Grenoble Cedex
France

ABSTRACT

The production of new silicon rich silicon dioxide films doped with fluorine, deposited by PECVD at low substrate temperatures is reported. Films deposited with substrate temperatures lower than 200°C exhibit a multiphase structure consisting of Si microcrystallites (60 Å-100 Å) embedded in an amorphous insulating tissue, whereas in films produced with substrate temperatures between 200°C and 300°C this microcrystalline phase is replaced by an a-Si:O:H:F zone and oxygen bubbles.

INTRODUCTION

Structural studies of the commonly called "semi-insulating polycrystalline silicon films" (S.I.P.O.S.) have received considerable attention. Films produced at temperatures ~ 700°C were composed of a two phase silicon and silicon oxide (SiO_x) network, the formation of crystallites ~ 100 Å being promoted after high temperatures annealing around 1000°C[1,2]. A new plasma processing technique whereby fluorine is used to enhance the plasma oxidation and nitridation of silicon at substrate temperatures below 600°C has been described by Chang et al.[3] Recently Morita et al.[4] have also used gases containing fluorine to enhance the thermal oxidation of silicon.

In this communication we report the results of a study of the production of a new silicon rich silicon dioxide fluorine doped film deposited by plasma enhanced chemical vapour deposition (P.E.C.V.D.) at substrate temperatures lower than 200°C.

EXPERIMENT

The silicon rich silicon dioxide films doped with fluorine ($SiO_xH_yF_z$) were deposited at low substrate temperatures (T_s) in a PECVD apparatus described elsewhere[5]. Our capacitively coupled deposition system enables us to vary the silane (SiH_4) to fluorine (F_2) gas ratio ($r = F_2/SiH_4$) controlling the F_2 and the Argon diluted SiH_4 flowrates. The main deposition parameters are listed in Table I.

Samples produced without the presence of fluorine in the gaseous mixture (r=0) in conditions that yield microcrystalline silicon films, were oxygen free [6,7]. The presence of oxygen in the reaction mixture and its subsequent incorporation in the films produced with r=0 is discussed in ref.8.

X-ray diffraction was performed using CuK_α radiation at nearly grazing incidence (~0.3°). I.R. absorption spectra of samples deposited onto crystalline silicon substrates were recorded in the transmittance mode using a Perkin-Elmer 683 double beam spectrophotometer. Dark conductivity measurements as a function of reciprocal temperature were performed on films deposited during the same run onto glass substrates using a coplanar configuration.

TABLE I - Deposition conditions

Total flowrate	: 20 s.c.c.m.	substrate temp.	: 100°C-300°C
Total pressure	: 3.8 torr	base pressure	: 10^{-5} torr
r.f. frequency	: 12.9 M Hz	power density	: 0.35 W/cm^3
$r=F_2/SiH_4$	: 0.3 ; 0.6	SiH_4/Ar	: 1:10

RESULTS AND DISCUSSION

A series of $SiO_xH_yF_z$ films were produced respectively with r=0.3 and r=0.6 varying the deposition temperature between 100°C and 300°C.

Typical examples of X-ray diffraction patterns of $SiO_xH_yF_z$ films are shown in fig. 1 for two samples deposited respectively at 100°C (r=0.3) and 300°C (r=0.6). The sample deposited at 100°C is a mixed phase amorphous-microcrystalline film with mean coherence length D ~ 90 Å. The amorphous nature of the 300°C sample can be distinguished by the high FWHM of the first maximum (which coincides with the (111) Bragg reflection of crystalline silicon) and the broad diffraction pattern in between the (200) and the (311) diffraction peaks[9]. In a general fashion the X-ray diffraction patterns of samples deposited at substrate temperatures below 200°C displayed microcrystalline patterns (with D between 60 Å and 100 Å), while samples deposited with T_s>200°C are amorphous, irrespective of the fluorine to silane gas ratio.

The corresponding I.R. spectra of samples deposited respectively with T_S=300°C (r=0.8) and T_S=100°C (r=0.3) are shown in fig. 2. The main features of these spectra are the strong absorption band in the frequency range 1200 - 1000 cm^{-1} ascribed of the Si-O-Si stretching modes from SiO_x complexes[10]. The amorphous nature to the T_S=300°C $SiO_xH_yF_z$ film and the observed shift in the frequency of the Si-O-Si stretching mode to

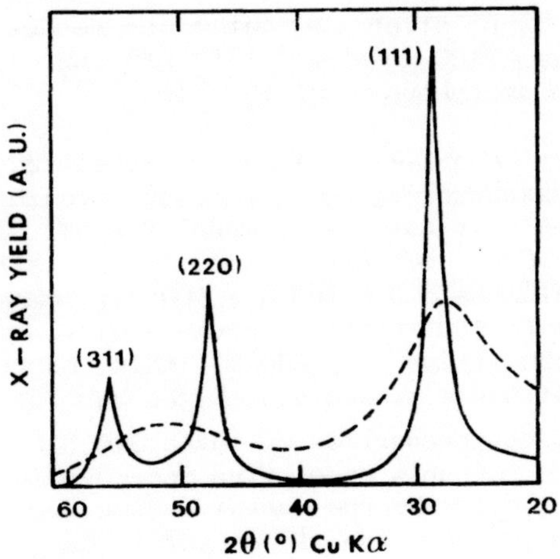

Fig. 1. X-ray diffraction patterns of $SiO_xH_yF_z$ films deposited respectively at 100°C (——) and 300°C (---).

frequencies higher than 1075 cm^{-1} [11] (in this case 1090 cm^{-1}) is indicative of the presence of a more electronegative atom (F) within the SiO_2 matrix[12]. The band located around 820 cm^{-1} in the I.R. spectra of these samples, attributed to a Si-F bond in an amorphous silicon matrix[13], suggests the incorporation of fluorine both in the insulating silicon dioxide phase, leading to the formation of SiO_3F complexes, and in the silicon rich semiconductiong a-Si:O:H:F tissue, competing with hydrogen as a dangling bond passivator. The departure from SiO_2 stoichiometry of the T_S =100°C sample may be evidenced by the growing high frequency shoulder of the Si-O-Si stretching mode[11] related to the presence of a silicon rich tissue in the low temperature samples.

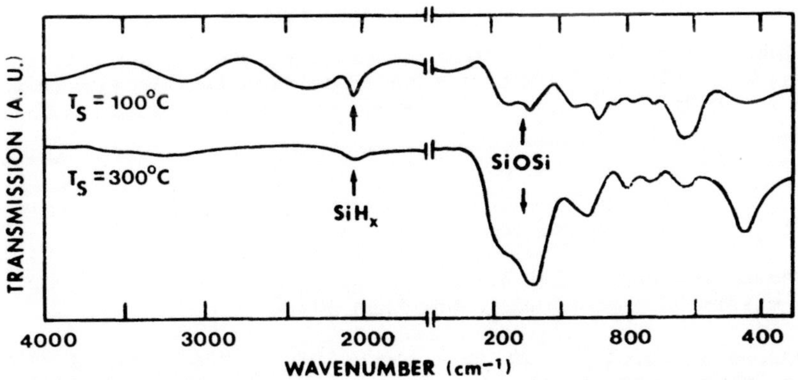

Fig. 2. I.R. absorption spectra of $SiO_xH_yF_z$ films deposited respectively at 100°C and 300°C.

The characteristic Si-H stretching vibration modes were also detected in the frequency range 2100-2000 cm^{-1} [14] of the I.R. spectra of both microcrystalline and amorphous $SiO_xH_yF_z$ films.

The above X-ray and I.R. data when correlated with previous Raman[15] and Rutherford backscattering spectroscopy studies[6], can be coherently interpreted in terms of a multi-zone model for our $SiO_xH_yF_z$ films. Samples deposited at $T_S < 200°C$ exhibit a multiphase structure consisting of silicon microcrystallites, $D \in |60$ Å ; 100 Å$|$, embedded in an amorphous Si:O:H:F insulating tissue (silicon rich SiO_x fluorine doped tissue). In the high temperature films ($T_S \geq 200°C$) this microcrystalline phase is replaced by an a-Si:O:H:F tissue and oxygen bubbles accomodated by a SiO_2 tissue[8]. The transport properties of these $SiO_xH_yF_z$ films was interpreted within the framework of this model. The decrease in the dark conductivity at room temperature (σ_{RT}) and the corresponding opposite dependence of the activation energy (ε_σ) upon the microcrystalline to amorphous transition (T_S~180°C) is indicative of a change in conduction mechanism detected in all samples under investigation (see fig. 3). Low substrate temperature samples having conductivity dominated by the less resistive c-Si phase display σ_{RT} and ε_σ values according to the predictions of classical models for polycrystalline semiconductors (σ_{RT} ~10^{-5} Ω^{-1} cm^{-1} and ε_σ~ 0.3 eV), while in the high substrate temperature samples these values are typical of a conductivity path ascribed to the amorphous silicon zone (σ_{RT} ~ 10^{-10} Ω^{-1} cm^{-1} and ε_σ~ 0.8 eV).

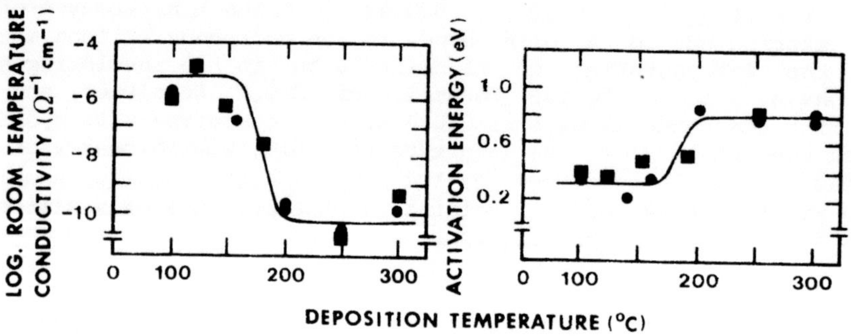

Fig. 3. Dark conductivity at room temperature and activation energy for various $SiO_x H_yF_z$ films versus deposition temperature (■ r = 0.6 ; • r = 0.3).

CONCLUSION

We have produced PECVD silicon rich silicon dioxide fluorine doped films at substrate temperatures lower than 300°C.

Based on X-ray, I.R. and dark conductivity measurements we derive a multi-zone model for these $SiO_xH_yF_z$ films consisting of silicon microscrystallites embedded in an amorphous insulating tissue which is replaced by an a-Si:O:H:F zone and oxygen bubbles accomodated by a SiO_2 tissue in films deposited at substrate temperatures higher than 200°C.

Fluorine is incorporated in the insulating silicon oxide tissue, mainly in SiO_3F complexes, and in the semiconducting amorphous phase.

ACKNOWLEDGEMENTS

We are indebted to Prof. Romão Dias (Centro de Quimica Estrutural da Universidade Técnica de Lisboa) for the I.R. spectrophotometer. Fruitful discussions with Drs. Y. Cros, J.C. Rostaing and D. Jousse from the "Laboratoire d'Etudes des Propriétés Electroniques de Solides-CNRS "are also acknowledged. This work was made possible by the attribution of a scholarship to A.G. Dias by the French M.R.E. and Embassy in Portugal.

REFERENCES

1. A. Harstein, J.C. Tsang, D.J. DiMaria and D.W. Dong, Appl. Phys. Lett. 36:836 (1980).
2. M. Hamasaki, T. Adachi, S. Wakayama and M. Kikuchi, J. Appl. Phys. 49:3987 (1978).
3. R.P.H. Chang, C.C. Chang and S. Darack, Appl. Phys. Lett. 36:999 (1980).
4. M. Morita, S. Aritome, M. Tsukuda, T. Murakawa and M. Hirose, Appl. Phys. Lett. 47:253 (1985).
5. R. Martins, A.G. Dias and L. Guimarães, J. Non-Cryst. Sol. 57:9 (1983).
6. R.C. da Silva, A.G. Dias, L. Guimarães and M.F. da Silva, Proc. "2° Simposio Ibérico de Fisica da Matéria Condensada" Sevilla (1986).
7. A.G. Dias, Proc. "Colloque Int. sur la Science des Matériaux pour l'Energie", Trieste (1986).
8. A.G. Dias, E. Bustarret and R.C. Silva, these proceedings.
9. S. Veprek, Z. Iqbal and F.A. Sarott, Phil. Mag. B, 45:137 (1982).
10. S.S. Chao, G. Lucovsky, D.V. Tsu, S.Y. Lin, P.D. Richard, Y. Takagi, P. Pai, J.E. Keem and J.E. Taylor, J. Non-Cryst. Sol. 77/78:929 (1985).
11. Y. Cros and J.C. Rostaing, Proc. "MRS - Symposium D", Les Editions de Physique, Les Ulis (1986) p. 77.
12. G. Lucovsky, Solid State Comm. 29:571 (1979).
13. C.J. Fang, L. Ley, H.R. Shanks, K.J. Gruntz and M. Cardona, Phys. Rev. B 22:6140 (1980).
14. G. Lucovsky, J. Yang, S.S. Chao, J.E. Tyler and W. Czubatyj, Phys. Rev. B 28:3225 (1983).
15. A.G. Dias and E. Bustarret, Proc. "2° Simposio Ibérico de Fisica da Matéria Condensada" Sevilla (1986).

ACKNOWLEDGMENTS

We are grateful to Prof. [illegible] Centre de Grenoble [illegible] [illegible] [illegible]. [illegible] [illegible]

REFERENCES

1. [illegible] (1980).
2. [illegible], S. [illegible] and M. [illegible], J. Appl. Phys. [illegible]
3. [illegible] Phys. Lett. [illegible] (1980).
4. [illegible] Appl. Phys. Lett. [illegible] (1981).
5. [illegible] (1981).
6. [illegible] (1981).
7. [illegible] "Colloque [illegible] sur la [illegible]", [illegible] (1980).
8. [illegible], private communication.
9. [illegible] Phys. Rev. B [illegible] (1982).
10. [illegible] (1981).
11. [illegible] 1979 [illegible]
12. [illegible] (1979).
13. [illegible]

HIGH FIELD TRANSPORT IN SiO_2

D.K. Ferry

Center for Solid State Electronics Research
Arizone State University
Tempe, AZ 85287 USA

INTRODUCTION

For a great many years, the transport of electrons in amorphous silicon dioxide has been of interest to the electronics community. This interest stems from the importance of this material as an insulator for microelectronics. Primarily, silicon dioxide has been studied for its structure and conductivity, as well as breakdown strength, due to the high voltages applied across the thin layers found in gate oxides of MOS devices. For VLSI, and the new ULSI, fields in the insulator are expected to be on the order of a few MV/cm, and thus relatively close to electrical breakdown.

From the early studies of the transport through silicon dioxide, it was thought that the scattering of electrons was dominated by the polar optical phonon interaction[1-3]. Through the emission of optical phonons, electrons can lose the energy which they gain in the electric field, and it was felt that this interaction was sufficiently strong to prevent the electrons from suffering polar runaway. This would imply that the electron distribution was stabilized at an average energy below the optical phonon energy (0.15 eV). It is now known that this is not the case. From both[4,5] experimental and theoretical studies[6] it is now known that the polar phonons are unable to stabilize the distribution above 1.5-2.0 MV/cm. This field is well below that at which silicon dioxide is found to break down ; so that we are left with the question as to what then determines the stabilization of the electron distribution function to avoid breakdown ?

Fischetti [7] suggested that Umklapp processes involving the acoustic phonons provided an additional scattering mechanism, and that this mechanism was sufficiently strong to stabilize the distribution, with an average energy in the 2-5 eV range found experimentally. Porod and Ferry[8], on the other band, introduced scattering to satellite valleys of the conduction band, which in turn stabilized the distribution and could also fit the experimental data. The key factor in each of these approaches, studied by ensemble Monte Carlo calculations was the introduction of an additional large density of states, which causes the large scattering rates necessary to stabilize the distribution.

In this paper, the current understanding of the transport and scattering processes will be reviewed. In particular, the role of Umklapp processes and intervalley processes will be discussed in the next two sections, which deal with the general problems of the electron-phonon interaction. Impact ionization is treated in the following section. Finally, a short discussion of the modifications expected with quantization will be presented.

THE ELECTRON-PHONON INTERACTION

The general features of the electronic structure for various crystalline modifications of silicon dioxide are actually quite similar despite their different crystal structures. This tends to point out the absolute importance of the short range order in determining the essential features of the electronic structure. We can actually carry this forward to the amorphous phase as well. In fact, one of the interesting aspects of silicon dioxide is that the mobility of the electrons is quite large when compared to the expected values for insulators, and is actually much closer to that expected for a low mobility, wide band-gap semiconductor. Indeed, it now appears that the mean-free path for the electrons is of the order, or smaller than the grain size, or ordering length of the material. As a consequence we really expect the details of the electron-phonon interaction to be described by the general lattice theory of crystalline solids (but modified in detail appropriate to the lack of long-range order in the amorphous phase).

Earlier attempts to treat the scattering by LO phonons in silicon dioxide led to relatively good agreement between low field mobilities found experimentally and theoretically. The major scattering processes, at least at low electric field, were previously shown to be the LO modes of the lattice[1]. There are two principle modes that interact, at energies of 0.06 and 0.153 eV, and these split the total polar coupling given by the difference between the optical and static dielectric constants. However, there are many other scattering mechanisms that could play a role, especially at high electric fields. To this end, it is probably well worth while to review briefly the electron-phonon interaction generally.

The relatively high mobility of most semiconductors, and even to a major extent in a material such as silicon dioxide, indicates that the electron-phonon interaction can be treated in perturbation theory. The basis of this simplification is the adiabatic approximation in which the motion of the electrons and the lattice are separated. The interaction term may then be expanded into the various interaction terms[9]. In a primitive sense, we can write the interaction orders as :

$$M = A/q + Bq^0 + Cq + \dots, \tag{1}$$

i.e. as a series in q. In purely homopolar materials, such as Si and Ge, the leading term has A=0, since this term depends upon the dipole fields existing within the unit cell. Thus, only those crystals lacking an inversion center have this term present. Silicon dioxide certainly fits this mould. This first term gives rise to the longitudinal polar optical phonon scattering for the optical modes and to piezoelectric scattering for the acoustic modes. The former is the dominant scattering process usually in material in which it can occur, at least over a portion of the energy range spanned by the electrons. The latter is usually only found at lower temperatures, but can be significant as a scattering mechanism.

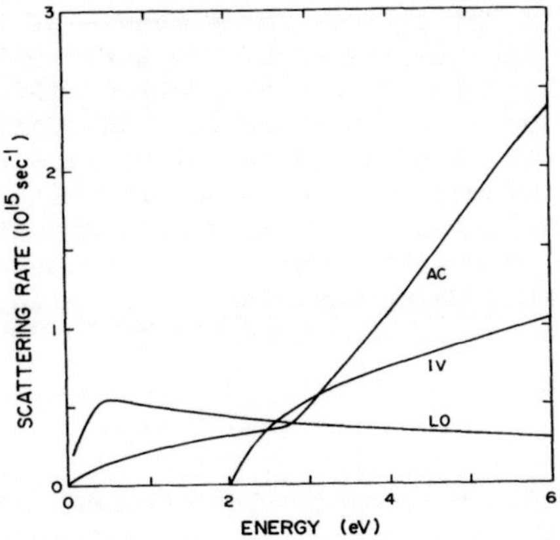

Fig. 1. Scattering rates used in the present study

The zero-order term always vanishes for the acoustic modes, as this amounts to a uniform displacement of the entire crystal. In the optical modes, however, it leads to non-polar optical phonon scattering, where the phonons arise near the zone center. It is also responsible for intervalley scattering, in which the phonon is a large momentum phonon, and is usually an umklapp process. In many cases, these interactions and the appropriate phonon branches must be carefully screened for the appropriate symmetry.

The first-order term gives rise to the deformation potential interaction for the acoustic modes. Of interest here is also the fact that umklapp processes have been used to treat the acoustic interaction at high energies as well[6]. This term, and the optical branch interaction as well, has also been shown to be important for intervalley processes in which the zero-order interaction is symmetry forbidden[10], such as in Si itself. The higher order terms give rise to multi-pole scattering terms, but are not thought to contribute significantly to scattering in semiconductors.

The form of the matrix elements and consequently the scattering rates, has been worked out for most of the lower order scattering processes for quite some time. They can be found e.g. in Conwell and Ridley[12]. What are not so well known are the values for the various "coupling constants" required to give numerical values to the scattering rates. In fig. 1, the scattering rates for silicon dioxide are plotted. All of the various emission and absorption processes for the two dominant polar modes are gathered together for the curve marked LO. The coupling constant for each mode is determined by the known split of the dielectric function between the two dominant modes and is not further adjusted. The acoustic curve (AC) includes piezoelectric, deformation potential and the umklapp process. Piezoelectric scattering is relatively weak, being only about 10 % of the acoustic mode at low energies. The acoustic deformation potential here is 10 eV. Throughout the present work, the electron mass is taken to be 0.7 m at low energies and 1.0 m at high energies, in keeping with the recent estimates of Fischetti et al.[13]. The strength of the umklapp process was adjusted to a value in keeping with the scattering strength used by the latter authors, but the form of the interaction differs somewhat to account for the actual threshold in this process. In addition, scattering by the nonpolar optical phonon to a second set of valleys has been included (IV). The details of this interaction are in keeping with the earlier reports of its use[8]. These scattering rates give good agreement with the low field mobility reported by Hughes[14]. Yet, it is important to note that almost none of the parameters, which describe the coupling between the electrons and the lattice have actually been experimentally determined.

HIGH FIELD STABILITY

The transport though silicon dioxide has been studied by the use of an ensemble Monte Carlo technique in which the scattering mechanisms of fig. 1 are incorporated. Typically, the ensemble contains about 2000 electrons, which is a large enough number to give meaningful ensemble averages for the kinetic coefficients, such as average energy. The electrons are treated as if the structure is a free electron band structure, with the sole exception of the varying effective mass, as discussed above. Still, this band structure introduces considerable multiplicity in the bands at higher energy, due to the pseudo-periodicity of the crystalline structure.

The significant factor in fig. 1 is the presence of the two scattering mechanisms which lead to very high scattering rates at high electron energies. In the absence of these scattering processes, the electron distribution function becomes unstable above 1.5-2.0 MV/cm. By unstable, we mean that the average energy of the carrier ensemble increases very rapidly. It is the presence of the additional (and large) scattering processes which are responsible for stabilizing the distribution. We can see this in fig. 2, which displays the average energy found for the electrons as a function of the electric field. The gray area is the range of average energies determined experimentally[4,5,15,16]. The curve marked 1 neglects the intervalley processes. We note that the scattering rate for the umklapp processes (fig. 1) is actually slightly weaker than that used by Fischetti[7], and the curve tends to begin to become unstable above about 8 MV/cm. While this curve lies at the bottom of the range found experimentally (up to 8 MV/cm) weakening the umklapp process would cause the entire distribution to become unstable. The curves marked 2 and 2' neglect the umklapp process, but do include the intervalley process at various strengths. Curve 2 increases the scattering by 50 % to overcome for the loss of the umklapp process, but clearly stabilizes the energy at a value well below that found experimentally. On the other hand, curve 2' weakens the intervalley process over that shown in fig. 1 by this same factor. In the latter case, the distribution becomes unstable already at 3-4 MV/cm. Curve 3 includes all of the various scattering processes, with the strengths shown in fig. 1. This curve lies somewhat below the experimental data at intermediate fields. Judicious adjustment of the various parameters, such as increasing the onset of intervalley scattering from 2 to 3 eV, allows us to actually put the entire curve within the experimental area.

By judicious adjustment of the coupling constants and the threshold energies for the high energy scattering processes, one can obtain relatively good agreement with the range of experimental data on the average energy of the electron distribution. However, what does this mean in light of the fact, mentioned above, that we really don't know the proper values for most of the coupling constants used in the theory. Indeed, the message from fig. 2 is that it is primarily the extra density of states

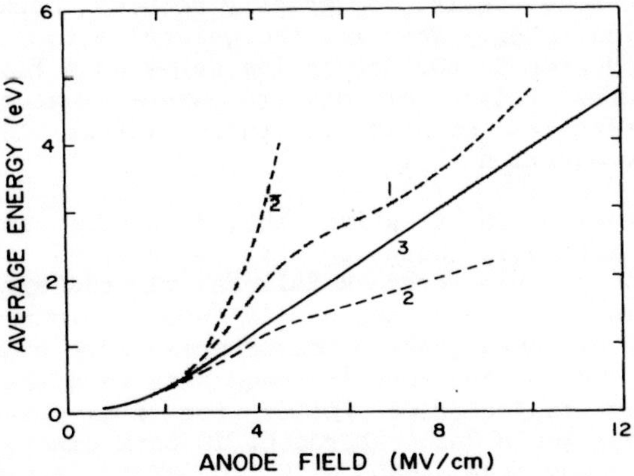

Fig. 2. The average energy calculated from the ensemble Monte Carlo model. The shaded area is the range of experimental curves, while the various curves are discussed in the text.

incorporated in the high energy scatterers, and not the details of the physics of these scatterers, that leads to the resulting stability of the electron distribution. What has been achieved is the essential proof that an additional scattering process, of relatively high scattering rate, can work to stabilize the distribution of carriers. Moreover, the average energy of the resulting distribution is not correlated to any particular parameter of the scattering process, in contrast to the result expected for polar scattering. On the other hand, this does not deter us from drawing some very important conclusions about breakdown from this mode, which will be discussed below.

QUANTUM EFFECTS

One needs, in principle, to consider the fact that the scattering indicated in fig. 1 is certainly not weak scattering, particularly at high electric fields, and high electron energies. The presence of such strong scattering means that we must replace the Fermi-golden-rule scattering rates shown in fig. 1 by those calculated using the proper self-energy calculated self-consistently. In essence, the energy is no longer a single-value given by the momentum, but these two quantities are related by a probability function, the joint spectral density function, such as the Lorentzian approximation

$$P(E) = \frac{1}{\pi}\left[\left(E - p^2/2m - \Sigma_r\right)^2 + \Sigma_i^2\right]^{-1}\Sigma_i , \tag{2}$$

where Σ is the proper self-energy (the imaginary part is the usual $h/\tau=\Gamma$). Equation (2) replaces the energy-conserving δ-function in the Fermi-golden rule, which in turn yields the self-energy Σ in the nearly free electron approximation. This expression is then solved self-consistently to give the new scattering rates, a procedure that has been carried out many times for semiconductors[17-20]. While most evaluate the expressions numerically, they can be solved in closed form in most cases. The largest changes are that sharp edges in the scattering rates, caused by the onset of new processes, are smoothed. In addition, the scattering rates are slightly enhanced in general, although strong peaks are greatly reduced back to the general background scattering rates. That is, the general effect is to reduce the effects of strong features in the scattering rates as a function of energy. These strong features relate closely to sharp features in the single particle density-of-states and it is these latter features which are strongly smoothed by equation (2).

The second problem in treating the transport in a fully quantum treatment is the transport equation. In principle, the motion of the carriers in the Monte Carlo process follows the classical trajectories. Fischetti and DiMaria[19] used a Feynmann path integral to replace the normal Monte Carlo process, although quantum trajectories were randomly generated. On the other hand, Porod and Ferry[20] used the fact that in homogeneous electric fields the trajectories follow the semi-classical paths and extended the classical Monte Carlo approach. In both cases, the results did not differ greatly from that found with the semi-classical approach, as depicted by fig. 2. Certainly, the values of the various coupling constants are not known well enough to be able to identify any significant changes in the transport.

A second quantum effect, whose role is totally unknown, is that of finite band-gap effects. Fischetti et al.[15] point out that some band structure calculations suggest that the lowest conduction band in silicon dioxide has a finite width of the order of 5.0-5.5 eV. If this were the case, and no other bands were present, there would be a sharp drop in the density of states at this energy level. If so, it would be quite difficult for electrons to be accelerated above this energy, which would provide an energy bottle-neck for the distribution. Indeed, if there were no scattering at all, the average energy would approach the half-band value, but the velocity would be zero[17]. While it is unlikely that such a minimum in the density-of-states exists, it is an intriguing possibility and one that would make most of the discussion about details of the scattering processes irrelevant.

IMPACT IONIZATION

The dielectric strength of amorphous silicon dioxide has an important bearing on the performance and reliability of MOS structures, and consequently has been of interest for quite some time. Dielectric breakdown has generally been attributed to thermal breakdown or velocity runaway, with subsequent impact ionization in the latter case. Carrier multiplication has been treated by many authors and was applied to the case of silicon dioxide by the present author[3]. There, it was felt that velocity runaway allowed the carriers to accelerate to a sufficient energy that they would undergo an ionizing collision. However, this earlier model did not fit accurately the transport of the carriers themselves, as is now well known. Moreover, it is now quite widely believed that no impact ionization is observed experimentally[15] for fields as high as 12 MV/cm.

To pursue this further, a model of the impact ionizing collision has been incorporated into the current ensemble Monte Carlo program. As Ridley [21] has pointed out, most early theories of impact ionization have tended to assume that once the threshold for ionization was reached by a carrier, impact ionization was a very probably process. This is termed a "hard" threshold, and was used in the earlier studies[3]. However, there is a different version, which holds that the ionization is a "soft" process with a collision probability of the same form as other collision mechanisms. This was shown to be the case in Monte Carlo studies of narrow-gap semiconductors by Curby and Ferry[22], and has recently been shown to be the case in moderate-gap semiconductors by Ridley[21].

The lowest energy which an electron needs to create an electron-hole pair is determined by the conservation of energy and momentum during the scattering process. These conservation conditions were studied for general energy bands extensively by Anderson and Crowell[23]. This provides a threshold energy, given by

$$E_i = E_G\ [(1 + 2\mu)/(1 + \mu)], \tag{3}$$

where μ is the ratio of the electron mass to the hole mass. Thus, we are somewhat stymied by the uncertainty in the value of the latter parameter. By assuming a Coulomb interaction between the incident, high-energy electron and the bound electron, one can the calculate the scattering cross-section and relaxation time. In the present study, we use the form calculated by Curby and Ferry[22] and by Ridley[21]. Once the scattering probability, as a function of energy, is known, this process can be incorporated into the Monte Carlo procedure. The number of ionizing collisions is monitored and the ionization rate α is calculated along with the other transport parameters. The ionization rate is plotted in fig. 3, for two different values of the hole mass. It is seen, from this figure, that α rises rather slowly with electric field, regardless of the mass. Moreover, the absolute value of α is not large.

Traditionally, one expects impact ionization to lead to avalanche multiplication and breakdown when $\alpha L=1$; that is, when the product of the ionization and the transport length becomes unity. At 12 MV/cm, and a hole mass of 5.0 m, α is only about 5000/cm, which requires an oxide thickness of 2 μm for breakdown by this process. Conversely, for an oxide thickness of 25 nm, we would require an ionization coefficient of 4 million per cm, which is more than two orders of magnitude larger than that found at this value of hole mass. It is more than an order of magnitude larger than observed with an infinite hole mass. These results seem to confirm the experimental conclusions that impact ionization by the electrons is not a significant factor in dielectric breakdown of silicon dioxide.

CONCLUSIONS

Progress in the understanding of electron transport in silicon dioxide over the past several years has led to the conclusion that the electron distribution cannot be stabilized by interaction with the longitudinal polar optical phonon modes above 1.5-2.0 MV/cm, which is well below the levels expected earlier. For fields above this level, the mean carrier energy is found experimentally to rise to a level of 2-4 eV and to be stable at this value up to a field of 10 MV/cm. The only way in which this can be understood on the basis of interactions between the electrons and the lattice is for the presence of a high energy scattering process to provide a very large density of final states. Both acoustic umklapp and intervalley processes have been suggested for this extra scattering mechanism. However, our working knowledge of the properties of silicon dioxide really prevents us from drawing any definitive conclusions about which of the processes, if either, is responsible for the stabilization of the electron distribution.

Breakdown of the silicon dioxide has also been a subject of much interest for many decades. Early studies suggested that avalanche breakdown, following impact ionization, was probably responsible, but this was not thought to be the case following recent experimental studies. In

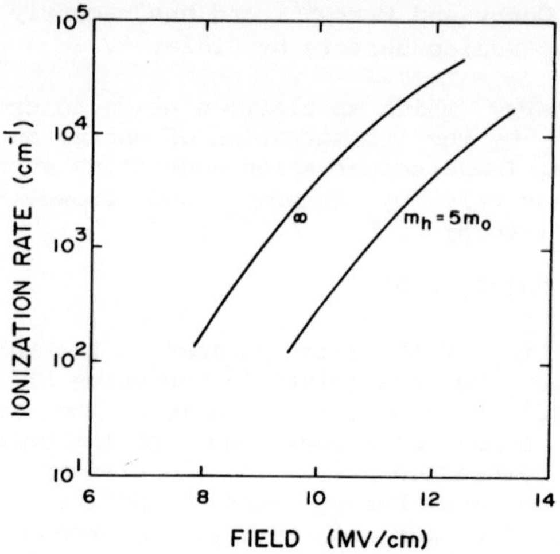

Fig. 3. The impact ionization coefficient for electron triggered ionization. The two curves are for different values of the hole mass.

this present effort, the impact ionization collision was included in the Monte Carlo program, in order to investigate such an effect. The calculated ionization coefficient is far too small to realistically lead to avalanche breakdown in thin dielectric layers. Thus, we may conclude that breakdown of silicon dioxide is not likely to be due to impact ionization by electrons, at least for fields up to about 15-20 MV/cm.

The author would like to express his appreciation to M. Fischetti (IBM) and W. Porod (Notre Dame) for many helpful discussions. This work was supported in part by SDI/IST through the Office of Naval Research.

REFERENCES

1. W.T. Lynch, J. Appl. Phys. 43:3274 (1972).
2. K.K. Thornber and R.P. Feynman, Phys. Rev. B 1:4099 (1970).
3. D.K. Ferry, J. Appl. Phys. 50:1422 (1979).
4. T.N. Theis, D.J. DiMaria, J.R. Kirtley, and D.W. Dong, Phys. Rev. Letts. 52:1445 (1984).
5. T.N. Theis, J.R. Kirtley, D.J. DiMaria and D.W. Dong, Phys. Rev. Letts. 50:750 (1983).
6. H.J. Fitting and J.U. Friemann, Phys. Stat. Sol. (a) 63:349 (1982).
7. M.V. Fischetti, Phys. Rev. Letts 53:1775 (1984).
8. W. Porod and D.K. Ferry, Phys. Rev. Letts. 54:1189 (1985).
9. P. Vogl, Phys. Rev. B 13:694 (1976).
10. D.K. Ferry, Phys. Rev. B 14:1605 (1976).
11. E.M. Conwell, "High Field Transport in Semiconductors", Academic Press, New York (1967).
12. B.K. Ridley, "Quantum Processes in Semiconductors", Clarendon Press, Oxford (1982).
13. M.V. Fischetti, D.J. DiMaria, L. Dori, J. Batey, E. Tierney and J. Stisiak, Phys. Rev. B 35:4404 (1987).
14. R.C. Hughes, Sol. State Electron, 21:251 (1978).

15. M.V. Fischetti, D.J. DiMaria, S.D. Brorson, T.N. Theis and J.R. Kirtley, Phys. Rev. B 31:8124 (1985).
16. D.J. DiMaria, M.V. Fischetti, J. Batey, L. Dori, E. Tierney and J. Stasiak, Phys. Rev. Letts. 55:2071 (1986).
17. R.K. Reich, R.O. Grondin, and D.K. Ferry, Phys. Rev. B 27:3483 (1983).
18. Y.C. Chang, Z.Y. Ting, J.Y. Tang and K. Hess, Appl. Phys. Letts. 42:76 (1983).
19. M.V. Fischetti and D.J. DiMaria, Phys. Rev. Letts. 55:2475 (1985).
20. W. Porod and D.K. Ferry, Physica 134B:137 (1985).
21. B.K. Ridley, Semicond. Sci. Technol. 2:116 (1987).
22. R.C. Curby and D.K. Ferry, Phys. Stat. Sol. (a) 15:319 (1973).
23. C.L. Anderson, C.R. Crowell, Phys. Rev. B 5:2267 (1972).

HOT ELECTRONS IN SiO_2 : BALLISTIC AND STEADY-STATE TRANSPORT

M.V. Fischetti and D.J. DiMaria

IBM Thomas J. Watson Research Center
Yorktown Heights, New York 10598

HOT ELECTRONS IN GENERAL AND IN SiO_2

One of the strongest impacts of the Very-Large-Scale-Integration (VLSI) on fundamental physics we have been witnessing in recent years is the revival of activities aimed to better understand the properties of electron transport in crystals. For many years the basic "semiconductor equations" (that is : the standard "Drift-Diffusion" coupled to the Poisson equation) have helped engineers to design semiconductor devices which have changed our lives in quite a remarkable way. Only in the past decade or so, with the shrinkage of the device dimensions towards a scale approaching the mean-free-path (or even De Broglie wavelength) of electrons in semiconductors, have we begun feeling the need for more realistic solutions of the Boltzmann-transport-equation (BTE). Hence, the present "golden age" of Monte Carlo simulations, of the "hot-electron", "high-field" problem, of the efforts to reach fully coupled an self-consistent Poisson-Schrödinger-Monte Carlo description of small devices which we see flooding the scientific literature today.

A remarkable exception during the dormant period of this field has always been the problem of transport in insulators. The "hot electron problem" has been connected, since the very beginning, to the phenomena of wear-out and dielectric breakdown of insulators, a problem of non-negligible importance in a number of applications. The high fields applied to these materials have made it clear that no simplified solution of the BTE assuming thermal carriers (as the Drift-Diffusion does) can help us to understand how breakdown occurs. Moreover, the small time- and length-scales at which interesting phenomena occur and the intrinsic theoretical difficulties in modeling the very-high-field electron transport have made this issue a very challenging (and controversial!) problem.

With the advent of silicon technology, SiO_2 has taken the place of the alkali-halides as the insulator. At present, it is probably safe to say that thin, amorphous films of thermally grown SiO_2 are the best known materials as far as hot-electron transport is concerned. Part of the reason obviously stems from its technological importance and the associated drive towards higher-quality material and the more ideal behavior observed in recent experiments. But, as the same reason could be invoked for Si and GaAs, there is more to it, as we shall discuss in the following.

In this paper, we wish to review the way which has been followed to achieve these results, and simultaneously understand why the issue has been (and, for obscure reason, perhaps still is) so controversial. There are two main considerations to keep in mind : the experimental information on the electronic properties of SiO_2 has often been indirect, incomplete, and, unfortunately, scarcely reproducible from one laboratory to another. Recently, a better standardization of the technology has improved this "irreproducibility" problem in a very concrete way. But old ideas remain hard to kill. Experiments on transport in thin SiO_2 films are also significantly more difficult to perform than in semiconductors. Since we are interested in high-field properties we are going to be plagued by breakdown events. There are no free electrons in the conduction band of an insulator, obviously. Thus, some effort is needed to inject them. We are also interested in thin films (<100 nm). Therefore, experiments like the time-of-flight determination of drift velocities enter the experimentally hard sub-picosecond time range.

The second thought concerns the winding path that our ideas are following. We have already said that high-field electron transport in SiO_2 has been regarded in the past as the constitutive problem of dielectric breakdown. Positive and negative oxide charging and the ultimate dielectric breakdown have been (very unfortunately in almost all instances in the past 15 years or so) the only tools available for inferring some kind of information about electron transport. No reliable direct measurement of, say, electron energies or distributions was available before the experiments performed in the early eighties (which we will refer to as to the "IBM-experiments"). Afterwards, we have looked at electron transport in SiO_2 as a problem worth investigating for its own importance and for its implications on the fundamental problems of high-field transport in solids. In so doing, we tried to free ourselves from the wealth of ideas (and misconceptions) crowding the breakdown issue, perhaps much too close to industrial applications. Once the transport properties are understood, only then can we revisit the problem of wear-out, charging, and breakdown. We shall say little about it. But earlier theoretical estimates[1] and more recent experimental evidence[2] are hinting quite strongly towards a correlation between electron heating and oxide degradation.

We divide our discussion into three main sections : first we shall review the experimental facts which have been associated (correctly and not) with the basic properties of transport. We do not imply any completeness in our discussion. As the high-field degradation of SiO_2 will be necessarily mentioned in discussing the early history, a look at Weinberg's[3] paper will be a useful and more complete reference. Afterwards, we shall move to some theory (but keeping theory completely separated from experiments will turn out to be an impossible task). Of course, we shall review and criticize old and new ideas, while emphasizing our own. In this case, a look at Ferry' paper[4] will have the overall effect of balancing our presentation (biased by definition). Finally, we shall present more recent results on transport in very thin films, which is the apex of our work. Details will be avoided throughout. We have published them already, many times. We shall concentrate mainly on the crux of the matter and regard this paper as a guided tour through the many references on the subject.

SOME HISTORY OF EXPERIMENTS

a. The Standard Model

Let's take an insulator. Most probably it is a highly ionic solid, as large bonding energies (thus, bandgaps) are obtained in polar materials. Let's apply a high field across some region of it by means of a couple of electrodes. The few electrons which we have succeeded in putting in the conduction band (by tunnel injection, internal photoemission, direct optical excitation, etc.) will be accelerated by the enormous field. The only means the electrons have for losing energy is by exciting the oscillatory modes of the polar molecules of the lattice (the so called Longitudinal-Optical - or LO - phonons. The transverse modes are associated with a field too low to bother considering them). At low electric fields, the electrons move at low speed and can lose energy efficiently, since the molecule can respond adiabatically to the incoming electron. The average electron remains "cool", at an energy of the order of 0.1 eV (i.e., the typical LO-phonon energy). As the electron speeds up while the external field is increased, the molecule will find it harder and harder to respond to the fast moving carrier. The mechanism for energy loss weakens. If now the energy the electrons gain from the field between two successive electron-lattice collisions, $eF\nu\tau$, (e is the magnitude of the electron charge, F the electric field ; ν the electron group velocity, and τ the time between collisions) exceeds the energy lost in a single scattering event ($\hbar\omega_{LO}$, the LO-phonon energy), the electron can gain enough energy to impact ionize in the crystal.

This is the core of all ideas about electron transport in insulators : the electrons are in a steady state and at low energy below a critical value of the electric field ($F_c = \hbar\omega_{LO}/(e\nu\tau)$, of the order of 10^6V/cm in typical insulators), while velocity runaway occurs above it. It also leads naturally into the issue of dielectric breakdown. When proposing these ideas, Fröhlich had in mind the alkali-halides[5]. In this case, the whole electron energy distribution runs away from the LO-phonon energy loss at fields above F_c. Avalanche multiplication with its associated Joule heating was the mechanism he envisioned for breakdown. With some modifications, this picture still holds true[6].

The application of this idea to SiO_2 necessitated some extra changes, but relative only to the final stage leading to breakdown. Not simply an avalanche multiplication in the bulk, the ionization process must be regarded as the source of instabilities via a positive charge feedback. This does not require a "global" runaway of the entire distribution, but simply the presence of a few "lucky" carriers at energies high enough to ionize. In a capacitor, such as the metal-oxide-semiconductor (MOS) capacitors used in VLSI Field-Effect-Transistors (FETs), injection occurs unavoidably at high-fields because of the well known Fowler-Nordheim tunnel injection from the cathode. As electrons flow into the insulator and impact-ionize, the resulting holes will either create bulk positive charge, or will be pushed towards the cathode by the field. Some of them will recombine with the oxide. This will increase the injecting field. The current will sense this increase in an exponential fashion. More electrons will be injected and destructive Joule heating could result from this mechanism, hence breakdown. The 1937 Fröhlich idea found its way into SiO_2 in the early seventies. Osburn and Weitzman[7], Solomon and Klein[8], Shatzkes and Av-Ron[9], and DiStefano and Shatzkes[10] all contributed to this model. It explained on qualitative grounds the observation of positive oxide charging at high fields. It could be made to explain quantitatively the breakdown fields observed at that time (7 to 10 MV/cm) with some fitting parameters. The most important was the mean free path for the electron/LO-phonon collision which was somewhere in the range of 0.1 to 0.2 nm.

At the time, there were other pieces of evidence supporting directly the LO-phonon-based model : the low field ($\leq$ 0.8 MV/cm) drift velocities measured by Hughes[11] and the already mentioned breakdown fields were consistent with theoretical predictions based on the LO-phonon interaction [12-14]. Weinberg et al. had interpreted their carrier separation data[15] on the basis of the absence of electron heating. Although no connection was made to the "standard model", this fitted the picture.

b. The exceptions

There were, however, a few puzzling observations. Berglund and Powell had studied the mechanisms of internal photoinjection[16] : current was indeed flowing across a moderately biased ($\leq$ 2 MV/cm) MOS capacitor illuminated with 3 eV light ! If the electrons were indeed giving up 0.153 eV of energy to an LO-phonon every 0.1 or 0.2 nm, how could they ever overcome the electrode-SiO_2 potential barrier rounded off by the image-potential ? Actually, 3 nm or so was the typical scattering length which was extracted from their data.

Maserjian and co-workers came the same conclusion in completely different experiments[17,18]. Electrons injected into thin SiO_2 films (5 nm or less) will tunnel through the first 2 nm or so of insulator and will emerge in the conduction band. They will travel for about 2 nm as conduction carriers before entering the anode. If the electrons were all ballistic (i.e., didn't suffer any phase-destroying collision), one should obtain quantum-mechanical reflections of the electronic wavefunction at the abrupt potential step between SiO_2 and the anode. These would be observed as oscillations of the tunneling current. Inelastic scattering with LO-phonons will prevent this observation, as the phase coherence yielding the reflections would be destroyed. Again, if such a collision occured every 0.1 or 0.2 nm, no oscillations would be seen. But oscillations were seen ! The mean-free-path deduced from the amplitude of the oscillations was about 1 nm. (It is appropriate to recall at this point that while the photoemission experiments are a measure of the "energy autocorrelation length" - that is, how long the electrons must travel to find themselves at equilibrium with the field after a few collisions - the Maserjian oscillations are destroyed by a single inelastic collision and are a measure of the mean-free-path itself).

Finally, in 1978 Solomon attempted a very ambitious experiment : to measure the electron energy-distribution by a vacuum-emission technique[19]. We shall describe this technique in the next section. Here it is sufficient to say that the equivalent electron temperatures he observed were too high to be accounted for by the "standard model". This model required the electrons to stay almost thermal up to the "breakdown" field $\hbar\omega_{LO}/(e\nu\tau)$, and run away towards ionization energies ($\geq$ 10 eV) at higher fields without any gradual heating. The "warm" tails seen by Solomon found no place in the model. But the unknown effect of the thin metal gate the electrons had to go through before emerging into the vacuum was casting a big doubt on the credibility of the observations. A final blow came from the high-field and the long stressing times needed to collect data : dielectric breakdown was a killing experimental problem preventing Solomon from obtaining full energy-distribution curves.

c. The IBM experiments

Solomon's experiments were inconclusive but highly stimulating. The only way to reconcile the standard model to the puzzling objections we have discussed above was to keep trying and obtain, one way or another, additional information on the energy distribution of the electrons at high fields. The most important information was thought to lie in the high-energy tail. All ionization-recombination models relied on the presence of highly energetic carriers to trigger the necessary first ionization events. Solomon was looking for them and found some, but couldn't say how many.

In 1982 Theis, Kirtley, DiMaria and Dong employed a variation of the vacuum emission technique which we shall call the **electroluminescence** experiment[20]. (In this case, as in the following ones, a couple of key facts must be stressed to understand why these measurements hadn't been done before : in the first place, the oxide technology was constantly improving and the increased reliability allowed the oxides to withstand high-fields for times much longer than those which could be sustained in the seventies. Secondly, the use of Si-rich SiO_2 buffer layers in large-area samples helped in reducing both breakdown problems, and the noise due to low currents in small samples). Instead of looking at electrons in vacuum, they injected them across the SiO_2 into an intentionally roughened metal gate. The hot carriers will lose energy mainly to collective excitation of the valence-electron-gas of the thin metal (surface plasmons). In the presence of a roughened surface, these will decay radiatively. The spectral distribution of the emitted photons will provide some indication of the energy distribution of the carriers which stimulated the plasmons initially. Despite the extrapolation needed to extract the essential information, there was no doubt that something was not going according to the theory of the standard model. The electrons already at field strengths around 4 to 6 MV/cm were not quasithermal, as they should have been at fields below breakdown. But they were not at above-bandgap energies, as expected above breakdown. The data showed clear signs of a gradual heating to average energies of 2 to 3 eV, expected. But there must have been also some other mechanism preventing them from accelerating towards ionization energies.

This experimental technique was not tailored to look at electrons hot "on average". A better technique for this situation, the **carrier separation** technique, had been already invented by Ginover et al.[24] and Weinberg et al.[15]. It hadn't been used simply because negative results were expected. But after the results of the electroluminescence experiment, DiMaria et al. employed the technique (using p-channel FET's)[21]. Electrons are injected from the polysilicon gate into the gate oxide of a p-channel FET. As they exit the oxide and enter the silicon substrate, they will impact ionize in the silicon. Any electron will have at least 3 eV of kinetic energy above the bottom of the Si conduction band. So, it will ionize at least once. If there are electrons at even larger energies, two ionizations are possible. Three at even larger energies and so on. Thus, the ratio of the substrate current (i.e., the electron current measured at the substrate contact) to the channel current (i.e. the hole current measured at the source/drain contacts) should decrease from about 2 (one ionization) to about 1.5 (two ionizations) as the electrons heat up. An accurate theory is needed to rework the way up from current ratios to average energies[22]. Actually, a deconvolution is needed. This works only because by a lucky accident the theoretical electron/hole current ratio vs. electron energy decreases linearly with energy above about 1 eV from the bottom of the SiO_2 conduction band[21,22]. (At lower energies,though, electrons can enter the substrate only via direct tunneling through very thin SiO_2 layers and their energy distribution approaches a delta function[21,23]). The results didn't come as a surprise any longer ; they confirmed the preliminary data from the electroluminescence experiment. Anticipating here later results, the electrons appear to stay "cool" for fields below 1.5 to 2 MV/cm. At this critical field, they acquire suddenly 2 eV of energy. Above 2.5 MV/cm the heating is still substantial, but more gradual. We illustrate this in fig. 1, together with additional experimental and theoretical data which we'll discuss shortly. Jumping immediately to a fundamental physical quantity, the slope $\lambda = dE/edF$ of the average energy E vs. electric field F is the already described "energy autocorrelation length". Another physical way to look at this important quantity is to consider it as the distance it takes the electrons to lose memory of their initial condition (or, in terms of energy autocorrelation time, how long it takes for the electron energy at a given time to lose any correlation with the initial energy at t = 0). This length is seen from the data to be of the order of 3 nm, consistent with the internal photoemission data of Berglund and Powell.

The final "frosting on the cake" comes from the most spectacular experiment : the **vacuum emission** technique[25]. The improved oxide technology has been pushed to the limit at which the vacuum emission technique Solomon had to abandon in 1978 could be revived and employed successfully. The breakdown problem being solved (...almost. A lot of patience is still needed !), the problem of the effect of the gate material remains. In the vacuum emission experiments electrons injected from the substrate of an Al-gate MOS capacitor through the oxide reach the thin metal gate and must penetrate it. (Thicknesses of 12 to 25 nm are typically used, but gates as thick as 1 μm have been used to check the role of the gate). Most carriers lose energy in the metal and are collected as gate current. But some (one out of 10^5, typically), probably emerge close to the edge of microscopic voids in the gate. These find only a small thickness of metal to cross and are likely to reach the vacuum of the chamber in which the sample is

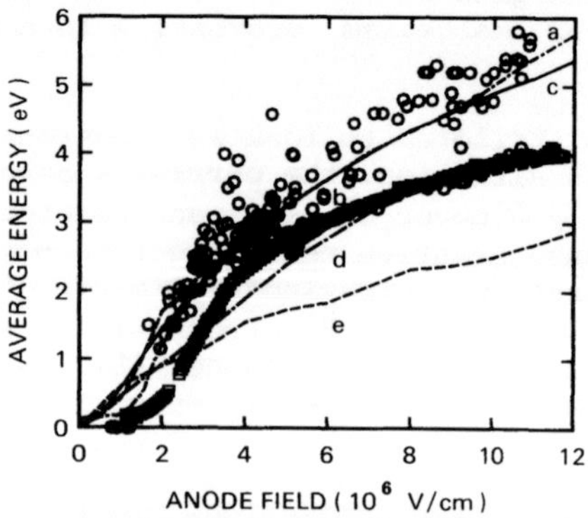

Fig. 1. Average energy of the electrons vs. anode field from experimental data (Squares : carrier séparation, Refs. 21, 28. Circles vacuum emission, Refs. 25, 28) and from various Monte Carlo simulations : (a) semi-classical (Refs. 40, 41) ; (b) overlapping the carrier-separation-data - with collisional-broadening (Ref. 41); (c) Quantum (Ref. 42) ; (d) semiclassical with intervalley scattering including collisional broadening and intracollisional -field effects (Ref. 47).

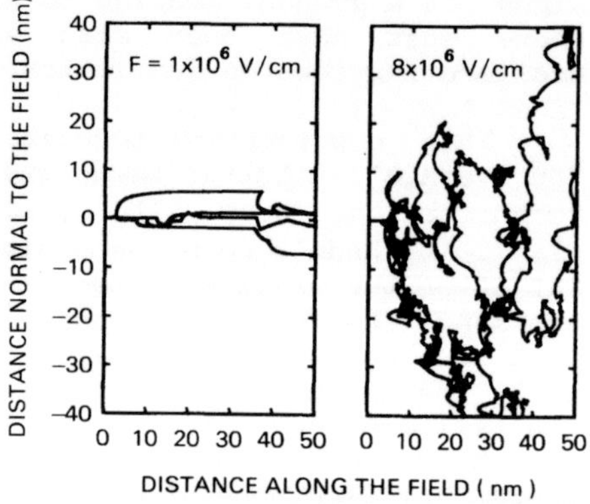

Fig. 2. Electron trajectories at 10^6V/cm (left) and $8x10^6$ V/cm (right) obtained from the semi-classical Monte Carlo simulation of Ref. 41. The low-field trajectories are controlled by the low angle polar scattering with LO-phonons. The high-field trajectories are diffusive, because of the large-angle (Normal and Umklapp) nonpolar scattering with acoustic and acoustic and zone-edge phonons.

mounted unaffected by the metal. A gold plate acts as a collecting electrode. Its potential can be varied to scan the energy distribution of the emitted electrons. Despite the confidence we have displayed in talking about the effect of the metal gate, we known nothing of what really happens. The picture we have presented above is suggested by many observations : 1. The effect of the metal thickness is much smaller than expected for an ideal, uniform, single crystal metal. No electrons should actually escape into the vacuum. Pinholes are known to be present in the metal films we have employed. 2. Changing the metal (Al, Cr, Au) has the obvious effect of shifting the electroncounts vs. applied bias curves, because of the different work functions, but has only relatively minor effects on the observed distributions : typically, some differences due to surface plasmons (characteristics of the metal) and on the high-energy tails (not always, though) can be observed. 3. The average energies obtained form the vacuum emission agree quite well those obtained from the carrier separation technique. This is an a posteriori justification, but it gives us confidence anyway. 4. Finally, the phonon replicas we have observed together with the ballistic carriers in thin films could not have been observed with a strongly scattering metal gate[26,27].

Leaving the observation of ballistic transport for the concluding section, we should now spell out explicitly some of the merits of these observations. In the first place, the set of three experiments, each with its own advantages and disadvantages, shows beyond any doubt that the polar electron-LO phonon interaction does not control the electron transport above 2 MV/cm (the "heating threshold" in fig. 1, discussed in Refs. 28 and 29). Secondly, and more importantly from a practical point of view, transport appears to occur at steady-state even at the highest fields achievable (up to 16 MV/cm). No signs of instability show up in films thinner than about 100 nm. It is probably easy to accept the basic consequences these experiments bear on electron transport, but it seems very hard to abandon some old ideas about ionization. Electron distributions obtained in SiO_2 films thinner than about 40 nm simply show a negligible number of carriers above a realistic ionization threshold (~ 12-14 eV). But even today we keep witnessing the ionization model being vigorously proposed as a viable mechanism for SiO_2 wear-out and breakdown even in these relatively thin films ! Definitely, we do observe large numbers of above-bandgap carriers in thicker (≤ 100 nm) films. [28]But while the practical importance of the high-field behavior of SiO_2 layers thicker than 50 nm is quite small, there isn't much evidence that ionization occurs even in these films. Actually, accepting with many "caveats" a self-consistent band structure of quartz as representative of what happens in thermal SiO_2, momentum, density of states, and matrix-element considerations let us guess that band-gap ionization is a very weak process with a relatively large effective threshold[30,31]. Having disposed once again of the impact-ionization model, we must say that dielectric breakdown remains a mystery to us. The good news : No intrinsic, unaviodable electronic processes seem to occur up to 14 MV/cm or so. This is the third, and most important, merit of these results. (A remark : since the electrons are "thermalized" in 3 nm or so, it will be the electric field in the last 3 nm from the anode which will fix the final distribution. This is why we plot average energies vs. anode field).

Then what was wrong with the standard model ? Made smarter by the experimental evidence, we may look for the most probable reasons in a few misleading theoretical estimates and in the scarce experimental information available. Having discarded for many years the puzzling data coming from internal photoemission, Maserjian's oscillations, and Solomon's vacuum emission, it was not recognized that the observation of positive charge generated by high-field injection into SiO_2 is by no means a proof that impact ionization occurs. Weinberg's paper[3] should give convincing evidence that the issue is by far more complex. Hughes' mobility data were limited to the low-field region in which the Fröhlich model still holds true (≤ 0.8 MV/cm). Other clear-cut experimental evidence was lacking. The standard model is by itself very elegant, appealing, and simple. It has been also very valuable as a first attempt to formalize with sound physics the degradation process of thermally grown SiO_2. Unfortunately, it is not consistent with the experimental data. And esthetical beauty by itself cannot be confused with truth.

SOME HISTORY OF THE THEORY

Having established that the standard model is inconsistent with the experimental data, the ball was passed to the theoreticians who had to explain what was going on. To pinpoint the problem : what is the additional mechanism which keeps the electrons hot, but at steady state, for fields than 2 MV/cm ?

THE FRÖHLICH MODEL AND SiO_2

We have briefly and qualitatively explained the Fröhlich model for breakdown in polar insulators. The basic idea lies in the fact that the polar electron-phonon interaction is the dominant scattering process, other mechanisms being too weak to matter. To state in a slightly more precise way the effect of this type of interaction, we shall recall a few simple considerations. The polar interaction (that is : the interaction between the moving electron and the dipole field associated with the ionic molecules) is controlled by a matrix element which, as explained for instance in Kittel's book[32], is of the form :

$$M_{polar}(q) = \frac{ie(\hbar\omega_{LO})^{1/2}}{2q}\left(\frac{1}{\epsilon_{high}} - \frac{1}{\epsilon_{low}}\right)^{1/2}\left(n_{LO} + \frac{1}{2} \pm \frac{1}{2}\right)^{1/2} \tag{1}$$

where q is the wavevector of the emitted/absorbed phonon, n_{LO} is the occupation number of the LO-phonon involved in the process, $\epsilon_{high}(\epsilon_{low})$ is the dielectric constant at a frenquency larger (smaller) than the phonon frequency ω_{LO}, and $\hbar$ the reduced Planck's constant. The key characteristics of this coupling - due to the Coulomb nature of the interaction - is the phonon wavevector in the denominator. This implies that the electron-lattice collisions will occur most likely at small angles (since $q = k_{before} - k_{after}$ is also the change in the electron wavevector k before and after the collision, small q's imply small changes of the electron direction). Moreover, even accounting for the widening of the region of the phase-space available to the final states as the electron energy increases, the coupling between electrons and dipole modes decreases. As we have mentioned before, this will lead eventually to the intrinsic instability associated with this interaction (runaway, breakdown etc..).

The properties of electron transport in SiO_2 have been calculated many times in the past, considering only the polar interaction whith the two LO phonons known in SiO_2 (of 0.063 and 0.153 eV of energy). Lynch[12] made use of a careful analysis of the frequency-dependent dielectric constant of SiO_2 to estimate the energy loss per unit length at a given field and therefore, to estimate the breakdown field. Unfortunately, his analysis was based on a calculation originally proposed by Fermi to evaluate the energy loss of heavy particles (α particles) in solids[33]. Electrons do not belong to the category of heavy particles and his conclusion (a breakdown field of 10 MV/cm) agrees with the experimental data of the time only by a fortuitous accident. Later, Ferry used an iterative solution of the Boltzmann equation[14] to reach the conclusion that a "stable" distribution can indeed exist even at high fields ; but for short times. His times were indeed very short (10^{-14} sec or so at 10 MV/cm). The electrons may not cross much more than a few monolayers during this time. Afterwards, they would run away. Ferry had also introduced the nonpolar scattering (which we shall discuss next), but only at low energy, in the deformation-potential approximation. This was necessary to fit Hughes' mobility data. As we will discuss later, this issue is still in a state of total confusion. Finally and more recently, Fitting and Friemann used a Monte Carlo technique to study the problem[34]. They found, quite consistently with the later IBM observations, that runaway was occuring at fields as low as 2 MV/cm. This result should have been a final blow to the standard model, but the paper passed unoticed. We should recall that the Monte Carlo calculations by Fitting and Friemann are not an isolated piece of work. Experimental and theoretical activity in this field had been going on and still continues at the University of Rostock[35,36]. We must also say that their experimental data appear to be somewhat inconsistent with ours. This may be due to the dubious "forming" procedure they employ to pre-stress thick oxides in order to obtain field-emission at very low fields[36].

NONPOLAR COLLISIONS

To the Fitting-Friemann work we should couple a few more ideas in order to come to the easy answer to our question relating to the missing mechanism. Ridley had suggested in 1976 that nonpolar scattering should also be considered[37]. Indeed the nonpolar coupling between the flying electron and the lattice ions displaced from their equilibrium position is described by the matrix element :

$$M_{nonpolar}(q) = \frac{i\hbar^{1/2} S_q}{(2\rho\omega_q)^{1/2}} \left(n_q + \frac{1}{2} \pm \frac{1}{2}\right)^{1/2} \qquad (2)$$

where ρ is the density of the material, and ω_q and n_q are the frequency and occupation number of the (acoustic or optical) phonons. Notice the difference between Eq. (1) and Eq. (2). In the latter the phonon wavevector q enters via the coupling constant S_q. At low energy (deformation-potential approximation[38]), it is customary to consider the electron as scattered by the energy-variation of the bottom of the conduction band as acoustic waves change the interionic distance. The shorter the wavelength, the larger the change of energy going from one unit cell to the other, so that $S_q \simeq Cq$, C being the deformation potential (i.e., S_q is the energy variation of the bottom of the conduction band as we change the lattice constant by a unit length). At higher energies, a similar form is taken for the coupling constant, assuming that displaced ions behave as incoherent scattering centers with a potential deep enough to scatter efficiently at short wavelengths.[39] In either case, the matrix element grows with q. Thus, the coupling is such that large-angle collisions are favored and the interaction gains strength as the electron energy (and k-vector) increases, as higher-q phonons will be emitted more and more efficiently, while the phase-space opens up.

Ridley suggested that a more careful look should have been given to these interactions, particularly for the case of SiO_2. Actually, these ideas were first applied to explain the frequency dependence of the laser-induced breakdown of alkali-halides[6].

Coupling the Fitting-Friemann results to Ridley's idea, in 1984 we added the electron-lattice interaction to the polar one and simulated the high-field electron transport with a Monte Carlo solution of the BTE.[40,41] All went well. As shown in fig. 1 (together with other simulations we'll discuss below), the results of the Monte Carlo simulation agree remarkably well with the experimental results. Perhaps, the best way to understand the physical significance of the nonpolar interaction is to look at fig. 2. We show the simulated trajectories of a few electrons at a field below the heating threshold (1 MV/cm in (a)) and a field above it (8 MV/cm in (b)). The "streaming" trajectories at low fields illustrate clearly the polar small-angle scattering. The electrons travel mostly in the "forward" direction, the mean-free-path and the mean-free-path projected in the direction of the applied field being practically identical. Not so above the heating threshold. The nonpolar scattering randomizes the motion via large-angle collisions. (Umklapp scattering - that is : scattering outside the first Brillouin Zone - is a key ingredient of this "randomization"). Now the electrons must travel in a diffusive fashion a much longer distance. In simple words, the acoustic phonons take away a little energy. But their main effect is to decrease the effective mean-free-path for polar collisions along the direction of the field. They really act as catalysts, triggering more frequent emissions of LO-phonons, thus effectively increasing the energy lost per unit length. Thus, a steady-state transport can be obtained up to very high fields.

To add weight to the value of this physical picture, we must add that the success does not stop at the correct prediction of the high-field average energy. The heating threshold is predicted almost at the correct field (more about the "almost" later).[28,29] The internal photoemission date have been simulated, obtaining an amazingly good agreement with Berglund and Powell's data.[41] Later, the Maserjian oscillations were modelled successfully.[26,27] So, everything fits...

Not quite. First, we have stressed several times some parameter-problems : a "guess" of 3.5 eV for the high-energy coupling constant C (defined after Eq. (2)) seems to work magnificently, but the value of the deformation potential at low electron energy does not. The drift velocity measured by Hughes is not explained.[41] (The inclusion of the Harrison nonpolar electron/TO-phonon interaction[35], the use of a very large deformation potential parameter[42], or of a very large effective mass[14] seem to fit the mobility data. But it is not clear which is the correct choice and whether any of these models would also explain the other properties of the transport). The heating threshold occurs at a field slightly lower than observed.[28,29] Also, we shall see that the ballistic-to-steady-state transition is softer than observed.[26,27] Clearly, a larger deformation potential is needed. We shall discuss later which experiments can be used to determine it.

Secondly, there are problems with the band-structure of the amorphous (!) SiO_2 films we simulate. Thirdly, more problems originate from the dubious validity of the semiclassical approximations which are behind the Monte Carlo simulation. This shows up dramatically in the high-energy tail of the electron energy distribution at high fields.[41] Given the importance of these tails, after some thoughts about the questionable

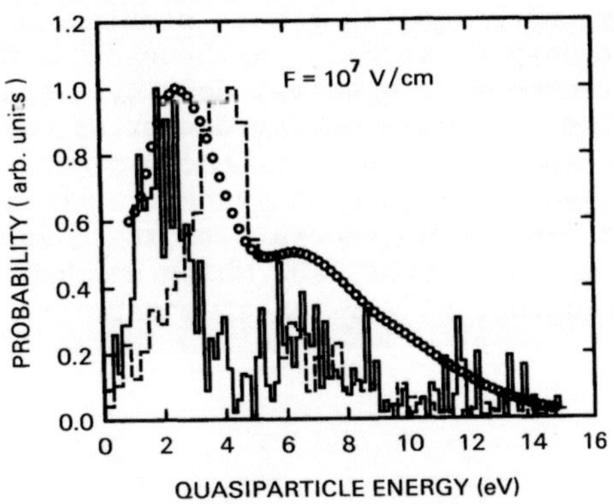

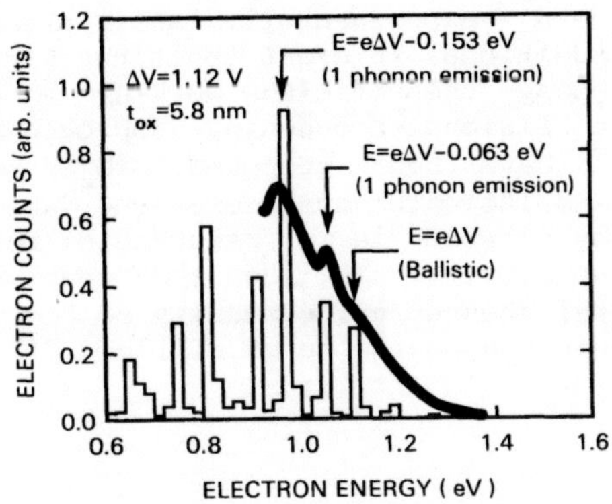

Fig. 3. (left). Electron energy distribution obtained from the vacuum-emission technique (symbols, Ref. 25) and compared to the semiclassical (dashed line, Ref. 41) and the Quantum (solid line, Ref. 45) Monte Carlo simulations.

Fig. 4. (right). Electron energy distribution from the vacuum-emission of a thin oxide (Refs. 26,27). The peaks correspond to the "phonon replicas" and are matched by semiclassical Monte Carlo simulations (histogram).

"collisional-broadening" trick proposed by Hess and co-workers[43] (fig. 1, curve (b)), we have attempted a move beyond the parabolic-band approximation and the semiclassical formulation of the transport problem by employing the experimentally observed [44] density of states (DOS) of thermally grown SiO_2.[45] We have computed the scattering rates beyond the Fermi golden rule from the Dyson equation (a formal resummation of the perturbative terms at all orders in the coupling constants) using the experimental DOS and accounting in an empirical way for the failure of the adiabatic approximation in (over) estimating the nonpolar interaction between the lattice and very high-energy electrons. Finally, we have modified the classical Monte Carlo procedure to evaluate the quantum mechanical probability-amplitude of the electron paths according to a Feynman-integral formulation. [46]This accounts for the phase interference in multiple collisions, for the effect of the high fields on the scattering rates, and for the dressing of the electron-phonon quasiparticle. The results enhance the high-energy tail, as shown in fig. 3, which is good. On the other side, we show in fig. 1, curve (c), how the quantum effects "wash-out" the heating-threshold which was so neatly predicted by the more naive semiclassical approach. And this is not so good.

To conclude this section on the theory, we must mention some alternative ideas. Porod and Ferry[42,47] have proposed, as a "missing mechanism", intervalley scattering. This gives the same effect, as far as large-angle scattering and randomization of the trajectories are concerned. It would be nice to know at which energy extra valleys actually come in to support this model, if they come in at all : Chelikowsky and Schlüter's band-structure calculation[48] do not seem to agree with Himpsel's inverse photoemission data. [44]The latter do not show any sign of extra DOS building up in the range of 2-4 eV, where additional valleys should enter, according to the theory. Notice also in fig. 1, curves (d) and (e), that the heating threshold is missing in these simulations, heating entering as a smooth rise and not as the steep run away from the polar scattering which is both seen experimentally and predicted by our semiclassical simulation. Additional interest about the high-field transport in SiO_2 has been aroused among theoreticians working on quantum transport : spectral-density[49] and collisional-broadening approaches (modified to ensure energy-conservation) [50] have been proposed. SiO_2, thanks to the IBM experiments and our early simulation, has become the "arena" of high-energy quantum transport. This by itself is a remarkable result. It may be appropriate to say that, perhaps, it is just the complexity of the physics of SiO_2 which has kept the theoretical community so "cold" and reluctant to work on this material. Let's hope the surge will not die.

BALLISTIC TRANSPORT IN THIN FILMS

We have already mentioned a couple of times the "energy autocorrelation distance" as the distance the electrons must travel to find themselves in the steady-state situation in the presence of the applied field. This distance is a function of the electric field and of temperature, but it is of the order of 3 nm at room temperature (increasing to 5 nm at T = 0, according to our Monte Carlo simulations[41]). If we now inject electrons in an oxide thinner than about 5 nm, we will see them tunneling through the first couple of nm or so (depending on the injecting field) and then travelling in the conduction band (CB) for a distance which is smaller than - or comparable to - the energy-autocorrelation length. If we dispose of inelastic tunneling as a small-probability process (those concerned about the details of tunneling are referred to Ref. 27 for some pessimistic observations), we should be able to observe initially the electrons flying ballistically through the CB when the distance is below about 2 nm. As the oxide is made thicker (or the applied field larger, so that a longer distance must be covered in the CB), the electrons should exhibit the streaming motion[51] typical of the polar interaction dominant at low energies. Since the electrons can lose energy only in discrete "chunks" (the almost dispersionless LO-phonons), we should see the ballistic peak reduce its strength as new peaks appear at energies shifted by the LO-phonon energies, corresponding to emission of one, two etc. LO-phonons of either type (the 0.063 and 0.153 eV, the latter being more tightly coupled to the electrons). As we bring the oxide thickness and/or the applied bias to even larger values, the nonpolar scattering should have the effect of washing out the peaks and push the electrons towards the steady-state situation.

Maserjian's oscillations which we have referred to in a previous section, are a consequence of the existence of ballistic electrons in SiO_2. We are now better off than Maserjian was in the mid-seventies: we can not only match the Monte Carlo simulations to the observed decay of the amplitude of the oscillations as the voltage is varied, but we can also measure the electron energy and energy-distribution with the carrier separation and the vacuum emission techniques.[26,27] At low bias, the nonpolar interaction (effective only at energies $\gtrsim 2$ eV) is pratically turned off and we can look only at the polar interaction. The amplitude of the oscillations in the I-V characteristics, of the ballistic peak and of its phonon-replicas seen in the vacuum emission - shown in fig. 4 - gives us a direct handle on the coupling constant between electrons and polar excitations of the lattice. All other parameters being fixed, we can play with the electron effective mass. A value of 0.7 m_{free}, m_{free} being the free electron mass, is consistent with the data. If we compare it with the value of 0.68 m_{free} which first-order perturbation theory would give us for the curvature mass of 0.5 m_{free} dressed by both LO-phonons (the "polaron" mass), we can be very satisfied. Actually, things seem to work so well, that we begin to believe strongly in the ordered structure of SiO_2 over the length scale of 3 nm[52]. The recent report that the first 5 or 6 monolayers of oxide may grow thermally in the crystalline tridymite form[53] reinforces this belief.

The spectacular results of fig. 4 can now be used - at higher biases and at lower temperatures - to look at the low-energy nonpolar interaction, having conclusively sorted out the polar interaction. The phonon replicas broaden at higher bias until, eventually, the featureless steady-state distributions are observed. They get sharper at 77 K. At this temperature a real wealth of structure appears in the vacuum emission experiments. This is mostly due to the action of the temperature-dependent acoustic-phonon population. Our value of 3.5 eV for the deformation potential does not do the job as neatly as we would like. A slightly larger value (about 5 eV) will probably do it. Measurements of low- and high-field drift velocities should be made available for the oxides of the eighties and provide independent information on the value of this constant. Experiments and simulations are now underway or are planned, and we'll get the missing information.

No other material has ever provided such direct and complete information about the electron-lattice coupling as SiO_2. GaAs is plagued by the electron-electron interaction in the "ballistic" regions.[54,55] Interesting physics, but it is so complicated that the electron-lattice coupling is obscured and buried under a zoo of 2-dimensional lattice/electron-gas excitations. Silicon is even worse off than III-V materials. With SiO_2, we have reached the point at which we can probe directly the wave-nature of the electrons, and observe electrons colliding only once with the lattice.

Oxides this thin are now of pratical interest in the VLSI industry. The recent observation that oxide-charging and Si-SiO_2 interface degradation on films nominally free of bulk water-related trapping sites is directly related to electron-heating[2] and the almost complete understanding of the transient transport in these layers should allow us to say something about the future of thin oxides. Our measurements on thick films have indicated that we should not be overly concerned about dielectric breakdown : nothing "intrinsic" - therefore unavoidable, such as the old ionization-recombination model - happens when we see SiO_2 layers breakdown destructively. We must "simply" (!) improve the technology. Similarly, if we can run our future devices under such bias conditions as to avoid electron heating, no destructive degradation should be observed.

ACKNOWLEDGEMENTS

Among the many contributors to the discovery of the electron heating, we would like to acknowledge T.N. Theis, J.R. Kirtley, and S.D. Brorson for the "early" experiments. J. Batey, E. Tierney, J. Stasiak, M. Arienzo, and L. Dori have helped us significantly with measurements and samples for the "ballistic" work. Finally, D.K. Ferry, A.B. Fowler, W. Porod, P. Price and K.K. Thornber have, at different times contributed to whatever we have understood theoretically.

REFERENCES

1. M.V. Fischetti, Phys. Rev. B 31:2099 (1984).
2. D.J. DiMaria, submitted to Appl. Phys. Lett.

3. Z.A. Weinberg, this volume.
4. D.K. Ferry, this volume.
5. H. Fröhlich, Proc. Royal Soc., A-160:230 (1937).
6. M. Sparks, D.L. Mills, R. Warren, T. Holstein, A.A. Maradudin, L.J. Sham, E. Loh and D.F. King, Phys. Rev., B 24:3519 (1981).
7. C.M. Osburn and E.J. Weitzman, J. Electrochem. Soc., 119:603 (1972).
8. P. Solomon and N. Klein, Solid. State Comm., 17:1397 (1975).
9. Morris Shatzkes and Moshed Av-Ron, J. Appl. Phys., 47:3192 (1976).
10. T.H. DiStefano and M. Shatzkes, Appl. Phys. Lett., 25:685 (1974).
11. R.C. Hughes, Phys. Rev. Lett., 30:1333 (1973).
12. W.T. Lynch, J. Appl. Phys., 43:3275 (1972).
13. D.K. Ferry, Appl. Phys. Lett., 27:689 (1975).
14. D.K. Ferry, J. Appl. Phys., 50:1422 (1978).
15. Z.A. Weinberg, W.C. Johnson and M.A. Lampert, Appl. Phys. Lett., 25:42 (1974).
16. C.N. Berglund and R.J. Powell, J. Appl. Phys., 42:573 (1971).
17. G. Lewicki and J. Maserjian, J. Appl. Phys., 46:3032 (1975).
18. J. Maserjian and N. Zamani, J. Appl. Phys., 53:559 (1982).
19. P.M. Solomon, in "The Physics of SiO_2 and Its Interfaces", Sokrates T. Pantelides ed., Pergamon, New York (1978).
20. T.N. Theis, J.R. Kirtley, D.J. DiMaria and D.W. Dong, Phys. Rev. Lett., 50:750 (1983).
21. D.J. DiMaria, T.N. Theis, J.R. Kirtley, F.L. Pesavento, D.W. Dong and S.D. Brorson, J. Appl. Phys. 57:1214 (1985).
22. R.C. Alig, S. Bloom and C.W. Struck, Phys. Rev. B 22:5565 (1980).
23. Chi Chang, Chenming Hu and Robert W. Brodersen, J. Appl. Phys., 57:302 (1985).
24. A.S. Ginover, V.A. Gritsenko and S.P. Sinitsa, Phys. Stat. Sol. (a), 26:489 (1984).
25. S.D. Brorson, D.J. DiMaria, M.V. Fischetti, F.L. Pesavento, P.M. Solomon and D.W. Dong, J. Appl. Phys., 58:1302 (1985).
26. D.J. DiMaria, M.V. Fischetti, J. Batey, L. Dori, E. Tierney and J. Stasiak, Phys. Rev. Lett., 57:3213 (1986).
27. M.V. Fischetti, D.J. DiMaria, L. Dori, J. Batey, E. Tierney and J. Stasiak, Phys. Rev. B 35:4404 (1987).
28. D.J. DiMaria, M.V. Fischetti, M. Arienzo and E. Tierney, J. Appl. Phys., 60:1719 (1986).
29. D.J. DiMaria, M.V. Fischetti, E. Tierney and S.D. Brorson, Phys. Rev. Lett., 56:1284 (1986).
30. Z.A. Weinberg, M.V. Fischetti and Y. Nissan-Cohen, J. Appl. Phys., 59:824 (1986).
31. Z.A. Weinberg, Appl. Phys. Lett., 27:437 (1975).
32. C. Kittel, "Quantum Theory of Solids", Wiley, New York (1963).
33. J.D. Jackson, "Classical Electrodynamics", Wiley, New York (1975).
34. H.J. Fitting and J.U. Friemann, Phys. Stat. Sol. (a), 69:349 (1982).
35. H. Köster and K. Hübner, Phys. Stat. Sol. (b), 118:293 (1983).
36. H.J. Fitting and A. Von Czarnowsky, Phys. Stat. Sol. (a), 93:385 (1986).
37. B.K. Ridley, J. Appl. Phys., 46:998 (1975).
38. J. Bardeen and W. Shockley, Phys. Rev., 80:72 (1950).
39. J.M. Ziman, "Principles of the Theory of Solids", Cambridge University Press, Cambridge, UK (1972).
40. M.V. Fischetti, Phys. Rev. Lett., 53:1755 (1984).
41. M.V. Fischetti, D.J. DiMaria, S.D. Brorson, T.N. Theis and J.R. Kirtley, Phys. Rev. B 31:8124 (1985).
42. W. Porod and D.K. Ferry, Phys. Rev. Lett., 54:1189 (1985).
43. Yia-Chung, D.Z.Y. Ting, J.Y. Tang and K. Hess, Appl. Phys. Lett., 42:76 (1983).
44. F.J. Himpsel, Th. Fauster and D. Straub, J. Lumin., 31-32:920 (1984).
45. M.V. Fischetti and D.J. DiMaria, Phys. Rev. Lett., 55:2475 (1985).

46. R.P. Feynman and A.R. Hibbs, "Quantum Mechanics and Path Integrals", McGraw-Hill, New York (1965).
47. W. Porod and D.K. Ferry, Physica B 134:137 (1985).
48. J.R. Chelikowsky and M. Schlüter, Phys. Rev. B 15:4020 (1977).
49. A.P. Jauho, lecture presented at the Workshop "Quantum Transport Theory with Application to Nanometer Electronics", S. Miniato, Italy (March 1987).
50. L. Reggiani, presentation given at the "Workshop on Small and Quantum-Structured Devices", Tempe, Arizona, December 1985.
51. Carlo Jacoboni and Lino Reggiani, Rev. Mod. Phys., 55:645 (1983).
52. Frank J. Feigl, Physics Today, 30:47 (1986).
53. A. Ourmazd, D.W. Taylor, J.A. Rentschler and J. Bevk, unpublished.
54. A.F. Levi, J.R. Hayes, P.M. Platzman and W. Weigman, Physica B 134:480 (1985).
55. M. Heiblum and M.V. Fischetti, in "The Physics of Quantum Devices", F. Capasso ed., Springer-Verlag, Heidelberg (1987).

ELECTRONIC CHARGE TRANSPORT IN THIN SiO_2 FILMS

D.R. Wolters, and A.T.A. Zegers-van Duynhoven

Philips Research Laboratories
5600JA Eindhoven, Netherlands

ABSTRACT

Some observations concerning the Fowler Nordheim conduction of electrons in SiO_2 are reviewed. The substrate hole currents accompanying the tunneling currents possibly originate from photon stimulated emission. Coulomb repulsion between adjacent sites inhibits trapping in the dielectric and realistic trapping kinetics have to account for this, as has been illustrated for high field and for avalanche injection.

It has been shown that small electron fluences of $\leq 10^{-6} C/cm^2$ are able to change the effective barrier height of the electrode-dielectric interface. The changes are attributed to the generation of interface traps.

The degradation of the dielectric by further injection is described in terms of the generation of low-ohmic paths which can explain a number of phenomena not previously understood.

INTRODUCTION

Electronic charge transport in thin silica films on silicon is governed by field emission at the negative electrode. The so-called **Fowler-Nordheim** conduction[1] consists of quantum mechanical tunneling of electrons through the energy barrier at the cathode-insulator interface, as has been used to describe field emission of electrons from cold cathodes into vacuum[2]. Using this concept of quantum mechanical tunneling of electrons through the potential barrier between the cathode and the dielectric Lenzlinger and Snow[3] have described current-voltage curves of capacitors of thermal SiO_2.

Weinberg[4] has investigated tunneling in unmetallised SiO_2 layers on silicon, measuring the decay of the surface potential with a Kelvin probe. The surface potential was brought out of equilibrium using a corona discharge. Measuring the potential decay curves over many orders of time Weinberg succeeded in finding very accurate Fowler-Nordheim parameters. He observed a difference between, the barrier heights calculated from current voltage curves and those predicted from optical measurements. The difference of about 0.3 eV is due to quantization of energy levels in the thin inversion layer.

Krieger and Swanson[5,6] assumed phonon-assisted tunneling to explain the differences in the barrier height of Si <111> and Si <100> interfaces. Weinberg[7], however, showed that this assumption is not consistent with tunneling from Si <100> interfaces. He showed that annealing treatments decrease differences between the orientations, and thus that surface chemistry largely affects the barrier height. It was concluded that, for tunneling electrons, there is probably no conservation of momentum in the direction parallel to the surface.

Eitan and Kolodny[8] observed a hole current in the substrate, related to the tunneling current. They attributed it to electron tunneling from the valence band of the silicon into the conduction band of oxide. Each tunneling electron leaves a hole in the valence band. Weinberg[9] showed this explanation to be inconsistent with intercepts and slopes of the Fowler-Nordheim curves. This point will be discussed below in more detail. Results of measurements interpreted in terms of Fowler Nordheim conduction are given in references[10-44].

A large number of papers[45-73] investigate space charge evolution during high-field injection and avalanche injection. There is, however, hardly any consistency between reported trapping site characteristics. Accounting for the appropriate trapping mechanism is very important for describing current transients.

Charge injection and transport through the dielectric largely affects the properties of a MOS- capacitor. Even small amounts may cause degradation, as will be shown below. In a large number of publications the generation of states and traps has been shown to occur[74-98]. This degradation determines trapping and interface state generation in severely stressed capacitors and eventually leads to breakdown.

We shall first discuss the **Fowler-Nordheim** mechanism and try to give some insight in the principal parameters.

In the third section a simple model explaining the origin of the **substrate current** accompanying the tunnel current is presented. It is suggested that it is caused by photon-stimulated emission. The effect of **Coulomb repulsion** on **trapping** leads to non-saturating space-charge evolution in dielectrics. Accounting for this effect leads to a more realistic description of the trapping process, as will be illustrated.

Dielectrics **change** when small amounts of charge are injected, as will be demonstrated in the fifth section. The effect of charge injection is seen in an early stage in the change of the slopes of the Fowler-Nordheim curves and the increase of dielectric losses (AC-losses). After further stress the degradation of the dielectric is readily seen from hysteresis of the CV curves and other instabilities. Finally, continued injection leads to breakdown[99,100]. We shall discuss these phenomena and explain them in terms of the growth of low-ohmic channels in the dielectric.

FOWLER-NORDHEIM TUNNELING

Rather than discussing details of the tunneling mechanism we shall indicate the relative importance of different terms in the Fowler-Nordheim expression which is given by :

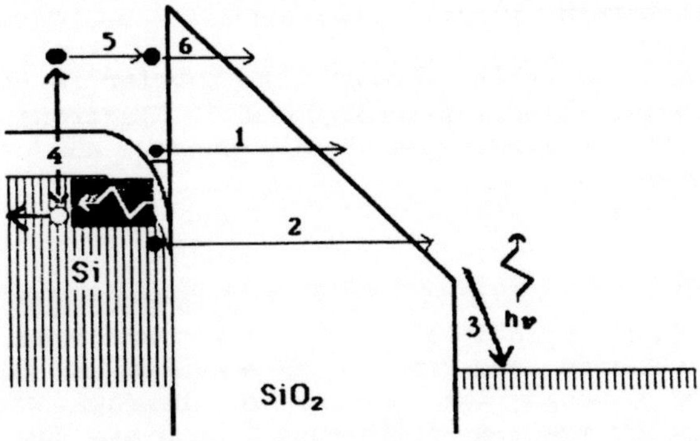

Fig. 1. Energy band diagram for an MOS-capacitor. Various paths for tunneling are indicated : 1) tunneling from a subband E_0. 2) tunneling from the top of the valence band. 3) Release of photon when electron loses conduction band energy. 4) Absorption of photon and generation of electron hole pair. 5) Transport to interface. 6) Tunneling from high energy level. Note that 5 and 6 only occur after process 1.

$$J = J_{max} \exp\left(-\frac{\beta}{F}\right) \quad [A/cm^2] \tag{1}$$

where

$$J_{max} = \frac{q^3 m F^2}{16\pi^2 \hbar m_{ox} \Phi} = \frac{3.10^6 F^2}{\Phi} \quad [A/cm^2], \tag{2}$$

and

$$\beta = \frac{4}{3} \frac{\sqrt{2m_{ox}}}{q\hbar} \Phi^{3/2} = 48.3\Phi^{3/2} \quad [MV/cm], \tag{3}$$

where J is the current density A/cm^2, F is the field at the injecting electrode in MV/cm, q is the electron charge, m is the mass of the electron in empty space, m_{ox} is the mass of the electron in the barrier ($m_{ox} \approx 2m$), $2\pi\hbar$ is Plancks constant and Φ is the barrier energy in eV. The significance of J_{max} and β will be discussed below.

Tunneling of electrons through the barrier at the interface of a p-type substrate is visualised in fig. 1.

If there is pronounced band bending, electrons are present in a sub-band at an energy $E_0 \approx 0.3$eV above the bottom of the conduction band. The effect is that electrons "see" only a barrier height of $\Phi - E_0$. This sub-band is caused by quantization of energy levels. According to the uncertainty principle the electrons cannot be confined to too small a location. On the same principle there is a definite probability that electrons can tunnel through the barrier to an equivalent position at the other side of the barrier.

a) The exponential term β

To illustrate that tunneling through the barrier is the critical step we may use the following illustrative argument [2]. Electrons close to the Fermi level have an indeterminacy in their energy and time given by the Heisenberg principle :

$$\Delta E.\Delta t \geq \frac{\hbar}{2} \tag{4}$$

The Heisenberg principle states that an electron wave has an indeterminacy in energy and time or in momentum and location. Whether an electron tunnels through the barrier or borrows the energy needed to surmount the barrier makes no difference : both are the consequence of its wave character. In order to overcome the energy barrier, Φ, the electron may borrow the energy but only for the time interval $\Delta t \approx \hbar/2\Phi$. Within this interval it can travel a distance $\Delta x = \Phi/qF$ with an extra velocity $v = \sqrt{2\Phi/m_{ox}}$. (It is assumed that the velocity in the ground state is negligible).

If $\Delta t = \Delta x/v$ is inserted in Eq. 4 the condition for tunneling can be calculated. Taking $\Delta E = \Phi$ in Eq. 4 yields :

$$\frac{\sqrt{2m_{ox}}}{q\hbar F}\Phi^{3/2} \leq 1 \tag{5}$$

This is about the same as the term β/F in Eq. 1. The factor 4/3 is obtained when the shape of the barrier is taken into account, and the tunneling probability is integrated over all energy levels in the Fermi-Dirac distribution. Eq. 5 illustrates the quantum-dynamical origin of β in Eq. 1.

b) The pre-exponential factor J_{max}

The pre-exponential may be regarded as the maximum current which would flow through the barrier if tunneling were not a restriction. The F^2 -dependence originates from the field dependence of the electron density and that of the energy level E_0 [9]. The supply of electrons can also be expressed by :

$$J_{max} = qN\nu \tag{6}$$

where $N = \epsilon F/q \approx 2.10^{13}cm^2$ is the number of electrons at the interface at F = 10MV/cm and ν is their attempt-to-escape frequency. For a value of E_0 = 0.3 eV, we get $\nu = E_0/h \approx 8.10^{13}sec^{-1}$ and $J_{max} \approx 2.6.10^8$ A/cm², which is close to the value according to Eq. 2 at a value of F = 10MV / cm [9]. The pre-exponential factor has been used by Weinberg[9] to discriminate between models for the origin of the substrate currents as will be discussed in the third section.

c) Resonant tunneling

Although resonant tunneling is bound to occur it has not yet been proved to be of practical importance in thin oxide structures. Hirose[10]

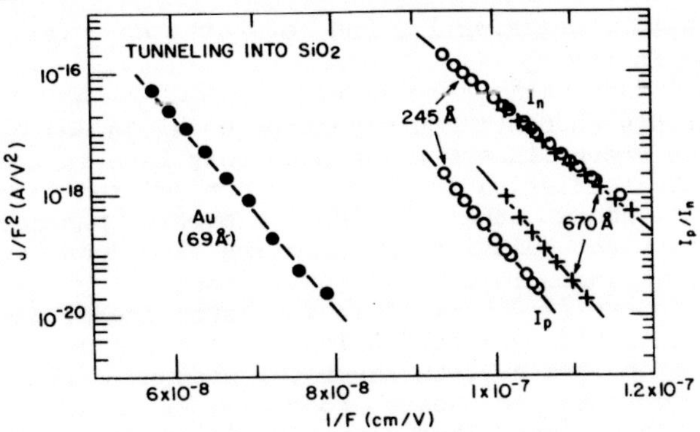

Fig. 2. Fowler Nordheim plot of the channel and substrate currents. After Weinberg[9].

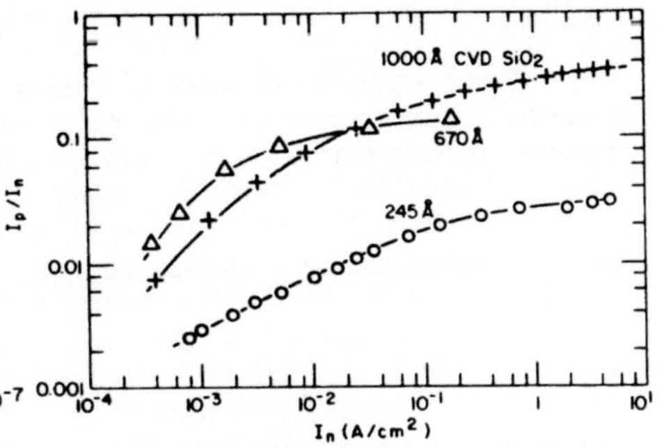

Fig. 3. The ratio of hole and tunnel current as a function of the tunnel-current. After Weinberg[9].

claims to have observed resonant tunneling in very thin sandwich layers of 4 nm thermal silica and polycrystalline silicon. His conclusions are, however, questioned by Ricco and Azbel[102], who state that the tunneling-transmission time constant τ_o depending on the barrier height. Φ, would be of the order of $\tau_o \approx 10^6$ sec. Resonant tunneling via an unfilled trap will have a large transmission coefficient but the event can only take place after τ_o which is large when the barrier energy is high[102].

Hartstein and Koch claim to have measured resonant tunneling in intentionaly contaminated MOS capacitors[103]. However the samples needed an electrical stress prior to the "resonant tunneling" condition and it is not quite sure whether the stress did not degrade the samples. IV curves with voltage peaks at regular points have not yet been observed in thick oxides. Perhaps that resonant tunneling in thick oxides will take place in the final stages of the formation of a low-ohmic path leading to breakdown[28].

SUBSTRATE CURRENTS

Eitan and Kolodny[8], separating electron and hole currents in p-type substrates by appropriate p-n structures, observes a substrate current due to holes which was closely related to the electron current tunneling through the energy barrier. The current is a fraction of the tunnel current but follows a Fowler-Nordheim curve corresponding to a barrier height of 4 eV. It was therefore attributed to electrons tunneling from the valence band into the conduction band of the oxide and leaving a hole in the valence band. The tunneling process is depicted in fig. 1 by step 2.

Weinberg[9] investigated this phenomenon in detail and found Eitan's model inconsistent with J_{max} values expected for tunneling through a 4 eV barrier. In fact, the observed value is orders of magnitude too high. This is shown in fig. 2 where the hole current I_p is compared with the normal tunneling current I_n and the tunneling current of a gold electrode which has $\Phi_{Au/SiO_2} = 4$ eV.

The ratio I_p/I_n increases with increasing I_n, and saturates at a value $I_p/I_n \approx 0.3 - 0.5$. This is shown in fig. 3.

The origin of these substrate currents could perhaps be explained by taking into account the surplus energy $\Delta E \approx \Phi$ which electrons have in the oxide conduction band and which is dissipated when they are emitted into the anode. See fig. 1, step 3. Among other dissipative events, photons with $h\nu \geq 3$ eV will be emitted. Light emission by tunneling electrons has been reported by Theis et al.[26]. There is only as small probability that these photons will be absorbed in the ($\approx$ 2.5 nm) thin inversion layer, due to the relatively large characteristic absorption length, λ = 250 nm for photons of 3 eV. In the region beyond the surface layer the photons will be absorbed, thereby generating electrons and holes (See fig. 1, step 4). However, the tunneling distance is so large that direct tunneling will not contribute to the current. The electrons and holes will recombine if either the electrons or the holes are not removed. But this will happen when electrons in the inversion layer tunnel through the barrier (step 1 in fig. 1). Their place will be taken by the photon-generated electrons (step 5) which, however, have a surplus energy of $h\nu - E_g = 3.2 - 1.1 = 2.1$ eV. Tunneling by the photon-generated electron is more probable than that of the directly tunneling electron (step 3 in figure 1). The point, however, is that it cannot occur unless normal tunneling takes place as well. The probability for tunneling of step 6 must therefore be multiplied by that of step 1.

a) Barrier energy

The barrier energy for hole currents is 4 eV. The barrier height for normal tunneling electrons is :

$$\Phi^* = \Phi - E_0 = 3.2 - 0.3 = 2.9 \text{ eV} \tag{7}$$

while the secondary electron, will see only an effective barrier of 1.1 eV. This is true when it conserves the energy it gets by excitation from a photon of 3.1. eV. The secondary electron cannot reach the interface under normal conditions except when the interface charge is lowered at that instant by a normal tunneling event. Therefore the tunneling probabilities for a photon-generated electron is less by a factor corresponding to normal tunneling. In first order approximation one expects :

$$\Phi_{second.elctr} = (1.1^{3/2} + 2.9^{3/2})^{2/3} \approx 3.3 \text{ eV}, \tag{8}$$

This is less than the observed 3.8 - 4 eV possibly due to the fact that the average electron does lose energy before it arrives at the interface.

b) Ratio I_p/I_n

The ratio I_p/I_n seems to saturate with increasing I_n, to a value above 0.4[9]. This is easily explained by the secondary photon-generated electron tunneling model. The tunneling probability for secondary electrons will, however, increase with the value of I_n, the maximum being reached when every electron is followed by a secondary electron. This leads to :

$$I \rightarrow 2I_n + I_p \qquad \text{if} \qquad I_n \rightarrow \infty \tag{9}$$

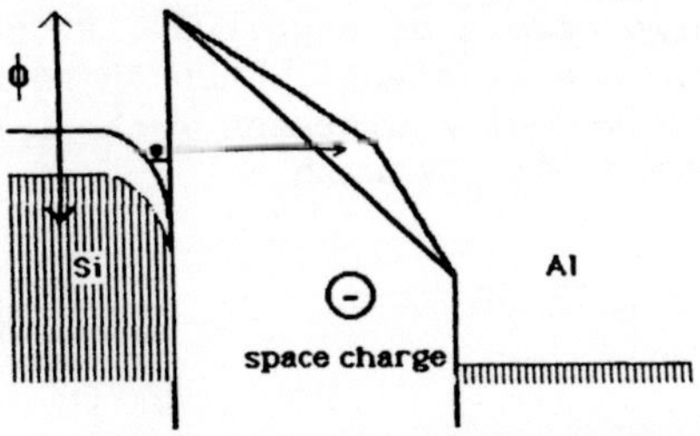

Fig. 4. Electrons tunneling through the energy barrier have to tunnel over a greater distance when negative space charge is present.

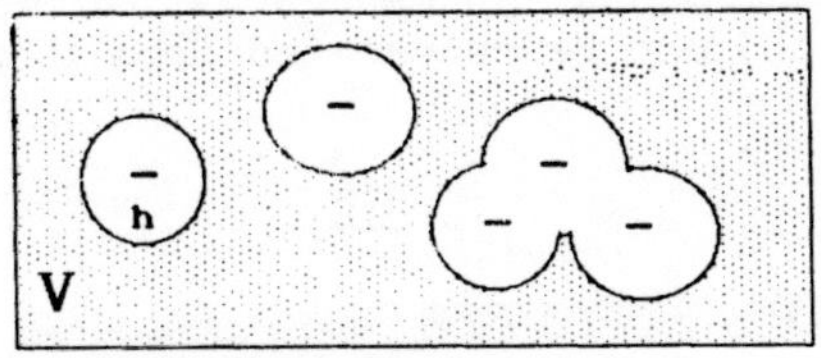

Fig. 5. A trapped electron repels adjacent trapping in a volume h, Note that these volumes can overlap.

The maximum value which the I_p/I_n ratio can have is 0.5.

c) Value of J_{max}

The value of J_{max} is about 4 orders of magnitude larger than that of the normal tunneling current. Although the maximum ratio $I_p/I_n = 0.5$ does not allow $I_p \geq I_n$, the **theoretical maximum** current density J_{max}^{holes} is orders of magnitude higher than that of primary electrons. Here again a simple explanation is possible, since photons with $h\nu = 3$ eV are absorbed in a layer about .25 μm thick. All the electrons less than 2 eV under the valence-band edge can be photon-stimulated to the conduction band and contribute to J_{max}^{holes}. From Eq.6 we calculate with $n_i \approx 10^{22} cm^{-3}$:

$$J_{max}^{holes} = \lambda nq\nu = 0.25x10^{-4}x10^{22}x1.6x10^{-19}x5x10^{13} = 2.10^{12} A/cm^2 \tag{10}$$

This is close to the observed value of $J_{max}^{holes} = 5.10^{11} A/cm^2$. The difference may be due to the fact that recombination can play a role even when the barrier is totally transparent.

TRAPPING KINETICS

The field at the injecting interface is the vital parameter for tunneling currents. This field is, however, not constant during stress. Space charge trapped in centers in the dielectric can distort the field at the interface, as has been depicted in fig. 4.

The effect of space charge is seen as a gradual change in the current during constant voltage stress or a change in voltage during constant current stress.

In order to avoid complex equations in the following derivations we shall start with a simplified form of the Fowler-Nordheim equation :

$$J = J_0 \exp\left(\frac{F}{F_0}\right), \tag{11}$$

which is a good approximation over one or two orders of magnitude. F is the actual field at the interface and F_0 and J_0 are constants. When there is a homogeneous space charge of density n centered at a distance x the actual field can be calculated from the applied field, F_{appl}, with :

$$F = F_{appl} - \frac{qn\bar{x}}{\epsilon}, \tag{12}$$

where ϵ is the dielectric constant of the dielectric. In order to account for the time evolution of n we must adopt a trapping model.

a) Trapping models

The literature is essentially concerned with three trapping models and these will be discussed breifly below.

First-order trapping

The trapping event depends solely on the availability of traps and the presence of electronic charge carriers. The trapping rate is given by :

$$\frac{dn}{dt} = (N - n)\, \frac{J}{q}\, \frac{\upsilon_{th}}{\upsilon_d}\, \sigma, \tag{13}$$

where υ_d and υ_{th} are the drift and thermal velocities of the electronic charge carriers, and σ is the capture cross section of the trap.

Trapping inhibited by Coulomb repulsion

Here the trapping rate is given by[104,35] :

$$\frac{dn}{dt} = (N - n)\exp\left(-\frac{nh}{V}\right)\frac{J}{q}\,\frac{\upsilon_{th}}{\upsilon_d}\,\sigma \tag{14}$$

The exponential term on the right comes from the assumption made that filling a trap adjacent to a filled trap is not possible when the latter is located within a distance $\bar{l}$. A space with a volume $h = 4\pi\bar{l}^{\,3}$ around the filled trap is not available for trapping (See fig. 5) V is the total volume of the dielectric. That we get an exponential term, exp(-nh/V) instead of a linear term (V - nh), is due to the fact that the spheres with volume h can overlap. It is easily shown that a single location can be overlapped by 8 adjacents spheres[105].

Trap generation

Following Badihi[106] we assume a generation rate for traps which is exponentially dependent on the electric field.

$$\frac{dn}{dt} = N_0 \exp\left(\frac{F}{F_0}\right)\frac{J}{q}\,\sigma \tag{15}$$

where F_0 and N_0 are constants. By inserting Eq.12 we can solve Eq.15.

Eqs. 13, 14 and 15 can all three be solved and their various solutions can be approximated within certain limits by :

$$n = N_{eff} \ln\left(\frac{Q}{q/\sigma^*} + 1\right) \tag{16}$$

where N_{eff} and σ^* are the effective density of traps and characteristic capture cross section, respectively, as specified in Table I. Note that Eq.16 is an approximation only for the first-order model. The differences between the models is predominantly the value of Q_{sat} relative to that of q/σ^*. A schematic diagram is given in figs. 6a, b and c.

Table I Expressions for N_{eff} and σ^* in Eq. 16 for the three trapping models.

Model	Density N_{eff}	cross section σ^*	Saturation Q_{sat}
First-order	N	$\sigma \dfrac{\upsilon_{th}}{\upsilon_d}$	$\dfrac{q}{\sigma^*}(e - 1)$
Coulomb rep.	V/h	$\dfrac{hN_0\upsilon_{th}}{Vq\upsilon_d}\sigma$	$\dfrac{q}{\sigma^*}(e^8 - 1)$
Trap generation	$\dfrac{\epsilon F_0}{q\overline{x}}$	$\dfrac{N_0 \overline{x}}{\epsilon F_0}\sigma \exp\left(\dfrac{FQ_{appl}}{F_0}\right)$	∞

Inserting Eq. 16 in Eqs. 11 and 12 we obtain the expression :

$$\ln\left(\frac{J}{J_0}\right) = \frac{F_{appl}}{F_0} - \frac{N_{eff} q\overline{x}}{\epsilon F_0} \ln\left(\frac{Q\sigma^*}{q} + 1\right) \tag{17}$$

Plotting log(J) vs log(Q) gives bent curves which incline to two asymptotes. The value of σ^* can be calculated from the intersection of the two asymptotes and the value of N_{eff} from their slopes. An example is given in fig. 7, while fig. 8 gives the values extracted vor σ and N_{eff}.

The most realistic model can be argued for as follows :

. **The first-order model** can be rejected since it predicts saturation of the curves at large Q_{inj} at a factor of 5 (=exp(e-1)) beyond the bending point, which saturation does not occur (cf. figs. 6 and 7).

. **Trap generation** is not appropriate since the value of σ^* should increase exponentially with the field F_{appl}. In practice, it decreases (cf. fig. 8).

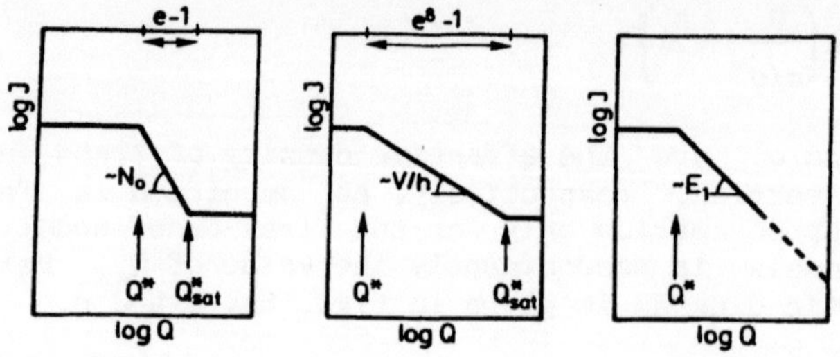

Fig. 6. 6a : First-order trapping : the capture is dominated by the number of empty traps. 6b : Trapping under Coulomb repulsion : More than one trap (all traps in volume h) are inactivated by a filled trap. Note that these volumes may overlap. 6c : Trap generation : The traps are generated by injection. Saturation need not necessarily take place.

Trapping inhibition by Coulomb repulsion appears the most realistic model here. Saturation is not observed but it should not be at a Q_{inj} less than a factor $e^8 - 1$ above q/σ^*. In reality the samples show a breakdown when measured above 1-10 C/cm^2 . This is below the point where saturation should occur.

The results show that Coulomb repulsion between adjacent traps is the most realistic model at this moment. Coulomb repulsion affects the trapping probability when traps are closer than 10-20 nm to each other which corresponds to a density of $N_{eff} \approx 10^{17}- 10^{18} cm^{-3}$. The same model (Eq. 17) has been used to analyse charge trapping at low fields with avalanche injection of hot electrons. As an example we have replotted data from Young[107] in fig. 9.

The $\ln(t + t_o)$ axis corresponds to a $\ln(Q\sigma^*/q + 1)$ axis. The extracted values for the capture cross section and density of traps are more consistent than those using the classical first-order model. The extracted data are given in the following table.

Table II : Trap characteristics of data of figure 9 using Eq. 17.

Conditions	$\sigma^* 10^{-18} cm^2$	$N_{eff} 10^{17} cm^{-3}$
No anneal	3.5	13
Form.gas 30 min	5.9	16.6
N2,30'+F.G 30'	3.3	8.6
N2,30 min	5.2	6.4
N2,60 min.	4.2	5.3

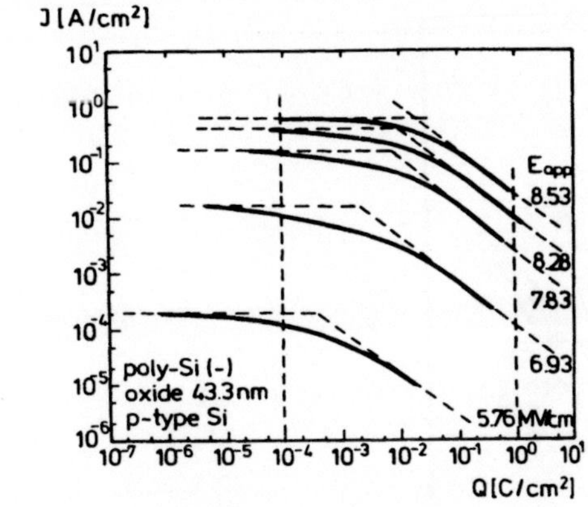

Fig. 7. Current transients on a double logarithmic scale. Note that the curves incline to two asymptotes.

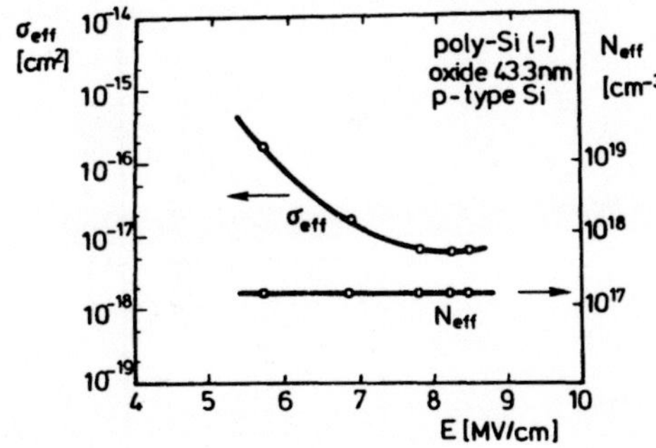

Fig. 8. The values of the effective cross section and effective density versus applied field calculated from curves in fig. 7.

The number of trap characteristics reduces by a factor of three compared to the first-order model which would have extracted about 15 different values. Also the consistency of the data is much better. The data for table II were taken from ref.107. A number of published data have been found to give consistent data using Eq.17[104,108].

THE EFFECT OF CHARGE INJECTION

The effect of charge injection can lead to severe degradation of the dielectric but even small amounts of charge may affect the properties of the MOS capacitor. First we shall discuss effects which, though small, are clearly detectable. Thse are the lowering of the effective tunneling barrier and the increase of dielectric losses. Then severe degradation will be examined.

a) Reduction of the Fowler-Nordheim barrier

In the following it will be shown that even at low fluences such as $Q_{inj} = 10^{-6} C/cm^2$ a dielectric can show degradation. Many papers have investigated changes in the space charge density due to stressing, and many have dealt with a systematic treatment of the Fowler-Nordheim curve. Not many have investigated the effect whereby the injection of charges degrades the barrier height[109].

The effect of degradation is illustrated in figs. 10a, b and c. Fig. 10a shows how the Fowler-Nordheim curves shift for a very small fluence of electrons. In fig. 10b the barrier height is plotted versus the fluence of electrons for two stressing conditions. As can be seen the degradation starts at a fluence as low as $10^{-6} C/cm^2$. Obviously, the

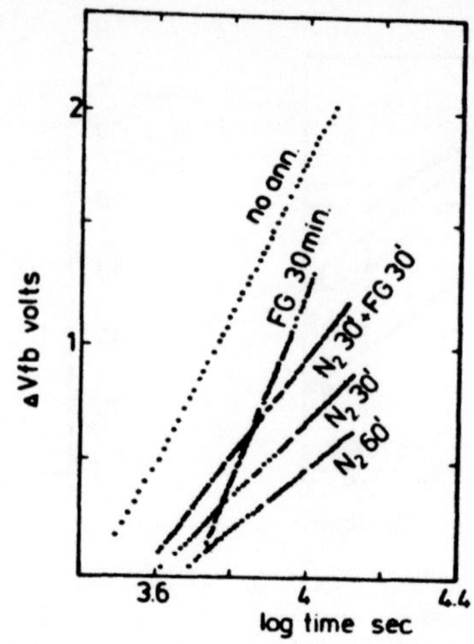

Fig. 9. Results of Young[107] replotted according to Eq.16. The trap characteristics are given in table II.

lowering of the barrier height is substantially larger when a stress current of $10^{-2}A/cm^2$ is used instead of $10^{-3}A/cm^2$. The critical charge needed for a shift is, however, practically the same. These shifts towards lower barrier height can be mistaken for an evolution of positive charge in the dielectric. As an example, the field at a fixed point of the current axis ($J = 10^{-5}A/cm^2$) is plotted versus Q_{inj} in fig. 10c.

Obviously, the applied field first decreases with injection as if there were a positive charge evolution. It will be clear that part, if not all, of the effect is due to barrier height changes and not to "hole trapping". The mechanism responsible for the lowering of the barrier is the generation of interface traps. These give rise to trap-assisted tunneling

b) The increase of losses

Many papers[51,53,66,110] have documented the generation of traps which can be measured by the distortion of CV curves. One of the most sensitive techniques consists in measuring the dielectric loss of a MOS capacitor. An example of losses generated by injected charge is given in fig. 11. As can be seen losses continuously increase due to charge injection. In most cases these losses are interpreted as interface states. The atomic picture of what is happening, however, is not at all clear[78].

a) Severe degradation

Properties of a dielectric change when a charge flux is sustained for some time. CV and IV analysis show that these changes comprise :

1. Change of the barrier height (interface traps)
2. Negative space charge (trapping)
3. Positive space charge (hole trapping or trap generation)

4. Increased AC losses (interface state generation ?)
5. Polarization effects (generation of slow states ?)
6. Dielectric breakdown.

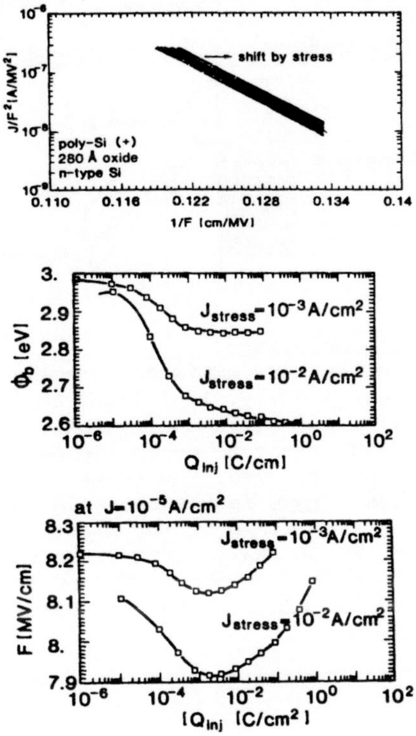

Fig. 10. 10a Fowler Nordheim plots of virgin and stressed capacitors. 10b. Fowler-Nordheim barrier height and 10c. The field at a current of $J = 10^{-5}$A/cm² as a function of charge injection.

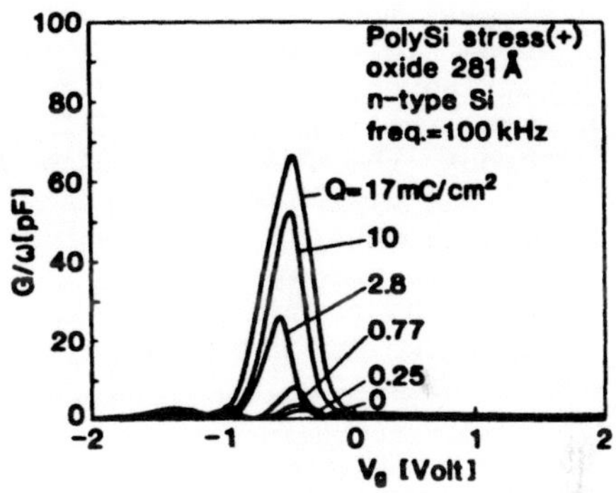

Fig. 11. The increase of the dielectric loss as a function of charge transport through the dielectric

As has been stated by several authors degradation due to charge injection is directly related to the eventual breakdown which occurs when sufficient charge has been injected[104]. The effect on the dielectric must therefore be related to the macroscopic features which form the low-ohmic path and finally cause breakdown. The generation of this path is caused by the energy dissipation of electrons losing their conduction-band energy at the anode. It is not vitally important to find out by what mechanism this transfer takes place[79]. It is clear, however, that 3-4 eV of energy is sufficient to break bonds at the interfaces. When a photon is generated its wavelength is about ten times the thickness of the dielectric. Therefore both the interfaces can absorb the energy. The result can be rupture of the network and hence formation of interface traps.

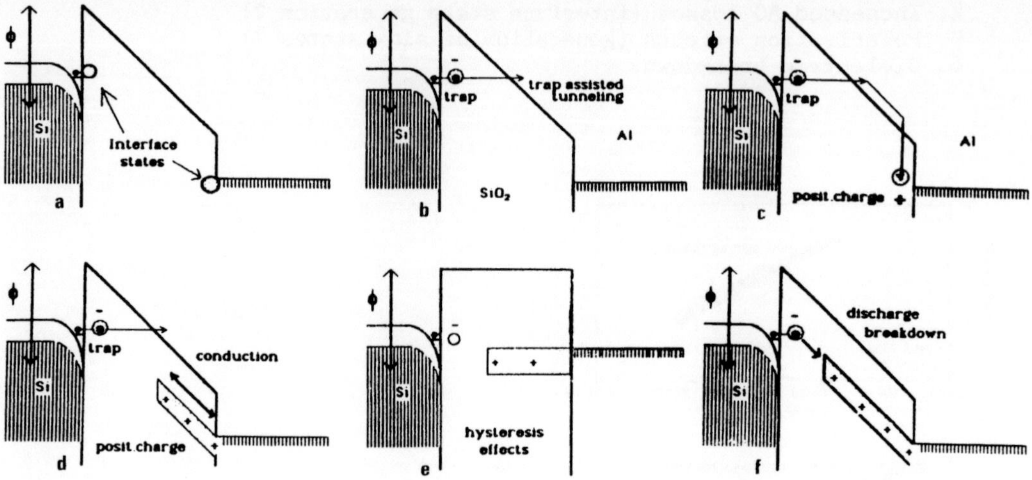

Fig. 12.

12a: The generation of broken bonds at the interface.
12b: Electrons can penetrate the barrier by trap assisted tunneling. Deeper than 3nm traps can be filled permanently. 12c: Traps at the anode interface will be charged positive and attract electrons. The confinement of energy dissipation can cause traps to be formed deeper into the dielectric. This generates a channel. 12d: Charge transport through the channels causes AC losses. 12e: Traps in the channels deep in the dielectric are only filled when the field of the electrodes overcompensates their mutual repulsion. Their charge content becomes field-dependent. Frozen in charge will cause hysteresis effects. 12f: When the channels grow and reach the region close to the cathode the capacitor may discharge its energy $1/2CV^2$.

The generation of the path is depicted in the following steps 1 to 5, which correspond with figs. 12a-f respectively.

1. When electrons cause broken bonds at the cathode the network can rupture over a region 1-2 nm into the dielectric. A state in this ruptured region may be observed from trap-assisted tunneling.

2. When traps are formed deeper than 3 nm from the cathode they may be filled permanently and form a negative space charge which locally inhibits further injection.

3. When a bond is broken at the anode, the interface trap can be charged positive. This charge attracts newly injected electrons which dissipate their conduction band energy close to the same location. This confines the energy dissipation to the damaged region. The region will protrude further into the dielectric and attract still more electrons, etc. If the region extends to locations more than 3 nm from the anode, a fixed positive charge may be formed. Charge can easily be exchanged with the electrode since the dielectric in between contains traps as well.

4. Charge transport via channels to places deeper in the dielectric gives rise to losses at high frequencies. If the channels increase in length the resonance frequency will decrease. Slow states are formed. The longer channels attract electrons more efficiently because of their

positively charged tips and this confines further energy dissipation and localizes the degradation of the dielectric.

5. When the channels, containing a large number of states, protude into the dielectric their charge content can increase in a strong field. The field is especially large when the channel is close to the counter-electrode. The saturation level of the traps thus becomes field-dependent and the space charge in the dielectric, which cannot be so easily exchanged any longer, causes hysteresis effects.

6. Finally, the channels reach the cathode within tunneling distance and the total energy of the biased capacitor $1/2CV^2$ can be discharged. This fast adiabatic process can burn a hole in the capacitor.

This simple macroscopic description of traps and losses may shed some light on a few hitherto unexplained observations[53,66,78,79].

Dynamic charging decharging

Dynamic trapping detrapping of charge in stressed dielectrics is caused by increasing and decreasing charge content of the channels.

Field-dependent saturation

The saturation level of the channels is dependent on the field. Together in one channel, charges exert a mutual repulsion and accumulation of charge can therefore only occur in a strong field. The stronger the field the more charge can be accomodated.

Centroid of charge

The centroid of positive charge is mainly at the anode for thin dielectrics. For reversed charge injection the centroid changes to the opposite electrode (the anode). This is consistent with the fact that the growing channels start at the anode.

Anode material

The evolution of positive space charge is independent of oxide quality by does depend on the anode work function. The energy loss at the anode is the driving force for channel growth, which will therefore be increased by a large work function.

Traps and states

There is a correlation between traps and interface states. This is due to the fact that traps in the channels close to the interface behave as states while those at some distance behave as traps. When channels are connected to the anode, traps behave as donors. In most cases, in fact donor-like centers are observed.

Charge and oxide thickness

Channels fill the volume of the dielectric to some extent. The thicker the dielectric the more positive charge[111].

Temperature and bias

Temperature- and bias-dependence are complex. Channels starting charge transport at a given temperature (e.g. 100 C) will release their positive charge and act as interface states. Generated at low temperatures, they may not conduct charge and act as states. When they are heated, they exchange charge with the anode and will contain positive charge (hole-trapping).

Breakdown depends on charge injection rather than on field

A breakdown is caused by injection of a certain amount of charge. This is due to the fact that all electrons losing their energy at the anode have at least the energy of its work function. The average additional energy gained in the field is small owing to the fact that electrons lose a lot of their energy by phonon interaction. Consequently, the quantity of electrons counts rather than their individual energy.

SUMMARY

A short description of the important parameters in the Fowler-Nordheim expression has been given. The tunneling probability and the maximum supply of electrons have been discussed. The substrate current, as a component of the tunneling current, possibly originates from photon stimulated emission. Trapping kinetics has been discussed using the probable inhibition of trapping by Coulomb forces. Degradation of MOS capacitors is of great importance in the interpretation of charge injection induced traps and states. It has been shown that fluences as small as $Q_{inj} \leq 10^{-6} C/cm^2$ are sufficient to change the dielectric's properties. Space charge and losses generated by charge injection are closely related to the low-ohmic paths growing from the anode to the cathode. Their existence has been used to model states and traps giving satisfactory explanations for phenomena hitherto not understood.

REFERENCES

1. R.H. Fowler and L. Nordheim, Proc. Roy. Soc. London, A119:173 (1928).
2. A.G.J. Van Oostrom, thesis, Amsterdam (1965).
3. M. Lenzlinger and E.H. Snow, J. Appl. Phys., 40:278 (1969).
4. Z.A. Weinberg, Solid State Elect., 20:11 (1977).
5. G. Krieger and R.M. Swanson, Appl. Phys. Lett., 39:818 (1981).
6. G. Krieger and R.M. Swanson, J. Appl. Phys., 52:5710 (1981).
7. Z.A. Weinberg, J. Appl. Phys., 53:5052 (1982).
8. B. Eitan and A. Kolodny, Appl. Phys. Lett., 43:106 (1983).
9. Z.A. Weinberg and M.V. Fischetti, J. Appl. Phys. 57:451 (1985).
10. V. Zekeriya and T. Ma., J. Appl. Phys., 56:1017 (1984).
11. K. Honda, A. Ohsawa and N. Toyokura, Appl. Phys. Lett., 45:270 (1984).
12. Z. Weinberg, D.R. Young, J. Calise, S. Cohen, J. Delucia and V. Deline, Appl. Phys. Lett.
13. G.G. Shirley, J. Electr. Chem. Soc., 132:488 (1985).
14. S. Jain and W. Dahlke, Solid State Elect., 29:597 (1986).
15. Z. Weinberg, M. Fischetti and Y. Nissan-Cohen, J. Appl. Phys., 59:824 (1986).
16. N. Matsukawa, S. Morita and H. Nozawa, Ext. Abstr. 16th Int. Conf. Solid. State Dev. Kobe 1984, 16:261 (1984).
17. H. Nozawa, N. Matsukawa and S. Morita, IEEE Trans. Elect. Dev., ED-33:275 (1986).
18. D.J. DiMaria, M.V. Fischetti and E. Tierney, Phys. Rev. Letts., 56:1284 (1986).
19. S. Horiguchi, T. Kobayashi, and K. Saito, J. Appl. Phys., 58:387 (1985).
20. S. Horiguchi and H. Yoshino, J. Appl. Phys., 58:1597 (1985).

21. A.K. Ghosh, T. Feng, J. Haberman and H.P. Maruska, J. Appl. Phys., 55:2990 (1984).
22. J. Halbritter, J. Appl. Phys., 58:1320 (1985).
23. P. Chattopadhyay and A.N. Daw., Solid. State Electr., 28:831 (1985).
24. Y. Nissan-Cohen, J. Shappir and D. Bentchkowsky, J. Appl. Phys., 58:2252 (1985).
25. E. Rosencher, R. Coppard and D. Bois, J. Appl. Phys., 57:2823 (1985).
26. T.N. Theis, D.J. DiMaria, J.R. Kirtley and D.W. Dong, Phys. Rev. Letts., 52:1445 (1984).
27. C.W. Jen, C.L. Lee and T.F. Lei, Solid. State Elect., 27:1 (1984).
28. P. Olivo, B. Ricco and E. Sangiorgi, J. Appl. Phys., 54:5267 (1983).
29. T. Changhua, H. Mingzhen and W. Yanguyan, J. Appl. Phys., 54:4398 (1983).
30. Y.K. Sharma, R.N. Sharma and V.S. Raghav, J. Appl. Phys., 54:4213 (1983).
31. J. Tang and K. Hess, J. Appl. Phys., 54:5145 (1983).
32. D. DiMaria, D. Dong, C. Falcony, T. Theis, J. Kirtley, J. Tsang, D. Young, F. Pesavento and S. Brorson, J. Appl. Phys., 54:5801 (1983).
33. A. Carim and A. Bhattacharyya, Appl. Phys. Lett., 46:872 (1985).
34. K. Nagai and U. Hayashi, Appl. Phys. Lett., 44:910 (1984).
35. D. Wolters and J. Van der Schoot, J. Appl. Phys., 58:831 (1985).
36. M. Garrigues, A. Pavlin and Y. Hellouin, J. Appl. Phys., 58:2672 (1985).
37. S. Brorson, D. DiMaria, M. Fischetti, F. Pesavento, P.M. Solomon and D. Dong, J. Appl. Phys., 58:1302 (1985).
38. D. DiMaria, T. Theis, J. Kirtley, F. Pesavento, D. Dong and S. Brorson, J. Appl. Phys., 57:1214 (1985).
39. A. Duong and A. Nassibian, J. Appl. Phys., 57:1256 (1985).
40. T. Hosoi, M. Azikawa and S. Matsumoto, J. Appl. Phys., 57:2072 (1985).
41. D. DiMaria, D. Dong, F. Pesavento, C. Lam and S. Brorson, J. Appl. Phys., 55:3000 (1984).
42. A. Khan, J. Woolam and Y. Chung, J. Appl. Phys., 55:4299 (1984).
43. C. Falcony and J. Helman, J. Appl. Phys., 54:442 (1983).
44. Z.A. Weinberg and A. Hartstein, J. Appl. Phys., 54:2517 (1983).
45. S. Kar, IEEE Trans. Elec. Dev., ED-34:420 (1987).
46. M. Aslam, R. Singh and P. Balk, Phys. Stat. Solidi a, 84:659 (1984).
47. I.P. Mikhailoxskii and A.E. Epov, Phys. Stat. Solidi a, 92:615 (1985).
48. R. Devine, J. Appl. Phys., 60:468 (1986).
49. D. Vuillaume and J. Bourgoin, J. Appl. Phys., 58:2077 (1985).
50. S. Lai, J. Appl. Phys. 54:2540 (1983).
51. S.K. Lai and D.R. Young, J. Appl. Phys., 52:6231 (1981).
52. C. Sah and J. Sun, J. Appl. Phys., 54:4378 (1983).
53. Y. Nissan-Cohe, J. Shappir and D. Frohman-Bentchowsky, J. Appl. Phys., 57:2830 (1985).
54. S. Manzini and F. Volonte, J. Appl. Phys., 58:4300 (1985).
55. E. Rosencher and R. Coppard, J. Appl. Phys., 55:971 (1984).
56. M. Johnson, D. Biegelsen, M. Moyer, S. Chang, E. Poindexter and P. Caplan, Appl. Phys. Lett., 43:563 (1983).
57. I. Strzalkowski, M. Marczewski and M. Kowalski, Thin Solid. Films, 99:331 (1983).
58. K. Agarwal and M. White F. Rhodes, Appl. Phys. Lett., 43:465 (1983).
59. S. Chang, H. Lam, B. Chen and J. Keenan, Appl. Phys. Lett., 48:663 (1986).
60. A. Stesman, Appl. Phys. Lett., 48:972 (1986).
61. N. Shiono, M. Shimaya and O. Nakajima, Appl. Phys. Lett., 48:1129 (1986).
62. C. Sah, W. Lin, C. Hsu and S. Pan, Appl. Phys. Lett., 48:1736 (1986).
63. S. Kar, D. Shanker and K. Chari, Appl. Phys. Lett., 47:1353 (1985).
64. S. Kar and S. Varma, J. Appl. Phys., 58:4256 (1985).
65. J. Tzou, J. Sun and C. Sah, Appl. Phys. Lett., 43:861 (1983).

66. Y. Nissan-Cohen, J. Shappir and D. Frohman-Bentchkowsky, J. Appl. Phys., 58:2252 (1985).
67. R. Devine, J. Appl. Phys., 58:716 (1985).
68. M. Heyns and R. de Keersmaecker, J. Appl. Phys., 58:3936.
69. W. Eades and R. Swanson, J. Appl. Phys., 58:4267 (1985).
70. U. Sharma and W. Dahlke, J. Appl. Phys., 57:1186 (1985).
71. C. Jorgensen, C. Svensson and K.H. Ryden, J. Appl. Phys., 56:1093 (1984).
72. C. Sah, J.C. Sun and J. Tzou, J. Appl. Phys., 55:1525 (1984).
73. M. Itsumu, J. Appl. Phys., 54:1930 (1983).
74. J.J. Tzou, C.C. Yao, R. Cheung and H. Chang, IEEE El. Dev. Letters, EDL-7:446 (1986).
75. G. Haller, N. Knoll, D. Braunig, F. Wulf and W. Fahrner, J. Appl. Phys., 56:1844 (1984).
76. N. Shiono, O. Nakajima and C. Hashimoto, J. Electrochem. Soci., 130:138 (1983).
77. C. Hsu, S. Pang and C.T. Sah, J. Appl. Phys., 58:1326 (1985).
78. M.V. Fischetti, Z.A. Weinberg and J.A. Calise, J. Appl. Phys., 57:418 (1985).
79. M.V. Fischetti, J. Appl. Phys., 57:2860 (1985).
80. M. Heyns, R. DeKeermaecker and M. Hillen, Appl. Phys. Lett., 44:202 (1984).
81. R. Devine and A. Golanski, J. Appl. Phys., 55:2738 (1984).
82. S. Hu, Appl. Phys. Lett., 43:449 (1983).
83. M. Jourdain, G. Salace, C. Petit, M. Favre and J. Despujols, Soli. State Elec., 26:251 (1983).
84. V. Zekeriya and T. Ma, Appl. Phys. Lett., 43:95 (1983).
85. W. Schmitz and D.R. Young, J. Appl. Phys., 54:6443 (1983).
86. P. Olivo, B. Ricco and E. Sangiorgi, Appl. Phys. Lett., 48:1135 (1986).
87. T. Hook and T. Ma, Appl. Phys. Lett., 48:1208 (1986).
88. P. Pan and C. Paquette, Appl. Phys. Lett., 47:473 (1985).
89. S. Haywood and R. de Keersmaecker, Appl. Phys. Lett., 47:381 (1985).
90. J. Wu, S. Lyon and W. Johnson, Appl. Phys. Lett., 42:585 (1983).
91. H. Boesch and F. McClean, J. Appl. Phys., 60:448.
92. C. Falcony and F. Salas, J. Appl. Phys., 59:3787 (1986).
93. C. Fiori, R. Devine and P. Meiland, J. Appl. Phys., 58:1058 (1985).
94. R. Pfeffer, J. Appl. Phys. 57:5176 (1985).
95. H. Ringel, M. Knoll, D. Braunig and W. Fahrner, J. Appl. Phys., 57:393 (1985).
96. J. Verdebout and R.L. Booker, J. Appl. Phys., 55:406 (1984).
97. J. Tzou, J. Sun and C. Sah, J. Appl. Phys., 55:846 (1984).
98. E. Takeda, T. Hagiwara and N. Suzuki, J. Appl. Phys., 55:3180 (1984).
99. D.R. Wolters, J.J. Van der Schoot and T. Poorter, Insulating films on Semiconductors, eds. J.F. Verwey and D. Wolters, North Holland, 3:256 (1983).
100. D.R. Wolters, Springer series in Electrophysics, eds. M. Schulz and G. Pensl, 7:180 (1981).
101. M. Hirose, M. Morita and Y. Osaka, Jap. J. of Appl. Phys., 16:561 (1977).
102. B. Ricco and M. Ya. Azbel, Phys. Rev. B 29:1970 (1984).
103. A. Hartstein and R.H. Kock, Surface Science, 170:391 (1986).
104. D.R. Wolters and J.F. Verwey, Springer series in Electrophysics, eds. M. Schulz and G. Pensl, 7:111 (1981).
105. D.R. Wolters and J.J. Van der Schoot, J. Appl. Phys. 58:831 (1985).
106. A. Badihi, B. Eitan, I. Cohen and J. Shappir, Appl. Phys. Lett., 40:396 (1982).
107. D.R. Young, E. Irene, D. DiMaria and R. Dekeersmaecker, J. Appl. Phys. 50:6366 (1979).
108. H. Bartholomae, Thesis, Erlangen, 202 (1986).

109. D.R. Wolters and H.L. Peek, Solid. State Elect., accepted for publ., (1987).
110. E. Harrari, J. Appl. Phys., 49:2478 (1978).
111. K.R. Hofman and G. Dorda, Springer series in Electrophysics, eds. M. Schulz and G. Pensl, 7:122 (1981).

THE ROLE OF HOLE TRAPS IN THE DEGRADATION OF THERMALLY GROWN SiO_2 LAYERS

M.M. Heyns and R.F. De Keersmaecker

Interuniversity Microelectronics Center (IMEC vzw)
Kapeldreef 75, B-3030 Leuven
Belgium

INTRODUCTION

Due to the high fields present in small-geometry metal-oxide-semiconductor (MOS) transistors charge injection into the gate oxide can readily occur leading to serious reliability problems related to the degradation of the SiO_2 layer and of the Si/SiO_2 interface. It is therefore extremely important to have a thorough understanding of the degradation mechanisms occurring upon charge injection.

The generation of fast states and of slow trapping instabilities during electron avalanche injection[1-7] or during electron injection using Fowler-Nordheim tunneling[8,9] has been demonstrated and the generation kinetics have been reported[10] ; their origin, however, is still poorly understood. In this work the close relationship between the slow trapping instabilities and the hole traps which are present near the Si/SiO_2 interface is demonstrated. The importance of the hole trap density in the degradation of SiO_2 is emphasized.

SLOW TRAPPING INSTABILITIES

Slow trapping instabilities (or slow states) have been observed in the form of anomalous positive charge near the Si/SiO_2 interface during electron injection. This charge responds to changes in the silicon surface potential and/or the internal field and, thereby, causes instabilities in the MOS-system. The time constant for charge exchange between these states and the silicon substrate typically ranges from seconds to several hours depending on temperature and on the history of the sample[3]. Several models based on a variety of physical mechanisms have been proposed to explain the build-up of this positive charge : the liberation of hydrogen from water-related trapping sites in the SiO_2 layer and the diffusion of this hydrogen to the Si/SiO_2 interface[4], the diffusion of neutral species (e.g. excitons[11]), the generation of hot holes at the anode as a consequence of the energy loss of the electrons when they exit from the oxide layer[12] (with the generation of surface plasmons as an intermediate step[13]), the generation of electron-hole pairs during high-field stressing via band-to-band impact-ionization[14-17] or trap-to-band impact ionization[18] and field-stripping of electrons from valence band orbitals[19].

When slow trapping instabilities are present during the injection experiment their density is usually derived from the generated positive interface charge[1,2,4-7]. It was demonstated[3], however, that due to the dynamic equilibrium between charging and discharging of these states only an unknown fraction of the total density is measured using this procedure.

Therefore, a more precise measurement technique based on the charging/discharging characteristics of these slow states as proposed in ref. 3 is used in this study. In this technique the slow state density is quantified by taking the difference, V_{ss}, between the midgap voltages

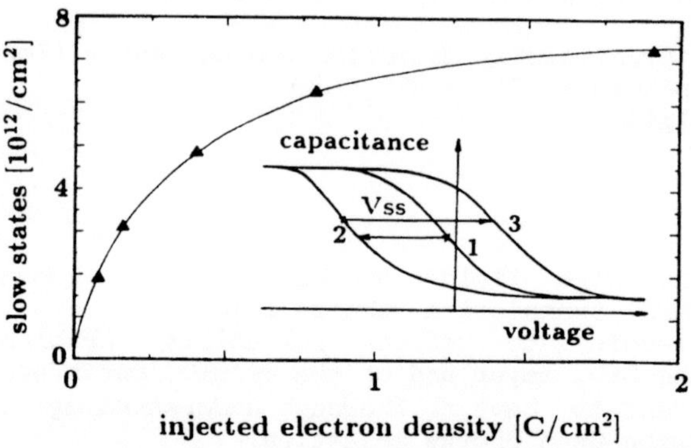

Fig. 1. Generated slow state density during avalanche injection of electrons as a function of the injected electron density for an Al-gate MOS sample with a 39 nm thick thermal oxide. The measurement procedure is shown schematically in the inset. Charging of the slow states is performed by applying a negative bias of -4MV/cm at 160°C for 10 min. This causes th high-frequency capacitance versus voltage (C-V) curve to shift from curve 1 (measured after charge injection) to curve 2. A positive bias of +4MV/cm at 160°C for 30 min. is used for neutralization of the slow states. The resulting C-V curve is given by curve 3. The midgap voltage shift between curve 2 and 3 (V_{ss}) is taken as a measure for the total amount of slow states present.

measured at room temperature and obtained after respectively charging and discharging (or neutralizing) all the slow states. This procedure is schematically demonstrated in the inset of fig. 1. The areal density of the slow states is calculated from V_{ss} by assuming that all slow states are located at (or at least very close to) the Si/SiO_2 interface. For every measurement a new capacitor is used. It was demonstrated that this procedure yields reproducible results and that only negligible annealing of the slow states occurs during the measurement procedure.

SAMPLE PREPARATION

Both p-type and n-type <100> silicon wafers with a doping level of more than $10^{17}cm^{-3}$ were used in order to allow laterally homogeneous avalanche injection of respectively electrons and holes. Oxidation was performed in a double-walled furnace tube[9] at 900°C in dry O_2 to a thickness of 39 nm. The wafers received a post-oxidation anneal for 5 min in N_2 at the same temperature. After the deposition of an aluminum layer using magnetron sputtering, capacitors of different areas were formed by standard wet lithography. Finally the samples received a post-metallization anneal at 435°C in forming gas for 20 min. The reported phenomena, however, were not critically dependent on processing details and are at least qualitatively typical for VLSI-grade SiO_2 layers.

BUILD-UP OF SLOW STATES

An identical saturation level for the generated slow state density is found whether the slow states are generated by means of avalanche injection of electrons (shown in fig. 1) or by high-field stressing with constant current density (shown in fig. 2). Only the generation rate is about 5 times larger during high-field stressing than during avalanche injection of electrons. During high-field stressing the generation rate is almost independent of the injection current density which was varied over 5 orders of magnitude and, rather, has a tendency to decrease with increasing injection current. The stress polarity influences the generation rate only at low injected charge where positive stress (injection from the silicon) is found to be more efficient in generating slow states than is

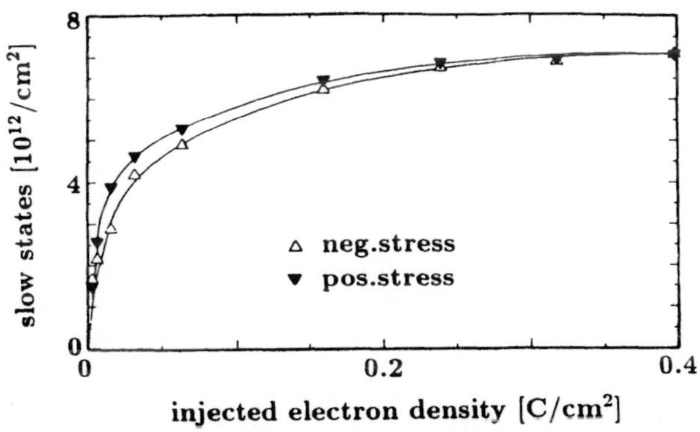

Fig. 2. Generated slow state density as a function of the injected electron density during constant current high-field injection with either negative or positive gate polarity for an Al-gate MOS-sample with a 39 nm thick thermal oxide.

negative stress (see fig. 2). The saturation level of the generated slow state density is independent of the injected current density and of the stress polarity.

These results indicate that there are a limited and fixed number of sites in the oxide network which can be converted into slow states, independent of the stress mode responsible for generating the slow states.

CORRELATION BETWEEN SLOW STATES AND HOLE TRAPS

The donor-type behaviour of the slow states suggests a possible correlation with hole traps. Therefore, the generation of slow states during avalanche injection of holes was studied. The generated slow state density is shown in fig. 3 together with the amount of trapped holes. In this experiment the quantification technique was slightly modified in that the order of the charging and discharging of the slow states was reversed. This is schematically shown in the inset of fig. 3. In the first measurement step the slow states are neutralized. This also results in a complete emptying of the charged hole traps and shifts the C-V curve from curve 1 (measured after avalanche hole injection) to curve 2. In the second step charging of the slow states is performed, shifting the C-V curve to curve 3. Only the positive charge that can be regenerated by this charging procedure under negative bias is measured and correlated with slow states. This charge density is calculated from the midgap voltage difference (V_{ss}) between curves 2 and 3.

This modified measurement procedure ensures that backtunneling of trapped holes to the silicon substrate during the neutralization step does not interfere with the measurement of the slow state density. The actual density of slow states may be underestimated due to an incomplete charging of the slow states after the neutralization procedure[3].

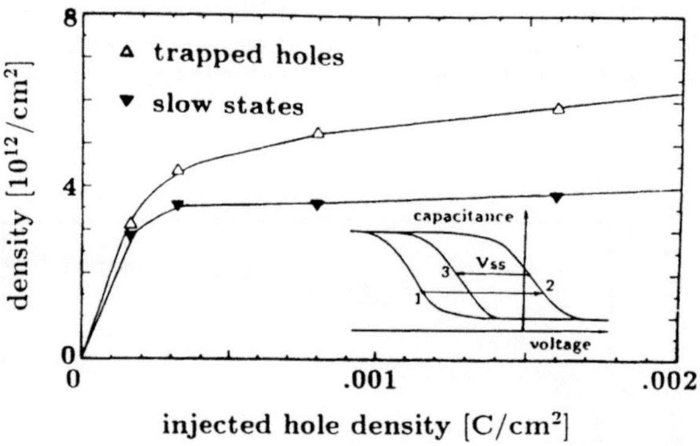

Fig. 3. Generated slow state density and trapped hole density during avalanche injection of holes as a function of the injected hole density for an Al-gate MOS sample with a 39 nm thick thermal oxide.

As shown in fig. 3, after hole injection and trapping the procedure for slow state measurement always yields a saturated slow state density equal to 60-70 % of the previously trapped hole density. The high efficiency for generating slow states found for avalanche hole injection is a clear demonstration of the role of hole trapping in the generation of slow states. In the experiment described hereafter it is demonstrated that the positive charge found in the slow states is located in hole traps already initially present.

When avalanche hole injection is performed on an MOS-structure in which positively charged slow states are first generated (e.g. by means of high field stressing) less holes can be trapped as compared to a virgin sample. This is shown in fig. 4a for both positive and negative high-field stressing. The saturation value of the midgap voltage shift during avalanche hole injection differs by an amount corresponding to the slow states which were positively charged at the start of the injection experiment. Therefore, when the absolute midgap voltage during hole injection is plotted for both samples, as shown in fig. 4b, the saturation

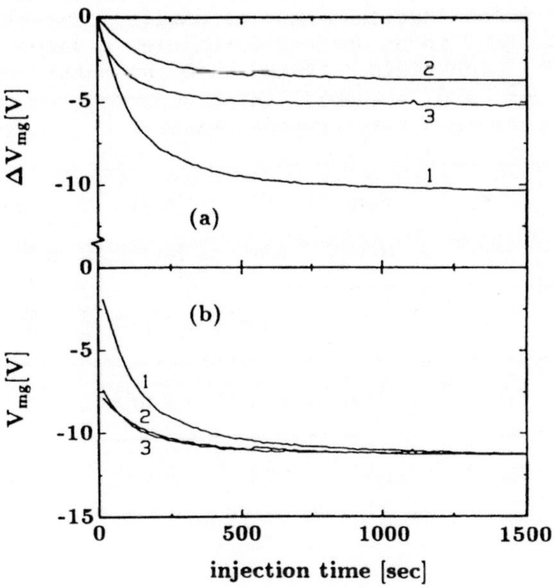

Fig. 4. Midgap voltage shift (a) and absolute midgap voltage (b) as a function of time during avalanche injection of holes with a constant current density of 6.37 x $10^{-7} A/cm^2$ on an Al-gate MOS sample with a 39 nm thick thermal oxide. Curve 1 was obtained on a virgin sample. Curves 2 and 3 were obtained on samples which respectively received a negative and a positive high-field stress with a total injected charge density of 0.159 C/cm^2 prior to the avalanche hole injection.

value and thus the total amount of positive charge which is built up in both samples is identical. This demonstrates that no new hole traps are generated during the stress and, therefore, is a strong indication that the positive charge present in the slow states is located in initially present hole traps.

It is important to remark that not all hole traps which are present in the oxide layer can be filled during avalanche injection of holes, due to the detrapping from hole traps which are located close to the Si/SiO_2 interface during the accumulation part of the avalanche injection pulse[20]. This, however, does not affect the conclusions made from this experiment because it influences the measured saturation level for both the stressed and the non-stressed samples to the same extent.

It could be argued that the saturation level found during avalanche injection of holes is not limited by the total number of hole traps present in the oxide but by the coulombic repulsion which limits the total charge in a given volume depending on the coulombic potential around the trapped hole. In this respect it must be noticed that during high-field stressing a much larger density of positive charge could be built up in the oxide layer (in spite of the larger field) which indicates that the saturation level is not imposed by coulombic repulsion in this experiment.

Table 1. Capture cross section and effective concentration of the hole traps present in an unstressed sample, a sample which underwent a negative high-field stress with a total charge density of 0.159 C/cm^2 and a sample which underwent a positive high-field stress up to this total charge density.

	σ_1 [cm^2]	N_{eff1} [cm^{-2}]	σ_2 [cm^2]	N_{eff2} [cm^{-2}]	σ_3 [cm^2]	N_{eff3} [cm^{-2}]
no stress	—	—	$2.3\ 10^{-15}$	$4.4\ 10^{12}$	$5.1\ 10^{-16}$	$1.3\ 10^{12}$
neg. stress	$6.7\ 10^{-15}$	$8.9\ 10^{11}$	$2.1\ 10^{-15}$	$1.9\ 10^{12}$	—	—
pos. stress	$8.2\ 10^{-15}$	$1.0\ 10^{12}$	$2.0\ 10^{-15}$	$1.8\ 10^{12}$	—	—

The study of the capture cross section of the hole trapping process before and after the generation of the slow states (shown in table 1) using a first order kinetic trapping model[21] reveals that hole trapping is not a fully reversible process i.e. the oxide has a memory for its history concerning hole trapping. It is seen from table 1 that although no new hole traps are generated during high field stressing (as demonstrated in fig. 4) a new capture cross section of 7-8 x $10^{-15} cm^2$ is found when avalanche hole injection is performed after high-field stressing and partial neutralization of the slow states. This new capture cross section is taken as evidence for a change in the nature of the hole traps concurrent with the generation of slow states. After the capture and detrapping of a hole the capture cross section of the center for recapturing a hole is strongly increased. It must be noted that on all three samples an identical injection current of 6.37 x 10^{-7} A/cm^2 was used (and thus an identical injection field) during the avalanche hole injection experiment. Therefore, even if the absolute values of the capture cross sections would be unreliable, it is excluded that this new capture cross section would be an artifact of a strong field dependence of the hole trapping process.

DISCUSSION

From these experiments it follows that slow states are generated from initially present hole traps. Based upon the recent observations and considerations of Grunthaner et al. concerning the structure of SiO_2 near the Si/SiO_2 interface[22] the following model is proposed. The capture of a hole (which is equivalent to the release of an electron) at a strained Si-O bond reduces the bond strength at this site which favors the rupture of this weakened bond. The cleavage of an Si-O bond results in the generation of two species : (i) a non-bridging oxygen (NBO) defect and (ii) a trivalent Si state bound to three oxygen atoms (O_3 - Si°). Due to the local strain in the SiO_2 network near the Si/SiO_2 interface[22] relaxation is likely to occur, preventing the direct recombination of the formed defects. The immobile O_3 - Si° defect will act as an efficient hole trap. This defect was already suggested as the possible origin for the slow trapping instability[5]. The local strain near the Si/SiO_2 interface and the presence of hydroxyl groups are the driving forces for the migration of the NBO defect towards the Si/SiO_2 interface where it forms a fast interface state. This explains the observed correlation between slow and fast states[6]. The generated fast interface states play a key role in the charging and discharging of the slow states[2,3] probably by facilitating the charge exchange between the slow state and the silicon substrate. The probability for this charge exchange determines whether the O_3 - Si° defect will be perceived as a hole trap or as a slow state. Wet oxides provide more OH groups, which facilitate the diffusion of the NBO defect towards the silicon and thereby contribute to preventing the recombination of the formed defects. Therefore, slow states are more easily generated in wet oxides[4].

The NBO defect will tend not to migrate until the trapped hole is neutralized because of its attraction to positive charge[22]. Therefore, both the fast[23] and the slow interface states are formed by a two-step process governed by the migration of the NBO defect after the neutralization of the trapped hole. This implies that in the case of avalanche injection of holes the actual slow state generation only occurs during the neutralization of the trapped holes in the first step of the slow state measurement procedure. The possible restoring of the bond, before the migration of the NBO stabilizes the defect, is the reason why only 60-70 % of the trapped holes effectively generate a slow state.

RELEVANCE OF THE HOLE TRAP DENSITY

The experiments described above demonstrate the importance of hole trapping in the degradation of the Si/SiO_2 system during charge injection. Hole trapping is also known to play an important role in the generation of fast surface states[23] and in the negative bias instability during bias-temperature stress measurements[24]. Recently it was also proposed that hole trapping plays an important role in the generation of electron traps [25]. In this respect it is important to notice that although hole trapping appears to play a more important role in oxide degradation than electron trapping, no evidence was found for the generation of extra hole traps during high field stressing neither at the Si/SiO_2 interface (as can be concluded from fig. 4) nor at the gate/SiO_2 interface, as was concluded from a combination of internal photoemission and avalanche hole injection experiments[20]. This despite the observation that for Al-gate capacitors a large amount of positive charge is built up under the Al-electrode after a positive high-field stress[26]. In contrast the generation of electron traps during high-field stressing has already been clearly demonstrated[26,27].

Although a large amount of positive charge is generated during electrical stressing in the pre-breakdown regime the existence of a relation between hole trapping and breakdown[28] has recently been seriously questioned[29,30]. The fact that all positive charge effects are due to initially present hole traps and that no new hole traps are created during high-field stressing is a further indication that hole traps are not the direct cause for breakdown. They could, however, affect the field distribution in the oxide during the stressing. It must be noticed that the role of shallow hole traps (which can not be filled at room temperature) on the degradation during high-field stressing has not been clarified so far and it is possible that they are generated in large amounts during electrical stressing.

Because of the evidence for the generation of electron traps during high-field stressing, it is tempting to identify them as the precursors for oxide breakdown. However, it has been shown that a saturation level exists for the total amount of electron traps generated during high-field stressing[20] whereas a continuous increase is expected for the defect which is to be the primary cause for breakdown. On the other hand, it must be noted that these deep electron traps may be amphoteric in nature and can also act as shallow hole traps. The existence of an amphoteric trap has already been suggested before[29,31,32]. The effect of these defects on the internal field distribution during the high-field stressing could, therefore, be rather complicated and new measurement methods and experiments are needed to complete the picture. The charge detection methods which have been applied so far either (i) remove the high field (such as C-V and internal photoemission measurements) and thereby may change the occupancy of the oxide defects, or (ii) measure at high-fields (such as the measurement of the voltage shift needed to sustain a constant current) but yield only limited information which is difficult to interpret.

CONCLUSION

The generation kinetics of the slow trapping instability during charge injection were compared. The saturation level of the generated slow states was independent of the charge injection method and the generation rate can be drastically increased when holes are directly introduced into the oxide. It was shown that the slow states originate from hole traps upon sequential trapping of a hole and an electron. The important role of the hole traps in the degradation of the SiO_2 layer during stressing was discussed. From this it can be concluded that in order to diminish the degradation of the Si/SiO_2 system during charge injection it is important to lower the density of hole traps in the oxide layer. It is, therefore, expected that the past efforts to improve the radiation hardness of the Si/SiO_2 system will also serve as a guideline for improving the resistance to hot carrier degradation in small geometry MOSFET's.

NOTE ADDED IN PROOF : The model proposed above to explain the transformation of hole traps into slow trapping instabilities and fast interface states is not unique. An alternative model based on the release and subsequent diffusion of hydrogen from Si-OH and/or Si-H bonds in the oxide layer[33,34] can be formulated[35] within which the experimental observations can equally be accounted for. Therefore, additional information is needed before a definitive detailed and microscopic model for hole trapping and slow state generation can be proposed. The study of the Si-OH and Si-H bond distribution in the SiO_2 layer and the technological dependence of both these distributions and the generation kinetics of the slow states must be combined with a detailed study of the characteristics of the generated defects in order to allow discrimination between the various models.

REFERENCES

1. D.R. Young, E.A. Irene, D.J. DiMaria, R.F. De Keersmaecker and H.Z. Massoud, J. Appl. Phys. 50:6366 (1979).
2. S.K. Lai and D.R. Young, J. Appl. Phys. 52:6321 (1981).
3. M.M. Heyns and R.F. De Keersmaecker "Characterisation of slow interface states created during avalanche injection in MOS devices", paper presented at the 6th Solid State Device Technol. Symp. (ESSDERC), Toulouse, France, Sept. 1981.
4. F.J. Feigl, D.R. Young, D.J. DiMaria, S.K. Lai and J.A. Calise, J. Appl. Phys. 52:5665 (1981).

5. M.V. Fischetti, R. Gastaldi, F. Maggioni and A. Modelli, J. Appl. Phys. 53:3120 (1982).
6. M.V. Fischetti, R. Gastaldi, F. Maggioni and A. Modelli, J. Appl. Phys. 53:3136 (1982).
7. C.T. Sah, J.Y. Sun and J.J. Tzou, J. Appl. Phys. 54:944 (1983).
8. K.R. Hofmann and G. Dorda in "Insulating Films on Semiconductors", Ed. M. Schulz, Springer, Berlin, (1981), p. 122.
9. M.W. Hillen, R.F. De Keersmaecker, M.M. Heyns, S.K. Haywood and I.S. Daraktchiev in "Insulating Films on Semiconductors", Eds. J.F. Verweij and D.R. Wolters, North-Holland, Amsterdam, (1983), p. 274.
10. M.M. Heyns and R.F. De Keersmaecker in "Dielectric layers in semiconductors : novel technologies and devices 1986", Eds. G. Bentini, E. Fogarassy and A. Golanski, Les Editions de Physique, Les Ulis Cedex, (1986), p. 303.
11. Z.A. Weinberg and G.W. Rubloff, Appl. Phys. Lett. 32:184 (1978).
12. Z.A. Weinberg, W.C. Johnson and M.A. Lampert, J. Appl. Phys. 47:248 (1976).
13. M.V. Fischetti, Phys. Rev. B 31:2099 (1985).
14. T.H. DiStefano and M. Shatzkes, Appl. Phys. Lett. 25:685 (1974).
15. M. Shatzkes and M. Av-Ron, J. Appl. Phys. 47:3192 (1976).
16. P. Solomon and N. Klein, Solid State Commun. 17:1937 (1979).
17. M. Knoll, D. Braunig and W.R. Fahrner, J. Appl. Phys. 53:6964 (1982).
18. Y. Nissan-Cohen, J. Shappir and D. Frohman-Bentchkowsky, J. Appl. Phys. 58:2252 (1985).
19. P. Olivo, B. Ricco and E. Sangiorgi, J. Appl. Phys. 54:5267 (1983).
20. M.M. Heyns, Ph. D.-thesis, K.U. Leuven (1986).
21. T.H. Ning and H.N. Yu, J. Appl. Phys. 45:5373 (1974).
22. F.J. Grunthaner and P.J. Grunthaner, Materials science reports, Vol. 1, Number 2,3, North-Holland, Amsterdam, (1986).
23. S.K. Lai, J. Appl. Phys. 54:2540 (1983).
24. S.K. Haywood and R.F. De Keersmaecker, Appl. Phys. Lett. 47:381 (1985).
25. I.C. Chen, S. Holland and C. Hu, J. Appl. Phys. 61:4544 (1987).
26. M.M. Heyns and R.F. De Keersmaecker, J. Appl. Phys. 58:3936 (1985).
27. M.M. Heyns, R.F. De Keersmaecker and M.W. Hillen, Appl. Phys. Lett. 44:202 (1984).
28. S. Holland and C. Hu, J. Electrochem. Soc. 133:1705 (1986).
29. Z.A. Weinberg and T.N. Nguyen, J. Appl. Phys. 61:1947 (1987).
30. J.J. Tzou, C.C. Yao, R. Cheung and H. Chan, IEEE Electron Device Lett. EDL-7:446 (1986).
31. M. Aslam and P. Balk in "Insulating Films on Semiconductors", Eds. J.F. Verwey and D.R. Wolters, North-Holland, Amsterdam, (1983), p. 103.
32. S.T. Chang, J.K. Wu and S.A. Lyon, Appl. Phys. Lett. 48:662 (1986).
33. C.M. Svensson in "The Physics of SiO_2 and its interfaces", Ed. S.T. Pantelides Pergamon, New York, (1978), p. 238.
34. D.L. Griscom, J. Appl. Phys. 58:2524 (1985).
35. D.L. Griscom, private communication.

5. [illegible], Phys. [illegible] (1965).
6. [illegible] and [illegible], [illegible] (1966).
7. G. [illegible] and [illegible], [illegible].
8. [illegible] and [illegible], "[illegible]", [illegible], [illegible].
9. [illegible], [illegible], [illegible] and [illegible], "[illegible]", [illegible] (1971).
10. [illegible] and [illegible], "Re [illegible]", [illegible] (1966).
11. [illegible] and [illegible], [illegible] (1965).
12. [illegible] and [illegible], [illegible] (1976).
13. [illegible], Phys. Rev. [illegible] (1970).
14. [illegible] and [illegible], Appl. Phys. Lett. [illegible].
15. [illegible] and [illegible], J. Appl. Phys. [illegible] (1976).
16. [illegible] and [illegible], Solid State Comm. [illegible] (1975).
17. [illegible], J. Phys. [illegible] (1968).
18. [illegible], [illegible] and [illegible], [illegible].
19. [illegible], [illegible] and [illegible], J. Appl. Phys. [illegible] (1993).
20. [illegible], [illegible] (1960).
21. [illegible] and [illegible], J. Appl. Phys. [illegible] (1975).
22. [illegible] and [illegible], Materials Science Series, Vol. 1, North-Holland, Amsterdam (1971).
23. [illegible], Phys. Rev. [illegible] (1953).
24. [illegible] and [illegible], [illegible] (1965).
25. [illegible], [illegible] and [illegible], J. Appl. Phys. [illegible] (1967).
26. [illegible] and [illegible], [illegible].
27. [illegible], [illegible] and [illegible], [illegible] (1968).
28. [illegible], [illegible] and [illegible], [illegible].
29. [illegible], [illegible] (1970) p. 231.
30. [illegible], [illegible].

THE INFLUENCE OF HIGH TEMPERATURE NITROGEN ANNEALING ON THE ELECTRICAL PROPERTIES OF PLASMA NITRIDED OXIDES

J. Camargo da Costa (1) and M. El-Sayed (2)

D. LETI/IRDI-Commissariat à l'Energie Atomique-CEN/G 85 X 38041 Grenoble, Cedex - France

ABSTRACT

The effect of nitrogen gas annealing on intrinsic electrical properties of thin plasma nitrided oxide films is studied. It is shown that nitridation related trap densities and other electrical properties have been considerably improved after a high temperature (1 000°C, 1 hour) nitrogen annealing.

INTRODUCTION

The development of submicron MOS (metal-oxide-semiconductor) integrated circuit technologies has led to the of use ultrathin (less than 20 nm) dielectric layers for the gate insulator[1,2].

Silicon dioxide, the currently employed insulator in MOS technologies, when used in ultrathin layers shows several limitations[3,4] : it tends to to react with gate electrode material, it is a poor barrier against impurity diffusion, it is susceptible to high-energy radiation damage, and it easily breaks down due to charge accumulation after ion implantation or reactive ion etching steps. As a consequence of these limitations, the need for a new high-quality insulator to replace silicon dioxide in ultrathin gate dielectric technologies has arisen and several alternative dielectric materials have recently been investigated[5,6].

Plasma-enhanced thermally nitrided silicon dioxide appears to be a prospective alternative to SiO_2, due to its good barrier properties against impurity diffusion[7], its higher dielectric constant[8], and its higher radiation resistance[9]. In this work, the effect of nitrogen gas annealing on intrinsic electrical properties of thin plasma nitrided oxide films is investigated.

---

(1) Depto. de Eng. Electrica-Univ. de Brasilia-BRASIL.
(2) Electrical Engineering Department, Alexandria University, EGYPT.

SAMPLE PREPARATION AND EXPERIMENTAL TECHNIQUES

SiO_2 films of 13 nm thickness were thermally grown on silicon(p-type) wafers of resistivity 6-8.5 Ω. cm and (100) surface orientation. The oxide films were nitrided in ammonia plasma (frequency : 13,56 MHz ; RF power : 400 W ; Pressure : 10 mTorr ; Temperature : 900°C ; Time : 2 hours). An increase in the thickness of the nitrided oxide films obtained was observed. Film thicknesses between 14 and 15 nm were in fact obtained from ellipsometry and C(V) measurements. Some samples were annealed at high temperature (1000°C, 1 hour) in N_2. The films were characterized using Auger depth profiling. The composition of a typical film is shown in fig. 1. Successful nitridation was achieved with nitrogen concentration in the film presenting two-distinct peaks ; at the exposed surface and at the insulator-silicon interface.

Aluminium metallization and postmetallization annealing were performed using standard techniques. The surface area of all devices was studied $4 \times 10^{-4} cm^2$. The electrical characterization consisted of I(V) and C(V) (1 MHz and quasistatic) measurements. High current electron injection was performed at constant current with current density varied between $25 \mu A/cm^2$ and $1 mA/cm^2$. The gate was always biased negatively (injection from Al-insulator interface). Insulator trap densities and their capture cross sections were therefore obtained.

A detailed description of the electrical characterization techniques employed in this investigation was presented elsewhere[10].

RESULTS

1. Device characteristics before high field injection

1.1. Current-voltage characteristics

Fig. 2 shows the I(V) characteristics presented as Fowler-Nordheim

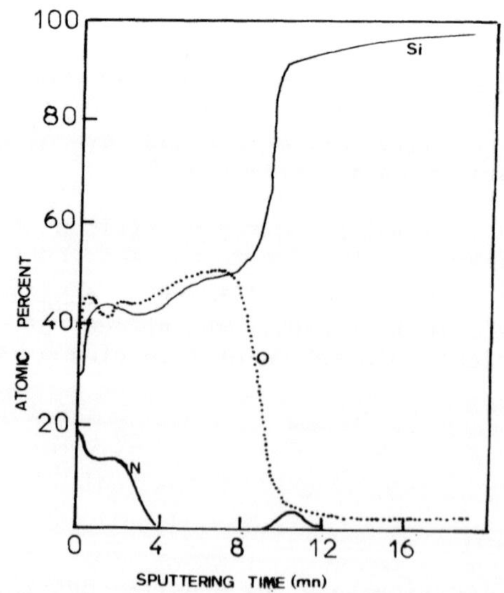

Fig. 1. Auger depth profile of a typical nitrided film.

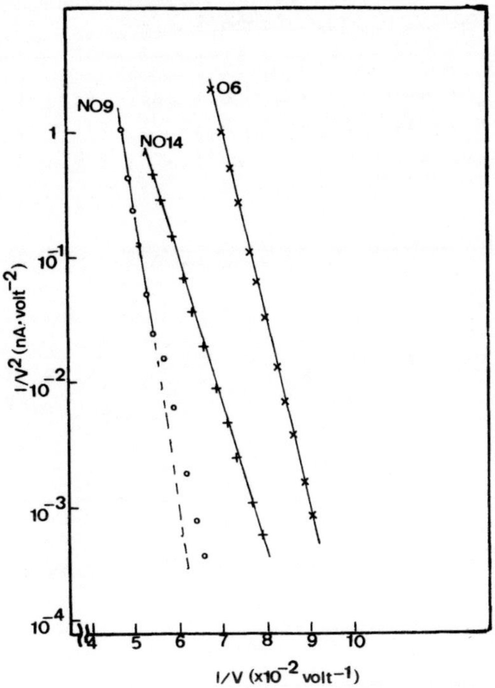

Fig. 2. Fowler-Nordheim plots for the basic oxide (O6), unannealed nitrided oxide (NO9) and annealed, nitrided oxide (NO14).

plots. The conduction mechanism in the original and nitrided oxide films is found to follow the Fowler-Nordheim injection mechanism except for relatively low voltages in the case of the unannealed nitrided oxide films (samples NO9).

1.2. Capacitance-voltage characteristics

From the 1 MHz C(V) characteristics, the original oxide has shown a negligible fixed oxide charge density while the nitrided oxides have shown a significant fixed positive charge density with values of about 1.1×10^{12} cm^{-2}. After an N2 annealing these features were reduced to 4.1×10^{11} cm^{-2}. The quasistatic capacitance measurements have revealed mid-gap interface state densities $D_{it}(E)$ of the order of $1 \times 10^{10} cm^{-2}.eV^{-1}$ even for the nitrided oxide samples.

2. Device characteristics after high field injection

Sample behaviour after injection is synthesized in Table I and in fig. 3. Comparison between nitrided oxides and the original oxide shows that the nitridation introduces electron traps in the insulator. A significant reduction in nitridation induced trap density is obtained after a high temperature nitrogen annealing.

DISCUSSION

The increase in the fixed charge density after nitridation of the silicon dioxide could be due to the incorporation of nitriding species, liberated in the ammonia plasma, near the Si-SiO_2 interface (formation of

TABLE I

	Sample	d_{ox} (nm) C-V measurements	V_{FB} (volt)	Mid-gap N_{ss} $(eV^{-1}cm^{-2})$	Effective Electron trap density (cm^{-2})	Capture cross-section (cm^2)
Original oxide	06	13	-0.85	$\sim 10^{10}$	-	-
Nitrided oxide	N09	16	-1.5	$\sim 10^{10}$	$Q_1=1.54\times10^{12}$ $Q_2=2\times10^{12}$	$\sigma_1=8.44\times10^{-18}$ $\sigma_2=6.36\times10^{-17}$
Annealed nitrided oxide	N014	15	-1.2	$\sim 10^{10}$	$Q=1.45\times10^{12}$	$\sigma=1.80\times10^{-17}$

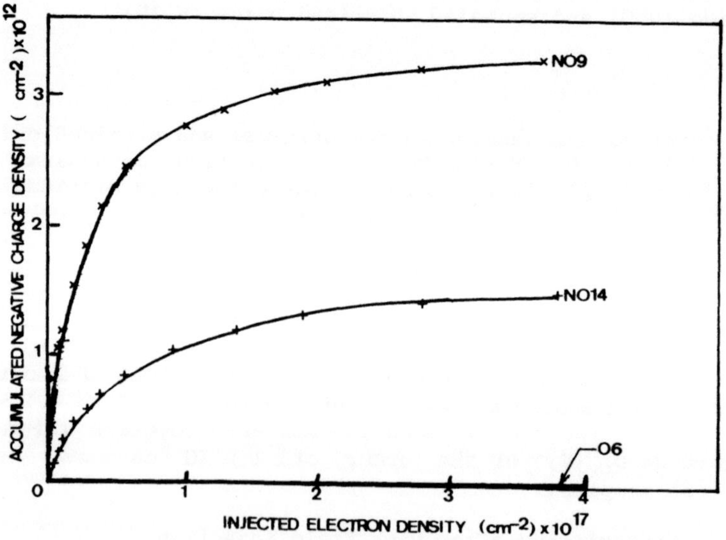

Fig.3

nitrogen-rich compounds) as shown in fig. 1. As a result, an increase in the trivalent silicon atom density near the interface would occur, as well as a deficiency of oxygen atoms in the same region. Since the surface nitrided region acts also as a diffusion barrier, excess oxygen atoms could not outdiffuse and an oxygen-rich region should appear in the oxide not far from the $Si\text{-}SiO_2$ interface.The annealing of the sample in nitrogen gas at a high temperature would stimulate the diffusion of excess oxygen atoms towards the interface region where they would react with trivalent silicon atoms and as consequence contribute to a decrease in the fixed charge density.The reduction in flatband voltage shift verified after high temperature nitrogen annealing (Table I) indicates that a decrease in fixed charge density, which seems in accordance with the considerations mentioned above, effectively occured.

CONCLUSION

A nitridation of thermally grown SiO_2 films in ammonia plasma, followed by a high temperature N_2 annealing can result in insulator films with good electrical properties. A reduction in the induced electron trap and fixed positive charge densities is obvious. Furthermore, an improvement in the insulating film reliability can be achevied.

REFERENCES

1. R.H. Dennard, F.H. Gaenssen, H.N. Yu, V.L. Rideout, E. Bassous and A.R. Le Blanc, IEEE J. Solid-State Circuits SC-9:256 (1978).
2. S.K. Lai, Extended Abstracts of the Spring Meeting of the Electrochemical Society (Electrochemical Society, Pennington, NJ. 1981) Vol. 81-1, abstract 214, p. 541.
3. J.A. Nemetz and R.E. Tressler, Solid State Technology. 26:79 (1983).
4. M.M. Moslehi and K.C. Saraswat, IEEE Trans. Electron Devices ED-32:106 (1985).
5. T. Ito. I. Nakamura and H. Ishikawa, IEEE Trans. Electron Devices ED-29:948 (1982)
6. S.S. Wong, C.G. Sodini, J.W. Ekstedt, H.R. Grinolds, K.H. Jackson, S.H. Kwan, and W.G. Oldham, J. Electrochem. Soc. 130:1139 (1983).
7. T. Ito, T. Nakamura and H. Ishikawa, IEEE Trans. Electron Devices ED-39:498 (1982).
8. T. Ito, T. Nozaki and H. Ishikawa, J. Electroch. Soc. 127:2053 (1980).
9. R.K. Pancholy and F.M. Erdmann, IEEE Trans. Nucl. Science, NS-30:4141 (1983).
10. M. El-Sayed and J. Camargo da Costa, "High Field Phenomena in Thin plasma Nitrided SiO_2 Films", presented at the International Conference on Insulating Films on Semiconductors-INFOS 87-Leuven, Belgium 1987.

ACKNOWLEDGMENTS

José Camargo da Costa would like to thank the Conselho Nacional de Desenvolvimento Cientifico e Tecnologico (CNPq-Brazil) for financial support.

[illegible]

CONCLUSION

[illegible]

REFERENCES

1. [illegible]
2. [illegible]
3. [illegible]
4. [illegible]
5. [illegible]
6. [illegible]
7. [illegible]
8. [illegible]
9. [illegible]

[illegible]

HIGH-FIELD POSITIVE-CHARGE GENERATION AND ITS

RELATION TO BREAKDOWN IN a-SiO_2

Zeev A. Weinberg

IBM Thomas J. Watson Research Center
Yorktown Heights
New York 10598, USA

INTRODUCTION

This paper reviews and discusses the issues of how electrically active defects are created in thermally grown SiO_2 during high-field stressing of MOS structures and their relation to dielectric breakdown. Positive charge, appearing in various electrical stress experiments, has been observed in early research. It is now clear that depending on the field strength, its duration, and oxide properties that positive charge, negative charge, fast interface states, slow states, amphoteric states, and creation of electron traps may all occur and furthermore they may be interrelated. The term "positive charge", therefore, is somewhat misleading as it does not convey the whole meaning of the phenomena expected to occur in the oxide under high-field injection. To avoid more confusion, however, I will use it throughout in the loose sense of lumping all the defects together.

The aim of this paper is to focus attention on the possible mechanisms that underlie or initiate the formation of the positive charge during high-field current injection. To keep this focus, I will exclude most of the radiation induced positive-charge literature because in that case the creation of holes by the radiation gives the positive charge a clear origin. The issues and literature on the nature of interface states or the microscopic bonding models for the various defects are also not reviewed.

Before proceeding with the review I would like to reemphasize several general points about the parameters of the defects, although most researchers have probably a good awareness of these issues. In modeling the dynamics of charge build-up it is necessary to use parameters such as cross-sections, trap densities, generation coefficients, and trap location in the oxide. There is, however, a general problem with this scheme. Much of the physics is in the question of what are the basic phenomena that are occuring. In this sense, the physics is in understanding the microscopic parameters rather than in fitting them to data. A dynamic curve of charging may be fitted, for example, by several traps with distinct fixed parameters or by one trap with variable parameters or by a dynamic trap model in which the nature of the defects and trapping keeps changing during the experiment. The cross-section for trapping an electron would be very dependent on the proximity of the trap to the interface and would be strongly limited by the tunneling of the electron out of the trap before it is stably trapped. This would of course depend also on the field polarity and strength. Because of these considerations the same traps may behave very differently in different experiments. Similar considerations complicate the physical understanding of the other parameters also.

The problem of the sensitivity of the parameters to the location of the defect is compounded by the fact that the defect location can also be subject to conflicting interpretations. For example, in experiments that sense the effect of charge at both interfaces the possibility of positive charges near both interfaces may be interpreted to give the centroid of the charge as being in the bulk of the oxide. Photo I-V experiments tend not to find any bulk charge but they, in turn, may miss some of it by erasing it very quickly. Finally, as a high field experiment is performed, the moment it is stopped for measuring C-V, photo I-V, etc. much of the charge may be altered by the change in the applied field and the conclusions as to what was occuring during the high field may not be correct.

BACKGROUND REVIEW

In the 1960s and early 1970s research in MOS technology was mainly concerned with certain oxide related instabilities. Besides the well known positive-ion drift and positive charge induced by radiation a third instability was identified which produced C-V shifts in opposite direction to that of ion drift. This instability occurred with gate voltages of both polarities but was known mainly as the negative voltage instability. Although this instability was produced by low or medium oxide fields where current injection by tunneling was negligible, and it was strongly temperature dependent, still it possessed certain similarities to the high field positive charge.

First, the end product of the instability was a defect which had the nature of positive charge but could slowly exchange electrons with the silicon. This behavior is known today as "slow states". The term, slow states, in connection with the oxide, appears already in 1959 in a paper by Attala[1]. In 1967, Deal et al.[2] mention slow states in conjunction with ionic contamination. Hofstein [3] describes a slow trapping instability in 1967, and in 1971 the term is used in conjunction with oxygen vacancy defects[4]. The term is absent from the terminology convention introduced by Deal et al.[5] in 1980, but is used again in 1981, this time to describe the behavior of the so called "anomalous positive charge"[6].

Second, there exists a substantial discussion in the literature on the possible mechanisms for the instability with some being mainly electronic and others based on chemistry, and without a final clear conclusion. Breed et al.[7-9] investigated the negative voltage instability extensively and tried to explain it by an electronic model of electrons in preexisting traps near the interface being thermally excited and emitted into the silicon conduction band. This was in contrast to the mechanism of thermal excitation and emission of holes from the silicon valence-band into the oxide[3]. On the other hand, chemical models such as bond breaking and diffusion[10] of OH, or oxygen vacancy production[11], were also proposed.

I will turn now to the literature on the high-field positive charge. In early work, the study of high field effects was mainly confined to breakdown studies performed by the self healing technique of Klein[12]. The study of the high-field conduction mechanisms in SiO_2 begins with the paper by Lenzlinger and Snow[13] who clearly identified the conduction as being due to Fowler-Nordheim (FN) tunneling. The authors observed also the well known slow decay of the current, which, for purposes of tunneling measurements, caused experimental difficulties only.

In 1972, Osburn and Weitzman[14] published their results on conduction and have described the same current decay. They mention that occasionally the current would, for positive bias, first increase with time and then decay. As this, in retrospect, is an important result, it is interesting to note that the origin of this effect was specified as unknown. The same paper showed that at high fields the current rises above the FN curve. The suggested explanation for this effect was that positive charge enhances the injection field. The positive charge was attributed to impact ionization of shallow traps by hot electrons at a distance of about 3 nm from the interface where electrons may have 2.7 eV energy. This idea of field enhancement due to positive charge was stated to be in qualitative agreement with models by O'Dwyer[15,16]. It is interesting to note that although O'Dwyer's original models were based mainly on hole production by collision ionization, he chose in 1973 to model SiO_2 by a two-carrier mechanism where hole injection occured at the anode, leading him to deduce an unrealistically small anode barrier[17]. Injection of holes by tunneling was also suggested by Williams and Woods[18] and by Rai et al.[19]

In 1974, Shatzkes et al.[20], proposed a model for positive charge production by hole generation caused by impact ionization in the SiO_2. The positive charge centroid was proposed to be stabilized by the combination of ionization-generation and the opposing mechanism of recombination. Their model depended on the idea that holes were relatively immobile in SiO_2, as was previously suggested by DiStefano and Eastman[21]. At about the same time, the model was mathematically established by DiStefano and Shatzkes[22] and more fully analyzed subsequently[23,24]. The model was based on electrons losing energy by LO phonons. To be compatible with the high breakdown fields, the authors deduced a mean free path which was unrealistically short and was not consistent with internal photoemission results. This problem was only recently clarified by Fischetti[25] who included the acoustic phonons that cause a strong spatial dispersion in the motion of the electrons. In 1976, Shatzkes et al.[26] published results in support of their model, where the centroid of the positive charge was deduced to be within the oxide bulk.

Solomon and Klein[27] used the idea of impact ionization to analyze the initial current increase and to calculate the impact ionization coefficient and its field dependence. A revised calculation of this coefficient was later given by Knoll et al.[28]. The model was extended to take electron trapping into account later[30]. Klein's model treats the electronic processes in a macroscopic way.

The model of positive charge production by impact ionization and the current runaway and breakdown it produces, were reviewed in 1977 by Solomon[31] who concluded that the impact-recombination model offers the best explanation for the intrinsic high-field breakdown mechanism.

The impact ionization model was not universally accepted, however, and difficulties with various aspects of this model appeared almost from its beginning. In 1975 the present author[32] described experiments with corona and with injection of holes by VUV photons (vacuum-ultra-violet photons with energy larger than the oxide bandgap). These experiments clearly showed that holes are quite mobile and, as expected, it was found that hole trapping near the Si-SiO_2 interface leads to an increase in the electron-injection current. However, the results of the very high fields attained by negative-corona charging were not consistent with the ideas of the impact ionization model. To explain the high-field positive charge, an alternative mechanism was suggested which was based on hot hole tunneling from the anode. This mechanism stipulated that hot holes created in the anode by the electrons from the oxide may tunnel back into the oxide and subsequently be trapped at the interfaces.

The creation of hot holes in the anode by the energy of the electrons falling into the anode was observed earlier[33]. The appearance of positive charge in corona experiments was observed also by Woods and Williams[34]. The corona work was summarized in a paper[35] published in 1976. The numerical analysis of the positive charge build-up yielded results which were more consistent with the idea that the positive charge was produced by a mechanism related to the electron current in the oxide and not by the independent tunneling of cold holes (thermal holes at the top of the silicon valence band).

Other flaws in the original impact ionization model were due to the relative fast motion of holes, as reported by Hughes[36], and to the fact that the location of the positive charge was found to be near the interface rather than in the bulk[37], which disagreed with Shatzkes et al[26].

Although the creation of positive charge by hot hole injection from the anode was an attractive mechanism, the present author had certain ideas about the possibility of neutral excitations which were not published. These ideas were reinforced by the experimental finding of positive charge during VUV illumination with negative gate bias[38]. In these experiments it was verified that the positive charge could not be directly created by holes, electrons, or by light reaching the Si-SiO_2 interface. The results were, however, more easily explained by invoking another possibility for a neutral excitation, namely, an exciton. Although the connection between this VUV effect and high-field positive charge was not established, it was nevertheless suggested later[39], and the energy required to produce the neutral excitations was proposed to be provided by the energy of hot electrons.

The VUV effect was combined with avalanche injection experiments and both were found to produce the anomalous positive charge which was strongly affected by water in the oxide[40]. This result reinforced the idea of a neutral species as the common mechanism for both, but added the possibility that the species was atomic hydrogen (H). The subjects of avalanche injection and H are discussed later in this review.

At this point, I will turn to the issue of high-field negative charge. While most early work focused on positive charge and its natural association with breakdown, negative charging was also evident in early work in the sense of the usual observation of decay in the current. In this review, I will emphasize the issue of high-field creation of negative charge via the generation of additional electron traps. This issue is separate from the issue of preexisting neutral electron traps, which are mostly water related, and which are revealed by the avalanche injection technique.

It is interesting to note that signs for the creation of extra negative charge already appear in the corona work of Lee (see the review by Johnson[41]) where positive corona charging which followed negative corona revealed that in addition to erasing the positive charge, extra negative charge was observed ; but the significance of this was not fully understood. Solomon[42], in 1977, observed a ledge in I-V curves which appeared at high fields. He associated this ledge with electron trapping on preexisting traps situated very close to the interface, within 1 nm. On the other hand, Harari[43], while studying breakdown in oxides which were thinner than was usually done before, postulated the creation of electron traps at high field and his explanation for breakdown was based on the increase in the field on the anode side and the possibility of bond rupture by this very high field.

Because the study of high-field negative charging had coincided with the larger effort on characterization of neutral traps by avalanche injection, it was assumed by most that the high-field negative charge was caused by the population of the same preexisting traps. The actual discussion of the generation of extra traps appears most clearly in a series of papers from the Hebrew University of Jerusalem which begins with the paper by Badihi et al.[44]. In this work, the injection of electrons was done by a dc technique in which hot electrons are supplied to the gate area from a nearby p/n junction. Because the supply of electrons can be independently controlled, this technique has the advantage of being able to separate the effects of the field from the effects of the current. The paper by Badihi et al. has shown that extra traps were generated upon stress. In a subsequent paper by Eitan et al.[45] it was argued that under high levels of injection, negative charge builds via direct electron tunneling into traps below the top of the barrier, without the need for conduction through the oxide.

The same dc hot electron injection technique was also used to study the positive charge formation by Nissan-Cohen et al.[46] who concluded that its formation and annihilation are consistent with the model of impact ionization and recombination of Klein[29]. In 1984 the same authors reported the first experiments of C-V shifts for both interfaces in oxides with polysilicon gates and concluded that the positive charge was located substantially away from the interface, in the oxide bulk. This information was used in a subsequent paper[47] to extend the impact-recombination model to take into account the spatial distribution of the charge and the field dependence of the parameters. The impact mechanism was stated as not identified but a tendency towards trap ionization was made. In their following paper[48] the build-up of either positive or negative charge was attributed to a dynamic model of trapping-detrapping. In this model a fairly high concentration of defects (about $10^{19} cm^{-3}$) was postulated to be distributed in the oxide bulk. These defects are, initially, mostly occupied with an electron. The filled traps may be charged positively by emitting their electron due to impact ionization of the trap by the hot conduction electrons. The empty traps may of course retrap an electron. The final level of the occupancy of the traps is determined by the field, and the action of trapping or detrapping is provided by the conduction electrons. The authors state that the oxide bulk charge should be distinguished from the effects of the interfacial slow states which appear mainly in oxides with Al gates and which mask the smaller effect of the bulk traps which are always present.

The model of Nissan-Cohen et al. has two major weaknesses. First, it requires a relatively large density of traps, occupied with an electron, whose presence is not substantiated. Second, the conclusion that the location of the charge is in the oxide bulk is in contradiction with some of the findings of De Keersmaecker and coworkers[49-53], which are described below (see also discussion in the previous section).

A different model of trapping-detrapping was proposed by Olivo et al.[54]. In this model, levels are present within the tunneling barrier and their occupancy depends on electrons tunneling into or out of them. These effects are independent from the injected current flowing in the oxide. This is not in agreement with the conclusion of Nissan-Cohen et al., as discussed above, that the charging effects are directly related to conduction electrons in the oxide. The subject of electron interaction with the tunnel barrier is discussed later.

Heyns and De Keersmaecker have done photo I-V experiments to find the location of the charges after various stress conditions. The photo I-V measurements were done for the first time for polysilicon gates and the work was carried out for both stress polarities and a comparison of Al and polysilicon gates was provided[52]. Their work was conveniently summarized in ref. 53. They emphazize that all charging effects are occuring at the interfaces and no electrically measurable effects occur in the bulk of the oxide. They find slow states at the Si-SiO_2 interface in all cases. They find generation of additional electron traps in the vicinity of the anode for both polarities of injection.

I will turn now to a discussion of the possible relation between the negative and positive charges. Such a relation may be inferred from the work of Lai[55,56]. In this work, holes were first injected from the substrate into the oxide by avalanche injection. The holes were then annihilated by photoinjection of electrons. Lai has found that a combination of these two steps was necessary to produce the peak in the fast interface states in the upper part of the silicon bandgap. These states are mostly charged negatively under conditions of positive gate bias. These findings were also confirmed by Haywood and De Keersmaecker[51]. The cause-and-effect relationship between positive charge and interface states appears also extensively in literature on radiation effects (this literature is not reviewed here, see refs. 57, 58, and references therein).

Although Lai's papers focus on the interface states, it is probably also true that the same mechanism will lead to the buid-up of negative charge in states somewhat away from the interface. This leads to the possible conclusion that slow states are generated by the combined action of first generating a positive charge and then trapping an electron on it. The relaxation that occurs and the energy released travel to the interface and result in the charge that becomes negative. One can find here consistency with the negative bias VUV generation of positive charge, which was described above. In Lai's experiment the energy of the recombining electron on positive charge creates fast states, some distance away, at the interface. In the VUV experiment the same energy of the recombining electron-hole pairs, created by the VUV photon absorption near the Al interface travels through the oxide to create the defects at the silicon interface[38]. This mechanism may even provide the explanation for the slow state build-up in avalanche injection. In this case the energy is provided by the electrons falling into deep traps.

The above discussion starts the subject of the connection between the positive charges produced by high field stress and by avalanche injection. A much more extensive discussion of the relation of these two experiments has been recently given by Fischetti[59]. A review of the avalanche literature will not be repeated here except for the following brief summary. The initial observation of the turn around effect and its explanation by formation of positive charge and interface states was first reported by Gdula[60]. It was later labelled as "anomalous positive charge" by Young who also found that it had the nature of slow states[61]. The strong dependence of this phenomenon on water in the oxide led to explanations based on release of hydrogen and its impact on interface properties. The release of hydrogen and its importance in understanding oxides were, of course, well known before (e.g. Nicollian[62], Revesz[63]). The effects of water and further discussion of the H model were provided by Feigl et al.[64] and later by Gale et al.[65] who found large H production. The mechanism of H serves as the basis for a rather extensive discussion by Sah et al.[66,70]

who provide an encyclopedic discussion of the many possibilities of H production by bond breaking and its subsequent effects at the interface.

A study of the kinetics of the slow states, produced by avalanche, was published by Fischetti et al.[71] in 1982. The current increase at high field was first thought to be not related to stable positive charge, Ricco and Fischetti[72], but this opinion was reversed and a connection was made to positive charge by proposing a model of holes emitted from the anode[73,74]. Similar to earlier suggestions by Weinberg[32], these holes were the holes produced by the energy lost by electrons falling from the oxide into the anode. This model was comprehensively analyzed by Fischetti[75] who developed a theory for holes being generated in the anode via the intermediate step of surface plasmons which are excited by the electrons. The avalanche and high-field effects were further discussed, as mentioned above, in a subsequent paper[59]. Within the model of Fischetti, the role of electron trapping is purely electrostatic in the sense that it serves to increase the anode field which enhances the injection of holes from the anode.

As difficult as it is to understand the positive charge formation mechanisms at room temperature, this task is not easier for lower temperatures. Breed[7] had already discussed two types of positive charge at 77 K and their association with interface states. Low temperature (90 K) effects have been studied by Lyon and coworkers[76-79]. The location of the positive charge was found to be very close to the interface (<2.7 nm)[77]. They find that positive charge is converted into amphoteric surface states after annihilation with photoinjected electrons[78] and after this conversion the defects can be cycled between surface states and negative charge[79]. These last characteristics of the charge have similarities to the room temperature slow states, the creation of which, at 77 K was studied also by Fischetti and Ricco[73].

The review, so far, has focused on the high-field charging issues. I have tried to follow certain themes and keep up some historical perspective, but in doing so, certain topics and some literature were omitted, and I have found it difficult to integrate the literature of the Far East and Eastern block. I will try to summarize these topics next.

For high positive gate field, in addition to the charging effects, another interesting effect occurs. This is the hole current in the silicon substrate which was first reported by Eitan and Kolodny[80] and investigated later by Weinberg and Fischetti[81,82]. The possible correlation of this effect with the positive charge is very important because it helps to eliminate certain mechanisms. This subject is treated more fully later.

The study of positive charge in very thin oxides should be also mentioned. Positive charge was observed in connection with tunnel oxides by Maserjian and Zamani[83,84]. The subject of tunnel oxides is deferred to the discussion.

An interesting correlation of the generation of interface traps and mechanical stress, in FN injection, was recently reported by Hook and Ma[85]. This follows Ma's earlier work on the effect of stresses produced by the mechanical properties of the gate material[86].

Nakagiri[87], in 1974, reported the formation of fast states and oxide defects after high field stress and these were related to bond breaking following the earlier suggestion of oxygen vacancy formation by Miura and Matukura[11]. Itsumi[88], in 1981, reported formation of both positive and negative charge induced by FN tunneling. He found positive charge for negative voltage stress and negative charge for positive stress. The generation rates of both charges were found to be proportional to the injected charge. He studied electron trapping as function of temperature[89] and reported shallow electron traps similar to findings of Balk et al.[90]. Hosoi et al.[91] studied FN stress on thin oxides and their results are modeled similarly to Nissan-Cohen's bulk positive charge. For thinner oxides there are several reports. Modeling of electron injection by impact and recombination was also given by Chen and Wu[92] and a model for breakdown by Chen and Cheng[93].

In Russian literature the same struggle of identifying the effects seems to take place. My files, undoubtedly, are not complete with regard to this literature. Nagin and Tyul'kin[94], in 1982, report on formation of positive charge at high fields and they emphasize that the charge is caused by the current and not by the action of the field alone. They also report the formation of neutral surface states. In 1985, they talk about the degradation of the barrier by interaction of free carriers with deep traps[95]. A model based only on the action of the field via field ionization of traps has been proposed by Mikhailovskii et al.[96,97]. Defect formation in the oxide in combination with MNOS structures was given by the previous authors[98] and by Efimov et al.[99] and Marczewski et al.[100].

It is interesting to note that literature on self trapped excitons also exists, see for example Trukhin and Plaudis[101] and Tanimura, Tanaka, and Itoh[102].

Hartman[103], in 1977, observed negative charge and positive charge with some discussion of possible mechanisms. Krause[104] summarized his work in 1985. He proposed the idea that electrons tunnel into clusters of traps near the injecting electrode and that this builds up fields which cause further trap induction.

So far, the main emphasis in this review was on the high-field charging effects. I will turn now to the issue of breakdown.

Naturally, in early work the occurence of charges in the oxide led to its association with breakdown via the field enhancement it provides, either on the cathode side for positive charge, or on the anode side for negative charge. Wolters and coworkers[105,106] have suggested a relation between the energy given off by the electrons to breakdown. A connection of breakdown with interface charges, especially on the anode side was also suggested by Hillen et al.[50].

The positive-charge feedback mechanism has recently enjoyed strong support from the work of Hu and coworkers. Their work begins[107] with the criticism of Harari's negative charge mechanism[43], they therefore chose the positive charge created by impact ionization as the basis for their modeling. They have included also negative charging in their model[108]. The breakdown was further correlated with positive charge build-up[109], and with process dependence of the positive charge[110]. The best proof of their model is given perhaps when the breakdown is tied to the substrate current which is seen to give a constant charge-to-breakdown in the substrate current[111]. This is interpreted as showing that a constant amount of holes has to flow to cause breakdown. This issue and the substrate current are treated further. Proponents of impact ionization are also Chen and Cheng[93].

Two papers critizing the idea of positive charge were recently given. One is by Tzou et al.[112] which is based on the temperature dependence being such that at higher temperatures the generation of positive charge decreases but the time-to-breakdown which should then increase, actually decreases.

Weinberg and Nguyen[113] find that large amounts of positive charge indeed lead to a large current enhancement, as predicted, but the current reverses and quickly falls. This was attributed to the annihilation of positive charge which then builds additional negative charge, therefore, the current runaway is prevented. It was also argued that, quantitatively, there is no relation between the fast initial development of positive charge and the much longer time usually observed for breakdown.

DISCUSSION

The conclusion of the review provided in the previous section is that there are diverse opinions in regard to the question of what is the mechanism that initiates the formation of charges in the SiO_2 during high-field current injection. The possible models can be broadly categorized into :

1. Trap ionization.
2. Electron-hole pair creation by impact ionization.
3. Injection of hot holes from the anode.
4. Interaction of electrons with states in the tunneling barrier.
5. Production of H, its diffusion, and interactions at the interfaces.
6. Various bond breaking mechanisms and migration of broken bonds.
7. Neutral species.
8. Other impurity or structural defects that become electrically active.

The various models given in the literature are usually forwarded to be applied to certain domains of experimental restrictions, such as, magnitude of injection, oxide processing, oxide thickness, or gate material. It is true in each restricted domain some models are very appealing. Because there are such strong similarities in the appearance of positive charge and its behavior, I prefer to search for a common mechanism for all cases. In view of this opinion I will try to analyze the problem according to the following scheme which, admittedly, is somewhat biased.

A. Is positive charge caused by the field or the current ?

A substantial body of experimental evidence exists now that connects the creation of charges to the conduction of current through the oxide. This has been best demonstrated by the experiments of Nissan-Cohen et al.[44,47,48] which showed that the action of the field alone is not sufficient and that the conduction of electrons is necessary. In most experiments, however, in addition to the conduction, there is also a flux of hot or thermal electrons impinging on the barrier which opens the possibility of effects caused by the tunneling of these electrons into the barrier. This issue is discussed more fully ahead.

I will continue here assuming that the effects are produced directly by electrons which are moving in the oxide conduction band.

B. The issue of electron trap generation

There seems to be now a consensus that, during high field injection, in addition to positive charge, there is also generation of negative charge by the generation of electron traps. These traps are not the positive charges themselves which of course act as coulombic traps for electrons. A simple recombination of electrons with these positive charges would just cause the positive charge generation to be limited.

Based on the results of Lai[56] and of Weinberg and Nguyen[113], my opinion now is that electron trap generation is a three step process. The first step is the creation of positive charge, the second step is an electron falling into it, and the third step is the formation of the acceptor like defect near the Si-SiO_2 interface which during positive-gate injection becomes negatively charged.

C. Relation of positive charge to substrate current

The issue of the substrate current was only briefly touched upon in the review. This current appears in the p substrate of n-channel FET's when the gate voltage is biased positively to draw an electron current into the oxide. During the experiment, the FET's drain and source are grounded. The substrate current is due to holes appearing in the silicon substrate which accompany the electron injection into the oxide. The substrate current is not due to any mechanism of generation in the silicon which is independent of the electron injection into the oxide.

It is reasonable to ask whether the positive charge and the substrate current are caused by the same mechanism. Weinberg and Fischetti have tried to investigate this correlation[81,82]. The time constants of the initial current increase, which are caused by the positive charge, were found to correlate both with the substrate current and the electron current. Of course, if the electron current is responsible for both then there should be such a correlation. This, however, does not rule out the possibility that the substrate current is caused by one mechanism (e.g. holes), while the positive charge is due to a different one (e.g. trap ionization).

Although the connection between the two effects has not been conclusively established, I will proceed with the assumption that the positive charge and substrate current are caused by the same mechanism. This is an important assumption. The ratio, substrate current/electron current, has been found to be temperature independent[81]. The ratio is also surprisingly large and can reach a few percents in 25 nm oxides. The substrate current appears as fast as the electron current and essentially follows it continuously[82]. The substrate current fluence until breakdown can be large (over 0.1 C/cm^2)[111].

The substrate current, therefore, needs a continuous supply and cannot be explained by mechanisms of atomic motion, such as H, or by trap ionization in the oxide. Due to its dependence on oxide thickness, and lack of sensitivity to interface-state generation, it seems that the substrate current is caused by an oxide conduction mechanism. It would be easy to explain the substrate current by holes and then invoke hole generation mechanisms. However there exists the difficulty with the temperature dependence, as hole conduction is strongly thermally activated.

At this point, the most important conclusion to be derived from the substrate current issue is that its basic cause cannot be explained by the release and migration of H, as discussed above.

D. The case for neutral species

My strongest argument for neutral species is based on my early observation of positive charge in the negative-bias VUV measurements, although this effect has not yet been verified by other researchers, it still is most simply explained by the motion of some kind of a neutral deformation. The VUV photons may, of course, cause release of H by bond breaking, but as stated above H cannot explain the substrate current. The idea of neutral species diffusion is consistent with the findings of Ma and coworkers[85,86] of the sensitivity to stress gradients, as neutral species will be affected by these stress gradients and will tend to diffuse towards the highest stress region.

A neutral species which has some attractive properties is the formation of non-bridging oxygen. Grunthaner[114,115] has proposed the idea that Si-O bonds may be broken and that the broken non-bridging bond may be mobile by bond forming and breaking. The migration of the broken bond is proposed to be aided by an H atom. As the broken bond arrives at the interface, surface states may form. This model has some difficulties with electron-spin-resonance identification of the defects[116,117].

I have favored for a long time a more vague idea of a neutral species which was only indirectly hinted at in my papers[38,39,81]. The idea assumes that electrons, when they lose energy, interact with the lattice in a non-linear way in addition to the phonon emission. The non-linear interaction produces lattice vibrations which carry a large localized deformation, perhaps in a solitonic manner. Notice that a broken or otherwise deformed bond may also be viewed as a strong localized deformation. This kind of excitation will carry also a large deformation in the local polarization. When the deformation arrives at the interface it can do several things. First, the polarization deformation will locally and momentarily deform the field at the interface and this will lead to a momentary enhancement in the injection of electrons both from the conduction and valence bands of the silicon. The enhanced tunneling from the valence band may explain the substrate current. Second, the deformation will create the interfacial defects which subsequently become positively or negatively charged.

In the model of neutral species, the role of water and H is secondary. Water-related defects may modify the generation and motion of the neutral species, and the release of H will certainly modify interfacial defects, however, the H is not the primary cause of the effects.

In this model a breakdown may be initiated when the enhanced injection and the neutral oscillations will become coherent so that a resonance will be established, or when enough damage is created locally at the interfaces.

The energy release required to produce the neutral species may result not only from hot electrons but also from the electrons falling into holes or into deep traps. In this sense a common thread is established for the effects of deep traps in avalanche, and the electron-hole recombination in the VUV case. In this model the role of trapping is not only electrostatic but also active in the sense of another channel for energy release.

E. Discussion of thin oxides

Most of the review and the discussion above have been devoted to effects occuring in oxides thick enough where injected electrons move a substantial distance in the conduction band. In such cases it appears that positive charge formation is definitely related to the conduction of electrons, hence all the ideas are related to various effects caused by these electrons. Defect formation occurs also in very thin oxides where electrons may tunnel directly through, or move only a short distance in the conduction band. In these cases it may be difficult to relate the defect formation to large energy loses. Evidence for defect formation in very thin oxides is given by Maserjian and Zamani[83,84] for 4-5 nm oxides, by Nagai and Hayashi for 3.5 nm oxides[118,119], and in slightly thicker oxides by several authors[82,120,91,121].

Fluctuations having a two level or a more complex nature have been observed for the current through ultra-thin oxides (1.6 nm), with an area of 1 μm^2, by Farmer, Rogers, and Buhrman[122]. The fluctuations are explained by "correlated, multielectron capture and emission in a strongly interacting cluster of localized trap states". The authors state also that at high fields the correlated switching of ensembles of traps can result in breaking of oxide bonds.

The observation that defect formation in very thin oxides leads to continuous increase in the current at lower gate voltages, after stress at higher gate voltage, and the implications of this finding to breakdown were recently reported by Olivo, Nguyen and Ricco[123,124].

One can raise here the question that if electrons can cause defects while tunneling through thin oxides, is it then perhaps possible that the same mechanism is responsible for the positive charge in all the other cases. It should be remembered here that in high field FN tunneling, in avalanche injection, and in the dc hot electron injection technique, electrons are impinging on the interface. The idea that these electrons would tunnel into levels in the barrier, when precise alignment of the levels with the energy of the electron occurs, and the subsequent occupancy of these levels, was the basis for the model of Olivo et al.[54,120].

This model, as the major explanation for the positive charge, cannot be accepted because it cannot explain the substrate current, the VUV effect, and it disagrees with the conclusions of Nissan-Cohen et al. in regard to the need for conduction electrons. Also, in thicker oxides the injection fluence needed to generate the positive charge is much smaller than in very thin oxides, and the positive charge builds up immediately at the beginning of injection.

As the attention turns now to very thin oxides, I cannot attempt to explain the phenomenon of defect generation in such oxides and one should look to interactions of electrons with defects that cause chemical reactions and which are triggered perhaps by smaller energy exchanges than in thicker oxides. As stated in ref. 122, smaller energies may be sufficient because the electrons may affect the lattice by multielectron correlated switching.

F. The breakdown question

In my opinion, the strongest evidence in the literature for the idea that positive charge causes breakdown is the finding of constant charge-to-breakdown in the substrate current (Ref. 111). In our laboratory, however, we have not been able to confirm this conclusion and in fact we have observed the charge-to-breakdown of the substrate current to increase with the field for oxides of 22 nm thickness[125]. We think that the field dependence of the charge-to-breakdown simply reflects the field dependence of the ratio of the two currents.

Although I have argued against the impact ionization model for a long time, still I have accepted the idea of current runaway by the feedback from the positive charge as very reasonable. I have understood only recently the fact that current-runaway is actually prevented when doing the experiments published by Weinberg and Nguyen[113]. In these experiments it was shown that the current increase is reversed by an intrinsic mechanism that removes the effectiveness of the positive charge by some dynamic changes that occur upon the trapping of an electron by the positive charge. This explains also why oxides can last a long time to breakdown while the positive charge phenomenon reaches its full development after a relatively short time.

The ultimate mechanism that initiates breakdown remains unresolved. It is suggested that a resonance between oscillations or deformations and the enhanced injection they cause may trigger the breakdown. The deterioration of the tunnel barrier by the accumulation of defects, which invariably accompany current injection, is also a good candidate for triggering the breakdown.

REFERENCES

1. M. M. Atalla, E. Tannenbaum, and E.J. Scheibner, Bell System Technical Journal 38:749 (1959).
2. B.E. Deal, M. Sklar, A.S. Grove, and E.H. Snow, J. Electrochem. Soc. 114:266 (1967).
3. S.R. Hofstein, Solid-State Elect 10:657 (1967).
4. A.J. Bennett and L.M. Roth, J. Phys. Chem. Solids 32:1251 (1971).
5. B.E. Deal, IEEE Trans. Elect. Dev. ED-27:606 (1980).
6. S.K. Lai and D.R. Young, J. Appl. Phys. 52:6231 (1981).
7. D.J. Breed, Solid-State Electr. 17:1229 (1974).
8. D.J. Breed, Appl. Phys. Lett. 26:166 (1975).
9. D.J. Breed and R.P. Kramer, Solid-State Electr. 19:897 (1976).
10. K.O. Jeppson and C.M. Svensson, J. Appl. Phys. 48:2004 (1977).
11. Y. Miura and Y. Matukura, Japan J. Appl. Phys. 5:180 (1966).
12. N. Klein, Advances in Physics 21:605 (1972).
13. M. Lenzlinger and E.H. Snow, J. Appl. Phys. 40:278 (1969).
14. C.M. Osburn and E.J. Weitzman, J. Electrochem. Soc. 119:603 (1972).
15. J.J. O'Dwyer, J. Appl. Phys. 39:4356 (1968).
16. J.J. O'Dwyer, J. Appl. Phys. 40:3887 (1969).
17. J.J. O'Dwyer, J. Appl. Phys. 44:5438 (1973).
18. R. Williams and M.H. Woods, J. Appl. Phys. 44:1026 (1973).
19. B.P. Rai, K. Singh, and R.S. Srivistava, Phys. stat. sol. (a) 36:591 (1976).
20. M. Shatzkes, M. Av-Ron, and R.M. Anderson, J. Appl. Phys. 45:2065 (1974).
21. T.H. DiStefano and D.E. Easteman, Phys. Rev. Lett. 27:1560 (1971).
22. T.H. DiStefano and M. Shatzkes, Appl. Phys. Lett. 25:685 (1974).
23. T.H. DiStefano and M. Shatzkes, J. Vac. Sci. Technol. 12:37 (1975).
24. T.H. DiStefano and M. Shatzkes, J. Vac. Sci. Technol. 13:50 (1976).
25. M.V. Fischetti, D.J. DiMaria, S.D. Brorson, T.N. Theis , and J.R. Kirtley, Phys. Rev. B 31:8124 (1985).
26. M. Shatzkes and M. Av-Ron, J. Appl. Phys. 47:3192 (1976).
27. P. Solomon and N. Klein, Solid State Commun. 17:1397 (1975).
28. M. Knoll, D. Braunig and W.R. Fahrner, J. Appl. Phys. 53:6496 (1982).
29. N. Klein and P. Solomon, J. Appl. Phys. 47:4364 (1976).
30. N. Klein, J. Appl. Phys. 53:5828 (1982).
31. P. Solomon, J. Vac. Sci. Technol. 14:1122 (1977).
32. Z.A. Weinberg, Appl. Phys. Lett. 27:437 (1975).
33. Z.A. Weinberg, W.C. Johnson, and M.A. Lampert, Appl. Phys. Lett. 25:42 (1974).
34. M.H. Woods and R. Williams, J. Appl. Phys. 47:1082 (1976).
35. Z.A. Weinberg, W.C. Johnson, and M.A. Lampert, J. Appl. Phys. 47:248 (1976).
36. R.C. Hughes, Phys. Rev. B15:2012 (1977).
37. D.J. DiMaria, Z.A. Weinberg and J.M. Aitken, J. Appl. Phys. 48:898 (1977).
38. Z.A. Weinberg and G.W. Rubloff, Appl. Phys. Lett. 32:184 (1978).
39. Z.A. Weinberg and G.W. Rubloff, in The Physics of SiO_2 and its interfaces, edited by S.T. Pantelides Pergamon, New York, (1978), p. 24.

40. Z.A. Weinberg, D.R. Young, D.J. DiMaria, and G.W. Rubloff, J. Appl. Phys. 50:5757 (1979).
41. W.C. Johnson, IEEE Trans. Sci. NS-22:2144 (1975).
42. P. Solomon, J. Appl. Phys. 48:3843 (1977).
43. E. Harari, J. Appl. Phys. 49:2479 (1978).
44. A. Badihi, B. Eitan, Y.N. Cohen and J. Shappir, Appl. Phys. Lett. 40:396 (1982).
45. B. Eitan, D. Frohman-Bentchkowsky, J. Shappir and M. Balog, Appl. Phys. Lett. 40:523 (1982).
46. Y. Nissan-Cohen, J. Shappir and D. Frohman-Bentchkowsky, J. Appl. Phys. 54:5793 (1983).
47. Y. Nissan-Cohen, J. Shappir and D. Frohman-Bentchkowsky, J. Appl. Phys. 57:2830 (1985).
48. Y. Nissan-Cohen, J. Shappir and D. Frohman-Bentchkowsky, J. Appl. Phys. 58:2252 (1985).
49. M.M. Heyns, R.F. De Keersmaecker, and M.W. Hillen, Appl. Phys. Lett. 44:202 (1984).
50. M.W. Hillen, R.F. De Keersmaecker, M.M. Heyns, S.K. Haywood, and I.S. Darakchiev, IEEE Trans. Elect. Insul. EI-19:245 (1984).
51. S.K. Haywood and R.F. De Keersmaecker, Appl. Phys. Lett. 47:381 (1985).
52. M.M. Heyns and R.F. De Keersmaecker, J. Appl. Phys. 58:3936 (1985).
53. M.M. Heyns and R.F. De Keersmaecker, MRS-Europe meeting, symposium on dielectric layers in semiconductors, Strasbourg, France, June 17-20, 1986.
54. P. Olivo, B. Ricco and E. Sangiorgy, J. Appl. Phys. 54:5267 (1983).
55. S.K. Lai, Appl. Phys. Lett. 39:58 (1981).
56. S.K. Lai, J. Appl. Phys. 54:2540 (1983).
57. P.S. Winokur, H.E. Boesch Jr., J.M. McGarrity and F.B. McLean, J. Appl. Phys. 50:3492 (1983).
58. G.J. Hu and W.C. Johnson, J. Appl. Phys. 54:1441 (1983).
59. M.V. Fischetti, J. Appl. Phys. 57:2860 (1985).
60. R.A. Gdula, J. Electrochem. Soc. 123:42 (1976).
61. D.R. Young, E.A. Irene, D.J. DiMaria, R.F. De Keersmaecker and H.Z. Massoud, J. Appl. Phys. 50:6366 (1979).
62. E.H. Nicollian and C.N. Berglund, J. Appl. Phys. 42:5654 (1971).
63. A.G. Revesz, J. Electrochem. Soc. 126:122 (1979).
64. F.J. Feigl, D.R. Young, D.J. DiMaria, S. Lai and J. Calise, J. Appl. Phys. 52:5665 (1981).
65. R. Gale, F.J. Feigl, C.W. Magee and D.R. Young, J. Appl. Phys. 54:6938 (1983).
66. C.T. Sah, J.Y.C. Sun and J.J.T. Tzou, J. Appl. Phys. 54:5864 (1983).
67. C.T. Sah, J.Y.C. Sun and J.J.T. Tzou, J. Appl. Phys. 54:2547 (1983).
68. C.T. Sah, J.Y.C. Sun and J.J.T. Tzou, J. Appl. Phys. 55:1525 (1984).
69. C.C.H. Hsu, S.C.S. Pan and C.T. Sah, J. Appl. Phys. 58:1326 (1985).
70. C.T. Sah, W.W.L. Lin, C.C.H. Hsu and S.C.S. Pan, Appl. Phys. Lett. 48:1736 (1986).
71. M.V. Fischetti, R. Gastaldi, F. Maggioni and A. Modelli, J. Appl. Phys. 53:3136 (1982).
72. B. Ricco and M.V. Fischetti, J. Appl. Phys. 55:4322 (1984).
73. M.V. Fischetti and B. Ricco, J. Appl. Phys. 57:2854 (1985).
74. M.V. Fischetti, Z.A. Weinberg, and J.A. Calise, J. Appl. Phys. 57:418 (1985).
75. M.V. Fischetti, Phys. Rev. B 31:2099 (1985).
76. J.K. Wu, S.A. Lyon and W.C. Johnson, Appl. Phys. Lett. 42:585 (1983).
77. S.T. Chang and S.A. Lyon, Appl. Phys. Lett. 48:136 (1986).
78. S.T. Chang, J.K. Wus and S.A. Lyon, Appl. Phys. Lett. 48:662 (1986).
79. J.M. Sung and S.A. Lyon, Appl. Phys. Lett. 50:1152 (1987).
80. B. Eitan and A. Kolodny, Appl. Phys. Lett. 43:106 (1983).
81. Z.A. Weinberg and M.V. Fischetti, J. Appl. Phys. 57:443 (1985).
82. Z.A. Weinberg, M.V. Fischetti and Y. Nissan-Cohen, J. Appl. Phys. 59:824 (1986).

83. J. Maserjian and N. Zamani, J. Appl. Phys. 53:559 (1982).
84. J. Maserjian and N. Zamani, J. Vac. Sci. Technol. 20:743 (1982).
85. T.B. Hook and T.P. Ma, Appl. Phys. Lett. 48:1208 (1986).
86. V. Zekeriya ant T.P. Ma, Appl. Phys. Lett. 47:54 (1985).
87. M. Nakagiri, Japan J. Appl. Phys. 13:1610 (1974).
88. M. Itsumi, J. Appl. Phys. 52:3491 (1981).
89. M. Itsumi, J. Appl. Phys. 54:1930 (1983).
90. M. Aslam, P. Balk and D.R. Young, Solid-State Elect. 272:709 (1984).
91. T. Hosoi, M. Akizawa and S. Matsumoto, J. Appl. Phys. 57:2072 (1985).
92. C.F. Chen and C.Y. Wu, Solid-State Elect. 29:1059 (1986).
93. D.N. Chen and Y.C. Cheng, J. Appl. Phys. 61:1592 (1987).
94. A.P. Nagin and V.M. Tyul'kin, Sov. Techn. Phys. Lett. 9:611 (1982).
95. A.P. Nagin and V.M. Tyul'kin, Phys. Chem. Mech. Surfaces 2:2383 (1985).
96. I.P. Mikhailovskii and A.E. Epov, Sov. Tech. Phys. Lett. 10:174 (1984).
97. I.P. Mikhailovskii and A.E. Epov, Phys. stat. sol. (a) 92:615 (1985).
98. I.P. Mikailovskii, V.N. Ovsyuk and A.E. Epov, Sov. Tech. Phys. Lett. 9:452 (1983).
99. V.M. Efimov, E.E. Meerson and A.A. Evtukh, Phys. stat. sol. (a) 91:693 (1985).
100. J. Marczewski, R. Nitecki, and M. Liberadzka, Phys. stat. sol. (a) 70:555 (1982).
101. A.N. Trukhin and A.E. Plaudis, Sov. Phys. Solid. State 21:644 (1979).
102. K. Tanimura, T. Tanaka and N. Itoh, Phys. Rev. Lett. 51:423 (1983).
103. U. Hartman, Phys. Stat. sol. (a) 43:103 (1977).
104. H. Krause, Phys. stat. sol. (a) 89:353 (1985).
105. D.R. Wolters and J.J. Van der Schoot, in Proceedings of the Insulating Films on Semiconductors conf. (INFOS-1985), edited by J.J. Simonne and J. Buxo Elsevier, Amsterdam, (1986), p. 145.
106. J.F. Verwey and D.R. Wolters, in Proceedings of the Insulating Films on Semiconductors conf. (INFOS-1985), edited by J.J. Simonne and J. Buxo Elsevier, Amsterdam, (1986), p. 125.
107. S. Holland, I.C. Chen, T.P. Man and C. Hu, IEEE Elect. Dev. Lett. EDL-5:302 (1984).
108. I.C. Chen, S.E. Holland and C. Hu, IEEE Trans. Elect. Dev. ED-32:413 (1985).
109. I.C. Chen, S. Holland and C. Hu, IEEE Elect. Dev. Lett. EDL-7:164 (1986).
110. S. Holland and C. Hu, J. Electrochem. Soc. 133:1705 (1986).
111. I.C. Chen, S. Holland, K.K. Young, C. Chang and C. Hu, Appl. Phys. Lett. 49:669 (1986).
112. J.J. Tzou, C.C. Yao, R. Cheung and H. Chan, IEEE Elect. Dev. Lett. EDL-7:446 (1986).
113. Z.A. Weinberg and T.N. Nguyen, J. Appl. Phys. 56:1947 (1987).
114. F.J. Grunthaner, P.J. Grunthaner and J. Maserjian, IEEE Trans.Nucl. Sci. NS-29:1462 (1982).
115. F.J. Grunthaner and P.J. Grunthaner, Materials Science Reports 1:65 (1986).
116. R.E. Mikawa and P.M. Lenahan, J. Appl. Phys. 59:2054 (1986).
117. W.L. Warren and P.M. Lenahan, Appl. Phys. Lett. 49:1296 (1986).
118. K. Nagai and Y. Hayashi, IEEE Trans. Elect. Dev. ED-30:715 (1983).
119. K. Nagai and Y. Hayashi, Appl. Phys. Lett. 44:910 (1984).
120. P. Olivo, B. Ricco and E. Sangiorgi, Appl. Phys. Lett. 48:1135 (1986).
121. S. Horiguchi, T. Kobayashi and K. Saito, J. Appl. Phys. 58:387 (1985).
122. K.R. Farmer, C.T. Rogers and R.A. Buhrman, Phys. Rev. Lett. 58:2255 (1987).
123. P. Olivo, T.N. Nguyen and B. Ricco, to be published.
124. T.N. Nguyen, P. Olivo and B. Ricco, Proceedings of the 25th International Reliability Symposium:ehp1, San Diego, 1987.
125. P. Olivo and Z.A. Weinberg, unpublished.

BREAKDOWN MECHANISMS OF THERMALLY GROWN SILICON DIOXIDE AT HIGH ELECTRIC FIELDS

N. Klein

Solid State Institute and Department of Electrical Engineering Technion
Israel Institute of Technology, Haifa 32000
Israel

ABSTRACT

Mechanisms of high field electrical breakdown have been widely investigated in recent years in thermally grown silicon dioxide films at fields larger than 7 MV/cm. We find for oxide films thicker than 10 nm that constant voltage and constant current tests produce breakdown by different mechanisms. Specifically, the fast breakdowns of constant voltage tests can be explained by the IIR (Impact ionization-recombination) breakdown model by the growth of positive charge at the cathode at fields larger than a critical field, F_r. The slow breakdowns in constant current tests cannot be explained by the IIR mode, since growth of electron trapping during a test keeps the critical field increasing. It appears that breakdown in constant current tests may be explained as due to the effects of the generation of a very large density of trap states at the injection barrier. The density of trap states generated may grow to $10^{19}/cm^3$. Such change in the oxide may produce current instability by diverse processes : barrier lowering, resonant tunneling, or transition of the oxide from an insulating to a conducting state. The mechanism of these breakdown processes has not yet been identified. Breakdown by effects of defect generation present novel models of insulator breakdown, and it is of interest to explore their range of validity for SiO_2 and also for other insulators.

INTRODUCTION

Electrical breakdown processes in thermally grown silicon dioxide films have been widely investigated in recent years at fields larger than 7 MV/cm [1]. Testing for breakdown was performed by ramp tests, or by the application of constant voltage[2,5], or constant current to a sample[6,7].

Comparison of experimental results with quantitative models showed that the breakdown mechanism can be fairly well explained in constant voltage tests[2,5]. The mechanism is little understood though in constant current tests[6,7] and the relationship between constant voltage and constant current tests is not known. To improve the understanding of breakdown in SiO_2 films at high fields we studied the breakdown mechanisms in constant voltage and in constant current tests in oxides thicker than 10 nm. We present here briefly the main results of this study. We start by recapitulating the properties of current instability that arises due to the rapid growth of positive charge at the cathode in constant voltage tests. Instability in constant current tests is discussed next. This, we suggest may result from a large generation of electrically active point defects in the energy barrier at the cathode during the test.

THE IIR (IMPACT IONIZATION-RECOMBINATION) BREAKDOWN MECHANISM AND CONSTANT VOLTAGE TESTS

Test results obtained on the application of constant voltage showed that breakdown can be interpreted with effects of impact ionization and recombination processes in fair agreement with a quantitative IIR breakdown model[2-5].

Properties of an IIR breakdown model can be well illustrated with j-F characteristics of the oxide sample j denoting the current density and F the constant applied field. Such a characteristic is schematically presented in fig. 1. Current in the lower part of this curve, below a field F_1 is controlled by Fowler-Nordheim tunneling through the cathode barrier.

Effects of impact ionization and recombination become significant above a field F_1. Holes generated in this field range by ionization in the bulk, or at the anode can be rapidly and deeply trapped. Such positive charges enhance the cathode field and the current injection so that pair generation and hole trapping are increased. The effect of this regenerative process on the positive charge is balanced by hole-electron recombination below a critical field[3] marked F_r in fig. 1. The regenerative effect cannot be balanced at fields F larger than F_r and current instability results in breakdown.

For fields F larger than the critical field F_r the time to breakdown, t_r, is found to decrease rapidly with increase of field[3]. These breakdown events are much faster than breakdowns in constant current tests and

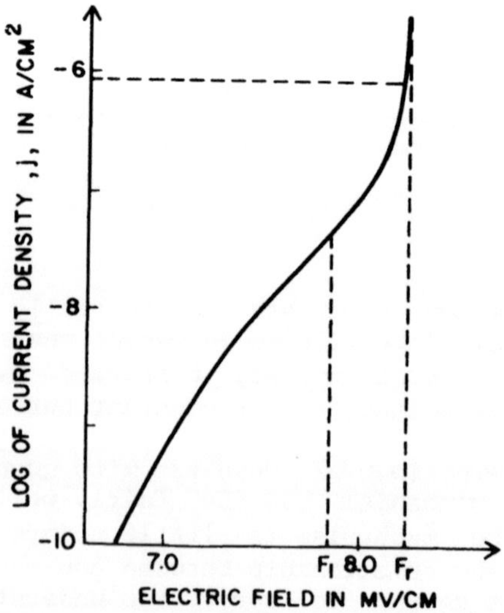

Fig. 1. Current density versus mean electric field characteristic.

effects of electron trapping, recombination and trap generation are usually negligible during t_r. Summing up, we find that the IIR model explains fast breakdown events at constant fields larger than the critical field F_r, for current runaway.

THE BREAKDOWN MECHANISM IN CONSTANT CURRENT TESTS

Figure 1 shows that breakdown is not expected to occur according to the IIR model when the applied field, F, is smaller than the critical field, F_r. Breakdown is however observed in constant current tests also at fields below F_r, but only after the passage of a considerable electron charge, Q_{bd} through the oxide of the order of 0.1 to 100 C/cm^2 [7,8]. These are breakdown events of longer duration than events according to the IIR model. Significant changes are observed to take place in the oxide during the constant current test prior to breakdown.

It is suggested that the mechanism of such breakdown is connected with the changes produced in the oxide during a test. Changes can be very complex and diverse and may degrade, but also increase the breakdown voltage. We will restrict here the discussion to effects of relatively simple changes in the oxide. Examination of test results indicates that important changes on a test can be (a) the growing trapping of electrons of a mean density, n_t, in the oxide, and (b) the continued generation[9-13] of defect-related electron traps of a mean density N. Origin and properties of the generated point defects are discussed elsewhere and we note here only that such electron traps may be produced at hole-electron recombination sites in the oxide[3,14,15].

To assess the effect of growing electron trapping during constant current tests, the critical field for breakdown, F_r, was calculated with the IIR model as function of n_t[16]. Calculations resulted in a simple, approximate relation for F_r,[16],

$$\sigma_h n_t \cong \alpha_{ar} \cong \alpha_o \exp(-H/F_r), \tag{1}$$

where σ_h is the hole-electron recombination cross section, α_{ar} a coefficient of impact ionization when $F = F_r$, and α_o and H are constants. Eq. (1) demonstrates that the critical field F_r grows during a constant current test with the increase of the trapped electron density, n_t. A numerical example showed that F_r grew by 11 % when n_t increased from 0 to $1.3 \times 10^{18}/cm^3$. This is an important finding on the effect of electron trapping. It is in contrast to the widely cited, qualitative assumption that breakdown is due to the growth of the density of trapped electrons[6].

Since the growth of n_t does not explain breakdown in constant current tests by the IIR mode, we turn to examine briefly the effect of the generation of new electron trap states on breakdown.

The density of generated traps N is found to grow linearly with the total charge crossing the oxide during a test[9-13]. Breakdown is observed after the passage of a considerable charge Q_{bd} through the oxide. The density of generated traps, N[6,10,13], is found to increase during a breakdown test to a large value assumedly of the order of $10^{19}cm^{-3}$. To study these breakdowns we consider figure 2 showing schematically the potential energy of the injection barrier a few nanometer thick at fields close to the breakdown range. Fig. 2 includes a potential well at a native, or at a generated defect related electron state. Effects of defect related states in the barrier have been the subject of recent investigations[17-19]. These showed that a defect related electron state can greatly increase the local current by resonant tunneling. It has been suggested[17] that such locally enhanced currents through a point defect may trigger breakdown.

Alternative models based on the effect of growth of defect states N in or near the barrier may explain breakdown with processes such as the degradation of the barrier height[20], or a transition from the insulating to a conducting state. Properties of such breakdown processes are hardly known though and models for the breakdown mechanism as function of the defect state density N have not yet been elaborated. Breakdown by effects of defect generation may present novel models of insulator breakdown and it seems of interest to explore their mechanism and range of validity for SiO_2 and also for other insulators.

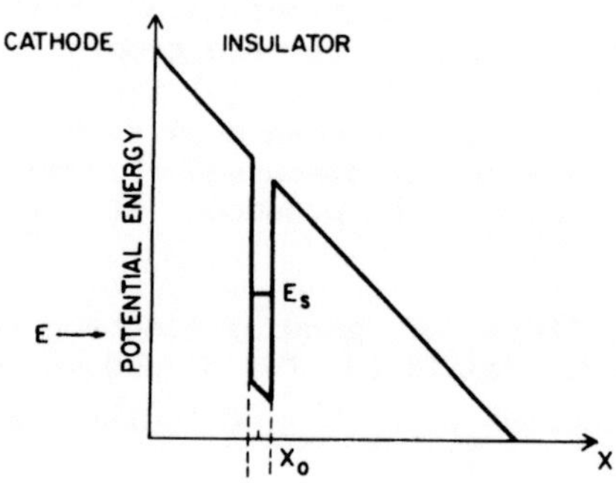

Fig. 2. Schematic representation of the potential energy of a barrier including a defect related state at high fields.

CONCLUSIONS

Studies of high field electrical breakdown in silicon dioxide films thicker than 10 nm reveal a complex picture. We find that constant voltage and constant current tests produce breakdown by different mechanisms. The fast breakdowns of constant voltage tests can be explained with the IIR (Impact Ionization-Recombination) breakdown model by the growth of positive charge at the cathode at fields, F, larger than a critical field, F_r.

The slow breakdowns on constant current tests cannot be explained by the IIR model, since $F < F_r$ and growing electron trapping during a test keeps the critical breakdown field increasing. It is suggested that breakdown in constant current tests results from changes produced in the oxide prior to the breakdown , by the generation of a large density of electron trap states, of the order of $10^{19}cm^{-3}$. The mechanism of such breakdowns has not been identified yet. Resonant tunneling via trap states near the cathode, or degradation of the barrier, or transition of the oxide from an insulating to a conducting state may be processes producing breakdown in constant current tests.

ACKNOWLEDGMENTS

I gratefully acknowledge valuable discussions of this publication with S.S. Cohen and K. Weiser.

REFERENCES

1. P. Solomon, J. Vac. Sci. Technol. 14:1122 (1977).
2. T.H. DiStefano and M. Shatzkes, Appl. Phys. Lett. 25:685 (1974) ; J. Vac. Sci. Technol. 12:37 (1975) ; 13:50 (1976).
3. N. Klein and P. Solomon, J. Appl. Phys. 47:4364 (1976).
4. M. Knoll, D. Braunig and W.R. Fahrner, IEE Trans. Nuc. Sci. NS-29:1471 (1983) ; in : Insulating Films on Semiconductors. Eds. J.F. Verweij and D.R. Wolters, Elsevier Sci. Publishers, North Holland, (1983), p. 107.
5. I.C. Chen, S. Holland and Chenming Hu, IEEE Trans. El. Dev. EDL-32:413 (1985) ; IEEE El. Dev. Lett. EDL-7:164 (1986).
6. E. Harari, J. Appl. Phys. 49:2479 (1978).
7. D.R. Wolters, J.J. van der Schoot and T. Poorter, in : Int. Conf. on Insulating Films and Semiconductors. Eds. F.J. Verweij and D.R. Wolters Elsevier Sci. Publishers, North Holland, (1983), p. 256.
8. A. Modelli and B. Ricco, IEEE Int. Electron Devices Meet. Washington, Techn. Digest (1984) 148.
9. Ching S. Jenq, T.R. Ranganath, Cheng H. Huang, H. Stanley Jones and Thomas, T.L. Chang, IEEE Int. Electron Devices Meet. Washington, Techn. Digest (1981) 381.
10. M.M. Heynes, R.F. de Keersmaecker and M.W. Hillen, Appl. Phys. Lett. 44:202 (1984).
11. A. Badihi, B. Eitan, I. Cohen and J. Shappir, Appl. Phys. Lett. 40:396 (1982).
12. Y. Nissan-Cohen, J. Shappir and D. Frohman-Bentchkowsky, J. Appl. Phys. 54:5793 (1983) ; 57:2830 (1985) ; 58:2252 (1985).
13. B. Balland, C. Plossu and S. Bardy, Thin Solid Films, 148:149 (1987).
14. S.K. Lai, Appl. Phys. Lett. 39:58 (1981).
15. K.R. Hofman, C. Werner, W. Weber and G. Dorda, IEEE Trans. El. Dev. ED-32:691 (1985).
16. N. Klein, J. Appl. Phys. 53:5828 (1982).
17. B. Ricco, M. Ya. Azbel and M.H. Brodsky, Phys. Rev. Lett. 51:1795 (1983).
18. R.H. Koch and A. Hartstein, Phys. Rev. Lett. 54:1848 (1985).
19. J. Halbritter, J. Appl. Phys. 58:1320 (1985).
20. Z.A. Weinberg and T.N. Nguyen, J. Appl. Phys. 69:1947 (1987).

FIELD DEPENDENCE OF TIME-TO-BREAKDOWN DISTRIBUTION OF THIN OXIDES

P. Olivo*[,a], Thao N. Nguyen*, and B. Ricco**

***IBM T.J. Watson Research Center**
Yorktown Heights 10589 N.Y.
****DEIS, University of Bologna**
Viale Risorgimento 2, 40136 Bologna
Italy

The accurate prediction of thin insulator reliability is of significant importance to the development of MOS VLSI technologies. In most reliability studies[1,5], the time-to-failure t_{bd} and/or the total injected charge prior to breakdown Q_{bd} are measured under high field stresses typically greater than or equal to 8 MV/cm. These data are then extrapolated down to normal operating fields to give a prediction of device wear-out. The important assumptions of this procedure are : a) the phenomena that take place at high fields and eventually lead to oxide failure also occur at the operating field and b) no additional failure mechanisms exist at low fields. These assumptions have been used routinely but never been proven to be valid to our knowledge.

In this work we report experimental data from high-quality devices suggesting that the above assumptions may not be valid. In particular, it will be shown that the spread in the t_{bd} distributions changes significantly with the stress field : under high fields these tend to be very tight, similar to intrinsic breakdown of very uniform oxides, while at lower fields they tend to broaden because of increasing domination of premature failures. This difference seems to suggest that time-dependent breakdown is initiated by localized defects that require a certain time to develop and evolve to a destructive failure and hence, may be difficult to reveal under very high field stresses. The implication of our findings is that accelerated testing may not detect all defects and consequently overestimate oxide reliability.

Finally, the role of Q_{bd} and electron trapping in oxide breakdown will be discussed.

EXPERIMENTS AND RESULTS

The devices used in this study were n^+ polysilicon-gate MOS capacitors with an oxide of 10.3 nm thick on 0.5 Ω-cm <100> n-type silicon substrate. The gate oxide was grown in O_2 + 4.5 % HCl at 800°C and subsequently

(a) present address : DEIS - University of Bologna - Viale Risorgimento 2 40136 Bologna - Italy.

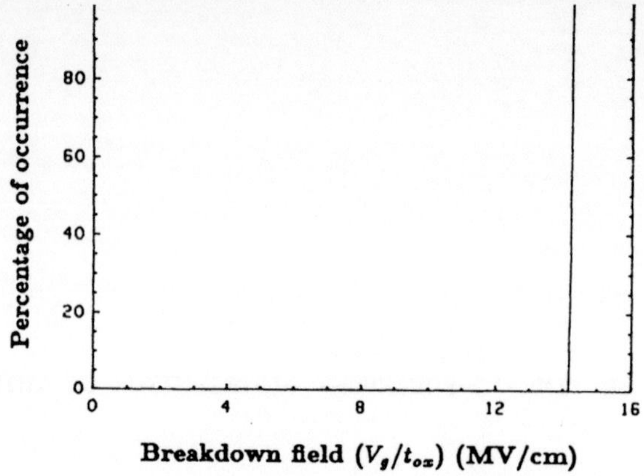

Fig. 1. Cumulative distribution of time-zero-breakdown

annealed in Ar ambient. The capacitor area was varied from 5.8×10^{-5} to 10^{-2} cm^2. Contacts to the polysilicon gate and the wafer backside were made by aluminum evaporation with a standard post-metal anneal.

Time-zero breakdown (TZB) tests were carried out by means of a HP4145A semiconductor parameter analyzer. A current ramp was applied to the device under test and the breakdown voltage is determined by detecting a large and sudden drop in the gate voltage. Time-dependent breakdown (TDB) test employed a Keithley 220 as current source and a Keithley 619 electrometer to monitor the voltage necessary to maintain a constant current through the insulator. The constant current stress is preferred in this work because it simplifies the determination of Q_{bd}. The time resolution of the TDB tests was 40 msec throughout the whole measurement. All measurements were automated and controlled by computers.

Figure 1 shows the cumulative distribution of breakdown field from TZB tests. The high average value of breakdown field and its small spread indicate that the oxide is uniform and of high quality. Figure 2 plots the cumulative distributions of t_{bd} at different levels of constant current stress from TDB tests. The data have been normalized to the median of t_{bd} (i.e. the t_{bd} corresponding to a cumulative failure of 50 %).

A suitable normalization is very important for a useful comparison between data varying by orders of magnitude (from about 8 sec with $I = 2$ A/cm^2 to 5×10^3 sec with $I = 10$ mA/cm^2)[6]. In particular, the use of the maximum or the mean of t_{bd} as normalization factor (the former strongly affected by a single, "lucky" device with a particularly high t_{bd} the latter by several small t_{bd} due to premature breakdown) would obscure the main results of the experiments.

It can be seen from fig. 2 that the t_{bd} distribution at the highest current level (corresponding to an initial average field (V_G/t_{ox}) of about 12 MV/cm) is very tight with more than 95 % of the devices between 0.8 and 1.2 x the mean-time-to-failure. As the stress level decreases, the distribution tends to broaden, especially its lower tail, indicating a larger percentage of premature breakdown. It should be pointed out that these data were obtained from the same wafer and the capacitors from each group were randomly distributed on the wafer. As a result, the different groups of devices should be reasonably identical.

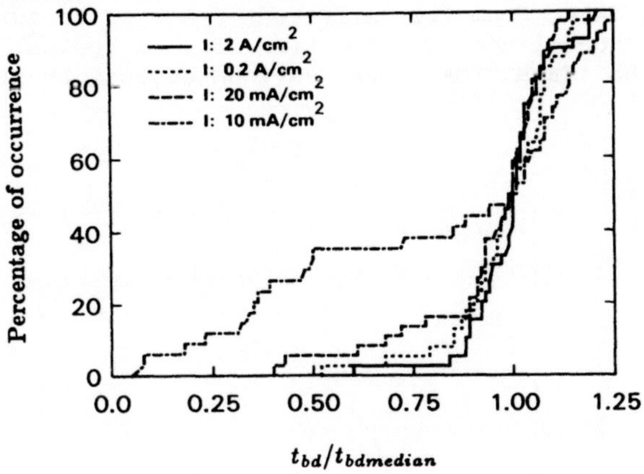

Fig. 2. Cumulative distribution for time-dependent-breakdown at different current levels.

An important factor that must be considered in generating the data of fig. 2 is the dependence of the t_{bd} distribution on capacitor area. Large area capacitors tend to produce a very broad distribution, as illustrated in fig. 3, which can easily conceal the effect of stress level. The previous results were obtained only with small area ($\leq 10^{-4} cm^2$).

DISCUSSIONS

The main result from the previous data concerns the different behavior of the t_{bd} distribution at high and low current levels (i.e. at high and low applied fields). These differences regard only the region of premature breakdown. For $t_{bd} \geq t_{bdmedian}$ the cumulative distributions are indistinguishable. This is in contrast with the results normally obtained from low-medium quality devices, when the number of premature defect-related failures increases with the applied fields.

The quality of the oxide used in the present work was very high, as demonstrated by the cumulative distribution of F_{bd} in fig 1 and by the high values of Q_{bd}[4]. In addition, the devices broken prematurely were not clustered in particular regions but scattered throughout the whole wafer.

As a consequence, the results to be discussed seem to be due to the intrinsic, physical nature of breakdown and not to low quality devices.

The premature breakdowns are due to particular behavior at localized spots, as demonstrated by the results of fig. 4. This figure shows the Fowler-Nordheim voltage shift ΔV required to keep $I = 10 mA/cm^2$ constant during the TDB experiments. For convenience ΔV is taken with respect to the voltage V_{ta} corresponding to the turn-around point of the $V_G$vs. time characteristic (i.e. that where the curve derivative goes to zero) always present in this type of experiments[7,8] and, because of the large initial transient, more easily determined that the initial V_G. In addition, the data have also been normalized with respect to V_{ta}. In this way, ΔV is relatively independent of the insulator thickness because the strong dependence of V_G and V_{ta} on the oxide field is essentially the same. The figure contains data from two different types of measurements. The curve represents the normalized ΔV vs. time for a typical device while the squares mark the data taken on various capacitors at the occurence of breakdown.

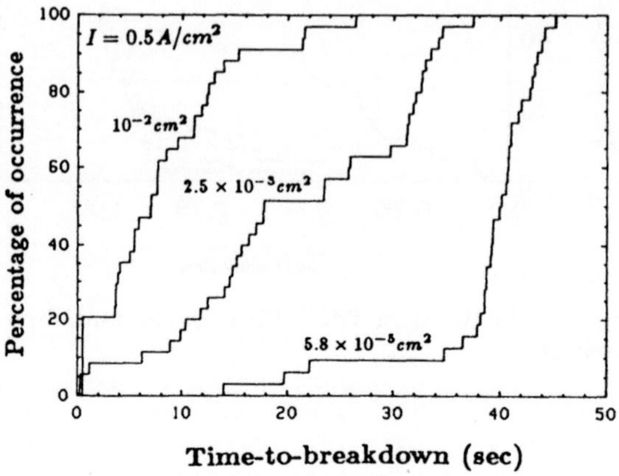

Fig. 3. Cumulative distribution for time-dependent-breakdown for different device area.

The fact that all the data points fall on the same curve indicates that the trapping characteristics of all the devices are essentially identical in spite of large differences in t_{bd}. This implies that the oxide failures, at least for what concerns premature breakdown, are not directly related to the collective effects of traps and defects uniformly distributed over the device area but, rather, to the individual behavior of localized sites providing a neglegible contribution to the measurable charge build-up.

As for the real cause of the difference between average and premature breakdown, unfortunately, there is not enough evidence to understand

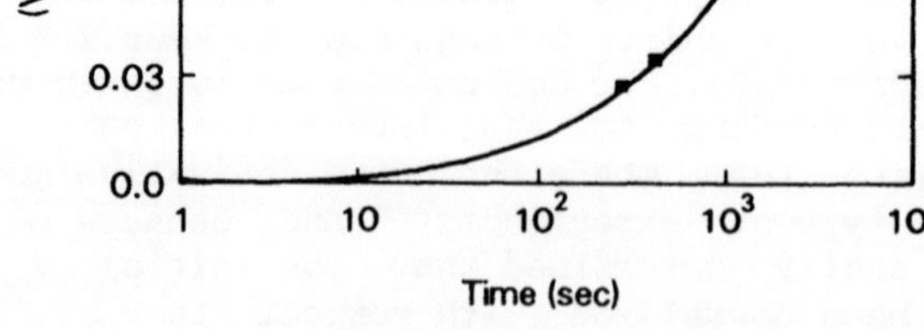

Fig. 4. Normalized Fowler-Nordheim voltage shift vs. time for a typical device. The squares represent the data at breakdown for various capacitors.

whether the two phenomena are due to really different mechanisms or simply to defects with substantially different time requirements.

In any case, the defects producing premature failure are evidently not active in the devices that break at high field and in most of the capacitors stressed at low fields. This can be either due to the fact that they are created during the stress only in devices presenting premature failures or that they are already present (or created) in all the devices but they require a certain amount of time or particular electrical conditions to develop up to a destructive stage.

Several models have been presented in the literature to describe the relationship between oxide degradation and the creation of defects[9,10] at the Si-SiO_2 interface, or to illustrate the need for particular electrical conditions within the oxide to trigger breakdown[11] by resonant tunneling (these conditions are also time-dependent).

In any case, regardless of the actual physical mechanism responsible for premature breakdown in high quality devices, our results rule-out the possibility of establishing meaningful relationships between breakdown and some measured entities, such as, in particular, Q_{bd}[2,12,13]. More specifically, however, they do not exclude that breakdown can be triggered by the amount of charge injected through a weak region, even though this latter can be quite independent of the total charge flown throughout the entire device.

The same considerations also apply to the trapping conditions. As already observed, until breakdown, all devices follow the same global trapping characteristic, though this does not necessarily apply to the spot responsible for breakdown.

Finally, it is worth mentioning that no signatures of premature failures have been noticed from the initial behaviour of the V_G vs. time characteristics. In particular, no differences have been noticed in V_{ta}, that depend on the creation of positive charge within the oxide at the initial stage.

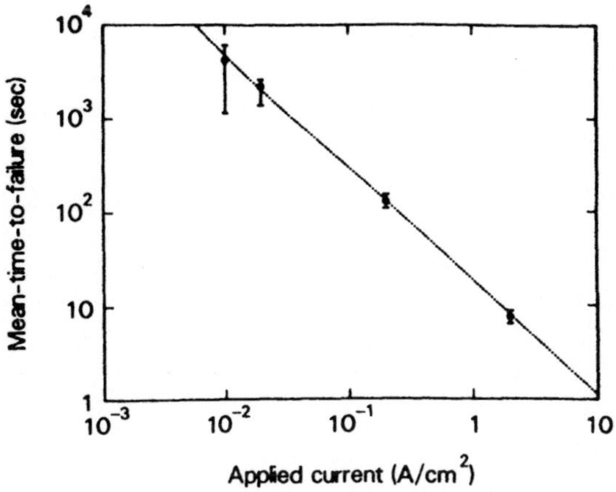

Fig. 5. Mean-time-to-failure vs. constant current level. The errors bars do not consider the best and the worst 10 % of the devices.

CONCLUSIONS

Time-dependent breakdown tests on high quality devices at different current levels show that the relative number of devices presenting a premature oxide failure increases for decreasing current levels. As a consequence, extra care is required when field accelerated tests are used to extrapolate reliability under usual operating conditions (see fig. 5).

Premature failures are due to strongly localized effects so that the relationships between breakdown and some measurable, i.e. distributed, entities can no longer be used to derive even empirical correlations among the main breakdown parameters.

REFERENCES

1. Y.P. Han, J. Mize, T. Mozdzen, T. O'Keepe, J. Pinto, and R. Worley, IEEE Tech. Dig. Int. Electron Device Meeting (1982).
2. I.C. Chen, S. Holland, and C. Hu, IEEE Elect. Dev. Letts., EDL-7:164 (1986).
3. I.C. Chen, S. Holland, K.K. Young, C. Chang, and C. Hu, Appl. Phys. Letts., 49:669 (1986).
4. M.L. Liong and J.Y. Choi, Appl. Phys. Letts., 50:104 (1987).
5. I.C. Chen and C. Hu, IEEE Elect. Dev. Letts., EDL-8:140 (1987).
6. In this test, a minimum current level of 10 mA/cm^2 has been used instead of 2 mA/cm^2 (i.e. one order of magnitude lower than the previous one). In that case, the average t_{bd} ($\simeq 40x10^3$ sec) is too high to allow a sufficiently large set of data.
7. P. Olivo, B. Ricco, and E. Sangiorgi, J. Appl. Phys., 54:5267 (1983).
8. Z.A. Weinberg and T.N. Nguyen, J. Appl. Phys., 61:1947 (1987).
9. E. Harari, J. Appl. Phys., 49:2478 (1978).
10. F.J. Grunthaner and P.J. Grunthaner, Material Science Reports, 1:65 (1986).
11. B. Ricco, M. Ya Azbel, and M.H. Brodsky, Phys. Rev. Letts., 51:1975 (1983).
12. J.J. Van der Schoot and D.R. Wolters, in Insulating Films on Semiconductors, Proceedings of the 3rd International Conference, J.F. Verweij and D.R. Wolters ed., North-Holland, Amsterdam (1983).
13. A. Modelli and B. Ricco, in IEEE Tech. Dig. Int. Electron Device meeting (1984).

RADIATION EFFECTS IN MOS VLSI STRUCTURES

H.L. Hughes

Naval Research Laboratory
Washington, D.C. 20375-5000

INTRODUCTION

Ionizing radiation is known to cause changes in the electrical properties of MOS devices[1]. Radiation effects in various oxide structures such as gate and field oxides have been previously studied[2,3]. As MOS technology now progresses to smaller feature sizes-one micron and less-new types of oxide regions have been added to VLSI circuits for device isolation and improved reliability. These new VLSI oxide structures, in turn, introduce new radiation-induced failure modes limiting circuit functionality and performance.

Buried-oxides, lightly-doped-drain (LDD) oxide spacers, and new lateral oxide-isolation structures are among the new VLSI oxide regions of concern. (See figure 1 for a graphic delineation of these regions).

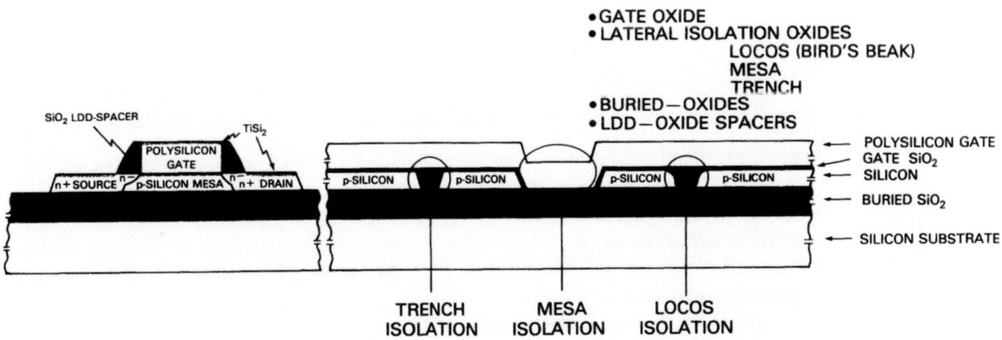

Fig. 1. VLSI Oxide regions susceptible to radiation effects

Buried-oxides, fabricated using various silicon-on-insulator (SOI) techniques [4] are being used to eliminate the possibilities of latch-up in CMOS circuits, and to reduce single-event-upset (SEU) and dose-rate upset sensitivities. Buried-oxides are also important to vertically isolate various active layers in future three-dimensional circuits.

Lightly-doped drain structures (LDD) are used to reduce reliability-limiting hot-carrier injection by reducing the electric field at the drain junction. In fabricating the n^- extension of the drain for N-channel transistors (see figure 1), the n^+ ion implantation is masked using a spacer created previously with an isotropic reactive-ion-etch (RIE). Silicon dioxide is normally used as a material for these spacers ; in some cases polysilicon has been used. However, for self-aligned silicide processes (such as $TiSi_2$ shown in figure 1) polysilicon can not be used, since the non-reactivity nature of the spacer is required to provide lift-off of the metal between the gate and source-drain contacts. Thus, deposited silicon dioxide has become the conventional spacer material for high speed VLSI circuits.

Lateral isolation is also an area where new radiation-sensitive oxide/silicon structures appear. (See figure 1). Most present integrated circuits use the local-oxidation-of-silicon (LOCOS) technique to laterally isolate adjacent active regions. However, due to deleterious encroachment (caused by the bird's beak extension) and concomitant radiation-induced leakage currents due to charge build-up in this region, other isolation techniques are now being pursued. Mesa and trench approaches fabricated by RIE techniques are being evaluated as alternative lateral isolation structures.

Furthermore, the gate oxide region, once assumed to be inherently harder as oxide thickness was reduced, [5] is now again a major concern in ultra-thin oxides, due to radiation-induced interface states (N_{IT}). The trapped oxide charge (N_{OT}) scales inversely with thickness cubed, whereas N_{IT} scales sublinearly[6].

The threshold-voltage shift (ΔV_{TH}) in MOS devices is caused by N_{OT} and N_{IT} charges in the gate-oxide and interface regions according to the following :

$$\Delta V_{TH} = \Delta V_{OT} + \Delta V_{IT} \tag{1}$$

where

$$\Delta V_{OT} = \frac{q\Delta N_{OT}}{C_{OX}} \tag{2}$$

$$\Delta V_{IT} = \frac{q\Delta N_{IT}}{C_{OX}} \tag{3}$$

C_{OX} = gate-oxide capacitance
q = electronic charge
ΔV_{OT} = usually negative in value for silicon dioxide gate-insulators
ΔV_{IT} ≥ 0 for N-channel devices, and ≤ 0 for P-channel devices

N_{IT} is a particular problem for N-channel devices where it causes threshold-voltage to shift to values greater than initial values, a phenomenon called rebound. This causes N-channel drive-current, and thus speed, to be reduced[7]. Speed is also affected by mobility reductions induced by increased N_{IT}[8].

$$T_D \sim \frac{1}{\mu(V_G - V_T)} \quad (4)$$

where

μ = channel mobility
V_G = applied voltage
V_T = threshold voltage

MESA-EDGE

Device isolation can be implemented by etching silicon islands or mesa structures on insulating substrates. Presently, these structures are defined using reactive ion etching (RIE) techniques. The vertical wall of the mesa is not orthogonal to the surface, but varies from 60° to 80° along the edge (see figure 2). Thus, there are a number of different vicinal planes along the edge. Also, as can be seen from figure 2, the oxide thickness varies greatly (2x) along the edge. Since it is known that MOS radiation-sensitivity varies with silicon orientation[10] and oxide

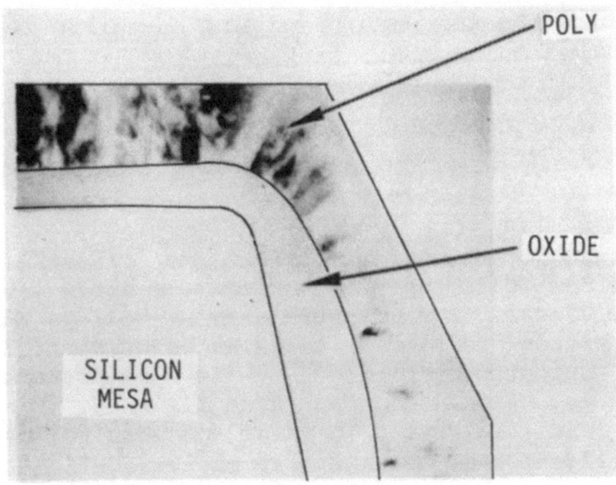

Fig. 2. Cross-sectional transmission electron micrograph showing the delineation of a silicon mesa edge.

thickness[6], these are most likely the bases for increased radiation-sensitivity of the mesa edge compared to that of the less sensitive gate MOS structure. Figure 3 shows a schematic of a silicon mesa on an SiO_2 substrate and the edge regions where the main radiation effect occurs. Radiation-induced N_{OT} causes an inversion layer to be generated along the p-type mesa-edge region from drain to source. Figure 4 presents the gate-transfer current-voltage characteristics versus various radiation-doses for an N-channel mesa-isolated transistor on an SiO_2-layer created by the SIMOX process (separation by the ion-implantation of oxygen)[11]. The edge leakage is seen as a large hump in the subthreshold region of the characteristics. At high doses, such as 10^7 rads, the build-up of N_{IT} reverses the trend by reducing mobility along the edge, and reversing the edge threshold voltage shift to low negative values.

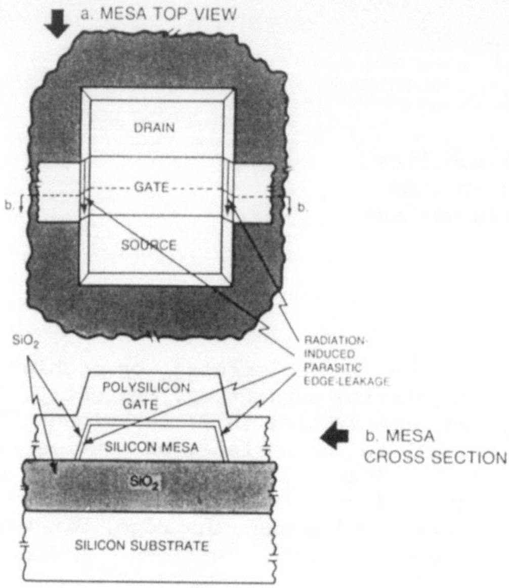

Fig. 3. a) Top view, and b) cross-sectional view of a silicon mesa over buried-oxide showing regions causing parasitic leakage current.

TRENCH-ISOLATION

Like the mesa structure, the trench has sidewalls created by RIE. Usually these silicon walls are oxidized and the whole trench filled with a deposited glass. Figure 5 presents the current-voltage characteristics for an N-channel transistor isolated by means of an oxide-filled trench region. This device was fabricated, otherwise, by the same process as the mesa-edge device in figure 4. As can be seen from figure 5, the radiation-induced edge leakage current is greater in value and less voltage controlled than that of the mesa-edge ; only curve 2 (for an irradiation of 10^5 rads) shows the large edge leakage hump similar to the mesa-edge. The reduced gate control of the edge is due to the larger distance of the poly-gate electrode from the edge-sidewall of the trench where the leakage occurs. Techniques to radiation harden trench structures have been reported[12].

BURIED OXIDES

MOS devices fabricated in thin films of silicon over buried oxide regions (as shown in fig. 1) experience modified electrical performance due to radiation-induced charges, N_{OT} and N_{IT}, in the buried-oxide structure. Utilizing test transistors without edges (enclosed configuration), the effects of back-channel regions can be isolated and studied. Figure 6 presents the current-voltage characteristics on such an enclosed N-channel MOS transistor versus various cobalt-60 radiation-doses. The back-gate region was irradiated with a negative bias (-5 volts) on the

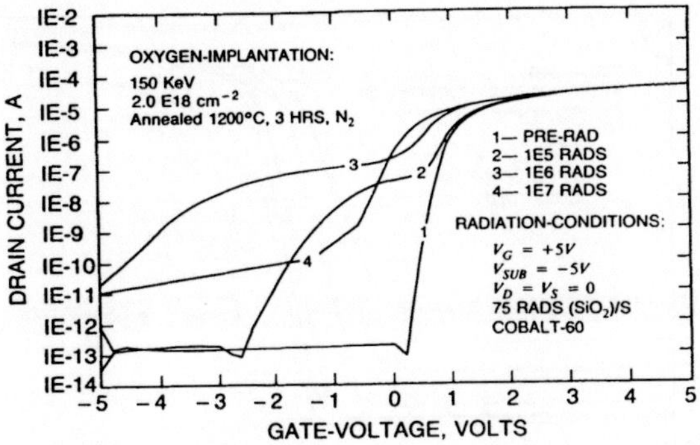

Fig. 4. Current-voltage characteristics versus radiation-dose for a mesa-isolated SIMOX N-channel MOS transistor.

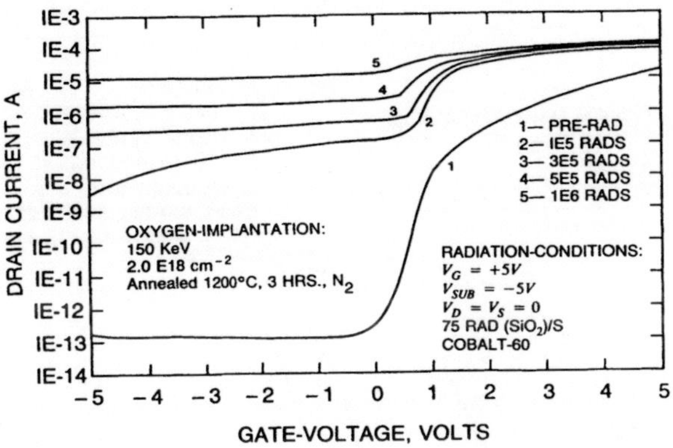

Fig. 5. Current-voltage characteristics versus radiation-dose for an oxide-filled-trench isolated SIMOX N-channel MOS transistor.

substrate to reduce the build-up of N_{OT} at the back-silicon/silicon dioxide interface[4]. The current-voltage characteristics of fig. 6 are for an active device built in the top region of a thin silicon layer affected by charge-build-up in the oxide region below the silicon layer ; essentially a back-channel effect.

Curves 1 and 3 are the "before" and "after" irradiation characteristics for a 1200 C, three hour nitrogen anneal of the high dose oxygen implantation (2.0×10^{18} per cm^2 at 150 KeV), and curves 2 and 4 for the 1100 C, three hour nitrogen anneal for the same oxygen implantation conditions. These results point to a striking difference from that observed for the nitrogen annealing of thermally-grown oxides, where high temperature annealing increases the charge build-up due to irradiation. As can be seen from fig. 7, the 1200 C nitrogen-annealed sample shifts substantially less than that observed for the 1100C nitrogen-annealed

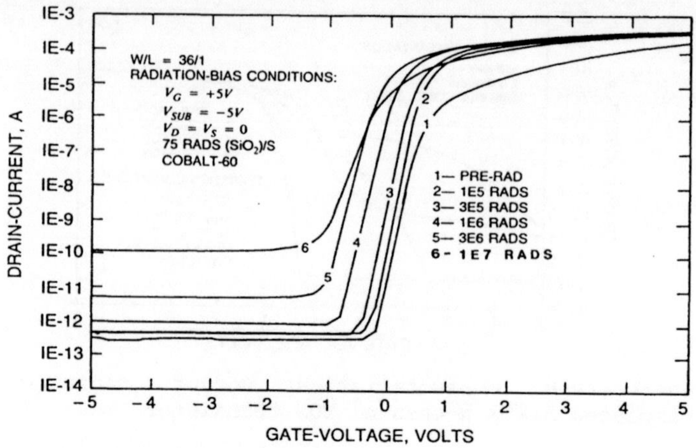

Fig. 6. Current-voltage characteristics versus radiation-dose for an enclosed SIMOX N-channel MOS transistor.

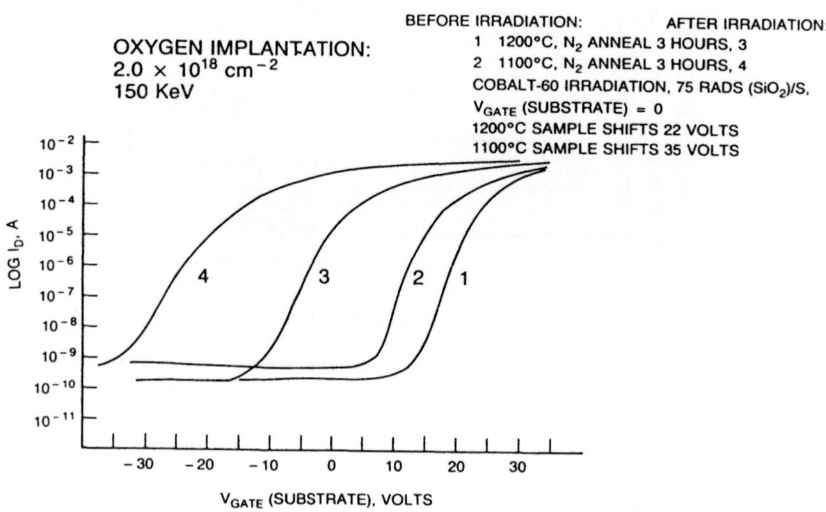

Fig. 7. Back-gate current-voltage characteristics for SIMOX N-channel MOS transistors.

sample. From figure 8 it can be seen that the nitrogen annealing of a thermally grown oxide greatly increases the radiation-induced shift (from one volt, for the unannealed control sample, to 25 volts for the sample annealed one hour at 1200 C in nitrogen). From secondary ion mass spectrometry analyses (SIMS) it was found that the nitrogen content at the silicon/silicon dioxide interface is also greatly increased due to the anneal process. It is possible that the increased radiation sensitivity is due to nitrogen related oxide centers serving as hole traps[13]. Performing a similar SIMS analysis on SIMOX samples annealed at both 1100 C and 1200 C revealed that the nitrogen content did not increase as a result of the annealing, and thus possible nitrogen-related hole traps (causing N_{OT}) would not have been enhanced in concentration. The superficial silicon

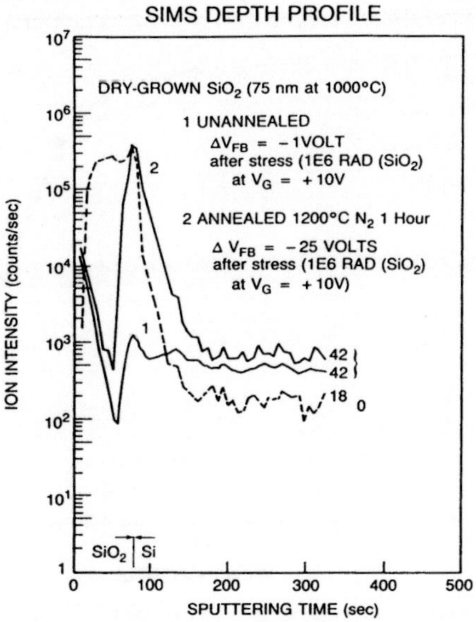

Fig. 8. Secondary-ion mass spectrograph and radiation-stress results for unannealed and annealed silicon/silicon dioxide structures.

layer resulting from the high-dose, high energy oxygen implantation serves as a mask against contamination during the high temperature processing.

LIGHTLY-DOPED DRAIN (LDD)

Oxide spacers are used to fabricate the n^-/n^+ drain regions needed to reduce hot-carrier effects by reducing the electric field at the drain junction. These regions are created by the anisotropic RIE of deposited oxide films. However, with the build-up of radiation-induced N_{OT} in the deposited oxide spacer, the n^- regions are converted to n^+ drain extensions, thus defeating the original purpose of reducing the electric field at the drain junction. Furthermore, hot electron trapping will also be enhanced through the build-up of N_{IT} with irradiation.

CIRCUIT SPEED DEGRADATION

The main cause for circuit speed reduction is the build-up of radiation-induced N_{IT}. Figure 9 presents the operating frequency of a CMOS ring oscillator versus cobalt-60 radiation dose. For instance, the speed falls off by a factor of 10 at a dose of 7.5 Mrad (SiO_2). From fig. 10, where the threshold-voltages (N- and P-channel) and mobility (for the N-channel transistor) are plotted versus cobalt-60 radiation-dose, it can be seen that at 7.5 Mrad (SiO_2) the mobility has fallen-off to a value 53 % of the pre-irradiation value and the N-channel threshold voltage has increased to a value of 4.2 volts. Using equation 4 the factor of 10 increase in propagation-delay (T_D) can be substantiated ; thus an order of magnitude reduction in speed is caused by mobility and threshold voltage changes.

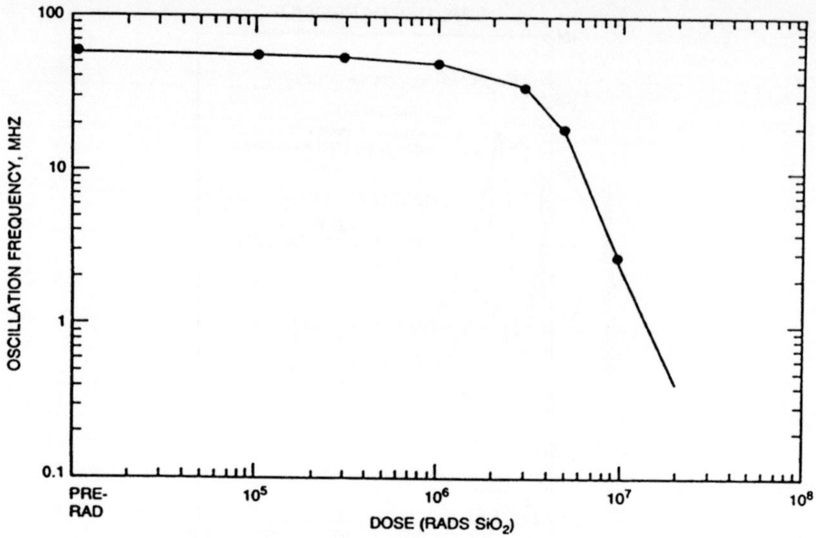

Fig. 9. Ring oscillator (21-stage inverter chain) frequency versus radiation-dose.

Thus far, the basic understanding of the mechanism for the radiation-induced build-up of N_{IT} includes two basic parameters : the P_b center at the silicon/silicon dioxide interface and interfacial hydrogen[14]. Techniques to reduce the radiation-induced P_b -center density in complex integrated circuits have yet to be developed. Furthermore, the hydrogen content in such circuits is very difficult to reduce since most chemical vapor deposition (CVD) fabrication processes include processing with hydrogen-containing compounds. Perhaps Si-MBE techniques for silicon deposition in ultra-high vacuum (hydrogen-free) will provide a way to reduce N_{IT}.

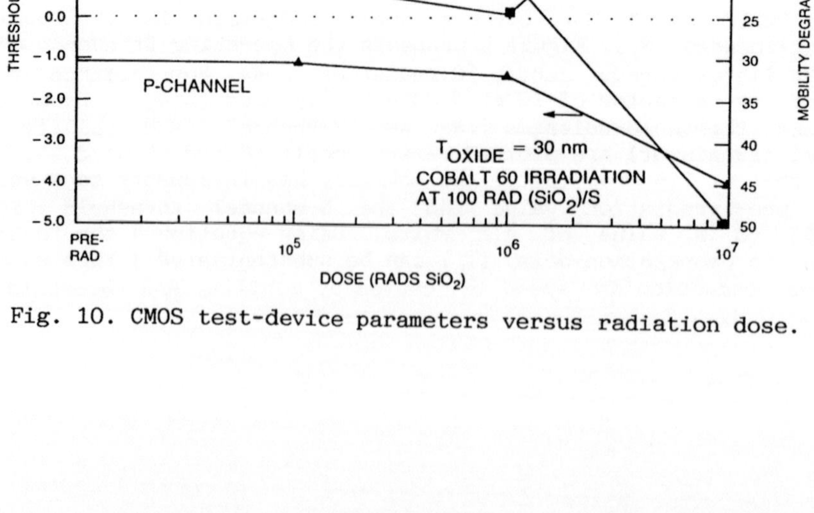

Fig. 10. CMOS test-device parameters versus radiation dose.

CONCLUSION

Scaling VLSI circuits to submicron geometries requires the incorporation of additional oxide structures for performance and reliability purposes. Included are : buried-oxides, lateral isolation oxides and LDD-spacers, in which radiation-induced charges, N_{OT} and N_{IT}, cause leakage currents and performance degradation. As gate-oxide migrate to thinner values, the key parameter causing speed degradation is N_{IT}. New processing techniques will need to be developed to alleviate these effects.

ACKNOWLEDGEMENTS

The device measurements provided by W. Jenkins, R. Hevey, R. Lawrence, and K. Maley are gratefully acknowledged.

REFERENCES

1. H.L. Hughes and R.A. Giroux, Electronics, Dec., 58 (1964).
2. G.F. Derbenwick and B.L. Gregory, IEEE Trans. Nuc. Sci., NS-22:2151 (1975).
3. K. Watanabe, M. Kato, T. Okabe, and M. Nagata, IEEE Trans. Nuc. Sci. NS-32:3971 (1985).
4. G.E. Davis, H.L. Hughes, and T.I. Kamins, IEEE Trans. Nuc. Sci., NS-29:1685 (1982).
5. J.M. McGarrity, IEEE Trans. Nuc. Sci., NS-27:1739 (1980).
6. K. Naruke, M. Yoshida, K. Maeguchi, and H. Tango, IEEE Trans. Nuc. Sci., NS-30:4054 (1983).
7. J.R. Burns, RCA Review, 627 (1964).
8. K.F. Galloway, M. Gaitan, and T.J. Russel, IEEE Trans. Nuc . Sci. NS-31:1497 (1984).
9. A.H. Johnston, IEEE Trans. Nuc. Sci., NS-31:1427 (1984).
10. Unpublished data, A.G. Revesz, G. Brown, and H.L. Hughes.
11. P.L.F Hemment, Semiconductor-On-Insulator and Thin Film Transistors Technology, Edited by A. Chiang, M.W. Geis, and L. Pfeiffer, MRS, 207 (1986).
12. K. Kasama, F. Toyokawa, M. Sakamoto, and K. Kobayashi, IEEE Trans. Nucl. Sci., NS-32:3965 (1985).
13. D.L. Griscom, J. Non-Crystalline Solids 31:241 (1978).
14. D.L. Griscom, J. Appl. Phys. 58:2524 (1985).

RELATIONSHIP BETWEEN HOLE TRAPPING AND INTERFACE STATE GENERATION IN THE Si/SiO_2 SYSTEM

S.J. Wang, J.M. Sung and S.A. Lyon

Department of Electrical Engineering
Princeton University
Princeton, NJ 08544

ABSTRACT

We have discovered a general relationship between the distance from the interface at which holes are trapped, and the subsequent generation of interface states. With photon assisted tunneling it has been previously established that there are two types of trapped holes near the Si/SiO_2 interface after high-field stress (Fowler-Nordheim tunneling) and radiation damage. These types are distinguished by their location and behavior upon electron capture. The first type ("near-interfacial trapped holes") are located between 20 and 70 Å from the interface and completely disappear upon electron capture. The second type ("interfacial trapped holes") lie within about 15 Å of the interface, and immediately become interface states when they capture an electron. The experiments show that these two types of interface states are not independent, but rather holes are first trapped in the near-interfacial sites, and then are converted to interfacial trapped holes by thermal energy or very high fields.

INTRODUCTION

Radiation damage is one of the oldest unsolved problems in metal-oxide-silicon (MOS) device physics. It has been known for some time that the high energy radiation generates electron-hole pairs in the SiO_2, and these holes play a major role in the subsequent degradation[1-4]. Not only are holes trapped in the oxide, but they also trigger the generation of interface states. Some experiments show a clear one-to-one correlation between the trapped holes and interface defects[5-7]. It has also been observed that no interface states form when the trapped holes produced by X-irradiation are annihilated with photoinjected electrons at low temperature[7]. However, in other experiments, the annihilation of the holes by electrons causes interface states to form[8]. Thus, while it is generally agreed that holes play a major role in the formation of interface states, the detailed nature of this role is not clear.

In this paper we present the results of experiments which reconcile some of the seeming contradictory observations mentioned above. Our studies are based upon the hole-profiling experiments of Chang and Lyon[9]. They found two different trapping positions for holes introduced by X-irradiation and Fowler-Nordheim tunneling. The two types of trapped holes could also be distinguished by their response to photoinjected electrons, as discussed above. In the present experiments we show that these two trapping sites are not unrelated defects, but rather they are two different stages in the evolution of a trapped hole. In particular, we find that the one-to-one correlation of interface states with trapped holes, the suppression of interface state formation by annihilation of the trapped holes, and the generation of interface states by hole annihilation are all limiting cases of a more complicated behavior. Even after reaching the vicinity of the Si/SiO_2 interface, radiation produced holes must pass through several stages before becoming an interface state. First the hole is trapped in "near-interfacial hole traps", located between 20 and 70 Å from the interface. Next, the hole is transferred to "interfacial hole traps" located within 14 Å of the interface. This transfer requires either very high fields (as in the Fowler-Nordheim experiments) or thermal energy. Finally, these interfacial trapped holes are transformed into interface states by the capture of an electron.

EXPERIMENT

All the samples used in these studies were similar in that they were MOS capacitors with thin semitransparent Al gates (about 130 Å) and oxide thicknesses of about 1000 Å. Both radiation hard and soft oxides have been investigated. One key ingredient of these experiments is the use of low temperature (90 K) to "freeze" the holes in some configuration, so that they remain stationary during the profiling. In some of the work, the irradiations have been performed while holding the MOS capacitor at low temperature. The holes are thought to form small polarons and are not mobile at 90 K[10]. Thus these holes are distributed throughout the bulk of the oxide. In order to study interface state formation, we must first move the holes to the vicinity of the Si/SiO_2 interface. There are several ways to accomplish this, including "pushing" them with a large (>4MV/cm) gate-positive field, applying a moderate gate-positive field (1MV/cm) and raising the temperature above about 130-150 K (but remaining below about 220 K, where interface states begin to form[11]), and applying a moderate gate-positive field while shining visible light on the sample[12]. We have investigated these three techniques, and find that the final distribution of trapped holes is similar for each. In all the low temperature irradiation experiments discussed in this paper, we used a 5MV/cm gate-positive field to push the holes. The field was applied for 8 hours, long enough to saturate the hole motion, as monitored by the shifts of the capacitance-voltage (C-V) curve. Typically, between 50 and 90 % of the holes (assuming they were initially distributed uniformly in the oxide) are lost during the pushing. The difference in low temperature hole retention between hard and soft oxides is only about factor of two, even though the room temperature hole trapping efficiency may differ by orders of magnitude. The origin of this difference will be discussed below.

The location of the trapped holes is determined with the hole-profiling technique introduced by Chang[9]. This technique is based upon the experimental observation that electrons do not tunnel directly from the Si into the trapped holes, even when assisted by a photon. Rather, electrons must first enter the oxide conduction band via photon-assisted tunneling, and they are then captured by the trapped holes. By varying the oxide field the position at which the electrons enter the oxide conduction band is changed, and thus holes at different distances from the interface can capture the electrons. Experimentally one starts with a low field, which implies a long tunneling distance for the electrons. Electrons are tunneled at this field until the CV curve stops shifting, which means that all the holes which are far from the interface have been annihilated. The field is then raised, and the process repeated. At each higher field, a new group of holes, each lying successively closer to the interface, can capture electrons. The total number of holes at a given distance is obtained from the shift of the CV curve at the corresponding field.

In the case of Fowler-Nordheim tunneling, Chang found that the holes trap between 15 and 17 Å from the interface and that these holes immediately become interface states when they capture an electron. In the present experiments we find similar trapped holes, however, in some samples they appear to lie a short distance from the interface, as seen by Chang, while in other samples they appear to be at the interface. It is not clear whether these differences correspond to real variations in the hole position, or whether some other effect changes the photon-assisted tunneling in such a way as to make it appear that there is a shift of a few Angstroms in the hole location. In all cases, how ever, we find that these holes lie within 15-20 Å of the interface, and immediately produce an interface state upon electron capture. We will use the term "interfacial trapped holes" for all of these holes, whether they appear to be at the interface or a few Angstroms away from it.

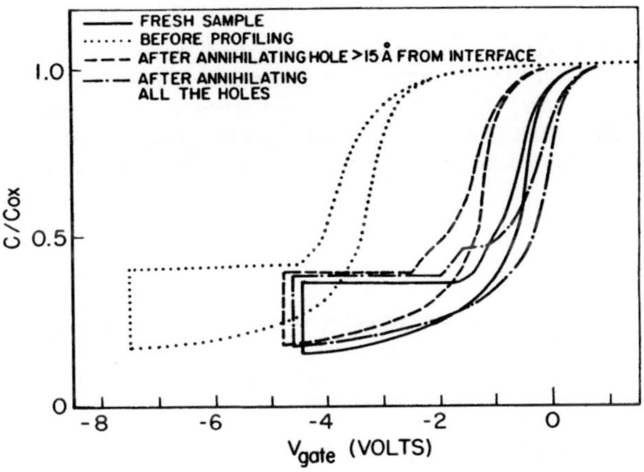

Fig. 1. Low temperature C-V curves showing several stages in the evaluation of hole trapping. Solid curve : fresh sample. Dotted curve : after low temperature irradiation, pushing holes to the interface, and holding the sample at room temperature for 3 days. Dashed curve : after using photon-assisted tunneling to annihilate all holes lying farther than 15 Å from the Si/SiO_2 interface. Dot-dash : after photoinjecting electrons from the metal gate and annihilating all remaining holes.

In Fig. 1 we show the low temperature C-V curves for a soft oxide. This sample had an oxide thickness of 1300 Å. The solid curve is for a fresh sample, and we see a small ledge indicating the presence of a few interface states[13]. The sample was X-irradiated at 90 K and the holes were pushed to the interface with a 5 MV/cm field as described above. For this sample, about 80 % of the holes were lost during the pushing. If electrons were photoinjected at this point, the C-V curve would return to the initial curve, and few if any interface states would be seen after warming the sample. Instead, the sample was warmed to room temperature for 3 days, and then cooled to 90 K to obtain the dotted curve. About 15-20 % of the holes were lost during this period. Next, photon-assisted tunneling was used to profile the remaining holes. The dashed curve was taken after all holes farther than 15 Å from the interface had been annihilated. The major effect has been to shift the curve back towards the initial one, though a more noticeable ledge, indicating interface states, has appeared. Finally, electrons were photoinjected from the metal gate in order to annihilate all the remaining holes. The dotdash curve in fig. 1 was taken at this point. These remaining holes have been tranformed into interface states, as shown by the increase in the breadth of the ledge.

In fig. 2 we show a histogram of the trapped hole density vs. the distance from the Si/SiO_2 interface obtained from profiling the sample whose C-V curves are shown in fig. 1. There are two distinct groups of holes in fig. 2, those very near the interface (hatched region), and those located between 20 and 70 Å from the interface. The large group of holes situated beyond 20 Å is the only part of the hole distribution which is seen if the sample had not been warmed to room temperature, and in that case the holes disappear upon electron capture[9]. Again, in the present experiments we find that these holes simply disappear when they capture an electron as shown by the dashed curve in fig. 1. Thus their behavior has not been changed by the warming. We will call this type of hole a near-interfacial trapped hole.

The group of holes located within 15 Å of the interface has not been previously observed after radiation damage. However, Chang saw a similar distribution of holes after Fowler-Nordheim tunneling at 90 K, which immediately became interface states upon electron capture, as in the present experiments (see dot-dash curve in fig. 1). We call this type

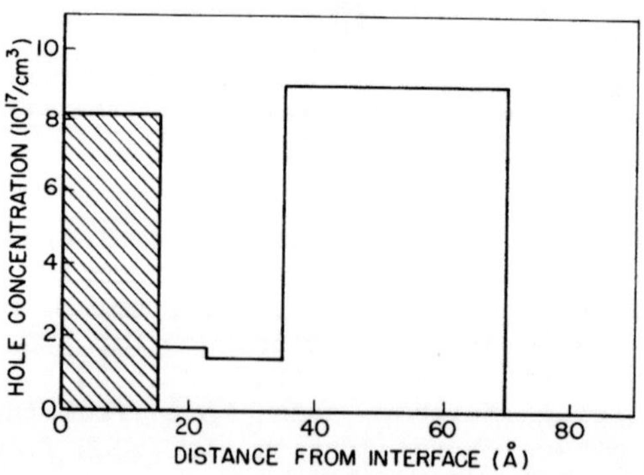

Figure 2. Histogram showing the density of trapped holes as a function of the distance from the Si/SiO_2 interface. This distribution was obtained after irradiating a MOS capacitor at low temperature, pushing the holes to the interface, and holding the sample at room temperature for 3 days. The hatched region shows those holes which became interface states when they captured an electron. The rest of the holes simply disappeared upon electron capture.

interfacial trapped holes. Warming the sample to room temperature has had the effect of converting some of the holes from near-interfacial to interfacial. After reaching the vicinity of the interface (near-interfacial sites) a hole still must first transfer to the interfacial sites, and then capture an electron in order to become an interface state. These processes require thermal energy. For this sample, about one fourth of the trapped holes were transferred in the 3 days at room temperature.

In addition to the relatively soft oxide and low temperature irradiation discussed above, we have investigated other oxides and procedures. In particular we have studied several radiation hard oxides. As mentioned above, the hard and soft oxides behave in a similar manner during the low temperature irradiation and while pushing the holes to the interface. Typically about 10 % of the holes are captured in the hard oxides, and if the sample is kept at low temperature they have the same distribution and behave like the near-interfacial trapped holes we found in soft oxides. However, there is a large difference when the samples are warmed. While 10-20 % of the holes in the soft oxide were lost after 3 days at room temperature, 90 % of the holes in the hard oxide escaped after a half day at room temperature, and those that remained had not transferred to the interfacial sites. From Fowler-Nordheim tunneling experiments we have found that holes can be trapped at the interfacial sites, and once there behave like interfacial trapped holes in soft oxides. Apparently, one effect of radiation-hardening an oxide is to decrease the depth of the near-interfacial traps.

We have also investigated the effect of different irradiation procedures on hole trapping in soft oxides. If the irradiation is done at low temperature, but the holes are not pushed to the interface, we still find both types of trapped holes after holding the sample at room temperature. However, the rate of transferring holes to the interfacial traps is significantly increased. After one hour at room temperature, about 20 % of the holes were found in the interfacial traps for a sample which had not been subjected to the 5MV/cm "pushing". Similarly, in the case of room temperature irradiations, if the sample is cooled to low temperature some time after the irradiation, both types of trapped holes are found. Again we find that the rate of conversion from near-interfacial to interfacial trapped holes is more rapid than when the holes were pushed to the interface at low temperature. Pushing the holes to the interface at 90 K apparently impedes the transfer to the interfacial sites, but we do not understand the reason for this effect.

DISCUSSION

In summary, we have found that it is possible to relate the formation of interface states to the physical location of radiation induced trapped holes in the oxide of MOS capacitors. When holes are first trapped in the vincinity of the interface, they are what we have called near-interfacial trapped holes. Upon warming, these holes can be converted to interfacial trapped holes, which are the precursors of the interface states. The interfacial trapped holes immediately become interface states, even at low temperature (as low as 30 K), when they capture an electron.

These results allow us to understand the previously mentioned, seemingly contradictory experiments. For example, in one experiment it was shown that the annihilation of radiation generated holes with injected electrons prevents the formation of interface states[7], while in another experiment the injection of electrons triggered the formation of interface states after avalanche injection of holes into the SiO_2 [8]. We can now say that these two results are not contradictory, since the first experiment was performed at low temperature, and holes could be trapped only at the near-interfacial sites, while the second experiment was done at room temperature so that both interfacial and near-interfacial trapped holes were present. Similarly, some results have shown that every trapped hole will eventually produce an interface state[5], but this is not always observed. We can now conclude that every interfacial trapped hole will eventually become an interface state, but holes can trap at the near-interfacial position without necessarily becoming an interface state. The holes studied in ref. 5 were introduced by Fowler-Nordheim tunneling, which we have found produces only interfacial trapped holes. Thus, we would now expect complete conversion to interface states in those experiments.

We have also found qualitative differences between radiation hard and soft oxides. While both types of oxides appear to have comparable numbers of near-interfacial traps (at least the low-temperature trapping efficiency is similar), these traps are deeper in the soft oxides. Thus the holes in the hard oxides escape from the traps at lower temperatures than those needed to transfer them to the interfacial sites. In soft oxides, on the other hand, the holes do not escape, and can move to the interface and eventually trigger the formation of interface states. For a room temperature irradiation of a hard oxide, one would expect very few of the holes to remain trapped long enough to be converted into interfacial trapped holes, and therefore very few interface states to form, as observed. In contrast, we find both types of trapped hole after a room temperature irradiation of a soft oxide.

The movement of holes from the near-interfacial to the interfacial traps appears to be a rather complicated process. It is not a simple release and recapture, since then we would expect to see some holes trap at the interfacial position initially. That is not seen. In fact, in a soft oxide the transfer process is relatively efficient, with only 10 to 20 % of the holes being lost. This fact would imply that a large number of holes should be found in the interfacial traps immediately after the "pushing". It seems more likely that the transfer involves the migration of the trapped hole defect to the interface. This picture is similar in spirit to some aspects of the Bond-Strain-Gradient mode,[14] though the present experiments do not support the details of that model. It is also not clear why trapping the hole at the near-interfacial site at low temperature would impede the migration as compared to the room temperature irradiation without pushing the holes. Possibly the energy released by the hole capture helps initiate the migration, though this explanation would seem to imply that the migration would either occur immediately upon hole capture, or be very slow. This does not appear to be the case. A definitive answer to these questions awaits further experiments.

The authors would like to thank P. Dressendorfer and R. Smeltzer for supplying samples for these experiments. This work was supported in part by grants from the GE foundation, RCA, Bellcore, and the National Science Foundation through the Presidential Young Investigator program (Grant No. ECS8351620).

REFERENCES

1. E.H. Snow, A.S. Grove and D.J. Fitzgerald, Proc. IEEE 55:1168 (1967).
2. K.H. Zaininger and A.G. Holmes-Sidele, RCA Rev. 28:208 (1967).
3. R.J. Powell and G.F. Derbenwick, IEEE Trans. Nucl. Sci. NS-18:99 (1971).
4. P.S. Winokur and M.M. Sokoloski, Appl. Phys. Lett. 28:627 (1976).
5. G.J. Hu and W.C. Johnson, Appl. Phys. Lett. 36:590 (1980).
6. S.T. Chang, J.K. Wu, and S.A. Lyon, Appl. Phys. Lett. 48:662 (1986).
7. G.J. Hu and W.C. Johnson, J. Appl. Phys. 54:1441 (1983).

8. S.K. Lai, Appl. Phys. Lett. 39:58 (1981).
9. S.T. Chang and S.A. Lyon, Appl. Phys. Lett. 48:136 (1986).
10. R.C. Hughes, Phys. Rev. B15:2012 (1977).
11. J.K. Wu, S.A. Lyon and W.C. Johnson, Appl. Phys. Lett. 42:585 (1983).
12. E. Harari, S. Wang, and B.S.H. Royce, J. Appl. Phys. 46:1310 (1975).
13. Short descriptions of the interpretation of low temperature C-V curves are given in ref. 5 and 6. A more complete discussion can be found in C.S. Jeng, Ph. D. Dissertation, Princeton University (Unpublished, available from University Microfilms International, P.O. Box 1764, Ann Arbor, MI 48106).
14. F.J. Grunthaner, P.J. Grunthaner, and J. Maserjian, IEEE Trans. Nucl. Sci. NS-29:1462 (1982).

RADIATION-INDUCED CONDUCTIVITY OF THIN SILICON DIOXIDE FILMS ON SILICON

V.A. Gurtov and A.I. Nazarov

Petrozavodsk University
Petrozavodsk, USSR

INTRODUCTION

The use of electron and X-ray lithography in the production of semiconductor devices poses some problems related to the study of changes arising in dielectric films on the surface of semiconductors due to their interaction with ionizing radiation[1]. Of special interest in this respect is silicon dioxide which is most commonly used in microelectronics. Some aspects of radiation induced positive charge formation in silicon dioxide are well understood[2-4]. However, some aspects of charge transport in silicon dioxide films of MOS structures due to irradiation of the gate are still poorly studied.

Charge transport in silicon dioxide subjected to X-ray photons was investigated in reference 5, and an assumption was made as to the role of a radiation induced space charge in this process. The relationship between the kinetics of radiation-induced current and the amount of stored space charge was not revealed in detail. Hughes[6] found that radiation resulted in saturation of drift rate of electrons and determined that the mobility of electrons $\mu_n = 20$ cm^2/V.s and their lifetime $\tau_n = 1{,}2\times10^{-8}$s. The energy value required for the generation of one electron-hole pair in silicon dioxide was estimated[7] to be 17 eV.

The present paper investigates the current flow mechanism in MOS structures which differ in silicon dioxide thickness and are continually irradiated with low energy electrons. The experimental results obtained are analysed using the model described in references 2 and 4. According to the model, the irradiation of polarized MOS structures results in the separation of electron-hole pairs and hole capturing at traps in the nearboundary areas of SiO_2.

MEASURING TECHNIQUE

MOS structures on silicon substrate with thermal silicon dioxide, SiO_2 as thick as $d_{ox} = 26 - 180$ nm were studied. Electron-transparent aluminium electrodes (0,07 x 0,07) cm in size were used as a gate. The structures were placed in a thick chamber made of tin and lead. The chamber was fixed to a teflon disc and installed in the photographic camera of an electron microscope. A tungsten thermionic cathode and a microscopic imaging system

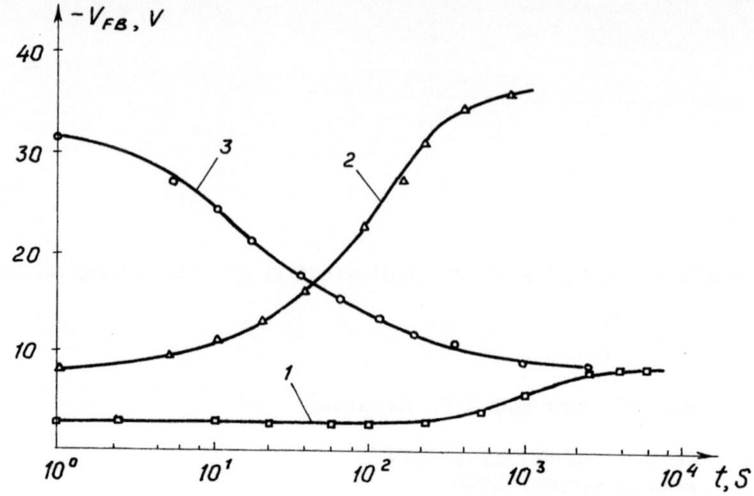

Fig. 1. Kinetics of radiation - induced flat band voltage, V_{FB} in the course of : 1,2 - space charge accumulation, 3 - "electronic" anneal d_{ox} = 60 nm ; (1) V_{FBO} = -2 V, V_G = 0 ; (2) V_{FBO} = -9 V, V_G = +30 V, (3) V_{FBO} = -30 V, V_G = 0.

were available. This made it possible to produce highly stable beams of electrons with energy E_o = 50 keV and to control their intensity.

To measure radiation-induced current (RIC), the silicon substrate was connected to an electrometric amplifier. Voltage was supplied from a stabilized feed source to the upper field electrode. Flat band voltage, V_{FB}, was measured with the diaphragm closed.

RIC measurements were made at a pressure P = 10^{-2} Pa and electron beam intensity n_e = $1,25 \times 10^{11}$ $cm^{-2}.s^{-1}$. The absorbed dose power calculated for silicon dioxide according to ref. 8 was D = $6,5 \times 10^3$ rad/s. It was shown in reference 5 that RIC is strongly dependent on the amount of space charge in the SiO_2 film, hence the kinetics of variation of the flat band voltage of MOS structures caused by an electron beam was studied prior to current measurement (fig. 1). It appears that, first, irradiation at gate voltage V_G = 0 causes the $|V_{FB}|$ value to increase to reach maximum $|V_{FBM}|$ (curve 1) ; second, subsequent irradiation at V_G= +30 V results in a markedly decreased $|V_{FB}|$ value (curve 2) ; and, third, further irradiation of MOS structures at V_G= 0 causes the $|V_{FB}|$ value to decline as low as $|V_{FBM}|$ value (curve 3). Therefore, subsequent irradiation at V_G = 0 results in the same final charge state of MOS structures (V_{FBM} value) regardless of their initial charge state (initial V_{FB} value). Hence, MOS structures must be in the same charge before measuring RIC. To this end, they were irradiated by an electron beam during a time required to achieve a steady-state V_{FBM} value.

EXPERIMENTAL RESULTS

An electron beam affects MOS structure by making transient RIC flow through silicon dioxide. The amount of the current was found to depend on many parameters such as beam intensity, voltage applied and the amount of the charge stored in SiO_2. The alteration pattern of a space charge is

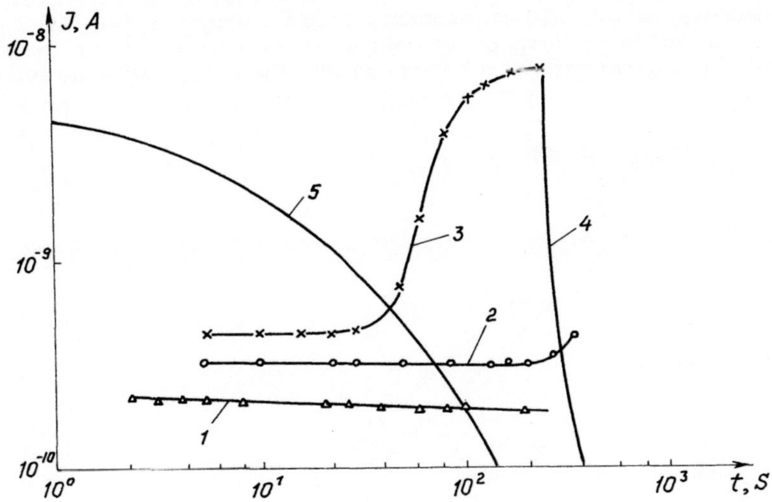

Fig. 2. Time-dependence of increase and decrease in RIC in MOS structures. d_{ox} = 60 nm ; V_G : (1)- 10V ; (2) - 20V ; (3)- 30V. (4)- = 30 V in the absence of electron beam, 5 - relation 4 reconstructed in short time range.

largely dependent on the amount of external voltage applied. Characteristic kinetic curves for different applied voltage (V_G) values are shown in fig. 2. At low gate voltage $V_G < 20$ V the current flowing through the structure decreased with irradiation time. However, application of gate voltage $V_G \geqslant 20$ V caused current to enhance with a subsequent saturation. The absorbed dose required for creating conditions under which current would start to increase declined with increasing applied field. Switching off an electron beam caused current to drop to zero values (fig. 2).

The dependence of RIC on reverse field value at the Si-SiO_2 contact at $V_G > 0$ is shown in fig. 3. In the course of irradiation of MOS structures here the centroid of space charge distribution in SiO_2 is within 10 nm of the boundary with the silicon[9]. Therefore, the field at the contact was calculated according to the relation :

$$E = (V_G + |V_{FB}|)/d_{ox} \tag{1}$$

It is seen from fig. 3 that relaxation of the current in the field range at the contact $E_G > 7{,}5$ MV/cm can be well described using the Fowler-Nordheim equation (2), the value of the constant $E_1 = 2{,}3 \times 10^8$ V/cm being close to that available in the relevant literature[10].

$$I \sim E^2 \exp(-E_1/E) \tag{2}$$

Now let us analyse the voltage-current characteristics measured in the absence of a considerable space charge in the SiO_2 film. In other words, RIC was fixed 5-10 seconds after voltage was supplied to the gate. In this case the change in the radiation-induced space charge was not large compared to the initial value $|\Delta V_{FB}|$ = (2-3)V for d_{ox}= 100 nm, and the relaxion processes related to the charging of the capacity were essentially finished.

The dependence of RIC on electric field intensity for MOS structures with different oxide thickness is shown in figure 4. It can be seen that the relations are symmetrical relative to the polarity of applied voltage,

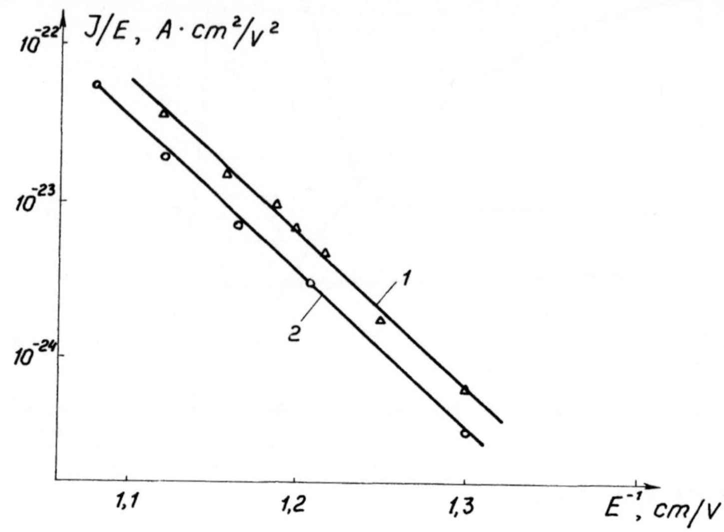

Fig. 3. Field dependence of RIC at the Si-SiO_2 contact, E_c, 1 - V_G = 30 V; V_{FB} (t) - space charge accumulation ; 2 - V_{FB} = -26 V ; V_G(t) - "electronic" anneal.

and do not depend on gate material (Al and Au). The RIC values for the samples with silicon dioxide as thick as d_{ox}= 100 nm and d_{ox} = 180 nm coincide.

A decrease in SiO_2 thickness from 100 nm to 25 nm causes the amount of RIC to diminish according to the law $J \sim d_{ox}^{1.3}$. The volt-ampere characteristics (fig. 4) can be divided into two regions : ohmic and sublinear. The transition of the dependence of RIC on field intensity in silicon dioxide E from the ohmic region to the sublinear region is not dependent on SiO_2 thickness d_{ox} and is observed when the electric field intensity E ≈ $2x10^5$V/cm.

The current integral $Q_1 = \int_0^t Idt$ during the time of RIC measurement was compared with the amount of the space charge Q_2 stored during the same time : $Q_2 = C_{ox}\Delta V_{FB}$. The amount of the charge Q_1 was found to be much greater than that of the trapped charge Q_2/Q_1 = 1/20.

DISCUSSION

It can be seen from fig. 1 that irradiation of MOS structures results in the formation of a positive space charge in SiO_2. The amount of the charge is dependent on the voltage applied to the structure in the course of irradiation and exposure time. The final charge state of the MOS structure does not depend on the initial state. Therefore, irradiation generates electron-hole pairs within SiO_2 , whereas the field and irradiation time are responsible for their spatial division and for the equilibrium concentration of the holes captured by the trapping centers in

SiO_2. Electron capturing by traps may be disregarded because of the small nonequilibrium electron concentration in conduction SiO_2 band[4].

Variation in flat band voltage (radiation space charge) resulting from irradiation causes the redistribution of electric field both within the

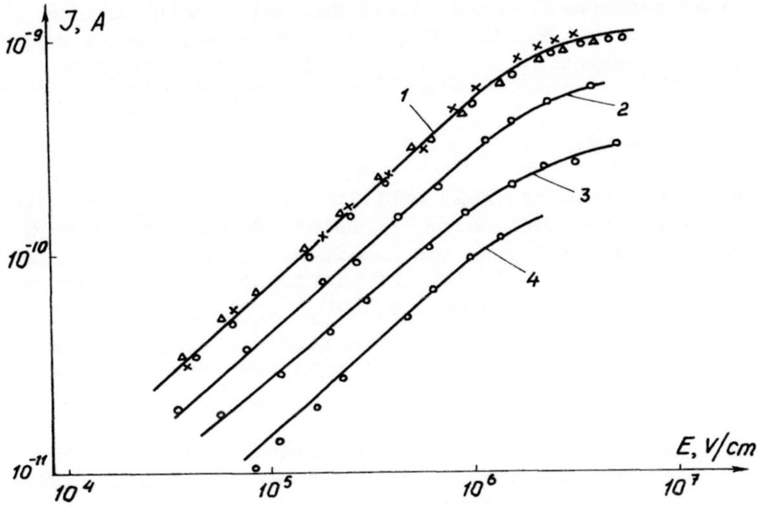

Fig. 4. Electron field intensity (E) dependence of RIC (J) in MOS structures differing in silicon dioxide thickness, d_{ox}.
1) o d_{ox} = 100 nm, $V_G > 0$, x d_{ox} = 100 nm, $V_G < 0$ Δ d_{ox} = 180 nm, $V_G > 0$; 2) d_{ox}= 70 nm $V_G > 0$; 3) d_{ox}= 40 nm, $V_G > 0$; 4) d_{ox}= 26 nm, $V_G > 0$.

the SiO_2 and at the SiO_2-Si boundary and affects the relaxation of the current measured (fig. 2). Current relaxation in the electric field intensity range at the contact $E_c < 7{,}5$ MV/cm is dependent on the mechanism of tunnel electron injection from Si to SiO_2. Thermalized charge carriers are most likely to be injected as the shape of the characteristics measured (fig. 3) concides with that of the voltage-current characteristics measured for MOS structures kept in the dark[10].

The relaxation of RIC in the field range $E_c < 7{,}5$ MV/cm seems to arise from electric field relaxation within SiO_2. The electric field within SiO_2 varies negligibly

$$E_{ox} = (V_G - |V_{FB}|.d_c/d_{ox})/d_{ox}, \tag{3}$$

where $d_c/d_{ox} \lesssim 0{,}15$ for d_{ox}= 60 nm, hence the current variation resulting from relaxation is also small. A decrease in RIC with increasing exposure time for $E_{ox} < 7.5$ MV/cm is due to positive charge formation in the Si-SiO_2 interface region for $V_G > 0$.

The RIC integral Q_1 may exceed the positive SiO_2 - trapped space charge value for two reasons: either carriers are injected from contacts or the dielectric-captured charge value is small. The latter condition implies that the electron and hole constituents of RIC are roughly equal, and only a small proportion of holes are captured on SiO_2 traps. The efficiency of carrier capture by the traps with cross-section close to the magnitude $\sigma \approx 10^{-13}cm^2$ was estimated in reference 11 and was less than

10 %. Consequently, the second reason for a large Q_1/Q_2 value seems quite probable.

Furthermore, the ohmic region of conductivity was observed in the field region $E < 2x10^5$ V/cm, the amount of RIC is not dependent on the polarity of applied voltage and gate electrode material, but is dependent on SiO_2 layer thickness for $d_{ox} < 100$ nm. All this is due to the bulk properties of silicon dioxide (fig. 4), i.e. radiation is responsible for certain carrier concentration in SiO_2 bands which determines the amount of current measured

$$J = qE(\mu_n n + \mu_p p), \tag{4}$$

where μ_n, μ_p is electron and hole mobility in SiO_2 and n, p is electron and hole concentration in the energy bands of SiO_2. The transition from the ohmic region of conductivity to the sublinear region seems to be due to the saturation of the drift velocity of electrons due to their relaxation by the optical and acoustic phonons of SiO_2[12]. Indeed, a transition to the sublinear region (fig. 4) is observed at electric field intensity $E \approx 2x10^5$ V/cm, which corresponds to the drift velocity of electrons close to the thermal velocity $\mu.E \approx 10^7$ cm/s.

RIC was observed to be field-dependent in the oxide. This can be alternatively explained using a model of current from a limited source[7], i.e. with decreasing dielectric thickness all carriers (electrons) leave SiO_2 without recombining. However, in this case a transition of RIC from ohmic to sublinear dependence with an increase in SiO_2 thickness would be observed at higher electric field values in the oxide, which was not the case in the experiment. A decrease in RIC, which depends on oxide thickness, is probably due to the energy dissipation mechanisms of incident electrons in SiO_2. At low d_{ox} values secondary electrons have no time to thermalize (dissipate their energy) and to generate electron-hole pairs in SiO_2 film.

CONCLUSIONS

The radiation - induced conductivity of MOS structures in the field region at the Si-SiO_2 contact $E_c < 7$ MV/cm is dependent on the bulk properties of silicon dioxide. Irradiation of MOS structures at $V_G > 0$ forms a positive space charge near the Si-SiO_2 boundary and, starting at fields $E_c > 7.5$ MV/cm, causes tunnel injection of thermalized electrons from silicon to silicon dioxide. Radiation - induced current in MOS structures consists of electron and hole constituents. A transition from the ohmic region of RIC conductivity to the sublinear region seems to result from the saturation of the drifting rate of electrons in silicon dioxide, whereas the dependence of RIC on SiO_2 thickness is due to electron energy dissipation in silicon dioxide film.

REFERENCES

1. Electron Beam Technology in Microelectronic Fabrication, edited by G.R. Brewer, Academic Press, N.Y., (1980).
2. J.N. Churchill, F.E. Holmstron and T.W. Collins, Adv. Elect. Elect. Phys., 58:1 (1982).
3. E.A. Anufriev and V.A. Gurtov, Phys. St. Sol. (a), 73:107 (1983).
4. E.M. Nicollian and J.A. Brews, "MOS physics and technology", N.Y. (1982).

5. J.M. Farmer and R.S. Lee, J. Appl. Phys. 46:2710 (1975).
6. R.S. Hughes, Phys. Rev. Lett. 30:1333 (1973).
7. J.R. Srour, O.L. Gurtis and K.Y. Chui, IEEE Trans. Nucl. Sci. 21:73 (1974).
8. J.P. Mitchell and D.K. Wilson, Bell Syst. Techn. J. 46:1 (1967).
9. D.J. DiMaria, Z.A. Weinberg and L.M. Aitken, J. Appl. Phys. 48:898 (1977).
10. M. Lenzlinger and E.A. Snow, J. Appl. Phys., 40:278 (1969).
11. T.H. Ning, C.M. Osburn and H.N. Yu, Appl. Phys. Lett. 26:248 (1975).
12. M.V. Fischetti, D.J. DiMaria, S.D. Brorson, T.N. Thies and J.R. Kirkfield, Phys. Rev. B31:8124 (1985).

INTERFACE DEGRADATION IN SHORT-CHANNEL MOSFETs

COMPARISON BETWEEN THE EFFECTS OF RADIATION AND HOT CARRIER INJECTION

H. Haddara, S.Cristoloveanu, B. Boukriss
A. Chovet and P. Jarron*

Laboratoire de Physique des Composants à Semiconducteurs
INPG, ENSERG, 23 rue des Martyrs, Grenoble 38031, France
***European Laboratory for Particle Physics, CERN**
EF Division, CH-1211, Geneva 23, Switzerland

INTRODUCTION

The degradation in performance of short-channel MOSFETs induced by ionizing radiation is a crucial problem for space applications. According to the radiation type, dose and energy, various defects are formed at the interface, in the oxide and in the substrate. This defect generation is almost homogeneous along the channel and is, therefore, different from the damage induced by hot carrier injection into the gate oxide. Indeed, the aging of short-channel devices after electrical stress consists in the progressive formation of a greatly localized defective region close to the drain.

In this paper, we present the evolution of MOSFET characteristics after different periods of exposure to radiation. The interface degradation is characterized using sensitive techniques such as charge pumping, dynamic transconductance and low frequency noise measurements. The results obtained by these methods are correlated and discussed. Similar analysis is performed on devices submitted to various conditions of electrical stress. Finally, the radiation-induced degradation is qualitatively compared to hot electron induced aging.

RADIATION-INDUCED DEGRADATION

Special N and P-channel MOSFETs with large widths have been fabricated at the MIETEC (Belgium) using normal CMOS process. Devices were irradiated in the European CERN synchroton. The particle flux was mainly composed of pions and background neutrons accelerated up to 300 GeV. The samples were classified into four groups according to the radiation dose (0,2,25 and 70 krad).

The deterioration of the static characteristics after exposure is analyzed in terms of : (a) extrapolated threshold voltage V_T, (2) subthreshold swing ($S = dV_G/d\log(I_D)$) and (c) maximum of the transconductance g_m. Figure 1 is a correlation diagram in which the rate of degradation of the maximum transconductance and subthreshold swing are plotted versus the threshold voltage shift, for both N and P-channel MOSFETs. As in the case of hot electron induced aging, the correlation curves consist of two parts. The initial regime is dominated by mobility (transconductance) degradation due to enhanced surface scattering and potential fluctuation[1] ; in the final regime, the total number of defects is large enough to introduce a significant threshold voltage shift through the modification of the flat-band voltage. It is observed that for P-channel devices, the threshold voltage shift is the dominant deterioration and reaches 0.4 V for 70 krad. In the case of N-channel devices, however, only a small shift (less than 0.1 V) is observed while the subthreshold swing and the maximum transconductance are greatly degraded. From these results, one can conclude that the created defects are most probably donor like interface traps.

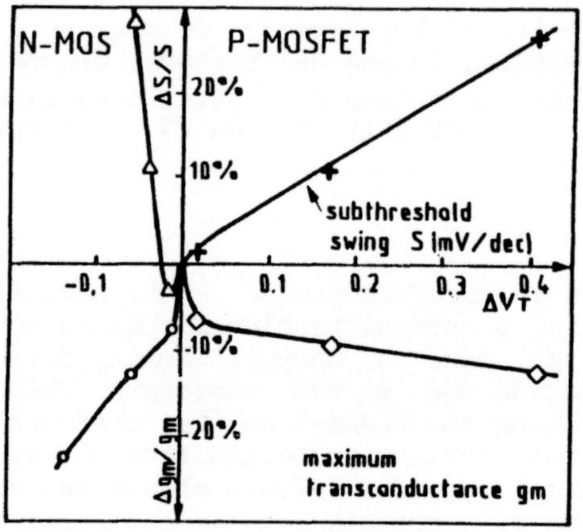

Fig. 1. Correlation diagram between the threshold voltage shift ΔV_T and the fractional degradation of the maximum transconductance g_m and subthreshold swing S, for N and P-channel MOSFETs irradiated at various doses.

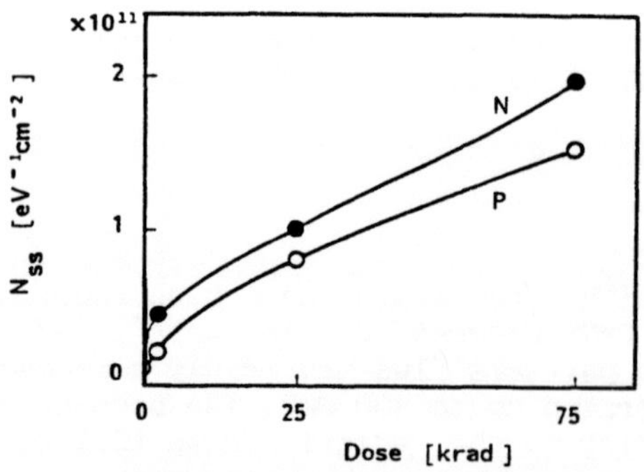

Fig. 2. Interface trap density N_{ss} against radiation dose as obtained from charge pumping measurements.

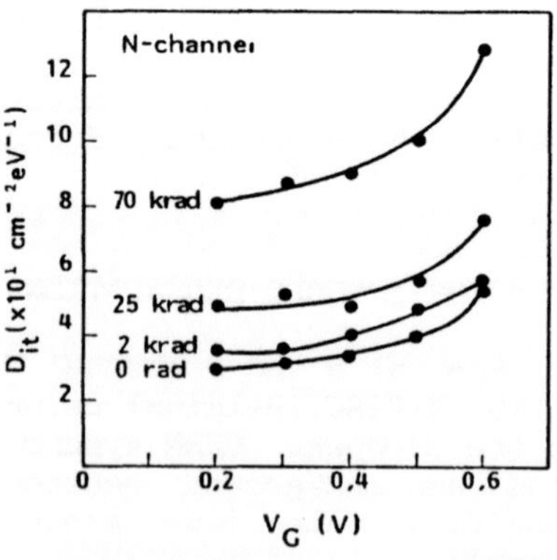

Fig. 3. Interface trap density versus gate voltage as obtained using the dynamic transconductance method in N and P-channel MOSFETs irradiated at various doses.

The density of interface traps N_{ss} as a function of exposure dose has been measured using the charge pumping technique[2] and the dynamic transconductance method[3] for both N and P-channel devices. Figure 2 shows the average value of N_{ss} versus radiation dose as obtained from charge pumping measurements. The experiment consists in measuring the substrate recombination current while continuously pulsing the gate from accumulation to strong inversion. Fig. 3 gives the interface trap density versus gate voltage as obtained from dynamic transconductance measurements. Note the qualitative agreement between charge pumping and dynamic transconductance results.

Low frequency noise is also a useful tool to characterize degradation processes dues to radiation effects as well as electrical aging. Indeed, the noise observed in MOS transistors at low frequency (<500 Hz), in weak or strong inversion regimes, is 1/f noise due to interface and oxide traps which induce fluctuations in the number of free carriers in the channel. Degradations able to create new traps or defects acting as carrier traps can increase the 1/f noise (or would exhibit supplementary noise components, like generation-recombination noise if created traps have identical kinetic characteristics). We find that irradiation at 70 krad increases by one order of magnitude the 1/f noise index measured as the **relative** noise power spectral density S_I/I^2 in the drain current fluctuations. Figure 4 shows that the trap generation due to irradiation is almost homogeneous along the MOSFET channel since the noise level is nearly constant with the drain voltage V_D. Figure 4 also allows comparing this degradation process with the electrical aging produced **homogeneously** with a strong gate bias V_G. It appears that, in this case, the 1/f noise increase is not very significant. In the next section, it will be shown that electrical stress produced in usual conditions of MOSFET operation (V_{DS} = 0) is also able to cause a strong noise increase, although the defect generation is no longer homogeneous.

HOT ELECTRON INDUCED DEGRADATION

In short-channel MOSFETs, electrons gain enough energy to be injected into the gate oxide where they create additional interface states and/or get trapped in already existing ones. A specific aspect of hot electron induced aging is that the created damage is extremely localized near the drain junction. This makes both the modeling and characterization of the resulting degradation rather difficult in comparison with the case of homogeneously distributed defects. Typical problems arise from : (1) the localization of defects in the high field zone near the drain which is a region of 2-D potential distribution, (2) the difficulty in determining the length ΔL of the defective region and (3) the interaction between the damaged and damage-free parts of the channel. This interaction has been shown to be responsible for a number of interesting and unexpected results[4] : (a) attenuation of the defect induced mobility reduction due to the presence of a transconductance overshoot, (b) modification of the subthreshold slope of the $I_D(V_G)$ characteristics in the case of localized fixed oxide charge and (c) attenuated influence of localized interface states in weak inversion.

Figure 5 shows the simulated correlation diagram between the transconductance degradation and the threshold voltage shift for the case of localized acceptor traps. The complex behaviour of these curves compared to those of fig. 1 is not only due to different successive regimes or mechanims of damage creation but also to the interaction between the damaged and non-damaged channel regions. Correlation diagrams strongly depend on stressing bias, channel length, temperature and device processing parameters. They provide a "signature" of the stressing conditions.

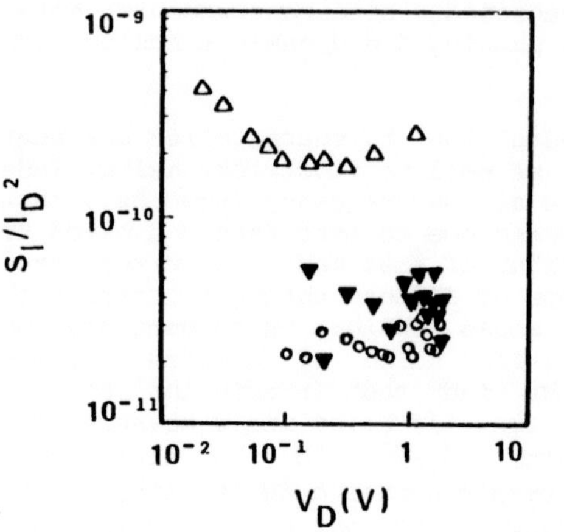

Fig. 4. 1/f noise spectral density S_I/I^2 (at 10 Hz) for similar N-type MOSFETs in weak inversion, for different degradations :
- virgin device.
- uniform electrical aging (V_D = 0 V, V_G = 25 V, 45 min).
- after 70 krad irradiation.

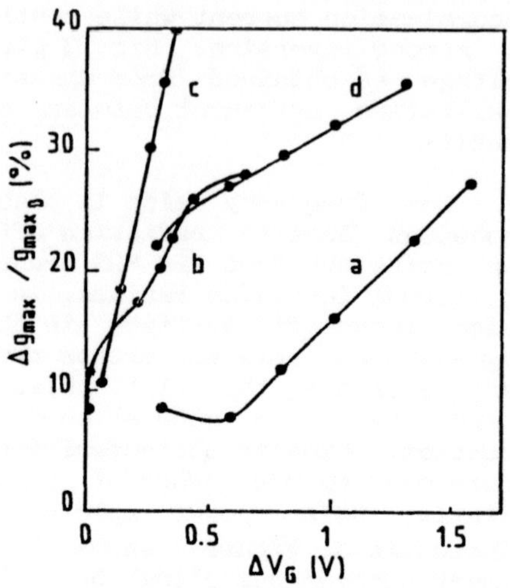

Fig. 5. Simulated correlation between threshold voltage shift and degradation of maximum transconductance for the case of localized acceptor states. The four curves represent : (a) L = 1 µm ; ΔL = 0.15 µm, (b) L = 1 µm ; ΔL = 0.05 µm, (c) L = 1 µm ; ΔL = L ; (d) L = 0.5 µm ; ΔL = 0.075 µm.

Several experimental methods have been proposed for the determination of ΔL. Most of them consist of gradually increasing the drain voltage until the entire zone of defects lies within the space charge region of the drain junction. This moment is determined by monitoring the evolution of a certain quantity (low frequency noise[5], charge pumping current[6], transconductance[2], etc...) as a function of the drain voltage until the initial value, obtained before aging, is recovered. The length ΔL is then taken to be equal to the width of the drain junction space charge region at this moment. Figures 6 and 7 show two examples of this method. Figure 6 makes use of the transconductance before and after aging in the normal and inverse (source and drain inter-exchanged) modes of operation to determine ΔL. As explained above, ΔL is obtained by calculating the extension of the drain junction space charge region at which the transconductance curve after aging rejoins that before aging. The second example (fig. 7) makes use of the measured 1/f noise versus drain voltage to determine ΔL.

After ΔL was found, the density of created defects can be determined from charge pumping or low frequency noise measurements. Figure 8 shows the interface trap density profile before and after electrical stress (V_D = V_G

= 9 V ; 2 hrs.) for a transistor of effective channel length L_{eff} = 2,4 µm. Notice the pronounced increase of trap density near the minority carrier (conduction) band edge.

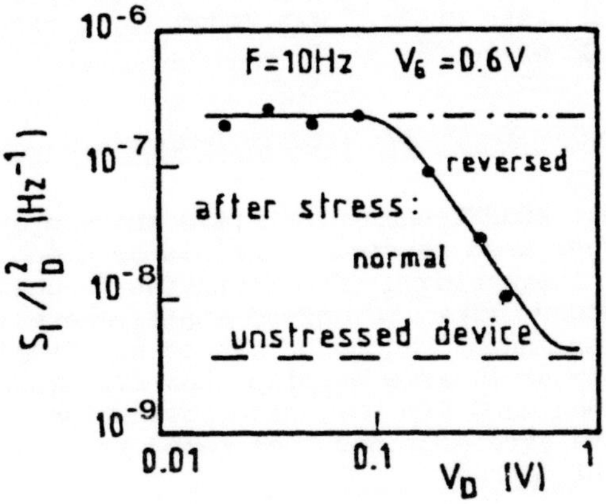

Fig. 6. Transconductance versus drain voltage in saturation region for an electrically stressed MOSFET (V_G=2 V, L_{eff}=2.4µm).

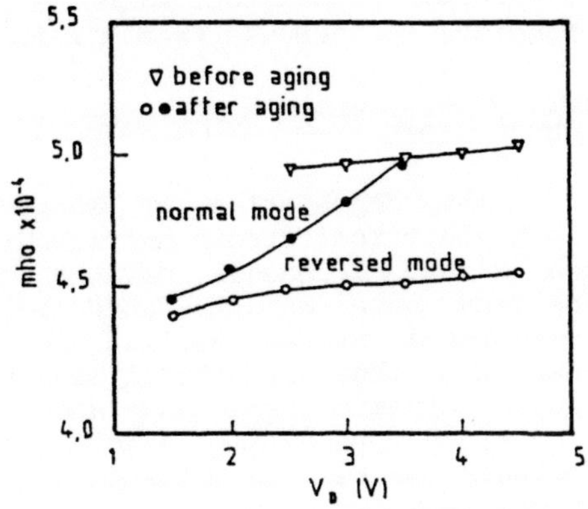

Fig. 7. 1/f noise magnitude in weak inversion versus drain voltage (N-type MOSFET with L=0.5µm). Before aging, the noise is the same in the normal and reversed modes of operation. After aging (V_G 10V, V_D=5V, 15hours), the 1/f noise is greatly increased ; in the reversed mode, the noise is constant with V_D, while in the normal mode it decreases to its initial level, due to progressive screening of created defects as the drain space charge region grows.

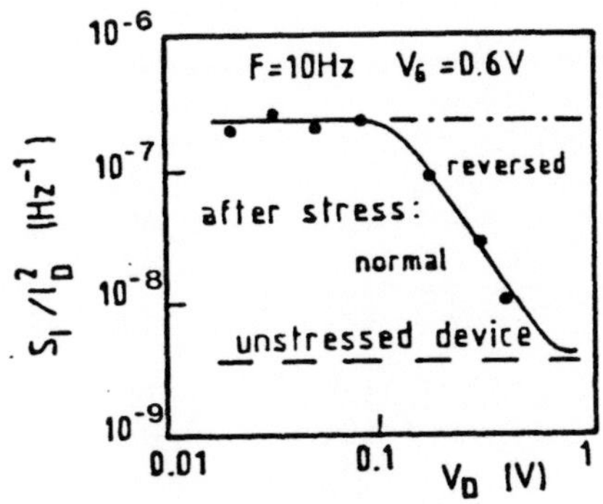

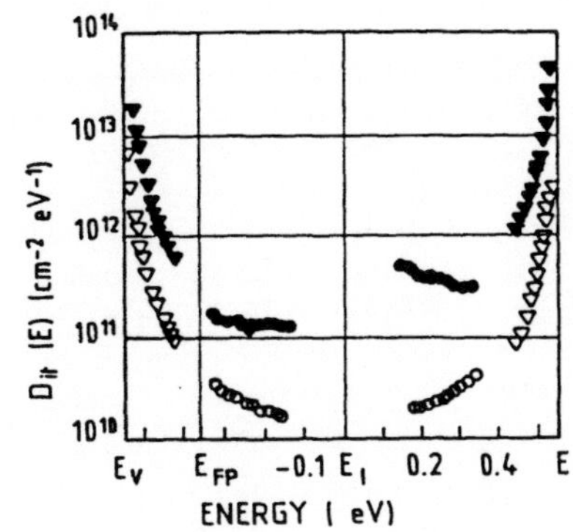

Fig. 8. Energy distribution of interface state density before and after electrical stress obtained by charge pumping technique for an N-channel MOSFET of L_{eff} = 2.4 µm.

As already indicated (in fig. 4), in order to make easier the comparison with the radiation induced damage, we have performed electrical stress with the source and drain grounded while applying a high gate voltage. This provides homogeneous degradation of the interface. After 45 minutes of stress at $V_G = 25$ V, the interface trap density in an N-channel transistor, with 39 nm thick gate oxide, was found (by charge pumping) to increase from 3×10^{10} to $10^{11} eV^{-1} cm^{-2}$.

CONCLUSIONS

The degradation of short-channel MOSFET characteristics induced by both electrical stress and irradiation has been studied. This has been done on N and P-channel devices having large widths. The radiation induced defects were mainly found to be donor like interface states whereas electrical stress resulted in the creation of acceptor states and fixed negative charges. Several sensitive methods (charge pumping, low-frequency noise, dynamic transconductance) have been used for the characterization of interface properties. A good qualitative agreement has been found to exist between the results obtained by different methods.

ACKNOWLEDGEMENTS

Our colleagues from MIETEC (Leuven) are sincerely thanked for device processing and interest shown in this work. Thanks are also due to Dr. E. Heijne (CERN) and Miss M. Gri (LPCS, Grenoble).

REFERENCES

1. B. Cabon, G. Ghibaudo, Rev. Phys. Appl. 21:305 (1986).
2. H. Haddara, S. Cristoloveanu, Solid. St. Electron., 29:767 (1986).
3. H. Haddara, G. Ghibaudo, ESSDERC 86, Cambridge, U.K.
4. H. Haddara, S. Cristoloveanu, IEEE Trans. Elect. Dev., ED-34:378 (1987).
5. Z. Fang, S. Cristoloveanu, A. Chovet, in Proc. Int. Conf. in Noise in Phys. Syst. Roma (Sept. 1985).
6. H. Maes, G. Groeseneken, Electron. Lett., 18:372 (1982).

SYNTHESIS OF BURIED DIELECTRIC LAYERS IN SILICON BY ION IMPLANTATION

Ian H. Wilson

Department of Electronic & Electrical Engineering
University of Surrey, Guildford, Surrey GU2 5XH, England

INTRODUCTION

Some of the ideas presented here are expanded from a review of the synthesis of SiO_2 by ion implantation written for the second radiation effects in insulators conference[1]. Since 1983 there have been many developments confirming (and a few contradicting) the theses and hypotheses contained therein. One major development has been the upsurge in interest in synthesis of buried layers of the nitrides and carbides of silicon and so these have been included here. This paper is an attempt to address the fundamental processes that occur when silicon is implanted with high doses of carbon, nitrogen and oxygen.

The order of topics is chosen to progressively build-up an understanding of the processes by which the microstructures described in the last section are created. First the basic chemistry is dealt with, then the physics of damage and mass transport, finally material science aspects of precipitation and microstructure.

Chronology

I will not attempt a complete historical review, but will pick out significant developments, with apologies for any omissions. The topic of synthesis of dielectric layers in semiconductors by ion implantation is as old as that of doping by the same technique, and originated from the same research groups. They had their beginnings in the mid 1960s. The first synthesis of SiO_2 and determination of electrical properties was by Watanabe and Tooi[2] in 1966. At the same time Pavlov and co-workers[3] commenced studying nitrogen implantation, his work culminated ten years later in a full description of the spherulitic growth of nitride precipitates[4]. By 1968 the work of several groups on the range profile of nitrogen in silicon had been reported[5,6]. Also at this time Schuttke and co-workers reported the formation of <u>buried</u> insulating layers by implantation of 2 MeV nitrogen, oxygen and carbon[7,8,9] and Freeman and co-workers identified formation of SiO_2, SiC and Si_3N_4 by infra-red transmission measurements[10,11]. The electrical properties and annealing of SiO_2 layers received further attention in the mid 1970s[12-15]. Also at this time Gill and Wilson[16] used Rutherford backscattering (RBS) to investigate the composition of oxygen implanted silicon and discovered saturation at the stoichiometric composition of SiO_2 in the as-implanted state. The first devices grown on buried SiO_2 layers were reported by Izumi, Doken and Ariyoshi[17] in 1978. They fabricated a high performance 19 stage CMOS ring oscillator in an epitaxial layer grown using the implanted silicon surface as a seed. They used high dose (1.2 x 10^{18} ions cm^{-2}) high energy (150 keV) $^{16}O^{+}$ implants with a high beam flux to heat the wafer in order to retain crystallinity of the silicon overlayer. A two hour anneal at 1150°C was found necessary to obtain good dielectric isolation. After this there was a rapid increase in the interest in buried dielectric layers because of the promise they held and still hold as a strong competitor with silicon on sapphire and other silicon on insulator technologies for high speed VLSI circuits. The development in 1980 of cross-sectional transmission electron microscopy (XTEM) by Booker and co-workers[18] yielded much more information on the microstructure of buried oxide layers. At the same time the application of dynamic secondary ion mass-spectrometry (SIMS)[19,20] to these layers highlighted the redistribution occuring as a result of annealing. More recently Kilner and co-workers have used SIMS in some interesting experiments to investigate mass transport using isotope markers[21], and have tested their results against the theoretical models of Maydell and Wilson[22] and Jäger[23]. Current developments include ultra high temperature annealing using lamp heating to melt the back surface of the wafer by Celler and co-workers[24], and the use of lamp heating during implantation, which results in quite different microstructures[25,26,27].

CHEMISTRY

In order to understand the synthesis of new phases by implantation one must first consider the incorporation of carbon, nitrogen and oxygen into the silicon lattice. These atoms are all chemically reactive and will therefore tend to form bonds with silicon, perhaps in defect complexes, with impurities or by interaction with point defects. One is interested in similarities and differences between the three implanted species.

Energetics

C, N and O are small, highly electronegative atoms which will readily be incorporated without imposing a large strain on the lattice, especially in the presence of defects. The topological constraint of the silicon lattice, which acts on larger atoms such as the traditional dopants boron and phosphorous, to impose tetrahedral bonding, will be weak.

The high electronegativity of C, N and O indicates that bonds with Si will have a considerable ionic character.

Defects will have the effect of lowering the potential barrier between an initial metastable state and the most energetically favourable state, thus increasing the probability of incorporation of the implanted species. Oxygen and, to a lesser degree, carbon exist as impurities in CZ silicon. The high strength of the C-O, C-N and N-O bonds indicate that O and C as impurities may play a key role in the nucleation of a second phase. For example it is well known[28] that carbon can influence the nucleation of oxide precipitates in CZ silicon. However available bonding sites at impurities will soon be exhausted during the high dose implants required to grow a second phase. C-Si and N-Si are rather intermediate in bonding energy, C-C and N-N bonds would be favoured, especially when the concentration of implant is greater than the stoichiometric ratio. The bond energy in the stoichiometric compounds shows a considerable difference however with SiC being weakly bound in contrast with Si_3N_4. In the case of oxygen the O-Si bonds are much stronger than O-O bonds.

From these simple energy considerations one can conclude that : 1) The phases Si_3N_4 and SiO_2 could be formed during implantation, provided the energy barrier to the stable state could be surmounted by thermal activation or by defect barrier lowering. 2) N-N (nitrogen gas) and C-C (graphite ?) bonds will be favoured at high concentrations, 3) O and C the "as grown" impurities could play an important role in nucleation of a second phase.

However one must exercise caution in carrying these conclusions too far as stability of a particular bonding pair depends on the depth of the potential well rather than the energy saved in creation of the bond.

Bonds

I will now turn from simple energy considerations to compare the nature of the bonds formed, and incorporation of the implanted atoms in the Si lattice.

Carbon like silicon, is isovalent (valency 4) with valence electrons $2s^2 2p^2$. In common with the other group IV elements four hybrid sp^3 states are formed by promotion of one 2s electron to 2p in order to achieve four unpaired spins. All group IV elements have a similar promotion energy (-400 K J mol^{-1}) which is retrieved by the energy saved in formation of the covalent bonds.

The features described above generally hold for group IV elements, but one must remember that the second period elements are rather different because of small size and high electronegativity. Carbon for example can form double or triple bonds, structures such as the benzene ring, and also the graphite sheet structure as well as the tetrahedral diamond structure. However one can speculate that carbon will be stable in a substitutional (tetrahedral) site in the silicon lattice, although lattice distortion and the presence of defect complexes may lead to other configurations being favoured.

Nitrogen has the valence electron configuration $2s^2$ or $2p^3$ (valency 5 or 3). When the silicon lattice can impose a topological constraint hybrid sp^3 orbitals can be formed with one s electron in an antibonding state. The promotion energy is only slightly greater than for group IV elements therefore the substitutional position in a tetrahedral configuration is stable. Disorder, or a high concentration of nitrogen will reduce the topological constraint, thus reducing the substitutional fraction and increasing the proportion of nitrogen in a three fold ($2p^3$) configuration with 90° bond angles, possibly forming puckered sheet structures.

Oxygen has the valence electron configuration $2s^2$ $2p^4$. The s electron energy level is lower than for groups IV or V, therefore the promotion energy to sp^3 with s^2 in antibonding states is high also two electrons have to be transferred across the energy gap. Thus bonding in a tetrahedral substitutional site is most unlikely. It is much more probable that two s spins are paired and also two p spins are paired forming the so-called "lone pair", leaving two p orbitals for forming bonds. Two bonds with ~100° bonding angle are favoured, and this is the configuration of the well known oxygen interstitial (O_i) situated between, and bonded to, two silicon atoms. This is illustrated in figure 1. This configuration has been found to be stable to high anneal temperatures where oxygen-defect complexes tend to disappear.

Defects and defect complexes created during implantation can profoundly affect the manner of incorporation of impurities in the silicon lattice and in general impurities will be most stable in sites where the number of nearest neighbours is equivalent to the natural valence. For example a silicon vacancy (Si_v) adjacent to a substitutional site satisfies an impurity seeking a three fold configuration such as N or O^+. It is worth noting here that the number of electrons available for bonding can depend on the charge state of the defect in silicon which in turn depends on the position of the Fermi energy.

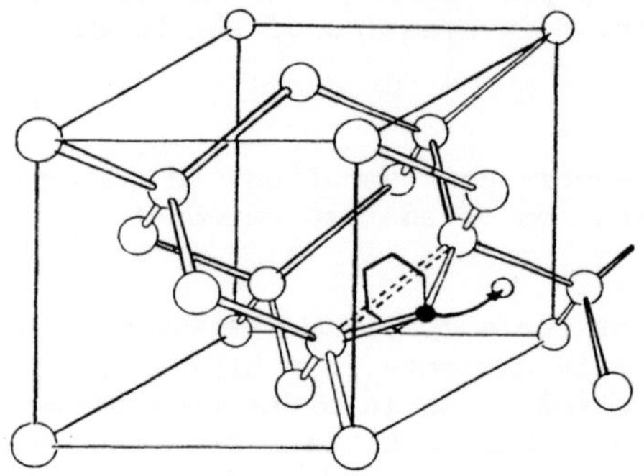

Fig. 1. The bonded oxygen interstitial in silicon showing the six equivalent orientations and a jump path to a neighborouring site.

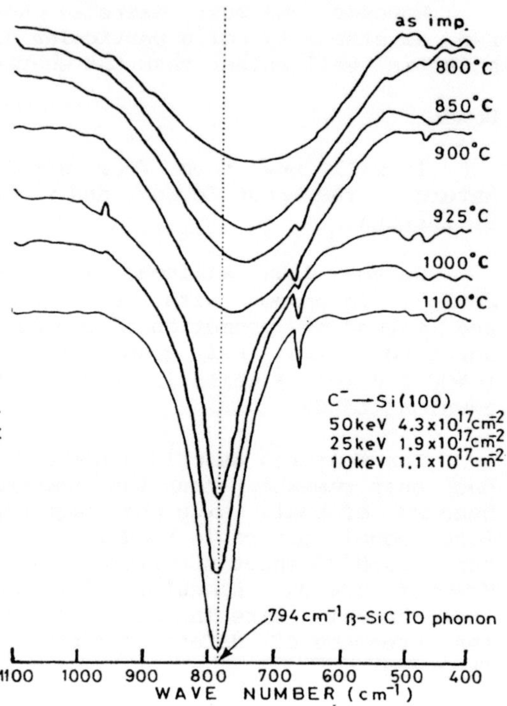

Fig. 2. Infra-red transmission of carbon implanted silicon for various anneal temperatures.

Formation of a second phase

At this stage one can speculate as to the structure of implantation synthesised phases. In the as-implanted state it is most likely that amorphous layers are formed especially at low implant temperatures (T_I) where it is known[29] that SiO_2 and Si_3N_4 are amorphised by ion impact. If conditions are favourable to crystal growth it seems possible that either α or β-SiC could be formed. The former, because of the polymorphic nature, could accommodate radiation damage but may then tend to amorphicity. The cubic form, β-SiC may be favoured, especially at high T_I, resulting from a topotactic transformation imposed by the silicon lattice. A topotactic transformation is a solid state transformation where the product is structurally and orientationally related to the parent phase (epitaxy is a two dimensional form of topotaxy).

The high $\alpha \rightarrow \beta$ transition temperature for Si_3N_4 indicates that only the lower nitride, α-Si_3N_4 could be formed by implantation. The possibility of only one form would greatly favour growth of the crystalline phase when high implant temperatures are used.

Crystallisation of SiO_2 would appear very unlikely in view of the polymorphic nature, the tendency to form a random network and the large thermal coefficient of expansion of the crystalline form. However, the cubic form, β-cristobalite, may be favoured in very small precipitates because of a topotactic relationship with the silicon lattice.

Experimental determination of the chemical state of implanted atoms

Evidence of the way in which implanted ions are incorporated into the silicon lattice comes chiefly from infra-red transmission (IRT) measurements. From three studies of carbon implanted silicon[30,31,32] on can conclude that Si-C bonds are formed in the as-implanted state forming microregions of SiC which may be crystallites of α-SiC or regions of truly amorphous SiC. It is difficult to distinguish between the two because of the large number of α-SiC polymorphs, the large percentage of C at Si/SiC interfaces, and radiation damage. At a critical anneal temperature of 800 to 900°C (depending on carbon concentration, T_I and silicon orientation) there is a transformation to cubic β-SiC. This is illustrated in fig. 2. One can speculate that this transformation is topotactic in nature resulting partly from the crystallisation of the Si host partly from misfit strain, and partly from phase stability considerations.

Borders and Beezhold[30] found that much longer times at higher temperatures (15 hours at 1000°C) are needed to grow crystalline Si_3N_4 in nitrogen implanted silicon. This was confirmed by Bourguet et al.[33] who found a broad band of Si-N stretching modes indicative of amorphous Si_3N_4 up to T_A= 900°C. At T_A= 1200°C a complex absorption spectrum appears and X ray diffraction lines for polycrystalline α-Si_3N_4 are seen, along with those for polycrystalline and single crystal silicon.

Stein[34] has investigated IRT of Si implanted with low doses of N and O (2×10^{15} ions cm^{-2}, 100 and 200 keV) at a low implant temperature (T_I = 50°C). He finds two peaks (767 and 963 cm^{-1}) for N implants, either two modes of Si-N bonds in the same defect or two defects with similar kinetics. These lines grow with increasing T_A up to 600°C and then decay. Presumably the growth arises from reduction in the quenching of the modes by other defects and the decay arises from removal of the defect complexes associated with the absorption peaks. At 750°C new bands appear, some associated with Si_3N_4 and some which have been found to be associated with N doped amorphous Si. All bands disappear at 900°C when the relevant defect complexes disappear.

Electrical measurements can also contribute clues as to the way in which nitrogen is incorporated in the Si lattice. For example recent measurements of carrier concentration in a buried layer N implant reveal a conducting layer near the buried Si_3N_4 layer[35]. It is concluded that the doping arises from complex formation (presumably analogous to the oxygen donor complex). The nitrogen complexes must be very stable as the conducting layer is only removed by annealing at over 1300°C.

It is convenient to separate the measurements on oxygen implanted silicon into those where the concentration of oxygen is low and interaction of isolated oxygen atoms with the lattice, defects, and small precipitates can be studied, and those where the oxygen concentration approaches that necessary to synthesise oxide layers. Oxygen dissolved in silicon during crystal growth has long been of interest[36]. After an anneal long enough for equilibrium to be established oxygen is in the form of bound interstitials (O_i-fig. 1) and in silica precipitates.

Stein[34], studying low dose oxygen implanted silicon, finds two lines in the as implanted state, shown in fig. 3. The line at 830 cm^{-1} is due to a neutral $O\text{-}Si_v$ complex (the A center shown in fig. 4) and that at 1103 cm^{-1} is due to O_i (fig 1). After annealing at 300°C a line at 889 cm^{-1} appears, the negative $O\text{-}Si_v$ complex with a complimentary reduction in the neutral $O\text{-}Si_v$ line. After annealing at 400°C only the unchanged O_i line remains, which is removed by annealing at 600°C. One presumes that at higher temperatures the oxygen is associated with more complex defects, and only at much higher doses or very long anneal times will the bands due to silica formation be seen.

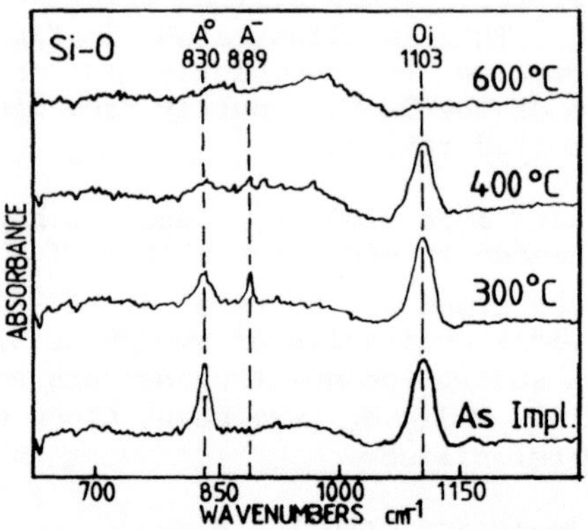

Fig. 3. Annealing of absorption bands of Si-O in oxygen implanted crystalline Si, $1x10^{15}$ion cm^{-2} at 60, 100 and 180 keV[34].

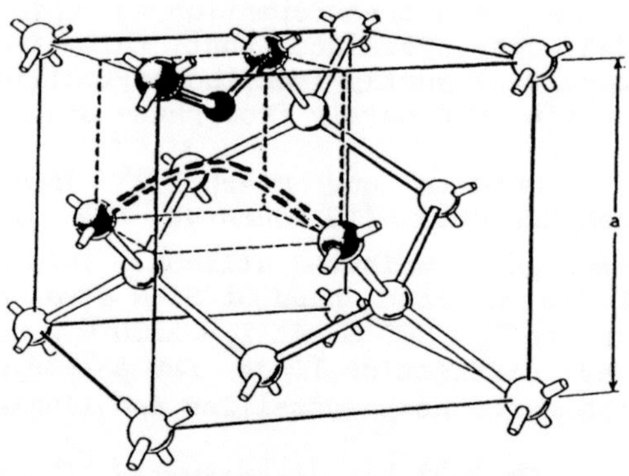

Fig. 4. The A-centre (vacancy plus an oxygen interstitial) in silicon.

As in the case of nitrogen implantation, electrical measurements can give clues regarding incorporation of oxygen in the silicon lattice. Hwang et al.[37] have used deep level transient spectroscopy (DLTS) to investigate buried low lifetime regions created by 400 keV oxygen ions. At low doses they find two electron traps typical of the precipitation of O_i in CZ Si. For higher doses (10^{16} ions cm^{-2}) the lines are obscured by additional defect levels and there is a basic change in the carrier generation mechanism. An increase in dislocation loop density and the appearance of small precipitates (presumably silica) is seen in the transmission electron microscope (TEM) at the same dose. The density of states in the band gap at the Si/SiO_2 precipitate interface has been determined and is shown in fig. 5. Defects associated with this interface have also been detected by electron paramagnetic resonance (EPR)[38]. The annealing of features in the EPR spectrum of high dose oxygen implanted wafers is shown in fig. 6. The center labelled T0 is similar to the E_1' center in damaged SiO_2, T2 is the amorphous silicon center, whilst T1 is the center most probably associated with dangling bonds at Si/SiO_2 (precipitate) interfaces. The structure on the annealing curve for T1 results from discontinuous changes in precipitate surface area due to growth and dissolution.

Conventional IRT measurements[13,39,40] of high dose oxygen implanted silicon show a very broad absorption band which for lower doses ($\leq 10^{17}$ ions cm^{-2}) is in the region of 10 μm. This narrows and shifts to 9.6 μm for high doses (10^{18} ions cm^{-2}). On annealing at 800 to 900°C (for doses $\geq 10^{17}$ ions cm^{-2}) the band narrows further and shifts to a value very similar to that for thermally grown SiO_2 at 9.2 to 9.3 μm. These shifts in position of the absorption peak are attributed to relief of bond strain in highly defective SiO_2. One presumes that one is observing compaction and ordering to form a continuous random network. This can be demonstrated in reverse by oxygen bombardment of thermally grown SiO_2[13] where the absorption peak broadens and shifts from 9.3 to 9.6 μm.

One can summarise the results on high dose implantation briefly as follows :

1) Amorphous SiO_2 precipitates form during implantation, presumably when a critical oxygen concentration is reached. Annealing at 800 to 900°C allows removal of defects in the SiO_2 and creation of a continuous random network.

2) Amorphous SiC is formed during implantation which transforms to β-SiC after annealing at 800 to 900°C possibly by a topotactic transformation.

3) Amorphous Si_3N_4 is formed during implantation which transforms after annealing at 1200°C to α-Si_3N_4 possibly by a topotactic transformation. There is some evidence for heterogeneous nucleation of Si_3N_4 at defect complexes after annealing at 750°C but these regions disappear when the complexes are removed at 900°C.

DEFECTS

One of the most important phenomena that distinguishes synthesis of dielectric layers in Si by ion implantation from other methods is that a large amount of radiation damage is created by the process. As has been already stated, defects can play a dominant role in determining the manner in which the implanted species is incorporated, and the kinetics of nucleation and growth of a second phase.

At low implant temperatures (T_I), and low doses, the damage is in the form of point defects and small clusters. Amorphisation occurs when a critical density of damage is created. At high T_I amorphisation can be avoided but large damage complexes can be formed which often collapse to form stacking faults, dislocation loops and voids.

This section is divided into four topics : The intrinsic or grown-in defects in CZ and FZ Si, radiation damage in the form of point defects, amorphisation and finally dislocations.

Intrinsic defects

The main impurities one expects to see in Si wafers are carbon in float-zone (FZ) silicon, and oxygen and carbon in Czochralski (CZ) silicon. The segregation coefficient for carbon to the solid is quite small

(6×10^{-2}) but it is found at concentrations equivalent to the solid solubility close to the melting point of silicon ($4 \times 10^{16} cm^{-3}$). It is known that carbon aids in the nucleation of oxide precipitates, and that stacking faults are formed at high concentrations (presumably generated by the strain caused by mis-match of carbide precipitates with the Si lattice). The segregation coefficient for oxygen is high (1.3) and concentrations in excess of the solid solubility ($5 \times 10^{17} cm^{-3}$ at 1200°C) are found in CZ wafers. The presence of oxygen, often in the form of precipitates has the beneficial effects of gettering impurities to the core of the wafer, and of pinning dislocations, thus bestowing mechanical strength. Oxygen exists in silicon chiefly as a bound interstitial O_i as described previously (fig 1). The minimum energy configuration has oxygen displaced from the <111> Si-Si bond axis in order to decrease the bond angle from 180° towards the preferred bond angle of 100°. When the silicon vacancy (Si_v) is mobile it can be trapped by O_i to form the vacancy oxygen complex illustrated in fig. 4. This is the A-center, where the oxygen atom is slightly displaced off the tetrahedral substitutional position. This center can re-orient and diffuse at low temperatures and is the type of defect complex leading to the oxygen donor effect after heat treatment at temperatures around 400°C. At lower temperatures Si_v diffusion is too slow for the oxygen donor effect to occur and at higher temperatures more

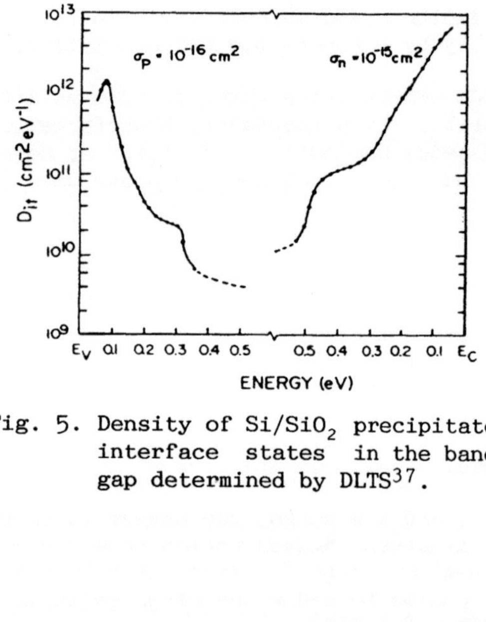

Fig. 5. Density of Si/SiO_2 precipitate interface states in the band gap determined by DLTS[37].

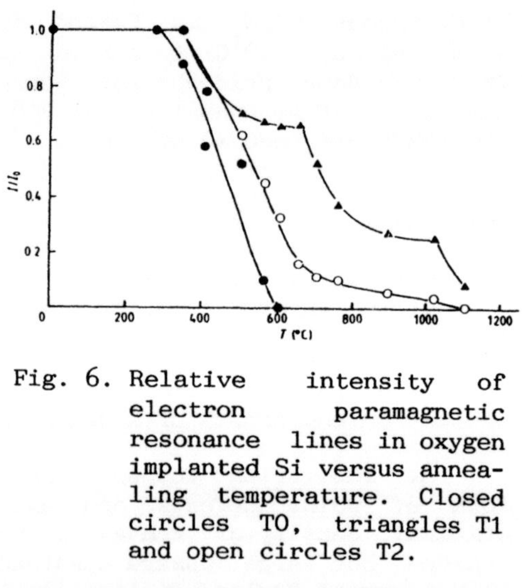

Fig. 6. Relative intensity of electron paramagnetic resonance lines in oxygen implanted Si versus annealing temperature. Closed circles T0, triangles T1 and open circles T2.

complex defect structures are formed which do not contribute electrons to the conduction band. Heat treatment of CZ wafers at 700°C leads to homogeneous nucleation of SiO_2 precipitates which grow at 800°C. At 900°C there is heterogeneous nucleation of precipitates at stacking faults and dislocation loops generated by condensation of Si_i which are emitted by the growing precipitates in order to relieve stress created by the volume expansion required on conversion of Si to SiO_2. At 1110°C the oxygen goes into solution (presumably as O_i).

Nitrogen has a very small segregation coefficient (7×10^{-4}) and a low solid solubility ($4 \times 10^{15} cm^{-3}$) in silicon and is not seen as an impurity,

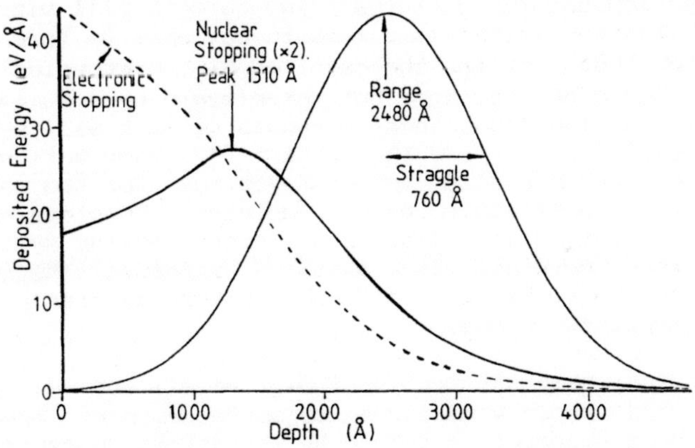

Fig. 7. Stopping of 100 keV nitrogen ions in silicon. Projected range, nuclear and electronic energy deposition. 31 keV lost in nuclear stopping, 69 keV lost in electronic stopping.

although when implanted it has been found to be more effective than oxygen in pinning dislocations[41]. Nitrogen has low electrical activity (~ 1 %). the activity arises not from occupation of substitutional sites but from defect complexes[35] in a way that is analogous to the oxygen donor effect.

Radiation Damage : point defects and defect complexes

In order to illustrate the effects of radiation damage in the case of light, high energy projectiles I have chosen 100 keV N^+ in Si as an example. Implant and energy deposition profiles, calculated using an analytical method[42] are shown in fig. 7. Electronic stopping accounts for 69 % of the total energy loss and diminishes monotonically with increasing depth. The balance of energy loss (31 %) arises from nuclear stopping, which in electrical conductors is responsible for the bulk of the displacement damage. Nuclear stopping peaks at just over half the mean projected range of the implanted nitrogen. This has the important consequence that the peak in the damage profile is shallower than the peak in the implant profile and therefore the upper and lower borders of synthesised buried layers will be in quite different damage environments. It is worth pointing out here that the peak in the damage profile will tend to be deeper than the peak in nuclear stopping. This difference arises from anisotropy of momentum in the collision cascade. Momentum is directed preferentially into the surface because the more energetic recoils will be so directed[43,44].

The mean number of displaced atoms $\langle \upsilon \rangle$ resulting from the impact of one ion can be estimated from the modified Kinchin-Pease formula[45] : $\langle \upsilon \rangle = kE_D/2E_d$. For 100 keV N^+ assuming that energy lost in Si in nuclear stopping E_D= 31 keV, k = 0.8 and the displacement energy E_d= 15 eV the formula predicts that $\langle \upsilon \rangle$ = 830 displacements. Many close interstitial-vacancy pairs will annihilate, but this can be accommodated by introducing a temperature dependent displacement energy and is seen experimentally as a reduction in the damage rate.

However, for low implant temperatures where the defects are relatively immobile large numbers of point defects will be created. As will be seen in the next section, when the concentration of point defects reaches a critical value a crystalline→amorphous phase transition takes place.

However at high implant temperatures point defects will migrate to sinks or coalesce into clusters which would tend to collapse to form stacking faults and dislocation loops or in the case of Si_v microvoids[46]. In this way crystallinity would be retained but there would be a build-up of extended defects with increasing dose. These extended defects will be stable to high temperatures (~900°C) and are the reason for poor activation of dopant implants into Si at elevated temperatures and for the reverse annealing effect for room temperature boron implants. Complete amorphisation is required for good activation of dopants. However during synthesis of buried dielectric layers crystallinity must be retained. The extended defects created will act as sinks, fast diffusion paths and precipitate nucleation sites for the implanted atoms.

It has often been assumed that Si_i the silicon self interstitial, is highly mobile and rapidly diffuses to sinks such as the surface or the substrate at room temperature driven by the defect concentration gradient. However both strain measurements[47] and optical absorption measurements[48] indicate the importance of silicon interstitials in the formation of defect complexes.

Radiation damage : the crystalline-amorphous transition

When radiation damage and the stress builds up to a critical value a crystalline-amorphous phase change occurs resulting in a change in volume and relaxation of stress.

There is evidence in the literature that high concentrations of dopant can alter the critical amorphisation temperature. Measurements of amorphisation dose versus temperature for the conventional dopants in silicon[49] indicate that, for low mass ions such as C, N and O, silicon

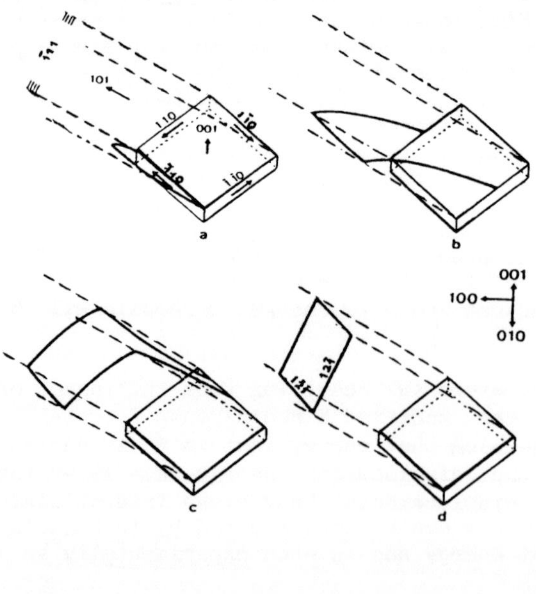

Fig. 8. Generation of prismatic punched dislocation loops by platelike SiO_2 precipitates.

should remain crystalline for $T_I > 100°C$. However amorphous regions each side of a buried oxide layer are seen for $T_I = 400°C$[50]. When the concentration of implant approaches ~ 1 % the defect properties must be greatly affected by the change in the chemical nature of the matrix and this poses greater problems in predicting phase stability.

Dislocations and stacking faults

For oxygen implanted silicon there are at least two other mechanisms that lead to formation of extended defects. During implantation to high doses precipitates of SiO_2 are formed with a considerable change in volume, and therefore Si_i are emitted in order to relieve stress. These will condense to form stacking faults and dislocation loops. This process also occurs during annealing. For temperatures between 900°C and 1000°C plate-like precipitates can also relieve stress by direct generation of prismatic dislocation loops as illustrated in fig. 8[51].

An understanding of the formation, interaction and evolution of dislocations and stacking faults during implantation of Si with C, N and O at elevated temperatures seems crucial in determining the nature of the as-implanted state and therefore the manner and form in which equilibrium is established during annealing. However little is yet known.

MASS TRANSPORT

In this section the means by which atoms migrate in C, N and O implanted Si will be discussed. One is interested in mass transport during implantation at temperatures between 400°C and 800°C and during annealing at temperatures between 1000°C and 1400°C in both defective silicon and in the synthesised layers.

Driving forces

The atomic flux $\underline{J}_i$ in a concentration gradient (C_i) of the ith impurity in a solid depends not only on diffusivity D, but also on the driving force $\underline{F}_i$. Thus the flux can be given by[52]: $\underline{J}_i = C_i\ (D/kT)\ \underline{F}_i$. the driving force may arise from the concentration gradient $\left[\underline{F}_i = -\nabla C_i\,(kT/C_i)\right]$, a chemical energy gradient $[\nabla\{\mu_i\ (C_i)\}]$ or external constraints such as stress $(-\sigma\,\underline{\epsilon}_i)$ or an electric field $(-q_i\underline{E})$. All four driving forces may well operate in the case of high dose implants of reactive ions, however as later sections will reveal the first of these, perhaps the most familiar, mass transport to reduce a concentration gradient, is relatively unimportant in spite of the presence of high concentration of defects. Instead the chemical free energy drives the system into segregation against the concentration gradient.

Chemical segregation

A second phase is stable if $\delta_{\mu i}/\delta C_i > 0$ where μ_i is the chemical potential in this case. Murch[54] has considered diffusion in a highly defective solid with a single mobile sub-lattice, for example diffusion of oxygen in an oxide.

The chemical diffusion coefficient $\tilde{D}$ is given by $\tilde{D} = D^*\ (\delta\ \ln A/\delta\ \ln C)$ where A is the thermodynamic activity, C is the mole fraction and D^* is the tracer diffusion coefficient. Close to stoichiometric composition A varies strongly with composition giving rise to an exceedingly large thermodynamic driving force and a large value of $\tilde{D}$.

The tracer diffusion coefficient is given by $D^* = (½)\ \Gamma\lambda^2 f$ (for one type of defect) where λ is the jump distance, Γ is the atomic jump frequency and f is the tracer correlation factor due to defect induced

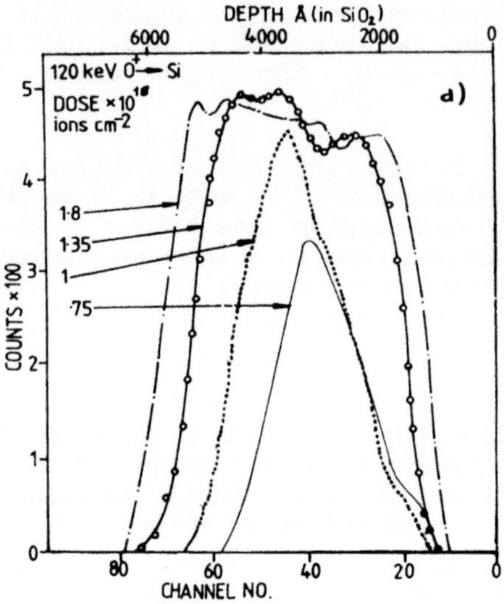

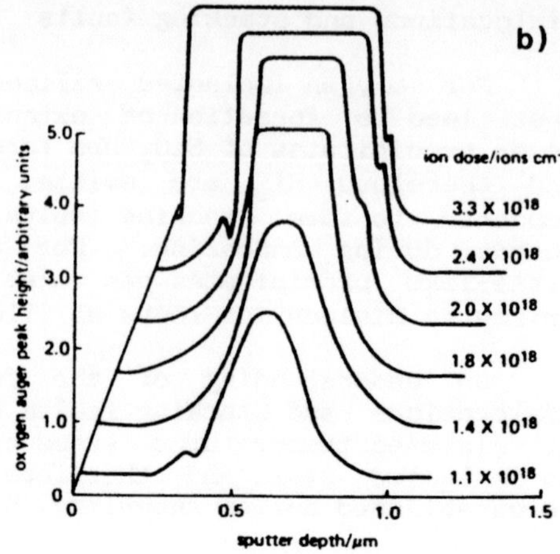

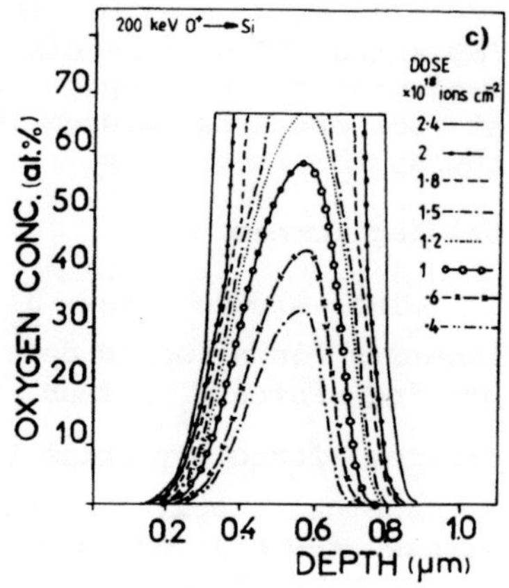

Fig. 9. Variation of buried oxide concentration profile with ion dose.
a) From Rutherford Backscattering[53] (120 keV $O^+ \rightarrow Si$)
b) From Auger depth profiling (200 keV $O^+ \rightarrow Si$)
c) From the theory of Maydell[22] (200 keV $O^+ \rightarrow Si$)

non-random motion of the tracer atom. Both Γ and f are strongly dependent on defect concentration.

If one wishes to apply the theory of diffusion in highly defective solids to the case of implantation synthesised layers there are several complicating factors. Defect structures are complex and largely unknown, there may be several species of mobile defect, and the high degree of disorder may lead to modification of the jump distance λ. At present the above theory cannot be applied quantitatively to the cases of interest here. The experimental results provide ample evidence for a strong driving force for segregation for all three cases of interest.

Chemical segregation operates during implantation in the case of oxygen bombardment of silicon. A flat topped profile is formed once stoichiometric compositon is achieved. As evidence for this an RBS profile from early work on this phenomenon is shown in fig. 9a[53], and a series of Auger depth profiles shown in fig. 9b.

There is a considerable oxygen concentration distribution on both sides of the stoichiometric layer, particularly between the layer and the surface. For implant temperatures in the region of 400°C conditions appear to be favourable for precipitate nucleation (this topic will be dealt with in more detail). Auger electron spectroscopy (AES) combined with cross-sectional transmission electron microscopy (XTEM) have been used to investigate segregation and precipitate growth during annealing for a low value of T_I, (400°C)[50]. AES oxygen profiles for various anneal temperatures are shown in fig. 10. Oxygen segregates to the implanted layer, the silicon surface and to a precipitate layer (seen as an oxygen spike). At T_I = 400°C an amorphous Si layer forms both sides of the oxide layer which transforms to polycrystalline on annealing, the oxygen spike forms at the interface between this layer and the crystalline (but heavily dislocated) silicon. Diffusion of oxygen will be rapid in this region and favourable sites will exist for nucleation of oxide precipitates. Depletion of oxygen caused by mass transport driven by the chemical segregation leads to formation of a hump in the oxygen profile between the spike and the surface. Precipitates nucleate at dislocation tangles in this region. This in turn sets up a chemical potential, oxygen segregates, and the hump grows.

At higher values of T_I precipitate nucleation can be suppressed and segregation is only to the stoichiometric layer (and perhaps the surface if an oxide cap is used). An XTEM micrograph of an oxide layer grown with T_I = 600°C and T_A= 1405°C is shown in fig. 11[54]. Segregation is almost complete, with atomically sharp amorphous SiO_2/crystalline Si interfaces. There are some spherical Si islands in the SiO_2 layer near the deep interface. These most probably arise from irregularities originally existing at the deep interface in the as implanted state due to statistical fluctuations of the collision cascade density. When interface smoothing occurs, in order to reduce surface energy during the anneal, Si islands are trapped in the SiO_2. In this case anneal temperatures have been taken to

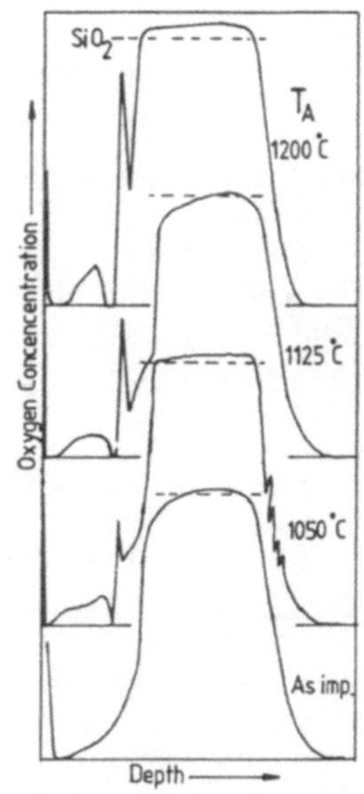

Fig. 10. Auger depth profiles for oxygen implanted buried layers with a low implant temperature (T_I = 400°C) for various anneal temperatures (2 hr.) (2.15×10^{18} ions cm^{-2} 200 keV)[50]

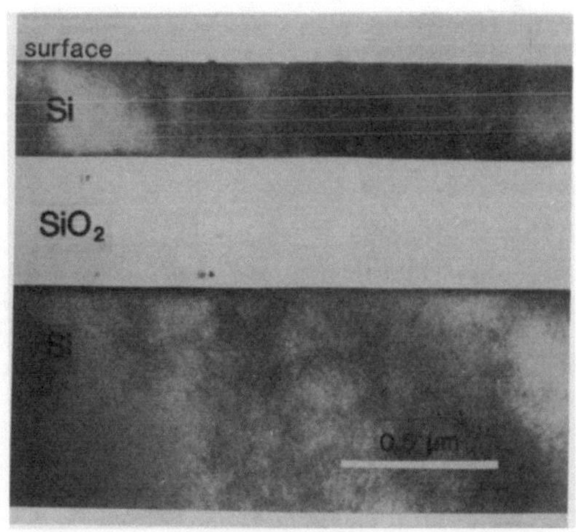

Fig. 11. Cross-sectional transmission micrograph of a layer formed with T_I= 500°C and 30 min. lamp anneal at 1405°C[54]. (2.15×10^{18} oxygen ions cm^{-2}, 200 keV).

the limit by heating (with lamps) the back of the wafer to some 7°C below the Si melting point[24]. RBS spectra, fig. 12, reveal perfect crystallinity in the Si overlayer, and all oxygen segregated to the SiO_2 layer.

High anneal temperatures are needed to remove extended defects and defect complexes by diffusion of Si_i and Si_v. Ion beam heating, used to obtain the high values of T_I discussed so far, inevitably means that T_I is low for the first minute or so of implantation. Damage formed then can be stable necessitating a very high T_A to remove. Independent in-situ heating can result in a much less defective Si overlayer in the as-implanted state. Holland et al.[25] have demonstrated this (fig. 13) with an optimum temperature of 475°C + beam heating (T_I - 800°C). At higher values of T_I there is rapid oxygen diffusion and growth of precipitates near the surface.

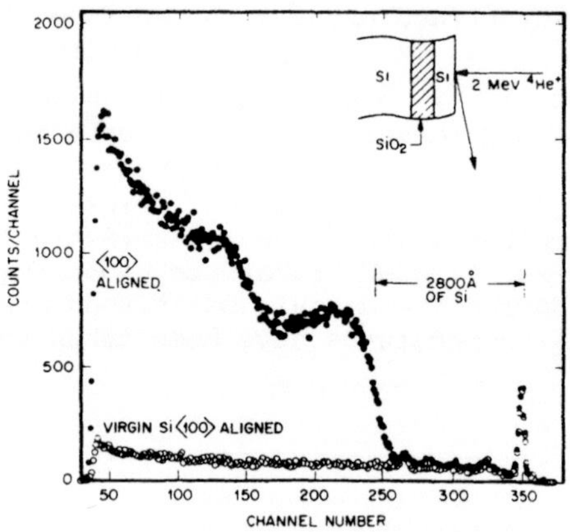

Fig. 12. High resolution RBS spectrum from a sample similar to that shown in fig. 11 compared to the yield from virgin silicon[24].

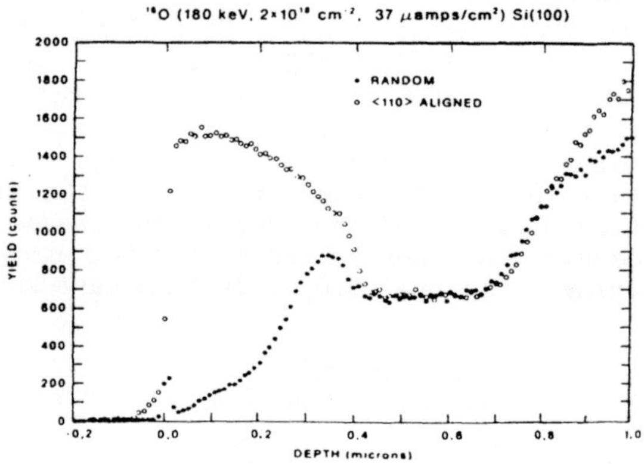

Fig. 13. RBS spectrum of the as-implanted state of a sample where lamp heating was used during implantation T_I = 800°C (475°C + beam heating)[25].

Nitrogen <u>does not</u> segregate during implantation, the skewed Gaussian profile increases in magnitude to peak values well above the stoichiometric ratio for Si_3N_4. Annealing at 1200°C, however, results in strong segregation as shown in the SIMS profiles in fig. 14[55]. Implants where the peak concentration is below or above the stoichiometric ratio both segregate to form Si_3N_4 layers, with strong nitrogen depletion in the wings. The layer formed from the high dose implant shows a central dip caused by sectioning through a nitrogen bubble layer. Apparently implantation synthesised Si_3N_4 is a good diffusion barrier to nitrogen in contrast to synthesised SiO_2 where there is rapid oxygen transport.

In the case of carbon implantation there is no macroscopic segregation either during implantation or during annealing, although the extremes of temperatures are yet to be explored. The IRT results (Section 2) show the existence of SiC, so one assumes that segregation operates on a microscopic scale to form very small SiC precipitates.

Stress and other effects

Stress must be considered when one is discussing mass transport in the presence of chemical segregation if a large change in volume occurs as in the case of synthesis of SiO_2. A stress field such as will exist at the temperatures used during implantation, will lead to enhanced diffusion (gettering) of oxygen to growing precipitates. Implantation damage, which also results in a volume change, is known to getter defects to the damage peak in silicon on sapphire wafers[56] and oxygen to the damage peak in CZ wafers[57,58]. Stress is relieved by emission of Si_i by growing precipitates and by changes in precipitate shape. The excess interstitials condense to form stacking faults and dislocations which will greatly enhance mass transport by providing the means for pipe diffusion.

Experiments on mass transport

Marker layers have been used to investigate the growth of SiO_2 layers during implantation[59,60]. An Si_3N_4 layer implanted at a greater depth

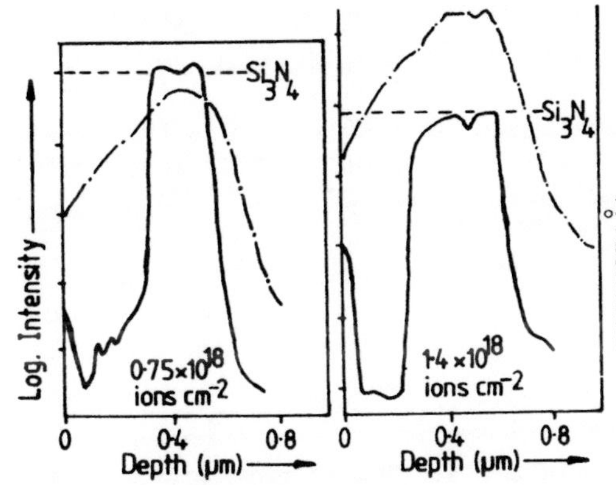

Fig. 14. Nitrogen profiles (SIMS), as implanted with 200 keV N^+ (dot-dash line) and after a 2 hr anneal at 1200°C (continuous line)[55].

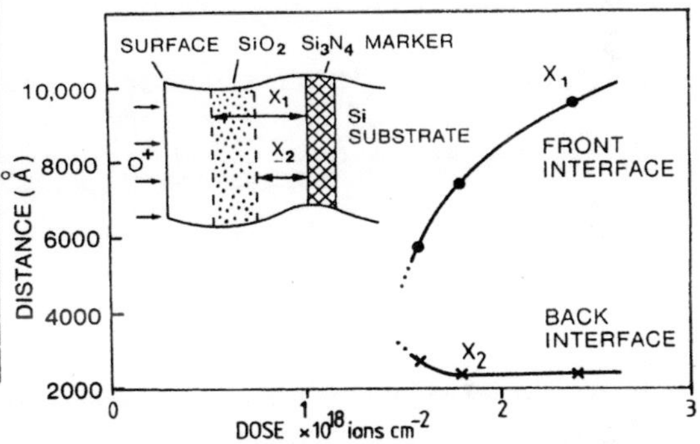

Fig. 15. Use of an implanted Si_3N_4 marker to follow the growth of an implanted SiO_2 layer by RBS[59].
Nearly all growth is at the front interface once SiO_2 is formed.
(200 keV O^+).

than the subsequently synthesised SiO_2 layer has been used as a marker. The results of interface depth versus ion dose determined by RBS are shown in fig. 15. The layer grows at the shallow interface. This is near the peak of the damage profile where there exists a favourable environment for the formation of the tetrahedral SiO_4 bonding configuration. The deep interface is formed as soon as stoichiometric composition is achieved and progresses no further at higher doses. Also the availability of the free surface as a sink for Si_i emitted by the shallow interface facilitates oxide growth.

Implantation of small doses of ^{18}O (2×10^{17} ions cm^{-2}) prior and subsequent to implantation of ^{16}O (2×10^{18} ions cm^{-2}) has been used to investigate mass transport in the growing SiO_2 layers[60,61]. SIMS profiles for ^{16}O are shown in figs. 16a (^{18}O before ^{16}O) and 16b (^{16}O before ^{18}O). (The apparent oxygen peak near the deep interface is a SIMS artefact[61]).

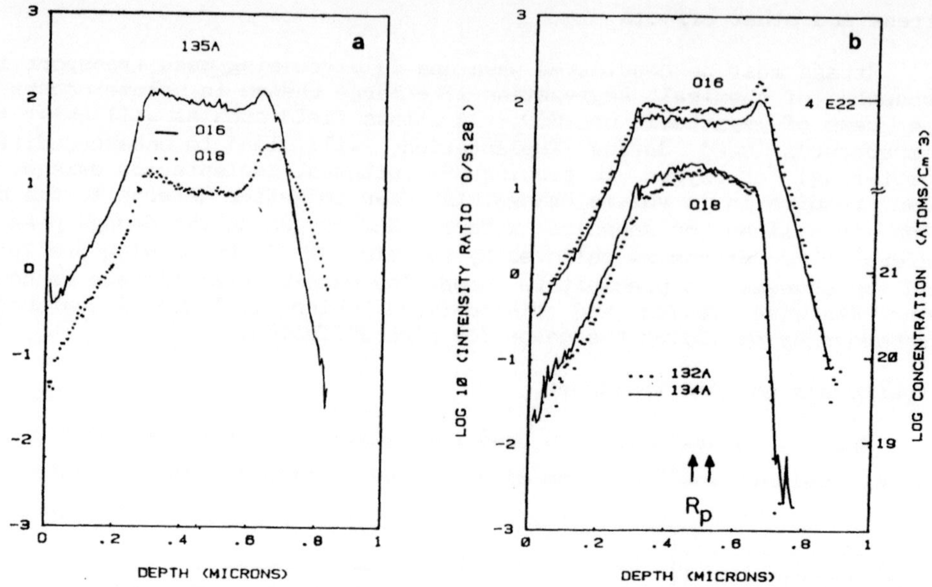

Fig. 16. Use of ^{18}O as a tracer to investigate by SIMS analysis the growth of buried oxide layers. a) A small dose of ^{18}O ($4 \times 10^{17} cm^{-2}$) <u>before</u> $1.8 \times 10^{18} cm^{-2}$ ^{16}O (200 keV). b) $2 \times 10^{18} cm^{-2}$ ^{16}O <u>followed by</u> $3.5 \times 10^{18} cm^{-2}$ ^{18}O (references 60, 61).

From this data one sees that the deep interface is formed very early during implantation. When ^{18}O is implanted first it goes deeper the ^{16}O and than vice-versa when ^{16}O is implanted first. SiO_2 has a higher stopping power than Si, this will reduce the rate of accumulation of oxygen at the deep interface once an SiO_2 layer is established. Also we know from earlier sections that because of the statistical nature of the collision cascade the deep interface is very irregular and isolated precipitates are formed which cannot be reached by rapid oxygen diffusion in the SiO_2 layer.

It is apparent from the ^{18}O SIMS profiles that there is extensive diffusion and exchange with the matrix oxygen. This is not direct radiation enhanced diffusion as, with the exception of the immediate back interface, oxygen diffuses equally throughout the layer. The diffusion rate is very rapid, some four orders of magnitude greater than for oxygen in vitreous silica. It was proposed by Kilner et al.[61] that this can be explained by the disordered nature of synthesised SiO_2 with a reduction in jump distance to roughly the 0-0 separation of 3 Å. However the diffusion mechanism, oxygen incorporation by exchange, is similar to that for oxygen in vitreous silica, and quite unlike that in thermally grown SiO_2 where short circuit diffusion is the dominant mechanism [62]. If nitrogen is implanted into implantation synthesised oxide layers it is rapidly transported to the Si/SiO_2 interfaces indicating a short circuit path in this case[63]. Implantation of ^{18}O tracer into <u>annealed</u> implantation synthesised SiO_2 layers has shown that oxygen is incorporated into the SiO_2 layer by exchange in the same way as during synthesis, but the diffusion rate is reduced, approaching that for oxygen in vitreous silica[85,86]. The difference in rate is easily explained by a reduction in unsaturated bonds due to the establishment of a continuous random network. The difference in mechanism between amorphous and thermal oxide may be structural in origin (channels, density or defects) or may be due to the lack of time for exchange between interstitial O_2 and network O during oxidation experiments.

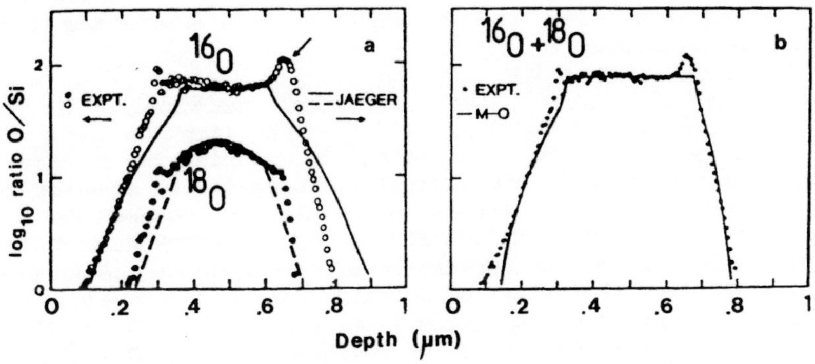

Fig. 17. Fit of theoretical models to SIMS results[60].
a) Jäger's model fits well to the ^{18}O tracer profile but not to the ^{16}O layer interfaces.
b) Maydell's model fits the total profile

Recent measurements have identified a rapid exchange of oxygen between an ^{16}O cap on the surface of a specimen with an ^{18}O buried layer during a very high temperature anneal[59]. ^{16}O is found at the shallow interface in the buried layer and ^{18}O is found near the SiO_2/Si interface of the ^{16}O cap. This clearly indicates that there is rapid oxygen diffusion throughout the system and that the existence of a cap can influence the development of a buried oxide layer.

Tracer experiments have also been reported for nitrogen implants using ^{14}N and ^{15}N (actually $^{28}Si^{14}N$ and $^{28}Si^{15}N$) SIMS profiles (tracer after main implant) for the as-implanted state and after annealing[64]. Like the case of oxygen, the tracer implanted after the main implant does not reach the deep interface. Nitrogen appears to be mobile in Si during the anneal, but only goes to the growing layer. There is no interaction with residual damage to form precipitates, as occurs during annealing of oxygen implanted Si. However it is worth pointing out that the point defect population is quite different, as oxide formation involves a much larger change in volume necessitating the emission of large numbers of Si_i.

Models of mass transport

Several models of oxygen distribution resulting from high dose oxygen implantation have been proposed[65,66]. The two most recent models[22,23] have in common an attempt to allow for the change in stopping power and therefore projected range as the oxygen concentration increases, also they account for expansion on forming SiO_2 and the sputtering of the silicon surface. Maydell and Wilson[22] used experimentally determined values of Si overlay, SiO_2 layer thickness (RBS and XTEM), and surface step height to determine the Si sputtering yield for 200 keV oxygen at high doses (0.1 atoms per ion) and the density of the SiO_2 layer (2.4 gm cm^{-3}). The model is a finite element simulation which builds up the profile by increments taking into account the factors already mentioned, and also the assymetry in the implant profile and the oxygen diffusion length assuming that this is inversely proportional to Si_i (and therefore oxygen) concentration. Once the local oxygen concentration exceeds that for stoichiometric SiO_2 the excess oxygen is added to the wings of the oxygen distribution. The simulated variation of oxygen profile with ion dose is shown in fig. 9 . A good fit is obtained to SIMS data for the data ^{16}O+ ^{18}O profile shown in fig. 17[60], and the model is very successful in predicting the thickness of the silicon overlayer.

The model of Jäger[23] uses analytical expressions in the form of partial differential equations to describe the build-up of oxygen concentration and a diffusion equation to describe the flux of oxygen in the SiO_2 layer driven by the concentration gradient, assuming a constant diffusion coefficient independent of radiation damage and a constant reaction rate at the Si/SiO_2 interfaces. An arbitrary assumption that the oxygen excess is zero at the boundaries of the layer and never exceeds 10 % at the center fixes the values of reaction rate and diffusion coefficient. The Si sputtering yield chosen (0.25 atoms per ion) is at variance with the experimentally determined value of Maydell and Wilson. Jäger uses data relevant to low dose bombardment of Si where the surface binding energy is unperturbed by the presence of large quantities of oxygen. The Jäger theory obtains a reasonable fit to experimental values of Si overlay thickness versus ion dose and also a good fit to the ^{18}O tracer profile (fig. 17). However the fit to the interfaces of the total ^{16}O + ^{18}O profile is poor. One can summarise the results of the two approaches in the statement that the Jäger model successfully describes tracer oxygen diffusion in the SiO_2 layer, and the Maydell-Wilson model successfully describes mass transport during synthesis of the buried layer.

PRECIPITATION

As has been shown previously the solubility of C, N and O in Si is very limited, the equilibrium form for concentrations above the solubility limit is as a precipitate of the natural valency compound. The driving force is chemical segregation and the kinetics are governed by defect enhanced diffusivity of the solute.

Sites for heterogeneous nucleation of precipitates exist in abundance in implanted wafers. Introduced by ion bombardment one has point defects, defect clusters, stacking faults and dislocation loops. This adds to the nucleation sites that exist in as-grown CZ wafers such as oxygen and carbon in solution and as SiO_2 and SiC precipitates.

Oxide precipitates

Bourret and co-workers have used lattice imaging (HREM) to observe oxygen segregation to dislocations in CZ silicon[67,68,69]. They conclude that in a cylindrical region around the dislocation core the strong elastic energy creates a quasi-equilibrium state for oxide clusters which are stable even when the cluster size and therefore the chemical potential is small. The pre-cursor to precipitate formation seems to be a small oxygen complex, similar (or identical) to that responsible for the thermal donor effect. The precipitates so nucleated will grow at low temperatures (450 to 650°C) along the dislocation core, the kinetics being controlled by the emission of Si_i by pipe diffusion along the dislocation with active sites at the ends of the ribbon-like precipitates.

Carbon in solution is attractive to oxygen as the lattice relaxation volumes are opposite in sign. Precipitates nucleated at carbon-oxygen complexes in CZ, wafers grow, after long heat treatment, to take form of amorphous SiO_2 platelets. They are square, 10 to 40 Å thick by 300 to 500 Å wide, and lie in the (100) plane[70]. These are generally seen to be the dominant form of precipitate after long heat treatments in the temperature range 650 to 950°C but have been seen at lower temperature[67].

The plate-like form in this temperature range is what may be expected from precipitation under the strain-field generated by the volume change on forming SiO_2. The shape is a compromise between surface energy and strain energy, and a disk is the shape of minimum strain energy independent of volume. The growth of disk precipitates has been considered by Hu[71] who concludes that the growth mechanism is a combination of incorporation of oxygen over the whole disk surface and diffusion over the disk surface to the most active site at the rim. The platelets are square rather than disk-like because of the anisotropy in surface free energy in crystalline Si.

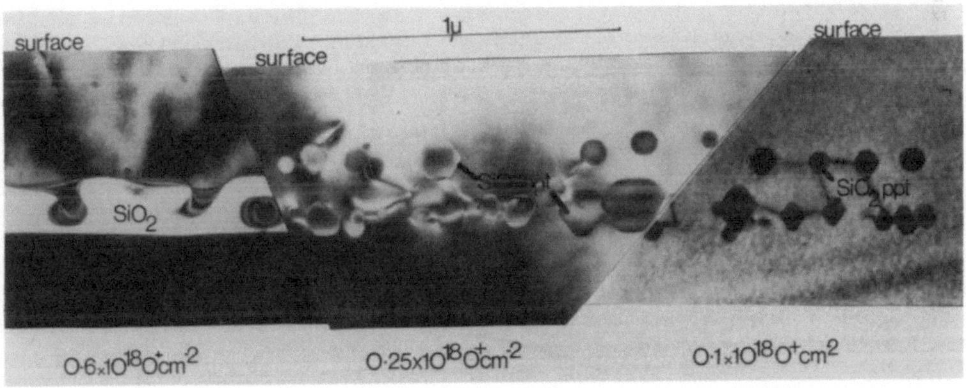

Fig. 18. Cross-sectional TEM of SiO_2 precipitates formed after a high temperature anneal (1405°C, 30 min) oxygen implants (200 keV)[54].

Large plate-like precipitates aligned along the (100) plane have been seen close to implantation synthesised layers[18]. The oxide "spike" discussed in section 4 arises from coalescence of such precipitates. Another example is the lamellae seen in Si close to the deep SiO_2/Si interface. It is these that create the discontinuity in tracer oxygen diffusion at this interface. It is clear from the work on CZ silicon that this form of precipitate will be favoured at implant temperatures of 600°C or so.

An interesting case of what appears to be homogeneous nucleation of SiO_2 precipitates has been reported by Van Ommen and co-workers[27]. Lamp heating during oxygen implantation to T_I = 500°C + beam heating was used with careful control over T_I and ion beam current in order that the Si_v and Si_i concentrations were kept stable. After 2.5 x 10^{18} oxygen ions cm^{-2} at 300 keV a regular structure of precipitates 20 Å diameter with 40 Å spacing on a simple cubic lattice in a matrix of dislocation free silicon was seen. This super-lattice was situated between a surface layer of largely dislocation free silicon (with a few small precipitates) and a layer close to the buried oxide containing a random distribution of larger precipitates. Their hypothesis is that Si_i are removed to the surface sink, and the excess Si_v allow a random distribution of homogeneously nucleated precipitates to form. These are ordered by long range elastic strain, in a similar manner to void lattices, with the symmetry determined by the elastic anisotropy of silicon. Regular structures under similar experimental conditions (lamp heating, T_I = 435°C + beam heating) have been seen before for oxygen implanted silicon. These appeared between the surface and buried oxide layer[25], but were identified as (100) lamellae. The difference may arise because the precipitate lattice could not be resolved in this case, or conditions varied during implantation allowing disk like precipitates to form from the (100) planes of the precipitate super-lattice.

For temperatures >1000°C typical of the anneal temperatures used for annealing implanted layers, the equilibrium form of the oxide precipitates seen in CZ silicon is a polyhedron[68,72]. At these temperatures Si_i and Si_v are highly mobile and so stress is rapidly relieved. Precipitate shape is determined by minimisation of surface energy. In an isotropic medium one would expect perfectly spherical precipitates, however the surface energy of crystalline silicon is anisotropic and developments of the Wolf theorem [73,74] show that a polyhedron is the equilibrium form of a bubble or small particle for a medium where there is anistropy in surface energy. Small (100 Å diameter) polyhedral precipitates have been seen in both wings of the implanted oxygen distribution close to the buried SiO_2 layer after annealing for 2 hours at 1150°C[26]. Recently, low dose oxygen implants (2.5×10^{17} ions cm^{-2} at 200 keV) have been subjected to very high temperature annealing (1405°C, 30 minutes). Cross-sectional micrographs (XTEM)[54] reveal complete segregation but not to a continuous layer. Polyhedral amorphous SiO_2 precipitates, 1000 Å in diameter, are formed near the peak in the projected range. Rounded precipitates are formed near the peak in the damage distribution where anisotropy of surface energy may be reduced by a high concentration of defects. A micrograph of both types of precipitate is shown in fig. 18.

The equilibrium precipitate radius for diffusion (O_i) limited precipitation in dislocation free silicon has been determined using kinetics, neutron scattering, etch pit density and optical scattering[104]. The particle radius increases with increasing temperature, the activation energy being that for diffusion of O_i. It is interesting to note that the equilibrium precipitate radius at 600°C is 40 to 90 Å similar to that described above for the homogeneously nucleated precipitate super-lattice in oxygen implanted silicon. The equilibrium radius at 1150°C, a typical anneal temperature, is 1000 to 3000 Å. This is in the same range as the large polyhedral precipitates shown in Figure 18.

The oxygen concentration and density of nucleation sites in oxygen implanted silicon is considerably higher than in dislocation free CZ silicon. One may therefore expect a smaller precipitate size, but defect enhanced diffusivity of oxygen must counterbalance this. All these factors will combine to increase the rate of approach to equilibrium.

It has been pointed out many times that oxide precipitate growth must be accompanied with emission of Si_i or absortpion of Si_v. Any perturbation of the point defect population will have a commensurate effect on precipitate growth. The formation of new nuclei will depress precipitate growth by increasing the population of Si_i. A change in SiO_2 polymorph may also result in an increase in Si_i[76]. Polymorphic changes may follow many routes, but typical changes with increasing temperature are : thermal donor → amorphous platelet + stress → polyhedron. Small precipitates, with a large surface to volume ratio and therefore high surface energy are more vulnerable to fluctuations in the point defect population and will tend to dissolve and disappear when polymorphic changes occur. This will have the effect of increasing the minimum precipitate size as temperature increases. The silicon surface can be a sink or source of point defects[77] as the annealing environment may place the surface under stress, particularly when a "cap" such as a deposited SiO_2 layer is used. This results in injection of point defects into the surface to relieve stress. If a thermal oxide grows during an uncapped anneal the surface acts as a sink for Si_v and source of Si_i thus inhibiting precipitate growth but increasing nucleation density by creating extrinsic stacking faults. A surface oxide will act as an infinite radius precipitate which will grow in a surpersaturation of Si_i at the expense of precipitates close to the surface, creating a zone denuded of oxygen[78]. A similar phenomenon must be expected during the annealing of buried oxide layers, indeed a large reduction in surface oxygen concentration after annealing at 1150°C is seen in SIMS profiles of buried oxide layers[79].

Nitride precipitates

The nucleation of Si_3N_4 precipitates in nitrogen implanted silicon is less well understood than in the case of oxide precipitates, but some clues exist in the literature. The same nucleation centers, point defect complexes, stacking faults and dislocation loops exist after implantation as do intrinsic defects such as carbide and oxide precipitates. The wealth of evidence is that Si_3N_4 precipitates only nucleate during annealing at 1200°C at the peak in the concentration of implanted nitrogen[80,81,82]. At low doses (2 x 10^{17}ions cm^{-2}, 40 keV) Pavlov[4] has seen finely dispersed precipitates decorating dislocation loops, formed by condensation of point defects, in the as-implanted state. Stein[34], in the IRT work reported in section 2, at lower doses observes formation of Si_3N_4 at 750°C, but at 900°C the signal disappears. It seems possible therefore that nitride precipitates can nucleate at extended defects but the precipitates are dispersed when the defects are removed by annealing.

Annealing low dose (2.5 x 10^{17}, 200 keV) samples at 1200°C results in the formation of a discontinuous layer of polycrystalline α-Si_3N_4 at the peak of the implant profile[80]. During the anneal α-Si_3N_4 crystallites grow radially parallel to the surface confined by the peak in the nitrogen concentration. Growth is dendritic forming spherulitic grains with nuclei 5 to 10 µm apart[4,80,82] for doses in the range 3.5 to 7.5 x 10^{17} ions cm^{-2} (150 keV, T_I 400°C, T_A 1200°C). The thickness of the crystalline layer is in the region of 1000 Å. A TEM micrograph from Te Kaat and Belz[82] of an isolated precipitate is shown in fig. 19a. The dendritic nature of the growth can clearly be seen. They have considered the growth of these precipitates from a nucleus, radius r which collects nitrogen atoms from a volume radius $r + \Delta$. They assume a maximum value for $\Delta = \Delta_c$ determined by the rate of diffusion of nitrogen to the growing precipitate. They have seen clear zones of crystalline silicon bridged by mismatch dislocations between heavily dislocated silicon and the growing precipitate. Assuming that these clear zones are depleted of nitrogen this gives $\Delta_c \sim 1500$ to 2000 Å. They estimate that the maximum radius of a stable spherical precipitate must be 2000 to 2500 Å, in agreement with their experimental results. Further growth must be dendritic to leave intermediate regions of crystalline silicon, whilst accumulating nitrogen. The nitrogen depleted zones are relatively defect free because the strong chemical potential opens bonds at the Si_3N_4/Si interface which allows for rearrangement to remove defects. A TEM micrograph of a continuous spherulitic layer formed at higher nitrogen doses (1 x 10^{18} ions cm^{-2}, 150 keV, T_I 400°C, T_A 1200°C) is shown in fig. 19b.

Fig. 19. TEM of Si_3N_4 precipitates in silicon, formed by implantation of 150 keV N^+ and annealing at 1200°C (2 to 15 hours)[82].
a) Isolated spherulite formed at low doses (3.5 x 10^{17} ions cm^{-2}).
b) Continuous structured polycrystalline Si_3N_4 layer formed at high doses (1 x 10^{18} cm^{-2}).

It is evident from the results presented here that nucleation of α-Si_3N_4 precipitates occurs during the high temperature anneal. The fact that this always occurs in the region of maximum nitrogen concentration indicates that this is homogeneous nucleation. Growth is controlled by the rate of nitrogen diffusion, which is low. This low rate of diffusion leads to instabilities in the growth front and confinement of the growing precipitates to the peak in nitrogen concentration.

There is some evidence for heterogeneous nucleation of Si_3N_4 during implantation but these nuclei are probably dissolved when extended defects are removed by annealing.

MICROSTRUCTURE

Much information on microstructure, gained chiefly by cross-sectional transmission electron microscopy (XTEM) has recently become available [24-27,50,59,64,80-85]. In this concluding section this will be pieced

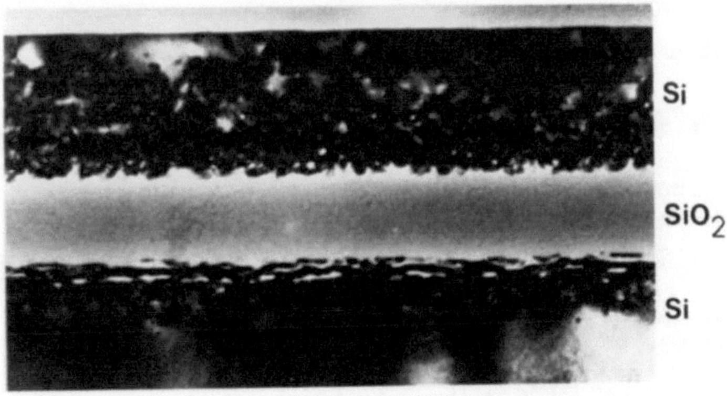

Fig. 20 a
1.8×10^{18}
O^+ at 200 keV
T_I 550°C
annealing for
2 hours at
1150°C.

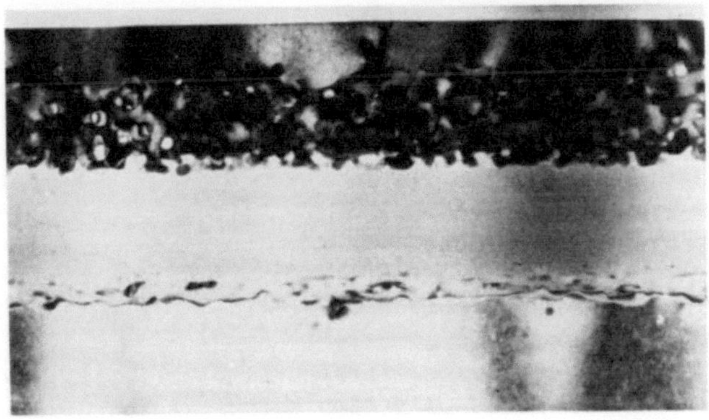

Fig. 20b
As for fig. 20a
but annealed
for 2 hours at
1200°C.

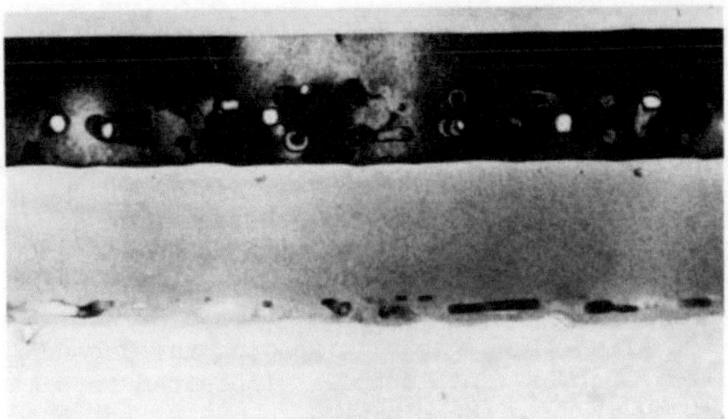

Fig. 20c
As for fig. 20a
but annealed
for 2 hours at
1250°C

together to form as complete a picture as possible of the structure resulting from the mechanisms discussed in sections 2 to 5. It will be demonstrated that structure depends critically on the implant dose, temperature and manner of heating, and on the annealing temperature and ambient.

Oxygen implantation : the as-implanted state

With the exception of implants where lamp heating has been used [25,26,27,86] the as-implanted state has not been as much studied by XTEM as the annealed structures. The information that is available shows that when the dose is sufficient to create stoichiometric SiO_2 a relatively featureless amorphous SiO_2 layer is formed with irregular mixed phase (Si + SiO_2) boundaries[85]. For $T_I > 450°C$ (beam heating) the regions both sides of this consist of heavily damaged crystalline silicon. The damage must arise from dislocation loops and stacking faults created by the collapse of point defect clusters.

For $T_I < 450°C$ (beam heating) an amorphous Si layer is formed between the SiO_2 layer and the damaged crystalline Si[50]. On annealing the amorphous layers nucleate independently of the neighbouring crystalline Si to form polycrystalline Si layers.

When lamp heating is employed[25,26,27,86] much less damage is seen in the as-implanted state. Evidently the shorter time taken to reach an equilibrium value for T_I prevents the formation of stable defect clusters. Structures similar to those seen after beam heating and annealing at ~ 1100°C are seen. Substrates were heated to temperatures around 500°C by lamp heating, but the true value of T_I, which will be increased by ion beam heating during implantation, has not been measured and unfortunately estimates seem to vary widely. At the optimum temperature (~ 475°C + beam heating) the immediate surface layer is largely dislocation free (Jassaud et al.[26] saw some small loops and rods) with very small SiO_2 precipitates [26,27]. Further from the surface regular arrays of precipitates are seen arranged in one case on a simple cubic super-lattice[27] or in another case as a regular laminar structure[25]. Details of these structures are discussed in the preceeding section. Closer to the SiO_2 layer larger SiO_2 precipitates are seen[27] arranged in a random distribution. The Si/SiO_2 layer interfaces are atomically sharp with the shallow interface being less convoluted than the deeper one, consistent with what is known about the statistical nature of the damage distribution, and the change in range during oxygen build-up.

Oxygen implantation : after annealing (beam heated specimens)

After annealing an oxygen denuded zone is formed near the silicon surface. This is largely precipitate and dislocation free with a number of threading dislocations which reduces from ~ $10^{10} cm^{-2}$ for a two hour furnace anneal at 1150°C [26] to ~ 10^9 cm^{-2} for a 30 min lamp anneal at 1405°C[24]. A series of XTEM micrographs[54] showing the dependence of microstructure on anneal temperature for samples implanted to a high dose (1.8×10^{18}ions cm^{-2}) at a high T_I (beam heated to 600°C) are shown in fig. 20. After a two hour furnace anneal at 1150°C (fig. 20) dislocation networks and precipitates, sometimes polyhedral[26], can be seen between the denuded zone and the SiO_2 layer. Plate-like precipitates can be seen at the deep interface of the layer. An increase of 50°C in anneal temperature (fig. 20b) results in a large reduction in damage and in interface area. Precipitates start to change to spherical or polyhedral shape. Damage is completely annealed at 1250°C, (fig. 20c) leaving only large polyhedral precipitates in the Si above the SiO_2 layer and rounded Si precipitates in the SiO_2 layer close to the deep interface. The layer interfaces are almost flat. As has been already shown (fig. 11) lamp annealing at 1405°C results in the removal of the polyhedral SiO_2 precipitates, only a few rounded Si precipitates remain in the SiO_2 layer.

Implanting at lower values of T_I can result in the formation of the secondary layers of SiO_2 that have been named hump and spike (fig. 10) [50,54]. Increase of annealing temperature from 1050°C to 1200°C results in the growth of these features according to Tuppen et al.[50], but Marsh[54] finds that at 1200°C the hump, seen after annealing at 1150°C, disappears. Obviously the details of these microstructures, which represent a non equilibrium thermodynamic state of the system, will depend sensitively on implant and anneal parameters. This is a technologically important issue as these features may be very important in obtaining desirable device characteristics.

Some early Russian work[87] reports the formation of polycrystalline SiO_2 (cristobalite) in regions away from the peak in the damage profile. There is corroboration of this from recent work where a three step annealing cycle is used (peak temperature 1250°C). Usually amorphous SiO_2 is seen both in the as-implanted state and after annealing, although no structure has yet been reported for the super-lattice precipitates seen by Van Ommen et al.[27]. It is interesting to speculate that these are small enough to relate topotactically to the host lattice and therefore be crystalline in nature, as are the cristobalite precipitates mentioned above.

Nitrogen Implantation : as-implanted

The structure of nitrogen implanted layers has been discussed by several authors[59,61,80,83]. Some results, particularly on the spherulitic growth of α-Si_3N_4 at the peak in the implant profile, have been described in the previous section. The work of Meekison and Booker[81] is the most recent and the most complete. They used XTEM to investigate 200 keV implants (T_I = 500°C) for doses of 0.25, 0.75 and 1.4 x 10^{18} nitrogen ions cm^{-2}. Their results are shown diagramatically in fig. 21 for the as-implanted state and after a two hour furnace anneal at 1200°C.

An amorphous layer is formed at the peak of the implant profile (not at the peak of the damage profile as has been reported in earlier work[82]). This increases in width with increasing dose until the peak nitrogen concentration becomes over stoichiometric when a layer of nitrogen bubbles is formed. The solubility and diffusion rate of nitrogen in Si_3N_4 is low, favouring coalescence rather than mass transport to free Si. A region of highly damaged silicon is seen near the surface which grows with increasing nitrogen implant dose consuming a region of good silicon below until at the highest dose both layers are partially consumed by the expanded amorphous layer. The reason for the existence of a damage free region at low doses is obscure at present. Within a certain range of nitrogen concentration the migration of point defects to form stable complexes must be inhibited, or Si_i-Si_v recombination enhanced.

The damage below the amorphous layer shows diffraction effects typical of Moiré fringes generated by a slight misorientation in crystalline Si, possibly due to strain.

Nitrogen implantation : annealed structures

After annealing there is almost complete segregation to an α-Si_3N_4 layer. In the low dose case this grows spherulitically with nuclei 10 μm apart. For all doses the position of the α-Si_3N_4 layer is centered on the amorphous layer seen in the as-implanted case, and is some 2/3 of the thickness of the amorphous layer. The α-Si_3N_4 layer is discontinuous in the low dose case, has rough interfaces (100 to 200 Å irregularity) at the medium dose and has smooth and atomically sharp interfaces at the high dose.

For the medium dose the Si_3N_4 layer occasionally contains regions of amorphous Si_3N_4 as well polycrystalline α-Si_3N_4. It is known that oxygen as an impurity inhibits crystallization of CVD Si_3N_4 and that oxygen implantation can have a similar effect on implanted layers[88]. It would appear therefore that oxygen as an impurity can prevent crystallization in certain regions by forming random oxy-nitride networks. Islands of silicon within the nitride layer are also seen in the medium dose case and the density of these correlates with high electrical leakage. It is evident that leakage occurs by electron tunnelling across the thinner regions of nitride.

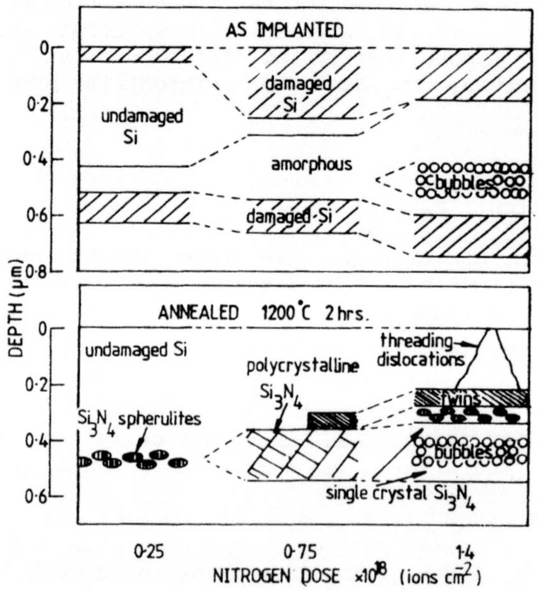

Fig. 21. Diagram of the structures seen in XTEM of nitrogen implanted silicon (200 keV, T_I = 500°C)[81].

Annealed high dose specimens have a complex structure. Nearest the shallow interface of the continuous nitride layer is a mixed layer of nitride spherulites and silicon, rather similar to that at the peak concentration in the low dose example. Both sides of the bubble layer formed at the peak of the implant distribution are regions of single crystal α-Si_3N_4. It is proposed that in this region the nitrogen concentration is already stoichiometric, so that crystal growth is achieved by small and rapid atomic rearrangements and is not limited by diffusion. There is a gradual variation of orientation in the single crystal regions arising from low angle grain boundaries. The bubble layer decouples the two single crystal layers which are not oriented with each other or with the silicon host.

The dislocation count in the nitrogen depleted Si surface is very low, $< 10^7 cm^{-2}$ for the low and medium dose specimens where the defect (and strain free) silicon layer can act as a seed during regrowth. In the case of the highest dose this layer does not exist and $10^9 cm^{-2}$ threading dislocations are seen. A discontinuous layer of twinned silicon at the shallow Si/Si_3N_4 interface occurs after annealing the medium dose specimen and this is continuous for the annealed high dose specimen. These could arise from mis-match or stress, but contrary to what one would expect, twins are not seen at the deep interface.

It is evident that silicon overlayers of better quality than for the oxygen implantation case can be achieved, but the optimum dose to achieve a good dielectric nitride layer is far more critical.

Carbon implantation

Very little information is available in the literature on the microstructure of carbon implanted silicon. Very recently however, XTEM has been used to explore high dose carbon implants, both before and after annealing at 1200°C[84].

There is a heavily damaged single crystal Si overlayer and a buried amorphous layer. Within the buried amorphous layer is a polycrystalline band, identified as β-SiC by selected area diffraction. Laminar structures are seen within the crystalline silicon overlayer close to the interface with the buried amorphous layer. These lead to size-effect elongation of the silicon diffraction spots.

After annealing the polycrystalline β-SiC layer has extended up to the interface with the Si overlayer.

The laminar defect structures in the crystalline Si overlayer disappear on annealing, and many carbide precipitates are seen throughout the overlayer. These are $\leqq$ 120 Å in diameter, consist of β-SiC and are aligned with the lattice of the Si host. This is taken as evidence for carbon occupying a substitutional site in the silicon lattice. However, as discussed in the case of oxide precipitates, the symmetry of the host imposes a constraint on small precipitates which therefore have a structure related topotactically to the host. In this case cubic β-SiC appears to be the equilibrium structure independent of the constraints imposed by the Si host. Topotaxy ensures that the precipitates are coherent with the Si lattice.

It has been pointed out that as the volume expansion on forming SiC is very small, the precipitates observed are too small to nucleate dislocations. Unlike the case of oxide precipitation no extra defects in the Si overlayer will be created by carbide precipitation. However at the anneal temperatures so far used the Si overlayer still contains a high concentration of defects compared with the best achieved by oxygen or nitrogen implantation.

The microstructure of carbon implanted silicon has similarities to nitrogen implanted silicon in that polycrystalline layers are formed, and to oxygen implanted silicon in that heterogeneous nucleation of precipitates in the defective Si overlayer is observed. This last observation indicates that SiC precipitates readily nucleate perhaps at oxygen-defect complexes. An explanation for the lack of large scale segregation to the buried layer during annealing at 1200°C is that the carbon diffuses to and is fixed in small immobile SiC precipitates.

REFERENCES

1. I.H. Wilson, Nucl. Instrum. and Meth. B1:331 (1984).
2. M. Wantanabe and A. Tooi, Japan, J. Appl. Phys. 5:737 (1966).
3. P.V. Pavlov, E.I. Zorin, D.I. Telel'baum and Yu.S. Popov, Soviet Phys. Doklady 10:786 (1966).
4. P.V. Pavlov, T.A. Kruze, D.I. Telel'baum, E.I. Zorin, E.W. Shitova and N.V. Godkova, Phys. Stat. Solidi A36:81 (1976).
5. F. Cianfrone, U. Fasoli and P. Mazzoldi, Nuovo Cimento B57:534 (1968).
6. W.J. Kleinfelder, W.S. Johnson and J.F. Gibbons, Can. J. Phys. 46:597 (1968).
7. G.H. Schwuttke, K. Brack, E.E. Gardner and H.M. De Angelis, "Radiation Effects in Semiconductors" F.L. Vook ed, p. 406, Plenum Press, New York (1968).
8. U. Bonse, M. Hart and G.H. Schwuttke, Phys. Stat. Solidi 33:361 (1969).
9. G.H. Schwuttke and K. Brack, Acta Crystallog. A25:S43 (1969) 8th Int. Cong. of Int. Union of Crystallography, AIP, Buffalo, Stony Brook and Upton, NY, USA (1969).
10. J.H. Freeman, Proc. Int. Conf. "Applications of Ion Beams to Semiconductor Technology", Editions Ophrys, Grenoble (1967), pp. 75, 669.
11. J.H. Freeman, G.A. Gard, D.J. Mazey, J.H. Stephen and F.B. Whiting, European Conf. on Ion Implantation, Reading, U.K., Peregrinus, Hitchin, (1970), p. 74.
12. J. Dylewski and M.C. Joshi, Thin Solid Films 35:327 (1976).
13. J. Dylewski and M.C. Joshi, Thin Solid Films 35:241 (1976).
14. J. Dylewski and M.C. Joshi, Thin Solid Films 42:227 (1977).
15. M. H. Badawi and K.V. Anand, J. Phys. D 10:1931 (1977).
16. S.S. Gill and I.H. Wilson, Thin Solid Films, 55:435 (1978).
17. K. Izumi, M. Doken and H. Ariyoshi, Electron. Lett. 14:593 (1978).
18. K. Das, J.B. Butcher, M.C. Wilson, G.R. Booker, D.W. Wellby, P.L.F. Hemment and K.V. Anand, Inst. Phys. Conf. Ser., 60:307 (1981).
19 H. Koyama, J. Appl. Phys. 51:3202 (1980).
20. Y. Homma, M. Oshima and T. Hasyashi, Japan, J. Appl. Phys. 21:890 (1982).
21. J.A. Kilner, R.J. Chater, P.L.F Hemment, R.F. Peart, E.A. Maydell-Ondrusz, M.R. Taylor and R.P. Arrowsmith, Nucl. Instrum. and Meth., B7/8:293 (1985).
22. E.A. Maydell-Ondrusz and I.H. Wilson, Thin Solid Films, 114:357 (1984).
23. H.V. Jäger, Nucl. Instrum. and Meth. (1986).
24. G.K. Celler, P.L.F. Hemment, K.W. West and J.M. Gibson, Appl. Phys. Lett. 48:532 (1986).
25. O.W. Holland, T.P. Sjooreen, D. Fathy and J. Narayan, Appl. Phys. Lett. 45:1081 (1984).
26. C. Jaussaud, J. Stoemenos, J. Margail, M. Dupuy, B. Blanchard and M. Bruel, Appl. Phys. Lett. 46:1064 (1985).
27. A.H. Van Ommen, B.H. Koek and M.P.A. Viergers, Appl. Phys. Lett..
28. A. Bourret, J. Thibault-Desseaux and D.N. Seidman, J. Appl. Phys. 55:825 (1984).
29. R. Kelly, Radiat. Effects, 64:205 (1982).

30. J.A. Borders and W. Beezhold, "Ion Implantation in Semiconductors" eds. I. Ruge and J. Graul, Springer-Verlag, Berlin (1971), p. 241.
31. J. Ishikawa, K. Miyata, H. Tsuji, T. Taya and T. Takagi, Proc. 9th Symp. ISIAT Tokyo (1985), p. 503.
32. I.P. Akimcherko, K.V. Kisseleva, V.V. Krasnopevtsev, A.G. Tourganski and V.S. Vavilov, Radiat. Effects 48:7 (1980).
33. P. Bourguet and J.M. Dupart, Rev. Phys. Appl., 15:547 (1980).
34. H.J. Stein, J. Electrochem. Soc., 132:668 (1985).
35. D.E. Davies, J.A. Adamski and E.F. Kennedy, Appl. Phys. Lett. 48:347 (1986).
36. J.G. Wilkes, J. Cryst. Growth, 65:214 (1983).
37. J.M. Hwang, D.K. Schroder and A.M. Goodman, IEEE Elect. Dev. Letts. EDL-7:172 (1986).
38. R.C. Barklie, A. Hobbs, P.L.F. Hemment and K. Reeson, J. Phys. C19:6417 (1986).
39. M.H. Badawi and K.V. Anand, J. Phys. D 10:1931 (1977).
40. S.S. Gill, Thesis, University of Surrey (1980).
41. J.A. Keenan and G.B. Larrabee, "VLSI Electronics, Impurities in Silicon", Vol. 6, Ch. 1.
42. J.P. Biersack and J.F. Ziegler, Nucl. Instrum. and Meth. 194:93 (1982).
43. O.S. Oen, Nucl. Instrum and Meth., B13:495 (1986).
44. I.H. Wilson, S. Chereckdjian and R.P. Webb, Nucl. Instrum. and Meth. B7/8:735 (1985).
45. M.J. Norgett, M.T. Robinson and I.M. Torrens, Nucl. Eng. Des. 33:50 (1975).
46. I.H. Wilson, J. Appl. Phys. 53:1698 (1982).
47. I.N. Smirnov, V.V. Konyshev and T.G. Alksnis, Soviet Phys. Solid. State 19:1401 (1977).
48. G. Davies, A.S. Oates, R.C. Newman, R. Woolley, E.C. Lightowlers, M.J. Binns and J.C. Wilkes, J. Phys. C.19:841 (1986).
49. H. Ryssel and I. Ruge, "Ionenimplantation", Teubner-Verlag, Stuttgart (1978).
50. C.G. Tuppen, M.R. Taylor, P.L.F. Hemment and R.P. Arrowsmith, Thin Solid. Films 131:233 (1985).
51. C. Claeys and H. Bender, Inst. Phys. Conf. Ser. No. 76:451 (1985).
52. P.S. Ho, Thin Solid. Films, 96:301 (1982).
53. S.S. Gill and I.H. Wilson, Mat. Res. Soc. Symp. Proc. 27:275 (1984).
54. G.E. Murch, "Atomic Diffusion Theory in Highly Defective Solids", Trans. Tech. SA Aedermannsdorf, Switzerland (1980).
55. C.D. Marsh, University of Oxford, in preparation (1986).
56. P.L.F. Hemment, K. Reeson, R.F. Peart, C.D. Meekinson, C. Marsh, G.R. Booker, R.J. Chater and J.A. Kilner, Rad. Eff.
57. Y. Yamamoto and I.H. Wilson, Appl. Phys. Lett. 34:403 (1979).
58. T.J. Magee, C. Leung, H. Kawayoshi, B.K. Furman, C.A. Evans Jr., and D.S. Day, Appl. Phys. Lett. 39:260 (1981).
59. I.J. Magee, C. Leung, H. Kawayoshi, R. Ormond, B.K. Furman, C.A. Evans Jr., and D.S. Day, Appl. Phys. Lett. 39:413 (1981) 413.
60. K.J. Reeson, Int. Conf. Ion Beam Modification of Materials, Catania (1986), to be published in Nucl. Instrum. and Meth. (1987).
61. R.J. Chater, J.A. Kilner, P.L.F. Hemment, K.J. Reeson and R.F. Peart, Mat. Res. Soc. Symp. Proc. (1985).
62. J.A. Kilner, R.J. Chater, P.L.F. Hemment, R.F. Peart, E.A. Maydell-Ondrusz, M.R. Taylor and R.P. Arrowsmith, Nucl. Instrum. and Meth. B7/8:293 (1985).
63. A.G. Revesz and H.A. Schaeffer, J. Electrochem. Soc. 129:357 (1982).
64. J.A. Kilner, Imperial College, London, in preparation (1986).
65. J.A. Kilner, R.J. Chater, P.L.F. Hemment, R.F. Peart, K.J. Reeson, R.P. Arrowsmith and J. Davis, Nucl. Instrum. and Meth..
66. T. Hayashi, H. Okamoto and Y. Homma, Inst. Phys. Conf. Series No. 59:559 (1981).

67. K. Ohwada, K. Izumi and T. Hayashi, Japan, Ann. Rev. Electron. Telecom. Semicond. Technol. 21:25 (1982).
68. A. Bourret and C. Colliex, Ultramiscroscopy 9:193 (1982).
69. A. Bourret, C. Colliex and P. Trebbia, J. Physique Lett. 44:L-33 (1983).
70. A. Bourret, J. Thibault-Desseaux and D.N. Seidman, J. Appl. Phys. 55:925 (1984).
71. W. Bergholz, J.L. Hutchinson and P. Pirouz, Inst. Phys. Conf. Ser. No. 76:11 (1985).
72. S.M. Hu, Appl. Phys. Lett. 48:115 (1986).
73. F.A. Ponce, ibid:1 (1985).
74. M. Von Laue, Z. Kristallog., 105:124 (1943).
75. W.K. Burton, N. Cabrera and F.C. Frank, Phil. Trans. A243:40 (1951).
76. M.J. Binns, W.P. Brown, J.G. Wilkes, R.C. Newman, F.M. Livingston, S. Messolovas and R.J. Stewart, Appl. Phys. Lett. 42:525 (1983).
77. T.Y. Tan and C.Y. Kung, J. Appl. Phys. 59:917 (1986).
78. R.B. Fair, J. Appl. Phys. 54:388 (1983).
79. J.G. Wilkes, Journal of Cryst. Growth, 65:214 (1983).
80. P.L.F. Hemment, E.A. Maydell-Ondrusz, K.G. Stephens, J.A. Kilner and J.B. Butcher, Vacuum 34:203(1984).
81. C.D. Meekison, G.R. Booker, K.J. Reeson, P.L.F. Hemment, R.J. Chater, J.A. Kilner and R.P. Arrowsmith, Inst. Phys. Conf. Ser. 76:489 (1985).
82. C.D. Meekison and G.R. Booker, in preparation (1986).
83. E.H. Te Kaat and J. Belz, Proc. 8th European Congress on Electron Microscopy, eds. A. Csanad, P. Röhlich and D. Szabo, Budapest (1985), Vol. 2, p. 967.
84. K.J. Reeson, P.L.F. Hemment, R.F. Peart, C.D. Meekison, G.R. Booker and J. Davis, Electron. Letts. 22:467 (1986).
85. K.J. Reeson, P.L.F. Hemment, J. Stoemenos, J. Davies and G.K. Celler, Proc. Int. Conf. "Microscopy of Semiconducting Materials, Oxford (1987), paper p. 2-4.
86. D. Fathy, O.C. Krivanek, R.W. Carpenter and S.R. Wilson, Proc. Conf. Microscopy of Semicond. Mat., Oxford (1983).
87. R. Yankov and I.H. Wilson, Univ. Surrey report (1985).
88. K.J. Reeson, private communication (1986).

ELECTRICAL PROPERTIES OF SIMOX MATERIAL AND DEVICES

Sorin Cristoloveanu

Institut National Polytechnique de Grenoble
Laboratoire de Physique des Composants à Semiconducteurs
ENSERG, 23 rue des Martyrs, Grenoble 38031, France

INTRODUCTION

The interest in Silicon On Insulator (SOI) structures as advanced substrates for integrated circuits has been generated by the inability of standard bulk Si processing to overcome some major VLSI limitations (parasitic capacitances affecting speed and power, intolerance to radiation effects, poor lateral isolation, etc.). Several technological approaches are under very active and competitive research : silicon hetero-epitaxy on various insulators (sapphire, zirconia), silicon deposition and subsequent recrystallization (laser, lamp, electron beam, etc.), oxidation of porous silicon, wafer bonding and, finally, deep implantation of insulator-forming ions into silicon (SIMOX). In spite of its relatively recent development SIMOX is now in a privileged position. The silicon overlayer is indeed a wafer scale monocrystal of high quality and offers the best perspective for microelectronics.

The characterization work supporting progress in SIMOX or in other SOI technologies was based initially on microscopic examination. Almost simultaneously, integrated circuits were fabricated on SOI and their performance compared with that of bulk Si components. This very crude comparison is a key element to decide on the interest for a deeper investigation and optimization of the material. Very little is known, at this early stage of maturity, about the basic electrical properties of the structure. The gap existing between the microscopic data and the integrated circuit cannot be totally attributed to a (reasonable) pragmatism dictated by expected economical impacts. The electrical characterization is also made difficult by the configuration of SOI materials (very thin Si overlay, multi-interface topology, inhomogeneous properties across the film, various contamination sources).

The aim of this paper is to answer these analysis problems and to provide a rather complete description of the SIMOX properties. The work is based on the experience accumulated on SIMOX at the PCS Laboratory ; relevant results will be compared with data produced by other groups. The discussion proceeds from the global transport parameters of the as-formed (unprocessed) material and includes the analysis of in-depth profiles of transport and optical properties in the Si overlayer and buried oxide. The formation of oxygen-induced thermal donors is shown to be a key aspect in SIMOX. The interface properties are revealed by the operation of MOS integrated devices. The static and dynamic characteristics, recombination parameters and low-frequency noise are reviewed. Adequate characterization methods, adapted from bulk Si or especially developed, are proposed. Finally, a new MOS-SIMOX device with greatly enhanced performance is described.

The electronic properties under discussion will be systematically cross referenced to processing conditions. The SIMOX material supporting this analysis has been fabricated at the University of Surrey and at the LETI, using low current implanters (50-200 μA). The implantation energy was 200 keV, the doses from 1.4 to $2.2 \times 10^{18} O^+/cm^2$ and the implantation temperature range was 400-650°C.

Two types of post-implantation annealing were used to perfect the process. After an anneal at low temperature (LTA : 1150-1200°C), many defects and SiO_2 precipitates subsist in a greatly disordered interfacial region, the composition of which depends upon the implantation temperature and dose[1,3]. Only the top layer, 0.1 μm thick, is almost free of defects.

High temperature annealing (HTA) above 1300°C causes the high quality region to extend over the whole film (0.2-0.3 μm thick), from the top surface to the interface[4]. This annealing is performed up to 1350°C in a poly-Si furnace[5] or at 1405°C with focused lamp irradiation on the back of the wafer to achieve partial surface melting[6]. During the HTA process most of the SiO_2 precipitates are dissolved and the residual oxygen concentration is reduced below $10^{18} cm^{-3}$ by the pumping of oxygen towards the oxide[7]. However, the density of dislocations which traverse the film is not significantly lowered (10^7-$10^9 cm^{-2}$)[5,8].

RESISTIVITY MEASUREMENTS AND IMPURITY ACTIVATION

Four point probe measurements are used on as-formed SIMOX material to control the degree of contamination, homogeneity across the wafer and activation of dopant species[9,10]. This experiment is routine in bulk Si but must be carefully reconsidered in very thin SOI layers. A convenient compromise is indeed necessary to achieve good ohmic contacts and small probe penetration (probe pressure below 20-30 g). On the other hand the extent of the depletion region induced by the metal-semiconductor contact or by interfacial charges could, in some cases, prevent the current injection and, therefore, the measurement feasibility ; an intentional doping of the sample is then helpful. Finally, the sheet resistance of a film 0.1 μm thick of resistivity 10^2-$10^4 \Omega cm$ is very large and a detection system with high impedance is required. The output voltage was measured with a differential voltmeter connected to each terminal via a Keithley electrometer.

Typical curves related to the activation of As dopant in LTA-SIMOX are shown in Figure 1. After implantation of As, the specimens were successively annealed from 300°C to 700°C for 30 minutes. The sheet resistance presents a steep decrease between 350°C and 500°C, towards a minimum level which correlates well with the dose. This behaviour is similar to that known for bulk Si and suggests thereby an easy control of this important step of the integration process on SIMOX substrates. In contrast, in SOI material synthesized by nitrogen implantation, arsenic is activated at a lower degree, probably as a result of partial precipitation.

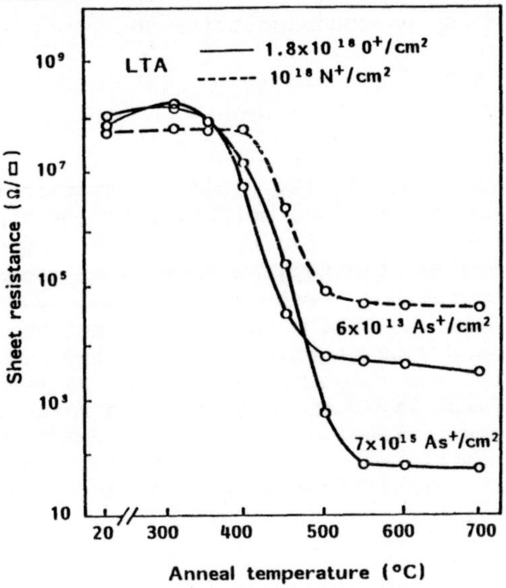

Fig. 1. Sheet resistance versus anneal temperature in LTA-SIMOX layers implanted with arsenic.

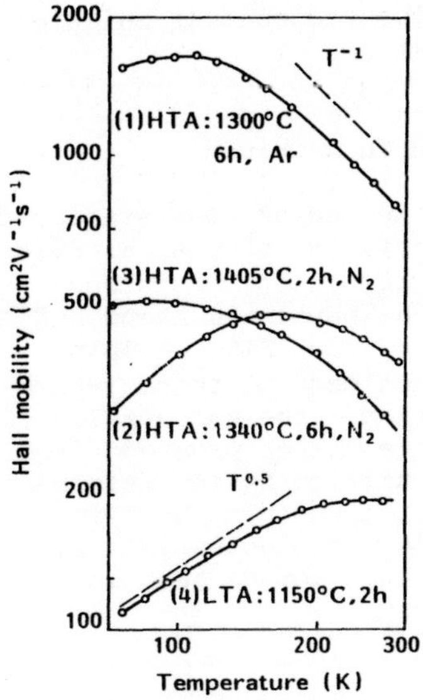

Fig. 2. Hall mobility against temperature in SIMOX layers ($1.8x10^{18}cm^{-2}$, 200 keV) annealed under various conditions

HALL EFFECT

The Hall effect permits a more detailed inspection of SIMOX properties in terms of carrier mobility and concentration but requires, in turn, ohmic contacts to be formed by masking, additional implantation and subsequent annealing. In Figure 2 are compared the variations of the Hall mobility at low temperatures for several Van der Pauw samples. Specimen 1 illustrates the beneficial effect of high temperature annealing (1300°C, 6 hours) in inert Ar ambient[4,7]. The mobility is high, presents a maximum at about 100 K and is dominated above by acoustic phonon scattering. This behaviour demonstrates a crystal quality comparable with that of bulk Si. The activation energy (50 meV) of the carrier concentration suggests an over-compensation of the initially p-type Si wafer (17-23 Ωcm) by thermal donors ($< 10^{16}cm^{-3}$) accidentally generated during the process.

In sample 2, oxygen donors are almost masked by a heavier contamination ($2 \times 10^{16}cm^{-3}$) during annealing in nitrogen ambient. This causes more intensive ionized scattering to occur and results in a lower mobility and a shift of the mobility maximum to higher temperatures[7]. Contamination with nitrogen and lower mobility are observed in sample 3 as well, annealed at 1405°C and doped with As ($3 \times 10^{17}cm^{-3}$). Sample 4, annealed at 1150°C only, shows a more dramatic mobility degradation due to crystalline defects in the interface region which dominate the electron scattering even at room temperature[9]. These electrically active defects are almost eliminated after HTA. Nevertheless, the control of the ambient and the purity of the capping layer (if any) appear as crucial elements in HTA process.

Lower temperature measurements give evidence for a rapid mobility decrease below 77 K. This is explained by an accentuated influence of dislocations[11] and by a transition to hopping conduction[12]. Evidence for electron hopping between impurities was also given by the increase in the carrier concentration below 40 K and by photoconductivity and photo-Hall effect.

IN-DEPTH PROFILES

Spreading resistance measurements are normally used to profile the resistivity of the Si film. The sample is bevelled uniformly at low angle and the sheet resistance ($<10^9\Omega$) is measured by two probes which progress on the bevel in steps of a few microns. The top surface is in general marked by SiO_2 capping. These operations represent supplementary limitations to those which are inherent to the 4-point probe method. In addition, the calculation of the conductivity profile from the as-measured average values proceeds from the first layer close to the buried insulator. The difficulty in determining this first layer in very thin and resistive films represents the main source of computational errors. Furthermore, the carrier concentration profile is sometimes abusively extrapolated by assuming homogeneous mobility and same mobility-doping laws as for bulk silicon.

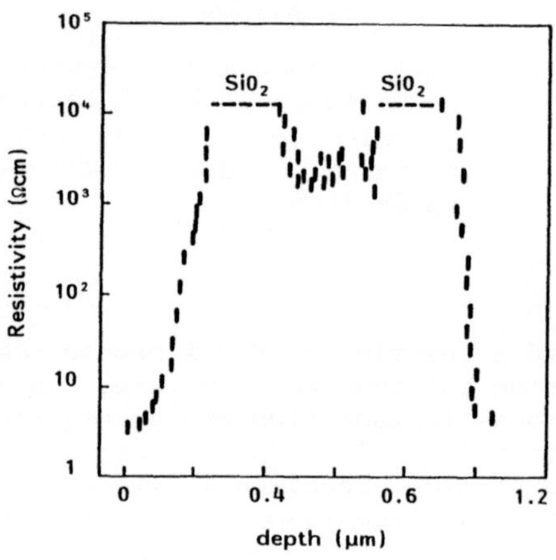

Fig. 3. Resistivity profiles deduced from spreading resistance in a SIMOX structure synthesized by double oxygen implantation.

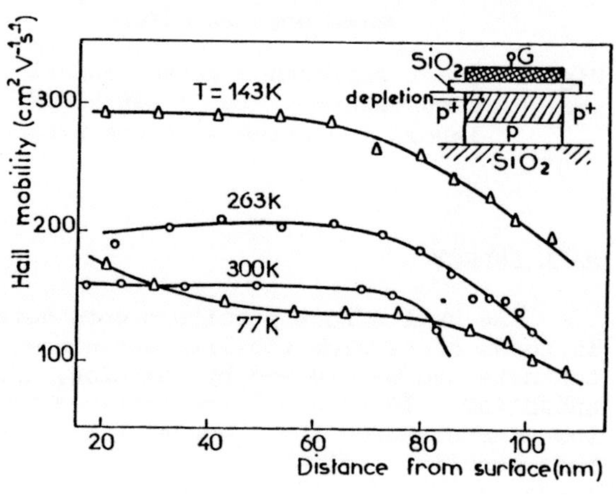

Fig. 4. Hall mobility profiles for holes in LTA-SIMOX ($N_A = 10^{17}cm^{-3}$).

In spite of these limitations, spreading resistance has proved to be a valuable method to verify the vertical configuration of SIMOX and the thickness of different layers[9,13]. In Figure 3, for instance, is shown the interesting case of a double SIMOX structure, formed by two successive oxygen implantations at different energies[14]. Two distinct buried oxides, 0.2 µm thick, are indeed visible ; the sandwiched Si layer has poorer quality and higher resistivity than the Si overlayer. Spreading resistance was also able to reveal that thermal donors are generated at 450°C not only in the Si film but also in the substrate close to the buried oxide.

Accurate profiles of the transport parameters along the vertical direction were determined in LTA-SIMOX using depletion-mode MOS-Hall devices[15]. The Hall effect was measured while gradually varying the gate voltage to reduce the thickness of the conductive region of the SIMOX film (fig. 4). An adequate differentiation technique provides the local values of carrier concentration and mobility at each point. MOS capacitance measurements are simultaneously performed to find the relation between gate voltage and depletion depth. The experiment shows that in the upper 100 nm of the film the mobility is high, nearly constant and normal scattering mechanisms occur. Closer to the synthesized oxide, the mobility falls rapidly.

At low temperature a discrepancy is observed between the profile of ionized impurities (deduced from capacitance curves) and that of the free carriers provided by the Hall effect. The explanation is based on the existence of small potential fluctuations (< 10 meV) induced by crystal imperfections which increase towards the buried oxide[15].

Although these measurements have not yet been repeated on HTA-SIMOX, it is expected, that there will be a quasi-flat profile. This anticipation is supported by the very large value of the average mobility and by the small difference (< 30 %) between the carrier mobilities in the inversion layers situated at the front and back interfaces[7].

The homogeneity of the buried oxide has been inspected by differential ellipsometry and in situ comparison with a thermal oxide[16,17]. The method (fig. 5) consisted in the thermal oxidation of the Si overlayer to obtain superposition of both oxides. The reflection angles were then measured during a progressive step by step oxide etching. The experiment does not reveal any differences between the optical properties of the thermal and synthesized oxides. The absence of any discontinuity at the boundary of the two oxides demonstrates experimentally that the buried oxide index is constant and identical to that of the thermal oxide.

A salient feature of SIMOX is the presence of a transition layer, 10-25 nm thick, at the buried oxide-bulk Si substrate[16] interface. This layer is neither pure SiO_2 since it is not consumed by the etchant, nor pure Si because it is measurable by ellipsometry. Its refractive index appears to be higher (2.5-3) than for SiO_2 and inhomogeneous. It must be underlined that the transition layer makes inappropriate the use of standard ellipsometry tables (calculated for the system thermal SiO_2-Si) in the analysis of the experimental data on SIMOX and can be responsible for significant errors.

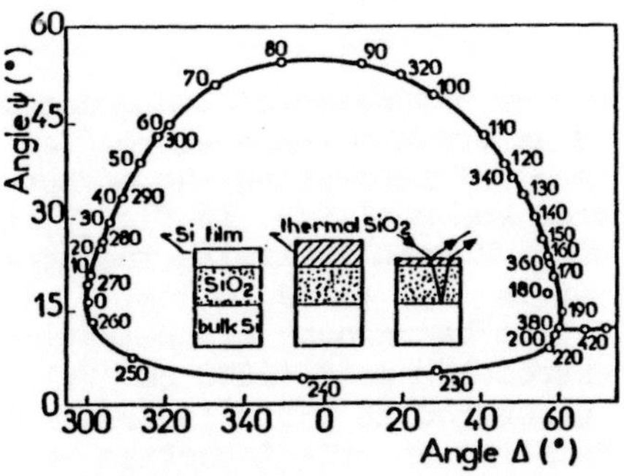

Fig. 5. Experimental ellipsometry points after various etching times ; the final points indicate the presence of a transition layer (after Levy et al.[16]).

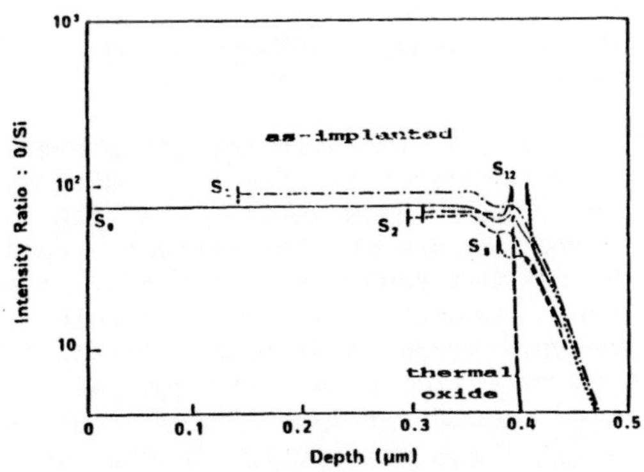

Fig. 6. SIMS oxygen ion profiles in the buried oxide of as-implanted SIMOX; subscripts correspond to etching time (minutes) (after Chater et al.[18]).

Further investigation of the transition layer was made by correlating differential ellipsometry, TEM and SIMS measurements[18]. Several samples were prepared by removing the Si overlay first and etching the buried oxide for different periods of time. The compositional profile obtained by SIMS (fig. 6) shows an interfacial region composed of two layers : the first of silicon with SiO_2 precipitates and the second of entirely SiO_2. This second layer, 20-30 nm thick and almost completely isolated from the buried oxide, corresponds well to the transition region revealed by ellipsometry. From the SIMS point of view the annealed interface layer is comparable with the thermal oxide-bulk silicon interface but twice as large. Prior to annealing the interface region is broader (30-50 nm) and consists of interpenetrating fingers of Si and SiO_2 also observed by TEM[18].

OXYGEN DONORS

Oxygen becomes electrically active in Si, after annealing at 450-750°C, and forms a number of donor species. As these free carriers remain mobile even at room temperature, the intentional doping level is modified and the device performance degraded. This problem is vital in SIMOX due to the high dose of deposited oxygen. It has been shown by spreading and 4-point probe analysis that in LTA-SIMOX, short additional anneals (30 minutes) are able to lower the wafer resistivity by orders of magnitudes. Two types of donors are distinguished[10] : thermal donors, related to the formation of more or less complicated oxygen-silicon aggregates between 450-550°C, and new donors at 750°C. At 650°C and above 900°C these donors are destroyed and the sample recovers its initial resistivity. This "normal" behaviour is modified by the presence of carbon : after carbon implantation in SIMOX the new donor formation is greatly enhanced.

Another important parameter is the implant temperature which governs the residual oxygen concentration in the SIMOX film. Due to inhomogeneous distribution of the temperature across the wafer, the density of oxygen-related donors ($1\text{-}2\times10^{16}cm^{-3}$), which are unintentionally formed during the processing of integrated circuits, increases from the edges of the wafer to the center[19]. This is demonstrated by shifts in the threshold voltages which correspond to an increase of both the overdoping of p-channel MOSFETs and the compensation of n-channel MOSFETs from the edge to the center. The device transconductance is also affected via a doping-induced (inhomogeneous) modification of the effective channel length.

Recent analysis has been carried out on HTA material[7]. Samples were prepared for Hall effect and then annealed at various temperatures for one hour. This measurement has the advantage of separating the effects of oxygen donors on the carrier concentration and mobility. In fig. 7 it is shown that annealing at 1350°C in Ar ambient hopefully limits the thermal donor generation at a reasonable level of a few times $10^{15}cm^{-3}$. After neutralization of thermal donors at 650°C, the background contamination is very low ($10^{15}cm^{-3}$) and the mobility extremely good (1250 cm^2/Vs). The large number of new donors ($10^{17}cm^{-3}$) generated at 750°C is presumed to result from positively charged states existing at the interface between small SiO_2 precipitates (< 5 nm) and the Si matrix. As a consequence, small accumulation layers are formed and electron transport occurs by percolation. The main portion of these new donors are probably situated in a narrow region close to the buried oxide[11]. This model is supported by the variation of the Hall mobility at low temperature which shows interface-like scattering processes above 100 K (fig. 9) and degenerated behaviour below 50 K.

Further analysis is necessary to complete our understanding of oxygen-related donors. Indeed, in other HTA materials examined the density of new donors is lower ; an argument, to be verified by SIMS, could be the existence of a lower degree of contamination with carbon, during the oxygen implant. The implantation of higher doses of oxygen seems to reduce paradoxically the residual oxygen concentration in the film, attenuating the donor activation as well.

MOS CAPACITANCE MEASUREMENTS

There are three alternatives to analyse the MOS capacitance and conductance in SIMOX film : between gate and substrate, gate and film or substrate and film. Typical examples are shown in figure 6 for capacitance measured between one contact and the two others (interconnected). Moreover, the third terminal can be independently biased to impose carrier inversion or accumulation at an interface. These experiments provide a significant amount data, very useful to characterize the three interfaces of the structure, the thickness of the oxides and the doping. It has been, however, demonstrated that each type of measurement involves simultaneous contributions from all of the interfaces, oxides and depletion layers.

This convolution explains the unusual variation of the high frequency capacitance in SIMOX (fig. 8) compared to standard C-V curves in bulk Si. For instance the gate-film capacitance $C_{G/FS}$, is normally measured to characterize the front interface, but internal capacitative links between the substrate, gate and film make this experimental curve very different from the pure curve C_1. Such errors depend upon geometrical dimensions, doping and oxide thicknesses and can become prohibitive as in fig. 8. It is clear that the experimental data must be corrected, using an appropriate equivalent circuit, in order to avoid the extraction of very erroneous parameters for interfaces and oxides.

A theoretical treatment has been recently proposed[20] and the corrected C-V curves are plotted in fig. 8 : C_1 for the front interface and $C_{2,3}$ for the two buried interfaces (involved in the capacitance between substrate and film). The extracted thicknesses of the gate and buried oxides show now good agreement with the processing conditions. These corrected curves, together with measurements of MOS conductance $G(\omega)$[17], have served to determine interface parameters which are summarized in fig. 10. Nevertheless the correction procedure does not hold for totally depleted films, where numerical simulations are required to analyse the correlation between I_1 and I_2.

Another major limitation of C-V measurements in SOI is due to the high and non-constant value of series resistances. This problem is alleviated using interdigitated capacitors or by removing the Si overlayer by plasma etching. In this case, an Al gate is evaporated onto the buried oxide and the properties of the interface I_3 are inferred from capacitance-voltage and transient capacitance curves[13]. The sum of these difficulties explains why in SIMOX, unlike in bulk Si, the capacitance and conductance techniques have been the object of limited interest. The same conclusion holds for the DLTS method which has been only tentatively applied. To simplify the experiment the thickness of the Si overlay was significantly increased to a few microns by over-epitaxy[21]. Unfortunately, the relevant DLTS data applies to the epitaxial layer, not to the original SIMOX film.

MOS TRANSISTORS

Interface characterization

The very good performance of MOSFETs fabricated on SIMOX[17,22,23] was demonstrated early on. Static I(V) characteristics from weak to strong inversion show that the carrier mobility in the inversion layer and the defect concentration are only a few percent inferior to those obtained in control transistors fabricated on bulk Si wafers in the same batch. Therefore, the implantation-induced damage at the front interface has been almost totally erased during the anneal. It follows that the properties of this interface cannot be significantly improved after HTA process (fig. 9).

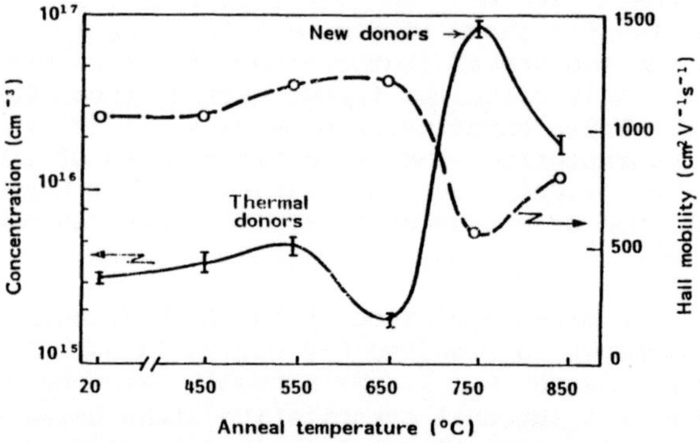

Fig. 7. Carrier concentration and Hall mobility versus temperature for additional anneals in HTA-SIMOX samples containing oxygen donors.

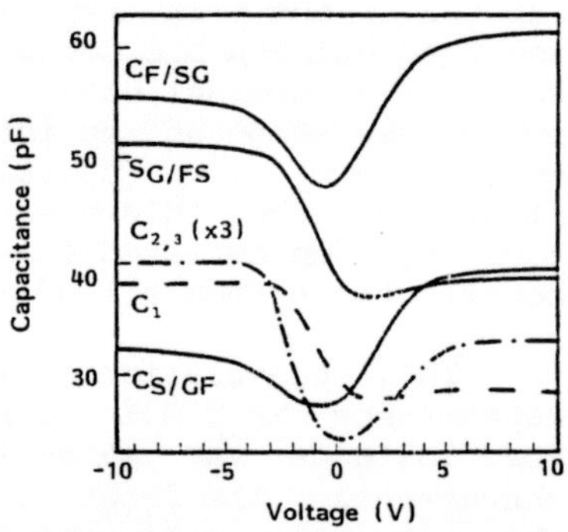

Fig. 8. Typical C-V curves measured at 100 kHz in LTA-SIMOX (full lines) and corrected curves[20] (dotted lines).

The existence of floating islands makes the operation of MOS transistors somewhat different in SIMOX and bulk Si. Substrate biasing is helpful to force the buried interface I_2 into accumulation, depletion or inversion regimes. This can be used to control the onset of the kink effect as well as leakage currents flowing mainly at the Si island edges. Indeed in edgeless transistors with close geometry the leakage is very small (< 10 pA/μm for $L_{eff} = 0.6$ μm and $V_D = 5$ V). On the other hand, the subthreshold swing improves from accumulation to depletion being a minimum for total depletion[7].

For the purpose of characterization, the roles of gate and substrate can be interchanged. The substrate becomes then a secondary gate which operates the inversion channel formed at the buried interface I_2. The static characteristics of the back channel transistor greatly depend on the annealing temperature (fig. 9). After LTA the density of interfacial defects exceeds $10^{13}cm^{-2}$, while the mobility is very poor and decreases further with lowering implant temperature and dose[23]. The improvement of the back interface induced by HTA is emphasized by the low density of fast interface states ($< 10^{12}cm^{-2}eV^{-1}$) ; as a consequence, the slope in weak inversion is increased and the threshold voltage is reduced from 50 V to 5-10 V. Figure 9 shows that the back channel mobility now becomes comparable with the front channel mobility, not only in magnitude (10-30 % lower) but also in dependence on temperature[7].

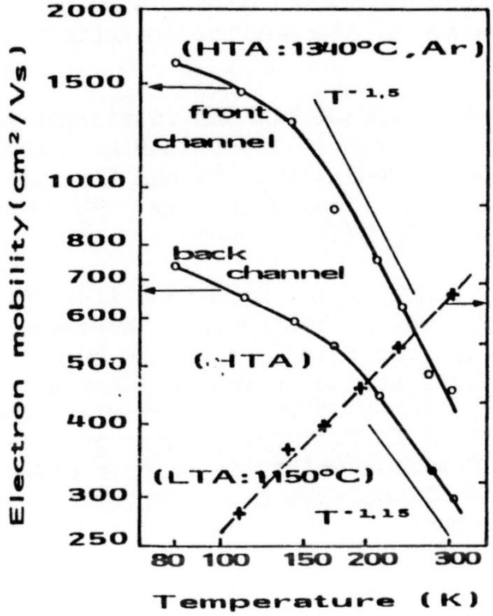

Fig. 9. Electron mobility versus temperature for inversion channels situated at the front and buried interfaces (after Cristoloveanu et al.[7])

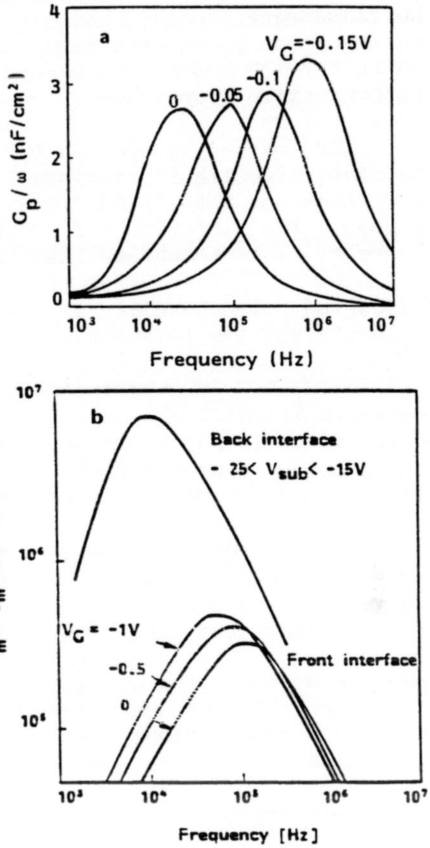

Fig. 10. $G_p(\omega)/\omega$ versus frequency deduced from dynamic transconductance measurements in enhancement (a) and depletion-mode (b) SIMOX-MOSFETs.

The **dynamic transconductance** technique has been recently proposed for accurate characterization of interface states in bulk Si MOSFETs[24]. A small frequency-dependent signal is used to modulate the gate voltage in weak inversion regime. The imaginary part of the dynamic transconductance provides the same (but greatly amplified) information as that found from conductance measurements $G(\omega)/\omega$ on MOS capacitors. The method has been adapted to short-channel SIMOX-MOSFETs and typical bell-shaped curves are plotted in fig. 10a. The magnitude and position of the peak are related, respectively, to the density and time constant of interface states. Figure 10a is used to determine the energy distribution of fast states in the gap. In some cases, two peaks are observed : one is frequency and voltage dependent and corresponds to interface states, while the other is almost constant and was associated with the presence of bulk traps in SIMOX implanted at low temperatures[25].

Results obtained by generalizing the dynamic transconductance method to depletion-mode transistors are shown in fig. 10b. An average state density of $10^{11}cm^{-2}eV^{-1}$ was deduced for the front interface. The peak corresponding to the back interface does not change either in position or in magnitude being presumably dominated by bulk traps.

Recombination Properties

The method proposed by Vu and Pfister[26] to measure the minority carrier lifetime in recrystallized SOI requires depletion-mode transistors and films not totally depleted. The gate being pulsed in the deep depletion region, the lifetime is determined by monitoring the transient drain current. This technique has been applied to thin SIMOX films highly doped to avoid depletion from one interface to another. A further improvement, discussed elsewhere[25], allows determination of the surface recombination velocity and the lifetime profile.

Lifetime is short in LTA material : 10-100 ns, as in silicon on sapphire films[27]. The superior quality of HTA-SIMOX is emphasized by a very high value (2 μs), also supported by photo-conductivity[12] and transient capacitance measurements[13]. The recombination velocity at the front interface has a normal value for the system SiO_2-Si.

These recombination properties in SIMOX have been qualitatively confirmed using double-injecting P^+NN^+ diodes. The forward current is governed by electron-hole recombination in the film and at the back and front interfaces. Applying a magnetic field shifts carriers towards one interface : the recombination rate is increased at this interface and reduced at the opposite interface. The current is modified according to the direction of the magnetic field. The magnetosensitivity of this device called a **magnetodiode** is proportional to the difference in recombination velocities at the front and buried interfaces[27].

Noise Measurements

Low frequency noise in bulk Si MOSFETs was demonstrated to provide valuable information about the interface quality in terms of fast and slow states. Presented here are recent noise experiments in SIMOX[28]. The fluctuations of the drain voltage are measured in the range 0.1 Hz-100kHz using a digital spectrum analyser and then converted in a "normalized" drain current spectral density S_I/I^2. Front and back channels are independently investigated by creating an accumulation layer at the opposite interface.

Typical spectral densities are plotted in fig. 11. In LTA material, the noise exhibits a 1/f spectrum in agreement with the theory[29]. For the weak inversion region, S_I/I^2 is almost independent of gate and drain bias, which demonstrates the lateral homogeneity of each interface. A quadratic decrease with gate voltage is found from weak to strong inversion. The noise magnitude is proportional to the concentration of slow states. At the back interface, it greatly exceeds that at the front interface (fig. 11a), although this difference is attenuated in edgeless transistors. It has been found that a smooth increase of the implantation temperature (from 490°C to 505°C) reduces the noise by one order of magnitude. This beneficial influence is also visible by TEM[30] and correlates well with the increase in the back channel mobility observed earlier[23].

The noise behaviour is far more surprising in HTA-SIMOX, where a $1/f^2$ spectrum is found at low frequency followed by a "normal" 1/f dependence (fig. 11b). The $1/f^2$ spectrum corresponds to a strongly emerging burst noise which is probably induced by dislocations. These dislocations traverse the film completely and generate, therefore, the same noise behaviour at both interfaces.

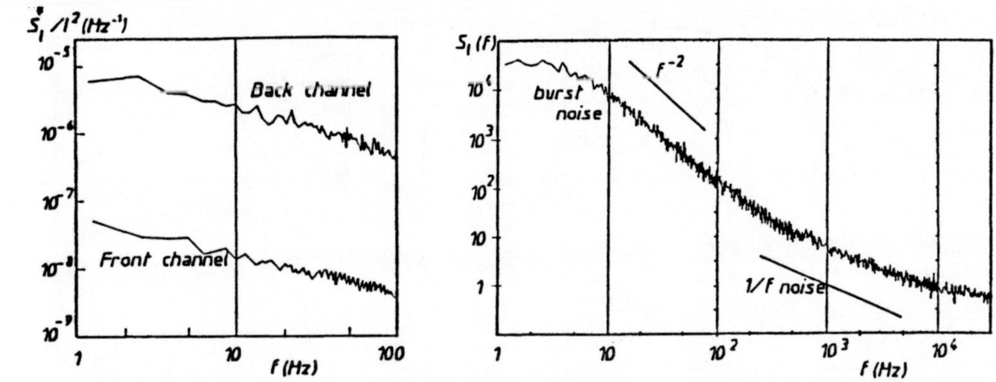

Fig. 11. Subthreshold drain current spectral density in SIMOX-MOSFETs annealed at low (a) and high (b) temperatures.

Volume Inversion MOS Transistor (VI-MOSFET)

The special multi-interface structure of SIMOX substrates gives the exciting opportunity for new devices to be conceived. The VI-MOSFET has a normal transistor configuration but is operated with simultaneous bias on both gates : $V_{G1} = KV_{G2}$, where K accounts for the difference in thickness between the gate oxide and buried oxide[31]. In thick films, the two inversion channels grow independently, whilst the middle of the film remains undepleted and governed by majority carriers : The behaviour is different in low-doped SIMOX films, 0.1-0.2 µm thick, where the two interfaces interact. The inversion channels make the whole film volume to go into strong inversion as well due to continuity.

The fact that in VI-MOSFETs minority carriers are no more confined to an interface offers many advantages[31]. In fig. 12 are compared the experimental characteristics of a VI-MOSFET (K = 10) with those of a normally operated transistor (K = 0). The greatly increased number of carriers allows higher current in VI-MOSFETs. Furthermore, the reduced influence of interfacial defects improves the subthreshold swing from 70 mV/decade (for K = 0) to 29 mV/decade and suggests a spectcaular switching capability (fig. 12a). The carrier mobility being much higher in the volume than at the interface, the transconductance of the VI-MOSFET is enhanced by 80 % (fig. 12b).

Similar improvements have been obtained for the case of depletion-mode transistors with volume accumulation. The double SIMOX structure shown in fig. 3 could be used to form independent secondary gates for each transistor. The Si layer lying between the two buried oxides acts as a gate and can be divided up by oxidation of the unused portion of this film.

CONCLUSION

In this review of the electrical properties of SIMOX, a number of key aspects have been identified. The top region of the Si overlay is always of high quality (good mobility, low density of defects) and accepts integrated-circuit processing. In LTA-SIMOX, the crystallinity drops rapidly in the region of the buried oxide (which becomes a "shielding" layer[32]). The main consequences of HTA are : the existence of a homogeneous

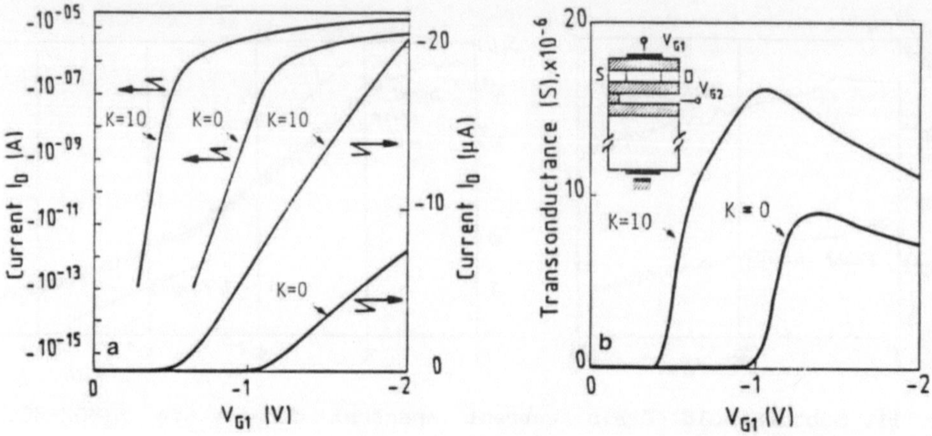

Fig. 12. Experimental I-V characteristic (a) and transconductance (b) for a p-type volume-inversion VI-MOSFET fabricated on SIMOX (after Balestra et al.[31]).

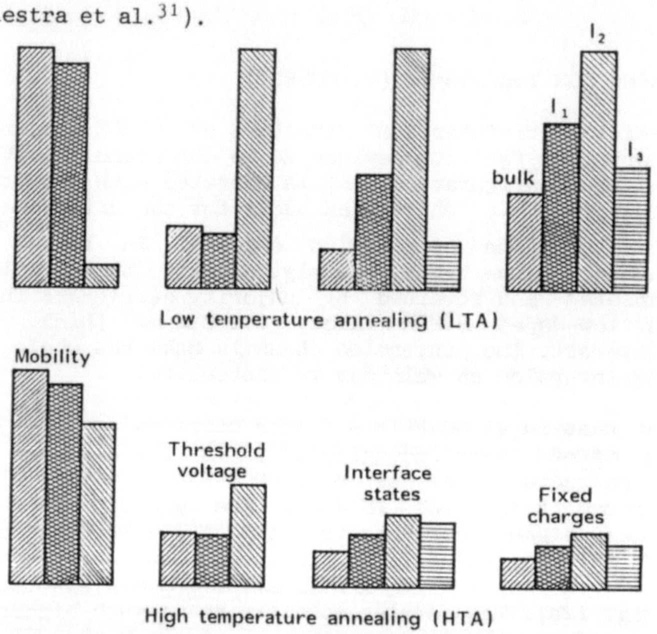

Fig. 13. Schematic comparison of the properties of Si-SiO_2 interfaces in bulk Si and SIMOX (I_1 : gate oxide-film, I_2 : film-buried oxide, I_3 : buried oxide-substrate) after low and high temperature anneals.

film with a low concentration of defects[33,34], very high carrier mobilities in the volume (1250 cm^2/Vs) and at both interfaces (fig. 13), reduced contamination and enhanced carrier lifetime. Residual dislocations have not been observed to influence the electrical properties other than by the sharp mobility decrease below 100 K. and the emergence of $1/f^2$ noise. Nevertheless, it is not obvious at all that this "ideal" HTA material fits all applications. At least, the construction of power devices[32] and sensors will exploit the opportunity for **interface engineering** offered by the lowering of the anneal temperature.

The activation of thermal donors can be greatly reduced by HTA, to levels that no longer interfere with intentional doping. Questions still exist, however, regarding the origin of new donors and their relation to very small SiO_2 precipitates[35] and carbon. The buried oxide has normal and homogeneous optical properties except at the boundary with the Si substrate where a thin transition layer has been observed.

Although only a few possibilities have yet been explored, SIMOX already appears to be a versatile structure allowing the conception of new devices with very high performance (VI-MOSFET). The state of the art of SIMOX technology offers comforting promise for its future role in microelectronics.

ACKNOWLEDGEMENTS

The author is pleased to thank his colleagues (F. Balestra, M. Benachir, B. Boukriss, J. Brini, A. Chovet, T. Elewa, S. Gardner, M. Gri, H. Haddara) involved in the electrical characterization of SIMOX at the PCS Laboratory. Gratitude for scientific interest, sample preparation and joint research is adressed to P. Hemment and K. Reeson from the University of Surrey (Guildford), C. Jaussaud, M. Bruel, J. Margail and A. Auberton from LETI (Grenoble), J. Davis and P. Arrowsmith from British Telecom, J. Kilner and R. Chatter from Imperial College (London), G. Celler from Bell Laboratories and I. Stoimenos from the University of Thessaloniki.

REFERENCES

1. P.L.F. Hemment, E. Maydell-Ondruesz, K.G. Stepens, J.A. Kilner, and J. Butcher, Vacuum, 34:203 (1984).
2. C.G. Tuppen, M.R. Taylor, P.L.F. Hemment, and R.P. Arrowsmith, Thin Solid Films, 131:233 (1985).
3. P.L.F. Hemment, in : "Semiconductor On Insulator and Thin Film Transistor Technology", A. Chiang, M.W. Geis, and L. Pfeiffer, eds. Mat. Res. Soc. Pittsburgh, 207 (1986).
4. J. Margail, I. Stoemenos, C. Jaussaud, M. Dupuy, P. Martin, B. Blanchard, and M. Bruel, in : "Energy Beam Solid Interactions and Transient Thermal Processing", V.T. Nguyen and A.G. Cullis, eds. Les Editions de Physique, Les Ulis, 519 (1986).
5. J. Stoemenos, C. Jaussaud, M. Bruel, and J. Margail, J. Crystal Growth, 73:546 (1985).
6. G.K. Celler, P.L.F. Hemment, K.W. West, and J.M. Gibson, Appl. Phys. Lett. 48:532 (1986).
7. S. Cristoloveanu, S. Gardner, C. Jaussaud J. Margail, A.J. Auberton-Hervé, and M. Bruel, J. Appl. Phys., in press (1987).
8. A. Mogro-Camperao, R.P. Love, N. Lewis, E.L. Hall, and M.D. McConnel, J. Appl. Phys., 60:2103 (1986).
9. S. Cristoloveanu, J. Wyncoll, P. Spinelli, P.L.F. Hemment, and R.P. Arrowsmith, Physica, 129B:249 (1985).
10. S. Cristoloveanu, J. Pumfrey, E. Scheid, P.L.F. Hemment, and R.P. Arrowsmith, Electron. Lett. 21:18 (1985).
11. F. Vettesse, J. Sicart, and J.L. Robert, Internal GCIS Colloquium, Toulouse, France (1987).
12. G. Papaioannou, unpublished work (1987).
13. A. Mogro-Campero, J. Appl. Phys. 61:639 (1987).
14. G.K. Celler, J.L. Batstone, K.W. West, P.L.F. Hemment, and K.J. Reeson, IEEE SOS/SOI, Technology Workshop, Captiva, Florida (1986).

15. S. Cristoloveanu, J.H. Lee, J. Pumfrey, J.R. Davis, R.P. Arrowsmith, and P.L.F. Hemment, J. Appl. Phys., 60:3199 (1986).
16. M. Levy, E. Scheid, G. Kamarinos, J.R. Davis, R.P. Arrowsmith, M.R. Taylor, and P.L.F. Hemment, in "Energy Beam Solid Interactions and Transient Thermal Processing", V.T. Nguyen and A.G. Cullis, eds. Les Editions de Physique, Les Ullis, 495 (1986).
17. R.J. Chater, J.A. Kilner, E. Scheid, S. Cristoloveanu, P.L.F. Hemment, and K.J. Reeson, Proc. Int. Conf. Insulationg Films on Semiconductors, Leuven, Belgium (1987).
18. J. Davis, M. Taylor, G. Spiller, P. Skevington, and P. Hemment, Appl. Phys. Lett. 48:1279 (1986).
19. J.H. Lee, and S. Cristoloveanu, IEEE Elect. Dev. Lett. EDL-7:537 (1986).
20. P.K. McLarty, J.W. Cole, J.F. Galloway, D.E. Ioannou, and S.E. Bernaki, IEEE SOS/SOI, Technology Workshop, Captiva, Florida (1986).
21. J.P. Collinge, K. Hashimoto, T. Kamins, S.Y. Chiang, E.D. Liu, S. Peng, and P. Rissman, IEEE Elect. Dev. Lett., EDL-7:279 (1986).
22. D.J. Foster, A.L. Butler, P.H. Bolbot, and J.C. Alderman, IEEE Trans. Elect. Dev., ED-33:354 (1986).
23. H. Haddara, and G. Ghibaudo, European Solid State Device Res. Conf. Cambridge, UK (1986).
24. T. Elewa, H. Haddara, and S. Cristoloveanu, this proceedings.
25. D.P. Vu, and J.C. Pfister, Appl. Phys. Lett., 47:950 (1985).
26. S. Cristoloveanu, A. Chovet, and G. Kamarinos, Solid State Electron., 21:1563 (1978).
27. A. Chovet, B. Boukriss, T. Elewa, and S. Cristoloveanu, Proc. of 9th Conf. "Noise in Physical Systems", Montreal (1987).
28. G. Reimbold, IEEE Trans. Electr. Dev., ED-31:1190 (1984).
29. A.E. White, K.T. Shart, J.L. Batstone, D.C. Jacobson, J.M. Poate, and K.W. West, Appl. Phys. Lett., 50:19 (1987).
30. F. Balestra, S. Christoloveanu, M. Benachir, J. Brini, and T. Elewa, IEEE Elect. Dev. Lett. in press (1987).
31. S. Nakashima, Y. Maeda, and M. Akiya, IEEE Trans. Elect. Dev., ED-33:126 (1986).
32. B.Y. Mao, P.H. Chang, H.W. Lam, B.W. Shen, and J.A. Keenan, Appl. Phys. Lett., 48:794 (1986).
33. P.H. Chang, and B.Y. Mao, Appl. Phys. Lett., 50:152 (1987).
34. A.H. Van Ommen, B.H. Koek, and M.P.A. Viegers, Appl. Phys. Lett., 49:1062 (1986).

FORMATION MECHANISMS AND STRUCTURES OF THIN BURIED LAYERS OF SiO_2 FABRICATED USING ION BEAM SYNTHESIS

K.J. Reeson, P.L.F. Hemment, R.J. Chater[1], J.A. Kilner[1]
C.D. Marsh[2], G.R. Booker[2], J.R. Davis[3] and G.K. Celler[4]

Department of Electronic and Electrical Engineering
University of Surrey, Guildford, Surrey, U.K.
[1]Department of Materials, Imperial College, London, U.K.
[2]Department of Metallurgy and the Science of Materials
University of Oxford, Oxford, U.K.
[3]BTRL, Martlesham Heath, Ipswich, U.K.
[4]AT&T Bell Laboratories, Murrey Hill, New Jersey, U.S.A.

ABSTRACT

The effects of variations in the implanted dose Φ_i and the annealing temperature T_A during the formation of buried layers of SiO_2 produced by ion beam synthesis (IBS) are discussed. It is found that when specimens implanted with oxygen doses below $0.3 \times 10^{18} O^+ cm^{-2}$ are annealed at 1405°C for 30 minutes a buried layer consisting of two discontinuous rows of SiO_2 precipitates is produced. For higher doses these two layers coalesce to give a single continuous layer of SiO_2. The interfacial irregularity of this layer is found to decrease with increasing dose such that for doses of $\geq 1.8 \times 10^{18} O^+ cm^{-2}$ the interfaces are almost planar. Studies of these higher dose specimens annealed in the temperature range 1280-1300°C for times of 2-20 hours show that the crystal quality of the silicon overlay improves with increasing temperature and time. The oxygen concentration in the silicon overlayer also appears to increase with annealing time and mechanisms are proposed to explain the observed phenomena.

INTRODUCTION

In recent years, silicon on insulator (SOI) structures produced by ion beam synthesis (IBS) have become increasingly attractive as substrates for VLSI device applications[1]. This is because they offer increased operating speeds, low electrical leakage, a decrease in the number of processing steps and immunity from parasitic transistor action (latch-up)[2]. In addition recent studies have also pointed to increased radiation tolerance[3]. These SOI structures are fabricated by implanting high doses of energetic oxygen ions into silicon substrates maintained at a temperature of approximately 550°C. After implantation the specimens are annealed at high temperatures, so that the resulting structure consists of a buried layer of SiO_2, with abrupt interfaces overlaid by high quality single crystal silicon.

A major breakthrough, which enabled buried oxide structures formed by IBS, to become one of the leading SOI technologies, came with the work of Margail et al.[4] at LETI in 1985. They showed that by increasing the annealing temperature and time from 1150°C for two hours to 1300°C for six hours it was possible to remove most of the implanted oxygen from the superficial silicon layer into the buried oxide layer. This phenomenon was shown, by experiment, to occur via a process of diffusion controlled precipitate growth. Later work confirmed that similar results could be obtained by annealing the specimens at higher temperatures for shorter times for example 1405°C for thirty minutes[5]. Test devices (CMOS) fabricated on this type of structure gave extremely encouraging results[6], which confirmed both the quality of the silicon overlayer and the suitability of this type of substrate for future generations of small geometry integrated circuits.

In this paper we report experiments which look at the evolution of the buried SiO_2 layer with increasing dose, anneal temperature and time. This is in order to identify firstly, the processing conditions which will allow very thin (< 1000 Å) planar layers of SiO_2 to be formed and secondly to establish the optimum conditions for annealing these substrates at high temperatures.

EXPERIMENTAL

100 device quality single crystal silicon wafers were implanted with 200 keV oxygen ions to doses in the range $0.1 \times 10^{18} O^+ cm^{-2}$ to $1.8 \times 10^{18} O^+ cm^{-2}$. During implantation the specimens were maintained at a temperature of between 540°C and 560°C using ion beam heating. After implantation the wafers were capped with 3000 Å of SiO_2 and annealed at temperatures in the range 1280-1405°C for times of 0.5-20 hours. All the specimens were annealed in a conventional nitrogen flow furnace with the exception of those annealed at 1405°C where a incoherent halogen lamp annealer was used[5]. Details of the implantation and annealing conditions are listed in Table 1.

TABLE 1

Specimen	Implantation: Dose $O^+ cm^{-2}$	Energy keV	Temperature °C	Annealing: Temperature °C	Time hrs
1	0.25×10^{18}	200	540	1405	0.5
2	0.35×10^{18}	200	550	1405	0.5
3	0.50×10^{18}	200	540	1405	0.5
4	0.60×10^{18}	200	560	1405	0.5
5	0.90×10^{18}	200	550	1405	0.5
6	1.8×10^{18}	200	540	1405	0.5
7a	1.8×10^{18}	200	540	1280	2
7b	1.8×10^{18}	200	540	1300	2
7c	1.8×10^{18}	200	540	1300	20

Specimens were analysed using Rutherford backscattering (RBS) and ion channelling to assess the impurity depth distribution and the crystallinity of the silicon overlayer. Secondary ion mass spectrometry (SIMS) provided a quantitative analysis of the oxygen depth distribution and cross sectional transmission electron microscopy (XTEM) provided information about the composition and stoichiometry of the buried layer and the nature of the defects in the silicon overlayer.

RESULTS AND DISCUSSION

Dose Dependence of Structure

Figure 1 shows the non channelled (fig. 1a) and channelled (fig. 1b) RBS spectra, for specimens 1,2,3 and 5 after annealing at 1405°C for 0.5 Hours. The low dechannelled fraction between channels 230 and 280 in the channelled spectra (fig. 1b) indicates that the silicon overlayer is composed throughout its thickness of high quality single crystal silicon. In the non-channelled spectra (fig. 1a) the dip between channels 190 and 220 gives a qualitative determination of the oxygen depth distribution. For the two highest doses the dip is steep sided and saturates at a depth commensurate with stoichiometric SiO_2. For the lower doses the dip is more triangular in profile and does not appear to saturate. this indicates a gradation in the oxygen concentration within the buried layer, suggesting that the buried layer may be discontinuous.

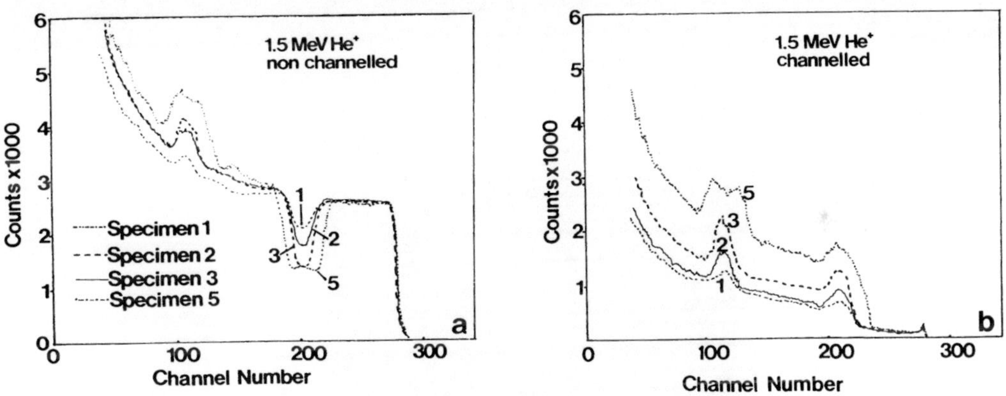

Fig. 1.
Part a. Channelled RBS spectra for specimens 1,2,3 and 5 after annealing at 1405°C for 0.5 hours.
Part b. Non channelled RBS spectra for specimens 1, 2, 3 and 5 after annealing at 1405°C for 0.5 hours.

Figure 2 shows XTEM micrographs for specimens 1,2,3,4 and 6 after annealing at 1405°C for 0.5 hours. For the lowest dose, two layers of SiO_2 precipitates are observed. These occur at depths of 3000 Å and 4900 Å, which are respectively commensurate with the calculated depths for peaks in the damage and impurity range profiles, for the implanted ions[7]. The precipitates associated with the damage distribution (layer A) are smaller and more spheroidal in shape than those associated with the range profile (layer B). The latter are usually polyhedral and some display an octahedral morphology similar to that observed in Czochralski silicon after prolonged annealing[8]. The precipitates are linked via threading dislocations which run both within and between the layers of precipitates.

As the dose is increased the two layers of precipitates coalesce firstly to form a discontinuous oxide layer and then a continuous layer with irregular interfaces[9]. The irregularity of the interfaces is a legacy of the two sites of nucleation involved in the formation of the structure. With increasing dose the interfacial irregularity diminishes such that for doses in excess of $1.8 \times 10^{18} O^{+} cm^{-2}$, at 200 keV, the interfaces are atomically abrupt with the only vestige of the lower dose irregularity being manifested by the inclusion of small silicon islands within the buried oxide. These islands are found to be slightly misorientated with respect to the silicon matrix[10]. It is postulated that this is due to viscous flow of the buried SiO_2 layer during annealing at 1405°C for 0.5 hours.

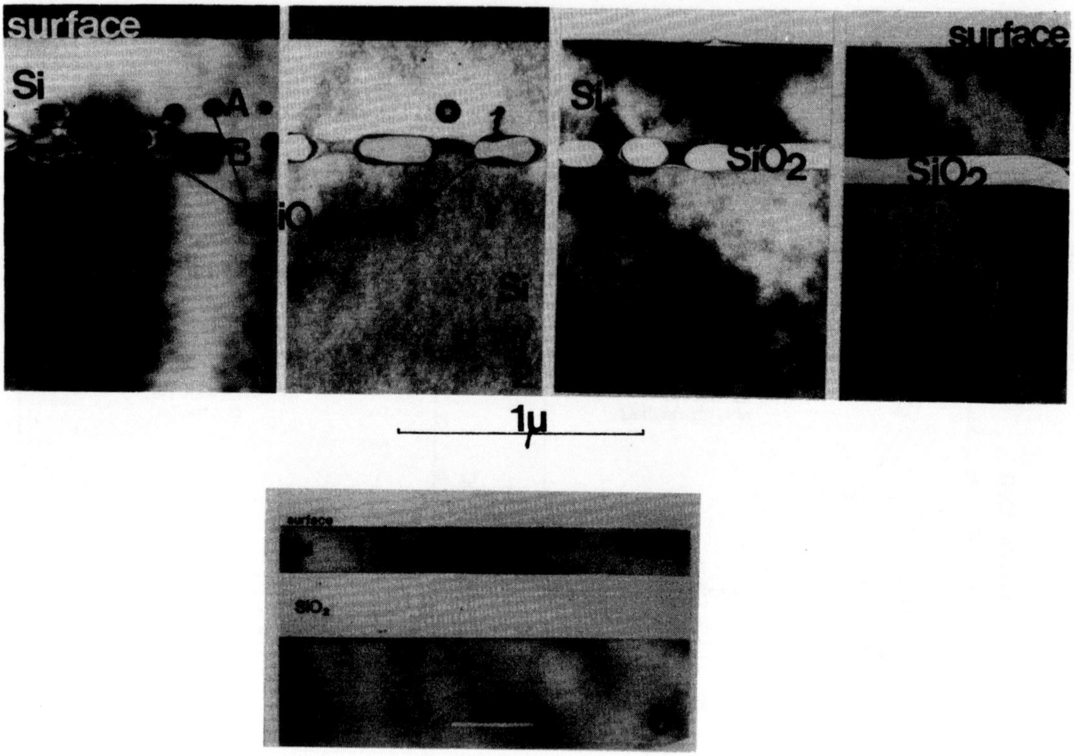

Fig. 2. XTEM micrographs showing the development of the buried oxide structure with increasing dose. Micrographs refer to specimens 1,2,3,4 and 6 after annealing at 1405°C for 0.5 hours.

Dependence of Structure on Annealing Temperature

Figure 3 shows the results of ion channelling studies in the silicon overlayer of parts a, b and c of Specimen 7. These show that as the annealing temperature and time are increased the crystal quality of the silicon overlayer improves, as demonstrated by specimens a)-c). In all three cases the silicon surface region is composed of high quality single crystal material $\chi_{min} = 4$ %. At greater depths, however, the dechannelled fraction increases indicating that the silicon overlayer is becoming progressively more defective. The increased dechannelling rate with depth is most apparent for specimen a) which was annealed at the lowest temperature and least apparent for specimen c) which was annealed at the highest temperature for the longest time.

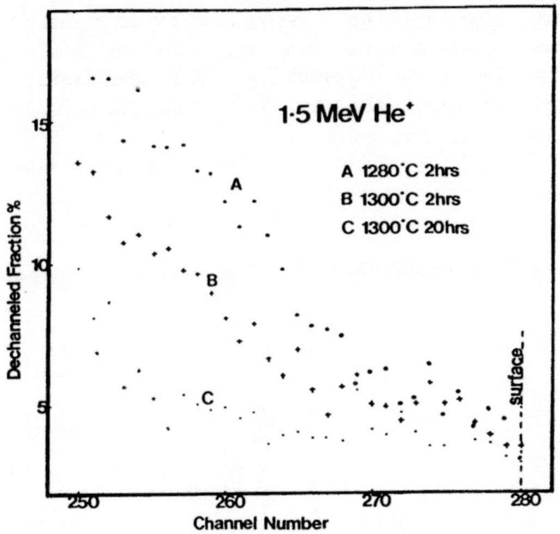

Fig. 3. RBS results for specimen 7 parts a), b) and c). Data shows the dechannelled fraction in the silicon overlay for anneals carried out at : a) 1280°C for 2 hours, b) 1300°C for 2 hours and c) 1300°C for 20 hours.

These results can be explained by examining the processes which occur during the annealing of high dose oxygen implanted specimens. In the as implanted specimen the silicon overlayer is composed of highly defective, single crystal silicon, containing small precipitates of SiO_2 the size and density of which increase with depth. As the annealing temperature increases the smaller precipitates become thermodynamically unstable and dissolve[11]. The oxygen released by their dissolution is then gettered by larger more stable precipitates. At even higher temperatures these precipitates, in turn, become unstable and release oxygen into the matrix. At sufficiently high temperatures and long enough times the buried layer will become the most stable precipitate and the majority of the oxygen will migrate towards it[12].

SIMS analysis of specimens b) and c) are displayed in fig. 4. These show that as the annealing time is increased, from two to twenty hours, the volume concentration of oxygen in the silicon overlay increases. At first sight these results appear to be at odds with the arguments advanced above. XTEM micrographs of specimens b) and c) (not shown) confirm the RBS results by showing a decrease in both the dislocation density and the number of SiO_2 precipitates in the silicon overlayer with increasing anneal time.

In order to explain these results it is necessary to consider the processes which occur during the high temperature anneal. Previous studies involving a buried layer of $Si^{18}O_2$ with a capping layer of $Si^{16}O_2$ have shown that isotopic interchange occurs during high temperature annealing[7]. It is postulated that this process is facilitated by both oxygen diffusion within the bulk and by enhanced oxygen diffusion along threading dislocations. The latter process is thought to be analogous to the pipe diffusion of arsenic observed in silicon on sapphire (SOS) material[13]. A similar mechanism may be evoked to explain the results described in fig. 4. During the long (20 hours) high temperature anneal any remaining precipitates of SiO_2 in the silicon overlayer will dissolve. A dynamic equilibrium will then be established with oxygen migrating between the capping layer and the buried layer via threading dislocations. After annealing when the specimen is cooled, the migrating oxygen will become effectively frozen in the threading dislocations, where it will be undetectable by conventional RBS and XTEM analysis. Electrical results tend to confirm this hypothesis, as the defects in the silicon overlayer are found to be electrically inactive. As would be the case if they were filled with oxygen.

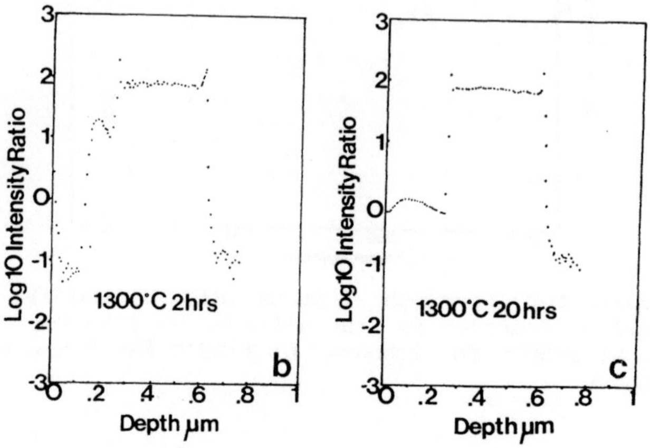

Fig. 4. SIMS profiles for specimen 7 parts b) and c). Showing the changes in oxygen concentration with depth for anneals carried out at 1300°C for b) 2 hours and c) 20 hours.

CONCLUSIONS

It is demonstrated that thin buried layers of SiO_2 can be fabricated in single crystal silicon substrates using IBS. For doses below 0.3×10^{18} $O^{+}cm^{-2}$ two layers of SiO_2 precipates are observed suggesting that the buried oxide nucleates at the peaks of the damage and range profiles. For slightly higher doses $> 0.5 \times 10^{18}$ O^{+} cm^{-2} the two layers of precipitates coalesce to form firstly a discountinous layer and then a continuous layer with irregular interfaces. At even higher doses ($1.8 \times 10^{18} O^{+}cm^{-2}$) the interfaces become abrupt. Annealing these specimens for long times at high temperatures (1300°C 20 hours) leads to a high oxygen concentration in the silicon overlay. This is thought to be due to migrating oxygen being trapped in threading disslocations.

The results of this study serve to illustrate the versatility of the technique of IBS. This allows a variety of structures to be fabricated in clean environments without a significant change in the operating parameters. It also allows the implantation conditions to be specifically tailored to produce the optimum structure for each application.

ACKNOWLEDGEMENTS

The authors would like to thank J.E. Mynard, J.F. Brown, M.D. Browton, M.K. Chapman, C.G.D. Knowler and R.R. Watt for their technical assistance during ion implantation and RBS analysis. They would also like to thank Mrs. B. Doré for her patience in typing this manuscript. The work is supported in part by the U.K. Science and Engineering Research Council (SERC).

REFERENCES

1. P.L.F. Hemment, MRS European Meeting, Strasbourg, May (1985).
2. J. Davis, IEE Tutorial Meeting on SOI current status and future potential, London, March (1987);
3. G. Davis, IEEE SOS/SOI Technology Workshop, Parks City, Utah, Oct. (1985).
4. J. Margail, I. Stoemenos, C. Jassaud, M. Dupuy, P. Marhin, B. Blanchard and M. Bruel, MRS European Meeting, Strasbourg, May (1985).
5. G.K. Celler, P.L.H. Hemment, K.W. West and J.M. Gibson, Appl. Phys. Lett. 48:532 (1986).
6. J.R. Davis, K.J. Reeson, P.L.F. Hemment and C.D. Marsh to be published.
7. K.J. Reeson, Nucl. Instrum. and Methods B19/20:269 (1987).
8. A. Bourret, Semiconductors, August (1979).
9. P.L.F. Hemment, K.J. Reeson, J.A. Kilner, R.J. Chater, C.D. Marsh, G.R. Booker, G.K. Celler and J. Stoemenos, Vacuum 36:877 (1986).
10. C.D. Marsh, J.L. Hutchison, G.R. Booker, K.J. Reeson, P.L.F. Hemment and G.K. Celler, Microscopy of Semiconducting Materials V, Oxford, April (1987).
11. R.C. Newman, M.J. Binns, W.P. Brown, F.M. Livingston, S. Merooloras, R.J. Steward and J.G. Wilkes, Physica 116B:264 (1983).
12. J. Stoemenos and J. Margail, Thin Solid Films 135:115 (1986).
13. D.J. Godfrey, SOS/SOI Technology Workshop, Parks City, Utah, Oct. (1985).

LOW-TEMPERATURE ESR STUDY OF SIMOX STRUCTURES

A. Stesmans* and G. Van Gorp

Departement Natuurkunde, Katholieke Universiteit Leuven
3030 Leuven, Belgium

X- and K-band ESR data are presented of a SIMOX structure fabricated by implanting [001] n-type c-Si with 1.7 $\times 10^{18} O^{+} cm^{-2}$ at an energy of 170 keV. The report focusses on three signals. A first one, with $g \simeq 2.070$ is due to interstitial Fe impurities probably introduced as a result of O^{+} implantation. A second anisotropic signal of $g(\bar{B} \| [001] ; 4.3\ K) = 1.99963$ reveals the presence of P piling up ; c-Si regions of effective P concentration $\sim 1.2 \times 10^{19} cm^{-3}$ are present in the bulk of the SIMOX sample. A third isotropic signal of $g = 1.9983$ is tentatively identified as a thermal donor. The observations are discussed in the light of the formation of internal Si/SiO_2 structures.

INTRODUCTION

The creation of SIMOX (Separation by IMplanted OXygen) structures by high dose oxygen ion implantation is a very promising technology[1,2], which may offer many attractive advantages for Si-on-insulator devices, e.g. small active p-n junction area, high packing density, fast switching rates, simple design, etc... In this method, a high dose (> 10^{18} cm^{-2}) of oxygen ions is implanted deep into a c-Si substrate using energies > 150 keV ; projected ranges R_p vary from 1000 - 3000 Å. During implantation, the substrate is kept at a temperature T > 500°C to continuously anneal the damage induced in the top Si layer by the impinging ions. At places where the O concentration exceeds the stoichiometric value [O] ~ $4.6 \times 10^{22} cm^{-3}$, SiO_2 is formed. The critical dose[3] to obtain SiO_2 formation is about $1.2 \times 10^{18} O^{+} cm^{-2}$ for 150 keV implantation. However, to grow an homogeneous buried SiO_2 layer with sharp Si/SiO_2 boundaries, a prolonged (~ 4-9 h) post-implantation high-T anneal (at $T = T_A$) is necessary[4,5]. The closer T_A approaches the Si melting temperature the better this is achieved[5]. Depending on the particular anneal treatment (i.e., T_A, anneal time, ambient) various types of Si oxide precipitates of different shapes (e.g. rods, plates, polyhedra) are formed at those Si regions where the O solubility in Si has been exceeded. Also, various types of dislocations and defects may remain. Even a 1405°C anneal for 0.5 h, while apparently eliminating all SiO_x precipitates in Si and creating

*Supported by the National Fund for Scientific Research, Belgium.

smooth SiO_2/Si interfaces, leaves some faceted Si crystallites embedded in the buried SiO_2 layer at the backside interface[6]. Hence the continuous interest in the detailed analysis of the SIMOX structure. Many studies of the morphology, structure and chemical composition of the prevailing precipitates and defects have been reported using various techniques[1], e.g., high-resolution transmission electron microscopy (HTEM), Rutherford backscattering (RBS) and secondary-ion mass spectroscopy (SIMS). Apart form the structural analysis of the SIMOX structure itself the study of the O-in-Si behaviour is of great interest because of the various beneficial effects O-in-Si may exhibit, e.g. wafer hardening and gettering of metallic impurities and defects.

In view of its intrinsically high sensitivity and outstanding ability to discriminate between the various types of spin-active centers, electron spin resonance (ESR) may add valuable information to the former structural analyses. Also, ESR is very sensitive to the local structure of the spin centers studied. ESR has been applied to SIMOX structures[7] and several defects have been identified, i.e., the amorphous E' center (zero crossing $g \simeq 2.0003$), the dangling-bond (DB) center ($g \simeq 2.0055$) and the anisotropic $P_{bo}Si/SiO_2$ interface defect. So far these observations have been confined to X-band (9 GHz) studies at 300 K. The present work extends these observations to lower T's and higher frequencies where increased sensitivity may be expected. It appears that one of the main problems left with SIMOX fabrication is possible contamination during ion implantation[1]; in this respect the utility of ESR analysis is evident.

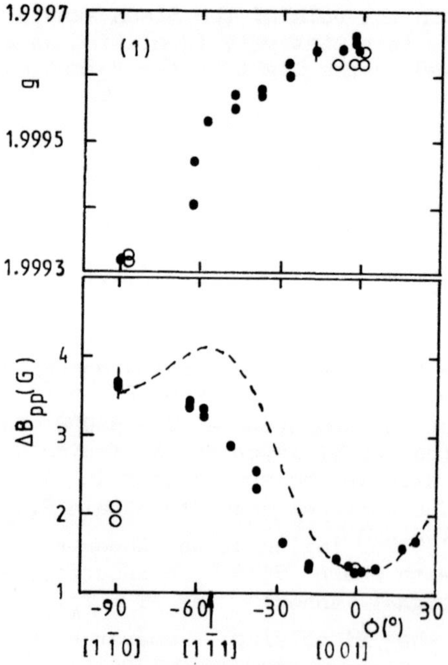

Fig. 1. Plot of g and linewidth anisotropy of signal A measured at T = 4.3 K for $\bar{B}$ varying in the (110) plane ; • : K band ; ○ : X band. The dashed line is a simulation based on a strain model.

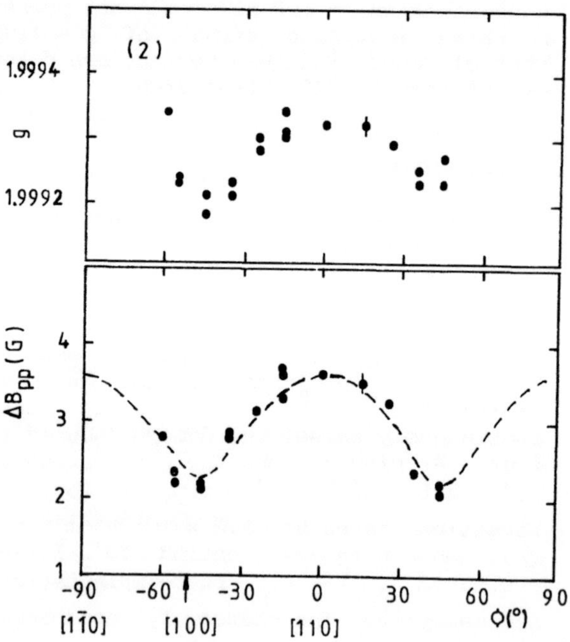

Fig. 2. As fig. 1, but for $\bar{B} \in (001)$ plane.

EXPERIMENTAL DETAILS

The samples were taken from a Cz-grown P-doped [001] Si wafer ($\rho_{RT} \simeq 5\Omega cm$) measuring 547 μm thick*. This wafer was ion implanted** with 150 keV oxygen ions to a dose of ~ $1.7 \times 10^{18} cm^{-2}$ and subsequently annealed at 1250°C in N_2 ambient for 8h upon which a SIMOX structure is formed. Using a diamond saw platelets 2 x 9 x 0.547 mm^3 were cut which had their 9 mm edge along a [110] direction. To increase the surface-to-volume ratio, these slices were thinned down at the backside using aplanar etch to a final thickness of 176 ± 10 μm. This thorough etching also eliminated at once any cracking-induced Si DB signal[8]. To analyse defect localization the

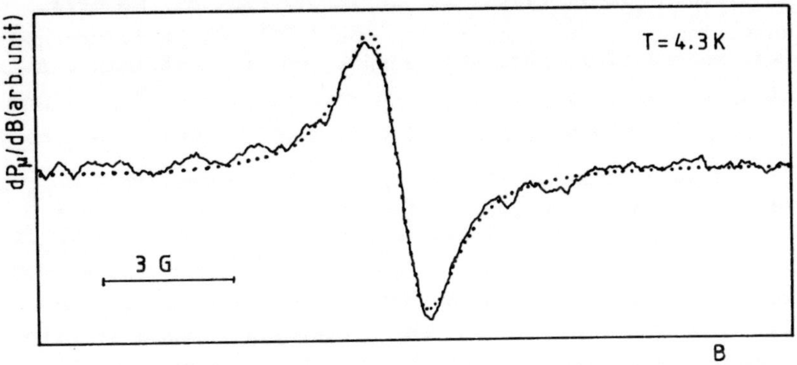

Fig. 3. K band ESR spectrum (full curve) of signal A observed at T = 4.3 K with P_μ = 1 mW. The dots correspond to a calculated Lorentzian line normalized to the experimental one at the peak positions.

superficial c-Si layer was etched in boiling (T ≈ 65°C) NH_4OH (26 vol %) during 8 min ; this took off a layer ~ 720 Å.

ESR experiments were carried out routinely at K band (20.9 GHz) in the range $2.4 \lesssim T \lesssim 34$ K ; some X band (9.0 GHz) data are reported too. Magnetic field modulation and phase-sensitive detection was applied so that absorption-derivative dP_μ/dB spectra were recorded ; P_μ and $\overline{B}$ represent the microwave power in the cylindrical TE_{011} ($Q_L \simeq 4000$) cavity and externally-applied magnetic field, respectively. During measurements the direction of $\overline{B}$ was varied either in the (001) or (110) plane ; in these planes the direction of $\overline{B}$ is specified by the angle Φ which $\overline{B}$ makes with the [110] and [001] directions respectively.

EXPERIMENTAL RESULTS

The low-T ESR spectrum exhibits various signals. First of all, there is the strong P hyperfine doublet[9] ($g \simeq 1.9984$ and hyperfine splitting a = 42.1 ± 0.5 G) due to donor electrons localized on ^{31}P (S = ½ ; 100 % natural abundance) atoms. This signal is anticipated[10] for Si:P doped to concentrations $\leqslant 10^{18} cm^{-3}$. Further, there are some signs of strongly-anisotropic signals for which the weak intensity, prevented their characterization. Presumably, these are due to the P_{bo} Si/SiO_2 interface defects reported previously[7,11] ; in view of the applied high-T anneal

---

*Obtained from Wacker-Chemitronic, Burghausen, Germany.

**Eaton NV-200 implanter.

these signals are expected to have largely disappeared[6,11]. The present work concentrates mainly on three other signals which are believed to relate closely to the SIMOX fabrication technique.

A first signal, labeled A, shows up at K band with $g = 1.99963 \pm 0.00003$ and $\Delta B_{pp}(K) = 1.36 \pm 0.02$ G for $\overline{B} \parallel [001]$ and T = 4.3 K. It could be observed from 2.3 to 34 K. This response behaves anisotropically as may be seen from figs. 1 and 2, in which g-vs-Φ is plotted for B rotating in the (110) and (001) planes, respectively. As may be seen from fig. 1 g decreases substantially from 1.99963($\overline{B} \parallel [001]$) to 1.99927 ± 0.00005 ($\overline{B} \parallel [1\overline{1}0]$) for $\overline{B} \in (100)$. Fig. 2, however, shows that the g anisotropy for $\overline{B}$ varying in the (001) plane is more vague, or at least less substantial. Somewhat surprising is that this g anisotropy is accompanied by a significant linewidth anisotropy (cf. fig. 1 and 2). For $\overline{B} \in (100)$, ΔB_{pp} increases from 1.36 ± 0.04 G ($\overline{B} \parallel [001]$) to $3.6. \pm 0.2$ G ($\overline{B} \parallel [1\overline{1}0]$). Measurements at X band for $\overline{B} \in (110)$ reveal that ΔB_{pp} grows more inhomogeneous with increasing Φ (cf. fig. 1), namely, for $\Phi = 0$, the X and K band data are equal within experimental error, referring to an homogeneous line for $\overline{B} \parallel [001]$; for $\overline{B} \parallel [1\overline{1}0]$, $\Delta B_{pp}(X) = 2.00 + 0.06$ G, which is much lower than the K band value. The inhomogeneous linewidth character is confirmed by line shape analysis. For $\overline{B} \parallel [1\overline{1}0]$ the line shape is almost Lorentzian as may be seen from figure 3. With increasing Φ, the shape gradually gets more Gaussian character. For $\overline{B} \parallel [1\overline{1}0]$ the data could be consistently computer simulated by Voigt profiles, which represent the convolution of Lorentzian (of width ΔB^L_{pp}) and Gaussian (of width ΔB^G_{pp}) broadening functions. This resulted in $\Delta B^G_{pp}(X) = 1.25 \pm 0.05$ G and $\Delta B^G_{pp}(K) = 2.90 \pm 0.05$ G, while $\Delta B^L_{pp}(X) = \Delta B^L_{pp}(K) = 1.34$ G. Notice that, as expected, $\Delta B^G_{pp}(K) = (20.9/9)\ \Delta B^G_{pp}(X)$, suggesting a g-distribution broadening. Also, the value $\Delta B^L_{pp}(\overline{B} \parallel [110]) = 1.34$ G is in agreement with $\Delta B_{pp}(\overline{B} \parallel [001])$ for which field direction no inhomogeneous broadening is observed.

The T dependence has been measured at K band for $\overline{B} \parallel [001]$; ΔB_{pp} is T dependent and shows a (slight) minimum of $\Delta B_{pp}(K) = 1.25 \pm 0.03$ G at 10 K and increases gradually towards $\Delta B_{pp} = 2.1 \pm 0.2.$ G at 34 K. Regarding g,

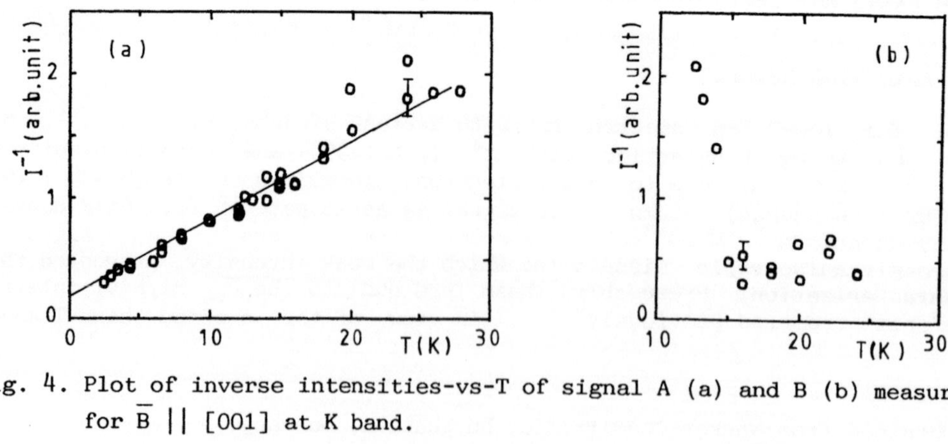

Fig. 4. Plot of inverse intensities-vs-T of signal A (a) and B (b) measured for $\overline{B} \parallel [001]$ at K band.

while it is found constant, i.e., $g = 1.99963 \pm 0.00003$ for $T < 22$ K, it decreases to the value $g = 1.99946 \pm 0.00005$ for T approaching 34 K. The inverse of the signal intensity I (area under absorption curves) is plotted versus T in fig. 4(a). As may be seen, the data are fairly well described by the relation $I \propto (T + \theta_a)^{-1}$, where θ_a is an asymptotic Curie-Weiss temperature given as $\theta_a = 2.81 \pm 0.3$ K. The spin density N_S^A has been determined for $\bar{B} \parallel [001]$. Relying on the observed Lorentzian line shape and Curie-Weiss behaviour of I, this gave $N_S^A = 3.4 \times 10^{11} cm^{-2}$ (± 20 % ; S = ½) or, $2.0 \times 10^{13} cm^{-3}$.

A second signal (B) only becomes observable at temperatures above 12 K. It exhibits a linewith $\Delta B_{pp} = 2.12 \pm 0.08$ G and $g = 1.99833 \pm 0.00003$, both parameters being T and Φ independent within experimental scatter. The line shape is found Lorentzian. The reason why this signal could only be observed for $T > 12$ K may be seen in the plot of I^{-1}-vs-T (cf. fig. 4(b)). As T decreases below ~ 16 K the signal strongly decreases so as to become unobservable below 12 K. The decrease is rather drastic and would seem the behaviour of a spin system going through an antiferromagnetic phase transition. At present the experimental accuracy is still poor because of the low signal intensity. However, one might try to deduce an order of magnitude for the activation energy E_a involved. Plotting the data as ln (IT)-vs-T^{-1} indicates an activated behaviour indeed giving $E_a = 5.8 \pm 0.5$ meV. The signal's spin density is found to be $N_S^B = 3.5$ (± 20 %) $\times 10^{11} cm^{-2}$ ($2.1 \times 10^{13} cm^{-3}$).

Both signals discussed so far appear rather insensitive to saturation, which is rather unexpected for isolated spins in a Si/SiO_2 structure. At K band and $T = 4.3$ K saturation effects in signal A become observable at $P_\mu \sim 1$ mW, which corresponds to $H_1 = 0.09$ G ; H_1 is the amplitude of the circularly polarized hf field at the sample site. A third signal (C) is observed at $g = 2.070 \pm 0.001$ with concentration $N_S^C > 3.6 \times 10^{10} cm^{-2}$ ($2.14 \times 10^{12} cm^{-3}$). The signal is undoubtedly to be identified with neutral Fe placed on the T_d interstitial site. It has been well studied[12] previously and was found to exhibit a "peculiar" line shape[9], sensitive to saturation. The latter effects suggest that the quoted N_S^C value is to be seen as a strict lower limit rather than a correct value.

Etching the top c-Si layer had no detectable influence on any of the signals discussed. This indicates that all three types of centers reside merely in the bulk of the SIMOX structure probably near to the buried Si/SiO_2 interfaces. Finally we remark that, as a matter of control, the basic unimplanted Si has been studied too ; none of the three signals could be traced thus confirming that the corresponding centers were introduced by the O^+ implantation (and/or annealing).

DISCUSSION

The $g = 2.070$ signal being identified with interstitial Fe impurities there remains the origin of signals A and B. Regarding signal A there is clear evidence that it originates from P-rich Si regions, i.e., there are small regions or globules in the SIMOX structure in which P has piled up to concentrations $> 10^{18} cm^{-3}$, far above the homogeneous dopant concentration N_p ($\sim 8 \times 10^{14} cm^{-3}$). Various features of the ESR response indicate this. A first stems from its saturability ; the signal is difficult to saturate and it almost behaves as a metallic or conduction electron resonance. It is well documented that donor electrons in Si:P of $N_D > 10^{18}cm^{-3}$ are strongly delocalized and behave in this manner[9,10,13]. Further, their g value, i.e., $g_p \simeq 1.9986$, is close to that of signal A. Also, the ΔB_{pp}-vs-T behaviour of signal A is typical[10] for Si:P of $N_D > 10^{18}cm^{-3}$. From the 4.3 K linewidth, the effective concentration N_e of the P-rich regions is estimated[10] to be $2 \times 10^{19}cm^{-3}$. In such Si:P the Fermi level has moved into the conduction band (CB) and Si:P behaves metallic. The observed decrease of g with increasing T for $\bar{B} \| [001]$ and the Lorentzian line shape are also as expected for Si:P doped to such concentrations[10,13]. The final identification concerns the observed anisotropies of ΔB_{pp} and g. To start with, one could try to fit the $g(\Phi)$ data with a tensor and extract the principal values. However, because of the particular symmetry shown, the data do not fit a physical g tensor ; an additional effect needs to be involved. The donor electrons in Si:P moving in one valley are known to have an axial symmetric tensor[14,15], i.e., $g_\| (\bar{B} \| \langle 001 \rangle) = 1.9995$ and $g_\perp = 1.9983$. Generally, since there are six equivalent valleys in Si centered around the <001> axes, this anisotropy never shows up in a strain-free sample. One always measures the isotropic average value[14,16] $g_0 = (g_\| + 2g_\perp)/3 = 1.9986 \pm 0.0001$ at $T \simeq 10$ K. However, if strain is present in c-Si, effects of the anisotropic g tensor become visible macroscopically due to valley repopulation and intra-valley g shifts. These, in general, have the form

$$g - g_0 = \frac{1}{6}(g_\| - g_\perp)\left(1 - \frac{3}{2}\sin^2\theta\right)(1-f) \quad (1)$$

where θ is the angle between $\bar{B}$ and the strain axis and f is a function of the applied stress. An additional indication of stress influence is the pronounced ΔB_{pp} anisotropy. It is well known that stress, or local lattice distortions in general, will induce a larger spread in $g_\perp$ than $g_\|$ - to first order no spread in $g_\|$ should occur[17]. The angular variation of ΔB_{pp} is then explained as due to a strain broadening of $g_\perp$. Unless the actual shape of the Si-rich regions and their strain fields are accurately known, the simulation in this model of the observed $g(\Phi)$ and $\Delta B_{pp}(\Phi)$ data would seem a formidable task in view of the many parameters involved. TEM studies, however, report that in SIMOX structures annealed at T > 1250°C for prolonged periods, there are small Si precipitates left in the buried SiO_2 layer, near to the back Si/SiO_2 interface[1,6] - in fact these appear to be the only precipitates remaining. The Si precipitates are faceted and of polyhedral size. Starting from these observations and in view of the cubic structure of Si, the shape of a regular truncated octahedron centred along the Si <001> axes was accepted and the strain was supposed to be perpendicular to the border planes. Indeed, the general features of the $\Delta B_{pp}(\Phi)$ curves could be simulated from this shape, as shown by the dashed lines in figs 1 and 2. However, to reproduce the observed $g(\Phi)$ data the model clearly needs further refinement. Also, the experimental data needs to be improved and extended.

The identification of signal B seems more cumbersome. A remarkable property of this signal is the activated behaviour of its intensity with $E_a = 5.8 \pm 0.5$ meV. This behaviour is very similar to the one of high-concentration thermal donors[17] (TD) created in O-rich Si by prolonged thermal annealing at 450°C. Also, their linewidths and g values largely agree, i.e., for TD, one has $g \sim 1.9985$ and $\Delta B_{pp} \sim 3$ G. Hence, in view of these close similarities and considering that signal B is likely to be related to the O-in-Si situation also, we tentatively identify it as a TD.

CONCLUSION

The implantation of $1.7 \times 10^{18} cm^{-2}$ $O^+ cm^{-2}$ in Si to create a SIMOX structure introduces Fe impurities, or better stated, the implantation followed by a thermal treatment has put an amount of Fe in interstitial ESR active positions at least 25 times more than previously present. In the SIMOX structure small "regions" are created containing an effective P concentration $N_e \sim 1.2 \times 10^{19} cm^{-3}$, these regions (precipitates) are located in the bulk of the structure - not at the air / Si interfaces. The P piling up is a natural consequence of Si/SiO_2 structure formation[13] and may occur even in p-type Si where P impurities form a minority. A third ESR signal is identified with a kind of TD. This, of course, relates to the question whether such a TD may be expected to survive a prolonged annealing at T = 1250°C. It is realized that this identification is rather vague since a detailed microscopic model of TD's is still lacking, although they are generally believed to be related to SiO_x complexes[17]. However, recently it has been suggested that much of the thermal and new donor story is a consequence of internal oxidation concurrent with inhomogeneous (band-bending effect) regions[18]. Consequently, the ESR spectra should mainly refer to the Si/SiO_2 interface properties. In fact, signal A seems to support this idea and seems suggestive that many TD and ND signals could in fact be traced back to P donor signals.

K. Yallup, IMEC, University of Leuven, is acknowledged for his cooperation on sample preparation.

REFERENCES

1. See, e.g., G.K. Celler, Solid State Technol. 30:93 (1987), and references therein.
2. P.L.F. Hemment, Applied Surf. Sci. (1987), unpublished.
3. P.L.F. Hemment, K.J. Reeson, J.A. Kilner, R.J. Chater, C.D. Marsh, G.R. Booker, J.A. Davis, and G.K. Celler, Proc. Ion Implantation Techn. Conf., San Francisco (1986).
4. C. Jaussaud and, J. Stoemenos, J. Margail, M. Dupuy, B. Blanchard and M. Bruel, Appl. Phys. Lett., 46:1064 (1985).
5. G.K. Celler, P.L.F. Hemment, K.W. West, and J.M. Gibson, Appl. Phys. Lett., 48:532 (1986).
6. C.D. Marsh, G.R. Booker, D.J. Reeson, P.L.F. Hemment, R.J. Chater, J.A. Killner, J.A. Alderman, and G.K. Celler, Proc. European MRS Les Editions de Physique, Les Ulis (1986).
7. R.C. Barklie, A. Hobbs, P.L.F. Hemment and K. Reeson, J. Phys. C. 19: 6417 (1986) ; in Proc. 14th Int. Conf. on Defects in Semiconductors, Paris, 1986, (unpublished) ; T. Makino and J. Takahaski, Appl. Phys. Lett., 50:267 1987).
8. See, e.g., B.P. Lemke and D. Haneman, Phys. Rev. B17:1893 (1978).
9. G. Feher, Phys. Rev. 114:1219 (1959).
10. J.D. Quirt and J.R. Marko, Phys. Rev. 5:1716 (1972) ; 7:3842 (1973).
11. E.H. Poindexter and P.J. Caplan, Prog. Surf. Sci. 14:201 (1983).
12. See, e.g., H.H. Woodbury and G.W. Ludwig, Phys. Rev. 117:102 (1960) ; E. Weber and H.G. Riotte, J. Appl. Phys. 51:1484 (1980).
13. See, e.g., A. Stesmans and J. Braet, Surf. Sci. 172:398 (1986) and references therein.
14. D.K. Wilson and G. Feher, Phys. Rev. 124:1068 (1961).
15. Y. Yafet, Solid. State Phys. 14:1 (1963).
16. A. Stesmans and G. De Vos, Phys. Rev. 34:6499 (1986).
17. M. Suezawa, K. Sumino, and M. Iwaizumi, J. Appl. Phys. 54:6594 (1983).
18. A. Henry, J.L. Pautrat, and K. Saminadayar, J. Appl. Phys. 60:3192 (1986).

DEFECTS IN SILICON-ON-INSULATOR STRUCTURES FORMED BY O^+ IMPLANTATION : THEIR DEPENDENCE ON IMPLANTATION TEMPERATURE

T.J. Ennis*, R.C. Barklie*, K. Reeson+ and P.L.F. Hemment+

*Physics Department, Trinity College, Dublin 2, Ireland
+Department of Electronic and Electrical Engineering
University of Surrey, Guildford, Surrey GU2 5XH, UK

EPR measurements have been made on silicon-on-silicon dioxide samples produced by implanting n type (100) silicon wafers with a dose of 1.4×10^{18} O^+cm^{-2} using 200 keV $^{16}O^+$ ions and an implantation temperature of 250, 350, 450 and 600°C. The EPR spectra reveal the presence of E_1', P_{bo} and amorphous silicon centres. The dependence of the concentrations of these defects on implantation and annealing temperatures is reported.

INTRODUCTION

A promising method for the production of device worthy silicon-on-insulator (SOI) material is the implantation into silicon of a dose of oxygen sufficiently high to form a buried oxide layer, followed by a high temperature anneal both to sharpen the interfaces of the buried layer and to greatly reduce SiO_2 precipitates and defects within the top silicon layer[1,2]. It is clearly desirable to find implantation and annealing conditions which lead to the best quality silicon layer. One of the most important implantation parameters is the implantation temperature, T_i. This determines whether or not the level of radiation damage is kept below that at which amorphous silicon layers are formed[3-5] and, in addition, it affects the extent of SiO_2 precipitate nucleation and growth as both of these are temperature dependent processes[6,7]. The morphology of the as-implanted state is therefore expected to depend on T_i and this is found to be the case[3,4]. It is also found that even after high temperature annealing the structure of the implanted region and oxygen distribution within it both depend on T_i[3,4].

Electron paramagnetic resonance measurements have already provided information about the paramagnetic defects produced by high dose oxygen implantation at T_i of about 500°C[8,10]. The aims of the present work are to determine how a variation of T_i in the range 250°C-600°C affects the type and/or concentration of defects produced and to find the extent to which, for each value of T_i, the defects can be removed by annealing. In this paper we concentrate on those defects associated with SiO_2.

Table 1. Sample implantation conditions.
All wafers are (100) n-type silicon

Sample	Ion	Beam Current (μA)	Implantation Temperature (°C)	Dose ($10^{18} cm^{-2}$)	Ion Energy (keV)
S020a	$^{16}O^{+}$	50±1	250	1.4	200
S021a	$^{16}O^{+}$	70±1	350	1.4	200
S023a	$^{16}O^{+}$	90±1	450	1.4	200
S033a	$^{16}O^{+}$	123±3	600	1.4	200

SAMPLE PREPARATION AND EXPERIMENTAL TECHNIQUES

The samples were prepared at the University of Surrey by implanting 3" n type (100) silicon wafers in a non-channeling direction with 200 keV $^{16}O^{+}$. Table 1 lists the implantation conditions used. Conductive heat losses were greatly reduced by supporting the wafers on silicon tips and hence the implantation temperature could be varied by changing the beam current. As only the central 2.5 x 2.5 cm portion of each wafer was implanted an approximately radial temperature profile existed and it has been found that for T_i in the range 400-550°C the temperature of the outer edge of the implanted[1] region is about 20°C less than that at the center.

At 200 keV a dose of $1.4 \times 10^{18} O^{+} cm^{-2}$ has been found to be the minimum required for the peak oxygen concentration, which for this energy occurs at a depth of approximately 0.5 μm, to reach that corresponding to SiO_2[12].

EPR measurements were made at Trinity College using a Bruker EPR spectrometer with 100 kHz field modulation ; all results were obtained at room temperature and with a microwave frequency of about 9.6 GHz.

RESULTS AND DISCUSSION

Fig. 1 shows the EPR spectra, with magnetic field $\underset{\sim}{B}$ ||(100), for the as-implanted state of each of the samples listed in table 1. In each case the spectrum consists of two feature A and C. For $\underset{\sim}{B}$ ||(100) feature A has a zero crossing g value $\simeq$ 2.0057 and peak to peak width, ΔB_{pp}, of about 0.45 mT but both its position and shape depend on orientation though this becomes less marked as T_i is decreased. On the other hand feature C is independent of orientation and fits to a slightly asymmetric line with zero crossing g = 2.0006 ± 0.0002 and ΔB_{pp} = 0.33 ± 0.02 mT. A very similar EPR spectrum was observed for sample 167 (n type (100) Si implanted at T_i = 520°C with a dose of $1.8 \times 10^{18} O^{+} cm^{-2}$ using 400 keV $^{32}O_2^{+}$) ; for this sample it was shown that feature C is most probably associated with E_1' centers, predominantly in the buried SiO_2 layer. Feature A is the superposition of spectra due to amorphous silicon centers (a-Si), which may be associated with oxygen, and also P_{bo} like centers primarily present at Si/SiO_2 interfaces of SiO_2 precipitates. The similarity of the spectra for the $^{16}O^{+}$ samples to that for sample 167, together with the annealing behaviour discussed later, strongly suggest that the same centers are present in the former samples and in particular that feature C is again due to E_1' centers. It is clear from fig. 1 that the ratio (intensity of feature C)/(intensity of feature A) increases as T_i increases from 250°C to 600°C. From a comparison of the intensities of computed fits to these spectra with the intensity of a standard sample of pitch in KCl we find the defect concentrations, N, to be $N(E_1') = 1.5 \times 10^{14}$, 2.0×10^{14}, 2.0×10^{14} and $2.3 \times 10^{14} cm^{-2}$, N(feature A) = 1.5×10^{15}, 1.4×10^{15}, 1.2×10^{15} and $1.1 \times 10^{15} cm^{-2}$ for T_i = 250, 350, 450 and 600°C respectively. In evaluating N account was taken of changes in cavity Q value and filling factor. As with most absolute spin concentrations these values may differ from the true values by up to a factor of 2 but their relative values should be accurate to about 10 %.

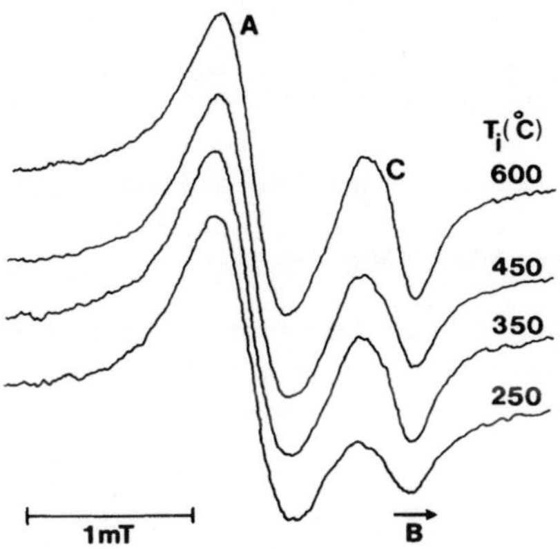

Fig. 1. EPR spectra of the samples, listed in Table 1, in the as - implanted state. $\tilde{B}||[100]$. The gains are different for each spectrum.

One striking feature of these results is that the E_1' concentration <u>increases</u> as T_i is increased from 250 to 600°C. Indeed it may be thought surprising that any E_1' centers are observed at all after an implantation at 600°C since these centers in ion implanted amorphous SiO_2 usually start to anneal out at about 100°C[13]. However, two points need to be born in mind : firstly, it has been shown that irreversible annealing of E_1' centers only begins at about 500°C[14] and, secondly, the annealing induced by the ion beam will always be accompanied by the creation of new E_1' centers so that their population will equilibriate at the value at which their creation and annihilation rates are equal. This however, does not explain our results as this equilibrium population would be expected to become smaller as T_i is raised. One additional piece of information is that E' centers can occur in SiO_2 precipitates. Although we concluded that for sample 167 the majority of the E_1' centers are in the buried SiO_2 layer we have found that the EPR spectrum, for $\tilde{B}$ ||[100], of a sample implanted at $T_i \simeq$ 570°C with a dose of $0.75 \times 10^{18} O^+ cm^{-2}$ 200 keV $^{16}O^+$ is very similar to those shown in fig. 1.

Since this dose is too low to form an SiO_2 layer in the as-implanted state [12] the presence of feature C implies that precipitates can also contain E_1' centers. Furthermore, although the oxygen concentration as a function of depth, in the as-implanted state, has been found[4] to be independent of T_i at least in the range 450 - 750°C, it is known that raising the temperature favours precipitate nucleation and growth[6,7,17]. The explanation for the increase in E_1' concentration as T_i increases may therefore simply be that the total volume of SiO_2, both as precipitates and buried layer, available for E_1' formation increases as T_i increases. Other factors which may also have affected the E_1' concentration are the changes in dose rate and possibly changes in the amount, if any, of SiO_x with $x < 2$.

It is difficult, for the as-implanted samples, to separate the contributions to feature A from the P_{bo}-like and a-Si centers ; however, our results indicate that as T_i increases the concentration of the former increases but that this increase is more than compensated by a reduction in the concentration of the a-Si centers so that the overall intensity decreases as T_i increases. The increase in P_{bo} concentration is consistent with the increase in total SiO_2 surface area which might reasonably be expected to accompany the increase in total SiO_2 volume and the decrease in a-Si concentration is very likely to be related to the reduction in width and eventual disappearance of amorphous silicon layers as T_i is raised[3]. It should be noted that the presence of a-Si centers does not require the presence of a continuous amorphous layer.

We have also annealed all but one (with T_i = 350°C) of the samples in nitrogen for ten minute intervals at successively higher temperatures from room temperature up to 1095°C in about 50°C steps. After each anneal the

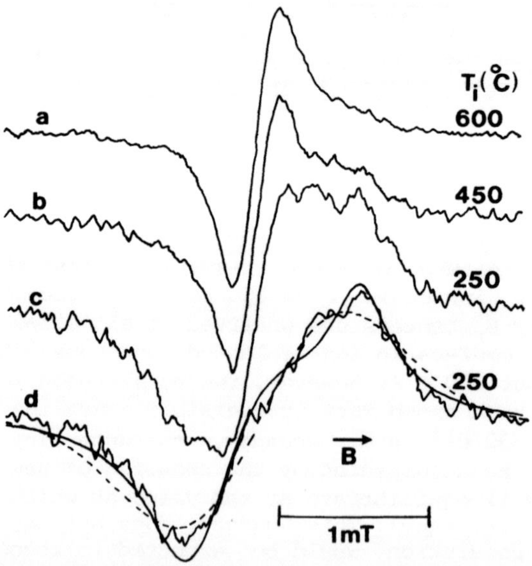

Fig. 2.a,b,c. EPR spectra of the samples with T_i = 600, 450 and 250°C respectively after step annealing to 1095°C. $\underset{\sim}{B} \parallel [100]$. Spectrum d is obtained by subtracting from spectrum c an appropriate amount of spectrum a. The dashed and smooth lines are fitted single Gaussian and powder pattern spectra with the parameters given in the text.

sample was quenched to room temperature and its EPR spectrum recorded. For each sample the E_1' signal commenced to fall at an anneal temperature, T_A, of about 450°C and had disappeared for $T_A \simeq 650$°C. It is not clear why the signal remains constant well above the value of about 100°C at which it usually starts to decrease in ion implanted amorphous SiO_2[13] but this behaviour is similar to that already reported[8] for the E_1' centers in sample 167. For the $T_i = 600$°C sample the T_A dependence of the shape and intensity of feature A closely follows that observed for sample 167 and can therefore be interpreted, as before[8], as arising from the more rapid annealing of the a-Si centers compared to the P_{bo} centers. The corresponding results for the sample with $T_i = 450$°C are similar though we find that, for any T_A value, the ratio of the intensity of the a-Si center signal to that of the P_{bo} centers has increased. At the higher T_A values however, it becomes apparent, as can be seen in fig. 2b, that an additional spectrum is present. This new feature is more clearly seen for the sample with $T_i = 250$°C as shown in fig. 2c. The spectrum in fig. 2a, which is strongly angular dependent, arises almost entirely from P_{bo} centers. Such a spectrum is present, but much less intense, for $T_i = 250$°C and when it is substracted the resultant spectrum, representing the new feature, is as shown in fig. 2d. It appears to be isotropic, and can be quite well fitted by a single Gaussian line with $g = 2.006$, $\Delta B_{pp} = 1.1$ mT. It can also be fitted, but not so well, by a powder pattern lineshape generated using the parameters $g_{\parallel} = 2.0026$, $g_{\perp} = 2.0088$ and with linewidths given by

$$\Delta B(\theta) = \Delta B_{\parallel} \mathrm{Cos}^2\theta + \Delta B_{\perp} \mathrm{Sin}^2\theta$$

with $\Delta B_{\parallel} = 0.30$ mT, $\Delta B_{\perp} = 0.53$ mT ; θ is the angle between $\underset{\sim}{B}$ and the axis of the g tensor. These principal g values are close to those of $g_{\parallel} = 2.0013(2)$, $g_{\perp} = 2.0082(2)$ which we measured[8] for the P_{bo} like centers in sample 167 and which we have also found, within experimental error, to be the values of the P_{bo} like centers in the present set of samples. In addition the inequality $\Delta B_{\perp} > \Delta B_{\parallel}$ is what we observe[8] for the P_{bo} like centers and almost certainly arises from a greater spread in the $g_{\perp}$ than in the $g_{\parallel}$ value. These similarities make it likely that the new feature comes from the same type of P_{bo} like centers at the interfaces of SiO_2 precipitates located in amorphous or polycrystalline regions of silicon. This is consistent with the fact that amorphous silicon layers, in this type of high dose sample, become polycrystalline rather than crystalline after high temperature annealing[3].

REFERENCES

1. I.H. Wilson, Nucl. Instr. and Meth. B1:331 (1984).
2. P.L.F. Hemment, Semiconductor-on-Insulator and Thin Film Transistor Technology Symposium., Boston MA. 1985 : Mater. Res. Soc., Pittsburg, (1986), p. 207.
3. C.G. Tuppen, M.R. Taylor, P.L.F. Hemment and R.P. Arrowsmith. Appl. Phys. Lett. 45:57 (1984).
4. S. Maillet, R. Stuck, J.J. Grob, A. Golanski, R. Pantel and A. Perio, Nucl. Instr. and Meth. B19/20:294 (1987).
5. R.F. Pinizzotto, J. Cryst. Growth. 63:559 (1983).
6. A. Bourret, J. Thibault-Desseaux and D.N. Seidman, J. Appl. Phys. 55: 825 (1984).
7. S.M. Hu, Appl. Phys. Lett. 48:115 (1986).
8. R.C. Barklie, A. Hobbs, P.L.F. Hemment and K. Reeson, J. Phys. C19:6417 (1986).

9. T. Makino and J. Takahashi, Appl. Phys. Lett. 50:267 (1987).
10. W.E. Carlos, Appl. Phys. Lett. 50:1450 (1987).
11. J.R. Davis, M.R. Taylor, G.D.T. Spiller, P.J. Skevington and P.L.F. Hemment, Appl. Phys. Lett. 48:1279 (1986).
12. P.L.F. Hemment, E. Maydell-Ondrusz, K.G. Stephens, J. Butcher, D. Ioannou and J. Alderman, Nucl. Instr. and Meth. 209/210:157 (1983).
13. R.A.B. Devine, Nucl. Instr. and Meth. B1:378 (1984).
14. A. Golanski, R.A.B. Devine and J.C. Oberlin, J. Appl. Phys. 56:1572 (1984).

INTERFACE PROPERTIES AND RECOMBINATION MECHANISMS IN SIMOX STRUCTURES

T. Elewa, H. Haddara and S. Cristoloveanu

Institut National Polytechnique de Grenoble
Laboratoire de Physique des Composants à Semiconducteurs
ENSERG, 23 rue des Martyrs, Grenoble 39031, France

INTRODUCTION

In this paper, we present new methods and results related to the characterization of silicon on insulator material fabricated by deep oxygen implantation (SIMOX). The minority carrier lifetime as well as the surface recombination velocity are obtained using depletion-mode MOSFETs. This is done by monitoring the drain current while the gate is being pulsed into deep depletion.

In addition, interface trap properties are determined using the dynamic transconductance technique[1], which is applied for the first time to enhancement SIMOX MOSFETs. This method gives the interface trap parallel conductance G_p as a function of frequency from which the density of interface traps as well as their time constants can be deduced.

Measurements have been carried out on SIMOX devices fabricated at the University of Surrey and LETI (Grenoble) under different processing conditions. The implantation energy was 200 keV and the dose 1.6×10^{18} O^{+}/cm^{2}. In particular, the influence of implantation and annealing temperatures is closely studied.

DYNAMIC TRANSCONDUCTANCE

Theory

The characterization of interface trap properties in VLSI MOSFETs is a difficult task. This is mainly due to the very small gate area of such devices which makes conventional capacitance and conductance measurements impossible. Recently, a new technique base on the measurement of the imaginary part of the inverse of the dynamic transconductance $g_m(\omega)$ has been proposed for the characterization of interface traps[1]. This technique, which benefits from the transistor current amplification, qualifies to be much more sensitive than usual conductance and capacitance methods.

Starting from the transconductance definition for enhancement MOSFETs operating in the ohmic region and making use of the instantaneous charge conservation relation, it is easily demonstrated that $g_m(\omega)$ is given by[1] :

$$g_m(\omega) = \frac{\mu_{eff}^2}{\mu_o} \left[\frac{Z}{L} V_d\right] \frac{C_{ox}\, C_{inv}}{C_{ox} + C_{inv} + C_d + Y_{it}/(j\omega)} \qquad (1)$$

with μ_{eff} : effective carrier mobility, μ_o : low field mobility, Z : channel width, L : channel length, V_d : drain voltage, C_{ox} : gate oxide capacitance, C_{inv} : inversion layer capacitance, C_d : depletion layer capacitance, Y_{it} : interface trap parallel admittance[2] ($Y_{it} = G_p + j\omega C_p$), and ω : angular frequency of the applied gate signal.

In weak inversion, the effective mobility can be approximated by μ_o and C_{inv} by $Q_{inv}/(kT/q)$ (Q_{inv} being the absolute inversion layer charge) so that $g_m(\omega)$ becomes :

$$g_m(\omega) = \mu_o\, V_d \frac{Z}{L} \frac{Q_{inv}}{(kT/q)} \frac{C_{ox}}{C_{ox} + C_d + C_{inv} + Y_{it}/(j\omega)} \qquad (2)$$

It is clear that G_p/ω is readily obtained from the imaginary part of $1/g_m(\omega)$ as follows :

$$\frac{G_p}{\omega} = \frac{C_{ox}\, I_d}{kT/q} \mathrm{Im}[1/g_m(\omega)] \qquad (3)$$

where I_d is the drain current.

Similar to the well known conductance technique[2], the interface trap density at a given gate bias is obtained from the magnitude of the maximum of G_p/ω versus frequency whereas the interface trap time constant is deduced from the position of this maximum.

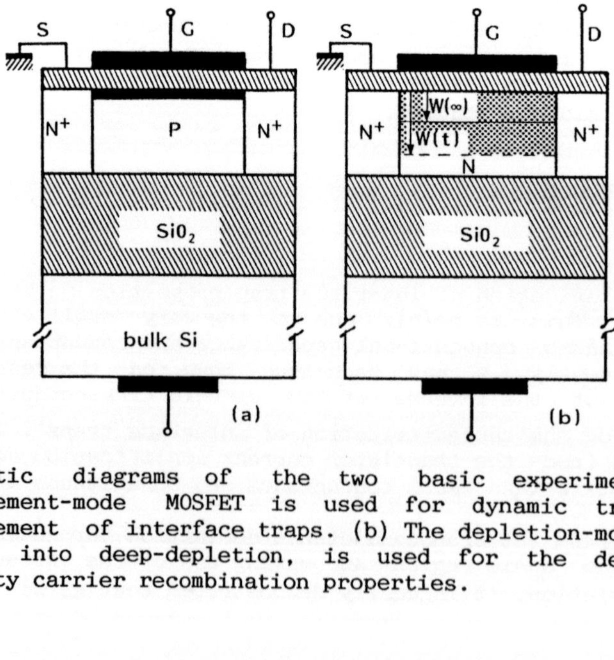

Fig. 1. Schematic diagrams of the two basic experiments. (a) The enhancement-mode MOSFET is used for dynamic transconductance measurement of interface traps. (b) The depletion-mode transistor, pulsed into deep-depletion, is used for the determination of minority carrier recombination properties.

EXPERIMENT

Dynamic transconductance measurements were carried out on enhancement-mode MOS transistors on SIMOX annealed and implanted at different temperatures. A schematic diagram of the measuring circuit is presented in fig. 1a. The gate voltage was modulated by a small signal (50 mV) and the a.c. drain current was measured in phase and magnitude.

Fig. 2 shows G_p/ω versus frequency, with gate voltage V_g as a parameter, for a transistor on a SIMOX substrate implanted at 515°C and annealed at 1405°C for 30 minutes. An average interface trap density D_{it} of $2 \times 10^{11} eV^{-1} cm^{-2}$ is deduced. On the other hand, fig. 3 shows the same curves for a transistor on a SIMOX substrate implanted at the same temperature but annealed at a lower temperature (1200°C) for 2 hours. By comparing the two cases of figs. 2 and 3, it clearly appears that the interface trap density is reduced by a factor 2 for substrates subjected to high temperature annealing.

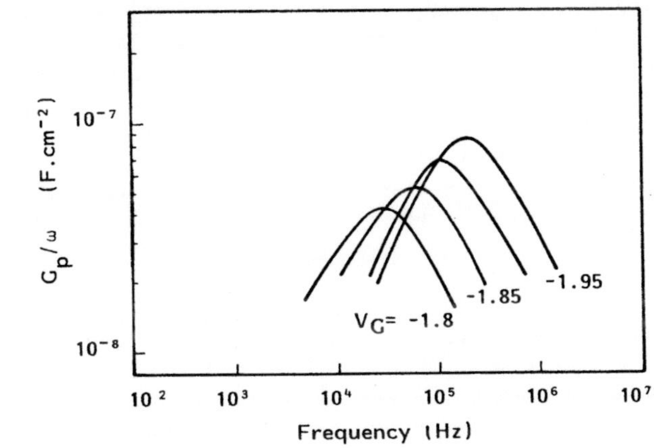

Fig. 2. G_p/ω versus frequency for an enhancement MOSFET ; the SIMOX substrate was implanted at 515°C and annealed at very high temperature (1405°C).

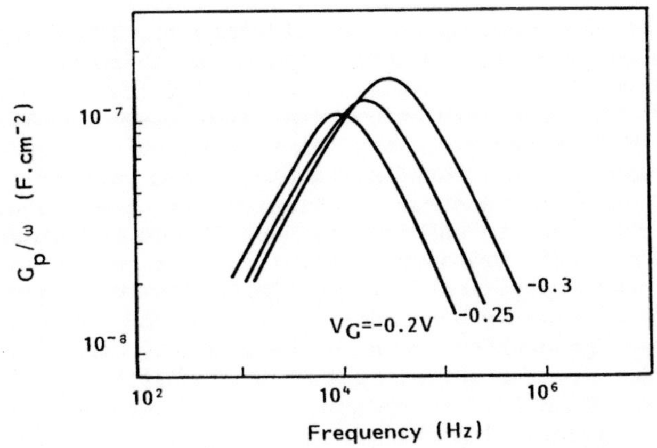

Fig. 3. G_p/ω versus frequency as obtained from dynamic transconductance measurements in a SIMOX-MOSFET (oxygen implantation at 515°C and annealing at 1200°C).

Finally, the effect of reducing the implantation temperature during manufacture of the SIMOX substrate is illustrated in fig. 4. In addition to the interface trap peak which shifts towards higher frequencies with increasing V_g, another peak is observed at a relatively low frequency (3 kHz). This second peak does not change either in position or in magnitude with increasing V_g and is therefore attributed to bulk traps[2]. Due to the complexity of the equivalent circuit including bulk traps, no quantitative information about them will be given here. Indeed, the extraction of G_p/ω for bulk traps from the plot $\mathrm{Im}[1/g_m(\omega)]$ versus frequency of fig. 4 requires a more detailed analysis.

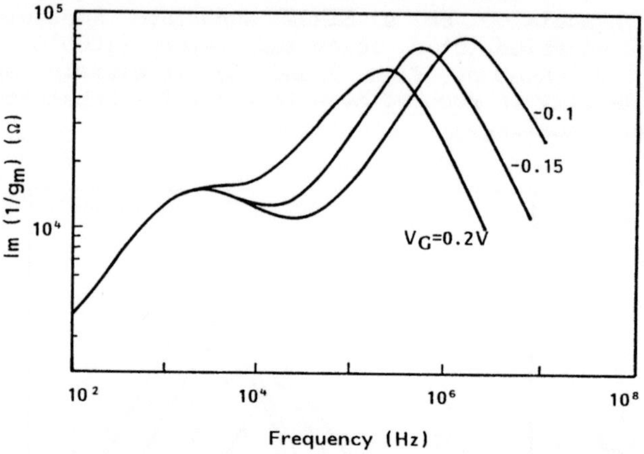

Fig. 4. Variation of the imaginary part of $1/g_m$ as a function of frequency for a SIMOX-MOSFET (oxygen implantation at 480°C and annealing at 1200°C). A constant peak appears at a relatively low frequency (3 kHz) and is related to bulk traps.

RECOMBINATION PROPERTIES

Model

One of the most important parameters in characterizing the quality of silicon material is the minority carrier lifetime τ_g. The Zerbst method[3] is commonly used to measure τ_g and the surface recombination velocity s_o by monitoring the high frequency capacitance during the transient of an MOS capacitor pulsed in deep depletion. In the case of MOS capacitors fabricated on SIMOX materials, the extraction of τ_g and s_o is not straightforward due to the errors introduced by the parasitic capacitances and series resistances[4]. Another difficulty is the requirement for a relatively large gate area to enable easy capacitance measurement. A method based on measuring the transient drain current of a depletion-mode transistor during the transition from deep depletion to steady state strong inversion has been recently proposed[5]. This method does not take into account the surface generation under the gate and/or over the sidewalls of the transistor. We present an exact analysis, similar to that of Zerbst[3] but transposed to MOSFETs, to evaluate the influence of the surface generation on the measured lifetime.

We start from the relation of charge conservation in an MOS system :

$$Q_{inv} + Q_{it} = C_{ox}(V_g - V_{FB} - \psi_s) - Q_d \tag{4}$$

with Q_d : depletion layer charge, ψ_s : the surface potential, V_{FB} : the flat band voltage and Q_{it} : the interface trap charge. The depletion charge Q_d and the surface potential ψ_s are given by :

$$Q_d = q N_d W \quad ; \quad \psi_s = \frac{q N_d W^2}{2 \epsilon_s} \tag{5}$$

with N_d : the film doping, ϵ_s : the silicon permitivity, q : the electron charge and W : the depletion layer width given as a function of time by :

$$W(t) = W_f - \frac{L}{\sigma Z V_d} I_d(t) \tag{6}$$

where W_f is the film thickness, σ : the film conductivity and $I_d(t)$: the transient drain current. Making use of 4, 5 and 6, we can write the rate of change of the sum of the inversion and interface trap charges as follows :

$$\frac{d}{dt}(Q_{it} + Q_{inv}) = -\frac{qN_dC_{ox}}{2\epsilon_s}\frac{d}{dt}\left[\left(W_f + \frac{\epsilon_s}{C_{ox}}\right) - KI_d(t)\right]^2 \tag{7}$$

where $K = L/(\sigma Z V_d)$. The total rate of generation in the silicon film is expressed as :

$$g = \frac{qn_i\,[W(t) - W(\infty)]}{\tau_g} + \frac{qn_i s_w A_w}{A_g} + qn_i s_g \tag{8}$$

where $A_g = Z.L$ is the gate area, s_g : the surface recombination velocity under the gate, $A_w = 2L[W(t) - W(\infty)]$: the depleted sidewall area and s_w: the surface recombination velocity over the sidewalls. The rate of change of interface trap and inversion charges is equal to the total generation rate g. Using 5, 7 and 8 we get :

$$-\frac{N_dC_{ox}}{2n_i\epsilon_s}\frac{d}{dt}\left[W_f + \frac{\epsilon_s}{C_{ox}} - KI_d(t)\right]^2 = \frac{K}{\tau_{eff}}[I_d(\infty) - I_d(t)] + s_g \tag{9}$$

where $1/\tau_{eff} = 1/\tau_g + (2s_w)/Z$. From (9) it is clear that by plotting the left hand side versus $[I_d(\infty) - I_d(t)]$ a straight line is obtained ; the effective lifetime and s_g can be determined from the slope and the intercept respectively.

RESULTS

Measurements were performed on the same samples as in the previous section, but with depletion-mode transistors (fig. 1b). The transient drain current was monitored as a function of time using a semiconductor parameter analyzer (HP 4145 B). Transistors with different dimensions were used to distinguish the generation in the bulk from that over the sidewall. Fig. 5 shows the transient behaviour of an N^+NN^+ transistor ($ZxL = 5 \times 5\ \mu m^2$) on a SIMOX substrate annealed at a high temperature (1405°C). The values of the τ_g(2.5 µs), s_g(0.2 cm/s) and s_w(260 cm/s) were obtained by comparison between two transistors having different channel widths (W = 5 and 10 µm) as shown in fig. 6. Transient drain current measurement was also conducted on transistors annealed at low temperatures. The low values obtained for the lifetime (10-100 ns) indicate the improvement achieved using high temperature annealing of the SIMOX substrate.

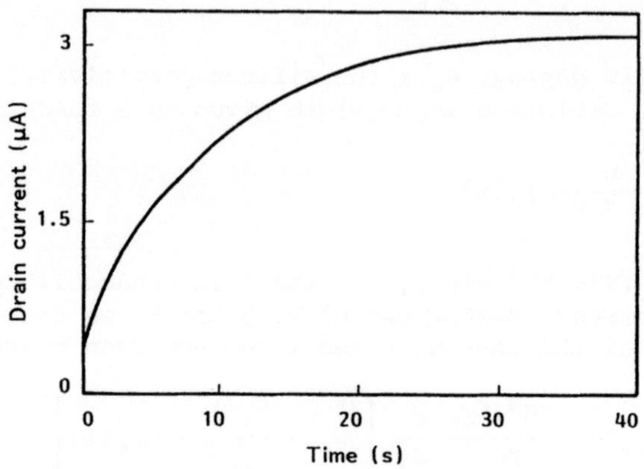

Fig. 5. Drain current as a function of time for depletion-mode transistor pulsed in deep-depletion (HTA-SIMOX, V_d= 50 mV, V_g = -5V).

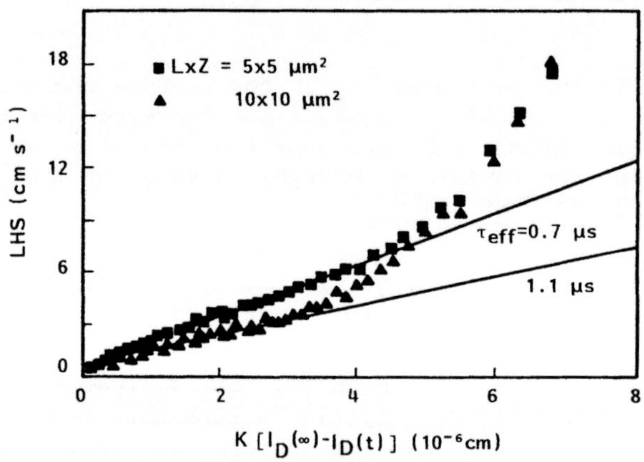

Fig. 6. Comparison of the transient behaviour of two transistors with different dimensions (HTA-SIMOX).

CONCLUSIONS

We have presented new methods for the characterization of interface traps and bulk recombination properties in SIMOX materials from measurements on enhancement and depletion-mode transistors. The effect of the implantation and annealing temperatures has been investigated. Increasing the implantation temperature eliminates bulk traps which could appear below 500°C. On the other hand, the importance of a high temperature anneal has been demonstrated ; this anneal improves the lifetime and reduces the interface trap density.

REFERENCES

1. H. Haddara, G. Ghibaudo, Europhy. Conf. Abstracts 10G, 47, ESSDERC'86, Cambridge, UK.
2. E. Nicollian, J. Brews, in "MOS Physics and Technology", Wiley, New York (1982).
3. M. Zerbst, Z. Angew. Phys., 22:30 (1966).
4. J.H. Lee, S. Cristoloveanu, IEEE Elect. Dev. Lett., EDL-7:537 (1986).
5. D.P. Vu, J.C. Pfister, Appl. Phys. Lett., 9:951 (1985).

POROUS SILICA SOL-GEL COATINGS FOR Nd : GLASS HIGH-POWER PULSED LASER USES

Hervé Floch, Jean-Jacques Priotton, and Ian Malvil Thomas*

Commissariat à l'Energie Atomique
Centre d'Etudes de Limeil-Valenton
B.P. 27 - 94190 Villeneuve Saint Georges

ABSTRACT

A method of forming a laser damage resistant antireflective (AR) coating on fused silica, borosilicate and phosphate glasses, and even onto KDP crystals has been used. The single-layer porous coating is applied by dip or spin from a specific silica colloidal suspension in ethanol. This sol-gel coating deposit is processed at room temperature and requires no subsequent treatment. The coated surfaces work near 100 % transmittance and exhibit laser-induced damage thresholds, measured with 1.06 µm, 1-ns pulses, two to three times greater than levels of widely used PVD (Physical Vapor Deposition) silica-titania multilayer[1], anti-reflection films.

INTRODUCTION

The output power of solid-state lasers is frequently limited by laser-induced damage to high-reflection (HR) or antireflection (AR) coatings. Thus, to minimize the cost of our large Nd : glass laser Phebus (20 TW - 1 ns) installed at Limeil, it was necessary to utilize various types of optical coatings to reduce optical losses and accommodate the highest possible power densities on all components. Antireflection (AR) coatings present the greatest difficulty in these lasers because they typically receive the highest-fluence loading of all surfaces in the laser amplifier chain and they have the lowest damage threshold. Concentrated efforts to increase the damage levels of thin-film AR coatings by the use of alternative materials, coating designs and processing methods have failed to achieve significant improvement[2].

Fraunhofer [3] and Lord Rayleigh[4] observed almost a century ago that the surface reflection of glass can be reduced by chemical treatment of the surface. Later, requirements for low-cost broadband AR surfaces for solar cell applications renewed interest in this etch/leach process, and such surfaces have been produced by acid leaching of phase-separated[5] and nonphase-separated[6] borosilicate glass. The latter process acquired for our own use from the Lawrence Livermore National Laboratory and called the

* Ian. M. Thomas is a chemist at Lawrence Livermore National Laboratory in Livermore, Calif.

Neutral-Solution Process, is currently used on BK-7 and BSC-7 optics for the 20-kJ Phebus laser [7].

More recently, some authors have investigated and reported the use of porous silica as an AR coating which can be applied to a wide variety of optical and nonlinear optical materials not treatable by the earlier processes[8-12]. Perhaps the most promising is the low density colloidal silica antireflection coating developed by Thomas at LLNL[7-14] deposited from suspension which can reduce Fresnel losses in laser components to $\ll 1$ % per surface with damage thresholds as high as 16 J/cm^2 for nano-second pulses at wavelengths around 1000 nm. In this paper we describe our experience in the fabrication and damage testing of colloidal amorphous silica coatings at the Limeil-Valenton CEA Laboratory.

SOL-GEL ANTIREFLECTIVE COATINGS

The method for preparing damage-resistant antireflection silica coatings is a particular application of the sol-gel process, a very versatile method for making metal oxide coatings. Because of the special nature of coatings required for laser applications, particularly damage resistance, in this investigation, the silica source was tetraethyl orthosilicate, $Si(OC_2H_5)_4$. This volatile metal-organic compound boils at a temperature of 167°C and is readily purified by fractional distillation, thus giving very pure silica. Hydrolysis requires a catalyst that may be either acidic or basic. The pathway along which the reaction proceeds depends on the choice of catalyst, but the end product in either case is amorphous silica (See fig. 1).

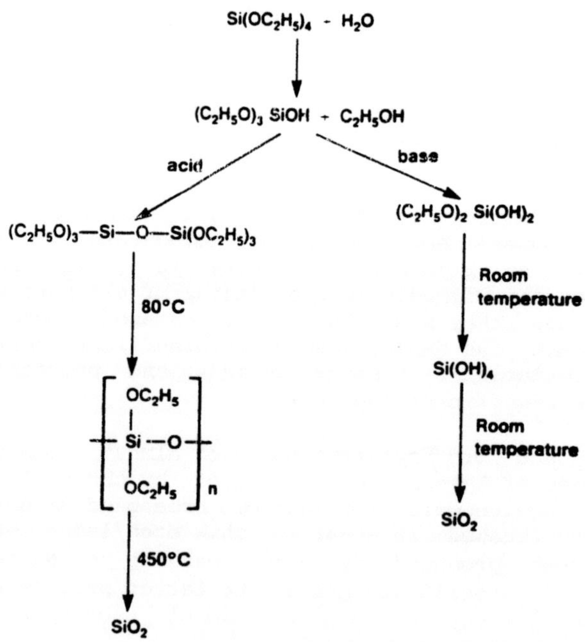

Fig. 1. Alternative reaction pathways in the hydrolysis of TEOS to silica.

Firstly, the acid-catalysed system was investigated. This consisted of applying a soluble ethoxysiloxane polymer to the substrate and converted it to silica by thermal decomposition at 450°C. Then, the porosity of the silica coating was increased by etching it in a dilute solution of hydrofluoric acid[11], to achieve a porosity of about 55 %. Although the optical properties of the resulting coatings were quite satisfactory, some of the coatings failed dramatically under irradiation with laser light. The cause of these failures was probably incomplete thermal elimination of the organic components in the original polysiloxane. This would leave carbonaceous entities that would absorb laser energy, causing destructive hot spots in the coating.

To avoid this source of trouble, Thomas[14] turned the attention to the base-catalyzed system, which produces a colloidal suspension of pure silica particles in ethanol. The preparation of this organosol is carried out at room temperature and is completed in ~ 72 h. The silica particles are roughly spherical and can be made in any diameter from about 5 nm to 2 μm by varying the reaction conditions. After preparation of the base sol with silica particles of the right size (~ 20 nm) the coating is applied simply by spreading the sol evenly on the substrate and allowing the ethanol to evaporate at room temperature. This eliminates high temperature baking and with it the possibility of unwanted organic residues. The porosity (~ 50 %) of the coating was such that an overall refractive index of ~ 1.22 was obtained.

Experimental procedure

Throughout the investigation, ammonia was used as the catalyst causing the formation of spherical particles. A typical coating solution is prepared by the hydrolysis of the distilled product by a procedure similar to that described by Stober et al[15]. Near theoretical quantities of water and high ammonia concentrations are used in the reaction. The chemical reaction is carried out by mixing concentrated ammonia solution (~ 28 % analytical reagent, 9.6 g), anhydrous ethanol (260 g) and tetraethyl orthosilicate (31.2 g) at room temperature.

Under these conditions, with gentle magnetic strirring, the sol formation is completed in three days.

After an invisible hydrolytic reaction forming silicic acid, the condensation of the supersaturated silicic acid occurs and is indicated by an increasing opalescence of the mixture. Finally, a turbid suspension is obtained, consisting very small SiO_2 particles ~ 20 nm diameter, at a concentration of 3.0 % by weight. Transmission electron microscopy indicated all silica particles were approximately spherical with a diameter averaging about 20 nm. Particle size can be varied from ~ 5 nm to ~ 2 μm by changing the quantity of water used and the amount of basic catalyst. High water and low base concentration favour large particle size and vice versa. In addition, some sols were prepared in methanol and reaction rates were faster than with ethanol. The performance of all these sols was similar.

Coating are dipped (fig. 2) or spun (fig. 3) onto optical surfaces from suspensions filtered through a 0.2 micron teflon membrane filter, and all the coating operations are carried out under stringently clean conditions. For AR coatings optimized for 350 nm wavelength light, a withdrawal rate of ~ 4 cm/min is suitable for the dip process. Coatings are normally spun onto surfaces at a rotation speed of 375 rpm, requiring ethanol dilution of a 3 % silica coating sol to 0,75 %.

Thus, the correct quarterwave thickness is achieved by varying the concentration of the stock colloidal silica, the thickness being directly proportional to the concentration, and by multiapplication of several layers if thicker coatings are required.

The technique is made particularly simple by the absence of any special drying or baking step in the process. We have now thoroughly stabilized this base-catalyzed sol process and have scaled it up to more than 350 litres, enough to dip-coat the large aperture (~ 80 cm) Phebus lenses. Also, 27 cm square KDP crystals are now spin-coated routinely with excellent optical performances. KDP surfaces were spin-processed rather than dip-processed because we needed coatings of unequal thickness on the two sides for the laser frequency-converter.

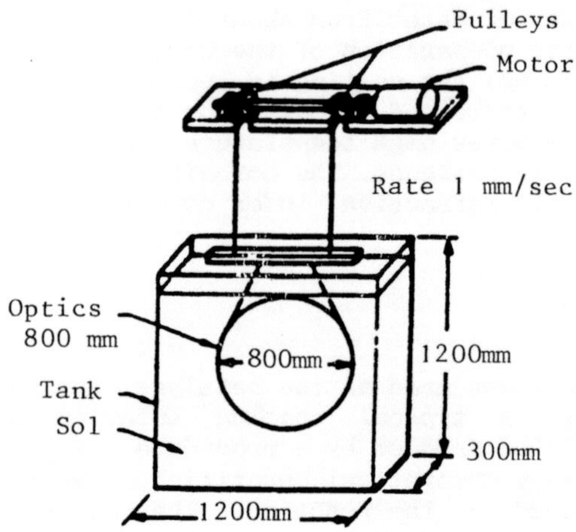

Fig. 2. Treatment of fused silica optics by dip-coating.

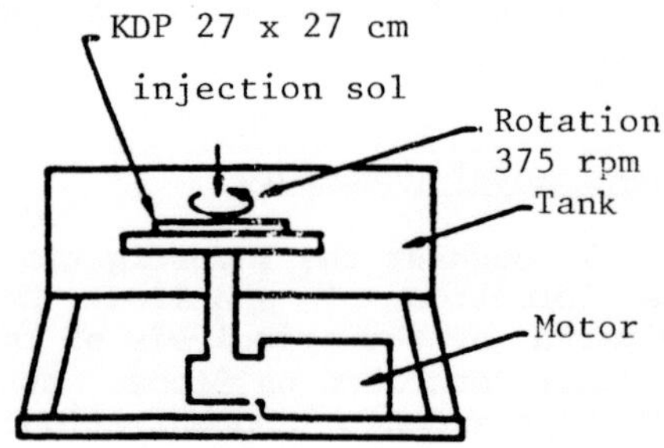

Fig. 3. Treatment of KDP by spin-coating.

Results and discussion

Figures 4 and 5 illustrate the transmission spectrum of successive layers of sol-gel coatings dip-processed on fused silica substrates. The amorphous silica film can be layered to produce antireflection properties at longer wavelengths. These transmission measurements indicated substantially 100 % transmission at wavelengths which were dependent on coating thickness. The latter could readily be varied by changing in the silica sol concentration and withdrawal or centrifuge rate. The effect of multicoats is particularily well illustrated in figure 4. This figure shows the transmission of the substrate when successive coats of a sol containing 2.24 % by weight of silica, at a withdrawal rate of 3 cm/min, are deposited. The coating with peak transmission at 550 nm had optical thickness of 137.5 nm which implies that ~ 46 nm of optical thickness was added during each coating application.

Surface effects caused problems in the dip-coating of large KDP samples. Therefore, it was decided to use a spin-coating process. Spinning has two principal advantages over dipping. In spinning, the two surfaces of a crystal are individually coated so that the coating applied to a given surface of the crystal can be tailored for the light that will pass through that surface. In addition, far less material is required for coating by

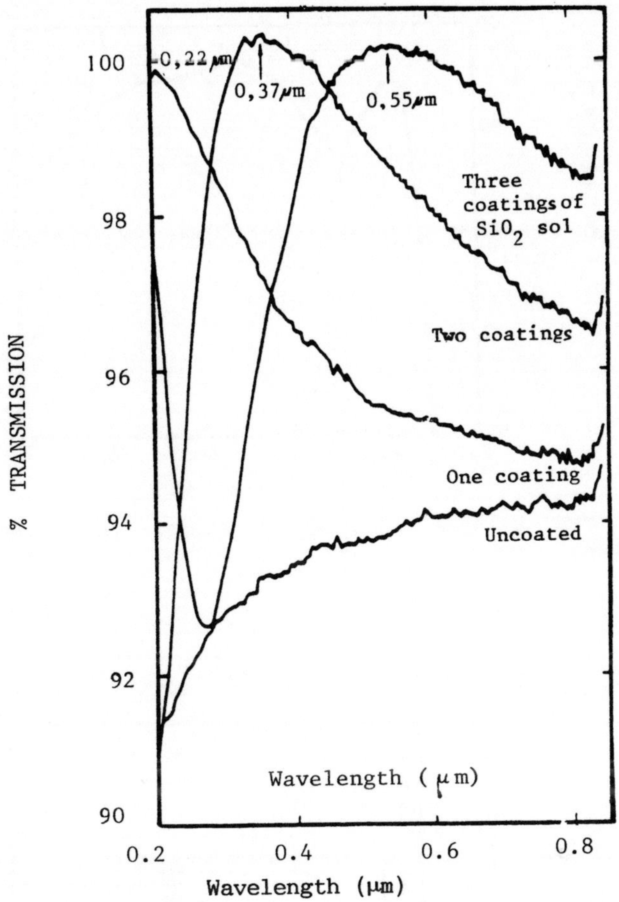

Fig. 4. Two-surface transmittance of a fused silica window with sol AR coatings.

spin than that for dip. The optical performances of the coated KDP crystals were similar to fused silica coated blanks.

Measurements of the laser damage threshold were made with 1ns, 1.06 μm pulses from a Nd : glass laser. The laser was focused with a 2.5 m focal length lens to a 2.0 mm spot diameter on the surface of the sample. After five shots by site, comparison of video photographs of the surface taken before and after irradiation allowed small changes in surface morphology caused by the laser irradiation to be reliably detected. Figure 6 shows the measured 1064 nm, 1 ns laser threshold of a series of sol-gel coated parts vs conventional multilayer antireflection films. The sol-gel coatings have two to three times the damage threshold of these conventional designs and make the use of coated fused silica and coated water sensitive materials like KDP possible. The use of sol-gel coatings on KDP has enabled us to eliminate not only the windows but also the index matching fluid of the classical frequency-converters. These fluid-filled cells are subject to damage initiated by the interaction of the high-intensity light with microscopic contaminants in the fluid. The light heats the particles and induces local decomposition of the fluid. Slight opacities then develop, which absorb energy on successive shots and grow progressively larger, eventually obscuring a large fraction of the aperture. At 350 nm, 1 ns the damage threshold of the sol-gel AR coating is in the range of 8.5 to 10 J/cm^2 on diamond-turned KDP substrates. Laser damage threshold at 248 nm, 15 ns pulse duration, was measured by Foltyn at Los Alamos National

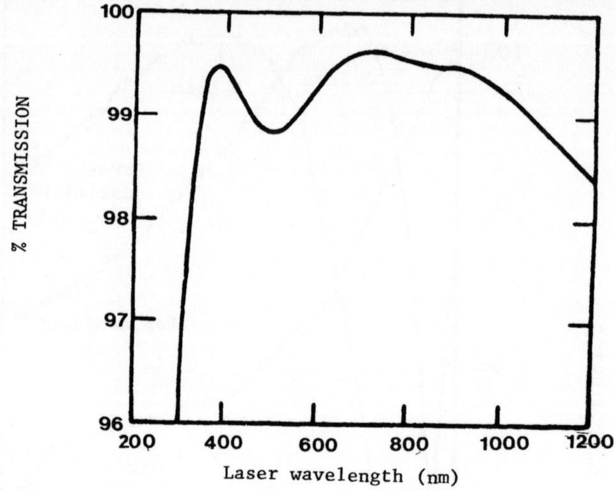

Fig. 5. AR porous sol-gel coating 3 successive silica layers 25°C processed.

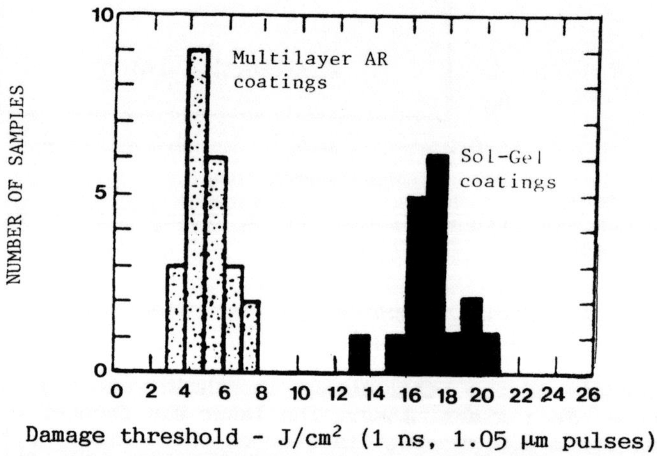

Fig. 6. Damage thresholds of sol-gel and PVD multilayer AR coatings.

Laboratory and gave for these AR coatings 4 to 5 J/cm^2. The two disadvantages of the porous sol-gel coating are its fragility and sensitivity to contamination in a vacuum system.

Its low abrasion resistance is in reality an advantage since sol-gel coated optics may be reprocessed after mechanical scrubbing (usually under alcohol) of the original coating.

CONCLUSION

By 1985 with the cooperation of the Lawrence Livermore National Laboratory the 20 KJ Phebus laser system became a reality at Limeil. This large optical instrument required new processes to withstand the high-power without any optical coating degradation. An important innovation in this way, has been the sol-gel process that is used to provide an antireflection coating on the KDP frequency-converter raising its ouput by 16 % and on the large focusing 'lenses, and optical filters. This new sol-gel method consists first of the preparation of a morphological catalyst with a high purity silica colloidal suspension in ethanol. Then, the silica sol is applied by a dipping or a spinning technique, as a thin liquid layer that dries to form the low packing density coating. The sol-gel process operates at room temperature and requires no further treatment. Without the development of this coating technology, it would have been impossible to meet the Phebus cost, performance, or schedule goals.

REFERENCES

The authors would like to express their gratitude and indebtedness to J.F. Mengue at Limeil for laser damage threshold measurements and to I.M. Thomas at LLNL for very helpful discussions and references.

1. B. Geenen, A. Malherbes, Private communication.
2. W.H. Lowdermilk, D. Milam and F. Rainer, in Laser-Induced Damage in Optical Materials : Natl. Bur. Stand. Sp. Publ., edited by A.J. Glass and A.E. Guenther (US. GPO, Washington, DC 1979).
3. J. Von Fraunhoefer, in Gesammelte Schriften, E. Lommel, Ed. (Verlagder K. Akademie, Munich 1888), p. 35
4. Lord Rayleigh, Proc. Roy. Soc. London 41:275 (1886).
5. M.J. Minot, J. Opt. Soc. Am. 66:515 (1976).
6. L.M. Cook, K.H. Mader, and R. Schnabel, "Integral Antireflective Surfaces on Silicate Glass" U.S. Patent Application 309, 149 (5/10/1981).
7. H. Floch, J.J. Priotton, "Neutral Solution and Sol-Gel Coatings for High Power Laser Applications", S.F.V. Proc. Int. Symp." On Trends and New Applications in Thin Films", 1, 1-7, (1987).
8. E.M. Pastirik and M.C. Keeling, "A Low Cost Durable Antireflective Film for Solar Collectors", Proc. Thirteenth IEEE Photog. Special. Conference, Washington, DC (June 1978).
9. H.R. Moulton, "Composition for Reducing the Reflection of Light", U.S. Patent 2, 601, 123 (1952).
10. H.R. Moulton, "Method of Producing Thin Microporous Silica Coatings Having Reflection Reducing Characteristics and the Articles so Coated", U.S. Patent 2, 474, 061 (1949).
11. B.E. Yoldas, D.P. Partlow, Appl. Opt. 23:1418 (1984).
12. H.L. Mc Collister and N.L. Boling, "Process of Making Glass Articles Having Antireflective Coatings and Product", U.S. Patent 4, 273, 826 (1981).
13. S.P. Mukherjee and W.H. Lowdermilk, Appl. Opt. 21:293 (1982).
14. I.M. Thomas, Appl. Opt. 21:293 (1982).
15. W. Stober, A. Fink, and E. Bohn, J. Colloid Interface Sci. 26:62 (1968).

VACUUM RE-EMISSION OF POSITRONS FROM a-SiO_2 LAYERS

R.I. Simpson, C.D. Beling and M. Charlton

Department of Physics and Astronomy
University College London
Gower Street, London WC1E 6BT

ABSTRACT

A study of vacuum re-emission of positrons from high electric field stressed a-SiO_2 layers is proposed. Measurements of positron re-emission energies and efficiencies would be of interest in understanding positron transport at high electric fields as well as offering the possibility of producing intense (up to $10^9 e^+/sec$) positron beams in the laboratory.

INTRODUCTION

Over the past 25 years positron research has been a rapidly growing field. Many different techniques have been developed to study the interaction of positrons with solid, liquid and gaseous media. Several general reviews have been published including those by Dupasquier and Zecca[1], Hautojärvi[2], Brandt and Dupasquier[3], and Griffith and Heyland[4]. One major advance has been the development of slow positron beams with intensities of up to $10^9 e^+/sec$.

There are many published studies on the behaviour of positrons in insulating and semiconducting materials[1-3]. The behaviour of both very fast and thermal positrons is now thought to be fairly well understood ; in particular with reference to positronium (electron-positron bound state) formation and defect trapping. Of particular relevance to the study proposed here is the work carried out by Mills and Crane[5] who used a tunable low energy positron beam to measure the energy distribution of positrons re-emitted from several materials including SiO_2. Some of their data is presented in figure 1. The significant feature of all the curves is the large spread in emission energies. The data was originally interpreted as positronium formed in the bulk diffusing to the crystal surface and breaking up to emit a positron with the electron becoming bound in a surface state. More recent work on solid rare gas films[6] suggests that this model is incorrect and that Mills and Crane were observing the diffusion of hot positrons rather than positronuim. The large energy of emission is presumably due to the slow energy-loss rate once the positron has "cooled" below the band-gap. This would seem to suggest that thc application of large electric fields might enable positrons to reach an equilibrium with large kinetic energies.

Little data exists for electric field transport of positrons in solid media, presumably because of difficulties in performing these measurements. A brief review of mobility measurement techniques for positrons is given by Beling et al[7]. Using an angular correlation method in which the angular shift from linearity of the two annihilation gamma rays is due to the field induced drift of the positron, Suoeka and Koide[8] have obtained upper limit on the low field mobilities of several insulators including SiO_2 for which a value of 15 $cm^2V^{-1}s^{-1}$ is quoted.

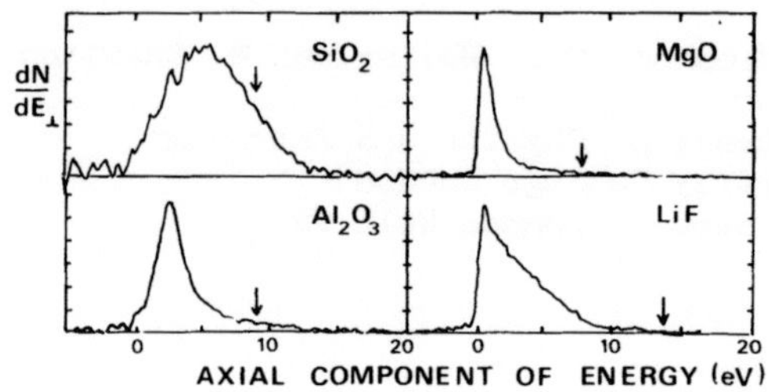

Figure 1. Positron emission spectra for SiO_2, Al_2O_3, MgO, and LiF single crystals. Positron implantation is at energy of 500 eV. Arrows indicate the position of the conduction band. Taken from the work of Mills and Crane[5].

In this paper we are proposing that a study of positron re-emission from a-SiO_2 layers would be of considerable interest in understanding positron tranport properties and may also be the basis of a field assisted positron moderator capable of producing intense beams of up to $10^9 e^+$/sec in the laboratory.

EXPERIMENTAL DETAILS

A schematic illustration of the experiment is shown in figure 2. The sample and channeltron (CEM) are mounted in a UHV chamber. A β^+ emitting radioactive source (^{22}Na) deposited on a gold pin is used to inject positrons into the sample. These rapidly lose energy by ionization, electron-hole pair creation and finally phonon excitation until, within 10^{-11}s, they reach an equilibrium with the applied electric field. They are then drifted towards a thin metal contact where, provided that they have sufficient energy to overcome the unfavourable SiO_2 positron work function and are not scattered in the contact, they are emitted into vacuum. They are guided by a weak (100 gauss) magnetic field to the channeltron detector. Channeltron pulses, after amplification and shaping, are fed to the start input of a time to amplitude converter (TAC). For a ^{22}Na source β^+ production is accompanied by the emission of a prompt 1.28 MeV gamma ray which is detected by a NaI-phototube arrangement. Phototube pulses are delayed so that they arrive at the TAC after the corresponding related channeltron pulse. This inversion is to ensure that the TAC is not overloaded by the high number of NaI- phototube pulses.

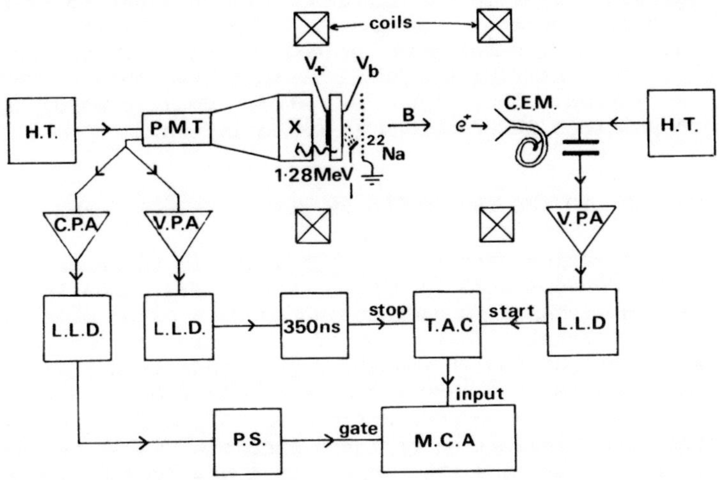

Fig. 2. Experimental arrangement at U.C.L. for the measurement of field emission of positrons from SiO_2. H.T. : High Tension supply, P.M.T. : Photomultiplier Tube, X : NaI crystal, V_b : Beam Voltage, C.E.M. : Channel Electron Multiplier, V.P.A. : Voltage Sensitive Preamplifier, C.P.A. : Charge Sensitive Preamplifier, L.L.D. : Lower Level Discriminator, T.A.C. Time to Amplitude Convertor, P.S. : Pulse Shaper, M.C.A. : Multi Channel Analyser.

Pulses from the TAC, proportional in height to the start-stop time differences, are stored in a multichannel analyser (MCA) according to their pulse height so that over a period of a few hours a time of flight spectrum is built up. Peaks corresponding to slow positron-gamma and fast positron or gamma-gamma prompt coincidences are observed superposed on an exponential background due to random coincidences. The peak separation gives the slow positron time of flight (TOF). Determination of the MCA time per channel by a subsiduary calibration experiment enables the positron energy distribution to be obtained.

The system has been calibrated using a well annealed tungsten foil which is a known low energy positron-remitter. By observing the singles count rate from the channeltron as a function of retarding voltage, the fast to slow conversion efficiency was found to be about 1×10^{-4}. TOF spectra taken, suggest that efficiencies 100 times smaller than this can quite easily be observed. 10 000 Å thick, wet grown oxide layers with 100 Å exit windows of either Sn or Al are currently being prepared for this study. About 1 % of the incident positrons can be expected to stop in this layer.

The energy resolution of the system is poor due to the slow response of the NaI scintillator and the short flight distance (about 10 cm). It is therefore hoped that a more sensitive study can be carried out using a low energy positron beam tunable between 0 and 100 keV. This would enable the mean implantation depth of the e^+ to be varied to suit the thickness of oxide being studied. Thus low energy e^+ of a selected energy can be implanted into the sample where, after slowing down, they drift in the bulk. A fraction of these e^+ (which diminishes as their impact energy increases) will drift back to the surface where they can be re-emitted. The re-emission energy spectra are measured by establishing a potential difference between a control grid placed in front of the sample, and the sample itself, and measuring the total annihilation rate as this potential difference is varied. This type of study would provide very detailed information on positron interactions in oxide layers.

APPLICATION OF SiO_2 TO THE PRODUCTION OF SLOW POSITRON BEAMS

The main motivation for these experiments is to produce intense slow positron beams. Conventional positron beams are usually produced by moderating fast positrons from radioactive sources using metal meshes, thin foils or single crystals. This type of moderator relies on diffusion to transport a small proportion of implanted positrons to a surface where they can be spontaneously emitted into vacuum with energies of a few eV. Fast to slow positron conversion efficiencies are $\lesssim 3 \times 10^{-3}$ [9]. It has been suggested that the application of large electric fields to semiconductors or insulators might provide the basis for a new type of moderator (field assisted) with an efficiency of up to 10 %[10]. In order to stop an appreciable proportion of fast positrons from the source, single crystals about 100 μm thick would have to be used necessitating large voltages to establish the required high electric fields. It is for this reason that we are proposing studying thin layers first before attempting to construct a moderator. In a preliminary experiment a 12 x.25 mm Z-cut quartz crystal with a 100 Å Al exit window surrounded by a 1000 Å Au guard ring was stressed by fields up to 0.8 $MVcm^{-1}$ at which point it broke down electrically. No positron re-emission was observed. These fields are substantially less than those at which electron emission has been observed[11].

Finally, it should be mentioned that other insulators may also be suitable as field assisted moderators. Epitaxial contacts can be grown on Si[12], GaAs[13], and many ionic crystals including LiF, NaCl, MgO and Al_2O_3[14] which would overcome problems of defects in the exit metal layer. LiF and MgO also have the advantage of having a negative positron work function[5]. However a-SiO_2 appears to be the best material for a first study because of a better knowledge of its properties.

ACKNOWLEDGEMENTS

We would like to thank Dr. F.M. Jacobsen for many useful discussions. We acknowledge SERC for funding this project and providing RIS with a Research Studentship. RIS would also like to thank the SERC and the conference organisers for providing funds to attend this conference.

REFERENCES

1. A. Dupasquier and A. Zecca, Rivista Del Nuovo Cimento 12:1 (1986).
2. "Positrons in Solids", (Topics in Current Physics vol. 12) P. Hautojarvi, ed., Springer Verlag, Berlin (1979).

3. "Positron Solid State Physics, Proc. S.I.F. Course LXXXIII", W. Brandt and A. Dupasquier, eds. North Holland., Amsterdam (1983).
4. T.C. Griffith and G.R. Heyland, Phys. Rep 39C:169 (1978).
5. A.P. Mills Jr., and W.S. Crane, Phys. Rev. Lett., 53:2165 (1984).
6. E.M. Gullikson and A.P. Mills Jr., Phys. Rev. Lett., 57:376 (1986).
7. C.D. Beling, R.I. Simpson, M.G. Stewart, Y.Y. Wang, S. Fung, J.C.H. Wai and T.N. Sun, to be published in Phys. Stat. Sol. (1987).
8. O. Sueoka and S. Koide, J. Phys. Soc., Japan, 41:116 (1976).
9. A. Vehanen, K.G. Lynn, P.J. Schultz and M. Eldrup, Appl. Phys. A32:163 (1983).
10. C.D. Beling, R.I. Simpson, M. Charlton, F.M. Jacobsen, T.C. Griffith and P. Moriarty, Appl. Phys. A42:111 (1987).
11. S.D. Brorson, D.J. DiMaria, M.V. Fischetti, F.L. Pesavento, P.M. Solomon, and D.W. Dong, J. Appl. Phys. 58:1302 (1985).
12. R.T. Tung, J.M. Poate, J.C. Bean, J.M. Gibson and D.C. Jacobson, Thin Solid Films, 93:77 (1982).
13. S.P. Svensson, G. Landgren, T.G. Andersson, J. Appl. Phys. 54:4474 (1983).
14. J. Chevallier, private communication.

AUTHOR INDEX

INDEX

MIX
Papier aus verantwortungsvollen Quellen
Paper from responsible sources
FSC® C105338

FSC
www.fsc.org

If you have any concerns about our products,
you can contact us on
ProductSafety@springernature.com

In case Publisher is established outside the EU,
the EU authorized representative is:
Springer Nature Customer Service Center GmbH
Europaplatz 3, 69115 Heidelberg, Germany

Printed by Libri Plureos GmbH
in Hamburg, Germany